BENZENE
AND ITS
INDUSTRIAL DERIVATIVES

CONTRIBUTORS

J. Alagy, Ingénieur Principal, Institut Français du Pétrole, Centre d'études et de développement industriels, Vernaison.

G. D. Allen, MA, Altecmar SA, Brussels.

J. R. Brooks, BPharm, PhD, MPS, ARIC, Manager, Medical & Technical Services, Glaxo (South Africa) (Pty) Ltd., Germiston, Transvaal.

C. Busson, Ingénieur de Recherche, Institut Français du Pétrole, Centre d'études et de développement industriels, Vernaison.

G. A. M. Diepen, dr ir, Professor of General and Inorganic Chemistry, University of Technology, Delft.

F. Eschard, PhD, Maître de Recherches, Direction Raffinage et Pétrochimie, Institut Français du Pétrole, Rueil Malmaison.

A. Forge, Sales Engineer, Engineering and Sales Division, Institut Français du Pétrole.

R. A. Fraser, BSc, CEng, FRIC, FInstGasEng, Technical Consultant.

K. H. V. French, MSc, FRIC, FInstPet, Technical Manager, Benzole Producers Ltd., London.

R. W. Gallant, BSChemE, Registered Professional Engineer, Production Manager, Dow Chemical Company, Plaquemine, La.

J. Habeshaw, BSc, ARIC, formerly of BP Chemicals International Ltd.

J. R. Horth, CEng, FIChemE, MInstF, Process Engineer.

F. A. Jackman, OBE, BSc, FRIC, FIGasE, CEng, MIL, independent consultant, chemical industry.

K. Jones, PhD, Port Sunlight Research Laboratory, Unilever Ltd.

R. O. Kerr, PhD, Senior Research Associate, Petro-Tex Chemical Corporation, Houston, Texas.

F. M. Konrad, PhD, Chemicals and Plastics Division, Union Carbide Corporation, Bound Brook, NJ.

R. Landau, PhD, President, Halcon International Corporation, Great Neck, NY.

R. E. Lidov, PhD, Chemical Consultant, Great Neck, NY.

J.-C. Marion, Ing-Chim, Altecmar SA, Brussels.

W. T. Reichle, PhD, Union Carbide Corporation, Bound Brook, NJ.

A. N. Roper, BSc, Manager, Technological Services Laboratory—Plastics, Shell Chemicals UK Ltd.

G. M. Rowell, Head, Petroleum Section, Process Department, Humphreys & Glasgow Ltd., London.

S. T. Swinfin, MSc, Senior Chemist, The Clayton Aniline Company Ltd.

D. Vu Quang, Senior Engineer, New Processes Development, Engineering Division, Institut Français du Pétrole.

M. A. de Zeeuw, ir, Lecturer, Department of General and Inorganic Chemistry, University of Technology, Delft.

BENZENE

AND ITS INDUSTRIAL DERIVATIVES

Edited by

E. G. HANCOCK

MA, MSC, FRIC, FInst Pet, FPI

UN Senior Adviser to the Chemical Industry Directorate-General in Indonesia

Formerly Shell International Chemical Company Ltd

A HALSTED PRESS BOOK

JOHN WILEY & SONS
New York

Published in the U.S.A. by Halstead Press,
a Division of John Wiley & Sons, Inc., New York

Library of Congress Cataloging in Publication Data
Main entry under title:

Benzene and its industrial derivatives

Includes bibliographies and indexes
1. Benzene. I. Hancock, E.G., ed.
TP248.B4B423 1975 661'.816 74-28074
ISBN 0-470-34780-5

Printed in Great Britain

PREFACE

LIKE ITS PREDECESSORS on Acetylene, Ethylene and Propylene this book was originally conceived by the late Dr S. A. Miller and its outline planned and some of the authors chosen before his untimely death. The present editor has tried to follow the lines laid down by Dr Miller and to fill up the gaps with new authors where necessary.

From being of industrial interest mainly as a constituent of motor fuels, a solvent and the raw material for the production of dyestuffs, pharmaceuticals and a range of fine chemicals, benzene has grown since World War II into a major bulk raw material for the production of plastics, detergents and synthetic fibres. Again before the forties nearly all benzene was a by-product in the carbonisation of coal but demand has now far outstripped any potential from this source and the production of benzene in hundreds of thousands of tons from petroleum has been one of the industrial achievements of the past thirty years.

This book therefore seeks to gather together up-to-date methods of production together with all the important end-uses of benzene, its physical and other properties as well as methods of testing, each the result of work by specialist authors.

As in previous books in the series we have endeavoured to keep a fair balance between scientific and technological information. Authors were asked to give important references, and while there has inevitably been some differences in the interpretation of this request we feel confident that the book will prove of value not only to research workers but also to people engaged in manufacture, planning, technologists of all types and to technical salesmen.

We would like to thank all authors for the trouble they have taken over their contributions and to apologise for any complications arising from the change of editorship.

Now one of the disadvantages of a book of this type, where the contributions come from a wide range of sources, lies in the difficulty of making the book appear up to date. The editor must therefore apologise in advance to the authors of many of the chapters who sent in their contributions promptly that the inevitable delays in publication of a multi-author volume has made the earlier chapters received not quite so up to the minute as the authors would have liked. Where possible this has been mitigated by introducing later statistics at the proof reading stage, but this of course does not entirely overcome the problem of some of the chapters having been completed three years before the book's appearance in print.

But at the present time this has been further complicated by the changing position in the benzene position itself. The setback in the demand for chemicals generally in the 1970 period led to a much reduced investment in capital projects and even though the situation changed during the boom period of late 1971 and 1972 it looked doubtful whether production of base chemicals and plastics would be adequate to meet the very rapidly expanding demand. Further, during 1971–72 a break was seen with the tradition of fixed currency ratios—the dollar was effectively devalued twice, and then allowed to float against the major European currencies, while the Japanese yen was first revalued and then devalued, the pound sterling and then the French franc allowed to float, mainly downwards, and so on. Some increase in prices took place, partly because of a desire for sellers to play safe with a floating currency, and partly due to increased costs arising chiefly from ever-increasing labour rates as workers demanded higher and higher wages to maintain and increase their standard of living against currencies which were steadily losing their purchasing power. Many countries, amongst them the UK and USA, attempted to check rising inflation by controlling prices and this, particularly when applied to base chemicals and plastics raw materials, inevitably leads to shortages, as no manufacturer can or will continue to produce with ever-diminishing returns. To cap all in 1973 the Arab countries began applying restrictions on the crude oil exported, and by insisting on a higher posted price on which to levy their taxes caused at the same time very substantial increases in the actual price of crude oil, which is of course the raw material for nearly all today's chemicals plastics and fibres.

Now, how does all this affect the benzene position? First we must separate the temporary from the long-term effects. Quantitative restrictions on oil are apparently dependent on a Middle East settlement, but even if this is delayed it is already becoming clear that some Middle East countries are beginning to ignore the restrictions, and in any case it is only being applied to certain countries alleged to favour a political opponent of the Middle East countries. We can therefore take this as being of merely temporary effect. But it is unlikely there will be any short-term reduction in prices; only when new oil fields are brought into production on a sufficient scale to produce a glut of oil will prices begin to slide downwards.

Since most benzene is produced from oil an increase in the price of this raw material can only lead to an increase in the price of benzene. Even if the price of coal is not correspondingly increased, and from the wage claims of miners in some countries a steady increase in this price might reasonably be expected, benzene from coal will fetch the same price as that attained by benzene from the major source of supply. How far this will affect the general downstream picture will depend on the elasticity of demand for the various products. This is an exercise being carried out in the economic departments of a great many industrial organisations at the present time, and we cannot go into it in detail here.

We had a discussion with the author of Chapter II in January 1974, when he forecast that supply and demand for benzene and hence its derivatives in Europe would be roughly in balance for some years to come, though there might be local shortages. In the USA it appears certain that many oil companies are purchasing, or considering purchasing, tracts of land from which coal will be available, intending to supplement their production of fuel and chemicals from oil sources by that from coal. Now the traditional method of producing benzene from the carbonisation of coal, even taking into account the availability of modern dealkylation techniques, does not yield more than about three gallons per ton, the benzene being just a by-product. Further, the recovery of benzene from the gas reduces the calorific value of the gas, and the benzene will need to fetch a price which would be equivalent, i.e. taking into account the cost of recovery and purification, to that it would realise as a fuel if left in the gas. In cases of absolute fuel shortage, e.g. in a steel works, it would probably not be recovered under any circumstances, now that modified designs have enabled steel works to use benzene-rich gas without any disadvantages. Natural gas contains some benzene—it varies according to the source, and there is certainly some in North Sea gas. Although small in quantity it might be worth recovering if the price it would fetch were high enough.

Liquid hydrocarbons can also be produced from coal by its low-temperature carbonisation, e.g. in the production of smokeless fuel, by the Fischer Tropsch process or modifications of this, and further by the hydrogenation of coal or coal products. Interest in these processes as sources of aromatics has faded in most countries due to the lower cost of hydrocarbons from petroleum sources, but they are almost certainly due for a new lease of life. Benzene is not directly produced in the pure state from any of them, including natural gas sources, for the benzene must be extracted and purified, but often the best way is to put the whole liquid fraction through a catalytic reformer which will greatly increase the proportion of aromatics in the fraction from which the benzene can be extracted, the quantity being further increased if necessary by dealkylation of the higher aromatics. The hydrogenation process is particularly suitable for this treatment, as the product is rich in naphthenes; it has been stated that by using this latter method as much as 15 gallons of benzene can be produced from a ton of coal, and work is in continual progress to increase this.

Another major source of many aromatic derivatives is the cyclohexane extracted from suitable feedstocks. So far all the work appears to have been carried out by superfractionation but any shortage of benzene will give a major spur to research on other techniques of separation.

One thing that is certain, quite apart from what has been said in the last paragraph, is that the present circumstances will provide the motive power for some most intensive research on the production of benzene from liquid hydrocarbons to increase the yields by more efficient catalysts, better designs of plant, including more efficient use of heat and more efficient separation and working-up methods. The same policy will be applied to the down-stream processes, as every additional ton of end product that can be squeezed out of 100 tons of benzene will in some way help to mitigate the higher cost, as well as the shortage of benzene.

Note in proof. At the time of going to press (January 1975) it is clear that major new plants are being planned or being erected and it is not going to be a question of benzene's availability but rather of its price that will be the determining factor in its use in the future. Due to the fantastic increases in the price of crude oil during 1974 the price of benzene is already two and a half times its level when the bulk of this book was being written and this will obviously affect the cost of all its derivatives and may reduce the demand for them particularly where alternative routes to the final product are available or even alternative final products exist. Without delaying the production of the book further we cannot however go into this subject more deeply.

Note on units

In this book unless, otherwise stated, degrees are expressed in centigrade units, gallons are imperial gallons and tons are metric tons.

E.G.H.

CONTENTS

CHAPTER 4

The Physical Properties of Benzene

F. A. Jackman, R. W. Gallant, G. A. M. Diepen and M. A. de Zeeuw

CHAPTER 5

Analysis and Handling

K. H. V. French

CHAPTER 6

Cyclohexane and its Derivatives

F. Eschard, A. Forge, D. vu Quang, J. Alagy, C. Busson, G. Allen and J.-C. Marion

CHAPTER 7

Ethylbenzene, Styrene and the Styrene Polymers

R. Landau, R. E. Lidov, J. Habeshaw and A. N. Roper

CHAPTER 8

Phenol and its Derivatives

W. T. Reichle, F. M. Konrad and J. R. Brooks

CHAPTER 9

Detergent Alkylate

K. Jones

CHAPTER 10

Nitration

S. T. Swinfin

CHAPTER 11

Maleic Anhydride

R. O. Kerr

CHAPTER 12

Other Industrial Uses of Benzene

R. A. Fraser

CHAPTER 8

Phenol and its Derivatives

W. T. Reichle, F. M. Konrad and J. R. Brooks

CHAPTER 9

Detergent Alkylate

K. Jones

CHAPTER 10

Nitration

S. T. Swinfin

CHAPTER 11

Maleic Anhydride

R. O. Kerr

CHAPTER 12

Other Industrial Uses of Benzene

R. A. Fraser

CHAPTER 1

THE HISTORY OF THE DISCOVERY AND DEVELOPMENT OF BENZENE

by F. A. Jackman

1.1. EARLY EXPERIMENTS IN CARBONISATION

IN many ways the later years of the eighteenth century were more fruitful of scientific thought and invention than any other period in history. Much of this was generated by societies, groups of men, drawn from various professional and business circles, who liked to meet and to discuss the newest ideas in science and technology purely for the love of such intellectual exercise. Today they would be described as enthusiastic amateurs but their like scarcely exists in the intensively specialised modern society in which we live.

One such group which has attained rather more than local fame was the 'Lunar Society' of Birmingham, which was active from about 1766–1800.

The practice of its members was to meet in turn at each other's houses in the afternoon of the Monday nearest to each full moon, to engage in scientific discussion and argument, and after dinner to return home assisted, in those days of unlit streets, by the light of the moon from whence the Society's name was derived.

Never large in numbers, its members and guests included such distinguished scientists and engineers as Dr Joseph Priestley, Matthew Boulton, James Watt, Josiah Wedgwood, Sir William Herschel, John Smeaton (who built the Eddystone lighthouse) and John Baskerville, a great printer.

Boulton and Watt were the proprietors of the Soho Ironworks, where many steam engines following Watt's ideas were built, and where also much British coinage was struck when the demand arose. A young apprentice at the Ironworks was William Murdock, who trained as a mechanical engineer, rose to be Works Superintendent and is frequently called 'the Father of the Gas Industry'. He too became a member of the Lunar Society and their philosophical discussions may well have led his thoughts into the study of the effects of heat upon carbonaceous matter.

The experiments he started in 1792 culminated in success five years later when he was able to light his workshop with coal gas.

It should perhaps be interposed that almost two centuries previously Turquet de Mayerne had obtained benzoic acid by heating benzoin, an Eastern spice, but it was not until 1834 that Mitscherlich discovered the effect of distilling the benzoic acid with lime. The resultant spirit, an impure benzene, thus came to light at approximately the same period of time as other workers were beginning to identify similar products made by heating various oils, both animal and vegetable, as well as the coal which was the subject of Murdock's experiments. Another scientist interested in the distillation of coal was G. T. Accum, who between 1803 and 1819 produced a variety of substances by this means. Some of his samples are still in existence in an Oxford museum and one, labelled 'Highly Rectified Essential Oil', was analysed recently by the modern gas-liquid chromatography technique, when a small benzene content was confirmed.[1]

Somewhat later a German chemist, A. W. Hofmann, became interested in the nature of the volatile products obtained by heating coal. Largely through the influence of the Prince Consort, he came to England in 1845, when only 27 years of age, to become Professor at the newly-created Royal College of Chemistry, where he remained for twenty years. Naturally he pursued his ideas about coal tar at the College and set some of his students the task of identifying the components. Among his pupils was Charles Mansfield, who studied at the Royal College of Chemistry from 1847–50 and is regarded as the founder of coal tar chemistry. He also followed with deep interest the researches of Michael Faraday, and was indeed associated with Faraday at the Royal Institution from 1849–53. Since 1825 Faraday had been working on the pyrolysis of fish oil and other similar materials, from which he actually isolated benzene, his 'bicarburet of hydrogen', and indeed it is generally accepted that he was the first worker to isolate benzene in a fairly pure form, to determine some of its physical properties and to express an opinion about its chemical analysis.[2] Faraday's interest, however, was that of the research scientist and wholly non-commercial. Mansfield on the other hand, although quite well-to-do, had a strong desire to convert his research into a business, and in this he was encouraged by his teacher Hofmann. Between them they devised methods for separating the components of coal tar, and Mansfield examined in great detail the light naphtha fraction obtained from tar distillation. The methods he employed for purification were undoubtedly the fore-runners of those still widely in use up to the present day, for he used successive washes of sulphuric acid, water, and caustic alkali in his process. His main difficulties occurred in distillation to separate the various hydrocarbons, because the apparatus available to him was comparatively inefficient. However, by repeated redistillations he managed to obtain benzene, toluene and xylene and also became aware of the presence of the C_9 hydrocarbons. He applied for a patent covering his process and this was granted in 1848, although patent law of the time gave insufficient protection and he spent much of his time subsequently battling with would-be competitors whom he accused of infringing his patent. There seems no doubt that some unscrupulous attempts were made to take advantage of his ideas. Another method of which he was a pioneer was the purification of benzene by crystallisation, another process which has returned to favour in modern times.[3]

Hofmann was always behind Mansfield, providing the driving force because he always believed that there would be a great wealth of chemical development springing from the first simple coal tar compounds. Thus they progressed to the manufacture of nitrobenzene and thence to aniline, which was to prove invaluable later on in the field of dyestuffs.

Once again Mansfield's desire for an opening into the commercial field came to the fore. He promoted the first use of benzene as a solvent for dry cleaning, and then had the idea that nitrobenzene, with its characteristic odour similar to bitter almonds, could be used as a perfuming agent and also for food flavouring. The latter proved inevitably disastrous because of the toxic nature of nitrobenzene.

It must be remembered that at the time when Mansfield was making those pioneer experiments in chemical manufacture he was only about 30 years old, and indeed he died tragically at 35 in a laboratory fire while preparing samples for the Paris Exhibition of 1855. In a few short years he made an indelible mark on the early history of coal tar chemistry.

It is perhaps worth recalling that the phase in the history of benzene from Murdock's first probings to the death of Mansfield covered a span of sixty years. The vital spur missing from the story, the incentive required to speed the research, was the lack of any large-scale commercial need for benzene as a chemical raw material. The early workers were primarily interested in the production of illuminating gases, and it was for this reason that Murdock tested a number of carbonaceous materials to see if they would yield such gases when heated.

It was then that he decided that coal was the most promising raw material upon which to work. Faraday on the other hand became involved in by-products from the gas made by the Portable Gas Company from whale and fish oils, and it was when he investigated the condensate deposited from such gases that he discovered the presence of benzene.

It was only when Hofmann and his pupil Mansfield began a systematic examination of coal tar that benzene emerged as an article of commerce, and even then although Hofmann was able to produce nitrobenzene and aniline as derivatives, the demand was to be measured in terms of only a few hundred gallons.

The vital point which so far had escaped notice was that the gases evolved themselves contained far more benzene than did the tar. There was to be a further considerable lapse of time before this fact emerged, and by then the stimulus of demand had arrived, mainly following the expansion of the new coal-based dyestuffs industry.

In parallel with this work in Britain, continuing over a long period of time, there was of course similar research being conducted on the Continent, particularly in France and Germany.

For example, Caspar Neumann reported as early as 1740 his observation that the distillation of 'fossil carbons' gave rise to a spirit which was no doubt a type of crude benzole,[4] and Macquer referred in his Dictionary of Chemistry in 1777 to a similar phenomenon.

It must also have been in Germany that A. W. Hofmann first became inspired to carry out his investigations, although it was not until he came to Britain in 1845 that he succeeded in concluding his work on the nature of coal tar.

Reference has already been made to the work of Mitscherlich, who produced benzene from benzoic acid, and carried out researches similar to those of Faraday from which he was able to determine a number of physical characteristics of his product. The results obtained by these two workers are extraordinarily similar.

1.2. THE FIRST INDUSTRIAL USES

Of these pioneer workers, therefore, it may be said that in roughly a hundred years between the mid-eighteenth and mid-nineteenth centuries, they had between them discovered that pyrolytic distillation of many carbonaceous materials, including coal and animal and vegetable oils, caused a breakdown of the original substance and the production of a spirit or tar, and that some distinct individual chemical compounds could be isolated.

A number of the hydrocarbons now known as 'aromatic' were thus obtained but most attention was paid to benzene, whose physical properties were examined and identified with some care.

Leading from this, Faraday showed the effect of treating benzene with nitric acid, and observed the characteristics of nitrobenzene. Subsequently Mansfield, who had been associated with Faraday, carried on the work and tried to promote nitrobenzene as a synthetic perfume.[5] An attempt to use it as a food flavouring failed because of the toxicity of nitrobenzene. However, Mansfield pursued a more profitable path in carrying out on a fairly large scale the reduction of nitrobenzene to aniline, a product of considerable interest to Hofmann.

It was while W. H. Perkin was studying under Hofmann at the Royal College of Science that his researches produced an aniline dyestuff with a beautiful purple colour. The main constituent is the compound known as mauveine, and Perkin seized the opportunity to enter into commercial production at Greenford, leaving his academic researches in spite of the entreaties of Hofmann to remain at the College.

And so it was, in 1856, that the great story of benzene as one of the world's most important chemical raw materials began in a small house on the western outskirts of London. Others had made a beginning in the commercial exploitation of benzene, albeit an impure variety, as a solvent somewhat earlier. Mansfield, for example, always hopeful of achieving some commercial success, introduced the use of benzene as a dry cleaning agent, and this many years before the discovery in 1859 of petroleum which eventually took its place. In fact until quite recently it was possible to buy bottled benzene for cleaning purposes in the home in spite of its pronounced unsuitability on the grounds both of inflammability and toxicity.

Another application of an aromatic hydrocarbon fraction which no doubt contained some benzene was commercialised in the early 1820's by Macintosh who had developed his process for the rubber-proofing of cotton sheeting. The rubber solvent he used was the light naphtha distilled from coal tar and in spite once again of its inflammability, which caused a number of disastrous fires, the product was a great success and the industry expanded enormously.

These were just typical examples of the way in which industry profited from the solvent qualities of benzene and the kindred hydrocarbons, but none of them were strictly chemical uses, and Perkin's synthetic dyestuff manufacture stands clearly alone as the first major use of benzene in the chemical industry.

The small yield obtainable by distilling coal tar has already been mentioned; the commercial exploitation of Perkin's process brought about an acute shortage of benzene, and it was fortunate that a far more fruitful source was at hand. During the 1860's several workers had tried the effect of different chemical reagents on coal gas, and they all had in common the observation that when the gas was bubbled through nitric acid the characteristic smell of nitrobenzene, likened to bitter almonds, was evident. G. E. Davis noted this in 1869,[6] but even earlier Dr Caro had made a similar discovery while working with Roberts, Dale & Co. This firm had not wished to pursue the study of this process and it was not until Dr Caro returned to Germany to work for the Badische company that he completed his investigations, with the result that he was able to take out a patent in 1869 covering the extraction of benzene from coal gas by oil scrubbing. From this point onwards the production of benzene expanded continuously as new chemical uses for it were developed.

In these earlier years there were perhaps two events, both linked with the medical world at the time, which eventually had greater impact upon benzene chemistry than any others. The first of these was the discovery by Lister in 1867 of the antiseptic properties of phenol. This once again was a product of coal tar refining which no doubt supplied adequate quantities at the beginning.

It had been known for some time that benzene could be reacted with sulphuric acid to form benzene sulphonic acid, and that the latter upon being fused with molten caustic soda would yield phenol.

This synthesis was not however employed commercially for some years, in spite of the discovery by Turpin in 1885 of the explosive nature of trinitrophenol (picric acid). It was only when the Boer War created a huge demand for picric acid that it became practicable to make phenol synthetically, and in 1899 Read Holliday & Co. commenced commercial production.[7] Thus began a synthesis which is now a cornerstone of the colossal world production of many synthetic fibres and plastic materials; the precise route has altered but the original starting point, benzene, still remains.

The other great discovery in the pharmaceutical field took place in 1899 when aspirin, acetyl-salicylic acid, was discovered in Germany and became the first genuine synthetic drug.[8] Still in its own right a vitally important and widely used analgesic drug, it has also proved to be the forerunner of a large family of kindred materials, many of which spring from benzene as a basic raw material. Included among these are such familiar names as phenacetin and sulphanilamide.

Thus by the turn of the century we find benzene firmly established in the world of chemistry, always derived from coal either through gas or tar, but as the gas lighting industry and the manufacture of coke for metallurgical purposes expanded, there was at the same time an ever-growing flow of crude material from which benzene could be extracted.

1.3. EARLY ACADEMIC WORK AND THE MOLECULAR STRUCTURE

In thus tracing the progress of benzene production as it followed the development of the organic chemical industry, it is appropriate to note the parallel academic work to ascertain the nature of the benzene molecule, a subject which generated at the time, a century ago, very considerable argument and dispute.

The foundation laid in 1825 by Faraday has already been mentioned. His part was to isolate the compound, to obtain it in a fairly pure form and to determine a number of its physical and chemical characteristics. Faraday's work led him to describe the new chemical which he had discovered as 'bicarburet of hydrogen' and it must be presumed that he was satisfied about the ratio of carbon to hydrogen in the molecule. It was however a few years later, in 1833, that Mitscherlich established the empirical formula for benzene as having six atoms each of carbon and hydrogen.[9] Following the establishment of this fact came much speculation about the manner in which these atoms were spaced in the molecule. In the aliphatic series a saturated hydrocarbon with six carbon atoms (hexane) contained also fourteen atoms of hydrogen. To devise a formation giving a stable six-carbon atom compound with only six atoms of hydrogen was much more difficult and a number of ingenious solutions have been proposed.

Most of these incorporate some sort of ring structure, and that suggested by Kekulé in 1865 is normally accepted today, although another worker, Loschmidt, suggested a similar form some four years earlier.

Kekulé put forward the theory that the six carbon atoms form a ring in which the alternate bonds are single and double, thus taking up three of the four valencies almost invariably attributed to carbon. The fourth valency for each carbon atom forms a link with an atom of hydrogen, thus satisfying the empirical formula C_6H_6.[10]

Kekulé carried out many painstaking experiments to prove his theory by preparing addition and substitution compounds, but these did not completely answer his critics and he was obliged to extend his theory.

In some cases, for example, if Kekulé's original proposals were accepted one would expect to find some extra isomeric substitution products, and his explanation for their non-existence is that the positions of the various bonds are not rigidly fixed within the molecule but are constantly changing. In view of modern theories, a century later, of the nature of atomic and molecular structure generally with the mobility conferred by the concept of the electron as a part of these structures, it may well be that Kekulé can be claimed as a brilliant forerunner of these present day ideas.

However, other workers propounded their different theories about the structure of the benzene molecule. Armstrong and Baeyer suggested a centric design where the fourth valency of each carbon atom was directed toward the centre of the ring, Ladenburg devised a structure in the form of a prism, Claus proposed a three-diagonal form and Dewar a rather similar idea but with only one diagonal. These various structures are illustrated in the following diagrams:

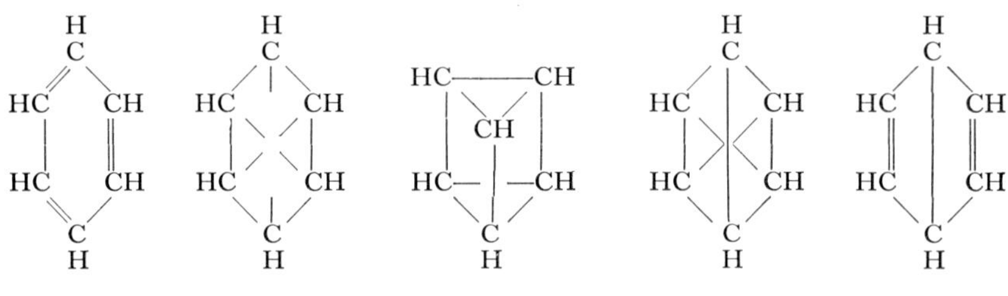

Kekulé | Armstrong and Baeyer | Ladenburg | Claus | Dewar

To each proposal there have been strong objections of one kind or another, but the original Kekulé form has best stood the test of time, is most generally accepted and is now supported by the modern views on atomic structure.

1.4. BENZENE FROM PETROLEUM

By about the year 1930 the production of benzene from coal gas had become a world-wide and high-tonnage industry, with a world total of refined benzene-rich fractions of about 1·5 million tons. Yet at that time the need for benzene as a chemical raw material represented only a small part of the potential available quantity, and the minds of the producers were directed far more towards the efficient use and marketing of benzene-rich distillates as components of motor fuel. It was only when, ten years later, World War II caused huge distortions in the supply/demand equation that the chemical industry turned its mind seriously to the production of benzene and its homologues from the obvious alternative source, petroleum.

Not unnaturally in war-time, the first acute shortage arose in the need for toluene for the manufacture of TNT and other explosives. In fact this situation had also been experienced during World War I, and at that time toluene had been extracted by fractional distillation from Borneo and other eastern petroleum, but the Japanese occupation in the Pacific prevented the exploitation of the same methods in 1939–45.

1.4.1. Catalytic Reforming

However, there had been many experimental processes in course of development in the years before 1939, mostly based upon the cracking of selected petroleum fractions, where the raw material is subjected to conditions of heat and pressure in the presence of catalysts. A pioneer in this field was Houdry, who had been seeking a method of improving the quality of motor fuels, and in 1927 devised a catalytic cracking process.[11] The first commercial plant based on this process was in operation ten years later and a new concept in petroleum refining had been initiated. Nevertheless although the improvement in motor fuel quality was due to the presence of hydrocarbons which differed from those in the original feedstock, and aromatics were among these, it was still technically difficult to extract the aromatics in pure form. It was to be a variant of the cracking principle which was eventually to provide a large part of the toluene needed for the American war effort.

Long years before, in 1911, N. D. Zelinski had discussed the possibility of dehydrogenating naphthenes, a series of ring compounds with carbon atoms in the ring but saturated by the presence of the full complement of hydrogen atoms. The basic member of the series is cyclohexane C_6H_{12} but others include methylcyclopentane $CH_3C_5H_9$ and methylcyclohexane $CH_3C_6H_{11}$.[12]

Zelinski's process consisted of exposing the chosen fraction to high pressure in a closed vessel under the influence of heat usually at about 500°C, and in the presence of a catalyst which could be platinum or palladium. These conditions cause hydrogen atoms to be detached from the naphthene ring so that cyclohexane, for example, is converted to benzene.

Another parallel effect occurred in the same process where some rearrangement of atoms within the molecule was also observed. Methylcyclopentane has of course a five-member ring of carbon atoms but still the empirical formula C_6H_{12}. Under the influence of the Zelinski process it was found not only that six atoms of hydrogen were detached, but the carbon atom in the methyl side chain entered the ring so that the final product was benzene, just as it was by the dehydrogenation of cyclohexane.

There seems to be a general consensus of opinion that the role of the platinum or palladium catalyst is to promote mobility of the hydrogen atoms within the naphthene molecule. Certainly this discovery marked the beginning of a new era in petroleum chemistry, with the possibility of transforming hydrocarbons of one series into another to meet chemical needs.

In a normal petroleum crude the naphthene series will at best only represent a proportion of the mixed hydrocarbons present, and usually the major part consists of straight-chain and branched-chain paraffins.

A further process to transform these into benzene and other aromatics has also been developed, and in this case the preferred catalyst is usually molybdenum or chromium in the form of the oxide, often coupled with aluminium. It is thus possible to convert normal hexane to benzene, and normal heptane to toluene, by this dehydrocyclisation reaction.

These processes are classed together under the generic name of catalytic reforming, and several variants have been offered to the petroleum industry under proprietary names, the best-known of which are Platforming, Hydroforming, Houdriforming and Catforming.

It is necessary to make a very careful choice of catalyst and conditions according to the type of feedstock available and the product required.

1.4.2. Purification by Solvent Extraction

Catalytic reforming usually gives rise to two main streams, a considerable amount of gas which is almost all hydrogen, and a liquid stream which is still a mixture of hydrocarbons.

The aromatics content will rise from an initially low level of less than 10 per cent in the feedstock to about 50 per cent in the product. Most of the naphthenes will have been dehydrogenated and straight-chain paraffins will have been transformed into ring compounds, but the branch-chain iso-paraffins are more slowly converted and comprise most of the non-aromatic part of the product. Conversion is not of course absolutely complete and small amounts of naphthenes and straight-chain paraffins will be found, but the main products are aromatics plus branch-chain paraffins.

It is now necessary to separate these if the aromatics are required in a pure quality, although many catalytic reformers are employed just to make a high-aromatic cut which is very suitable as a blending stock into high-grade motor gasoline.

There are numerous methods for the separation of the aromatics, usually by some type of solvent extraction.

The general principle underlying all of these is to make use of the differing rates of solubility of aromatics and paraffins in a chosen agent.

It is not usual to attain the desired degree of separation at a single mixing, and some sort of counter-current device is necessary. By using this the two streams, one the reformer product and the other the extractant, move in opposite directions so that the fresh extractant encounters reformer product from which one component, the aromatics has largely been removed by partly-spent extractant earlier in the contact process.

With all their modern refinements such solvent extraction units are remarkably efficient and the final aromatic and paraffinic streams are almost completely separated from each other. The aromatics contain only the smallest trace of paraffins, and the paraffin stream may well have a benzene content of less than 0·05 per cent.

Thus the way is clear for the manufacture of benzene of very high purity, a product of great value to the chemical industry.

The forerunner of solvent extraction methods was the Edeleanu process which used as its extractant liquid sulphur dioxide, which dissolves preferentially the aromatic components of the reformer product, leaving the paraffins behind. This process was introduced as long ago as 1907 for the refining of special grades of kerosene and is still used for this purpose, but for the production of aromatics on a large scale other types of solvent extraction are now preferred. The two most commonly used are the Udex process of Universal Oil Products which employs diethylene glycol as the extractant, and the Shell sulpholane process. Other methods have been designed by Lurgi in Germany, and by the Institut Français du Pétrole.

1.4.3. Other Petrochemical Methods for Benzene Production

A catalytic process of a different kind is the Catarole process developed by H. Steiner. Although Dr Steiner brought his process to Britain at the onset of World War II, it was not until 1949 that it went into commercial operation when a plant at Partington, near Manchester, was built by Petrochemicals Ltd. The essential difference attributable to the Catarole method is that there is a much higher proportion of gaseous product, up to about 50 per cent of the whole, and this gas consists not of hydrogen as in the classic catalytic reforming plants, but of ethylene and propylene. These gases can of course be separated and subsequently used for chemical synthesis.

The liquid product also differs from the catalytic reformer, mainly because it contains a higher proportion of benzene vis-à-vis the other hydrocarbons, and also quite a low percentage of the unwanted aliphatic material. Purification of the aromatics is effected by azeotropic distillation rather than by solvent extraction.

One of the problems which has beset the petroleum industry has been the balance of supply and demand for the various aromatic hydrocarbons which may be produced. Catalytic reforming tends to yield high proportions of the benzene homologues, toluene and the xylenes, and rather less of benzene itself.

Under present-day conditions the xylenes are finding a market in polyester fibre and phthalic anhydride manufacture, and benzene has a huge offtake into synthetic fibres and plastics, which will later be described in detail.

Toluene however is much less required by the chemical industry other than in time of war, when non-nuclear explosives of the TNT type are responsible for a much increased demand. On the other hand the production of toluene by catalytic reforming is quite high, even if there is careful selection of the charging stock fed into the reformer system. Much can be done in this way by pre-cutting the feed to give a preponderance of aliphatic hydrocarbons with the desired number of carbon atoms (isomeric hexanes, C_6, to give benzene, heptanes, C_7, to give toluene and so on), but inevitably there will be some overlapping and unwanted products will arise.

It was therefore of great benefit when another type of reforming process was invented, which would detach methyl groups from benzene homologues and reduce them down to the basic benzene. Thus toluene, $C_6H_5CH_3$ and xylene, $C_6H_4\,(CH_3)_2$ can be dealkylated to give benzene C_6H_6.

The dealkylation process has been developed by a number of designers, the two most often used being the Houdry Detol method and the Hydeal plant offered by Universal Oil Products in conjunction with Ashland. Universal have also collaborated with Toray Industries of Japan and they have jointly developed another plant employing similar principles, known as the Tatoray process.

Having regard to the long history of benzene production, starting as it did from coal and expanding so prolifically in the petroleum field, it is perhaps appropriate to refer to a petroleum-type process which is actually being used to refine crude benzoles of coal origin from coke ovens, and making product benzene of the finest quality, comparable with any originating from the petroleum industry.

As far as Britain is concerned this modern and sophisticated technique, the Houdry Litol process, has transformed the production of coal-based benzene from the by-product industry which it was, into a purpose-built operation strictly on the same lines as those employed in oil refineries.[13]

The Litol process combines in a single plant a number of complementary operations designed to eliminate all unwanted products, leaving only the main stream of pure benzene. It consists of a catalytic reactor into which crude benzole from coke ovens, together with overhead fractions from tar distillation, is passed with an accompanying stream of hydrogen. The catalytic reaction takes place at temperatures around 600°C and pressures of about 700 lbs per square inch, and during its course the sulphur compounds are converted to hydrogen sulphide, non-aromatics are broken down to low-boiling hydrocarbons, and higher-boiling aromatics such as toluene and xylene are converted into benzene. This main product emerges with less than 1 p.p.m. of sulphur, over 99·9 per cent purity, virtually all toluene and paraffins eliminated, and its freezing point of 5·45°C or higher demonstrates beyond doubt that given the right refining methods coal-based benzene can match the quality of any petroleum-based material. The first British plant incorporating the Litol process commenced operations at Staveley, Chesterfield in 1969, and has a capacity of 40,000 tons per year.

1.5. DEVELOPMENT OF THE USES OF BENZENE

It is clear from the intervention of the petroleum industry into the world of aromatics that a massive increase in demand for benzene has come from the chemical manufacturers. In 1930 the total world production of pure and 90's benzene, covering chemical and solvent use, was about 50,000 tons per year, all derived from coal, and by the outbreak of World War II this output had increased only to about 60,000 tons. By contrast, by 1965 the total world consumption of benzene (excluding Eastern Europe) was calculated to be 5·0 million tons, and by 1970 the predicted total was 9·0 million tons or about 150 times the pre-war figure. On a world-wdie basis the demand from the chemical industry overtook the capacity of the coal-based by-products production by the early 1950's, and the petroleum-based catalytic processes started to supply the additional requirements so that they now have a dominating share of the trade.

There are three great outlets for benzene in the chemical industry which largely account for this very large tonnage of consumption, and some 80–90 per cent of all benzene used for chemical manufacture is absorbed in these three. They are the production of styrene, which is used either in its own polymers to make the well-known glass-like plastic material polystyrene, or in copolymerisation with butadiene to give a wide range of synthetic rubbers; cyclohexane, which is an essential intermediate step in the production of nylon; and cumene, equally important as an intermediate in the synthesis of phenol.

Other uses which make up the total of benzene consumed include nitration and subsequent reduction to aniline, the oldest of all ways of using benzene as a raw material and to which reference has already been made.

Benzene is also chlorinated to give either products where chlorine is substituted in the benzene ring, such as monochlorobenzene, or others where the chlorine gives rise to addition products like benzene hexachloride.

The development of synthetic detergents is known to all, and benzene is the starting point for the production of many of them, usually by the addition of alkyl groups to form a side-chain attached to the benzene ring.

Direct oxidation of benzene gives maleic anhydride, which is used for the production of alkyd resins for the surface coatings industry, particularly for automobile finishes. Other types of resin can also be produced in this way.

1.5.1. Styrene

From the very earliest days of the study of benzene chemistry, initiated by Faraday in 1825, it was realised that addition compounds with side-chains attached to the benzene molecule were possible, even though another forty years would elapse before Kekulé put forward his proposals concerning the structure of the benzene molecule. Among the derivatives made in those early experiments was ethylbenzene, and by 1831 Bonastre had isolated styrene.[14] The formation of polystyrene under the influence of sunlight was also observed very early in the history of benzene chemistry, but these discoveries remained as little more than academic curiosities for decades to come.[15] It was only after 1920 when the first commercial production of ethylene on a significant scale was initiated by the Linde Air Products Company in the United States, that attempts were made to develop the manufacture of styrene. This commenced in the USA in 1925 and by 1929, I G Farbenindustrie in Germany had built a plant to make 500 tons per annum.

Interest was now thoroughly awakened and in Germany in particular the possibility of making a type of synthetic rubber by the copolymerisation of styrene and butadiene was well appreciated. Although the quality of the first synthetic rubbers was markedly inferior to that of natural rubber, the Germans, spurred on by the need to conserve foreign currency, developed and improved their home-produced synthetic rubber, known as Buna-S, during the period of about five years before the outbreak of World War II. During that war they were almost entirely cut off from supplies of natural rubber and relied largely upon synthetic material.

Meanwhile in the United States, a parallel development of the industry was proceeding, and by 1935 Dow Chemical Company was in production of styrene.

Following the Japanese occupation of the main areas of natural rubber growing in the East Indies, the Americans also were compelled to rely very largely on synthetic material and by the end of the war were making it at the rate of about one million tons per year. In Canada also a large synthetic rubber plant was built by the Polymer Corporation, a Crown Corporation, at Sarnia, Ontario and came into operation in 1942.

In more recent years the synthetic rubbers have come into their own and it is now customary to use for automobile tyres a blend of natural rubber with one of the synthetic grades.

By so doing the makers have achieved results showing marked improvements in qualities of wear and resistance to attack by oil and other solvents.

The other important use for styrene is as a polymer, which has found a great variety of uses in the industrial and domestic fields. It is familiar throughout the civilised world, a clear transparent glass-like plastic material, giving off a characteristic metallic ring when tapped. It has excellent insulating properties in the electrical sense and is widely used in packaging.

The dominant position of styrene manufacturers as consumers of benzene is shown in the following table (Table I.1).

TABLE I.1

Consumption of Benzene in the Chemical Industry North America and Western Europe (Thousands of Tons)

	1965	*1970*	*1975 (Forecast)*
Cyclohexane	900	1,630	2,320
Styrene	1,610	2,760	3,830
Phenol	1,000	1,370	2,000
Miscellaneous	780	900	1,210
	4,290	6,660	9,360

1.5.2. Cyclohexane

Cyclohexane, a simple derivative of benzene, was, as in the case of styrene, well known quite early in the history of benzene chemistry, but equally failed to excite much interest as an article of commerce.

Eventually this interest emerged as a result of research work by W. H. Carothers in the laboratories of E. I. du Pont de Nemours in the United States, when he discovered the possibility of forming polymers from esters made by the conjunction of dialcohols with dicarboxylic acids.[16] For example, hexamethylene glycol forms an ester with adipic acid in the following manner:

$$\underset{\text{Hexamethylene Glycol}}{HO(CH_2)_6OH} + \underset{\text{Adipic acid}}{HOOC(CH_2)_4COOH}$$

$$\longrightarrow \underset{\text{Ester}}{HO(CH_2)_6O{\cdot}OC(CH_2)_4COOH}$$

This ester is susceptible to polymerisation and the product is a type of fibre.

Subsequently it was found that a much improved fibre could be obtained if the glycol were replaced by a diamine such as hexamethylene diamine. This was the beginning of the great nylon industry, and although there are several variants of this first truly synthetic fibre, they all have in common the basic principle of Carothers's discovery, and in each variant the most popular route for the synthesis starts from benzene. In the earlier methods the first step was to convert benzene into phenol and thence eventually to adipic acid, but more recently it has been preferred to make cyclohexane the first intermediate product.

In fact in some plants, including the modern coal-based unit of Staveley Chemicals at Chesterfield, the benzene produced passes immediately to the adjacent unit where it is hydrogenated to give cyclohexane as the main product.

Nylon has proved to be a most versatile plastic material and undoubtedly the first discovery in 1936 was most timely in the event, for World War II provided countless opportunities for its use, and gave a great stimulus to its development. It is well known that the fibres can be spun to a high degree of fineness and yet retain their tensile strength, thus being very suitable for textiles where light weight and strength are required. Not only does this apply in the familiar case of socks and stockings, but also for sheeting of various gauges. Another valuable property is the impermeability of nylon to water, and thus the material is suitable for many marine uses including ropes and sails. As an alternative to drawing nylon out into a fibre for spinning, it can also be moulded into solid shapes and used as an insulator, or can be machined and used for some types of bearings and other mechanical purposes.

It is therefore not surprising that the production of nylon has become an industry of considerable size, responsible for the use of over 1,600,000 tons of benzene per year in the Western World, or nearly a quarter of the total make. As new uses for the product, either in fibre or in solid form, are developed there will no doubt be a further substantial growth in its manufacture.

1.5.3. Phenol

Reference has already been made to the early experiments in synthesising phenol from benzene, and by the time of the Boer War in 1899, the demand for picric acid by the explosives industry was so great that the synthesis was operated on a commercial scale.

The plant was operated by Read, Holliday & Company who carried out the classic process of fusing benzene sulphonic acid with caustic soda, and then treating the sodium benzene sulphonate with sulphur dioxide.[7]

During World War I a somewhat similar situation arose and on this occasion phenol was synthesised on a much larger scale in a number of plants up and down the country.

These methods were necessarily much more expensive than the conventional extraction of phenol from coal tar, and the modest requirements for phenol in peacetime were easily met by the coal tar industry, so the synthesis fell once again into disuse. However after 1918 demand for phenol was stimulated by the development of early synthetic resins of the phenol-formaldehyde type, originally discovered in the United States by Baekeland in the years before the war, and usually known by the generic name 'Bakelite'.

During the past twenty years phenol production has joined styrene and cyclohexane in experiencing phenomenal growth. It has provided one of the routes to the production of adipic acid for nylon manufacture, and although it has been displaced by cyclohexane to a large extent for this purpose, there is still some nylon made by the phenol route.

Other uses include the still popular formaldehyde and similar synthetic resins, and a more recent bulk outlet into agricultural chemicals, particularly selective weed killers. Chlorinated phenols are also widely used as germicides.

In spite of its venerable history, the sulphonation method is still in use to convert benzene into phenol, although there have been considerable modifications in the detail. Other processes start by chlorinating benzene to chlorobenzene, but the most recent method proceeds by a route where cumene (isopropyl benzene) is the main intermediate.

Initially used during World War II as a high-octane fuel component, cumene is a product of the alkylation of benzene with propylene, and the next phase of the reaction where the cumene is oxidised to give phenol has been developed by the Distillers Co. Ltd., and Hercules Powder Co., so that this method is now widely used.

1.5.4. Other Uses for Benzene in the Chemical Industry

The very early recognition by Mansfield and others of the powerful solvent properties of benzene led to a long period when it was marketed, usually in the less pure form known as 90's Benzene (or Benzole), simply as an industrial solvent. It was particularly useful where natural rubber was required in solution and also in the production of some types of adhesive tapes. Its great disadvantage however is its toxicity, a condition of which users have become increasingly aware. Even when in comparatively low concentration, benzene vapour can if inhaled for any prolonged period damage the blood cells and can prove fatal. In very recent years the attention of health authorities has been drawn to this hazard particularly because it can be fatal even when benzene is present as a component of hydrocarbon solvents to an extent of as little as 5 per cent.

Some fatalities have occurred in the automobile industry where such solvents are used in adhesives formulation, the vapour has not been effectively removed but has been inhaled and the benzene has taken effect.

Consequently the use of benzene as a solvent has much diminished and it is now only acceptable when totally enclosed in properly designed vapour-tight chemical plant.

There are however still some of the earlier processes, for which benzene was employed, actively operating.

Perkin's first synthesis of a dyestuff opened up a great field of activity in the synthetic dye industry and this still accounts for much of the so-called 'Miscellaneous' use of benzene. It is a versatile chemical and still remarkably cheap—there can be few chemical raw materials where a comparable standard of purity, 99·9 per cent or better, is available at so low a price. In 1971 supplies have been available at well under 20 U S Cents per U S gallon or less than £25 per ton.

1.6. BENZENE AS A FUEL

In early years of development, benzene had little industrial use, the great bulk methods of production had not been discovered, and the internal combustion engine had not been invented. By the end of World War I much had changed, for the demand for toluene for explosives had caused large quantities of benzene to be made as a by-product. Methods of extracting the aromatic hydrocarbons from coal gas as well as from tar were fully understood and very rapidly it became clear that there was an urgent need for a market for benzene. At this time the motor car was becoming much more widely available and the era of the small low-priced family car had begun.

In fact benzene, usually not in its pure state but admixed with its homologues toluene and xylene and known as 'motor benzole', had been marketed as a motor fuel before World War I. As early as 1907 Messrs Wm. Butler & Co., of Bristol had set up a marketing company which sold motor benzole, various other refiners followed this lead and by February 1914 the journal *The Motor* was able to publish an article describing the sale of benzole as a motor fuel by Messrs. Thos. Crow & Sons of Barking.[17]

The main consideration in these early years was however one of price. Motor engines were designed with only comparatively low compression ratios and there was little need for fuels with anti-knock qualities such as are available today.

The post-1918 small cars were however obliged to derive their motive power from engines of low capacity but of ever increasing mechanical efficiency.

This was achieved largely by increasing compression ratio, a process which has continued steadily for fifty years.

Thus the small car of the early 1920's had a compression ratio of about 5:1, whereas now, in the 1970's, an engine of the same nominal cubic capacity could well have a ratio in excess of 9:1.

This development in engine design, and hence in fuel technology, created a new opportunity for benzole as a fuel, as its marked anti-knock properties became apparent. From this period of the early 1920's arose a specialised marketing organisation, the National Benzole Company, devoted to the distribution of benzole as a motor fuel. After the earliest experiments it became customary to sell benzole in the form of a mixture with petrol, usually from 15–40 per cent of benzole being mixed with petroleum spirit. During the 1930's a standard test engine came into universal use and the concept of the octane number as an index of anti-knock quality was accepted on a worldwide basis.

In Germany a very similar organisation was set up to market benzole mixtures, originally known as the Benzol-Verband and later as BV-Aral.

Thus the use of benzene and its homologues as motor fuels became accepted on a world-wide scale, and the test engines demonstrated the value of the aromatic hydrocarbons as an additive to petrol. It was found that toluene had an even higher value than benzene for this purpose, and xylene higher still (although with the latter some loss of volatility provided a slight disadvantage).

Meanwhile engine design had been advancing and creating an enormous demand for high-octane fuels. Once again the supply of benzene (or benzole) for this market could no longer be met from coal-based sources and the catalytic refining processes of the petroleum industry were used to satisfy the growing need.

By 1941 there were fourteen plants in the United States operating by the Houdry process. The use of these catalytic reformers to produce benzene and, particularly, toluene for war purposes has already been described but after the return of peace the use of the process developed on an enormous scale purely to provide aromatics for motor fuel blending.

Normally these aromatics are not extracted in pure form, just as in earlier days the coal-based motor benzole was a benzene-rich mixture rather than a pure hydrocarbon.

The reformate product is usually made in most petroleum refineries to give an aromatics content of 40–50 per cent, and it is this stream, blended with a number of other components, which gives the high-octane motor fuels of the present day.

1.7. BENZENE IN THE 1970's

Production figures for the East of Europe are not accurately known, but it must be certain that the world today is making over 10 million tons annually of benzene, of which 90 per cent or more is derived from petroleum. The story of this phenomenal growth of more than a hundredfold in thirty years could be said to epitomise the whole chemical industry. Time after time there has been a situation where a new demand has heavily out-weighed existing methods of supply, and resulting research and development work has thrown up new sources, new production processes to meet the demands on a scale more bountiful than ever before, and frequently, as a bonus, more cheaply also. The early painstaking experimental production of Mansfield and his contemporaries proved inadequate for the dyestuffs manufacturers led by Perkin in 1856, with the result that in the ensuing years the price rose fivefold, from three shillings to fifteen shillings per gallon. This stimulated the extraction of benzene from coal gas so that by the outbreak of war in 1914 the motor benzole offered to motorists was sold at little more than one shilling per gallon. After that war production was so prolific that distribution on a nation-wide scale of benzole mixtures as a motor fuel became practicable and economic.

Then once again industrial demand on a very large scale arose and overtook the capacity of the coal-based by-products refiners to meet it, thus providing the springboard for the great expansion of aromatics supply from petroleum. By the early 1950's the new source had taken the lead and by 1970 dominated the market with its huge total tonnage. This must indeed be a classic example of the interaction of supply and demand, when over-supply and low prices stimulate research into new uses, and equally shortage of supply concentrates attention on new methods.

In the latter part of 1970 and early 1971 over-supply has again been very evident and the price of benzene at the refineries has been little, if any, more than it was in 1914, in spite of continued inflation of the currency and a reduction of its purchasing power to about one-eighth of the 1914 level. This situation obviously cannot continue, and by all historical standards will not continue. The likelihood is that eventually new uses for benzene will emerge, and this aspect is dealt with in a later section of this chapter.

When, in about 1965, the market forecasts were compiled and the level of requirements for the next ten years was predicted, it was thought that by 1975 the demand would be more than double that of 1965. At that time the slackening in the pace of demand could not have been completely foreseen, a situation felt far beyond the world of benzene alone but throughout industry.

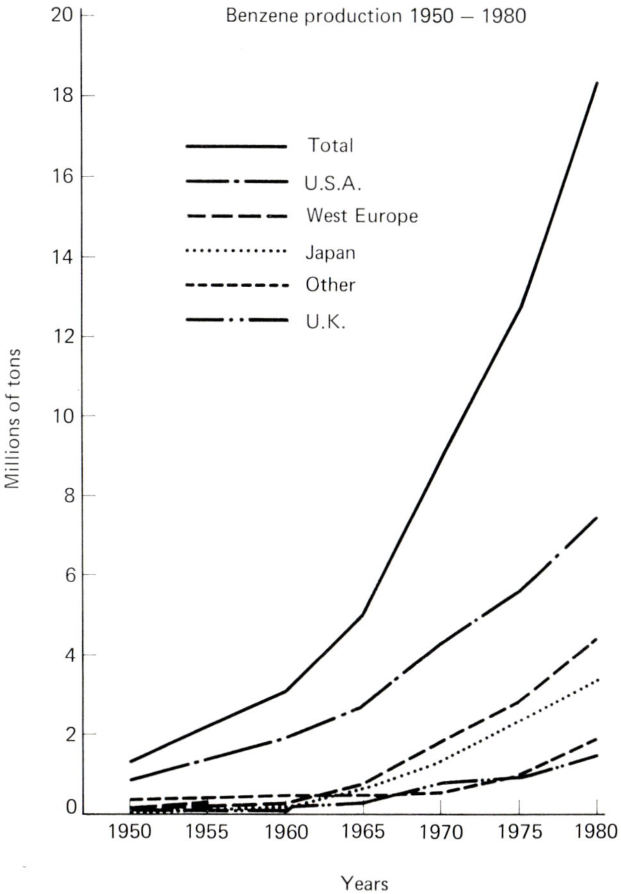

FIG. 1.1 *Benzene Production 1950–80*

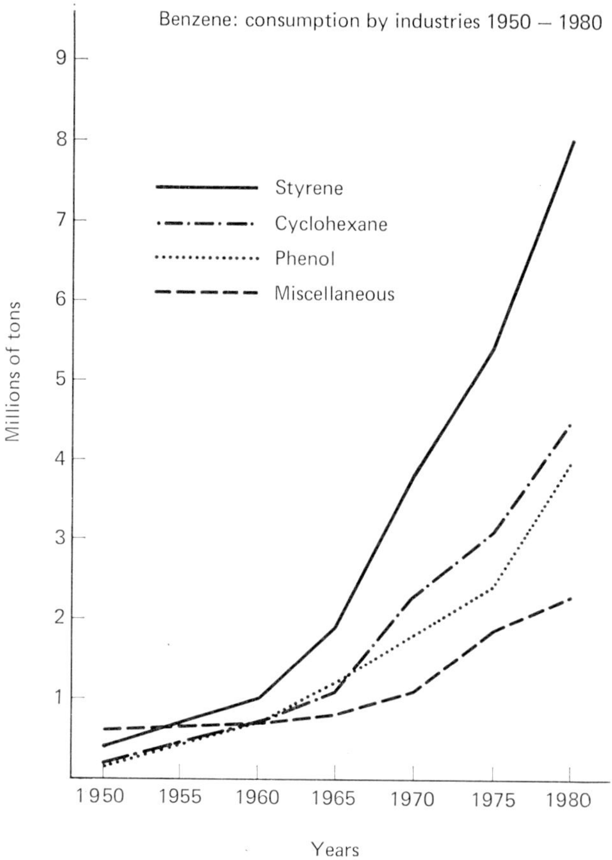

FIG. 1.2 *Benzene: Consumption by Industries, 1950–80*

Most benzene finds its way through its derivatives into consumer goods, either as fibres and dyestuffs for clothing, as packaging material or into specialised industrial outlets, and it is in these sectors of the chemical industry that demand is currently lacking. However, the forecasts predicted a growth rate of over 40 per cent in the five years 1970–5 for existing uses alone, and although there has been a hesitation in the growth over 1970–1, it seems probable that by 1972 consumption will again be rising with the result that while the five-year (1970–5) programme may be as much as two years late in being achieved, it will eventually come to fruition. It is therefore reasonable to expect that at some time in the next two or three years the level of demand will again reach that of capacity. Any new large-scale uses for benzene will of course accelerate this time-table.

The motor car industry accounts for a very substantial proportion of the benzene produced, quite apart from the fuel aspect. Styrene goes into synthetic rubber for tyres, nylon into fabrics and tyre cords, and maleic anhydride is the base for the production of alkyd resins for surface coating formulations.

Thus the lack of growth in motor car manufacture has a very definite influence on the consumption of benzene. However, particularly in motor tyres, recent developments have tended to make more use of synthetic materials derived from benzene. The blend of rubbers used now incorporates an ever-growing proportion of synthetic rubber which confers its own particular qualities of resistance to wear and to attack by oil and other solvents with which it comes into contact.

It will be in this section of the industry, where much lively research and development work is being carried out, that considerable increase will be evident.

The two graphs (Figs. 1.1 and 1.2) show respectively the rates of production of benzene, divided into geographical areas and covering the period 1950–80, and the consumption of benzene divided between the main types of outlet during the same period.

1.8. FUTURE PROSPECTS FOR BENZENE

In the historic past it has always occurred that high availability of benzene, as with many other chemical raw materials, and consequent depression of selling prices, has stimulated research into potential new uses which, when commercially exploited, have redressed the balance of supply and demand. In the early 1970's petroleum refineries all over the world find themselves with excessive capacity for petrochemical products, high stock inventories and sluggish growth rates. This is in other words the classic situation for the development of new discoveries, but unfortunately as far as benzene is concerned these new uses are simply not in evidence. Certainly this is true of the chemical industry, although on the fuel side there are possibilities which could explode into a vast increase in consumption of benzene. To understand this it is necessary to look back at the evolution of the motor car engine over the last fifty years. This evolution followed in Britain a markedly different path from that in the USA and Europe for many years, simply because of the eccentricities of the British taxation system. In recent years however, the old tax method having been abandoned, the development of engine design has been much more uniform throughout the world, although domestic tastes in the USA still dictate the need for cars with larger bodies and engines but rather lower compression ratios.

Reference has already been made to the changes in motor engine design which followed the introduction in the early 1920's of the low-priced family car. In Britain the motor taxes were based on a formula incorporating the bore size of the cylinders, so in order to keep the tax liability as low as possible, engines were designed with small bore but a long stroke to give the requisite cylinder capacity. This meant that the speed of rotation of the engine was slowed down and poor acceleration resulted. The cure for this problem was to increase the compression ratio of the cylinders, a process which was facilitated by great metallurgical improvements in the materials of construction.

Increased compression brought with it a need for fuels with superior anti-knock characteristics, and by the early 1930's a standard test engine, the Co-operative Fuel Research (CFR) engine built by the Waukesha Motor Company of Wisconsin, had been designed and was in use all over the world for the measurement of octane number, an index of anti-knock quality.

In the 1920's good quality motor fuels only had an octane number of about 70 and although the British type of small long-stroke engine was more exacting, it was recognised everywhere that an increase in compression ratio would confer improved efficiency, better fuel consumption and generally enhanced engine performance.

This was the situation which really brought recognition to the valuable qualities of benzene as a fuel and the sales of benzole mixtures increased.

Other methods of improving octane number also appeared on the market, however, the most important being the admixture into motor spirit of small amounts of an organo-metallic compound, tetra-ethyl lead (TEL).

The phenomenal qualities of TEL were discovered by Thomas Midgley in the Standard Oil Laboratories in 1921, when he was doing research work on the use of vaporising oil for farm tractors.[18] He recognised the enormous potential of the additive, the Ethyl Corporation was set up to manufacture and promote the use of TEL, and a completely new approach to motor fuel formulation was initiated. It has been universally so successful that at the present time, fifty years later, it is still a dominant feature of the motor fuel market and the Ethyl Corporation, its successors and affiliates have become a major element in the industry. The present day motor spirits with octane numbers ranging from 93 to 101 all contain some TEL, usually blended with a very complex mixture of hydrocarbons containing a substantial proportion of benzene and its homologues.

This outstanding success story is however marred by a serious doubt. It is well known that lead has some unpleasant toxic qualities, and there can be no doubt that all the TEL mixed into motor fuel eventually finds its way through the exhaust system of the car into the surrounding atmosphere, probably in the form of lead oxides and carbonate. In the UK it is believed that some 8,000–10,000 tons of lead are used annually for TEL manufacture, and of course in countries like the USA with their enormous consumption of motor fuels, the discharge of lead into the atmosphere must be proportionately larger.

Much of the lead discharged falls harmlessly to the ground as the final products are heavy, but there must be a suspicion that some are dispersed in the form of dust and may be inhaled.

In certain areas, particularly in the USA, there are appreciable problems in connection with air pollution, partly because of exceptional climatic and geographical conditions, but public opinion is now very sensitive on questions of pollution and it seems probable that some legislative steps will be taken to limit it as far as motor car exhaust fumes are concerned.

Another aspect of air pollution from motor engines which is relevant is that there is a desire to reduce the amounts of carbon monoxide and unburned hydrocarbons which are emitted.

One practical method of dealing with this problem is to fit the car with an exhaust after-burner, which often relies on a catalytic principle to convert the carbon monoxide and hydrocarbons to other less noxious products. Unfortunately the catalyst, which may be platinum or palladium, is rapidly poisoned by lead and rendered useless, thus providing another argument for those who advocate the elimination of TEL from motor fuels.

To do this however would bring in its train some further difficulties. If TEL is not permitted, the octane number of fuels will drop, possibly by an index of 10 to 15, so that the present ratings of 93 to 101 may drop to as low as 80. At this level many modern engines would be totally unserviceable unless modified, so it would be virtually essential to add some other component to restore the octane number. The only practical additive would be aromatics, and the normal fuel, with TEL removed, would require some 15 per cent of aromatics added to restore, even approximately, the lost quality.

If such a decision were taken the problems facing the petroleum industry would be extremely serious. Even given the time to build the catalytic reformers required, together with ancillary equipment, the additional catalyst needed, usually platinum, would be so large in quantity that the world's resources would be severely strained.

The total world production of platinum (excluding the Soviet block) is only about 2 million ounces per year and this is mostly already committed. The large additional demand for reforming just could not be met on a world-wide and continuous basis.

However, the likelihood is that some areas, parts of the U S A, and some countries in Europe, will gradually introduce restrictive legislation on motor car exhaust fumes, there will be some reduction in the use of TEL and a consequent increase in the need for aromatics for fuels. Thus the use of benzene will again show considerable growth, and this may well take place in the period up to 1975.

TABLE I.2

Thousands of (Metric) Tons

	1965		*1970*	
	Coal	*Petroleum*	*Coal*	*Petroleum*
Australia	30	—	30	25
Austria	14	—	15	—
Belgium	51	—	54	100
Bulgaria	—	—	—	25*
Canada	—	120	—	150
France	149	74	160	225
Germany (East)	—	—	25	28*
Germany (West)	248	70	260	220
India	20	—	20	50
Italy	35	130	35	460
Japan	204	180	250	1,250
Netherlands	33	—	—	60
Poland	30*	—	30*	166*
South Africa	5	—	8	—
Spain	8	—	8	60
United Kingdom	160	45	140	570
United States	406	2,353	380	3,750
USSR	400*	N/A	400*	N/A

* Estimated.

1.9. WORLD PRODUCTION, 1965–70

Table I.2 shows the main areas of production of benzene, together with an approximate evaluation of the annual totals for the two years 1965 and 1970. The figures have been derived from many sources, not all official and many somewhat conflicting, for the availability of such statistics varies considerably from one country to another. However, it will probably suffice to show the trends in the production pattern and to point to the areas where new plant has been installed, and also to a decline in the older methods in some parts of the world.

1.10. THE FUTURE OUTLOOK FOR BENZENE

It is a truism in the history of the chemical industry that the existence of plentiful supplies of a low-priced raw material invariably provides a powerful stimulus for the discovery and exploitation of new processes making use of such material.

The history of benzene itself gives ample illustrations of the point, yet although in peal terms the price of first-quality benzene is far lower than ever before, there has been little sign of the emergence of new outlets.

Research work to this end is undoubtedly being done and new developments will come along, but their nature is still unknown outside the immediate circle of the researchers. Meanwhile another change must take place because of the dramatic increase in the cost of the ultimate raw material, crude petroleum.

Following a sharp rise in 1971 in the price of all Middle East crudes, which will cause the cost at a U K port to increase by about 40 per cent over the levels of the previous year, there will be inevitably a comparable rise in the price of motor gasoline. It has been shown in an earlier section that benzene, albeit in a less refined form, is an all-important component of gasoline and that importance may be enhanced. Thus a very great base outlet for benzene exists, and a floor level for its value is determined. In the long run, unless the chemical grade of pure benzene shows a comparable price, making allowance for refining costs, it will not be made and the semi-refined product will be diverted into gasoline. It seems therefore quite certain that, whatever the market condition, benzene will be less cheap before many months have passed. It is to be hoped that any new processes to use it will not have to be abandoned on considerations of price.

However, the chemical industry of the present time is itself expected to raise its demands very substantially, and projections represented in the graphs in Section 6 predict that over the period from 1970 to 1980, the use of chemical grade benzene will be more than doubled. There is furthermore another consideration, although it may take a longer time to reach fulfilment. It has already been shown that a very high proportion of benzene-based chemical products lie in the field of man-made fibres and plastics, most of which are aimed at the domestic market.

It has been shown that the consumption of such fibres and plastics, in clothing and other personal uses, is very much a function of the per capita income level in the country concerned.[19]

The accompanying graph (Fig. 1.3) shows this very clearly, but it may be expected that as the standard of living rises in many countries, their use of such products as nylon and styrene will increase also. Thus an ever-growing market for these products will be generated, and while it would not be wise to expect immediate growth for this reason, it does seem most likely that a ten- to twenty-year view will need to take these factors into consideration.

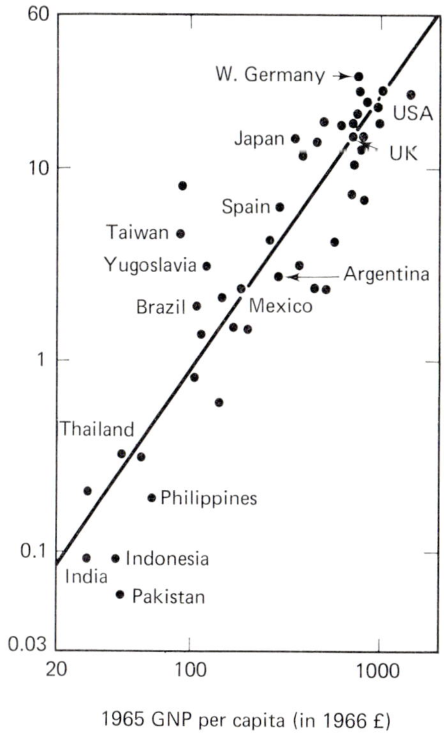

FIG. 1.3 *Plastics Consumption (kg per Capita in 1965)*

Meanwhile it is probable that the most rapid increase will be in the motor fuel market. The world is becoming more and more conscious of its environment and the effects of pollution, and although the influence of motor exhaust fumes is rather limited except in certain special areas, there is no doubt that motor car usage is going to grow at a phenomenal rate and with this will be a comparable rise in the emission of exhaust fumes. It is therefore logical to assume that in one country after another legislative restrictions will be imposed, and this all seems to point to a much greater use of aromatics in the formulation of motor fuels. There will surely be some difficulties because of the demand for precious metals as catalysts, which may well outstrip availability, but it is likely that newer methods, other than catalysts, will come along to supplement supplies. It must be remembered that catalytic reforming still represents only a small proportion, about 10 per cent, of the total tonnage of oil refined throughout the world.

Looking a little further into the future, therefore, perhaps as far as the end of the century, one can only see continued growth in the industry. A major change which is likely is that the producing units will be much wider spread. While today a very high percentage, certainly over 95 per cent, of world production of aromatics is concentrated in North America, Europe and Japan, it must be expected that future expansion will take place much more in countries less developed industrially. In almost every country in South America, and many in Asia, there is an interest in developing a petrochemical industry, and even if the market for products does not yet exist, there will certainly be expansion into these areas, one country after another. Unfortunately the petrochemical industry is not labour-intensive, although demanding high technical skills, and in the beginning will probably be launched by expatriate staff with capital provided internationally. Nevertheless the prizes by way of improved national prosperity are such that these developments must come. Another compelling reason is the incidence of freight costs as part of the total.

Benzene is in 1971, as has been shown, a very cheap chemical indeed, but the cost of trans-atlantic shipment can be as much as 20 per cent of the landed value of a cargo. This must also contribute to the trend towards local manufacture much nearer the points of consumption. Modern catalytic reforming plants commonly have a throughput capacity of the order of 500,000 tons per year, and this brings substantial economies of scale.

No developing country could contemplate building on such a scale, but it would not be surprising if some plants in the 100,000/200,000 ton range were built within the next ten years in such areas.

Finally, to return, as it were, to the point of origin, i.e., the coal-based production of benzene, this is of course linked with the manufacture of steel. This is still growing a world-wide basis, although once again it is the newer industrial countries such as Australia, India, Japan and South Africa that new steel production capacity is increasing substantially.

Thus we can foresee growth in every way; the main routes of consumption as a chemical and as a fuel, and the main sources of supply, petroleum- and coal-based alike. Of these two, the petroleum-based products are likely to be the fastest growing and in the long run coal, once the only source of benzene, will be unlikely to contribute more than about 5 per cent of the world's needs of this very important raw material.

REFERENCES

[1] Taylor, F. S., *Endeavour*, 1942, **1**, 67.
[2] Faraday, M., *Trans. Roy. Soc.* (*London*), 1825, **115**, 440.
[3] Mansfield, C. B., *J. Chem. Soc.*, 1849, **1**, 244.
[4] Lunge, G., '*Coal Tar and Ammonia*' (Gurney and Jackson, London, 1916), p. 224.
[5] Ward, E. R., 'Articles on the Synthetic Perfume Industry', *Chem. and Ind.*, 1961, 1663 and 1966, 815.
[6] Davis, G. E., *J. Soc. Chem. Ind.*, 1883, **2**, 515.
[7] Whittaker, C. M., *J. Soc. Dyers and Colourists*, 1956, **72**, 557.
[8] Claxton, G., '*Benzoles, Production and Uses*' (National Benzole and Allied Products Assoc., 1961), p. 749.

[9] Mitscherlich, E., *Ann. Physik*, 1833, **29**, 231.
[10] Kekulé, A., *Bull. Soc. chim.*, 1865, **3**, 98.
[11] Dickerman, W. C., Jr., *J. Inst. Pet.*, 1953, **39**, 7651.
[12] Zelinski, N. D., *Ber. deut. chem. Ges.*, 1911, **44**, 3121.
[13] Logwinuk, A. K. *et al.*, '*Upgrading Coke Oven Aromatics*' (Industrial and Engineering Chemistry, April 1964), p. 20 *et seq.*
[14] *J. Pharm.*, 1831, **17**, 338.
[15] Hofmann, A. W., *Ann.*, 1845, **53**, 289 and 311.
[16] Claxton, G., '*Benzoles, Production and Uses*' (National Benzole and Allied Products Assoc., 1961), p. 773.
[17] *The Motor*, 17 February 1914.
[18] Claxton, G., '*Benzoles, Production and Uses*' (National Benzole and Allied Products Assoc., 1961), p. 680.
[19] Richards, R. B., *Chem. and Ind.*, 21 November 1970, p. 1491.

CHAPTER 2

BENZENE FROM COAL

by K. H. V. French

2.1. INTRODUCTION

WHEN coal is carbonised, the main products are coke, gas and tar, but, in addition, between 0·5 and 2 gallons of benzene are produced per ton of coal. The exact quantity will depend primarily on the nature of the coal, and, to some extent, on the type of carbonising plant. The bulk of coal carbonisation in the world today is performed in coke ovens for the production of metallurgical coke. Because the temperature of carbonisation is high, few non-aromatic hydrocarbons are produced and the problem of refining the aromatic hydrocarbons is reduced to a minimum. The aromatic hydrocarbons are extracted from the gas as crude benzole and only a minor quantity of benzene is produced by the distillation of coal tar. The extraction process usually involves scrubbing the gas with oil and distilling the oil to recover the crude benzole. This product contains approximately 60 per cent benzene, together with toluene, xylenes, sulphur compounds and small quantities of saturated and unsaturated non-aromatic hydrocarbons. The refining of the crude benzole involves chemical or physical treatment to remove these impurities, and distillation to separate the benzene from the other hydrocarbons.

The quantities of benzene produced from coal carbonisation in 1970 are shown in Table II.1. Given modern refining equipment the quality of coal-based benzene is indistinguishable from that produced from petroleum.

TABLE II.1

Production of Benzene from Coal 1972

	Tons	
Austria	15,000	(nil)
Belgium	40,000	—
France	81,000	(302,000)
West Germany	262,000	(600,000)
Italy	22,000	(518,000)
Spain	6,000	(100,000)
United Kingdom	117,400	(556,000)
Japan	269,000	(1,588,000)
United States	268,000	(3,820,000)

The values in brackets are the corresponding production from petroleum.

FIG. 2.1 *Horizontal Retorts Charging Side*

2.2. COAL CARBONISATION PROCESSES

The carbonisation of coal involves heating coal, in the absence of air, in a suitable plant. During the process, tar, gas and water are evolved leaving a residue of coke. Today the coke is the most important product and the plant and conditions will be dependent upon the end use of the coke. Until 1955, in the United Kingdom, the quantity of coal carbonised primarily for gas production was of the same order as that carbonised for coke. With the introduction of oil-gas processes and, more important, the discovery and utilisation of natural gas from the North Sea, carbonisation of coal for gas manufacture has ceased in the United Kingdom. Carbonising processes which are designed for gas making will be considered, therefore, only briefly.

2.2.1. Carbonisation for Gas Production

2.2.1.1. Horizontal retorts The coal is heated in horizontal refractory D shaped tubes 18–20 ft. long, 18–25 in. wide and 13–16 in. deep fitted with doors at each end. (See Fig. 2.1.) A number of these retorts is arranged in a setting and gas heated so that the coke eventually reaches a temperature of 1000–1050°C. Coal charging and coke discharging are usually by machine and the gas offtake is from the ends of the retorts. The intermittent nature of the process and the small capacity of each retort (usually less than 1 ton) means the labour costs are high and the process is generally unattractive from an economic point of view. The crude benzole recovered from the gas and tar produced by this process is, however, of high grade with high benzene and low non-aromatic hydrocarbon content. The composition of typical horizontal retort crude benzoles is shown in Table II.2. The lower non-aromatic hydrocarbon content is due to the high temperature of carbonisation and the intimate contact of the gases with the hot top surface of the retorts.

TABLE II.2

Composition of Crude Benzoles from Gas-Making Processes

	Horizontal retorts			*Intermittent vertical retorts*		*Continuous vertical retorts*	
Sp. gr. 15·5°C/15·5°C *% vol/vol*	*0·878*	*0·869*	*0·854*	*0·861*	*0·851*	*0·837*	*0·826*
Benzene	61	54	48	54	52	42	27
Toluene	14	16	13	17	19	13	13
Xylenes	5	7	7	7	7	6	10
Higher aromatics	7	8	7	7	12	7	7
Saturated non-aromatics	8	10	17	8	7	19	25
Olefines	5	5	8	7	3	13	18
Saturated non-aromatics in Benzene boiling range	1·5	2	6	5	3	12	14

2.2.1.2. Intermittent vertical retorts These retorts hold 1–5 tons of coal. They are usually about 20 ft. high, 4–10 ft. broad and 8–12 in. wide and tapered to facilitate discharge from the bottom. A typical intermittent vertical retort is shown diagrammatically in Fig. 2.2. Analyses of crude benzoles from this process are also given in Table II.2.

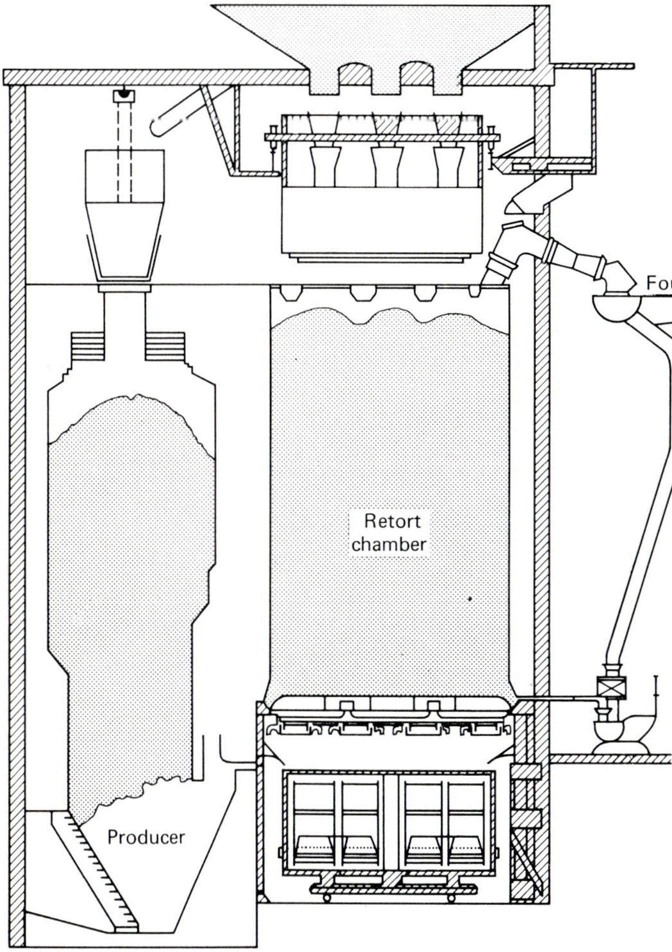

FIG. 2.2 *Intermittent Vertical Retort*

2.2.1.3. Continuous vertical retorts Fig. 2.3 shows a typical design of a unit of this type. The retorts are 25 ft. high and elliptical or rectangular in cross section 12–18 in. by 3–9 ft. and tapered from bottom (wide) to top. They are arranged in settings and

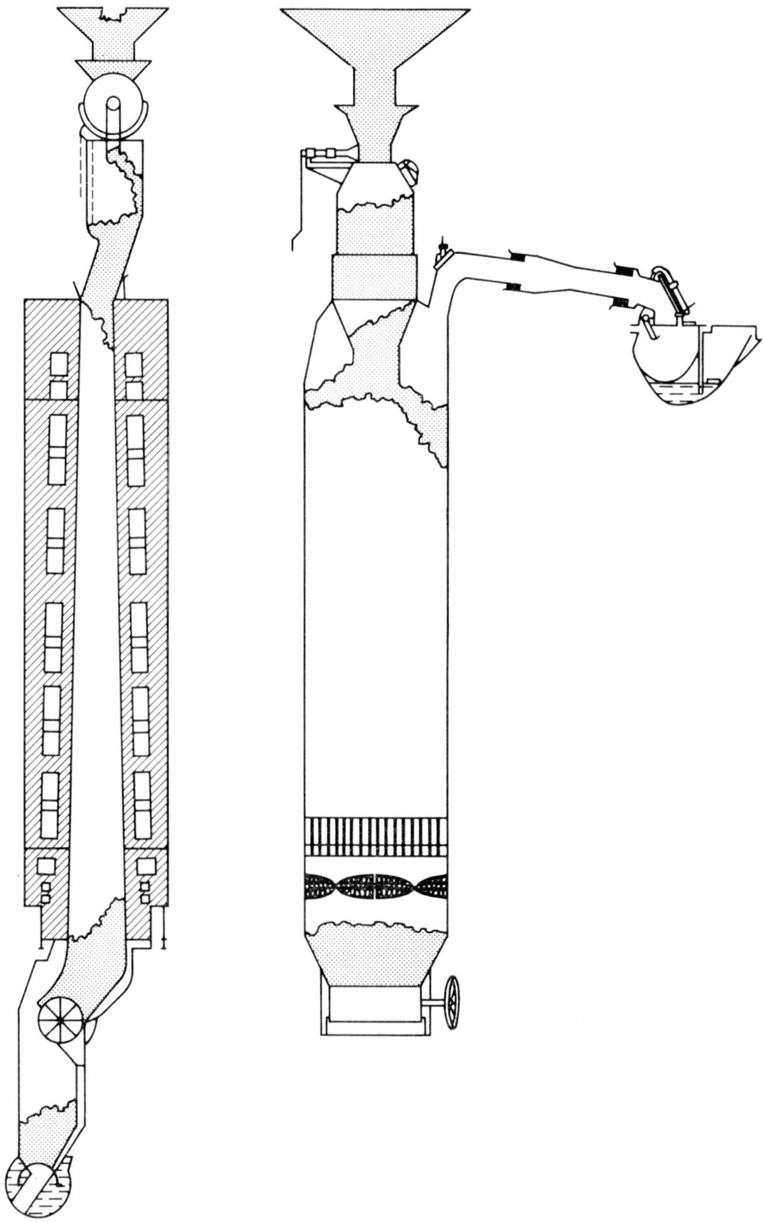

FIG. 2.3 *Continuous Vertical Retort*

gas heated. Coal falls from a hopper into the top of the retort at a speed set by the rate of mechanical withdrawal of coke from the bottom. The gases do not pass through the hottest part of the charge since they are removed at the cool top end of the retort. For this reason less cracking occurs and the benzole produced has a much lower benzene content and a higher non-aromatic hydrocarbon content than benzole produced by the other processes which have been described. Typical analyses are again given in Table II.2.

2.2.2. Carbonisation for Domestic Solid Fuels

The coke produced by the three gas-making processes described is suitable for domestic use. Coke ovens may also be operated under conditions which yield coke sufficiently active for domestic purposes, but the crude benzole produced will contain more non-aromatic hydrocarbons than that produced under the conditions required for blast furnace coke.

In the low temperature carbonisation process for the production of smokeless solid fuels, coal is carbonised at temperatures in the range 450–650°C. At these temperatures little cracking occurs and the light oil which can be extracted from the gas evolved is of no value for the production of benzene.

FIG. 2.4 *Coke-Oven Battery—Pusher Side*

2.2.3. Coke Oven Practice

A coke-oven is a rectangular chamber up to 40 ft. long, 14–20 ft. high and 14–18 in. wide. They are constructed in silica or special refractory bricks and are arranged side by side in long batteries of perhaps fifty ovens. Figs. 2.4 and 2.5 are photographs of a typical coke oven battery. The heating flues are in the walls between the ovens. The ends of each oven are closed by cast iron or steel doors having a firebrick lining. Coal, crushed to a convenient size is charged into the oven through filling holes in the oven roof. On completion of the carbonisation, the oven doors are removed and the coke is discharged on one side of the battery by means of a ram which enters the oven on the opposite side. The coke is collected in a coke car, in which it is conveyed to the water quencher.

FIG. 2.5 *Coke-Oven Battery—Coke Side*

During the carbonising period, the coal is heated to about 1000°C. Gas and tar vapours leave the oven through ascension pipes fitted at the top of the oven at one or both ends. The ascension pipes from all the ovens are connected to a collecting main from which the gas is conveyed for cooling and purification.

Each oven is charged with 10–20 tons of coal and carbonisation is completed in 14 to 25 hours. The exact temperature and time of carbonisation will depend on the type of coke being produced. Coke for use in blast-furnaces requires the highest temperatures and shortest carbonising times. Foundry coke requires lower temperatures and long carbonising times. The conditions required for domestic coke are intermediate between these. The composition of the crude benzoles produced reflect these differences in carbonising conditions as shown in Table II.3.

TABLE II.3

Composition of Coke Oven Crude Benzoles

Coke oven producing	*Benzene % wt/wt*	*Toluene % wt/wt*	*Non-aromatics in Benzene boiling range % wt/wt*
Blast-furnace coke	60–70	15–20	0·0–0·8
Domestic coke	60–70	10–20	0·4–1·25
Foundry coke	45–55	20–25	1·0–1·4

TABLE II.4

Effect of Carbonising Temperatures on Benzole Quality

	Lower temperature	*Higher temperature*
Inspection data on crude benzole		
Specific gravity 15·5°/15·5°C	0·872	0·881
Total sulphur % wt/wt	0·420	0·440
Sulphur as CS_2 % wt/wt	0·105	0·100
Inspection data on fraction (drop to 100°C) from a distillation through a 15-plate column		
% wt/wt of crude benzole	59·9	59·5
Specific gravity	0·872	0·875
Bromine number	10·8	12·1
*Cyclo*pentadiene % wt/wt	1·9	1·6
GLC analysis—% wt/wt of drop to 100°C fraction		
iso- and *n*-pentane	0·23	0·16
2,2-dimethylbutane	0·36	0·22
2-methylpentane	0·28	0·13
n-hexane	0·65	0·34
*2,4-dimethylpentane	0·04	0·04
*methyl*cyclo*pentane	0·24	0·18
*3-methylhexane	0·30	0·12
**cyclo*hexane	0·80	0·45
**n*-heptane	0·60	0·46
**cyclo*hexane	0·04	0·00
methyl*cyclo*hexane	0·21	0·08
benzene	92·1	93·2
toluene	1·15	2·2

* Undesirable non-aromatic hydrocarbons for benzene production.

Although the differences in the quantities of non-aromatic hydrocarbons shown in Table II.3 may appear to be small, they are paraffins and naphthenes that cannot be separated from benzene by normal distillation. Some of the non-aromatics form azeotropes with benzene, others have abnormal vapour/liquid equilibrium characteristics with benzene when the concentration of the non-aromatic hydrocarbon is low. The importance of the concentration of these compounds is such that this factor may dictate the type of refining process that is required.

Whilst the differences in composition shown in Table II.3 are mainly due to the different carbonising temperatures, the results did not come from a single coking plant and the design of the coke ovens could have influenced the benzole composition.

At one British coking unit a deliberate change in carbonising conditions was made which resulted in an increase of 100°C in the temperature in the free space above the coke oven charge. The effect of this change in carbonising conditions is shown in Table II.4. It may be seen that the increase in carbonising temperature has reduced substantially all the non-aromatic hydrocarbons.

There is little evidence that the type of coal carbonised has a significant effect on the hydrocarbon composition of the crude benzole. The nature of the coal, however, has a profound effect on the quantity and relative proportions of sulphur compounds present in the crude benzole. The main sulphur compounds occurring in coke oven crude benzoles produced from four different British coals are presented in Table II.5. These differences must be considered when selecting a refining procedure.

TABLE II.5

Comparison of Sulphur Compounds Occurring in Crude Benzole Fractions Below 160°C

	Source of coal			
Sulphur occurring as	*Glamorgan* % wt/wt	*Monmouth* % wt/wt	*S. Yorks* % wt/wt	*S. Yorks* % wt/wt
Carbon disulphide	0·035	0·279	0·474	0·372
Thiophen	0·040	0·109	0·364	0·350
Methylthiophens	0·019	0·025	0·077	0·181
Ethyl- and dimethyl-thiophens	0·006	0·004	0·038	0·084
Propylthiophens	0·000	0·001	0·003	0·010
Sulphides	0·011	0·000	0·014	0·017
Disulphides	0·016	0·010	0·016	0·072
Total	0·127	0·428	0·986	1·086

2.3. BENZOLE RECOVERY

2.3.1. Introduction

The theory and practice of benzole recovery from coal gas has been considered in detail by Claxton who also provides a bibliography of over 300 references on the subject.[1] Material has been freely drawn from that source which should be consulted if greater detail is required.

No matter what process is used for benzole recovery, some preliminary treatment of the gas is essential. It must be cooled to at least 20°C and freed from tar, tar fog and ammonia. The naphthalene content should be reduced and there are considerable advantages in removing the hydrogen sulphide and hydrogen cyanide before recovering the benzole.

As the gas passes from the carbonising plant through the raw gas main, its temperature will fall to about 80°C. A primary indirect water cooler will then lower the gas temperature to about 35°C before it enters an exhauster. After the exhauster, the gas will pass through a tar extractor, a direct cooler, a naphthalene scrubber and an ammonia scrubber before reaching the benzole recovery plant at a temperature of about 20°C. An electrostatic detarrer to remove tar fog and a unit for hydrogen sulphide removal may also be included.

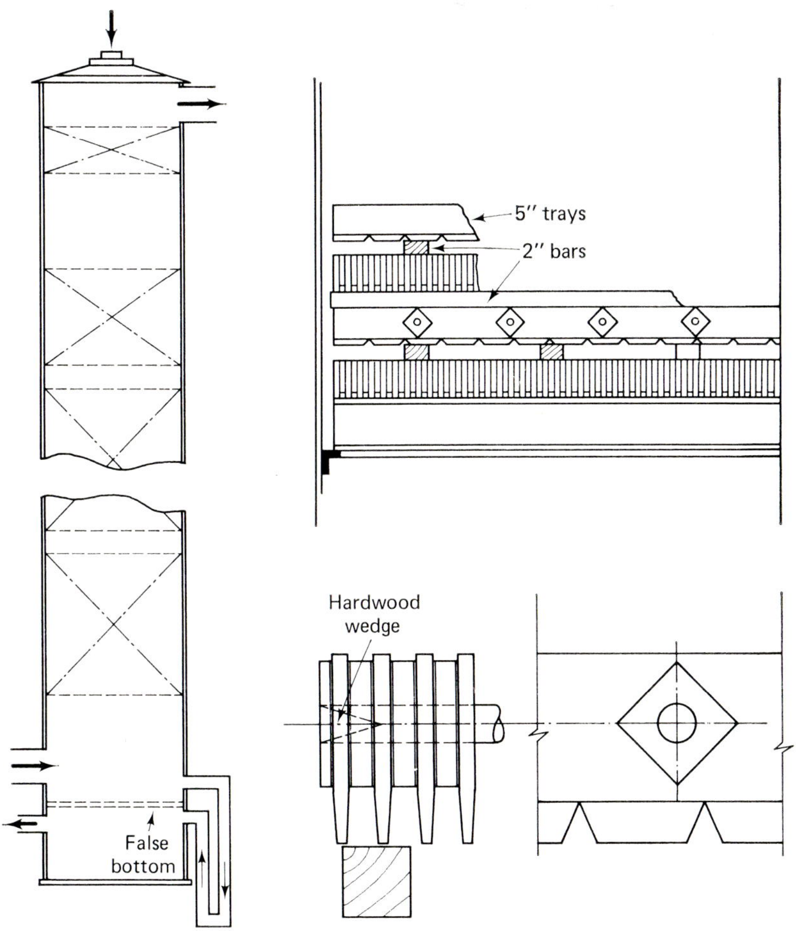

FIG. 2.6 *Board-Packed Tower and Tower Packing*

2.3.2. Oil Washing Process

On leaving the cooling and purification system, the gas passes to a scrubber where it is brought into intimate contact with a high-boiling oil which dissolves the greater part of the benzole from the gas. The benzolised oil passes to a still in which the benzole is removed by a continuous steam distillation process. The oil, which has been freed from benzole, is cooled and recycled to the scrubber.

The commonest type of benzole scrubber is a board-packed tower (Fig. 2.6). The operation is counter-current, the ascending gas meeting the descending flow of wash oil which presents a large surface area as it flows over trays of boards arranged

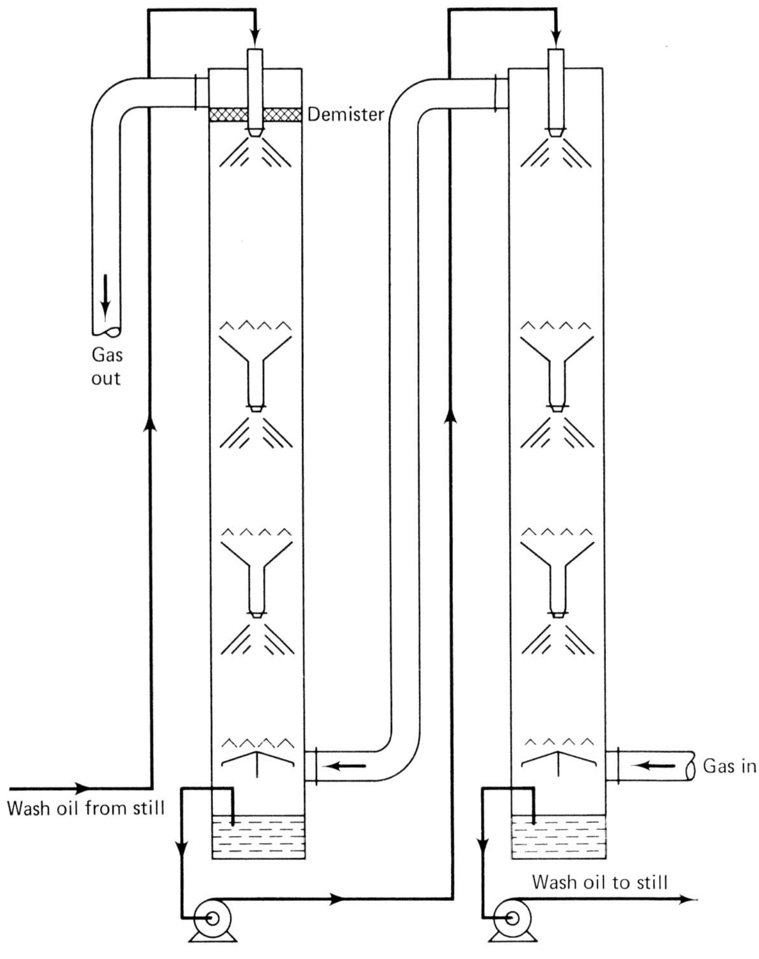

FIG. 2.7 *Willputte Benzole Scrubbers. Direct Spray Type*

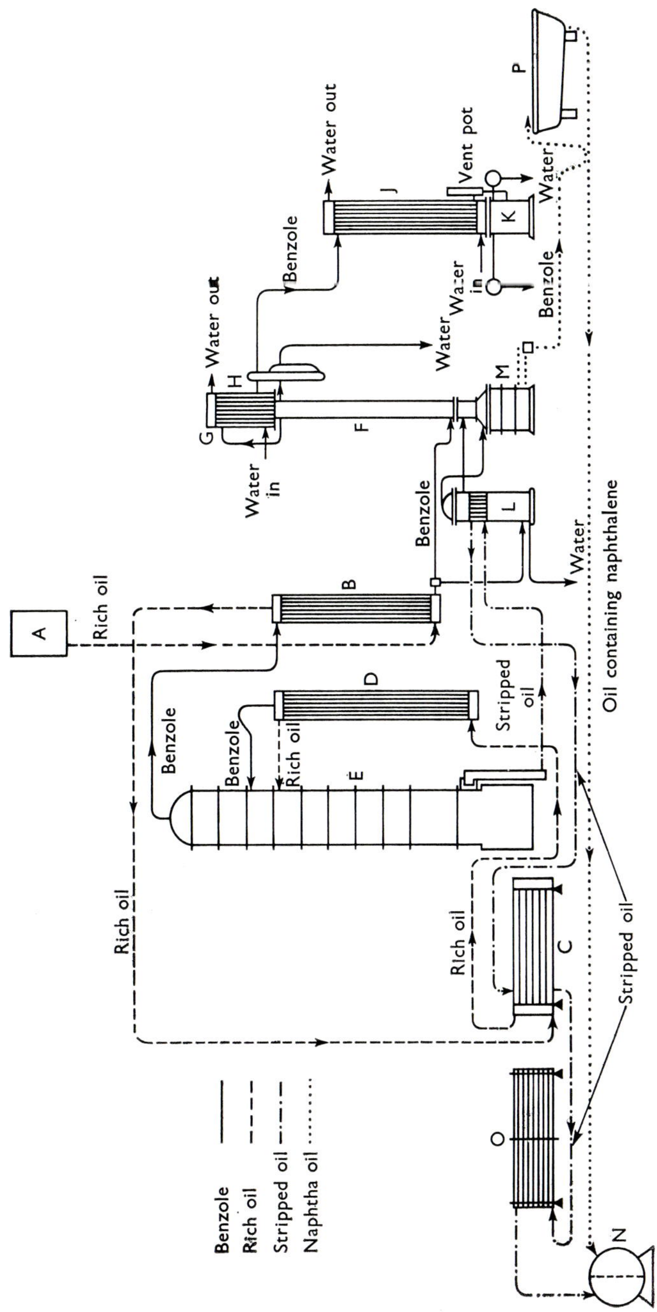

FIG. 2.8 *Woodhall-Duckham. Crude Benzole Plant*

on edge as shown in the figure. Alternate layers of boards are at right angles and redistribution devices are fitted at various levels in the tower. Many other packings have been used successfully in tower scrubbers—the essential features of a good packing are large surface area, low back pressure, good redistribution of oil and freedom from sludge hold-up.

Largely because of their low back pressure characteristics, towers without any packing are becoming popular. A tower of this type is shown in Fig. 2.7. The oil is sprayed into the top of the tower and collected and resprayed at intervals down the tower. The respraying stages may involve additional pumping externally to the tower or it may be achieved by gravity within the tower. Highly efficient self-cleaning jets are available for spraying the oil.

A number of scrubbers in series will usually be required to reduce the benzole in the gas to an acceptable level. The rich oil leaving the final scrubber will contain about 3 per cent of benzole and this must be recovered from the oil. A typical benzole recovery plant is shown diagrammatically in Fig. 2.8. The benzolised (rich) oil from the scrubbers passes to the buffer tank A and from thence through heat exchangers B and C, and steam preheater D. It then passes to the still E, where the benzole is distilled out of the hot benzolised oil by direct injection of steam. Some fractionation of the crude benzole is obtained in column F. The benzole vapours from the top of this column are condensed and taken to storage; a naphthalene rich oil is drawn from the bottom of the column F. The debenzolised oil from the still is cooled prior to its return to the scrubbers.

2.3.3. Benzole-Absorption Oils

The oils used for the benzole recovery process are of two types, petroleum oils and creosote oils. Properties of typical oils of each type are given in Table II.6.

TABLE II.6

Properties of Benzole Absorption Oils

	Petroleum oils		*Creosote oils*	
Specific gravity, 60°F	0·86	0·86	1·06	1·06
Distillation °C				
Drop	280	161	230	—
5%	295	220	—	235
95%	355	355	320	300
Dry	361	—	—	—
Viscosity, Redwood No. 1, sec at 70°F	50	40	40	39
Specific heat, Btu/lb/°C	0·83	0·83	0·61	0·61
Absorptive capacity g per 100 g solution	2·8	3·3	3·9	4·1

These physical properties indicate the suitability of the oil for benzole recovery and will be considered individually.

2.3.3.1. *Specific gravity* Emulsification of the oil with water can create serious problems and, if the oil has a specific gravity well removed from that of water, separation is rapid and the danger of emulsification is minimised. The pumping costs will be higher with higher specific gravity.

2.3.3.2. *Distillation* The distillation range is an indication of the volatility of the oil. Too great a volatility will result in excessive losses of oil to the gas in the scrubbing stage.

In the distillation stage a sharp separation between benzole and wash oil is desirable. For this reason the drop point of the wash oil should be preferably above 200°C.

A low molecular weight results in desirable viscosity and absorptive capacity characteristics, but undesirable volatility characteristics. The choice of boiling range must thus be a compromise.

2.3.3.3. *Viscosity* The viscosity should be as low as possible to minimise pumping costs.

2.3.3.4. *Specific heat* Because the overall process involves heating and cooling of the oil, a low specific heat is desirable to minimise heat requirements.

2.3.3.5. *Absorptive capacity* The absorptive capacity of an oil for benzene may be expressed as that concentration of benzene in the oil which has a partial pressure of 6·5 mm Hg. at 20°C. It is thus a measure of the relative amounts of benzene that different oils will absorb. The higher the absorptive capacity, the smaller the quantity of oil necessary to scrub a given volume of gas. Because the quantity of oil is thus reduced, there is a similar reduction in the size of plant and in the quantities of steam, cooling water and power. A low molecular weight results in the oil having a high absorptive capacity, but the minimum molecular weight will be restricted by volatility considerations.

From a knowledge of these properties, a choice of a benzole absorption oil can be made, but in practice other important factors are involved. Whilst it is true that the absorptive capacity of fresh petroleum wash oil is lower than that of a creosote wash oil, this difference becomes less as the oils age. The absorptive capacity of the petroleum oil rises as high boiling aromatic materials from the gas accumulate in it, whilst that of the creosote falls due to loss of the more volatile (lower molecular weight) constituents.

In use, petroleum wash oils tend to deposit sludges whilst creosote wash oils tend to thicken. These probably have the same cause, namely, the accumulation of polymerised material in the oil. This material is less soluble in the petroleum oil and therefore deposits as a sludge, but remains in solution in the creosote and increases its viscosity. Whilst the selection of the oils may reduce this problem, the real solution lies in an oil regeneration unit which removes the sludge-forming material from the oil.

2.3.4. Adsorption Processes

Adsorption processes for the recovery of benzole have been in use at a number of plants in the United Kingdom, Germany, France and the Netherlands. All were based on active carbon and none is still in use. The process is suitable for small production units providing the gas is purified before benzole recovery. Claxton[1] has included a detailed consideration of the process and alternative adsorbents.

2.3.5. The Economics of Benzole Recovery

The reasons for recovery of benzole from gas may be:

(1) to purify the gas and render it usable and saleable as a domestic or industrial fuel;

(2) to provide a source of revenue from the sale of benzene, toluene and xylene or motor benzole.

If benzole is not recovered from the gas, there are some limitations to the uses of the gas. The most serious is associated with the tendency of gas containing benzole to deposit carbon (soot) which generally renders the gas unsuitable for domestic purposes. The soot is deposited in particular in 'neat-flame' gas fires, on the elements of flame failure devices and in the heat exchangers of water heaters.[2] The practice of benzole recovery also reduces the organic sulphur content of the gas to a level that makes it acceptable for domestic purposes.

In the industrial uses of coal gas, both the aromatic content and the sulphur content can cause problems. The use of gas containing benzole for under-firing coke ovens can cause excessive carbon deposition.[3] 'Blooming' has also been reported when unstripped gas is used in sheet steel annealing ovens. Gardner and Kalushner have considered the use of plastic pipes for carrying gas and conclude that aromatic hydrocarbons are freely absorbed and cause softening of the pipes and loss of mechanical strength.[4] If benzole-rich gas is to be compressed, condensation of the benzole may occur. It may be calculated, for example, that if gas containing 2 gallons per 10,000 ft^3 is compressed to 100 p.s.i. at 40°F, approximately 1 gallon of benzole will be condensed.

On the credit side, leaving 1 gallon of benzole in 10,000 ft^3 of gas will result in a net increase in the calorific value of about 15 Btu. per ft^3 and an increase in volume of 40 ft^3.

When there was a ready domestic and industrial market for purified coke oven gas, benzole recovery was practiced as part of the purification of the gas and the products produced a useful revenue. When, however, in the United Kingdom and other European countries, natural gas became available, the markets for coke oven gas contracted and the economics of benzole recovery became less attractive and advantages and disadvantages of benzole recovery needed to be examined more closely.

Today in the United Kingdom it may be found that a benzole recovery plant exists alongside a coke oven unit, but there is no domestic market for coke oven gas. Two situations may then arise:

(a) The whole of the gas can be used on the plant (e.g. a steelworks) associated with the coke oven unit. If a demand exists for more fuel than that supplied by the whole of the coke oven gas, and benzolised gas is acceptable, then benzole left in the gas can be credited at fuel value. A comparison of this value with the realisation value of benzene, toluene and xylene etc., after due allowance for recovery and refining costs allows, a decision to be made on the economic advantage (or otherwise) in recovering benzole.

(b) Even when benzole is recovered, only part of the coke oven gas can be used and the surplus is flared-off, i.e. there is no demand for extra fuel. In these circumstances the value of benzole left in the gas will be zero or even negative if there are flaring charges. Recovery and refining of benzole then becomes desirable providing the realisation value of the products exceeds the process costs.

The value of the products and charges applicable for recovery and refining vary so widely from plant to plant and from year to year that it is impossible to give a generalised answer to the question 'is it profitable to recover and refine benzole'.

The production figures given in Table II.1 make it clear that in 1972 the majority of coke oven operators took the view that benzole recovery was worth while.

2.4. REFINING CRUDE BENZOLE TO SYNTHESIS BENZENE

The purity of synthesis grade benzene must be such that it contains only a few p.p.m. of sulphur compounds, 1000–3000 p.p.m. of aliphatic hydrocarbons and perhaps 50 p.p.m. of toluene. Specifications for the finished product are considered in detail in Chapter V. From what has already been said of the composition of crude benzoles, it is clear that the chief function of the refining process must be the virtual elimination of all sulphur compounds and the separation of the benzene from the bulk of the other hydrocarbons in the crude benzole.

2.4.1. Sulphur Removal

The sulphur compounds occurring in crude benzole include thiophen, carbon disulphide, sulphides, disulphides, mercaptans, hydrogen sulphide and free sulphur.

Distillation is effective in removing those sulphur compounds with boiling points remote from that of benzene; treatment with sulphuric acid will remove thiophen, sulphides and disulphides; hydrogenation will convert all sulphur compounds to hydrogen sulphide which can be eliminated subsequently by physical or chemical means; treatment with selective reagents will remove specific groups of sulphur compounds.

The various methods of sulphur removal will now be considered in detail.

A combination of defronting, distillation and thiophen removal by acid treatment can yield synthesis grade benzene with no other processing providing the non-aromatic hydrocarbon content of the crude benzole is low.

2.4.1.1. Carbon disulphide removal The low boiling point of carbon disulphide (46·25 °C) compared with that of benzene (80·1 °C) makes the removal of this sulphur compound by distillation a simple matter. Reasonably efficient equipment is needed, however, if the concentration of carbon disulphide is to be reduced to below 1 p.p.m. A continuous distillation unit is normally used for the process which is known as defronting. Low-boiling hydrocarbons present in the crude benzole and, of particular importance, cyclopentadiene, are also removed in the process.

A number of plants are available for this operation. The essential features are a fractionating column so designed that its performance is not impaired by the presence of water, and a partial condenser. There is invariably direct injection of steam into the base of the column. The column may be fitted with various types of bubble-cap trays or it may be packed with suitable distillation rings. Partial condensation is achieved by control of the temperature of the condenser cooling water.

An effective defronting unit designed by the APV Company Ltd. is shown diagrammatically in Fig. 2.9. The cold crude benzole is pumped to the feed plate at the top of the column and open steam is injected at the base of the column. This steam rises countercurrent to the descending benzole and strips the carbon disulphide and light hydrocarbons from the benzole. The vapours leaving the top of the column pass to a direct condenser which is fitted with a number of plates. The temperature of the water supplied to this condenser is carefully controlled so that the carbon disulphide leaves the top of the condenser as vapour, but the quantity of benzene associated with it is very small. The water and condensed hydrocarbons pass from the base of the condenser to a decanter as shown.

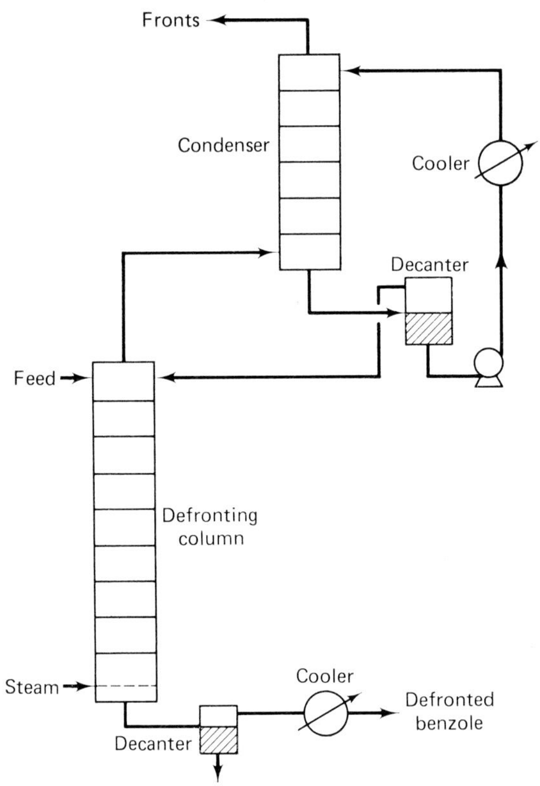

FIG. 2.9 *APV Continuous Defronting Column*

2.4.1.2. Thiophen removal The boiling point of thiophen is too close to that of benzene to permit substantial reduction of thiophen content by distillation. The traditional method for the removal of thiophen together with some other sulphur compounds from a crude benzole or benzene fraction is by treatment with sulphuric acid. Unsaturated hydrocarbons are also removed from the benzene in this process.

One of the more recent plant designs for continuous acid washing is that by APV Company Ltd. A flow diagram is given in Fig. 2.10. The acid washing unit is a pulsed column having slotted plates. Dispersion of the acid in the benzene is achieved by the action of the pulse pump which forces the liquid through the slots in the plates. A controlled quantity of 100 per cent sulphuric acid is pumped to the top of the washer and benzene is pumped into the lower part of the column via the pulse line. Settling zones at the top and bottom of the washer allow disengagement of the two phases. The rate of spend-acid run-off from the base of the column is governed by an interface controller. The washed benzene passes to neutralising and water washing vessels. Feedstocks containing 1800–2000 p.p.m. of sulphur as thiophen yield benzene containing less than 0·1 p.p.m. sulphur as thiophen with an acid consumption of 2–2·5 per cent v/v of 100 per cent sulphuric acid.

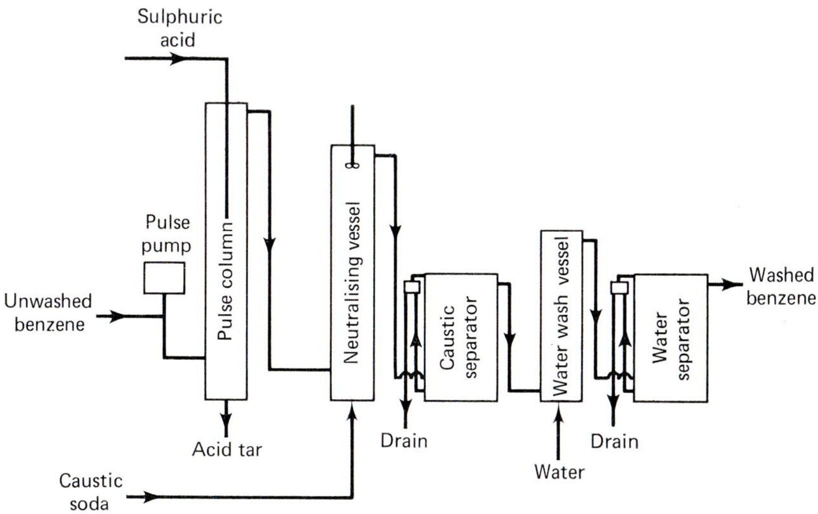

FIG. 2.10 *Continuous Acid Washing Plant*

Even in batch acid washers, benzene can be prepared with 0·1 to 1·0 p.p.m. sulphur as thiophen. This degree of thiophen removal became possible once the 'reversal' phenomenon was appreciated.[5] The reaction between thiophen and sulphuric acid is reversible:

$$H_2SO_4+C_4H_4S \rightleftharpoons C_4H_4S.SO_3H+H_2O$$

The position of the equilibrium will clearly depend on the concentration of water in the acid phase. The quantity of thiophen present is relatively small and little water will be formed from its reaction with sulphuric acid. Unfortunately the benzene that is present also sulphonates and sufficient water is formed to upset the reversible reaction with thiophen. Some thiophen is, in fact, regenerated and returns to the benzene phase.

Thus, if benzene containing thiophen is washed batch-wise with concentrated sulphuric acid, the thiophen content of the benzene will fall to a certain point and then rise again due to the reversal effect. To overcome this effect in a batch process, multistage washes must be used—when a minimum is reached, the used acid is withdrawn and the process continued with fresh acid to the next minimum and so on until the thiophen is reduced to an acceptable level.

2.4.2. Hydrorefining

An alternative approach to sulphur removal is by selective hydrogenation under conditions that will convert all the sulphur compounds in the crude benzole to hydrogen sulphide, but will not cause hydrogenation of aromatic hydrocarbons. Hydrorefining processes are available from most chemical engineering companies. They depend upon vapour phase hydrogenation over a catalyst of the cobalt-molybdenum type, either hydrogen or coke oven gas being used as the reducing agent. Catalyst temperatures are usually in the range 300–450°C and pressures between 30 and 50 atmospheres.

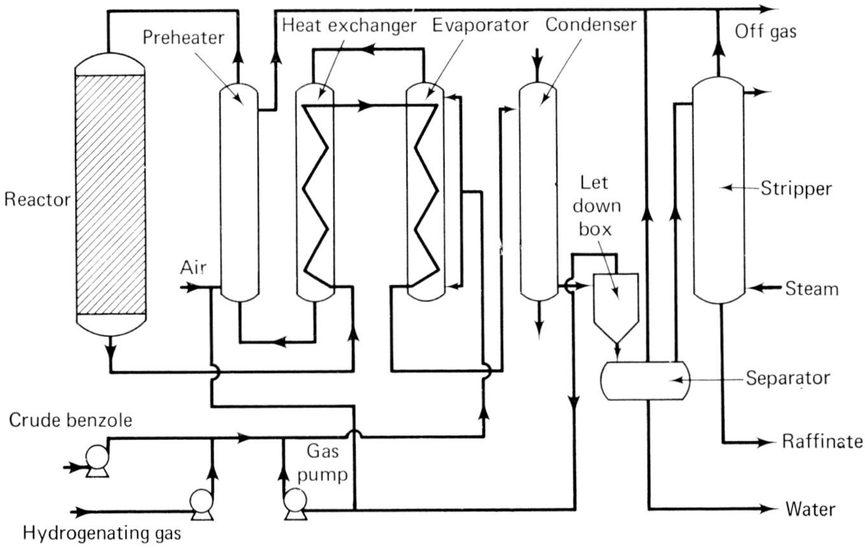

FIG. 2.11 *Lurgi Hydrorefining Process*

Two hydrorefining units were operating in the United Kingdom but they closed in 1970 and 1972. The last to close was a Lurgi plant operating on coke oven gas. The flow diagram is shown in Fig. 2.11. The crude benzole is vaporised into a stream of coke oven gas and the mixture is heated to the appropriate temperature before entering the reactor. In this vessel all sulphur compounds are reduced to hydrogen sulphide and unsaturated hydrocarbons are saturated. The process stream is cooled and the gases separated. The liquid product passes to a stripper where it is freed from hydrogen sulphide and then to a distillation unit for separation of benzene, toluene and xylene. Whilst hydrorefining processes of this type are capable of reducing the total sulphur content of benzene to less than 1 p.p.m. and of eliminating unsaturated hydrocarbons, problems may be created due to the formation of saturated non-aromatic hydrocarbons from the olefines.

2.4.3. Hydrodealkylation Processes

A major disadvantage of hydrorefining processes is the formation of saturated non-aromatic hydrocarbons that may prove difficult to separate from the benzene. This problem does not arise in hydrodealkylation processes which eliminate sulphur compounds and saturated and unsaturated hydrocarbons from a benzene rich feed in a single operation. The best-known processes of this type are the Houdry Litol Process[6] and the UOP Hydeal Process.[7] A simplified flow diagram of the Litol Process is shown in Fig. 2.12.

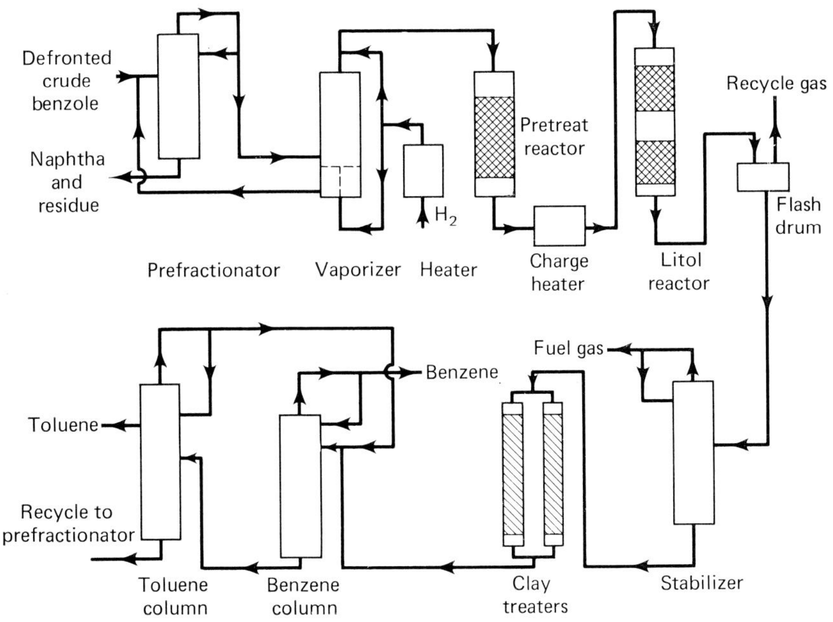

FIG. 2.12 *Simplified Litol Flow Diagram*

There are some limitations on the feedstock to the plant if it is to give trouble-free and economic operation:

(a) The feed should be defronted so that it is substantially free from carbon disulphide and cyclopentadiene. These compounds are hydrogen consumers and yield no useful products. The crude benzoles are often defronted at their production points and are satisfactory if they contain less than 200 mg/kg sulphur as carbon disulphide.

(b) The styrene and indene in the feed should not exceed 2·5 per cent. This is to avoid polymerisation troubles and a prefractionator can be provided that rejects these higher boiling materials.

(c) The total non-aromatic hydrocarbon content of the feed should not exceed 5 per cent. Non-aromatic contents above this value cause high reactor temperatures. Feedstocks containing up to 30 per cent non-aromatics can be treated by the similar Houdry Pyrotol process.

In the Litol Process the prepared feed is vaporised into a stream of hydrogen and passed into a pre-treatment reactor where olefines and other reactive materials are hydrogenated. The stream is then heated further before entering the Litol reactor. Here the main reactions occurring are typified by the following:

Hydrodesulphurisation

$$C_4H_4S + 4H_2 \longrightarrow C_4H_{10} + H_2S$$

Thiophen — Butane

$$CS_2 + 4H_2 \longrightarrow CH_4 + 2H_2S$$

Carbon disulphide — Methane

Removal of Paraffins and Naphthenes

$$C_6H_{12} + 3H_2 \longrightarrow 3C_2H_6$$

Cyclohexane — Ethane

$$CH_3CH(CH_3).CH_2.CH(CH_3).CH_3 + H_2 \longrightarrow C_3H_8 + C_4H_{10}$$

2,4-dimethylpentane — Propane — Butane

Removal of Unsaturated Hydrocarbons

$$CH_2 = CH.CH_2.CH_2.CH_2.CH_3 \longrightarrow C_6H_{14} \longrightarrow C_3H_8 \text{ etc.}$$

Hexene-1 — *n*-Hexane

$$C_6H_{10} \longrightarrow C_6H_6 + 2H_2$$

Cyclohexene — Benzene

Hydrodealkylation

$$C_7H_8 + H_2 \longrightarrow C_6H_6 + CH_4$$

Toluene — Benzene Methane

The product, consisting of benzene, some toluene and xylenes, gaseous and light aliphatic hydrocarbons and hydrogen sulphide passes to a stabiliser tower where the light material is removed. The last traces of unsaturated materials in the aromatics are eliminated by clay treaters prior to the distillation unit from which benzene and toluene products are obtained. The higher boiling material is recycled.

Under typical operational conditions 50 per cent of the toluene and 85 per cent of the xylenes in the feed is dealkylated. Thus the yield of benzene is substantially greater than that present in the feed. Typical analyses of benzene and toluene produced by the Houdry Litol Process are given in Table II.7.

TABLE II.7

Typical Analyses

Benzene		
Acid wash test		0·05
S as thiophen	mg/kg	< 0·005
S as CS_2	mg/kg	< 0·06
Free S		nil
Crystallising point	°C	5·50
Toluene	mg/kg	80
n-Hexane	mg/kg	20
Cyclohexane	mg/kg	280
Methylcyclopentane	mg/kg	220
Toluene		
Acid wash test		0·05
Total sulphur (Granatelli)	mg/kg	< 0·1
Benzene	mg/kg	30
Aliphatic hydrocarbons	mg/kg	10
C_8 aromatics	mg/kg	20

The hydrogenating gas used for the process should contain at least 78 per cent hydrogen and must contain less than 10 p.p.m. carbon dioxide and carbon monoxide.

2.4.4. Removal of Non-aromatic Hydrocarbons

Reference has already been made to the undesirability of non-aromatic hydrocarbons associated with benzene. A number of these hydrocarbons, and in particular those listed in Table II.8 form azeotropes with benzene or their vapour/liquid equilibria with benzene is such that their concentration cannot be reduced sufficiently by normal distillation.

TABLE II.8

Undesirable Non-aromatic Hydrocarbons that Interfere with Pure Benzene Production

		Azeotrope with benzene	
Hydrocarbon	*Boiling point* °C	*Boiling point* °C	*% wt/wt benzene*
Methylcyclopentane	72	71·4	11·7
2,4-dimethylpentane	80·6	75·2	48·3
Cyclohexane	80·6	77·4	49·7
3-methylhexane	91·85	—	—
n-heptane	98·4	no azeotrope	
Methylcyclohexane	100·95	no azeotrope	

No data are available on azeotropes between 3-methylhexane and benzene. If no azeotrope is formed, the 3-methylhexane would be removed with the benzene-toluene intermediates.

The first three hydrocarbons in Table II.8 can be removed by efficient separation of forerunnings. The shape of the vapour-liquid equilibrium curves for benzene/*n*-heptane and benzene/methylcyclohexane are such that great difficulty is experienced in removing the last of these compounds from benzene.

These undesirable saturated hydrocarbons may be present in the original crude benzole, particularly if it has been produced under milder carbonising conditions (see Tables II.3 and II.4). Their concentration will not be reduced by acid washing or distillation and more will be formed if the crude benzole is hydrorefined. The acceptable concentration of these hydrocarbons in the refined benzene is strictly limited by the crystallising clause in the specification.

Hydrodealkylation processes are most attractive because all the non-aromatic hydrocarbons are eliminated from the benzene whereas, with acid washing or hydrorefining processes an additional stage of refining is essential to remove the non-aromatics.

Processes which have been used for the separation of non-aromatic hydrocarbons from benzene are:

(1) pressure distillation.
(2) azeotropic distillation.
(3) extractive distillation.
(4) liquid/liquid extraction.
(5) crystallisation.

As these processes are also applicable to the production of benzene from petroleum sources where higher proportions of non-aromatics are to be separated (see Chapter 3), the present discussion will be limited to the consideration of the treatment of benzene fractions containing less than 5 per cent of non-aromatics, i.e. feed compositions typical of those from coal carbonisation sources.

2.4.4.1. Pressure distillation It has been found that an increase in the pressure of a distillation unit modifies the vapour/liquid equilibrium for aromatic/non-aromatic systems sufficiently to increase the separation that can be obtained. Heinrich Koppers GmbH have built plants at Osaka, Japan and Linz, Austria utilising this principle.[8] Both plants treat hydrorefined benzole and their feed rates are 1750 and 875 gal/h respectively. The benzene produced has a crystallising point of 5·4°C, but the toluene usually contains about 1 per cent of dimethylcyclohexane.

The unit consists of a series of continuous distillation columns. Column 1 is a defronter operating at atmospheric pressure, column 2 operates at about 6 atm and pure benzene is taken as the overhead fraction; *n*-heptane together with toluene leaves the column at its base. Column 3 is again at atmospheric pressure; the overhead is a benzene/toluene intermediate and contains 20 per cent of *n*-heptane. Toluene is taken as the overhead of column 4 which operates at about 5 atm. The paraffinic-rich materials produced as overheads in columns 1 and 2 may be passed to a reformer in which the non-aromatics are converted to gas. The aromatic portion is recycled.

The quantities of non-aromatics in the feed that can be handled successfully are limited and it appears that the highest grade toluene cannot be produced. These limitations no doubt account for the limited number of plants of this type that have been constructed.

2.4.4.2. Azeotropic distillation Azeotropic distillation is not a popular method for upgrading aromatic hydrocarbons. A plant has been operating, however, at the Spondon Works of British Celanese Ltd. since 1961.[9] It was built by the APV Co. Ltd. and is operated in conjunction with a continuous acid washing unit.

The feedstock is a hydrocarbon by-product from a naphtha cracking plant. It contains about 32 per cent benzene, 17 per cent toluene, 7 per cent xylenes and large quantities of saturated and unsaturated non-aromatic hydrocarbons. It is defronted and then separated into a benzene fraction, a toluene/xylene fraction and heavier materials in two continuous distillation columns. The benzene fraction passes to an azeotropic distillation column in which a 50 per cent acetone/water mixture is used as the entrainer. The benzene from the base of this column is acid-washed and then distilled in a blow-over unit. The product has a crystallising point of 5·5°C. The toluene/xylene fraction is treated similarly and then separated by a batch distillation.

The non-aromatics plus acetone are the overhead products from the azeotropic distillation columns. The non-aromatics are sprung by the addition of water and the water/acetone mixture is returned to the azeotropic columns.

The steam requirement is quoted as being 48 lb/gal of product. This high value is probably due, to some extent, to the high non-aromatic content of the feed, but even if the process were applied to a crude benzole containing only a few per cent of non-aromatic hydrocarbons, the steam costs are likely to be much higher than those for extractive distillation.

2.4.4.3. Extractive distillation In an extractive distillation (Fig. 2.13) a non-volatile solvent is introduced into the distillation column a few plates from the top of the column and above the feed point. The ratio of solvent to feed is in the range 2 : 1 to 5 : 1. The plates above the solvent feed point are merely to separate the solvent from the overhead product. In the remainder of the column the presence of the solvent on each tray modifies the relative volatilities of the aromatic and non-aromatic hydrocarbons so that an efficient separation can be obtained with only a small loss of aromatic hydrocarbon in the overhead fraction. The solvent and the aromatic hydrocarbon are separated in a second column and the solvent is recycled. By its very nature the process must be continuous.

The factors controlling the effectiveness of the process are the nature of the solvent, the ratio of solvent to feed, and the usual factors such as the column efficiency and the reflux ratio.

Although inexpensive solvents such as phenol and cresol may be used, materials such as sulpholane, N-methylpyrollidone (NMP) and N-formylmorpholene are so much more effective that their higher price is justified.

Three plants in the United Kingdom treat a benzene fraction using NMP as the extractive solvent. All three plants are based on a Lurgi design and are capable of producing benzene with a crystallising point of 5·5°C. Sulphur compounds are removed by acid treatment. In Germany an extractive distillation unit utilises N-formylmorpholene for the separation of non-aromatic hydrocarbons from a hydrorefined benzene fraction. The plant is designed by Heinrich Koppers GmbH of Essen and has a capacity of 170,000 tons per year. The feedstock to the hydrorefining stage is coke oven crude benzole and the benzene product normally has a crystallising point of 5·48°C although the plant is capable of producing material with a crystallising point of 5·53°C.

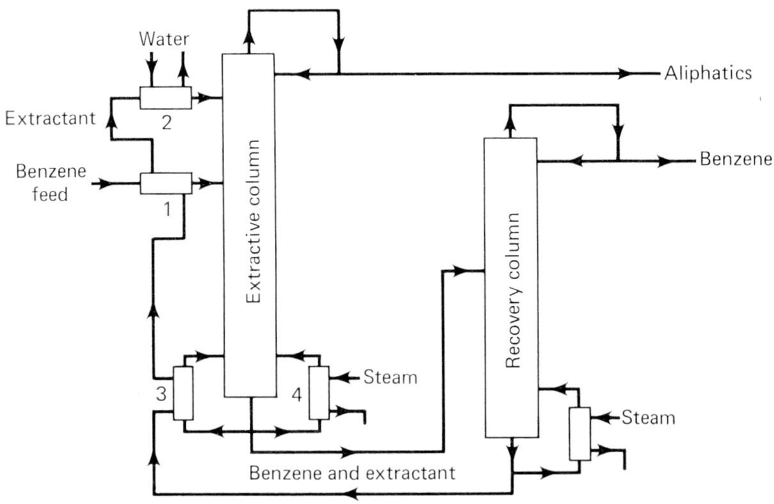

FIG. 2.13 *Extractive Distillation Unit*

2.4.4.4. Liquid/liquid extraction Liquid/liquid extraction processes, which are widely used for extracting aromatics from petroleum fractions, have not been applied to coke oven benzoles in Europe. The small scale of operation and the low non-aromatic content of the feed are the main reasons for the preference for extractive distillation. In the United States, however, the Udex process has been applied to hydrorefined crude benzole. This process is considered in detail in Chapter 3.

2.4.4.5. Crystallisation Crystallisation is a classical method of purification and the convenient crystallising point (5·53°C) of benzene should make processes based on this technique attractive. The aromatic and non-aromatic hydrocarbons associated with benzene all form eutectic systems with that hydrocarbon so that freezing produces crystals of pure benzene contaminated only by adherent or occluded mother-liquor. The product purity is determined by the efficiency of removal of the mother-liquor, rather than by inter-phase equilibria and is largely independent of the identity of the impurities.

Unfortunately, thiophen forms mixed crystals with benzene so that a number of fractional crystallisation stages would be required to effect a moderate separation. In the preparation of synthesis grade benzene from a coke oven crude benzene fraction a reduction of thiophen by a factor of about 20,000 is required which is unlikely to be obtained in any physical process. It may be concluded, therefore, that crystallisation processes will not provide adequate thiophen removal, but may be used in association with a desulphurisation process, to remove other impurities.

A crystallisation process was used for a time at the Carrington plant of Shell Chemicals (U K) Ltd. The feed was a benzene concentrate from pyrolysis gasoline. The process operated batch-wise and involved direct cooling of the feed with brine, separation of the benzene crystals by centrifuging and washing them to remove surface impurities.

Many other crystallisation processes have been proposed for the purification of benzene, but few have been operated beyond the pilot plant scale.

The Wilton Proabd Refiner which has proved successful in the purification of naphthalene has been offered in a modified form for benzene.[10] A benzene fraction is charged into a tank filled with finned tubes. A coolant is pumped through the tubes until all the benzene has crystallised. The temperature is then raised at a controlled rate and the thaw liquor allowed to drain. When the impurities have been drained off, the remaining benzene is melted and run off. Crystallising points up to 5·5 °C can be obtained.

Various crystallisers based on the principle described by Schildknecht[11, 12, 13] have been suggested for the purification of benzene. The principle is shown in Fig. 2.14.

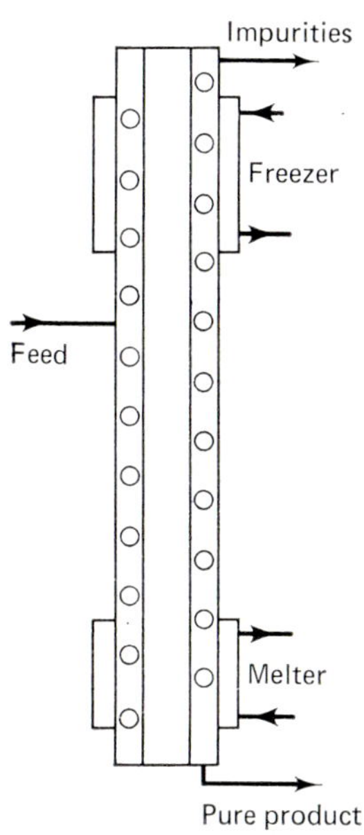

FIG. 2.14 *Rotating Spiral Crystalliser*

A close-fitting helix rotates in the annular space between the stationary core and the jacket. There is a cooling section at the top of the column and a melting section at the bottom. A benzene-rich feed is introduced near the centre of the column. Crystals formed in the cooled section are transported down the column by the spiral to be melted at the bottom where benzene is removed. Some benzene passes up the column thus washing the downward-moving crystals. The non-aromatic hydrocarbons remain in the liquid phase and their concentration increases to the top of the column where they are removed.

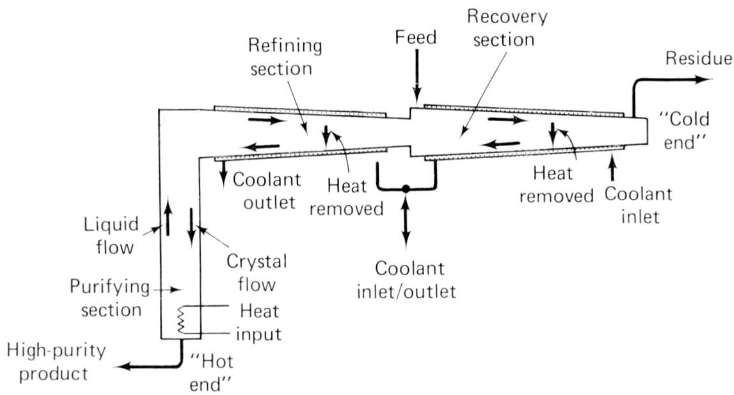

FIG. 2.15 *Typical Form of Brodie Purifier*

The Brodie Purifier[14] operates on a somewhat similar principle as may be seen from Fig. 2.15. The crystals are transported towards the high purity end by means of rotating blades and they are washed with the pure product as they pass counter current to it. Pilot plants runs on a feed containing 90·5 per cent wt/wt benzene, 8·42 per cent wt/wt toluene, 0·05 per cent wt/wt thiophen and 1·03 per cent wt/wt, other impurities yielded 99·97 per cent wt/wt purity benzene the only impurities being thiophen 0·01 per cent wt/wt and toluene 0·02 per cent wt/wt. Commercial plants of this type are in operation on products other than benzene.

2.5. FUTURE TRENDS

The demand for coke for iron making is likely to continue for several decades and, it may be assumed, the existing and new coke ovens will be the source of this coke. Changes in the design of coke ovens will occur and these will be directed towards a greater production of coke per oven per day. Most ovens in the UK and USA are 4–5 m high. Larger ovens (up to 7 m high and 50 ft long) are now being built and will operate at fast carbonising rates.[15] Similar yields of benzole per ton of coal may be expected from these plants and the quantity of non-aromatic hydrocarbons associated with the benzene will be low. Higher throughputs of coke may also be obtained by charging the coke ovens with dry or preheated coal. Reduction of the moisture content of the coal charged to the ovens from 10 per cent to 3 per cent can result in a 25 per cent increase in output for the same flue temperatures. The British Coke Research Association has investigated the effect of charging the coke ovens with dry coal heated to 300°C. and a full-scale plant is now operating in the United Kingdom. The coke output is increased by up to 50 per cent. Both these techniques result in a slightly higher yield of benzene per ton of coal and a substantial reduction in the non-aromatic hydrocarbons occurring in the benzene range. When preheated coal is used, the reduction in these undesirable hydrocarbons is so great (see Table II.9) that special refining processes to remove them become unnecessary.

TABLE II.9

Effect of Preheating Coal Charged to Coke Oven

Charge condition	*Normal charge*	*Preheated to 150°C*	*Preheated to 160°C*
Benzene % wt/wt	66	67	68
Crystallising point °C	5·38	5·52	5·52
Impurities: % wt/wt of benzene fraction:			
2,4-Dimethylpentane	0·01	below 0·01	below 0·01
Methylcyclopentane	0·04	below 0·01	below 0·01
3-Methylhexane	0·02	below 0·01	below 0·01
Cyclohexane	0·06	0·01	below 0·01
n-Heptane	0·10	0·01	below 0·01
Methylcyclohexane	0·04	0·01	below 0·01

Whilst these developments in the coking industry will tend to improve the quality of the feedstock for the production of benzene, future policy on recovery of benzole is difficult to forecast. Much will depend on the local competition from natural gas and oil and the outcome of investigations on underfiring with benzolised gas. Maloy[15] is of the opinion that the use of non-recovery ovens would be considerably less profitable than the installation of by-product ovens for coke making.

In a long term appraisal of coal as a source of benzene, consideration must be given to routes based on the hydrogenation of coal. During World War II coal hydrogenation plants for the production of gasoline were successfully operated in the UK, Germany and Japan. When petroleum gasoline became readily available after the war, the plants were closed for economic reasons. In 1968 the US Department of the Interior published a report on Chemical By-products from Coal[16] which included a detailed consideration of the availability of benzene by this route. The report concluded that up to 10 gallons of benzene per ton of coal could be obtained compared with a maximum of 2 gallons per ton by the coking route. Pulverised coal is hydrogenated to yield a synthetic crude which may then be subjected to hydrocracking and aromatics are extracted from the product. The higher aromatics can be dealkylated to increase the yield of benzene.

As the world reserves of coal are estimated to be sufficient for 1,600 years, whereas the corresponding figure for oil is less than 100 years, the day will surely come when coal is once again the main source of benzene.

REFERENCES

[1] Claxton, G., '*Benzoles, Production and Uses*' (National Benzole and Allied Products Assoc., London, 1961), pp. 157–276.
[2] Prigg, J. A. and Rooke, D. E., *Gas Council Research Communication*, 1962, GC.85.
[3] Woolhouse, T. G. (Dorman Long (Chemicals) Ltd.) (now British Steel Corporation), Private Communication.
[4] Gardner, W. E. and Kalushner, I., *J. Inst. Gas Engrs.*, 1962, **2**, 184.
[5] Hyslop, W. M. *et al.*, *Ind. Chem.*, 1961, **37**, 331.
[6] Lorz, W. *et al.*, *Erdol u. Kohle*, 1968, **21**, 610.
[7] Asselin, G. F. and Erickson, R. A., *Amer. Inst. Chem. Engrs.*, 46th National Meeting, 1962.
[8] Heinrich Koppers Gesellschaft, BP 949,046.
[9] Worrall, K. W. and Thwaites, J. M., *Brit. Chem. Eng.*, 1965, **10**, 158
[10] Molinari, J. G. D., *Ind. Chem.*, 1961, **37**, 323.
[11] Schildknecht, H. and Vetter, H., *Angew. Chem.*, 1961, **73**, 612.
[12] Schildknecht, H. and Maas, K., *Warme*, 1963, **67**, 121.
[13] Betts, W. D. *et al.*, *J. Appl. Chem.*, 1968, **17**, 180.
[14] Brodie, J. A., Institution of Engineers, Australia, *Mech. and Chem. Engr. Trans.*, May 1971, 37.
[15] Maloy, J., *Coke and Iron Making* (Iron and Steel Publication No. 127, 1970).
[16] *Chemical Products from Coal* (Office of Coal Research Report No. 43, 1968, US Dept. of the Interior; West, H. J., *Ind. Eng. Chem.*, 1970, **62**, 28.

CHAPTER 3

BENZENE FROM PETROLEUM

by J. R. Horth and G. M. Rowell

3.0. INTRODUCTION

BENZENE and its homologues (toluene, xylenes, ethyl benzene and, in short, all alkyl benzenes) are present in all crude oils. As an indication of its abundance, see Table III.1 below.

TABLE III.1

Benzene Content of Typical Crude Oils (Wt per cent)

Kirkuk (Iraq)	0·01
Kuwait (Persian Gulf)	0·08
Ponca City (USA)	0·08
Es Sider (Libya)	0·09
El Morgan (Egypt)	0·11
Qatar (Persian Gulf)	0·12
Gach Saran (Persia)	0·15
Agha Jari (Persia)	0·19
Abu Dhabi Land (Persian Gulf)	0·20
Abu Dhabi Marine (Persian Gulf)	0·25
Hassi Messaoud (Algeria)	0·39
Nahorkhatiyah (Assam)	0·75

These figures are only given to show the probable range of benzene contents of crude oils: of fields not mentioned, North Sea, Alaskan, West European, Canadian and Australian crudes are generally low in aromatics. West African, American and Russian crudes such as Romashkino, Tuimaza and Mukhanovskaia are intermediate, and South American, Mexican, Rumanian, Russian (Black Sea), Borneo and Indonesian crudes are usually high in aromatics. Benzene content is loosely related to toluene content (generally about two to four times benzene content) and C_8 aromatics content (three to ten times benzene content). Also, high aromatics contents are usually associated with high naphthene contents, and as discussed later, benzene is most easily made from higher aromatics and naphthenes.

Benzene can be obtained from crude oil by:

(1) Separation (e.g. distillation, plus solvent extraction and/or crystallisation or adsorption).

(2) Addition (polymerisation of lighter hydrocarbons).

(3) Subtraction (dealkylation of higher aromatics, dehydrogenation and dealkylation of naphthenes).

(4) Alteration (cracking of higher hydrocarbons, dehydrocyclisation of paraffins).

None of these methods alone will produce benzene from oil on a commercially viable basis; in combination, they make the great majority of world production, coal-produced benzene being perhaps one-fourth to one-fifth of the total and steadily declining, relatively to oil, in importance.

It is customary to treat benzene, toluene and xylenes as a single entity (called BTX for short) since similar technologies are used in producing all three, and one can readily be transformed, in good yield, into another.

Processes exist specifically for the production of BTX and hence benzene and for the production of other products, e.g. ethylene, where BTX is an important by-product.

3.1. BENZENE AS MAIN PRODUCT

3.1.1. Benzene from Catalytic Reforming

When it is required to produce benzene as a main product, the first step is almost always catalytic reforming of naphtha, i.e. petroleum distillate boiling between about 70°C and 190°C. (Below 70°C the predominant reaction is production of lower paraffins, while distillate boiling higher than 190°C reduces catalyst life: any intermediate cut can of course be reformed.)

Naphtha can be reformed either over a base-metal catalyst (typically 10 per cent MoO_3 on alumina) or a precious metal catalyst (typically 0·2–0·5 per cent Pt on a variety of supports, and often with a second precious metal such as rhenium). At the present time the operation is entirely, or almost entirely, carried out with precious metal catalysts.

Both base- and precious-metal reforming processes operate at similar pressures (10–35 atm) and temperatures (475–560°C) with large amounts of hydrogen recycle (approx. 3–10 mols hydrogen per mol oil) to suppress olefine formation and coking. Both catalysts are sensitive to sulphur, but coke laydown on molybdena catalysts is so much higher than on a platinum catalyst that the former must be regenerated during operation (either *in situ* or externally) so sulphur is burnt off at the same time as coke. Thus the feed to a base-metal reformer needs no pretreatment, while the feed to a platinum reformer must be desulphurised down to 5–10 p.p.m., almost universally by selective hydrogenation over cobalt-molybdenum catalyst (typically 15 wt per cent MoO_3 and 3 per cent CoO on an alumina support; nickel oxide (NiO) is sometimes substituted, in equal amount, for CoO). The usual operating conditions are: temperatures 340–425°C, pressures 20–35 atm, hydrogen/oil molecular ratio 0·5/1, space velocity 4–8 volumes of cold oil per volume of catalyst per hour. Hydrogen of any purity down to about 60 per cent is suitable provided the contaminants are paraffins or nitrogen; olefines or oxides of carbon are deleterious but permissible in small quantities (1–2 per cent).

A typical fixed-bed catalytic reformer with pretreating section is shown schematically in Fig. 3.1.

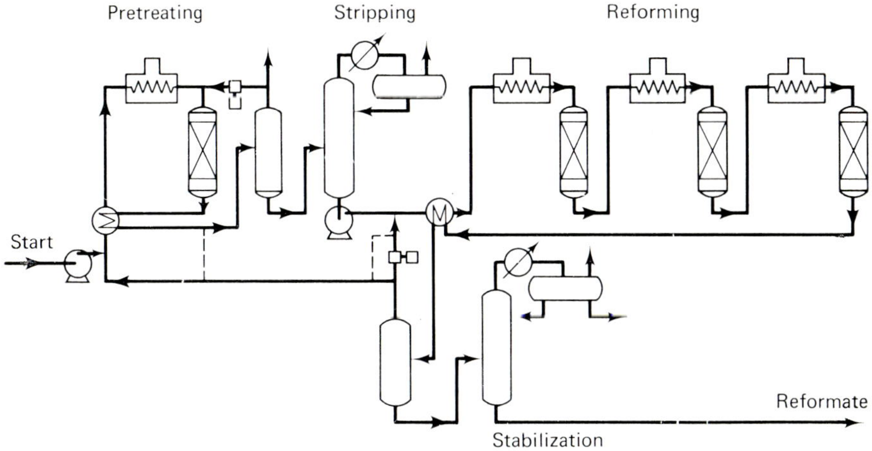

FIG. 3.1

The reforming reactions may be summarised briefly as follows:

(1) Naphthenes are hydrogenated to the corresponding aromatic, e.g. $C_6H_{12} \rightarrow C_6H_6 + 3H_2 - 53{,}010$ cal/gm mole (at 525°C). This class of reaction is quick and almost quantitative with practically all catalysts used for this purpose.

(2) Paraffins are cyclised to naphthenes, e.g. $C_6H_{14} \rightarrow C_6H_{12} + H_2 - 15{,}250$ cal/gm mole. This reaction is slow, not quantitative, and the yield is a good index of catalyst quality, with platinum much superior to base metals. Thereafter, reaction (1) takes place.

(3) Olefins are hydrogenated to paraffins followed by reactions (2) and (1). The reaction is strongly exothermic, so theoretically could be used to counterbalance the endothermic reactions above, but, in general, cracked feedstocks are unsuitable for aromatics production, quite apart from fetching a higher price (as motor fuel) than does ordinary naphtha.

(4) Alkylcyclopentanes are isomerised to cyclohexanes, or hydrogenated to lower paraffins.

(5) Aromatics pass through the catalyst unchanged, or slightly dealkylated.

The ideal reformer, then, would produce aromatics and hydrogen only. In practice, there is also the formation of coke and light paraffins (C_1 to C_5), and the desired reactions do not go to completion, either because of reaction equilibria or because it is uneconomic to run at very low space velocities. Since the number of moles leaving the reactor is markedly greater than that entering, the reaction is favoured by low pressure (see Table III.2 for effect of pressure on conversion to benzene at 482°C).

Low pressure, however, will have three undesirable effects: (a) lower catalyst life, (b) higher gas circulation and/or piping and equipment costs, and (c) higher losses of reformate in the hydrogen off-gas. Thus a compromise in pressure must be sought.

TABLE III.2

Effect of Pressure on Conversion to Benzene

Feedstock	*Partial pressure,* atm H_2	*Equilibrium conversion to benzene*
n-Hexane	40	3%
Methyl cyclopentane	40	40%
Cyclohexane	40	84%
n-Hexane	10	86%
Methyl cyclopentane	10	95%
Cyclohexane	10	99+%

3.1.2. Development of Reforming

Historically, the first commercial hydroformers were installed during World War II to make toluene and aviation gasoline components. The process consisted of two reactors in series, filled with molybdena on alumina catalyst, operating at 480–540°C and 11–22 atm, with intermediate reheat.[1] Catalyst cycle time before regeneration was only 8–16 hours, so reactors were set up in pairs, with two reforming while the other two were regenerated by controlled burning-off of carbon and sulphur. Yields of liquid products were low, of light paraffins, polymer and coke high. This was followed in 1949 by Platforming (a process using a platinum catalyst at higher pressure, usually 35 atm, developed by Universal Oil Products) and several other very similar platinum-catalyst processes, all with fixed-bed reactors.

The base-metal processes developed along two routes, both following catalytic-cracker technology. The first, fluid hydroforming, used two fluidised-bed reactors, in one of which naphtha was continuously reformed over 10 per cent MoO_3 on alumina, with the catalyst removed and regenerated by controlled combustion in a separate reactor. The second, Hyperforming and Thermofor Catalytic Reforming, used a moving bed of granular catalyst (cobalt-molybdena and chromia on alumina respectively) in a vertical reactor from which spent catalyst was removed from the bottom, regenerated and returned to the top again.

The latter processes are at present obsolete, due to development of better platinum catalysts and the safety hazards inherent in fluid hydroforming, but it is not impossible that better base-metal catalysts could alter the situation.

Platinum reforming has been intensively developed since its inception; operating pressure, formerly 35–40 atm, is now as low as 8 atm, catalyst life between regenerations in fixed-bed reformers is six months or more, and as we go to press two organisations (Universal Oil Products and Institut Français du Pétrole) have announced moving-bed reformers with intermediate reheating and external continuous regeneration, an important step forward in operating at maximum severity. (See Fig. 3.2.)

Yields also have increased; Table III.3 shows typical aromatics yields from modern fixed-bed catalytic reforming.

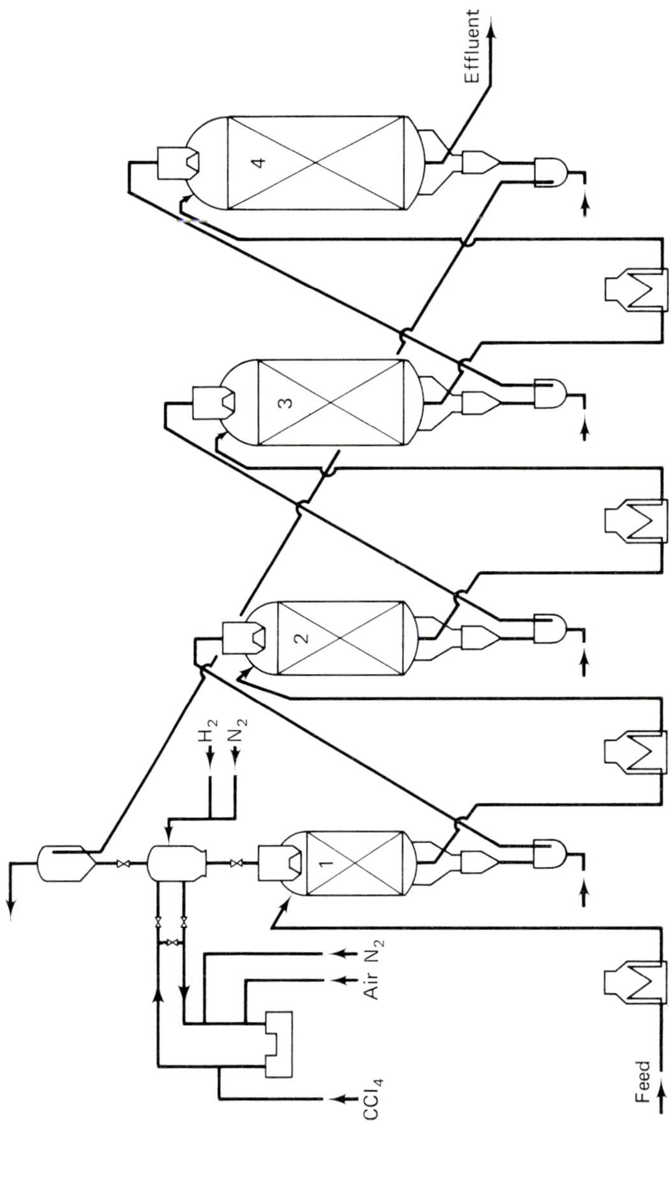

FIG. 3.2 *Regenerative Cat Reformer (IFP)*

TABLE III.3

Typical Yields from Fixed-Bed Catalytic Reforming[2]

Feed naphtha boiling range, °C	100–180	
Feed naphtha paraffins content, vol %	36·3	
Feed naphtha naphthenes content, vol %	59·0	
Feed naphtha aromatics content, vol %	4·7	
Estimated run length, months	6	
Reactor outlet pressure, atm	8	15
Hydrogen yields, vols at NTP/vol cold feed	320	297
Aromatics production wt % on feed		
C_6	1·6	1·5
C_7	18·6	17·6
C_8	28·4	27·1
C_9+	21·4	20·8
Total aromatics, wt %	70·0	67·0

Many variants of platinum reforming have been, and doubtless will be, proposed by the various licensors of their processes. These fall into three classes:

(1) The obvious, but previously untried. A good example is the use of a 'swing' reactor, freshly regenerated, and substitutable for all or some of the other reactors, so that operation can be continued, unimpeded by shutdowns for regeneration.

(2) Alternative approaches in which accepted thinking is discarded or modified.

(3) New process concepts such as bi-metallic catalysts to give longer life, or the above-mentioned continuous regeneration to maintain catalyst activity. These soon give rise to a flock of 'me too' processes, where every licensor tries to duplicate his rival's results, in one way or another, without infringing patent law.

A typical example of the class (1) approach is the Magnaforming process of Atlantic-Richfield, in which hydrogen partial pressure is increased, by increasing recycle gas rate, to further paraffin dehydrocyclisation in the last reactor(s). In the example given by the authors cited, the hydrogen to oil molal ratio in the first two reactors is 2·5/1, while in the last two it is 9/1.[3] Fig. 3.3 shows a typical schematic flow sheet. Reactor effluent is proportioned to the oil-gas and supplemental gas feed streams by valves V_1 and V_2, to get maximum heat recovery. Inlet temperature of the first reactor would be around 470°C, which is high enough to dehydrogenate naphthenes while still obtaining good catalyst life. Temperature of subsequent reactors would be higher, to promote aromatisation of paraffins, and to the last reactor might be 510°C; catalyst life in this case being maximised by the high hydrogen partial pressure.

The class (2) approach tries to get round the *status quo* by introducing another variable; a good example is the Mobil Selectoforming process,[4] in which a dual-functional zeolite-based catalyst is employed as an adjunct to reformate production or for high-aromatic stream upgrading. Main reaction is cracking of paraffins, and hydrogenation to propane. This process, available from UOP, or from California Research Corporation as part of their Rheniforming S process, could be attractive where LPG and aromatics only are required, and hydrogen is in excess. The paraffins are minimised by cracking, rather than aromatisation, but this may be equally acceptable to an aromatics producer. The above-mentioned description of the process is based on gasoline production, when the high-octane iso-paraffins are required to pass through the reaction zone unchanged, but it might well be possible to use the Selectoforming concept to eliminate iso-paraffins also and leave a high-aromatic concentrate. In any event, the aromatic-free raffinate from solvent extraction or distillation would have a substantially higher octane number, and value as a by-product, than otherwise.

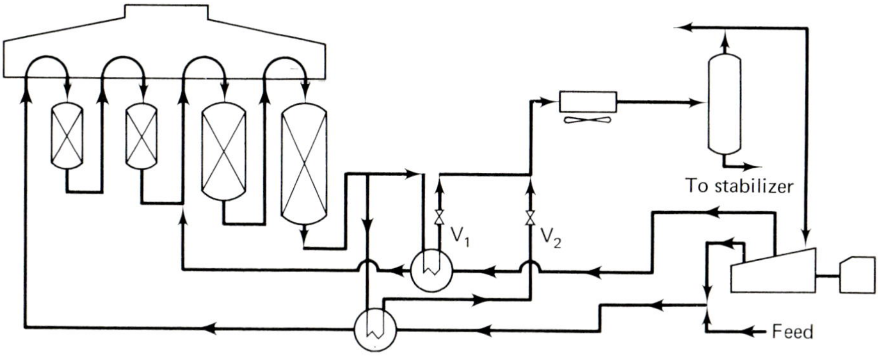

FIG. 3.3 *Magnaforming*

Operating conditions are similar to those used in catalytic reforming, and the last and/or penultimate reactors thereof often have reforming catalyst replaced by Selectoforming zeolite catalyst, in part at least.

3.1.3. Benzene from other processes

Other processes for aromatics production are (a) high temperature cracking and polymerisation of light paraffins, and (b) high-severity cracking of naphtha, kerosene or gas oil to yield olefines and gas; the aromatic content of the feed is more resistant to cracking and is either dealkylated or unchanged.

Neither process is currently used to make aromatics alone, as they both give low yields (typically around 20 per cent) contaminated by olefines and dienes. The former involves a two-stage reaction,[5, 6, 7] for instance:

(a) $C_2H_6 \longrightarrow C_2H_4 + H_2 + 32700$ cal/gm. mole.
(b) $3C_2H_4 \longrightarrow C_6H_6 + 3H_2 - 18600$ cal/gm. mole.

The 'Catarole' process is nominally a catalytic process cracking kerosine using copper or copper-iron catalyst at 2·5–4·5 atm and 700°C[8]; subsequent work showed that practically the same results could be obtained non-catalytically.

3.2. BENZENE AS BY-PRODUCT

Benzene and other aromatics are produced in large amounts as by-products from (a) pyrolysis of hydrocarbons for ethylene, and (b) gasification of hydrocarbons for town's gas, and in small amounts from other reactions such as pyrolysis to acetylene.

3.2.1. From Ethylene Crackers

Aromatics production as a by-product of ethylene varies widely with feedstock, process and severity of cracking. Some typical yields are given in Table III.4.

TABLE III.4

Yield of Benzene and Other Aromatics in Ethylene Cracking Processes

Type process	*Yield wt %*		*Reference*
Tubular cracking of naphtha to 25·5 wt % ethylene	Benzene	6·0	9
	Toluene	3·5	
	Xylene	1·5	
	Total C_6–200°C gasoline	19·6	
Cyclic refractory-bed cracking of naphtha to 36·5 wt % ethylene	Benzene	6·35	9
	Toluene	1·70	
	Xylene	0·40	
Hoechst continuous crude oil coker to 19·4 wt % ethylene	Benzene	4·25	10
	Toluene	3·55	
	Xylene	0·85	
	Total C_6–200°C gasoline	20·8	
Tsutsumi moving bed kerosine cracker to 34·2 % ethylene	Benzene	5·8	11
	Toluene	2·9	
	Xylene	2·9	
	Total C_6–200°C gasoline	18·6	
Tubular cracking of light gas oil to 23·0 wt % ethylene	Benzene, Toluene, Xylenes	10·6	12

In the absence of any yield data, and for preliminary estimates, BTX yield can be estimated for tubular crackers, from Table III.5, at normal severities, though these are average figures only.

TABLE III.5

BTX/Ethylene Yield from Tubular Crackers Using Different Feedstocks

Feedstock	*Wt ratio, BTX/ethylene*
Ethane	Negligible
Propane	0·10
Butane	0·11
Naphtha	0·45
Kerosine	0·50
Gas oil	0·50

3.2.2. From Acetylene Production

Yields of benzene from pyrolysis of acetylene are similar to, or lower than, those from ethylene. From electric-arc cracking Gladisch[13] gives 2–3 wt per cent from natural gas feedstock and 3·0 per cent from natural gas—LPG feed. Rosendahl[14] gives 3·0 per cent benzene from Wulff cyclic cracking of propane, while Tebboth[15] for the same process and feedstock gives 4·6–7·7 per cent, the difference probably being due to different severities or analytical errors.

A modern variation of the arc process, in which the product gas is quenched with naphtha, would give increased aromatics production, by a factor of about two.

As far as is known benzene is not recovered as such in any of the acetylene plants currently operating.

3.2.3. From Town's Gas Production

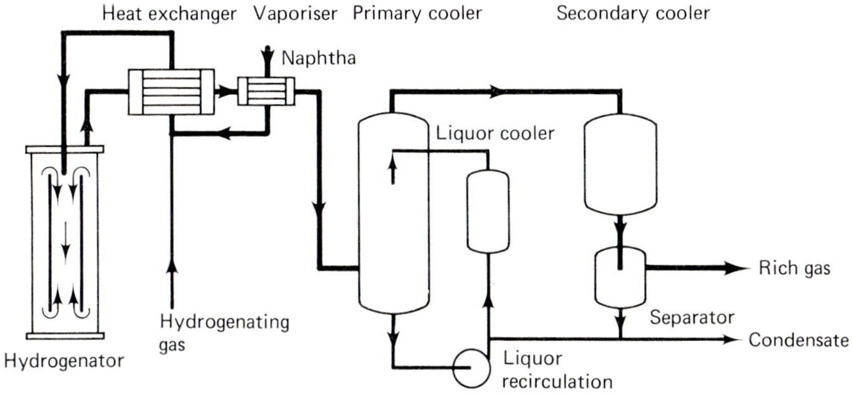

FIG. 3.4 *Simplified Flowsheet of Typical GRH Process*

Aromatics appear as a by-product of town's gas in the Gas Council gas recycle hydrogenation process, in which naphtha is gasified to raise the calorific value of a hydrogen-rich gas. Operating conditions are typically 20 atm and 750°C; the reaction is non-catalytic, and the heat of reaction is controlled by injecting the reactants into a central sleeve in the reactor, thereby inducing recycle of product gas to the reactor inlet via the outer annulus (Fig. 3.4). Principal reactions are hydrogenation of paraffins and naphthenes, and dealkylation of aromatics to benzene. A typical naphtha would yield 12·5 wt per cent of aromatic condensate, containing 82·9 per cent benzene, 8·2 per cent toluene and 2·9 per cent xylenes, together with 0·3 per cent paraffins, 0·9 per cent higher monocyclics, 3·3 per cent naphthalene and 1·5 per cent polycyclics.

Installed capacity of these plants could produce over 2,000,000 tons per annum of benzene if required; in most cases the cost of collection and refining does not warrant this step, and the aromatic concentrate is used as solvent, fuel or otherwise.

A similar Gas Council process uses a fluidised coke bed at 50–60 atm and 750°C, to gasify crude oil with hydrogen to make a methane-rich gas. Only one of these plants has so far been built, in Japan, by Humphreys & Glasgow Ltd., but the process has obvious promise, since conditions are such that the aromatic condensate is practically free of paraffins (<0·1 per cent). The Middle East crude yielded 12·55 per cent wt per cent BTX on crude, of which 93·45 wt per cent was benzene, 5·75 per cent toluene, and 0·8 per cent xylene.

3.3. INTERMEDIATE TREATMENT OF BTX FRACTION

BTX fractions from catalytic reforming can be separated as described in later sections of this chapter, without other treatment. BTX from other sources may contain impurities necessitating further treatment. Pyrolysis gasoline from ethylene production, in particular, will contain dienes, olefines and sulphur compounds. Dienes and olefines would tend to form gums which foul downstream equipment and increase solvent losses during extraction, while sulphur compounds tend to be extracted with aromatics and thus degrade purity. Oxygen and nitrogen compounds are usually present, in very small amounts, but are equally troublesome.

Olefines and sulphur compounds are commonly removed by cobalt molybdate desulphurisation, as for catalytic reformer pretreatment. Dienes are far more reactive, and hence it is usually necessary to carry out a preliminary selective hydrogenation, under milder conditions, to avoid gum and coke deposition.

3.3.1. Diene Elimination—Base-Metal Catalyst

Several processes are available using base-metal (usually nickel) catalysts, of which British Petroleum's is typical and widely used.[16]

Usual operating conditions are: pressure 20–60 atm, temperature 150–175°C, space velocity 2–3 vols cold feed/vol catalyst/hr, hydrogen consumption 25–60 vols at NTP/vol oil. The reaction is markedly exothermic ($\Delta T = 30$–70°C) and it may be necessary to moderate the temperature by recycling cold treated liquid to intermediate points in the reactor, or otherwise. Catalysts are generally not regenerable *in situ* and are discarded at end of life. The higher the pressure the higher the catalyst life and capital cost, which sets an optimum. Hydrogen from catalytic reforming or steam reforming of naphtha is of adequate purity. It is customary to redistil the product to remove small amounts of heavy polymer formed in the process. Little hydrogenation of mono-olefins takes place.

3.3.2. Diene Elimination—Precious-Metal Catalyst

An alternate method is the use of a precious-metal catalyst (typically palladium on a suitable support) and such processes are marketed by Engelhard, Institut Français du Pétrole (IFP) and Bayer AG. The advantage of a precious-metal catalyst is that, being more reactive it can operate at much lower temperature (e.g. 35–75°C) and therefore there is much less polymer formation. Catalyst space velocities are higher (5–8 vols feed/vol/hr) and the Bayer process uses a tubular reactor, somewhat like a fixed-tube-sheet heat exchanger, cooled on the shell side by a circulating stream of demineralised water. The disadvantage, of course, is the high catalyst cost, but the catalyst is regenerable by controlled burning, as for a platinum reforming catalyst, and licensors claim very high catalytic life. Pressure is around 30 atm, and if impure hydrogen is used enough must be circulated to ensure the partial pressure of hydrogen does not drop significantly.

Both with base- and precious-metal catalysts, the consumption of hydrogen is almost stoichiometric, apart from mechanical losses (say 4 vols H_2/vol cold oil) and in solution in the cold product, which is seldom worth recovering.

3.3.3. Clay Treating for Diolefin Removal

This process is obsolete, because of high losses and difficulty of disposing of spent reagent. There may occasionally be exceptional circumstances when its low capital cost and zero royalty could make it usable, by small producers at least, to remove diolefins. Typical conditions for ethylene pyrolysis gasoline are: pressure 4–5 atm, temperature 250°C (i.e. partially vaporised), catalyst 3–5 mm acid-treated clay, space velocity 1·6 vols gasoline/hr. Catalyst life is 3–4 weeks, whereon it is dumped. Two reactors are installed, one working, one loading or stand-by. Treated vapour and liquid are flashed into a fractionating column from which polymer is taken as bottom product, with a heavy distillate sidestream recycled to the charge heater to prevent gum laydown.[17, 18, 19]

3.3.4. Olefine and Sulphur Elimination

This is done with cobalt molybdate, very much like pretreating catalytic reformer feedstock as described above. There are two points of difference; first, saturation of olefines is much more exothermic than desulphurisation, so provision should be made for temperature control, usually by cold gas or oil injection at intermediate points in the reactor; second, while olefines, sulphur and nitrogen compounds must be hydrogenated, conditions must not be so severe that aromatics are hydrogenated to naphthenes. This calls for considerable skill and experience; lower temperatures and pressures, and higher catalyst volumes, than in virgin naphtha desulphurisation are usually employed.

3.4. BENZENE FROM HIGHER AROMATICS

3.4.1. Dealkylation

While BTX can be separated into its constituents by solvent extraction, crystallisation, distillation, etc., this is impractical where

(a) production is too small to support aromatics separation, i.e. where the BTX is low in toluene and xylenes;

(b) where there is a requirement for benzene only.

The most practical procedure in this situation is dealkylation. Catalytic dealkylation processes, such as the Houdry Litol and UOP Pyrotol processes mentioned in Chapter 2 are frequently used, but thermal dealkylation in the presence of hydrogen, as takes place with naphtha in the Gas Council recycle hydrogenator, is also possible. Yields are high (around 90–95 per cent of benzene rings in feed come out as benzene) and conversion in one pass substantially complete. Operating conditions are quoted by one licensor as: pressure 35–70 atm, temperature 600–750°C; high yields require high hydrogen rates (3–5 moles H_2/mole feed). The product (benzene, methane, ethane, hydrogen, etc.) is cooled to ambient temperature. The condensate is stabilised, clay-treated hot in liquid phase to remove tar, and redistilled, with product benzene overhead and toluene and heavier recycled. The gas phase is also recycled, hydrogen purity being maintained by purging a portion to fuel gas, or, where hydrogen is scarce, by removing methane and heavier cryogenically or otherwise. Benzene content of BTX is 98 wt per cent.

3.4.2. Toluene Disproportionation

At present and in the near future the BTX components most in demand are benzene (as general intermediate, especially cumene and cyclohexane) *p*-xylene (for *tere*-phthalic acid) and *o*-xylene (for phthalic anhydride). Any of the xylenes can be isomerised to give the equilibrium mixture (very roughly, 1 para, 2 meta, 1 ortho), so an excess of any xylene presents no problem. On the other hand, a large aromatics complex is likely to have an excess of toluene, which can be disproportionated to benzene and xylene almost stoichiometrically according to the equation $2C_7H_8 \rightleftharpoons C_6H_6+C_8H_{10}+400$ cal (at 530°C). (The xylenes are in the ratio given above.)

Three processes exist at time of writing:

(1) The 'Xylenes-plus' Atlantic-Richfield process (Engelhard licence).[20, 21]
(2) The 'Tatoray' Toyo Rayon process (UOP licence).[22, 23]
(3) The LTD Mobil Chemical Co. process.[24, 25]

The above processes can also be used where both toluene and benzene are in excess, to disproportionate trimethyl benzenes to xylenes; this is perhaps outside the scope of this chapter, but the facility is a useful insurance that the plant, if installed, can be operated profitably in one way or another.

The 'Xylenes-plus' process uses a base-metal catalyst in a fixed bed in the presence of hydrogen, with little or no hydrogen consumption. Coke laydown on catalyst necessitates fairly frequent regeneration by controlled combustion. The volumetric yields are shown in Table III.6.

Catalyst life is claimed to be four years plus, but a lot must depend on the care with which it is regenerated.

TABLE III.6

Effect of Varying Trimethyl Benzene/Toluene Ratios on the Products from the Xylenes-Plus Process

Feed, vol %			
Toluene	100	66·67	50
Trimethyl benzenes	0	33·33	50
Products, vol % on feed			
Light hydrocarbons (S.G. 0·386)	3·85	5·17	3·87
Benzene	37	12·5	8·63
Xylenes	55	83·5	87·25
Trimethyl benzenes and higher	5	—	—

The 'Xylenes-plus' process can also be operated under more severe conditions to carry out hydro-dealkylation concurrently with disproportionation, and thus increase benzene yield.

The 'Tatoray' process is understood to operate at the same, or less severe, conditions using a noble-metal or rare-earth catalyst, with higher aromatics yield and less frequent regeneration, but requires high hydrogen recycle rate (5–12 mols H_2 per mol of fresh feed) and has a net consumption of hydrogen. Catalyst life is claimed to be up to three years. Weight yields are given in Table III.7.

TABLE III.7

Typical Yields from Tatoray Process

Feed, wt per unit time	
Toluene	1000
Hydrogen	4
Products	
Benzene	414
Xylenes	561
Heavy ends (C_{10}+)	10
Off gas	19

The Mobil LTD (low temperature disproportionation) process operates in the liquid phase, using a special zeolite-base catalyst. No hydrogen is required. Weight hourly space velocity is 1·5 kg oil/kg catalyst. Operation starts at 260°C, and temperature is gradually raised, as catalyst loses activity, to 315°C, when it is regenerated

TABLE III.8

Composition of Stream at Various Stages of the Mobil LTD Process

Identification number *Component (wt/unit time)*	(*1*)	(*2*)	(*3*)	(*4*)	(*5*)	(*6*)	(*7*)
Non-aromatics	—	<0·1	0·2	0·2	<0·1	—	0·2
Benzene	—	0·1	43·9	2·1	41·7	—	43·8
Toluene	99·7	221·8	122·1	—	—	0·1	0·1
Ethylbenzene	0·1	0·1	0·3	—	—	0·2	0·2
p-Xylene	0·1	0·2	12·3	—	—	12·2	12·2
m-Xylene	0·1	0·2	27·4	—	—	27·3	27·3
o-Xylene	—	—	11·5	—	—	11·5	11·5
Trimethyl benzenes	—	—	4·0	—	—	4·0	4·0
Other C_9+ aromatics	—	—	0·7	—	—	0·7	0·7
Total	100·0	222·4	222·4	2·3	41·7	56·0	100·0

Identification number key
(*1*) = Fresh feed.
(*2*) = Fresh feed, plus recycle toluene, to reactor.
(*3*) = Reactor effluent.
(*4*) = Non-aromatics column overheads.
(*5*) = Benzene column overheads.
(*6*) = Toluene column bottoms.
(*7*) = Net product yield on feed.

by controlled burning of coke. Regeneration is said to be infrequent, and catalyst life at least one and a half years. As with the other disproportionation processes yield per pass is limited by equilibrium conditions, and only 45 per cent of toluene fed to the reactor is disproportionated, the remainder being recycled. A typical weight balance is given in Table III.8, in which the reactor is followed by three distillation columns, successively distilling overhead non-aromatics, benzene, and recycle toluene, and leaving as end-product a xylene and heavier stream for separation as described in later sections.

3.5. BENZENE FROM PETROLEUM: PURIFICATION METHODS

3.5.1. General

As has already been indicated earlier in this chapter, the principal petroleum sources of BTX aromatics are from hydrocarbon reforming or from the dealkylation of toluene or other aromatics, themselves originally obtained in the reforming operation.

Having obtained the BTX components by means of any of the processes already described they must be concentrated and then separated into benzene, toluene, xylenes, etc.

Benzene is a major chemical feedstock, used for a wide variety of applications which are dealt with elsewhere in this book. As a chemical feedstock product, quality and consistency of quality are prime considerations. Typical industrial grades might analyse as shown in Table III.9.

TABLE III.9

Typical Properties of Benzene Made to Various Grades

Grade	*ASTM nitration grade D-835*	*Average quality produced*	*Best obtainable in quantity*
Purity (wt %)	—	99+	99·99+
Freezing point, °C	4·85 °C min.	5·44 min.	5·56 min.
Initial boiling point, °C	(1 °C spread, including	79·5 min.	79·8 min.
Final boiling point, °C	80·1 °C)	80·1 max.	80·1 max.
Specific gravity (15·5/15·5 °C)	0·882–0·886	0·884	0·8850–0·8854

The aromatics present in the reformate must be separated from the paraffins and the residual naphthenes. Of the known recovery processes, liquid-liquid extraction, selective adsorption, extractive distillation, azeotropic distillation, and simple distillation, liquid-liquid extraction is by far the most prominent.

Using a solvent which selectively dissolves the aromatic hydrocarbons while rejecting the paraffinic and naphthenic ones, the extract (a solution of almost pure aromatic hydrocarbons in the solvent), is generally purified by extractive distillation, or back-washing with a non-aromatic of different boiling point to displace the impurities, followed by distillation; the aromatics are then separated from the solvent by distillation. The economics of such a process depend upon the solvent chosen, which will fix the flow rates used (power, equipment size), and the reflux ratios in the various distillations (heat energy).

Many solvents have been proposed or used for this purpose; the more important are given in Table III.10.

TABLE III.10

Typical Solvents Used in Soluent Extraction of Aromatics

Process	*Solvent*	*Licensor*	*Notes*
Edeleanu	Sulphur dioxide	None	1
Udex	Diethylene glycol	UOP-Dow	2
Sulfolane	Tetramethylenesulphone	UOP or Atlantic-Richfield	3
Arosolvan	*n*-methyl pyrrolidone (NMP)	Lurgi	4
Distapex	NMP and ethylene glycol	Lurgi	5
Mofex	Monoethyl methyl formamide	Leuna Werke	6
DMSO	Dimethylsulphoxide	Institut Français du Pétrole	7
Carmex	Methyl carbonate	ONIA-ERAP	8
SNAM	Morpholine	SNAM Progetti	9
REDEX	Dimethyl formamide	Marathon Oil Co.	10

Notes to Table III.10:

1. Used 1908 onwards, last plant installed around 1955, no longer competitive except in special circumstances. Disadvantages: high utility costs, toxicity, corrosion. Might yet be applied where other solvents unavailable.
2. Uses 5–10 per cent water to decrease viscosity and increase selectivity. Installed 1955–65, most plants now converted to sulfolane or other solvents, e.g. tetraethylene glycol/water (95/5), because of high (8–11/1) solvent/oil ratio.
3. Most widely used extraction process, developed by Shell, licensed through UOP. Main patents expired, hence designs available from other sources than UOP-Shell.
4. Some commercial application, mainly in Germany.
5. Uses an extractive distillation step; not likely to be widely applied.
6. Unstable solvent which can cause corrosion without expensive precautions; wide use unlikely.
7. Some commercial application, mainly in France.
8. Uses 10 per cent water as antisolvent, claimed to be operable in presence of mono-olefins; current application small.
9. Solvent recovery complicated by boiling point (128°C); later developments use formylmorpholine.
10. Currently applied to heavier aromatics, may be applicable to BTX. Uses back-extraction as in DMSO.

Of these processes, the following are described later in more detail:
Edeleanu (SO_2) extraction;
Sulfolane, developed by Shell and licensed by UOP;
UDEX, developed by Dow and licensed by UOP;
DMSO, developed and licensed by IFP.

The general principles are roughly similar, but the different solvents call for different operating conditions.

3.5.2. Solvent Extraction: General Principles

Solvent extraction has been used in the petroleum refining industry since 1908.[26] Originally, the upgrading of kerosene was the major use, but the improvement of lubricating oils is now of great importance. More recent applications include the upgrading of charge stocks for catalytic cracking, the separation of light aromatics from gasolines, and the preparation of jet fuel components.

Extraction is possible because different liquids have different affinities for each other. Some are completely miscible in all proportions. Some are completely immiscible. Components are removed from a feed liquid by means of the solvent action of another liquid. Counter-current processes involve two operations:

(1) The obtainment of equilibrium between feed and solvent; and

(2) the separation of the contacted materials.

Separation in absorption and fractionation operations is relatively simple, because one phase is a liquid, the other is a vapour. In solvent extraction, however, both phases are liquids, and the separation of these two liquids by gravity or centrifugal settling is a major design problem.

The Distribution Law (Henry's Law modified for two liquid-phases) states that ideally there is a constant ratio between the concentrations of a component that is distributed between two mutally insoluble liquid-phases:

$$C_1/C_2 = K$$

where: C_1 = concentration of solute in first liquid-phase; per cent wt
C_2 = concentration of solute in second liquid-phase; per cent wt
K = a constant depending only on temperature.

If the solute is a mixture, the law holds for each component of it. Modifications of this equation to cover either dissociation or association will be found in textbooks on physical chemistry. As in other separation processes, a single contacting stage does not normally suffice, and counter-current multi-stage units are needed. Refluxing may also be needed if the product is required to be extra-pure.

The following terms are frequently employed in solvent extraction operations:

Solvent: The extracting agent;
Solute: The readily-soluble material in the feed;
Extract: Material removed from the feed;
Raffinate: The non-dissolved product;
Reflux: Extract returned to the extractor;
Lean Solvent: Solvent as charged to the reactor;
Rich Solvent: The solvent-extract solution leaving the extractor.

Various methods may be used for contacting the feed with the solvent and in the design of extracting equipment the principal problem is to provide sufficient dispersal of one of the phases into the other, so that material may transfer between them. A wide variety of designs have been used, and an even wider variety proposed.[27, 28] In the petroleum industry the most popular types have been mixer-settler combinations, packed columns, sieve-tray, baffle columns. A 'Rotating Disc Contactor' (RDC) was put into commercial use about 1955. Pulsed columns and centrifugal separators have also been proposed.

The principal objection to mixer-settlers is the many pieces of equipment that are required. Usually four to eight counter-current stages are employed, but as many as fourteen have been used.[29] Each stage consists of a mixing chamber, a settling chamber, and a pump.

Packed columns have been used for the extraction of aromatics from lube oils and from kerosene.[27] One-inch Raschig rings are a popular packing. A packed height of between 20–50 ft is typical, although heights of 100 ft have been employed.[30] Tower sizing is described by Crawford and Wilke.[31] Typical commercial columns provide between 1·4 to 6 theoretical stages. In packed columns it is usually desirable to hold the interface between the raffinate and solvent phases near the bottom, because the packing is preferentially wetted by the solvent. But note that sulphur dioxide is an exception (with heavy feedstocks) because its low viscosity promotes transfer of material between phases.

According to Morello and Poffenberger,[27] of fixed equipment, sieve-tray columns appear to have been the most popular extracting devices, particularly for lube oils. Fifteen to twenty trays are commonly used. Mayfield and Church[32] describe tower sizing in this context.

Baffle columns, using simple horizontal baffles to force counter-current liquids to zig-zag are sometimes used. There are three types: 'side-to-side'; 'centre-to-side'; and 'disc and doughnut'. For each such baffle, contacting efficiency is low, but a reasonable column efficiency is obtained by short baffle spacing (4–6 in. say). Because the cost of dismantling is high, this type of column is not recommended where precipitates may form that are removeable only by mechanical means.

Because all the above column types rely on gravity to provide the impetus for mixing, their liquid-containing efficiencies are low. Rotating Disc Contactors (RDCs) employ mechanical power to augment the force of gravity.[29] The internals of an RDC are similar to disc and doughnut baffles except that they do not overlap, and that the discs are attached to a central rotating shaft. This rotation disperses one of the counter-flowing liquids into the other. The incoming streams are brought into the system tangentially in the direction of rotation. Because they are high-efficiency devices, RDCs are favoured by conditions requiring a considerable number of stages. When seven stages are required, it is claimed that the cost of a RDC is less than one-half of that of a mixer-settler system or of a packed column.[29]

In the design of extraction systems, using the Distribution Law mentioned above: $C_1/C_2 = K$, it can be shown, for multiple stage extractions[33] that if:

Y_o = wt solute in treated liquid before extraction
Y_n = wt solute in treated liquid after n^{th} extraction
P = volume of solute-free treated liquid
W = volume of solute-free treating solvent

then, for a single-stage operation:

$$\frac{Y_1/P}{(Y_o - Y_1)/W} = K$$

and

$$Y_1 = Y_o \left(\frac{KP}{KP+W}\right)^n$$

and for an operation involving n stages:

$$Y_n = Y_o \left(\frac{KP}{KP+W}\right)^n$$

From this general equation, it is evident that Y_o approaches Y_n only after an infinite number of extractions with a given volume of solvent. Fortunately it has been found that essentially complete removal of the solute is obtained by a relatively few extraction stages. For a given volume of solvent, about 94 per cent of the theoretical amount of solute extractable can be separated with the solvent divided into five portions.[34]

It is evident that more efficient extraction (low value of Y_n) is obtained as the volume of the solvent used (solvent-oil ratio) is made larger. Maximum efficiency is obtained when the solvent portions are of equal volume.[35]

The theory of solvent extraction has been discussed by many other authors.[36, 37]

3.5.2.1. Solvent extraction with liquid sulphur dioxide (*Edeleanu process*) In the early days of the petroleum industry, one of the principal products was kerosene, for lighting.

In kerosene, the presence of aromatic hydrocarbons increases the tendency to smoke when burned. Unsaturated compounds also tend to form gum and reduce the colour-stability of the product. The object of refining kerosene is to improve its quality by removing unsaturated and aromatic hydrocarbons, sulphur compounds, acidic and basic bodies, and any resinous or asphaltic impurities that may be carried over during distillation, and to produce a marketable kerosene to a given specification.

The traditional method (no longer used except in special circumstances) of refining was formerly by washing with sulphuric acid. Agitation with acid was followed by water, and caustic soda washes, and further water washes to remove any entrained caustic soda. This procedure was followed by acid-clay or bauxite treatment to ensure colour stability.

In 1908, British Patent No. 11,140 was granted to Edeleanu for removing the aromatic fractions from kerosene by means of liquid sulphur dioxide. Sulphur dioxide has a normal boiling point of -10°C, and at this temperature aromatic and unsaturated hydrocarbons are completely miscible in it, but paraffins and naphthenes are not. Sulphur dioxide also extracts cyclic sulphur compounds and nitrogen compounds, a property that make it useful in handling high-sulphur content kerosenes. The early work of Edeleanu showed that liquid SO_2 could profitably be used for this purpose. Furthermore, none of the oil was lost to sludge (as is the case with sulphuric acid treatment). The extracted fractions proved to be valuable by-products which could be utilised as solvents or as sources of aromatic compounds, so permitting a full utilisation of raw petroleum distillates such as had not before been achieved.[38]

The 'Edeleanu Process', and its successors are applicable to a wide variety of feedstocks, and all plants are similar except for the operating temperatures. Stocks in the naphtha range are extracted at about -30°C, lubricating oils at temperatures of 10–25°C, and intermediate stocks at intermediate temperatures.

Approximately one volume of solvent is circulated per volume of feedstock. When a high-purity aromatics product is desired, the process is modified by the addition of a wash-oil reflux step. Lubricating oils are processed by the addition of a second solvent (benzene) to increase the capacity of the solvent for dissolving heavy aromatics.

A typical sulphur dioxide extraction plant is schematically illustrated in Fig. 3.5. The charge will be petroleum fractions in the boiling range from kerosene to lubricating oils. The kerosene fractions (which are our concern) will yield illuminating oils with high smoke point. Due to their very low content of residual aromatics, kerosene raffinates find the most modern use as jet fuel components.

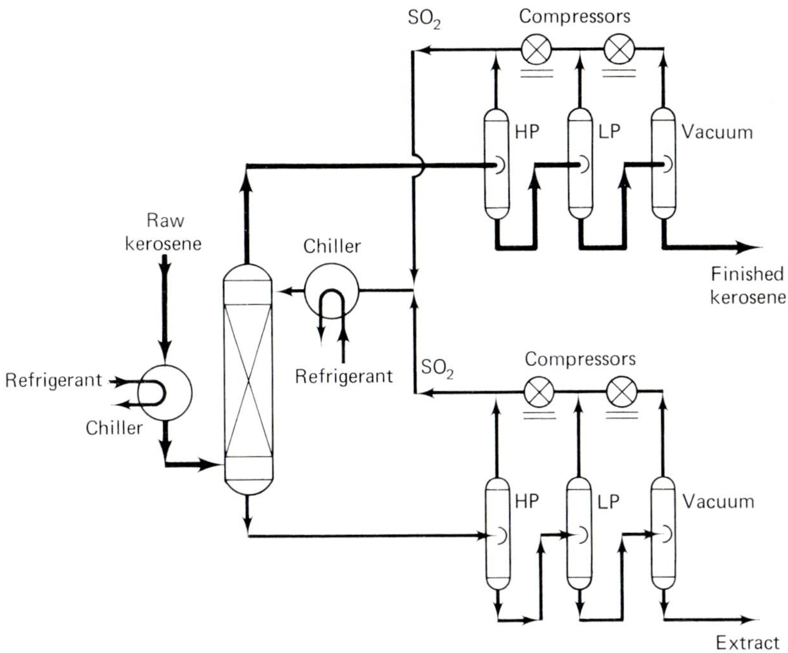

FIG. 3.5 *Schematic Diagram of SO_2 Extraction Plant*

After passing through a coalescer to remove any free water, the feed is dried (0·005 wt% max. H_2O over calcium chloride), cooled to extraction temperature by exchange with raffinate and with a refrigerant, and charged to the extraction column. Raffinate as top product passes to a set of three flash evaporators. Rich extract (containing the aromatics) is taken from the bottom of the column, and passes to a second set of three flash evaporators. The first evaporator in each set operates at 6 atm or higher—a pressure high enough to permit the vapours to be condensed without compression. Compression is required for the other two stages, the last of which operates at about 25 mm Hg abs. Two vacuum stages are used in order to reduce compression requirements, which would be high if all the solvent were evaporated at 25 mm Hg abs.

Because wet sulphur dioxide is corrosive, a slip-stream of circulating solvent is distilled to reject water as bottoms, and thus hold the water content of the circulating sulphur dioxide below 0·05 per cent. Thus, the whole equipment of the plant, except the SO_2 drying facility, can be made of carbon steel.

Some sulphur dioxide is lost to the products, to the reject water, and by leakage. Solvent losses may total about 0·5 kg/ton of charge.

A typical packed column for the extraction of light distillates contains about 30 ft of packing.[39]

With some naphtha feeds, solvent recovery can be accomplished in one stage and no compression is required.[40]

Raffinates from sulphur dioxide extraction contain about 1–5 per cent aromatics, extracts from 55–80 per cent. In the recovery of light aromatics, higher concentrations are usually required. These are obtained by displacing the diluent materials with a wash-oil, and then separating the wash-oil from the product aromatics by distillation. Some wash-oil passes into the raffinate, which is also separated by distillation.

In operations that include the use of wash-oil, 90 per cent benzene, nitration-grade toluene, and high-purity (99 per cent plus) xylenes are produced.[40] Recoveries are: for benzene 95 per cent; for toluene 98 per cent; for xylene 96 per cent.[41]

Benzene can be raised to nitration-grade by azeotropic distillation with methanol at an overall recovery of 90 per cent of the benzene in the charge for extraction.[40] A detailed description of a plant of this type has been published.[42]

Typical separations obtained are indicated in Tables III.11 and III.12.

TABLE III.11

Effect of Various Raw Materials and Extraction Conditions on Yield by Edeleanu Process

Extraction conditions	*Gas oil*	*Kerosene*		*Diesel*	*Naphthas*	
Solvent: feed, vol/vol	125	160	100	135	75	175
Wash oil: feed, vol/vol	0	0	0	0	0	75
Temperature: °C	0	−13	−18	−6	−29	−29
Extract: vol %	23	23·5	15	25	20·6	53·4
Aromatics content: vol %						
Feed	22	18	14	24	16·7	51
Extract	79·5	75	82	78	75·9	94
Raffinate	5	2	2	3	1·3	1·8

More recent data[43] give the following operating conditions and yields for kerosene processing.

TABLE III.12

Effect of Crude Oil Sources on Edeleanu Extraction Process

Crude sources	*Near East*	*Tia Juana*
Solvent/feed, vol %	100	50
Treatment temperature, °C	−10	−10
Raffinate, wt %	75	83·5
Extract, wt %	25	16·5
Sulphur, wt %: in feed	0·15	0·22
in raffinate	0·026	0·07
in extract	0·522	0·979
Smoke point, mm: feed	24	21
raffinate	45	30

The above table shows one disadvantage of solvent extraction, namely, that sulphur compounds are as soluble as aromatics and must be otherwise removed.

3.5.2.2. Sulfolane process The first reference to this process was by Deal, Evans, Oliver and Papadopoulos.[44] It is used for extraction of aromatics from catalytic reformates (or straight-run fractions) for octane improvement. It is also used to give nitration-grade aromatics. Basically similar to other extraction processes it differs in operating detail because of the unique properties of the solvent which it uses. Sulfolane is a five-member hetero-cyclic compound having the structural formula:

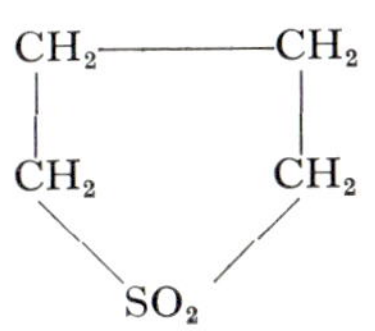

(tetra methylene sulphone or tetrahydrothiophen 1-1 dioxide)

some of its properties are as follows:

Boiling point (°C)	287
Freezing point (°C)	28
Sp. gr. at 100°C	1·20
Viscosity at 100°C cp	2·5
Sp. heat (cal/gm)	0·40
Interfacial tension at 100°C dynes/cm	3·0

The plant feed is charged to a rotating disc contactor (RDC) for counter-current extraction of the aromatic components with sulfolane. Rich solvent from the base of the RDC is flashed after heat exchange with stripper bottoms and further heating. The flash vapours are condensed and collected in a separator, from which the hydrocarbons are returned to the RDC as a 'back-wash', and water is sent to the stripper reboiler, for conversion to stripping steam. The flash liquid is charged to the top of the stripper, which produces the aromatic extract (as a top product) and lean solvent for recycle to the extractor. The raffinate from the RDC extractor is cooled, and, after separation of solvent, is charged to the raffinate washer (also an RDC) where the water from the extract accumulator and/or fresh process water, removes dissolved or entrained solvent from the raffinate, which then goes to storage for further treatment (distillation to separate the aromatic components). The water solution of solvent is charged to the stripper reboiler. The flow diagram (Fig. 3.6) illustrates this system.

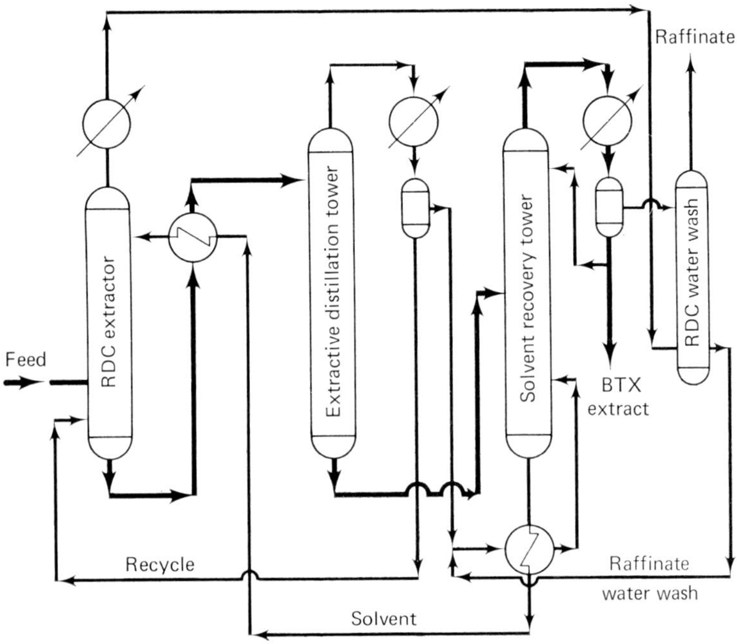

FIG. 3.6 *Shell Sulfolane Aromatics Extraction Plant*

Typical operating conditions have been indicated by the Shell Chemical Company as follows:

Solvent/feed ratio wt	6·8
Backwash/feed ratio wt	0·5
Stripping steam/feed ratio wt	0·13
Stripper base temp. (°C)	190
Extractor top temp. (°C)	100
Extractor pressure (atm)	2
Feed temperature (°C)	240
Water in lean solvent (wt per cent)	1·3

3.5.2.3. UDEX process This process was licensed by UOP for the production of high-octane motor fuel blending components, or, alternatively, for the separation of high-purity aromatics from mixtures with other hydrocarbons. The process is sufficiently flexible to take stocks of wide boiling range, and can provide (after suitable fractionation) nitration grade benzene and toluene, and a xylenes-ethylbenzene concentrate of high purity. If the boiling range of the feedstock is wide enough, it can also produce a concentrate of C_9 to C_{11} aromatics.

The process uses an efficient multistage counter-current extraction column of 'novel design developed especially for this purpose'—(understood to be a perforated-tray shower-deck contactor). A mixture of glycols and water is used as solvent. The feedstock is fed to a refluxed reactor.

Rich solvent from the extractor passes to a solvent stripper, from which the extract is taken overhead with the air of steam as a stripping agent, a portion of the extract being recycled to the extractor as reflux. Raffinate from the extractor is water-washed to recover dissolved glycol. The washings are combined with the solvent from the bottom of the solvent stripper, and the mixed stream is recycled to the extractor to complete the cycle.

Typical UDEX results when charging a catalytic reformate fraction (78–180°C) are given opposite in Table III.13.

The individual products shown in this Table were fractionated from the UDEX extract.

Fig. 3.7 is a schematic flow diagram of the process.

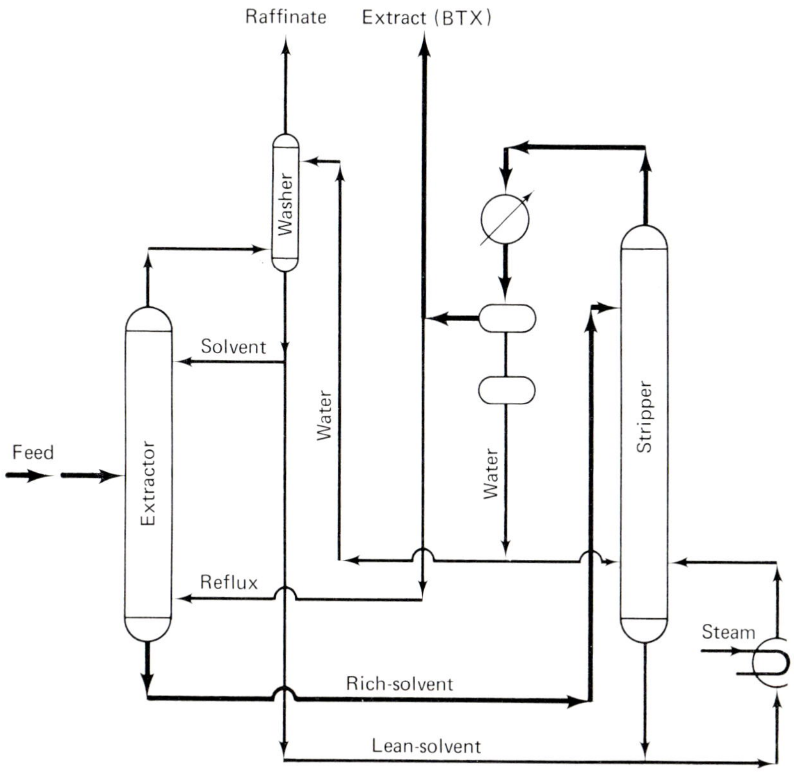

FIG. 3.7 *Elements of UOP UDEX Unit*

TABLE III.13

Products from a Typical UDEX Extraction on a Catalytic Reformate

Material	*Percent by volume of charge*	*Percent by volume of extract*	*Percent aromatics in fractionated product*
Benzene	4·6	8·3	99·9
Toluene	24·0	43·0	99·8
Xylenes + Ethylbenzene	20·8	37·2	99+
C_9 + aromatics	6·4	11·5	
Raffinate	44·2		

3.5.2.4. DMSO process This process has been developed, and is licensed by the Institut Français du Pétrole (IFP). The first reference appeared in 1965, and more details appeared in 1966. It is based on the utilisation of a commercially available solvent—dimethyl-sulphoxide (DMSO) having the structural formula:

$$CH_3—SO—CH_3$$

Some of its properties are listed below:

Sp. gr. (60/60°F)	1·1
Viscosity at 25°C, cp	2·7
Boiling point at 144 mm Hg abs (°C)	170
Freezing point, °C	18·4
Sp. heat at 20°C, cal/gm/°C	0·49
90% DMSO/ 10% H_2O, viscosity at 25°C, cp	2·3
90% DMSO/ 10% H_2O, freezing point, °C	−1·0

The solvent is claimed to have the following special characteristics:
(a) high aromatic solvent selectivity;
(b) high extraction power, giving low solvent circulation rates;
(c) good stability minimising degradation losses; and
(d) can treat any kind of aromatics containing feedstock.

A schematic flow diagram is given in Fig. 3.8. The process scheme differs somewhat from other liquid-liquid extraction processes. Feed and back-wash are charged counter-current to DMSO and small amounts of process water in a first reactor (rotating disc contactor (RDC)), at atmospheric temperature. The extract phase, rich in aromatics, is obtained at the base of the first RDC, and this is then counter-currently back-extracted in a second extractor (also an RDC), with a paraffinic hydrocarbon also at atmospheric temperature. This paraffin dissolves the aromatics out of the solvent phase. Regenerated DMSO is ready for re-use; the paraffinic solvent is separated from the aromatics by conventional distillation, and recycled to the second extractor. Raffinate and extract are then water-washed in order to recover the small amounts of dissolved DMSO (approx. 1 per cent of the entire DMSO stream). Water content of the solvent phase, and reflux (back-wash) rates are selected according to feed composition and required product purity for aromatics.

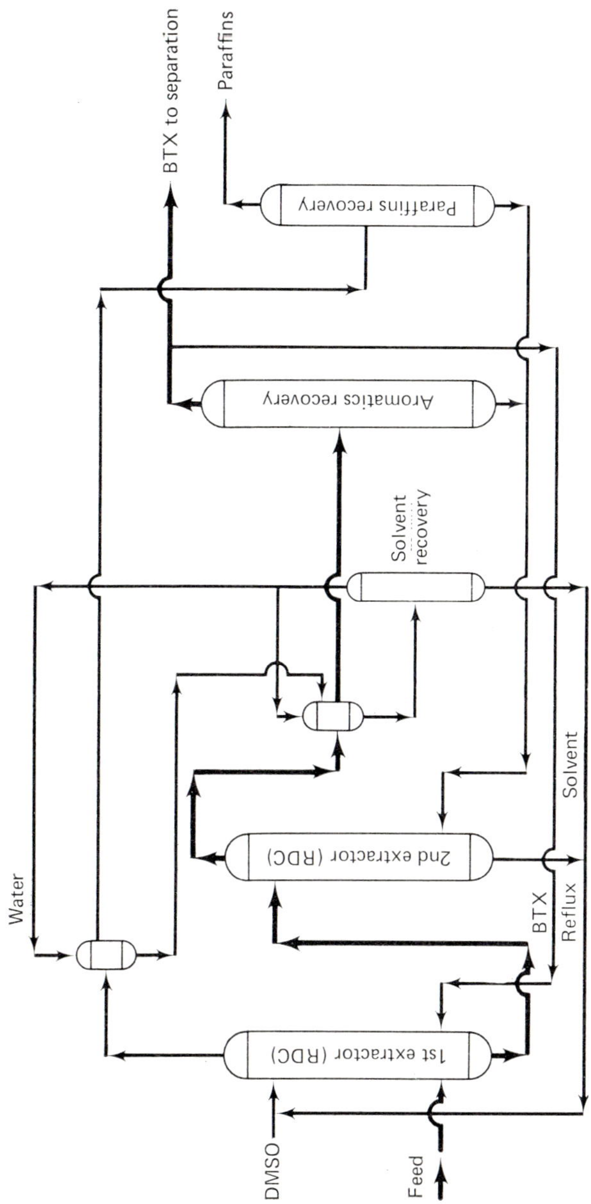

FIG. 3.8 *IFP Aromatics Extraction Process (Solvent = DMSO)*

Typically, the extraction of aromatics from a catalytic reformate containing 10 per cent benzene, 27 per cent toluene, and 18 per cent xylenes will give economic recoveries of benzene at 99·9 per cent purity; toluene at 98·5 per cent purity; and xylenes at 95 per cent purity.

3.5.3. Adsorption. General Principles

Adsorption is a physical process, in which a solid (porous) material is used to remove selectively one or more components from a liquid or gas stream. The required component(s) will concentrate on the surface of the solid, making separation possible. Being a surface phenomenon, any practical adsorbent must be highly porous. A typical surface area would be of the order of 700 sq. m/g.[45] Most adsorbents attract hydrocarbons in the following order: aromatics, olefins, naphthenes, and paraffins.[46] Sulphur-containing compounds are attracted about as strongly as aromatics.

Adsorbents in general use include silica gels, silica-alumina complexes, activated aluminas, and activated charcoals. The most highly selective are the silica gels, while the silica-alumina complexes are the least.[47] On the other hand, the activated charcoals have the greatest capacity while the silica-aluminas have the least. However, silica-alumina particles are more resistant to fracture in adsorption operations. Mention should also be made of a new class of adsorbents that are becoming increasingly important in petroleum refinery operations. These are synthetic zeolites ('molecular sieves'), having pore openings of a more or less uniform size: 3, 4 and 5 Angstrom units in diameter, a size small enough to preclude all but straight-chain hydrocarbons. Their main use at present is in the preparation of 'normal paraffins' of high purity as feedstock for detergent manufacture. Desirable qualities of an ideal adsorbent would be:

high capacity;
high selectivity;
high diffusivity;
high bulk density;
high strength;
low cost;
low pressure drop;
long life;
ease of desorption.

In all cases, practical adsorbents represent a compromise between desirable and undesirable qualities.

With liquid feeds, fixed-bed operation is used; the bed may be 'de-sorbed' by heating the rich adsorbent in the presence of a liquid strippant. Or an adsorbate may be displaced by a material that is equally attracted to the adsorbent. For example, benzene may be displaced by xylenes, and vice-versa. (This is the principle of the 'Arosorb' process described below.) With liquid hydrocarbons, impurities such as water, organic acids, and nitrogen bases, are difficult to de-sorb. For this reason the feedstock should be pre-treated in 'guard-chambers' to eliminate them.[48]

Many adsorption processes have been developed for general use in the petroleum industry, for example Hypersorption, Fluid-Char Adsorption, Cyclic Adsorption, and the 'Arosorb' process. Only the last two are normally applicable to BTX recovery.

A recent development, the UOP 'Parex' adsorption process, selectively removes *para*-xylene from the other xylenes, competitively, it is claimed, with crystallisation. It is too early yet to assess the commercial viability of this process, but it is possible that other BTX components may be similarly fractionated in the future with other selective adsorbents.

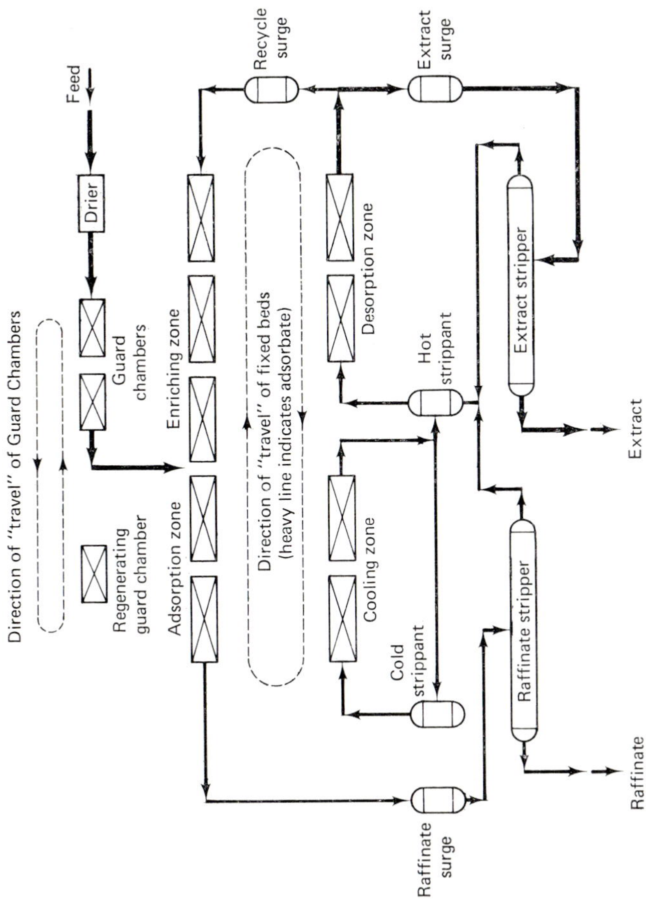

FIG. 3.9 *Schematic of Cyclic-Adsorption Unit*

3.5.3.1. Cyclic adsorption Cyclic adsorption separates liquids by chemical type.[46] It finds its major use in the separation of aromatics from hydrocarbon mixtures.

Feed is passed over fixed beds of adsorbent (in this case silica gel) and adsorbate is subsequently stripped from the adsorbent by a hot liquid paraffinic strippant (pentane or *iso*-octane).[49] In order to simulate counter-current flow of adsorbent to feed and strippant, each of the liquid streams passes through several beds of adsorbent in series. The adsorbent beds are so piped that they can be used on a cycle in any position. Hence the beds can be considered to 'move' counter-current to the flow of liquid feed and strippant.

Fig. 3.9 gives a schematic idea of the 'flow pattern' when separating aromatics. Regenerated adsorbent 'travels' counter-current to the feed in an adsorption zone, in which aromatics are adsorbed, along with small quantities of non-aromatics. The adsorbent 'carries' the adsorbate, along with the liquid trapped in the interstices of the bed, into an enriching zone. Here the adsorbate is contacted with a counter-current recycle stream, which is mainly of paraffinic nature (but contains some aromatics), and this displaces the non-aromatics in the bed. From the enriching zone, the rich adsorbent passes to a desorption zone; here it is contacted with hot paraffinic strippant liquid at 200–250°F, and the aromatics are desorbed. Immediately after transfer to the desorption zone, the first increment of effluent is rich in strippant, and this increment is recycled back to the enriching zone. Thereafter the remainder of the effluent is sent to an extract fractionator to separate the paraffin strippant from the aromatic product. Raffinate is also fractionated to recover the paraffinic strippant introduced via the recycle stream. The adsorbent cycle is completed by having the hot adsorbent beds (from the desorption zone) cooled by circulation of cold paraffin strippant; the first portion of such strippant emerging from any bed is at elevated temperature, and is sent to hot strippant storage. The adsorption beds, now cooled, 'pass' to the adsorption zone, and the cycle is re-started.

Materials such as water, organic acids, and nitrogen bases are difficult to desorb, and, in consequence of this, the feedstock is dried and then pretreated to remove

TABLE III.14

Effect of Cyclic Adsorption Processes on Various Feedstocks

Feedstock	*Catalytic reformate*	*Kerosene*	*Catalytic cycle oil*	*Cracked gasoline*
Aromatics, vol %	35·8	22·5	40	19
Bromine no. g/100 g	3	0	9	71
ASTM 10% point	96·7	202·2	240·6	110·0
ASTM 90% point	131·7	245·6	310·0	178·9
Strippant	Pentane	*i*-Octane	*i*-Octane	Pentane
Adsorbent	0·5–0·06 mm silica gel			
Feed/gel, wt/wt	0·3	0·3	0·6	0·6
Recycle/feed, vol/vol	2·27	1·02	1·74	1·28
Strippant/feed, vol/vol	4·7	2·6	5·5	4·9
Cycle time, hrs	1	3	0·75	0·75
Raffinate yield, vol %	65	76	59	78
Raffinate bromine no.	4	0	0	55
Extract yield, vol %	35	24	41	22
Extract bromine no.	1	0	20	80
Extract aromatics, vol %	99·8	92	95	87

such materials. This pretreatment consists of passing the feed over adsorbent guard chambers, which are periodically regenerated by displacement with ethyl alcohol. The alcohol itself is recovered by stripping the adsorbent with hot paraffinic vapours (pentane). Some typical operating results are shown in Table III.14.

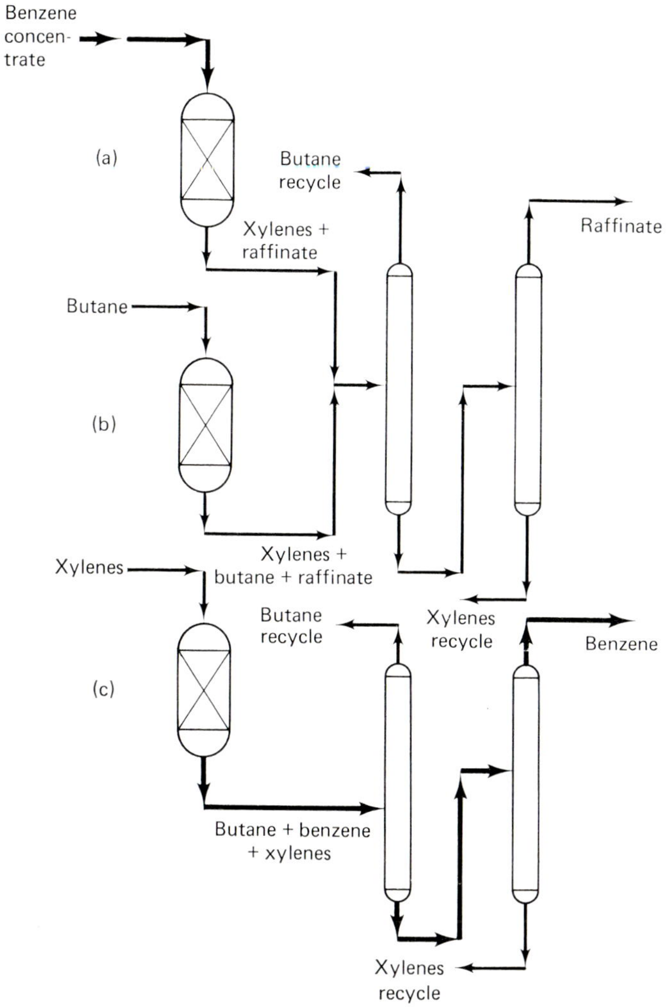

FIG. 3.10 *Production of Benzene by Arosorb Process*

(*a*) *1st step: As bed goes on stream it is saturated with xylenes. Benzene in feed replaces xylenes which pass out in the raffinate.*
(*b*) *2nd step: Butane displaces the benzene boiling range non-aromatics from bed interstices and from adsorbate, leaving butane, benzene and some xylenes in the bed.*
(*c*) *3rd step: Xylenes displace butane and benzene. Some xylenes pass over into the butane-benzene stream.*

3.5.3.2. Adsorption by the 'Arosorb' process This process, like that of cyclic adsorption, is used for separating aromatics from non-aromatics,[45] but it uses a different technique for desorption. Whereas cyclic adsorption uses a paraffinic displacing material, the Arosorb process uses another aromatic material with a boiling range differing from that of the feedstock. By way of illustration of the principle involved, consider the separation of benzene from a hexane stream, using xylenes as the desorbent. Fig. 3.10 and the following notes describe what happens:

(a) first step: as the section goes on stream, the silica gel is saturated with xylenes. Benzene in the feedstock replaces the xylenes, which pass out in the raffinate. The operation is continued until the benzene-xylenes interface approaches the outlet end, when the feedstock is diverted to the next bed.

(b) second step: butane, used as a 'push liquid', now displaces the non-aromatics in the benzene boiling range from the interstices of the bed and from the adsorbate, leaving butane, benzene and some xylenes in the section.

(c) third step: xylenes are now used to displace butane and benzene from the section. Some xylenes pass through into the butane-benzene stream.

The manner in which the effluent from an adsorbent bed changes is shown in Fig. 3.11.

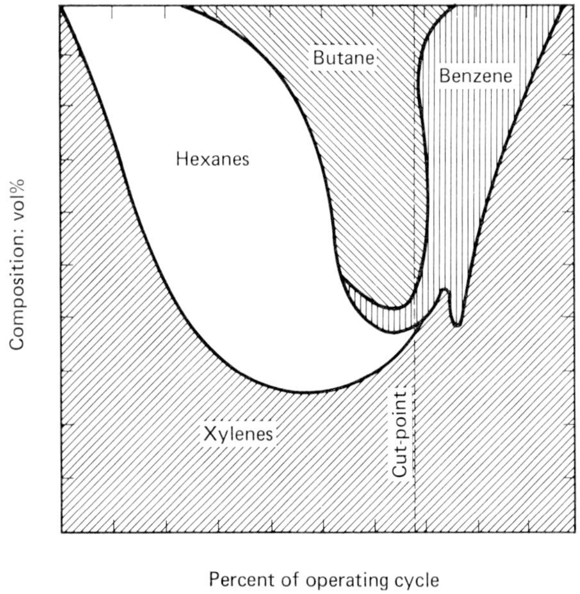

FIG. 3.11 *Effluent from a Typical 'Arosorb' Operation*

First a mixture of xylenes/hexanes come through. This is followed by a mixture of xylenes/hexanes/butanes plus a trace of benzene. Next, as the hexanes disappear the benzene concentration rises sharply, and by the correct choice of cut-point, the effluent can be divided into two streams, one containing most of the hexanes and one containing most of the benzene. (Both contain both 'push liquid' butanes, and xylenes, from which they are separated by fractionation.) To obtain maximum recovery and maximum purity of benzene, the cut-point can be tightened by recycling a 'cut-point fraction' back to the primary feed.

For sizing an Arosorb bed, a fresh feed superficial velocity of 1·5 cm/min. on cross section is recommended.[45]

3.5.4. Extractive and Azeotropic Distillation: General

Extractive and azeotropic distillation have the common feature that a substance not normally present in the mixture to be separated is deliberately introduced into the system in order to increase the difference in volatility of the most hard-to-separate components. Extractive distillation can be defined as distillation in the presence of a substance which is relatively non-volatile compared to the components to be separated, and which, therefore, is charged continuously near the top of the fractionating tower, so that an appreciable concentration is maintained on all plates in the tower below its entry. Azeotropic distillation can be defined as distillation in which the added substance forms an azeotrope with one or more of the components in the feed, and by virtue of this is present on most of the plates in the tower above its entry at an appreciable level of concentration.

These separation methods find their principal applications in the separation of mixtures whose components boil too close together for the economical use of simple fractionating equipment. These separation methods are particularly applicable when the components to be separated differ in chemical type. The theoretical principles involved are well documented,[50, 51, 52, 53] and will not be further considered here. The processes differ in the means used to maintain the desired solvent concentration on the plates of the tower. In extractive distillation the high concentration of solvent is maintained by virtue of its non-volatility, and by the fact that it is charged at a high point in the tower. The solvent is, necessarily, removed from the base of the principal tower. In azeotropic distillation, most of the solvent is taken off overhead, with relatively small amounts (ideally, none) drawn off with the bottoms.

Extractive distillation is generally more flexible than azeotropic distillation, a greater variety of solvents and a wider range of operating conditions are available; and the concentration of solvent may be controlled by heat and material balances rather than by the accident of azeotrope composition. Furthermore, since vaporisation of the solvent is not required, heat loads are usually considerably less. It has been mainly used for the separation of toluene, not benzene, but is mentioned here for the sake of completeness.

Thus the extractive distillation process using phenol, developed by the Shell Development Co.[54] has been widely employed for toluene purification. During World War II more toluene was produced in the USA by this process than by any other.

In a typical operation[55] a narrow-boiling fraction (95–120°C) hydroformate fraction containing 63·4 mol per cent toluene is charged 19 trays from the bottom of a 65-tray tower. Phenol is charged to this tower at a rate of 2·6 vols/vol of feed. Hydrocarbon reflux to the top tray is at a rate of 1·4 vol/vol of feed. The feed is split into almost equal volumetric proportions of overhead and bottoms. The overhead contains 14·3 mol per cent toluene. The bottoms (hydrocarbon product plus phenol) is fed to a 30-tray stripper, the overhead from which is 99·9 mol per cent toluene. This is finished to nitration grade toluene by a light acid treatment and re-running.

The above process is no longer competitive, but a recent development by Heinrich Koppers GmbH, the 'Morphylane' process uses morpholine derivatives as extractive agents and may be competitive if charged with high aromatic feedstocks. The Lurgi 'Distapex' process, using *n*-methyl pyrrolidone has also been offered recently.

3.5.4.1. Azeotropic distillation applied to BTX from petroleum The use of azeotropic distillation as a means of separation of BTX components from other non-aromatic hydrocarbons has been known and employed for some thirty years. Acetone is used as an entrainer to purify benzene from similar-boiling non-aromatic hydrocarbons.[56] Toluene can be separated by the use of either methanol,[57] or of methyl ethyl ketone.[58] Ethyl benzene may be separated from styrene either by iso-butanol,[59, 61] or by *l*-nitropropane.[60, 62]

In a 1966 review paper, further information was made available.[63] Fifty-eight possible entraining agents for separating ethyl benzene (bpt 136·2°C) and *para*-xylene (bpt 138·4°C) have been examined. It would appear that 2-methyl butanol is the most successful agent, requiring a column with only 48 per cent of the number of theoretical plates required if no entrainer were used.

The separation of *para*- and *meta*-xylenes (bpts 138·2 and 139·2°C respectively) is much more difficult. Of thirty-five entrainers examined the best appears to be 2-methylbutanol, but the change of relative volatility is only from 1·020 to 1·029, and hence it can be safely concluded that azeotropic distillation for the separation of the *para*- and *meta*-xylenes is not an economic proposition.

Last, and easiest of the C_8 aromatics is the separation of *meta*- from *ortho*-xylene (bpts 139·2 and 144·5°C respectively). Twenty-eight entrainers were examined, the best being formic acid, requiring a column containing only 70 per cent of the theoretical plates required if no entrainer is used.

Two commercial processes have been developed for the separation of toluene using azeotropic distillation; one using an aqueous solution of methyl ethyl ketone and water[64] and the other using methanol.[65] Both processes operate on a narrow boiling range concentrate. The equilibrium relationships are well established.[66] One commercial plant has been built using the methanol azeotrope which was developed by Socony-Vacuum Oil Co. Fig. 3.12 gives the toluene concentration step. The azeotropic system consists of two towers each containing 36 trays operating at 30 p.s.i.g. The toluene concentrate is mixed with 75 vol per cent of anhydrous methanol for a feed containing 25 vol per cent toluene, and charged to the first tower. Cold reflux of 0·6 to 0·7 vols/vol of feed is employed. Practically complete recovery of toluene is obtained in the form of a 70 vol per cent concentrate from the bottom methanol. Cold reflux of the order 4·0 to 4·5 vols/vol of feed is employed, and toluene recovery from this tower averages about 95 per cent, while overall yields of about 80 per cent of the toluene present in the initial feed are obtained. Losses are distributed mainly in the azeotrope overhead and in the acid treating and re-running stages.

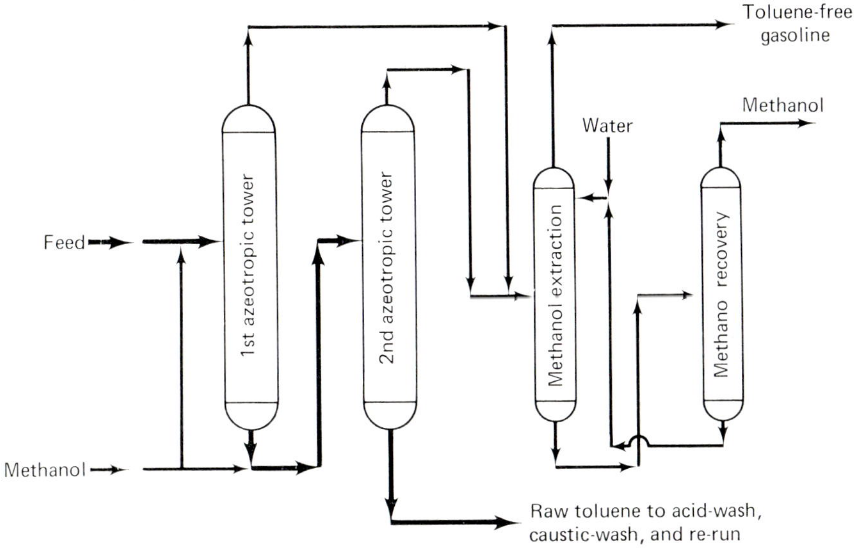

FIG. 3.12 *Concentration of Toluene Azeotropic Distillation with Methanol*

The MEK process, developed by Union Oil Company, has had one commercial plant operating, and handles a wide variety of straight-run and cracked stocks. A preconcentrate feedstock with a toluene content in the range 35–85 vol per cent is charged, together with the solvent (90 vol per cent MEK plus 10 vol per cent H_2O), to the 22nd tray of a 50-tray tower. The volumetric ratio of solvent + water/hydrocarbon charge varies between 2·0 to 3·0 to 1 depending upon the boiling range and type of non-aromatics present. Cold reflux in the ratio of 0 to 1 reflux/feed is employed. The azeotropic distillate is composed of about 75 vol per cent solvent. A small amount of water-rich phase separates from it in the reflux drain (80 vol per cent water, 20 vol per cent MEK) and is recycled back to the feed. 95–97 per cent of the toluene present in the feed is recovered as bottoms from the azeotropic tower as 99 per cent purity toluene on a solvent-free basis. The removal of MEK from the azeotropic distillate is accomplished by four stages of water-extraction thus reducing the ketone content to approximately 1 vol per cent. The washed material is then fractionated in an auxiliary tower to remove the remaining ketone. Recovery of MEK from the water extracts is done by distillation of the MEK-water azeotrope from the excess water in a tower with only a few trays and a low reflux ratio. The azeotrope thus obtained is then fed directly to the primary azeotrope tower together with the toluene pre-concentrate feed. The crude toluene so produced, after removal of MEK is treated with sulphuric acid, and followed by redistillation to produce nitration-grade toluene. Overall recovery of toluene is around 90 per cent.

3.5.5. Crystallisation: General Principles

Crystallisation is a classical technique for producing pure compounds, and its relevance to the separation of BTX can be best understood by reference to the freezing points of the basic aromatics:

TABLE III.15

Boiling and Freezing Points of the Principal Aromatics

Compound	*Boiling point* °C	*Freezing point* °C
Benzene	80·08	−5·56
Toluene	110·8	−95
o-Xylene	144	−25·2
m-Xylene	139·3	−47·9
p-Xylene	138·5	13·3

Most other hydrocarbons in this boiling range have very low freezing points (−50 to −100°C), but some exceptions may be noted. Among the *iso*-paraffins, 2,2,3-Trimethyl butane (Triptane) has a freezing point of −25°C and cyclohexane, has a freezing point of 6·7°C.

From the above table it is evident that:

(a) it is economic to separate benzene from non-aromatics by crystallisation;

(b) it is not economic to separate *meta*- and *para*-xylenes by distillation because of the closeness of their boiling points;

(c) it is possible to crystallise *para*-xylene from other C_8 aromatics because of its high freezing point.

Recent developments in crystallisation equipment have resulted in a number of successful commercial process, the purification of *para*-xylene being the most important application, though the technique has also been used for the separation of pure benzene, e.g. by Shell Chemical Co. at Carrington. Commercial separation and purification processes using crystallisation from a melt can be grouped according to three basic operating approaches:[67] 'Normal Freezing'; 'Zone Refining' and 'Column Crystallisation'. The term 'normal freezing' is applied to processes in which a crystal layer is deposited upon one which has been formed from the same melt. The freezing of a lake is an illustration of this principle, that is, it is the 'normal' or 'natural' manner for freezing to take place. Commercial production of naphthalene and benzene has been described in the literature.[68] In general terms, a large charge of impure material is introduced into the shell side of a shell and tube heat exchanger. Coolant is fed into the tube side at steadily decreasing temperatures until the entire charge is solidified. Then a heating fluid is circulated through the tube-bundle and the material is re-melted, drained from the unit and collected in fractions. The more impure material melts at lower temperatures and hence the fractions collected increase in purity. Cycle times of 24–48 hours are used.[69]

In many industrial applications, crystal growth occurs primarily on nuclei within the melt rather than on the crystal surfaces which are formed directly on the cooling surface of the equipment. In carrying out this type of operation it is often convenient to utilise direct contact cooling with an immiscible fluid, thereby eliminating the need for expensive heat-transfer equipment. This principle has been applied in the recovery of benzene from petrochemical by-products.[69]

Zone melting is a recent development.[70] The procedure differs from normal freezing in that a molten zone is passed through an ingot, instead of forming a solid ingot from an entirely molten initial charge. The most important commercial applications of zone refining are in the production of high-purity materials for semi-conductors.

Column crystallisation appears to offer the greatest potential for commercial application to hydrocarbon and other organic materials. In a column crystalliser, solid and melt are moved counter-currently in intimate contact. Advantages inherent in this method of processing are firstly, the efficient re-use of the energy of crystallisation, and secondly the possibility of operating a single crystallisation unit under steady-state, continuous-flow conditions.

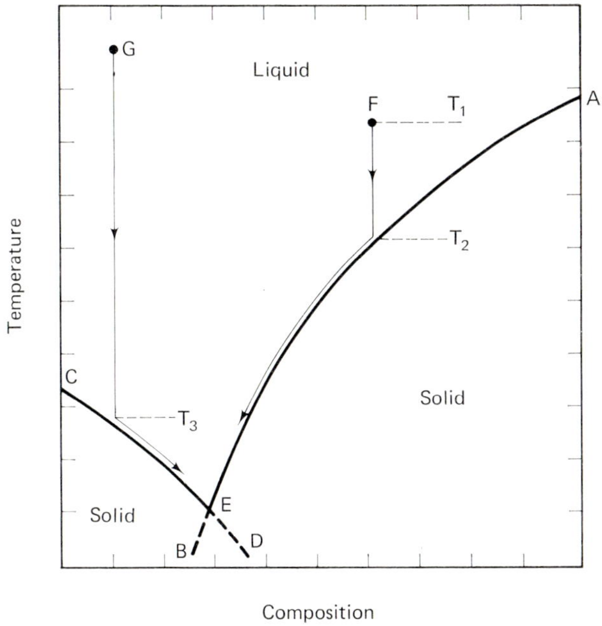

FIG. 3.13 *Typical Freezing-Point Diagram*

It is well known that the freezing point of a liquid is lowered by the addition of a second, miscible liquid. The system *p*-xylene/*m*-xylene is illustrated in Fig. 3.13. Pure *p*-xylene freezes at 13·3°C. Addition of *m*-xylene causes its freezing point to fall along the line AB of Fig. 3.13. Similarly, *m*-xylene freezes at −47·9°C, and addition of *p*-xylene causes its freezing point to fall as indicated by the line CD. The two lines AB and CD intersect at point E. At this point both xylenes crystallise out together and the crystals are the same composition as the liquid. It is clear from the diagram that only one material can be crystallised in pure form out of a given mixture, and the recovery of that component is limited by the composition of the eutectic point E. Consider the cooling of a solution of composition and temperature T_1 shown at point F in Fig. 3.13. Nothing happens until the temperatures reaches T_2 when *p*-xylene begins to freeze out. Further cooling of the mixture causes the temperature and composition of the remaining liquid to follow the line AB. Similarly, if starting at composition point G, freezing out of *m*-xylene starts at temperature T_3 and further cooling causes composition and temperature changes to follow the line CE.

Finally it should be noted that because the crystals formed in the cooling process trap liquid in their interstices, it is not possible to obtain a pure product in one stage. Melting and subsequent recrystallisation are frequently used for purification.

3.5.5.1. Benzene production by crystallisation Benzene with a crystallising point higher than 5·4°C can be produced from any suitable feedstock by the Newton Chambers Crystallisation Process.[71] This process will remove any impurity other than thiophen. Benzene yield is dependent upon feedstock quality, and yields in excess of 95 per cent product benzene with a crystallising point of 5·45°C can be obtained.

A Process Flow Diagram is shown in Fig. 3.14. Crude benzene feedstock (thiophen-free) is mixed with cold brine to cause the crystallisation of the majority of the benzene in the feed. This benzene/brine slurry is fed to a batch centrifuge, which separates benzene crystals from mother liquor plus brine. After the crystals have been spun, they are treated while still in the centrifuge with a wash liquor, and the organic layer removed at this stage is recycled to the feedstock in order to raise the overall yield from the process. The crystals are given a further centrifugal treatment before being discharged into a crystal melter. From here, pure product benzene is sent to storage.

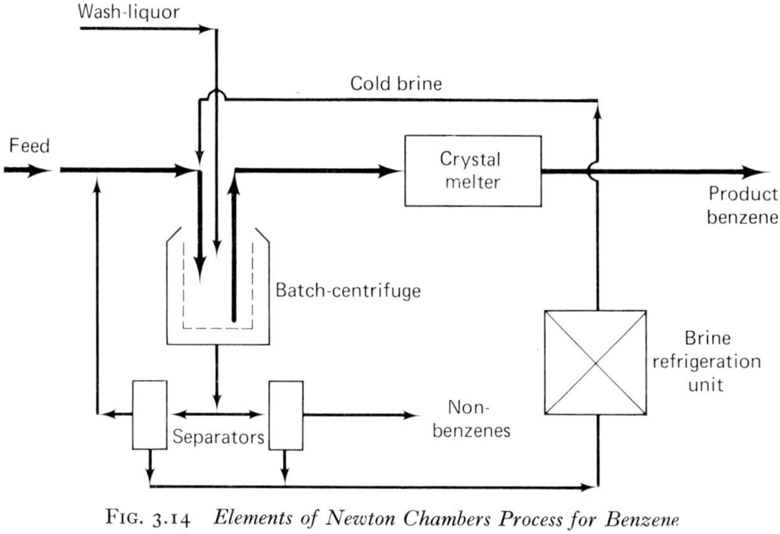

FIG. 3.14 *Elements of Newton Chambers Process for Benzene*

The process is not truly continuous, but is cyclic in operation, and being time-cycle controlled, supervision is held to a minimum.

3.5.5.2. p-xylene production by crystallisation This process is of interest to the benzene producer, first, as demonstrating an application of crystallisation, and second, because BTX production must be treated as a whole, with each component earning the maximum revenue to be profitable.

As already indicated in the section on 'Crystallisation: General Principles', the fractional crystallisation of a binary system which forms an eutectic can only produce one of the components in a pure form. However, if a third component is introduced into a binary system, thus altering the solid-liquid phase relationships, then it is possible by a series of crystallisation steps, to separate the desired two components in a pure form. The added third component, normally a liquid, is called the solvent, and the process is known by the general name of 'extractive crystallisation'. Such a process has been used to separate *meta-* and *para*-cresols.[72] The separation of *meta-* and *para*-xylenes by the use of *n*-heptane has been described.[73] No industrial applications for this process have as yet been reported.[74]

Counter-current fractional crystallisation has been successfully developed for the separation of *para*-xylene (an essential raw material for the manufacture of terephthalic acid) from a mixture of xylenes. This process has been developed by Phillips Petroleum Co. (USA), and a schematic diagram is given in Fig. 3.15. Feedstock containing about 15–20 per cent of *p*-xylene is pre-cooled to about −40°C, before being passed to a scraped-surface chiller, cooled with ethylene refrigeration. The emergent stream, at about −75°C, and containing a mass of tiny crystals of *p*-xylene, passes to a rotary filter where a filter-cake containing about 65 per cent of *p*-xylene is recovered. The mother-liquor, containing about 7 per cent of *p*-xylene and at a temperature of about −75°C is used to pre-cool the incoming feed, and is then recovered.

The filter-cake, of high *p*-xylene content, is melted, and passed to a second scraped surface exchanger, this time cooled with ammonia, and emerges as a crystalline slurry or magma at about −18 to −26°C. The crystals are extremely small. This slurry is then pulsed through a column containing a filter at the upper end, which removes mother-liquor (33–42 per cent *p*-xylene), and a melting section at the base. High-purity crystals (above 98 per cent) fall to the base of the tower, where they are melted; part of the melt is removed from the system as finished product, and the rest is forced back up the column counter-current to the falling crystals. The counter-current contacting of the warm, pure melt with the cold, impure crystals, results in the partial freezing of the melt, and the crystals are returned to the melting section. The overall result is the almost perfect separation of crystals and mother-liquor. Careful control of the temperature gradient across the column (−75 to +13°C) can produce a finished product containing 99·5 per cent of *p*-xylene. The mother-liquor contains about 5 per cent *p*-xylene.

The column itself is quite small, rarely exceeding 3 ft in height, and its diameter is small compared to its capacity; production rates from 2–6 m^3 hr/m^2 cross section have been reported.[74]

Further details of this have been given in the literature.[75, 76]

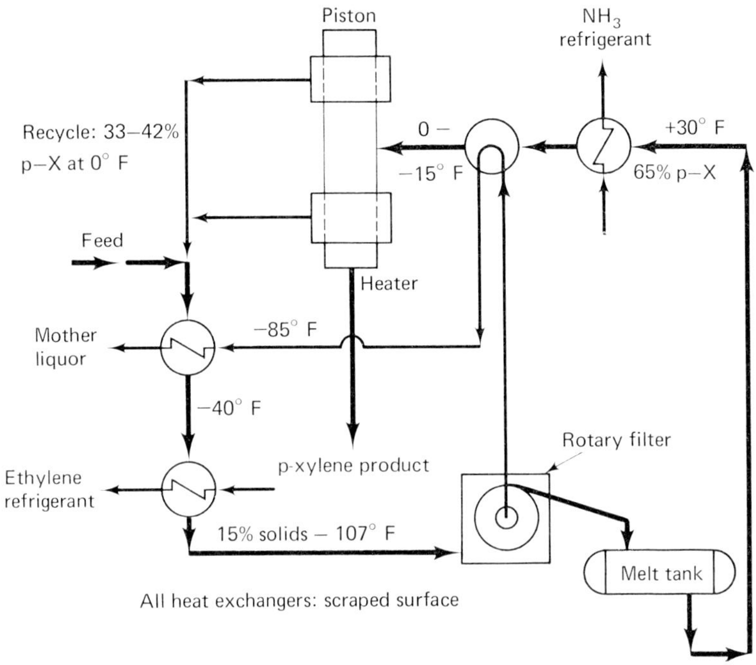

FIG. 3.15 *Elements of the Phillips* p-*Xylene Process*

3.5.6. Desulphurisation of BTX by Sodium

Petroleum fractions from certain sources are high in both aromatics and sulphur and in such cases it may prove desirable to remove the sulphur contaminants by treatment with metallic sodium. Finely divided sodium is a versatile treating agent for removing reactive compounds from petroleum stocks.[77] At about 20°C, it reacts only with compounds having active hydrogen, i.e. phenols and mercaptans. At temperatures of about 200°C sodium reacts with sulphur compounds to form sodium sulphide. While no commercial applications of sodium treating have been reported, the process is the subject of a US patent,[78] and has been described in the literature.[79]

Originally developed for desulphurising coke-oven light oils, it is equally applicable to appropriate petroleum stocks, as Table III.16 illustrates:[80]

US Industrial Chemicals Company (the process owners) consider that if less than 75 cu. m per day of BTX are being treated, a ribbon blender contactor is the most suitable device. If larger quantities are being treated, then a fluid bed contactor should be used as giving lower capital investment and total costs. (Private communication.)

If the crude BTX cut contains appreciable materials of CS_2, simple mercaptans and disulphides, light C_5 di-olefins, and heavier olefins of the coumarone-indene type, then these must be removed prior to sodium treatment, in order to avoid excessive sodium usage, as well as to minimise polymer formation.

TABLE III.16

Treatment of Various Feedstocks with Sodium

Material treated	*Temp.* °C	*Time* sec	*Recovery %*		Na used kg/ton	*Sulphur %*	
			Liquid	*Polymer*		*Feed*	*Product*
Light virgin naphtha	300	8	99	0·5	2·0	0·037	0·001
Light cat. naphtha	300	25	96	3·1	5·3	0·061	0·001
Heavy cat. naphtha	315	10	95	3·3	12·5	0·29	0·053
Benzene	314	6	98·5	1·0	14·2	0·7	Thiophen-free
Solvent hexane	250	11	96·3	0·6	2·2	Sour	Sweet
Solvent heptane	253	13	96·5	1·1	3·3	Sour	Sweet

The preparation of sodium dispersions is of the utmost importance. Sodium particles in the range 1 to 20 microns are small enough to react completely, using very nearly the theoretical amounts of sodium. The term High Surface Sodium (HSS) is applied to films of sodium approaching colloidal dimensions spread over inert solids of high surface area. Its preparation is accomplished simply by mixing molten sodium with suitable inert solid materials having very large surface areas. Recommended materials are blast furnace slag, silica sand, soda ash, and salt. HSS may be readily prepared by mixing molten sodium with any of the inert solid materials already mentioned. Usually, sodium spreads spontaneously over the surface of hot dry solids at temperatures between 100 and 200°C. In order to obtain maximum efficiency and to protect the reactivity of HSS it is necessary to carry out its preparation and subsequent use under an inert atmosphere. Likewise, whenever possible, it is desirable to prepare and use HSS in the same reaction vessel. Basic designs for such equipment are available and requirements for handling metallic sodium are completely detailed.[81]

The Mechanically Agitated System is illustrated in Fig. 3.16. A jacketed, ribbon blender, containing an inert sodium carrier, is used. The inert material in the blender is coated with metallic sodium at a level of 2–6 per cent sodium metal and is under nitrogen. The blender is heated by circulating hot oil. Metered benzene is vaporised into the ribbon blender, thereby eliminating thiophene and other sulphur compounds. Sulphur-free vapour is taken overhead and condensed to storage.

The Fluidised Solids System (for larger capacities) is illustrated in Fig. 3.17. The inert material is in the 80–200 mesh size, and a level of 2·5–3·0 per cent sodium is carried. With a vapour temperature of 225–315°F, and a contact time of 2–6 seconds, thiophene removal is completed.

In both cases, spent beds can be fed to a screw conveyor, and treated with steam to eliminate any last traces of active sodium. Investment for sodium desulphurisation plant is low. Both types of unit will reduce thiophene content from 1,000 p.p.m. to less than 1 p.p.m.

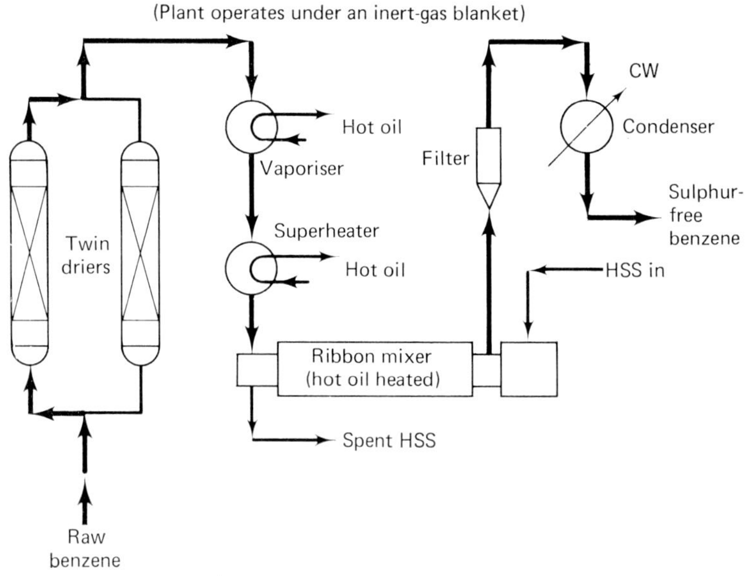

FIG. 3.16 *Desulphurising of Benzene with High Surface Sodium (HSS) Using a Mechanically Agitated Bed*

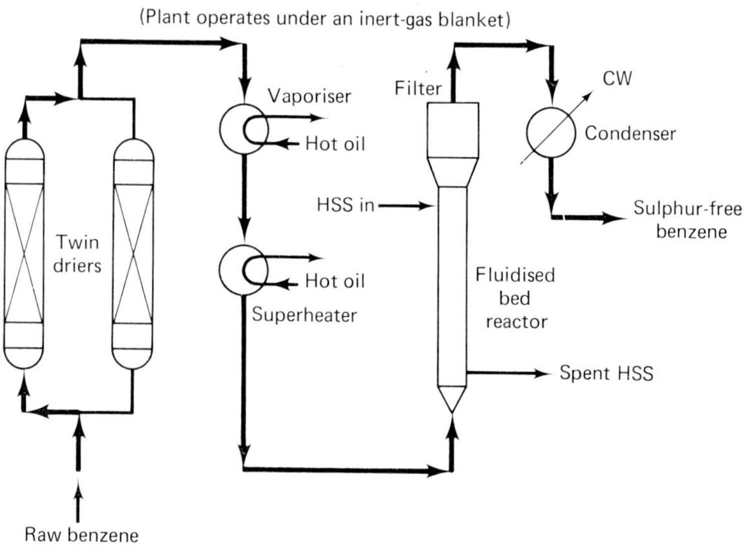

FIG. 3.17 *Desulphurising of Benzene with High Surface Sodium (HSS) Using a Fluidised Bed Reactor*

Note by Editor It was appreciated that many of the techniques for working up and refining benzene would be identical whether the benzene arose from coal or petroleum. Chapter 2 was written first and handed to the authors of Chapter 3 with the request that they would not duplicate description of processes dealt with in detail in Chapter 2. Readers interested in other methods of refining petroleum benzene should therefore refer to Chapter 2.

REFERENCES

[1] Hill, Vincent and Everett, *Trans. Am. Inst. Chem. Eng.*, 1946, **42**, 611.
[2] Jacobson and McCoy, *Hydrocarbon Processing*, 1970, **49**, 109.
[3] Kopf, Decker, Pfefferle, Dalson and Nevison, *Hydrocarbon Processing*, 1969, **48**, 111.
[4] Burd and Masiuk, *Hydrocarbon Processing*, 1972, **51**, 97.
[5] Dunstan, Hague and Wheeler, *Ind. Eng. Chem.*, 1934, **26**, 307.
[6] Frey and Hepp, *Ind. Eng. Chem.*, 1932, **24**, 282.
[7] Maisel, *Petroleum Processing*, August 1953, p. 1185.
[8] Swaminathan, *Petroleum Engineer*, June 1954, p. C7.
[9] Humphreys and Glasgow Ltd., Private Communication.
[10] Miller, S. A., '*Ethylene and its Industrial Derivatives*' (Ernest Benn, 1969), p. 82.
[11] Tsutsumi, *Petroleum Refiner*, 1957, **36**, 289.
[12] Brooks and Newman, Eighth World Petroleum Congress, Moscow, June 1971.
[13] Gladisch, *Hydrocarbon Processing*, 1962, **41**, 159.
[14] Rosendahl, *Chim. et Ind.*, 1960, **84**, 357.
[15] Tebboth, *Inst. Pet. Rev.*, October 1967, p. 347.
[16] Lester, *Hydrocarbon Processing*, 1961, **40**, 175.
[17] Kalichevsky and Kobe, '*Petroleum Refining with Chemicals*' (Elsevier, 1956).
[18] Nelson, '*Petroleum Refinery Engineering*' (McGraw-Hill, 1958).
[19] Bland and Davidson, '*Petroleum Processing Handbook*' (McGraw-Hill, 1967).
[20] Verdol, J. A., '*The Xylenes Plus Process*' (available from Engelhard).
[21] Leprince, Chauvel, Catry and Castex, '*Procedes de Petrochimie* (Editions Technip, Paris, 1971).
[22] Anon, *Hydrocarbon Processing*, 1971, **50**, 133.
[23] Otani, Matsuoka and Sato, *Japan Chemical Quarterly*, 1968, **4**, 16.
[24] American Chemical Society's 162nd Annual Meeting, 12 September 1971.
[25] Grandio, Schneider, Schwartz and Wise, *Oil and Gas J.*, 29 November 1971, p. 62.
[26] Edeleanu, USP 911,533; BP 11,140 (1908).
[27] Morello and Poffenburger, *Ind. Eng. Chem.*, 1950, **42**, 1021.
[28] Mumford, C. J., *Brit. Chem. Eng.*, 1968, **13**, 981.
[29] Reman and Olney, *Chem. Eng. Prog.*, 1955, **51**, 141.
[30] Reman and Van de Vusse, *Petroleum Refiner*, 1955, **34**, 129.
[31] Crawford and Wilke, *Chem. Eng. Prog.*, 1951, **47**, 423.
[32] Mayfield and Church, *Ind. Eng. Chem.*, 1952, **44**, 2253.
[33] Hunter and Nash, *Proc. World Petroleum Cong.*, 1933, II. 340.
[34] Evans, *Ind. Eng. Chem.*, 1934, **26**, 439.
[35] Gruse and Stevens, '*Chemical Technology of Petroleum*' (McGraw-Hill, 1942).
[36] Treybal, '*Liquid Extraction*' (McGraw-Hill, New York, 1951).
[37] Scheibel, *Ind. Eng. Chem.*, 1954, **46**, 16; *Petroleum Refiner*, 1959, **38**, 227; 1961, **40**, 189, and *Chem. Eng.*, November 1957.
[38] Kalichevsky and Stagner, '*Chemical Refining of Petroleum* (Reinhold, 1933).
[39] Gerster, *Advances in Chemistry*, 1951, **5**, 177.
[40] Wilkinson, Ghublikian and Obergfell, *Chem. Eng. Prog.*, 1953, **49**, 257.
[41] Anon, *Petroleum Refiner*, 1952, **31**, 99, 259.
[42] Ratliff and Strobel, *Petroleum Refiner*, 1954, 33, 151.

[43] Anon, *Hydrocarbon Processing*, 1966, **45** (2), 231.
[44] Deal, Evans, Oliver and Papadopoulos, *Petroleum Refiner*, 1959, **38**, 185.
[45] Shuman and Brace, *Petroleum Engineer*, 1953, **25**, C9.
[46] Hengstebeck, '*Petroleum Processing*' (McGraw-Hill, 1959), pp. 81, 85.
[47] Eagle and Scott, *Ind. Eng. Chem.*, 1950, **42**, 1287.
[48] Eagle and Rudy, *Ind. Eng. Chem.*, 1950, **42**, 1294.
[49] Eagle and Scott, *Petroleum Processing*, 1949, **4**, 881.
[50] Happel, '*Science of Petroleum*', Vol. 5, pt. 2, 1952.
[51] Benedict and Rubin, *Nat. Pet. News*, 1945, **37**, No. 36.
[52] Scheibel, *Chem. Eng. Prog.*, 1948, **44**, 927.
[53] Colburn and Schoenborn, *Trans. Am. Inst. Chem. Eng.*, 1950, **46**, 421.
[54] Dunn *et al.*, *Trans. Am. Inst. Chem. Eng.*, 1945, **41**, 631.
[55] Drickamer *et al.*, *Trans. Am. Inst. Chem. Eng.*, 1945, **41**, 555.
[56] Field, USP 2,212,810.
[57] Lee and Holm, USP 2,319,694.
[58] Clark, USP 2,388,040.
[59] Berg *et al.*, *Ind. Eng. Chem.*, 1946, **38**, 1149.
[60] Carlson, *JSCI*, 1944, 333.
[61] Lake, USP 2,375,478.
[62] Natta, USP 2,308,229.
[63] Berg, *Hydrocarbon Processing*, 1966, **45**, 103.
[64] Lake, *Trans. Am. Inst. Chem. Eng.*, 1945, **41**, 327.
[65] Foster, *Oil and Gas J.*, 1944, **42**, 130.
[66] Benedict and Rubin, *Trans. Am. Inst. Chem. Eng.*, 1945, **41**, 353.
[67] Powers, *Hydrocarbon Processing*, 1966, **45**, 97.
[68] Molinari, '*Symposium on the Less Common Means of Separation*' (Institution of Chemical Engineers, 1964), p. 63.
[69] Molinari, '*Fractional Solidification*' (Marcel Dekker Inc., 1967), p. 393.
[70] Pfann, '*Zone Melting*' (John Wiley, 1966).
[71] BP Appl. 26499/58.
[72] Chivate and Shah, *Chem. Eng. Sci.*, 1956, **5**, 232.
[73] Findlay and Weedman, '*Advances in Petroleum Chemistry and Refining*' (Interscience, 1958), Vol. I.
[74] Mullin, '*Industrial Crystallisation*' (Butterworth, 1961), p. 168.
[75] Weedman and Finlay, *Petroleum Refiner*, 1968, **37**, 195.
[76] McKay *et al.*, *Chem. Eng. Prog.*, 1967.
[77] Vanderbilt, *Ind. Eng. Chem.*, 1957, **49**, 697.
[78] USP 2,960,546.
[79] Hansley, Watson and Faye, *Hydrocarbon Processing*, 1961, **40**, 127.
[80] USI Bulletin, 'High Surface Sodium', 1953.
[81] USI Bulletin, 'Handling Metallic Sodium', 1959.

CHAPTER 4

THE PHYSICAL PROPERTIES OF BENZENE

by F. A. Jackman, R. W. Gallant, G. A. M. Diepen and M. A. de Zeeuw

4.1 GENERAL PROPERTIES

by F. A. Jackman

4.1.1. Introduction

BENZENE, as a chemical product and at the same time a raw material for many processes, finds itself in a rather special category, shared only by a small number of compounds mainly of petrochemical origin. Generally speaking its principal physical properties have been well known for many years, although the values quoted in reference books have varied somewhat. It is recognised that the reason for these variations is basically the question of the purity of the sample under examination. The tremendous surge in the use of benzene for chemical synthesis over recent years has provided the means for the production of benzene of ever-increasing purity, and thus the source of physical data of greater and greater accuracy. Although there have been many recent refinements in the design of laboratory apparatus for the determination of physical constants, it is most probable that a more important factor in the attainment of greater accuracy has been the availability of specimens of very high purity.

It is interesting to note in Table IV.1 how the recorded results for the freezing point of benzene have varied over the long years since it was first discovered.

TABLE IV.1

1825	Faraday	5·5°C
1833	Mitscherlich	7·0°
1914	Richards and Shipley	5·483°
1920	Sidgwick	5·7°
1946	Rossini	5·533°

In the petrochemical and carbochemical industries which produce benzene in very large quantities, it has long been a convention that the boiling point test, in which a boiling range is determined, has been the main criterion of quality, but in present-day circumstances this standard is rapidly being superseded because the quality of product is so high that simple apparatus as used in an industrial works laboratory is insufficiently precise to determine the minute traces of impurities which may be present. Modern thinking tends towards the opinion that a freezing point determination, combined with an analysis by gas-liquid chromatography, will provide all necessary controls for the supply and use of what is today a highly refined chemical, manufactured in very large quantities. In this chapter many of the physical properties of benzene are examined with regard for the historical, academic and industrial viewpoints.

Necessarily in a work of this nature the author must draw heavily from the publications of others, and he would wish to acknowledge with deep gratitude the outstanding contributions to the knowledge of the properties of benzene from two workers of world-wide reputation, F. D. Rossini and his colleagues in the United States, and G. Claxton and his colleagues in Britain. References to the original works have been given wherever possible.

4.1.2. Boiling Point

Laboratory apparatus for the determination of the boiling point of a single substance with extreme accuracy has been available for very many years. In the case of benzene, however, the obstacle has been, for this property, mainly that of ensuring the removal of every sort of contaminant which could cause a variation in the boiling point observed.

Until comparatively recent times most benzene for industrial use was made by the carbonisation of coal, and the removal of the last traces of some contaminants was almost impossible. Impurities such as carbon disulphide, pyridine, cyclohexane and paraffin hydrocarbons occur in the crude benzene, some of which have boiling points fairly close to 80°C so that separation by fractional distillation is difficult, and the older conventional refining methods could not effect total removal. The availability of bulk supplies of benzene of petroleum origin has reduced the number of trace impurities and the introduction of modern refinery techniques has resulted in a substantial improvement in the general standard of quality. Nevertheless for the most precise determinations of boiling point and other physical properties it is still necessary to use specimens of benzene prepared synthetically. The classic method of reacting benzoic acid with lime and distilling off benzene is commonly used.

There are two other sources of error in these determinations. One is the presence of traces of water, which is very slightly soluble in benzene, and this must be removed by intensive drying. When benzene is kept in contact with drying agents such as phosphorus pentoxide or sulphuric acid for long periods, another source of error may appear in the presence of an associated molecular form. Baker has found that this association proceeds at least to the double molecule,[1] and the effect is to raise the boiling point considerably, figures of 94°C and 106°C having been observed.

For pure uncontaminated benzene the generally accepted boiling point at 760 mm atmospheric pressure is 80·099°C.[2]

4.1.2.1. Effect of variations in atmospheric pressure upon boiling point Changes in atmospheric pressure cause variations in the boiling point of benzene in accordance with the normal pattern defined by the Clapeyron–Clausius equation. This is derived from the second law of thermodynamics and states the relationship between the liquid and vapour phases when they are in equilibrium (*v.* Section 4.3).

Table IV.2 gives a selection of boiling points of benzene which obtain at different atmospheric pressures.[3] These results are also shown on the accompanying graph (Fig. 4.1).

TABLE IV.2

Atmospheric pressure mm *of Hg.*	*Boiling point* °C
760	80·099
700	77·8
600	73·1
463	65
394	60
273	50
184	40
122	30
76	20
46	10
26	0

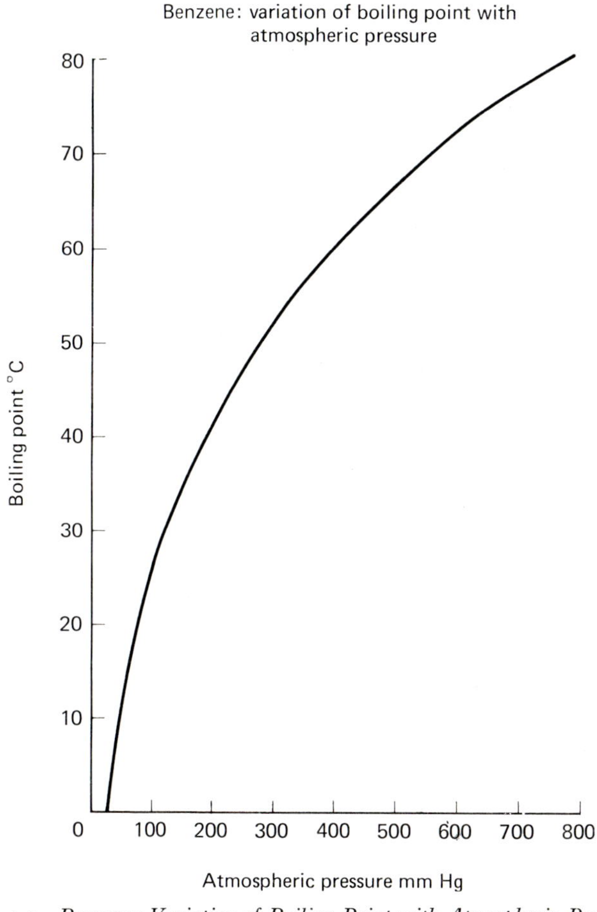

FIG. 4.1 *Benzene: Variation of Boiling Point with Atmospheric Pressure*

4.1.2.2. Effect of the presence of other substances upon the boiling point of benzene In his work '*The Constituents of Coal Tar*', P. E. Spielmann[4] cites a number of values for the boiling point of benzene, *viz.*, 80·08°C, 80·12°, 80·2° to 80·4°. These variations are almost certainly due to the presence of various impurities, and have considerable relevance in the design of refining and fractionation equipment for the industrial production of benzene.

For this purpose the most important hydrocarbon impurities are *n*-heptane (b.p. 98·427°), toluene (b.p. 110·626°), cyclohexane (b.p. 80·738°), methyl cyclohexane (b.p. 100·934°) and methyl cyclopentane (b.p. 71·812°). Many of the data are given in a publication by Benzole Producers Ltd.[5] and the subject is dealt with in very great detail by Claxton and his co-workers.[6]

Boiling point differences for many blends of organic liquids with benzene have been investigated very fully by J. S. Lewis.[7]

4.1.3. Freezing Point (Melting Point)

Under perfect test conditions and with a sample of complete purity, the freezing point and melting point should theoretically be identical. In practice there are usually minute differences, although once again the general availability of material of higher quality has reduced these. Two common impurities which have a discernible effect on the result are air and water, and it is quite usual in industrial practice to ensure that the sample under test is saturated with these two contaminants before taking the readings. Subsequently it is necessary to apply corrections to bring the result back to a dry, air-free basis. It is usually accepted that the most accurate determination of the freezing point of benzene of absolute purity gives a result of 5·533°C ± 0·010, and was carried out by F. D. Rossini and his co-workers.[8] The correction applied to the observed reading, according to the British Standard method, when the sample is saturated with water and the result for dry material is required, is to add 0·09°C.

4.1.3.1. Elevation and depression of freezing point As in the case of the boiling point, the two main features which may affect the freezing point of benzene are the atmospheric pressure to which the sample is subjected, and the presence in solution of other organic compounds.

Under the influence of increased atmospheric pressure, a corresponding rise in the freezing point is observed. Claxton records an average increase up to 500 atm of 0·0285°C per atmosphere.[9] Other hydrocarbons, such as may be found in benzene produced industrially on the large scale, cause depressions of the freezing point. These also are recorded by Claxton,[9] who refers to the detailed information given in the International Critical Tables.[10] One of the older refining techniques, now revived, is the use of fractional crystallisation for the production of very pure qualities of benzene, and the data mentioned are of value for the design of equipment for this process.

4.1.4. Density

In recent years, under the auspices of the International Standards Organization, there has been a growing trend towards the universal acceptance of standards of density which use as the reference datum the value for water at 4°C. This unit is by definition the mass of 1 ml of water at 4°C, one gram. Thus it is now usual to express density of a specimen, as at an ambient temperature of 20°C, by relation to this reference datum, and the value is usually written d_4^{20}. For benzene this figure is 0·8790. The ambient temperature of 20°C is acceptable throughout the temperate zones of the world, but in tropical climates, such as for example those in India, it is necessary to use conversion tables as the ambient temperatures are often substantially higher. Indian national standards use a reference datum of 30°C.

There is a further conflict in practical use because a large section of the petroleum industry use, for the purpose of calculation, specific gravity at 60°F (15·54°C), which uses as its reference datum the specific gravity of water at 60°F. This practice is also supported by Customs usage, and facilitates quick calculations based upon the premise that one Imperial Gallon of water weighs 10 lb avoir., and thus that the weight of a known volume of benzene, etc., can be obtained rapidly by the formula:

$$\text{Weight in lb} = \frac{\text{Volume in gallons} \times \text{Sp.Gr. at } 60^\circ\text{F}}{10}$$

The increasing use of metric measurement is likely to accelerate the change to the density system referred to water at 4°C.

Tables have been produced for the conversion of values from each system to the other. A convenient version is contained in the British Standard Specification for Benzene.[10]

4.1.4.1. Coefficient of expansion In the industrial production and use of benzene it is frequently necessary to make calculations for the conversion of density readings from one temperature to another, and these are based upon the coefficients of expansion for the product. Reliable tables have been drawn up and published to facilitate such calculations.[11] For benzene the coefficients are:

0·001209 at 20°C
0·001261 at 40°C

4.1.4.2. Molar volume The molar volume at any given temperature can be calculated by dividing the molecular weight in grams by the density at that temperature.

The molar volume for benzene is 88·9.

4.1.4.3. Mixtures of aromatic with non-aromatic hydrocarbons When benzene is used for chemical synthesis it is unusual to find it in mixtures with paraffin hydrocarbons, but this situation is of course normal in blends for motor fuels. It has been found that significant volume changes occur when such mixing takes place, and it is of some importance to oil companies, customs authorities and others involved in dealing with the mixtures that the volume changes should be predictable. G. Claxton[12] has evaluated these changes statistically, and has shown that the degree of change varies considerably with the individual hydrocarbons. As an example, whereas a benzene/heptane mixture with 65 per cent by weight of benzene shows a volume expansion of 0·49 per cent, a comparable toluene/heptane mixture with the same proportions only shows 0·15 per cent expansion and a xylene/heptane mixture shows only 0·04 per cent expansion. Tables giving the detailed statistical factors have been published by Kurtz and Sankin.[13]

4.1.5. Refractive Index

Observations of refractive index are sensitive to a number of external influences, which include temperature, atmospheric pressure, the wavelength of the light source and the air in which the test is made. To achieve an absolute standard the readings are usually conducted at 20°C, 760 mm Hg. barometric pressure, using the light of the sodium D line of the spectrum and in vacuum. Correction factors are available so that readings for benzene carried out under ambient conditions in air can be converted to the absolute standard.

The refractive index for benzene at 20°C and 760 mm Hg. pressure, using a sodium D light source and carrying out the test in dry air, is 1·50112. Values for tests performed in other conditions are given by G. Claxton[14] and Rossini, Mair and Streiff.[2]

4.1.6. Vapour Pressure

The fact that vapour pressure of liquids will vary with temperature is a fundamental law of physics which is discussed in detail in the section of this book which deals with phase equilibria. There are, however, some practical applications for which some knowledge of the factors involved is essential, and these occur equally in the refining, storage and use of benzene. For the design of distillation columns, the calculation of stresses in storage tanks, and the considerations applicable to carburation of motor fuels, it is required that vapour pressure changes which occur in different physical conditions shall be fully understood. For vapour pressures between 10 mm and 1500 mm Hg. the Antoine Equation is generally applied:

$$\log_{10} p = A - \frac{B}{C + t}$$

where p and t are respectively the pressure in mm Hg and the temperature in °C. A, B and C are constants, and have values as follows for benzene:[2]

$$A = 6{\cdot}91210$$
$$B = 1214{\cdot}645$$
$$C = 221{\cdot}205$$

4.1.7. Inflammability (Alternative: Flammability)

It is a matter of common knowledge that benzene is highly inflammable, and from the legal and safety viewpoints it is always so regarded. The relevance of this property is far greater when benzene is being used either as a solvent or as a fuel, and in both of these cases it is unusual to be dealing with the pure product. Solvent grades often contain some toluene and a little xylene, and motor fuels are almost invariably blends of benzene and other aromatic hydrocarbons with aliphatic and cyclic (naphthenic) materials. In the main uses of benzene as a chemical raw material, it is of course a matter of good works practice that its inflammability should be recognised and appropriate precautions taken, although it is frequently used in chemical processes in conjunction with other raw materials which may be equally or even more hazardous.

4.1.7.1. Flash point Basically, a test for flash point consists in placing the required sample in an enclosed cup-like container which is fitted with an aperture through which a small flame can be applied. The temperature of the sample is raised slowly and steadily, the flame is applied at regular intervals and eventually the vapour contained in the sample container ignites. This is known as the flash point, but it will be readily noted that the test is entirely empirical as it depends on a number of extraneous factors. These include the size of the sample, the volume of the vapour space above it, the frequency of application of the test flame, the size of the flame, and several others.

To achieve anything like repeatable results it is necessary to lay down a rigid specification for the dimensions of the Flash Point apparatus and its method of use. In the United Kingdom this is set out by law in the Petroleum (Consolidation) Act of 1928, and the method of test thus has legal force which is applicable in such matters as the licensing of storage premises and the construction of vessels and road vehicles. The apparatus used in the U K is usually known as the Abel Apparatus, so named after its inventor Sir Frederick Abel. In Germany the Abel-Pensky apparatus is in general use, and in the United States the Tag (Tagliabue) method is most usual. It is somewhat difficult to correlate results by these different methods.

Initially in the U K the Flash Point test was intended only to determine whether a hydrocarbon, or mixture containing hydrocarbons, had a flash point above or below 73°F, a limit which had legal significance in connection with storage or transport. Subsequently the same method was used for materials such as petroleum white spirit, coal tar naphtha and turpentine which all normally have flash points over 73°F. The method is really quite unsuitable for materials with low flash points which must be cooled to very low temperatures before the test can be carried out. Nevertheless the results of such tests are often quoted but must be regarded as unreliable.

In the case of benzene there are further difficulties because the flash point by the Abel apparatus is certainly well below the freezing point of the product. Thus in cooling the sample it will become solid and the vapour will be emitted less regularly. At an approximate estimate the observed flash point is about 8–10°F.

4.1.7.2. Spontaneous ignition temperature The phenomenon of spontaneous ignition is familiar to anybody who understands the internal combustion engine, whether it be the spark-ignition or the compression-ignition type. Hydrocarbon fuels will ignite and combustion will commence in certain conditions, when the appropriate limits of temperature and pressure are reached, without the necessity of applying a spark or other source of flame.

In fact the compression-ignition or Diesel engine relies upon this characteristic for its very existence.

Precise measurement of the point at which a hydrocarbon will ignite spontaneously is however not easily achieved. This is probably because of the nature of the combustion reaction which is still the subject of research and investigation. However, a number of methods has been used to examine benzene and to evaluate the property. They have in common the function of bringing benzene in air into contact with surfaces heated to a known temperature, which can be raised until the spontaneous ignition occurs. This can be done by using a surface such as a crucible, a tube, a bulb or a wire and various methods have been devised to introduce the sample of benzene. The results obtained by different methods vary greatly, but a very detailed investigation was carried out by H. Moore, one of the early research workers in hydrocarbon science.[15] Moore's results seem to lie in the range of approximately 600–650°C, with a median figure of 630°C.

4.1.7.3. Explosive range The fuel intake for an internal combustion engine must necessarily be composed of a mixture of the (hydrocarbon) fuel with oxygen or air so that the combustion process is supported. To optimise the design of fuel introduction systems it is essential to have available precise knowledge of the limits of composition of the fuel/air mixture, and much work has been done to accumulate this knowledge.

For combustion to occur, the proportions of the fuel and air must lie within certain defined limits; too much air, or too little, will produce mixtures which will not ignite.

The actual ignition and combustion phenomena must be understood before tests can be evaluated. It is generally thought that the application of the spark or other heat source only initially affects and ignites the fuel/air mixture in the immediate vicinity of the source. The heat generated at this point is then transferred to the next adjoining layer, and so on. Thus the tests which can be applied to any given hydrocarbon can only be empirical, as the results will vary with the size and shape of the container, with the position and direction of the spark, and general atmospheric conditions. In these circumstances it is perhaps remarkable that results of experiments by many different workers are as consistent as they are.

Typical results are reported by Zabetakis *et al.*,[16] and give for benzene/air the lower and upper limits of inflammability at 100°C as 1·33 per cent and 7·9 per cent benzene vol/vol. The effect of temperature on the results has been studied by Briand *et al*[17] who quote limits of inflammability for benzene at 100°C as 1·37 per cent to 5·32 per cent vol/vol.

4.1.8. Thermal Properties

It goes almost without saying that knowledge of calorific value latent heats of vaporisation and fusion, and of specific heat, is essential for the design of plant for the refining and chemical use of benzene.

4.1.8.1. Calorific value The calorific value of a fuel is defined as the amount of heat evolved when a unit weight of the fuel is completely burnt. In the case of hydrocarbons such as benzene, the products of complete combustion are carbon dioxide and water, and the calorific value can be calculated accurately from the molecular composition.

In Britain the unit of heat employed is the British Thermal Unit (BThU) which is the amount of heat required to raise 1 lb of water from 60°F to 61°F, i.e. through 1°F. In Continental Europe the unit is the gram-calorie, which is sometimes referred to a base-level of 15°C and sometimes to 20°C. The gram-calorie unit at 15°C (cal_{15}) is the amount of heat required to raise 1 g of water from 15°C to 16°C, and similarly the unit at 20°C (cal_{20}) is the amount of heat required to raise 1 g of water from 20°C to 21°C. The conversion from one system to the other is that:

$$\text{One } Cal_{15} \text{ per gram} = 1{\cdot}8 \text{ BThU per lb.}$$

There is a distinction which must be noted between the gross and the net calorific values. The testing is usually carried out in a bomb calorimeter, where a portion of the sample under test is introduced in liquid form, ignition takes place, the heat evolved is measured and the products of combustion are released at the ambient temperature. The effect of this is that included in the observed result are the loss of heat due to the vaporisation of the fuel, the gain due to the condensation of the products and the gain due to the cooling of the products down to the ambient temperature. The result before correcting for these errors is the gross calorific value, and the net calorific value is the figure obtained after the corrections are applied.

Table IV.3 gives the values for benzene as determined by Rossini.[18]

TABLE IV.3

	Calories/g	*B.Th.U/lb*
Gross calorific value at 25°C	10102·4	18172
Net calorific value at 25°C	9698·4	17446

4.1.8.2. Latent heat of vaporisation The latent heat of vaporisation of a given liquid is defined as the quantity of heat required to convert a unit mass of the liquid into the vapour phase. As the value for latent heat varies according to the temperature at which the test is carried out, it is necessary also to define the temperature. At normal atmospheric pressure this temperature is of course the boiling point of the compound, but as many of the operations for the refining and use of benzene take place at pressures higher or lower than normal, it is necessary to know the latent heat over a wide range of temperatures. J. H. Perry[19] gives a diagram from which these can be derived, and a selection from the range is given in Table IV.4 overleaf. Latent heats of vaporisation can also be calculated by the use of the appropriate Clapeyron-Clausius equation (*vide* the section dealing with phase equilibria).

4.1.8.3. Latent heat of fusion The latent heat of fusion is the quantity of heat required to convert a unit mass of solid into liquid at the same temperature (thus excluding the specific heat which is required to create a rise in temperature). This property can also be calculated from a Clapeyron-Clausius equation. The latent heat of fusion for benzene is given as 30·45 calories per g.[20] P. E. Spielmann quotes another value as 30·1 calories per g.[21]

TABLE IV.4

Latent Heat of Vaporisation of Benzene

Temperature °C	*Latent heat* calories/g
60	97·47
80	94·17
80·10	94·14
100	90·57
140	82·59
180	74·08
220	62·09
289·5	0

4.1.8.4. Specific heat The specific heat of a chemical compound is defined as the quantity of heat required to raise the temperature of a given mass of the compound through one unit of temperature. It will be appreciated that expressed in this form, three arbitrary but interlinking units are involved, those of heat measurement, mass and temperature. It is therefore more usual to express the specific heat, or heat capacity, in the form of an equation:

$$C = \frac{dQ}{dt}$$

where C is the heat capacity per mole

dQ is the heat applied to the substance (measured in calories)

and dt is the corresponding rise in temperature, measured in °C.

For benzene the heat capacity at 25°C is 0·2499 cal/g/°C.

TABLE IV.5

Thermal Conductivity of Liquid Benzene

Temperature °C	*Thermal conductivity* cals/sec/sq cm/°C per cm
0	$36{\cdot}8 \times 10^{-5}$
10	$35{\cdot}7 \times 10^{-5}$
20	$34{\cdot}8 \times 10^{-5}$
30	$33{\cdot}9 \times 10^{-5}$
40	$33{\cdot}0 \times 10^{-5}$
50	$32{\cdot}2 \times 10^{-5}$
60	$31{\cdot}4 \times 10^{-5}$
70	$30{\cdot}7 \times 10^{-5}$
80	$29{\cdot}9 \times 10^{-5}$
90	$29{\cdot}3 \times 10^{-5}$
100	$28{\cdot}7 \times 10^{-5}$
120	$27{\cdot}8 \times 10^{-5}$
140	$27{\cdot}2 \times 10^{-5}$

For thermal conductivity in the vapour phase see Section 4.2.2.2.

4.1.9. Thermal Conductivity

Thermal conductivity is a measure of the rate at which heat flows through the compound concerned and, where expressed as a physical factor, it is calculated on unitary basis of time, cross-sectional area and temperature difference per unit length.

Liquid thermal conductivities have been measured by numerous investigators. Horrocks and McLaughlin measured it from 20–75°C,[22] Venart from 0–80°C,[23] Scheffy and Johnson from 68–169°C,[24] Briggs at 20°C[25] and McAdams at 30° and 60°C.[26] The agreement between the investigators is generally within 5 per cent. Table IV.5 was developed by plotting the data and drawing the most representative curves.

4.1.10. Viscosity

The viscosity of a liquid is customarily expressed in one of two ways; one is absolute viscosity, for which the unit is the *poise*, and the other is kinematic viscosity, for which the unit is the *stokes*. The relation between the two can be expressed for a liquid at t°C as:

$$\text{Kinematic viscosity at t°C} = \frac{\text{Absolute viscosity at t°C}}{\text{Density at t°C}}$$

Viscometers used in laboratories are mostly of the flow type, where the liquid passes through an orifice or a capillary tube and such apparatus will measure kinematic viscosity.

The figures for benzene at different temperatures are given in Table IV.6.[27, 28, 29]

TABLE IV.6

Viscosity of Benzene at Various Temperatures

Temperature °C	*Absolute viscosity* centipoises	*Kinematic viscosity* centistokes
0	0·910	
10	0·771	
20	0·654	0·742
30	0·570	
40	0·506	
50	0·449	0·512
60	0·404	
70	0·361	
80	0·318	0·383
90	0·286	
100	0·261	
120	0·221	
160	0·163	
200	0·119	
280	0·062	

Note—Most of the figures in Tables IV.5 and IV.6 were supplied by Dr R. W. Gallant (Ed.).

4.1.11 Surface Tension

When benzene is used as a solvent, now an increasingly rare outlet mainly because of the toxicity hazard described in another section, it is sometimes necessary to have information about its surface tension, which has an important effect on the spreading properties. Another occasion when surface tension is relevant arises in fuel blending, when benzene is a component of the mixture, and the passage of the fuel blend through carburettor jets must be considered.

The surface tension is reduced as the temperature rises, and can be calculated from equations using the parachor. Table IV.7 gives a selection of values for surface tension at a range of temperatures but Claxton[30] quotes a larger number.

TABLE IV.7

Surface Tension of Benzene

Temperature °C	*Surface tension* dyne/cm
15	29·6
20	28·9
40	26·4
60	23·8
80	21·3

4.1.12 Solubility

Although benzene is only slightly soluble in water, and also so is water in benzene, these properties have relevance in the manufacture, storage and shipping of the product, particularly now that it is handled in such very large tonnages.

It is completely miscible with other liquid hydrocarbons of both the aromatic and aliphatic series, and with aliphatic alcohols, ketones, ether and carbon bisulphide. For further details, and references, see section 4.3, pages 117–186.

4.1.12.1. Solubility of benzene in water At 22°C the solubility of benzene in water is 0·082 per cent on a vol/vol basis. This is equivalent to 0·072 per cent on a wt/wt basis (see also page 118).

4.1.12.2. Solubility of water in benzene Almost all benzene supplied in bulk for industrial use must necessarily be saturated with the maximum amount of water which it will tolerate at the ambient temperature. This is of some importance commercially. The following Table IV.8 gives the solubility of water in benzene at different temperatures.[31]

TABLE IV.8

Solubility of Water in Benzene

Temperature °C	*Solubility* % wt/wt
10	0·030
20	0·0425–0·0445
26	0·054

4.1.12.3. Other solubility relationships The complete miscibility of benzene with many groups of oxygenated organic compounds has already been noted. A more complex situation arises with three component or ternary mixtures, especially when one of these is water. In fact the ternary system benzene-ethanol-water has attained considerable importance because of the use of benzene-ethanol-petrol blends for motor fuels, with the inevitable risk of contamination of such blends with water. Table IV.9 gives some examples at different temperatures of the solubility of benzene in ethanol-water mixtures[32] and this is also shown in graphical form in Fig.4.2.

Shepherd has given details of ternary blends of methanol-*n*-heptane-benzene[33] and Leibnitz has published his work on blends of benzene and ethylene glycol with paraffinic and naphthenic hydrocarbons.[34] The latter has considerable relevance in designing solvent extraction systems for the separation of aromatic from other hydrocarbons in refining processes (see also page 118).

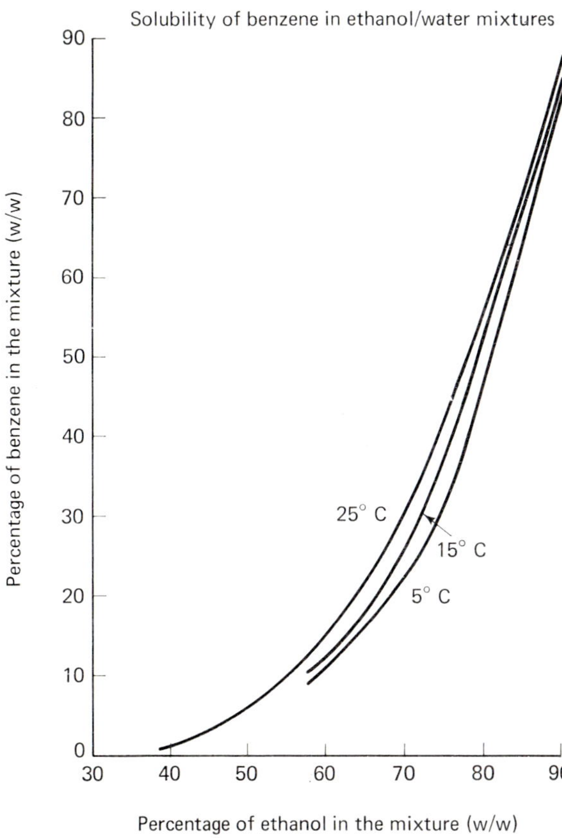

FIG. 4.2 *Solubility of Benzene in Ethanol/Water Mixtures*

TABLE IV.9

Solubility of Benzene in Ethanol—Water Mixtures

Ethanol, % wt/wt	90·06	80·10	69·08	57·66	39·0
Temperature, °C		*Percentage of Benzene*, wt/wt			
25	87·22	59·41	27·46	12·61	1·0
15	85·05	55·00	23·38	10·47	—
5	83·05	46·92	20·88	8·95	—

4.1.13 Critical Temperature, Pressure and Density

The properties relating to phase equilibria are described in detail in the appropriate part of this work, and it suffices to mention only briefly here the phenomena of critical temperature, pressure and density. In a liquid/vapour system the effect of increasing temperature on the liquid is to cause the vapour pressure to rise until eventually the boiling point is reached. On the other hand the effect of increasing pressure upon the vapour is to cause condensation to take place.

If the temperature and pressure are increased simultaneously, eventually the two effects will merge and a condition of equilibrium will exist. The points at which this occurs are known as the critical temperature and critical pressure. The density of the liquid/vapour system at this point is the critical density.

For benzene the values for these three properties are:

Critical temperature	289·5°C
Critical pressure	48·7 atm
Critical density	0·304 g/ml

4.1.14. Absorption

The production of crude benzene as a by-product of the carbonisation of coal frequently depends on the property of absorption. This is more generally true of older coking plants, as in modern units there is a tendency to extract the benzene by adsorption (*q.v.*).

The principle of absorption processes is that the gases containing benzene and other oily components are brought into contact with a flow of liquid so arranged that it takes the form of a thin film. This is usually in a scrubbing tower packed with ceramic or similar material where the liquid, a petroleum or creosote oil, flows over the surface of the packing. The solubility of benzene in oils has already been described and this assists the absorption process.

The distribution of the benzene between the gas and the scrubbing oil follows Henry's law, from which the relative proportions can be calculated. It was usual in scrubbing plants in coke ovens to allow about 25 per cent capacity above the theoretical level, but nevertheless this theoretical figure can be calculated with precision.

4.1.15. Adsorption

Adsorption processes, now largely used in benzene recovery at coking plants, rely on a theoretical principle which is not yet fully understood although certainly different from absorption, as may be deduced from the descriptive name. Whereas in absorption the benzene is actually dissolved by the scrubbing oil, in adsorption it is taken into and retained in the interstices of a porous solid, from whence it is subsequently retrieved by a recovery process. It is the mechanism of the retention within the adsorbing medium which is the subject of discussion.

Initially it was thought that the vapour adsorbed was condensed into liquid in the internal spaces of the adsorbent, and later expelled by re-vaporisation.

A newer theory put forward by Langmuir[35] is that an inter-molecular action takes place at the interface between the adsorbent solid and the adsorbed vapour so that some loose chemical link is created between them.

Langmuir's theory has perhaps tended to find favour in recent times, but Claxton[36] points out that the two views are not mutually exclusive, and that taken together they may provide a picture of the nature of the adsorption process.

It is certainly true that benzene readily conforms with these theoretical considerations and thus that the adsorption techniques have great value in its production.

4.1.16. Toxicity

For very many years it has been known that benzene vapour has a powerful narcotic effect, and accidents have been fairly common, particularly in refineries, where as the result of a spillage of benzene, plant operators have been affected and have lost consciousness. More recently there has been full appreciation of the chronic effect of benzene poisoning which can eventually destroy the red and white corpuscles.

A further development occurred in the late 1950's as a result of a fatal accident in a European motor car factory. Car bodies were sprayed with paint containing a solvent, part of which was a hydrocarbon mixture. These hydrocarbon solvents of petroleum origin commonly contained between 5 and 10 per cent of benzene, but the accident drew the attention of health authorities to the fact that the inhalation of fumes from such solvent-based paints could cause benzene poisoning by cumulative effect. Starting with Italy and France, and then spreading to most countries, there are now stringent regulations governing the benzene content of solvents used in this way.

At the 56th session of the International Labour Conference, Geneva, 1971, the Conference adopted Recommendation 144 which provides *inter alia* that benzene should not be used as a solvent or diluent nor should products be so used if their benzene content exceeds 1 per cent, and even where the benzene content is below 1 per cent, care must be taken to ensure that the atmosphere in workrooms must not have a benzene content greater than 25 parts per million.[37]

It should be added that it is possible to provide safe working conditions in such a way that the atmosphere of workrooms is not inhaled, and some countries (including the UK) prefer this approach. It is clear, however, that whichever method of control is adopted, the toxicity of benzene, and of solvents containing benzene, is a very real hazard.

4.1.17. Summary of Principal Physical Properties of Benzene

Boiling point at 760 mm pressure	80·099°C
Calorific value (gross) at 25°C	10102·4 cal/g
	18172 BThU/lb
Calorific value (net) at 25°C	9698·4 cal/g
	17446 BThU/lb
Coefficient of expansion at 20°C	0·001209
at 40°C	0·001261
Critical density	0·304 g/ml
Critical pressure	48·7 atm
Critical temperature	289·5°C
Density D_4^{20}	0·8790
Explosive range	1·33 per cent (lower) to 7·9 per cent (upper)
Flash point (approximate)	8 to 10°F, −13·3 to −12·2°C
Freezing point (dry benzene)	5·533°C
Latent heat of fusion	30·45 cal/g
Latent heat of vaporisation at 80·10°C	94·14 cal/g
Molar volume	88·9
Refractive index at 20°C and 760 mm pressure	1·50112
Specific gravity at 60°F (15·54°C)	0·8847
Specific heat (heat capacity) at 25°C	0·2499 cal/g/°C
Spontaneous ignition temperature (approximate)	630°C
Surface tension: at 15°C	29·6 dyne/cm
at 20°C	28·9 dyne/cm
Thermal conductivity: at 30°C	$33·9 \times 10^{-5}$ cal/sec/sq. cm/°C/cm
at 60°C	$31·4 \times 10^{-5}$ cal/sec/sq. cm/°C/cm
Viscosity at 20°C: Absolute	0·654 centipoise
Kinematic	0·742 centistoke

A new publication presents values in S.I. units of some thermodynamic properties of benzene for temperatures between 290 K and 650 K and for pressures up to 7 MN/m². The tables were correlated and calculated by Dr. I. J. Lawrenson of the National Physical Laboratory and are published by Engineering Sciences Data Unit, London, 1973 under reference number 73009.

4.2 SOME THERMODYNAMIC AND TRANSPORT PROPERTIES OF BENZENE

By R. W. Gallant

4.2.1. Thermodynamic Properties of Benzene Gas

4.2.1.1. Enthalpy, entropy and volume Table IV.10 presents the saturated liquid and vapour thermal properties of benzene. The data are based on the extensive investigations of Organick,[38] Gornowski[39] and Canjar.[40]

The superheated gas properties are tabulated in Table IV.11.[38, 39, 40]

TABLE IV.10

Thermal Properties of Saturated Benzene

Temp. °C	*Volume*, cc/mole *Liquid*	*Vapor*	*Enthalpy*, cal/mole *Liquid*	△H	*Vapor*	*Entropy*, cal/mole-°C *Liquid*	△S	*Vapor*
40	90·99	115,322	19,982	7873	27,855	42·41	25·15	67·56
60	93·33	55,198	20,478	7620	28,098	44·37	22·88	67·25
80	95·82	29,208	21,137	7347	28,484	46·55	20·78	67·33
100	98·55	16,452	21,853	7047	28,900	48·66	18·91	67·57
120	101·57	10,103	22,600	6739	29,339	50·77	17·19	67·96
140	104·94	6,577·9	23,381	6409	29,790	52·80	15·55	68·35
160	108·84	4,369·1	24,179	6049	30,228	54·91	13·98	68·89
180	113·37	3,028·1	24,986	5685	30,671	56·79	12·57	69·36
200	118·69	2,155·3	25,841	5272	31,113	58·50	11·10	69·60
220	125·37	1,537·5	26,731	4799	31,530	60·14	9·69	69·83
240	134·29	1,093·2	27,668	4257	31,925	61·86	8·28	70·14
260	147·41	759·2	28,688	3502	32,190	63·82	6·56	70·38
280	172·81	487·1	29,955	2308	32,263	65·92	4·07	69·99
289·5	262·39	262·4	31,443	0	31,443	68·66	0	68·66

TABLE IV.11

Thermal Properties of Superheated Benzene

Atm		*Temperature*, °C 60	80	100	150	200	300	400
·075	v	400,823	425,204	449,585	510,049	587,619	690,469	811,398
	h	28,106	28,566	29,048	30,376	31,873	35,271	39,137
	s	71·55	72·80	74·36	78·42	81·62	87·87	93·97
1·0	v	—	24,498	29,954	34,128	38,278	46,772	55,101
	h	—	28,484	28,979	30,332	31,838	35,262	39,133
	s	—	67·33	68·89	72·95	76·24	82·48	88·65
2·0	v	—	—	14,848	16,911	19,022	23,318	27,502
	h	—	—	28,887	30,267	31,812	35,232	39,115
	s	—	—	67·33	71·39	74·83	81·08	87·33
4·0	v	—	—	—	7997	7294	11,591	13,356
	h	—	—	—	30,133	31,699	35,158	39,076
	s	—	—	—	69·91	73·19	80·22	85·61
10	v	—	—	—	—	3364	4487	5427
	h	—	—	—	—	31,374	35,002	38,990
	s	—	—	—	—	70·85	77·64	83·19
20	v	—	—	—	—	—	2022	2570
	h	—	—	—	—	—	34,667	38,781
	s	—	—	—	—	—	75·77	82·25

Table 11 continued

Atm		*Temperature,* °C 500	600	700	800	900	1000
·075	v	932,327	1,052,771	1,174,189	1,295,606	1,413,610	1,530,639
	h	43,390	47,976	52,815	57,849	63,065	68,311
	s	99·82	105·4	110·6	115·6	120·3	124·6
1·0	v	63,439	71,680	79,872	88,113	96,207	104,155
	h	43,390	47,976	52,815	57,849	63,065	68,311
	s	94·51	100·1	105·3	110·3	114·9	119·3
2·0	v	31,690	35,845	39,926	44,042	47,928	51,980
	h	43,377	47,972	52,810	57,840	63,056	68,306
	s	93·19	98·81	104·0	109·0	113·6	117·9
4·0	v	15,779	17,866	19,997	21,709	24,093	26,122
	h	43,351	47,946	52,797	57,831	63,047	68,298
	s	91·70	97·40	102·6	107·5	112·0	116·5
10	v	6300	6924	7943	8802	9655	10,508
	h	43,272	47,894	52,750	57,783	63,030	68,528
	s	89·90	95·45	100·7	105·6	110·3	114·7
20	v	3038	3483	3921	4335	4779	5222
	h	43,138	47,785	52,676	57,740	63,004	68,489
	s	86·70	93·97	99·20	104·2	108·9	113·3

Atm		*Temperature,* °C 200	300	400	500	600	700	800	900	1000
40	v	—	751·4	1162	1453	1677	1846	2168	2385	2613
	h	—	33,787	38,352	42,869	47,590	52,528	57,623	62,908	68,415
	s	—	73·11	80·45	86·70	92·40	97·79	102·7	107·4	111·9
70	v	—	163·4	560·7	766·0	985·5	1080	1218	1357	1498
	h	—	30,809	37,623	42,474	47,321	52,342	57,469	62,782	68,311
	s	—	67·10	78·66	85·14	91·15	96·54	101·6	106·2	110·8
100	v	—	146·8	320·4	500·3	632·4	751·4	854·8	954·3	1046
	h	—	30,428	36,703	41,971	47,004	52,094	57,284	62,631	68,202
	s	—	66·39	76·63	83·97	90·06	95·53	100·7	105·4	109·9
130	v	—	143·8	237·0	373·0	482·7	574·9	658·8	736·3	814·3
	h	—	30,306	35,969	41,584	46,714	51,851	57,102	62,522	68,120
	s	—	66·08	75·30	83·03	89·28	94·85	100·1	104·9	109·4
160	v	—	134·6	200·9	300·4	393·5	472·0	542·2	605·1	666·6
	h	—	30,241	35,661	41,146	46,471	51,704	56,972	62,388	67,977
	s	—	65·77	74·52	82·25	88·65	94·36	99·51	104·4	108·9
200	v	—	129·2	175·5	246·2	320·4	385·7	442·3	494·9	549·5
	h	—	30·207	35,288	40,769	54,716	51,483	56,807	62,283	67,894
	s	—	65·46	73·74	81·31	87·87	93·73	98·89	103·9	108·4

v = volume, cc/mole h = enthalpy, cal/mole s = entropy, cal/mole-°C

4.2.1.2. Ideal gas properties Kobe has calculated the ideal gas enthalpy and heat capacity (Table IV.12).[41] The equation for the heat capacity is the following:

$$Cp = 17{\cdot}700 + {\cdot}0785t - 6{\cdot}022 \times 10^{-5}t^2 + 1{\cdot}854 \times 10^{-8}t^3$$

where heat capacity is cal/mole °C and temperature is °C.

TABLE IV.12

Ideal Gas Thermodynamic Properties of Benzene

°C	*Heat capacity* *calories* gm/mole-°C	*Enthalpy* *calories* gm/mole
0	17·70	0
25	19·52	465
100	24·93	2123
200	31·28	4964
300	36·52	8360
400	40·75	12,190
500	44·23	16,500
600	47·13	21,090
700	49·58	25,900
800	51·65	30,970
900	53·43	36,230
1000	54·94	41,650

Free Energy of Formation $= 30{,}989 \dfrac{\text{calories}}{\text{gm/mole}}$ at 25°C.

Heat of Formation $= 19{,}820 \dfrac{\text{calories}}{\text{gm/mole}}$ at 25°C.

4.2.1.3. Fugacity The fugacity-pressure ratios are shown in Table IV.13.[40]

TABLE IV.13

Fugacity Coefficients of Benzene

Temp. °C	*Saturated vapour*	*Pressure*, atm ·075	*1·0*	*4*	*10*	*40*	*100*	*200*
40	·985	·998	—	—	—	—	—	—
60	·976	·998	—	—	—	—	—	—
80	·962	·998	·962	—	—	—	—	—
100	·944	·998	·968	—	—	—	—	—
120	·926	·998	·972	—	—	—	—	—
140	·902	·999	·976	·882	—	—	—	—
160	·876	·999	·979	·929	—	—	—	—
180	·844	·999	·982	·940	·889	—	—	—
200	·816	·999	·984	·948	·871	—	—	—
220	·786	·999	·986	·956	·888	—	—	—
240	·752	1·000	·988	·962	·903	—	—	—
260	·717	1·000	·990	·966	·916	—	—	—
280	·679	1·000	·992	·971	·926	—	—	—
289·5	·659	1·000	·992	·973	·931	·727	·383	·254
400	—	1·000	·997	·985	·962	·858	·725	·503
500	—	1·000	·999	·991	·978	·915	·838	·684
600	—	1·000	·999	·994	·986	·945	·878	·805
700	—	1·000	1·000	·997	·990	·965	·924	·888
800	—	1·000	1·000	·998	·994	·976	·951	·937
900	—	1·000	1·000	·999	·996	·984	·969	·969
1000	—	1·000	1·000	1·000	·997	·990	·984	·993

4.2.1.4. Compressibility The compressibility factor is a measure of how a gas deviates from an ideal gas in its pressure-volume-temperature relationship. It is defined as the pressure times the volume divided by the gas constant times the temperature. Thus, it is dimensionless. The compressibility factors for saturated benzene vapour (Table IV.14) and superheated vapour (Table IV.15) have been determined by Gornowski and co-workers.[39] For benzene, the compressibility factor increases as the temperature increases, and decreases as the pressure increases.

Connolly and Kandalic report the specific volumes of liquid benzene (Table IV 16) from 5–180 atm.[42]

TABLE IV.14

Compressibility Factors for Saturated Vapour Benzene

Temperature °C	*Pressure* atm	*Compressibility factor*
240	25·6	0·667
250	28·4	0·640
260	33·6	0·585
270	38·3	0·537
280	43·4	0·466
285	46·1	0·407
289·5	48·7	0·276

TABLE IV.15

Compressibility Factors for Superheated Vapour Benzene

Temp. °C	*Pressure*, atm						
	30	*35*	*40*	*45*	*50*	*55*	*60*
260	·666	—	—	—	—	—	—
280	—	·660	·557	—	—	—	—
300	—	·716	·653	·582	·482	·306	—
320	—	·757	·711	·662	·608	·545	·471
340	—	·789	·753	·715	·675	·632	·590
360	—	·815	·787	·759	·728	·694	·660

TABLE IV.16

Specific Volumes for Liquid Benzene, cc/g

Temp. °C	*Pressure*, atm					
	5	*10*	*50*	*100*	*150*	*180*
130	1·3248	1·3224	1·3098	1·2952	1·2823	1·2756
150	—	1·3693	1·3637	1·3346	1·2863	1·2775
170	—	1·4236	1·4010	1·3775	1·3578	1·3477
200	—	—	1·4878	1·4521	1·4241	1·4107
220	—	—	1·5622	1·5122	1·4747	1·4574
240	—	—	1·6556	1·5804	1·5297	1·5073
270	—	—	1·9019	1·7167	1·6372	1·5984

4.2.2. Transport Properties of Benzene

4.2.2.1. *Viscosity* Thodos and Flynn have developed an equation for calculating the viscosity of aromatics at normal pressures.[43] This method has been used to calculate the viscosities from 0° to 1000°C (Table IV.17)

TABLE IV.17

Viscosity of Benzene Vapour

Temperature °C	*Absolute viscosity micropoise*
0	65·6
66	82·0
100	90·2
200	113
300	137
400	159
500	180
600	203
700	222
800	244
900	262
1000	279

4.2.2.2. Thermal conductivity Lambert has measured the vapour thermal conductivity as 32·2 cal/cm-sec-°C at 66°C.[44]

4.2.2.3. Diffusivity Rathburn and Babb report the self-diffusion coefficient of benzene to be ·0000214 cm^2/sec at 25°C.

4.3. PHASE EQUILIBRIA, INCLUDING SOLUBILITIES IN BINARY AND MULTI-COMPONENT SYSTEMS WITH BENZENE

by G. A. M. Diepen and M. A. de Zeeuw

4.3.1. Introduction

Phase equilibria are of great importance in designing chemical processes involving the separation of benzene in the presence of other reagents which may occur as main products, by-products or catalysts in a reaction or of making use of benzene to achieve an otherwise difficult separation. The number of determinations in this field are so large that it is not possible within the compass of this book to provide details of all the various known systems. Instead it has been decided to produce a bibliography of all the systems, details on which are available, giving the references to the original work at the same time indicating some specific data such as compounds formed, azeotropes, eutectics etc.

It is believed that information has never before been provided in just this form and it is hoped that it will prove of value both to chemical engineers and to research workers desiring information in this field.

The systems are arranged in the following way:

1st. Systems with inorganic compounds following the order used in Landolt-Börnstein: 'Zahlenwerte und Funktionen', which is indicated in Volume II 2a, Heidelberg 1960, page IX.

2nd. Systems with inorganic-organic compounds in the same way.

3rd. Systems with organic compounds which are arranged according to their functional groups. Compounds of the same kind are arranged after increasing number of C-atoms. Compounds with several ligands such as e.g. secondary amines are arranged after increasing number of C-atoms and nature of the ligands, aromatic after aliphatic.

4.3.2. Phase Equilibria in Binary Systems

Systems of a noble gas and benzene

Solubilities in benzene at different temperatures of He,[45–47] Ne,[45, 46] Ar,[45, 46] Kr,[45, 48, 49] Rn[50–52] and An (actinon).[53] Solid-liquid equilibria in the system Helium-Benzene at high pressures.[54] Solubility of benzene in compressed argon at various temperatures and pressures.[55]

Hydrogen—Benzene

Liquid-vapour equilibria at 25, 50, 70, 100 and 150° and pressures up to 3000 atm.[56] Solubility of hydrogen in benzene at 10° and 35°.[57] Further literature.[58–60]

Deuterium—Benzene

Solubility of deuterium in benzene at 10° and 35°.[57]

Oxygen—Benzene

Solubility of oxygen gas at different temperatures in benzene.[61–65] Solubility of benzene in compressed oxygen at various temperatures and pressures.[55]

Water—Benzene

Solubility of water in benzene.[66–67] Solubility of benzene in water.[68–72] At 20° immiscibility between $x = 0{\cdot}00038$ and $0{\cdot}9979$ mole fraction benzene.[73] Mutual solubility of water and benzene,[74–75] at temperatures to 300°.[76–80] at pressures to 350 atm and temperatures to 450°.[81] Liquid-vapour equilibria at 20°.[82] Three-phase equilibrium liquid-liquid-gas from 150° and 10·6 atm to 267·8° and 92·7 atm (critical endpoint.)[83] Critical curve between 260 and 386° up to 3700 bar.[84–86] Further literature.[87]

Hydrogen fluoride—Benzene

Melting equilibria from solid benzene in liquid hydrogen fluoride.[88–89]

Chlorine—Benzene

Solubility of chlorine gas in benzene.[90]

Hydrogen chloride—Benzene

Solubility of HCl gas in benzene at different temperatures.[91–96] Melting equilibria. A compound $HCl.C_6H_6$ is formed.[97]

Bromine—Benzene
Melting equilibria. A compound $C_6H_6.Br_2$ is formed with congruent melting point.[98]

Hydrogen bromide—Benzene
Melting equilibrium with eutectic at $-95°$ and 7·3 mole per cent benzene.[99] Solubility of HBr gas in benzene at 30 and 50°.[100]

Iodine—Benzene
Solubility of solid iodine in benzene at different temperatures.[101–103]

Sulphur—Benzene
Solubility of rhombic sulphur in benzene.[104]
This system shows immiscibility in the liquid phase at temperatures lower than 162°.[105–106] A lower critical solution point of sulphur polymer is reported at 230°.[107] Further literature.[108]

Hydrogen sulphide—Benzene
Solubility of H_2S gas at 20° in benzene.[91, 109] Melting equilibrium with eutectic at $-100·1°$, 7·4 mole per cent benzene.[110]

Sulphur dioxide—Benzene
Solubility of sulphur dioxide gas at different temperatures in benzene.[62, 111] Melting equilibrium with eutectic at $-78°$ and 6 weight per cent benzene.[112] A compound $C_6H_6.SO_2$ with incongruent melting point $-58°$ is reported.[113]

Nitrogen—Benzene
Solubility of nitrogen at different temperatures in liquid benzene.[48, 62, 114–115] Solubility of benzene in compressed nitrogen at 25–75° and pressures up to 300 atm.[116–117] Solid-liquid equilibrium at high pressures.[54]

Ammonia—Benzene
Solubility of ammonia gas in benzene at 20°.[91, 118]

Nitrous oxide—Benzene
Solubility of N_2O gas in benzene at different temperatures.[62] Solid-liquid equilibria at high pressures.[54]

Nitric oxide—Benzene
Solubility of NO gas in benzene at different temperatures.[119–120]

Nitrogen tetroxide—Benzene
There is a solid compound $C_6H_6.N_2O_4$ melting at $-70°$.[121]

Nitrosyl chloride—Benzene
Melting equilibrium with eutectic.[121]

Phosphorus—Benzene
Solubility of solid and liquid phosphorus in benzene.[122]

Phosphorus trioxide—Benzene
The solubility of P_2O_3 in benzene at 15° is 0·067 mole per cent.[123]

Arsenic tribromide—Benzene
Melting equilibrium[124] with eutectic at −15° and 60 mole per cent benzene.

Antimony trifluoride—Benzene
The solubility of SbF_3 in benzene at 25° is 0·55 mg SbF_3/100 ml solution.[125]

Antimony trichloride—Benzene
Melting equilibria.[126] There is a solid compound $C_6H_6.2\ SbCl_3$, melting point 79·5°. Eutectic $C_6H_6+C_6H_6.2\ SbCl_3$: 1° and 93·34 mole per cent benzene. Eutectic $C_6H_6.2\ SbCl_3+SbCl_3$: 62° and 14 mole per cent benzene.

Antimony tribromide—Benzene
Melting equilibria.[127–128] There is a solid compound $C_6H_6.2\ SbBr_3$, melting point 92·5°. Eutectic $C_6H_6+C_6H_6.2\ SbBr_3$: 4·5°, 98·07 mole per cent benzene. Eutectic $C_6H_6.2\ SbBr_3+SbBr_3$: 85°, 15·1 mole per cent benzene.

Carbon monoxide—Benzene
The solubility of carbon monoxide gas in benzene at different temperatures.[62, 115, 119, 129–131] Vapour-liquid equilibria at 120–300° and pressures up to 200 atm.[60]

Carbon dioxide—Benzene
The solubility of carbon dioxide in benzene at different temperatures at 1 atm[115, 119, 129, 132] and at different pressures up to 85 atm.[133–134] Solid-liquid equilibria at high pressures.[54]

Phosgene—Benzene
The solubility of phosgene gas at 20° and pressures from 74–760 mm in benzene.[135–136]

Carbon disulphide—Benzene
Vapour-liquid equilibria[137–141] at temperatures 0–80° and pressures from 26·9 to 2030 mm. The solubility of carbon disulphide gas at 25°.[62] Solid-liquid equilibria. Eutectic at −115·8°, 2 mole per cent benzene.[142]

Carbon oxysulphide—Benzene
Solubility of carbon oxysulphide in benzene.[143]

Hydrogen cyanide—Benzene
Melting equilibria of solid benzene in liquid mixtures of benzene and HCN.[144]

Trichlorosilane—Benzene
Total vapour pressure.[145]

Silicon tetrachloride—Benzene
Silicon tetrabromide—Benzene
Solid-liquid equilibria.[146]

Silicon tetraiodide—Benzene
Solid-liquid equilibria.[147]

Boron trifluoride—Benzene
Solubility of BF_3 in benzene at 1 atm and temperatures 22–47°[48]

Sodium azide—Benzene
Solubility of NaN_3 in benzene.[149]

Sodium cyanate—Benzene
Solubility of NaCNO in benzene.[149]

Sodium stearate—Benzene
The solubility of Na stearate[150] in benzene.

Sodium oleate—Benzene
At 20° 0·04 g Na oleate is soluble in 100 g benzene.[151]

Potassium azide—Benzene
Solubility of KN_3 in benzene.[149]

Potassium cyanate—Benzene
Solubility of KCNO in benzene.[149]

Potassium laurate—Benzene
At 25° 0·005 g K laurate is soluble in 100 g benzene.[152]

Potassium myristate—Benzene
Solubility of K myristate in benzene.[153]

Magnesium hydnocarpate—Benzene
The solubility of the Mg salt of hydnocarpic acid in benzene is at 15°, 30° and the normal boiling point resp. 0·02, 0·04 and 21·42 g/100 ml solution.[154]

Magnesium chaulmoograte—Benzene
The solubility of the Mg salt of chaulmoogric acid in benzene in g per 100 ml is 0·01 at 15°, 0·004 at 30° and 20·07 at the boiling point.[154]

Calcium oleate—Benzene
At 15° 3·44 g Ca oleate is soluble in 100 g benzene.[155]

Barium hydnocarpate—Benzene
Solubility of this salt in benzene at 15°, 30° and the boiling point is given as 0·025, 0·04 and 0·13 g/100 ml solvent resp.[154]

Barium chaulmoograte—Benzene
The solubility of this salt in benzene at 15°, 30° and the boiling point is resp. 0·02, 0·03 and 0·07 g/100 ml benzene.[154]

Cerium camphocarbonate—Benzene
More than 70 g Ce camphocarbonate ($Ce(C_{11}H_{15}O_3)_3$) are soluble in 100 ml benzene solution at 17°.[156]

Niobium pentachloride—Benzene
Tantalum pentachloride—Benzene
Solid-liquid equilibria.[157]

Chromium carbonyl—Benzene
Molybdenum carbonyl—Benzene
Tungsten carbonyl—Benzene
Manganese carbonyl—Benzene
Solubility of the carbonyls of these transition metals in benzene.[158]

Uranyl camphocarbonate—Benzene
At 7° the solubility of $UO_2\ (C_{11}H_{15}O_3)_2$ is 7·91 g/100 cm³ solution.[156]

Cobalt (II) chloride—Benzene
At 25° 0·02 g $CoCl_2$ is soluble in 100 g benzene solution.[159]

Cobalt (II) xanthate—Benzene
200 g of cobalt xanthate are soluble in 1 l benzene.[160]

Nickel(II) chloride—Benzene
At 25·0° 0·03 g $NiCl_2$ is soluble in 100 g benzene solution.[159]

Silver perchlorate—Benzene
Melting equilibrium.[161–163] There is a compound $AgClO_4.C_6H_6$ with incongruent melting point at 145·0°, 63·0 g $AgClO_4$/100 g solution.

Silver nitrate—Benzene
At 35° 0·22 g $AgNO_3$ is soluble in 100 g solution and at 40·5° 0·44 g.[164]

Zinc benzoate—Benzene
At 15° 4·2 g Zn benzoate is soluble in 100 g benzene solution.[155]

Cadmium iodide—Benzene
Solubility of CdI_2 in benzene.[164]

Mercury—Benzene
Solubility of mercury metal in benzene ($\simeq 10^{-6}$ mol/l).[165]

Mercury(II) chloride—Benzene; *Mercury(II) iodide—Benzene*
Solubility in benzene of $HgCl_2$[164, 166] and of HgI_2.[166]

Mercury benzoate—Benzene
At 15° 2·49 g Hg benzoate is soluble in 100 g solution.[155]

Mercury camphocarbonate—Benzene
At 15° 56·03 g Hg camphocarbonate ($Hg(C_{11}H_{15}O_3)_2$) is soluble in 100 cm³ benzene solution.[156]

Aluminum chloride—Benzene
Melting equilibria.[167–168]

Aluminum bromide—Benzene
Melting equilibria.[167] There is a solid compound $C_6H_6.AlBr_3$ with an incongruent melting point 37·0°, the liquid contains 74 mole per cent C_6H_6.

Aluminum iodide—Benzene
Melting equilibria.[167]

Aluminum stearate—Benzene
At 15° 0·18 g Al stearate is soluble in 100 g benzene.[155]

Aluminum oleate—Benzene
At 15° 0·58 g Al oleate is soluble in 100 g benzene.[155]

Gallium trichloride—Benzene
Melting equilibria.[169–170]

Gallium tribromide—Benzene
Melting equilibria.[170]

Indium halogenides—Benzene
Only traces of InF_3 and $InCl_3$ are soluble in benzene. At 20° 0·48 g $InBr_3$ and 6·27 g InI_3 are soluble in 100 g solution.[171]

Thallium(I) methylate—Benzene
At 25·0° 3·16 g CH_3OTl is soluble in 100 g benzene solution.[172]

Tin(IV) chloride—Benzene
Vapour-liquid equilibria at 100° and 125° with pressures up to 3 atm.[173]

Tin(IV) bromide—Benzene
Melting equilibria. Eutectic −9·9°, 65·9 mole per cent benzene.[174]

Tin(IV) iodide—Benzene
Solubility of SnI_4 in benzene at temperatures 10–40°.[175–176]

Lead laurate—Benzene
The solubility of Pb-laurate is 0·011 g in 100 g benzene.[177]

Lead myristate—Benzene
Solubility in benzene.[177]

Lead palmitate—Benzene
At 15° 0·009 g Pb palmitate is soluble in 100 g benzene.[177]

Lead stearate—Benzene
The solubility of Pb stearate in benzene is given[177] as 0·008 g/100 g benzene at 15°.

Lead chaulmoograte—Benzene
The solubility of this salt in benzene at 15°, 30° and the boiling point is resp. 0·05, 0·06 and 2·42 g/100 ml benzene.[154]

Tributyl phosphate—Benzene
Vapour pressure at 20–60°.[178]

Some systems of an organometallic compound and benzene

Vapour-liquid equilibria of the system *tetraethylsilicane—benzene* were measured at 20, 35 and 50°.[179]

The solubility of some organometallic compounds in benzene at 20° is given in Table IV.18.

The solubility of *tri-cyclo-hexyllead*, *triphenyllead* and *tri-p-tolyllead* in benzene at 30° is indicated[182] as resp. 2·28, 20·90 and 5·078 g/100 g benzene.

TABLE IV.18

Solubility of Some Organometallic Compounds in Benzene at 20°

Name of compound	*Solubility in g/l solution*
Triphenylboron[180]	175
Diethylmagnesium[181]	1·3
Diphenylmagnesium[181]	0·11
Diphenylzinc[181]	79
Diphenylcadmium[181]	16
Diphenylmercury[180]	143
Triphenylaluminum[180]	10·5
Triphenylgallium[180]	60·2
Triphenylindium[180]	49·0
Triphenylthallium[180]	63·7

Methane—Benzene
Vapour-liquid equilibria at 65·6° and pressures up to 325 atm.[183] Vapour-liquid equilibria from 40–150°.[184–186] Further literature.[187] Solubility of methane gas at different temperatures and 1 atm.[62, 188–189] Solid-liquid equilibrium at high pressures.[54]

Ethane—Benzene
Vapour-liquid equilibria from −1·3° to 288° and pressures up to 97 atm.[190] The solubility of ethane vapour in benzene at different temperatures.[62]

Ethylene—Benzene
The solubility of ethylene gas in benzene at 1 atm and different temperatures;[62, 188] at 20–150° and pressures up to 120 atm.[191–193] Critical points in the ethylene-benzene system.[192, 194] Pressure-temperature-density relationship at various temperatures, pressures and compositions.[192, 195] Solubility of benzene in compressed ethylene.[196]

Acetylene—Benzene
The solubility of acetylene in benzene at 1 atm and different temperatures.[62, 109, 197–198]

Propane—Benzene
Solubility of propane gas in benzene at 1 atm and 21·5°.[199] Vapour-liquid equilibria[200] at temperatures 37·8–204·4° and pressures up to 60 atm. Further literature.[201]

Propylene—Benzene
Solubility of propylene gas in benzene at 20–60° and pressures up to 7·6 atm.[193]

1,3-Butadiene—Benzene
Vapour-liquid equilibria.[202]

Vinylacetylene—Benzene
Vapour-liquid equilibria at 740 mm.[203]

Pentane—Benzene
Vapour-liquid equilibria.[204]

Hexane—Benzene
Vapour-liquid equilibria at 200, 300 and 400 mm,[205] at 735 mm,[204, 206–207] at pressures from 4–18 atm.[208–209] Further literature.[210–216]

1-Hexene—Benzene
Vapour-liquid equilibria at 25°.[217]

Heptane—Benzene
Vapour-liquid equilibria at pressures varying from 180–760 mm;[204, 218–226] at 75°.[227] Solid-liquid equilibria.[228–229] Eutectic temperature −91·275°.

2,4-Dimethylpentane—Benzene
Vapour-liquid equilibria. Azeotrope with minimum temperature at p = 757 mm, t = 75·2°, x = 0·545 mole fraction benzene.[230] Further literature.[231–232]

2,2,3-Trimethylbutane—Benzene
Vapour-liquid equilibria.[219, 233] Azeotrope with minimum temperature at p = 736 mm, t = 75·6°, x = 0·567 mole fraction benzene.

Octane—Benzene
Vapour-liquid equilibria at 1 atm,[219, 234] at 55, 65 and 75°.[235] Melting equilibria; eutectic at −58·4°, 8·2 weight per cent benzene.[236]

Isoöctane—Benzene
Vapour-liquid equilibria.[237]

2,2,4-Trimethylpentane—Benzene
Vapour-liquid equilibria at 1 atm.[219, 234]

Dodecane—Benzene
Melting equilibria.[236] Eutectic at 17·9°, 34·9 weight per cent benzene.

Hexadecane—Benzene
Melting equilibria.[236] Eutectic at −1·3°, 32·4 weight per cent benzene.

Heptadecane—Benzene
Melting equilibria.[236] Eutectic −1·9°, 37·2 weight per cent benzene.

Dotriacontane—Benzene
Melting equilibria.[236, 238–239]

cyclo-*Pentane—Benzene*
Vapour-liquid equilibria.[240–241]

cyclo-*Hexane—Benzene*
Liquid-vapour equilibria.[219, 230, 242–251] Azeotrope with temperature minimum at p = 759 mm, t = 77·4°, x = 0·515 mole fraction benzene. Change of azeotropic composition with temperature and pressure change (100–760 mm).[252] Vapour-liquid equilibria at pressures from 4–18 atm.[208–209] Melting equilibrium with eutectic −42·5°, 26·5 mole per cent benzene.[253–255] Solid-liquid and liquid-vapour equilibria.[256]

Methyl-cyclo-*pentane—Benzene*
Vapour-liquid equilibria.[240, 246, 257] Azeotrope with temperature minimum at p = 760 mm, t = 71·5°, x = 0·0995 mole fraction benzene.

cyclo-*Hexene—Benzene*
Vapour-liquid equilibria at 740 mm.[233] Azeotrope with temperature minimum at 78·9° and 0·657 mole fraction benzene.

Methyl-cyclo-*hexane—Benzene*
Vapour-liquid equilibria at 1 atm.[219, 240]

Camphene—Benzene
Melting equilibria.[258] Eutectic at −45°, 36·6 mole per cent benzene.

Toluene—Benzene
Liquid-vapour equilibria at temperatures 0–100° and pressures up to 45 atm.[139, 259–266] Melting equilibria;[267–271] eutectic at −100·5° and 14·0 mole per cent benzene.

o-*Xylene—Benzene*
Melting equilibria.[272] Eutectic at −43·5°, 40·7 mole per cent benzene.

m-*Xylene—Benzene*
Vapour-liquid equilibria at temperatures from 100–310° and pressures up to 44 atm.[259] Melting equilibria.[273–275] Eutectic at −59·5°, 27·6 mole per cent benzene.

p-*Xylene—Benzene*
Melting equilibria.[272, 276–277] Eutectic at −22·2°, 62·7 mole per cent benzene.

Ethylbenzene—Benzene
Melting equilibria.[272, 278] Eutectic at −98°, 10·8 mole per cent benzene. Vapour-liquid equilibria.[266]

Propylbenzene—Benzene
Vapour-liquid equilibria.[266]

Cumene—Benzene
Melting equilibria.[278]

Indene—Benzene
Melting equilibria.[279] Eutectic at −39·0°, 44·0 mole per cent benzene.

Butylbenzene—Benzene
Melting equilibria.[278]

Naphthalene—Benzene
Melting equilibria.[102, 280] Eutectic at $-3 \cdot 4°$, 86·6 mole per cent benzene.[267] Eutectic temperature rises from $-3 \cdot 5°$ at 1 atm to 20° at 1003 atm, the eutectic composition changing only from 87·5 to 88 mole per cent benzene.[281–284] Critical temperature and volume.[285]

Hexamethylbenzene—Benzene
Melting equilibria.[286]

Diphenyl—Benzene
Activity coefficient as function of the mole fraction benzene in the liquid phase at temperatures 15–80°.[287–288] Melting equilibria.[289] Solubility in benzene at 20°.[180] Solid-liquid and liquid-vapour equilibria.[290]

Diphenyl methane—Benzene
Vapour pressure and activity coefficient as function of the mole fraction benzene at temperatures from 25–75°.[288, 291] Solid-liquid equilibria.[292]

Fluorene—Benzene
Melting equilibria.[180, 293]

Dibenzyl—Benzene
Activity coefficient as function of the mole fraction benzene at temperatures from 25–75°.[288]

cis- *and* trans-*Stilbene—Benzene*
Solid-liquid equilibria.[294–295]

Tolan—Benzene
Solid-liquid equilibria. A compound $C_{14}H_{10} \cdot C_6H_6$ is formed.[296]

Anthracene—Benzene
Solubility of anthracene in benzene at temperatures 15–30°.[297–300] Critical temperature and volume.[285]

Phenanthrene—Benzene
Melting equilibria; eutectic at $-1 \cdot 6°$, 88·3 mole per cent benzene.[298, 300–302]

Fluoranthene—Benzene
Melting equilibria. Eutectic 1·2°, 86·33 mole per cent benzene.[303]

Triphenylmethane—Benzene
Melting equilibria.[304] There is a compound $C_6H_6 \cdot C_{19}H_{16}$ with melting point 78·2°. Eutectic $C_6H_6 \cdot C_{19}H_{16} + C_6H_6$ 4·2°, 98·75 mole per cent benzene. Eutectic $C_6H_6 \cdot C_{19}H_{16} + C_{19}H_{16}$ 74·0°, 31·0 mole per cent benzene.

Methyl chloride—Benzene
Solubility of methyl chloride gas in benzene at 1 atm and 20 and 25°.[62, 305]

Iodomethane—Benzene
Vapour-liquid equilibria at 35·0°.[306]

Chloroform—Benzene
Vapour-liquid equilibria at 1 atm.[307–308] Vapour pressures of liquid mixtures at temperatures varying from 0–100°.[139] Further literature.[309–310] Melting equilibrium with eutectic: −79°, 2·7 mole per cent benzene.[311]

Bromoform—Benzene
Melting equilibria.[312–313] Eutectic at −27°, 47·6 mole per cent benzene.

Carbon tetrachloride—Benzene
Vapour-liquid equilibria and activity coefficients at temperatures from 30–70°.[242, 314] Vapour-liquid equilibria at 1 atm.[220, 315–317] Vapour pressure of the liquid at temperatures from 0–70°.[139] Further literature.[247, 318–322] Melting equilibria.[254, 271, 323–324] There is a compound $C_6H_6.CCl_4$ with incongruent melting point −34·2°, 48 mole per cent benzene in the liquid phase. Eutectic CCl_4+ $CCl_4.C_6H_6$ −47°, 23 mole per cent benzene.

Monochlorotrifluoromethane ('Freon-13')—Benzene
Dichlorodifluoromethane ('Freon-12')—Benzene
The solubility of CF_3Cl and of CF_2Cl_2 in benzene at 1 atm partial pressure is resp. at 24·5°, 0·61 and 7·3 mole per cent and at 35·3°, 0·57 and 5·6 mole per cent of the gas.[325]

1,2-Dichloroethane—Benzene
Compositions of coexisting liquid and vapour phases as a function of the pressure at 49·99°[326] and at 39·99, 49·99, 60·00 and 70·00°.[327] Vapour-liquid equilibria at pressures 100–760 mm as function of temp.[328] Temperature as function of the mole fraction benzene in the liquid phase at 1 atm.[315] Mole fraction benzene in the gas phase as function of the mole fraction benzene in the liquid phase at 1 atm.[220] The vapour-liquid equilibrium of this system is nearly ideal.[329] Further literature.[318, 330] Melting equilibria, eutectic at −55·4°, 32 mole per cent benzene.[331–332]

Hexachloroethane—Benzene
Melting equilibria.[333] Eutectic $C_6H_6 + \gamma C_2Cl_6$ at −0·9° and 90·0 mole per cent benzene. Transition $\gamma C_2Cl_6 \rightarrow \beta C_2Cl_6$ at 42°, 72 mole per cent benzene. Transition $\beta C_2Cl_6 \rightarrow \alpha C_2Cl_6$ at 70·0° and 52 mole per cent benzene.

Bromoethane—Benzene
Vapour-liquid equilibria at 1 atm.[310] Melting equilibrium with eutectic at −120·5°, 4·5 mole per cent benzene.[334]

1,2-Dibromoethane—Benzene
Solid-liquid equilibria[332, 335] with an eutectic at −27·4° and 53·9 mole per cent benzene.[335] Vapour-liquid equilibria.[336]

Ethylene chloride—Benzene
Vapour-liquid equilibria at 760 and 900 mm.[337]

Trichloroethylene—Benzene
Vapour-liquid equilibria at 1 atm[220] and at 50, 70 and 80°.[338]

Tetrachloroethylene—Benzene
Vapour-liquid equilibria at 1 atm.[339]

Trifluorochloroethylene—Benzene
Solubility of $CF_2 = CFCl$ (Freon no. 1113) in benzene.[340]

Perfluoroheptane—Benzene
Immiscibility in the liquid phase.[341] Critical solution point at 113·5° and 21 mole per cent benzene.

1-Chlorododecane—Benzene
Melting equilibria.[342] Eutectic at −24·0°, 35·7 weight per cent benzene.

1-Iodododecane—Benzene
Melting equilibria.[342] Eutectic at −19°, 35·7 weight per cent benzene.

1-Bromotetradecane—Benzene
Melting equilibria.[342] Eutectic at −13°, 41·6 weight per cent benzene.

1-Iodohexadecane—Benzene
Melting equilibria.[342] Eutectic at −3°, 55·5 weight per cent benzene.

Perfluoromethyl-cyclo-*hexane—Benzene*
Immiscibility in the liquid phase. Critical solution point 85·3°, 73 mole per cent benzene.[343–344]

Perfluorodimethyl-cyclo-*hexane—Benzene*
Immiscibility in the liquid phase.[344]

Fluorobenzene—Benzene
Vapour-liquid equilibria.[345]

Pentafluorobenzene—Benzene
Solid-liquid equilibria.[346] Two compounds are formed with incongruent melting point $C_6H_6.C_6HF_5$ and $(C_6H_6)_2.(C_6HF_5)_3$.

Hexafluorobenzene—Benzene
Solid-liquid equilibria. There is a compound $C_6H_6.C_6F_6$ with congruent melting point[347] at 24·06°. Liquid-vapour equilibria. At 500 mm there is a maximum and a minimum azeotrope at 68·1° and 66·8°[348] respectively.

Chlorobenzene—Benzene
Vapour-liquid equilibrium at 70°.[349] The system behaves ideally. Further literature.[350] Melting equilibria.[271] Eutectic at −60·8°, 27·0 mole per cent benzene.

1,4-Dichlorobenzene—Benzene
Vapour liquid equilibrium at 70°.[351]

Pentachlorobenzene—Benzene
Melting equilibria.[286]

Hexachlorobenzene—Benzene
Melting equilibria.[286] Eutectic at −0·9°, 90 mole per cent benzene.[352]

Bromobenzene—Benzene
Vapour-liquid equilibrium at 70°.[349, 353]

Dibromobenzene—Benzene
Melting equilibria.[293, 300, 354–355]

p-*Chlorotoluene—Benzene*
Solid-liquid equilibria.[356] Eutectic at −26·8°, 56 mole per cent benzene.

Pentachlorotoluene—Benzene
Melting equilibria.[286]

p-*Bromotoluene—Benzene*
Melting equilibria.[276, 357] Eutectic at −19°, 66 mole per cent benzene.

4,5-Dichloro-o-xylene—Benzene
3,4,5-Trichloro-o-xylene—Benzene
Tetrachloro-o-xylene—Benzene
1,3,5-Trimethyl-2,4,6-Trichlorobenzene—Benzene
1,2,5-Trimethyl-3,4,6-Trichlorobenzene—Benzene
1,2,3,4-Tetramethyldichlorobenzene—Benzene
Melting equilibria.[286]

Perfluoronaphthalene—Benzene
Solid-liquid equilibria with a compound 1:1 congruently melting at 69·5°. The eutectic: solid compound-solid perfluoronaphthalene-liquid is reported at 63·4° and about 30 mole per cent C_6H_6.[358]

Pentamethylchlorobenzene—Benzene
Melting equilibria.[286]

Perfluorobiphenyl—Benzene
Solid-liquid equilibria.[358] A solid compound 1:1 is formed with a congruent melting point at 65·2°. The eutectic: solid compound-solid perfluorobiphenyl-liquid is reported at 56·1° and about 22 mole per cent benzene.

2,2-Bis-(4-chlorophenyl)-1,1,1-trichloroethane (DDT)—Benzene
The solubility of DDT in benzene is given[359] as 6·8, 44·0 and 57·8 weight per cent at 0, 24 and 48° respectively.

Methanol—Benzene
Vapour-liquid equilibrium at 20°, azeotrope with max. pressure p = 142·5 mm, x = 0·483 mole fraction benzene.[73] Vapour-liquid equilibrium at 1 atm. azeotrope with temperature minimum at 57·6° and x = 0·391 mole fraction benzene.[360] Vapour-liquid equilibria at 4·4; 7·7; 10·9; 14·3 and 17·7 atm.[361] Vapour-liquid critical points.[362] Vapour-liquid equilibria and critical locus.[403] Further literature.[139, 225, 363–369] Melting equilibria.[283, 370–373]

Ethanol—Benzene
Vapour-liquid equilibria at 1 atm, azeotrope with temperature minimum 67·9°, mole fraction benzene = 0·551.[360] Azeotrope as function (p, t, x).[374–376] Vapour-liquid equilibria at elevated pressures,[377] with critical locus.[403] Further literature.[222, 366, 373, 378–384] Melting equilibria. Eutectic at −115°, 1·3 mole per cent benzene.[283, 370, 385–386]

1-Propanol—Benzene
Vapour-liquid equilibrium at 40°, azeotrope with pressure maximum p = 196·3 mm, mole fraction benzene = 0·865.[365] Change of azeotropic composition with temperature and pressure (100–760 mm).[252] Vapour pressure of mixtures at temperatures 0–70°.[139] Vapour-liquid equilibria from 3–8 atm,[387] to 18 atm,[209] with critical locus.[403] Further literature.[212, 227, 366, 373, 388] Melting equilibria.[283]

2-Propanol—Benzene
Melting equilibria.[370, 389] Eutectic −88°, 1 mole per cent benzene. Vapour-liquid equilibria at 500 mm,[390] at 30–60°, azeotrope.[391] Change of azeotropic composition with temperature and pressure (100–760 mm).[252] Boiling point–composition diagram at 1 atm.[392]

Deuterated isopropyl alcohol $(CH_3)_2CHOD$—Benzene
Boiling point composition diagram at 1 atm.[392]

1-Butanol—Benzene
Vapour-liquid equilibrium at 25°.[393] Vapour-liquid equilibria at elevated pressures and critical locus.[403] The system is azeotropic only above 7·5 atm.[403] Further literature.[366, 373, 388, 394–396] Melting equilibria.[370]

2-Butanol(dl)—Benzene
Vapour-liquid equilibrium at 25°, azeotrope with max. pressure p = 96·4 mm, mole fraction benzene = 0·905.[393] Vapour-liquid equlibrium at 1 atm, azeotrope: mole fraction benzene = 0·847,[397] t = 78·36°.[245] Further literature.[249]

Isobutyl alcohol—Benzene
Vapour-liquid equilibrium at 25·0°.[393] Change of azeotropic composition with temperature and pressure (100–760 mm).[252]

tert. Butyl alcohol—Benzene
Vapour-liquid equilibrium at 25·0°. Azeotrope with max. pressure p = 106·5 mm, mole fraction benzene = 0·805.[393] Minimum boiling azeotrope at 1 atm: 72·6°, 63·5 mole per cent benzene.[398] Further literature.[399]

1-Pentanol—Benzene
Vapour-liquid equilibria.[366]

tert.-Amyl alcohol—Benzene
Vapour-liquid equilibrium.[400]

Methyl isobutyl carbinol—Benzene
Vapour-liquid equilibrium at 760 mm.[401]

1-Decanol—Benzene
Melting equilibria. Eutectic −7·5°, 62·0 weight per cent benzene.[402]

1-Dodecanol (Lauryl alcohol)—Benzene
Melting equilibria. Eutectic 2·5°, 76·1 weight per cent benzene.[402]

1-Tetradecanol (Myristic alcohol)—Benzene
Melting equilibria. Eutectic 5·2°, 93·6 weight per cent benzene.[402]

1-Hexadecanol (Cetyl alcohol)—Benzene
Melting equilibria. Eutectic 5·5°, 98·3 weight per cent benzene.[402]

1-Octadecanol—Benzene
Melting equilibria. Eutectic 5·5°, 99·2 weight per cent benzene.[402]

Phenol—Benzene
Benzyl alcohol—Benzene
Vapour-liquid equilibrium at 70·0°.[351]

m-*Cresol—Benzene*
Boiling temperature as function of the liquid composition at 1 atm.[404] Melting equilibria. Eutectic at −22·0°, 39 mole per cent benzene.[405]

2-Naphthol—Benzene
Melting equilibria.[406] Eutectic at 4·33°, 98·97 mole per cent benzene.

Glycol—Benzene
Liquid-liquid equilibria.[407]

Pyrocatechol—Benzene
Solubility of o-$C_6H_4(OH)_2$ in benzene at 20–100°.[408]

Resorcinol—Benzene
Melting equilibria.[408–410] At 96° equilibrium between solid m-$C_6H_4(OH)_2$ and two liquids resp. with 87 and 39 mole per cent benzene. Upper critical solution point 109·5°, 65 mole per cent benzene.

Hydroquinone—Benzene
Melting equilibria.[410] At 157·1° equilibrium between solid p-$C_6H_4(OH)_2$ and two liquids, containing 82 and 37 mole per cent benzene resp.

Estradiol—Benzene
Solubility of estradiol in benzene at 0–40°.[411]

Dimethyl ether—Benzene
The solubility of dimethyl ether gas in benzene at 25° and pressures from 93–1007 mm.[62]

Diethyl ether—Benzene
Vapour pressure of mixtures at temperatures from 0–70°.[139] Boiling temperature of mixtures at 761·8 mm.[364] Further literature.[209, 412] Melting equilibria. Eutectic at −126·5°, 5·2 mole per cent benzene.[283, 413]

Diisopropyl ether—Benzene
Vapour-liquid equilibria.[414]

Dimethoxy methane—Benzene
Vapour-liquid equilibrium at 35°.[306]

Anisole—Benzene
Vapour-liquid equilibrium at 70·0°.[349] Boiling temperature as function of the liquid composition at 1 atm.[404]

Veratrole—Benzene
Solubility of o-$C_6H_4(OCH_3)_2$ in benzene.[415]

2-Methoxy ethanol—Benzene
Vapour-liquid equilibrium at 1 atm.[243]

Diethylene glycol—Benzene
Immiscibility in the liquid phase. Upper critical solution point 88·8°, 55 weight per cent benzene[416] (82°, 50 weight per cent benzene).[417] Further literature.[407]

Triethylene glycol—Benzene
Liquid-liquid equilibria.[418] Critical solution temp. 42·5°.

Guaiacol—Benzene
Melting equilibrium with eutectic at −8°, 65 mole per cent benzene.[419]

Acetaldehyde—Benzene
Vapour-liquid equilibrium at 1 atm.[420]

Furfural—Benzene
Vapour-liquid equilibria at 1 atm.[243] Further literature.[237] Melting equilibria. Eutectic at −5·5°, 62·5 mole per cent benzene.[276, 421–423]

Systems of the hydroxybenzaldehydes with benzene
Melting equilibria of *ortho*-, *meta*- and *para*-hydroxybenzaldehyde with benzene.[424]

Systems of some hydroxy-methylbenzaldehydes with benzene
Melting equilibria of 2-hydroxy-5-methyl-, 4-hydroxy-2-methyl- and 4-hydroxy-3-methylbenzaldehyde with benzene.[424]

Acetone—Benzene
Vapour-liquid equilibrium at 25·0 and 35·0°[306] and at 45°.[425] Vapour-liquid equilibrium at 1 atm.[307–308, 420] Further literature.[140, 320, 364, 368] Melting equilibria.[413] Eutectic at −98°, 8 mole per cent benzene.

2-*Butanone—Benzene*
Vapour-liquid equilibria with an azeotrope with temperature minimum[388, 426] at 1 atm, t = 78·25° and mole fraction benzene = 0·550. Vapour-liquid equilibria at high pressures; the system becoming nonazeotropic above 8·5 atm.[209, 427] Solid-liquid and liquid-vapour equilibria.[256]

Methyl-isobutyl ketone—Benzene
Vapour-liquid equilibria at 80° and at 450, 600 and 760 mm pressure.[428] The system approaches ideality.

2-*Nonanone—Benzene*
Melting equilibria.[429] Eutectic at −26·9°, 42·9 weight per cent benzene.

2-*Tridecanone—Benzene*
Melting equilibria. Eutectic of $C_6H_6+\alpha C_{13}H_{26}O$ at −5·8°, 68·0 weight per cent benzene. Metastable eutectic of $C_6H_6+\beta C_{13}H_{26}O$ at −6·5°, 66·1 weight per cent benzene.[429]

2-Nonadecanone—Benzene
Melting equilibria. Eutectic at 5·2°, 93·7 weight per cent benzene. Metastable eutectic with $\beta C_{19}H_{38}O$ at 5·1°, 93·0 weight per cent benzene.[429]

10-Nonadecanone (Caprinone)—Benzene
Melting equilibria. Eutectic at 3·9°, 90·0 weight per cent benzene.[430]

12-Tricosanone—Benzene
Melting equilibria. Eutectic at 5·4°, 99·6 weight per cent benzene.[430]

14-Heptacosanone—Benzene
Solubility of $C_{27}H_{54}O$ in benzene at 10, 30 and 50°.[430]

16-Hentriacontanone (Palmitone)—Benzene
Solubility of $C_{31}H_{62}O$ in benzene at 30–68·8°.[430]

18-Pentatriacontanone (Stearone)—Benzene
Solubility of $C_{35}H_{70}O$ in benzene at 10–80·1°.[430]

Camphor—Benzene
Melting equilibria.[431] A compound $C_6H_6.C_{10}H_{16}O$ is reported with melting point −29·2°. Eutectic $C_6H_6+C_6H_6.C_{10}H_{16}O$ at −32·3°, 58 mole per cent benzene. Eutectic $C_6H_6.C_{10}H_{16}O+C_{10}H_{16}O$ at −32·2°, 47·1 mole per cent benzene. Mixed crystals up to 10 per cent $C_{10}H_{16}O$.

Acetophenone—Benzene
Melting equilibria.[432]

Benzophenone—Benzene
Melting equilibria. Eutectic at −10·5°, 75·5 mole per cent benzene.[433]

Anthraquinone—Benzene
Solubility of anthraquinone in benzene at temperatures from 0–80°.[299]

Formic acid—Benzene
Immiscibility in the liquid phase. Upper critical solution point 73·2°, 52 weight per cent benzene.[434–436] Vapour-liquid equilibria at 25·00 and 59·96°.[437] Hetero azeotrope with pressure maximum at 25·00°: $p = 124$ mm, $X_{L1} = 0·0909$, $X_{L2} = 0·8121$, $X_G = 0·5674$; at 59·96°: $p = 517·7$ mm, $X_{L1} = 0·1715$, $X_{L2} = 0·6763$, $X_G = 0·5617$ (X = mole fraction benzene).

Acetic acid—Benzene
Vapour pressures of mixtures at 10–70° indicate a max. pressure at X benzene = about 0·8.[139] A max. pressure was not found at 49·99°.[327] At 750 mm an azeotrope was found at X benzene = 0·9806.[438] Further literature.[315, 439–444] Melting equilibria.[445–446] Eutectic at −8·2°, 65·3 weight per cent benzene.

Acetic anhydride—Benzene
Liquid-vapour equilibrium at 1 atm.[447]

Some systems of an aliphatic monocarboxylic acid and benzene
Eutectic temperatures and compositions of the melting equilibria in systems of benzene with an aliphatic acid are given in Table IV.19.

TABLE IV.19

Eutectic Point of Some Binary Mixtures of a Monocarboxylic Acid and Benzene

Acid	*Eutectic temperature* °C	*Eutectic composition weight % benzene*
Caprylic acid[448]	−10·5	49·5
Pelargonic acid[448]	−13·1	45·9
Capric acid[448]	− 2·0	65·5
Undecanoic acid[448]	− 4·0	61·1
Lauric acid[448]	4·5	88·8
Tridecanoic acid[448]	3·7	85·4
Myristic acid[448, 449]	5·2	97·1
Pentadecanoic acid[448]	5·15	96·4
Palmitic acid[448, 449, 450]	5·4	99·8
Heptadecanoic acid[448]	5·35	99·6
Stearic acid[448, 450]	5·50	99·985
Oleic acid[451]	− 9·2	41·3
Linoleic acid[451]	−21·2	25·3

Benzoic acid—Benzene
Melting equilibria.[452, 453] Eutectic at 4·3°, 96·7 mole per cent benzene.

Systems of the hydroxybenzoic acids with benzene
Melting equilibria of *ortho-* (salicylic), *meta-* and *para*-hydroxybenzoic acid with benzene.[453]

Systems of the aldehydobenzoic acids with benzene
Melting equilibria of *ortho-*, *meta-* and *para*-aldehydobenzoic acid with benzene.[454]

Systems of some hydroxy-methylbenzoic acids with benzene
Melting equilibria of 2-hydroxy-3-methyl-, 2-hydroxy-4-methyl-, 2-hydroxy-5-methyl-, 3-hydroxy-4-methyl- and 4-hydroxy-3-methyl benzoic acid with benzene.[453]

Phenylacetic acid—Benzene
Melting equilibria.[453]

Hydrocinnamic acid—Benzene
Melting equilibria.[453]

Systems of some aliphatic dicarboxylic acids with benzene
The solubility of succinic acid, adipic acid, suberic acid, sebacic acid, dodecanedioic acid and hexadecanedioic acid (thapsic acid) in benzene at temperatures from 30–65°.[455]

Methyl acetate—Benzene
Vapour pressures of mixtures from 0–80°.[306]

Ethyl acetate—Benzene
Vapour pressure of mixtures at 0–60°.[139] Azeotrope with max. pressure at t = 20°: p max. = 107·3 mm, mole fraction benzene = 0·50. Further literature.[456]

n-*Butyl acetate—Benzene*
Vapour-liquid equilibria.[457–458]

Systems of some esters of aliphatic monocarboxylic acids and benzene
Eutectic temperatures and compositions of the melting equilibria in systems of some methyl esters of aliphatic monocarboxylic acids[459] with benzene are given in Table IV.20. The solubility of the methyl, ethyl, propyl and butyl ester of stearic acid was determined at temperatures from 10–30°.[459]

TABLE IV.20

Eutectics of the Methyl Ester of Some Monocarboxylic Acids with Benzene

Ester	*Eutectic temperature* °C	*Eutectic composition weight % benzene*
Methyl ester of caprylic acid	−42	27
Methyl ester of lauric acid	−14	50
Methyl ester of myristic acid	− 6	62
Methyl ester of palmitic acid	− 1	73

Salol (Phenyl salicylate)—Benzene
Melting equilibria.[433] Eutectic at −10° and 76 mole per cent benzene.

Systems of some triglycerides of aliphatic monocarboxylic acids and benzene
Melting equilibria of glycerol triacetate, -tributyrate, -tricaproate, -tricaprylate, -tricaprate, -trilaurate, -trimyristate, -tripalmitate and -tristearate with benzene.[460]

Cotton wax —Benzene
Solubility of $C_nH_mO_o$ in benzene.[461]

1,4-Dioxane—Benzene
Melting equilibria.[462] Eutectic at −26°, 56·5 mole per cent benzene. Vapour-liquid equilibria at 200–760 mm.[463–464]

Trioxane—Benzene
Vapour-liquid equilibria at 1 atm.[465]

2-Chloroethanol—Benzene
Vapour-liquid equilibrium at 1 atm.[466]

1,1,1,3,3,3-Hexafluoro-2-propanol—Benzene
Liquid-vapour equilibria at 25, 40 and 55°,[467] showing an azeotrope with max. pressure.

*1-Hydroperfluoro-*n*-heptane—Benzene*
Dependence of critical solution temperature on pressure.[468]

Systems of the chlorophenols with benzene
Melting equilibria.[469] Eutectic with *o*-chlorophenol at −19·5°, 38·42 weight per cent benzene. Eutectic with *m*-chlorophenol at −6·5°, 61·61 weight per cent benzene. Eutectic with *p*-chlorophenol at −5·5°, 62·5 weight per cent benzene.

Bis-β-chloro-ethyl ether—Benzene
Melting equilibria.[470] Eutectic at −60·0°, 29·0 mole per cent benzene.

α-Chloroacetophenone (phenacyl chloride)—Benzene
Melting equilibria.[432] Eutectic −1·6°, 77·1 weight per cent benzene.

Chloroacetic acid—Benzene
Melting equilibria.[471–472] Eutectic $C_6H_6+\alpha C_2H_3O_2Cl$ at 4·0°, 94·6 mole per cent benzene. Metastable eutectic $C_6H_6+\beta C_2H_3O_2Cl$ at 3·65°, 92·9 mole per cent benzene.

Trichloroacetic acid—Benzene
Melting equilibria.[472] Eutectic at −3·8°, 76·5 mole per cent benzene.

Systems of the chlorobenzoic acids with benzene
Melting equilibria of *ortho-*, *meta-* and *para-*chlorobenzoic acid with benzene.[453]

Benzoyl chloride—Benzene
Melting equilibria.[473] Eutectic at −26·8°, 36·4 per cent benzene.

Dimethyl sulphoxide—Benzene
Total vapour pressure at 25, 40, 55 and 70°.[474] Solid-liquid equilibria with eutectic at −6·59°, 64·5 mole per cent benzene.[475]

Thiophene—Benzene
There is complete miscibility in solid and liquid phases.[422, 476–478] Further literature.[479] Vapour-liquid equilibria.[480–481]

Systems of some naphthalenetri- and naphthalenetetrasulphonyl chlorides with benzene
The solubility of 1,3,5-, 1,3,6- and 1,3,7-naphthalenetrisulphonyl chloride and 1,3,5,7-naphthalenetetrasulphonyl chloride in benzene at 20 and 30°.[482]

Trimethyl amine—Benzene
The solubility of trimethylamine gas in benzene at 25° and 27·8–33·1 mm.[483]

Some systems of an aliphatic amine or an amine salt with benzene
Data about solid-liquid equilibria of these systems are collected in Table IV.21.

Aniline—Benzene
Vapour-liquid equilibria at 50, 70 and 90°[349, 494] and at 350 and 760 mm pressure.[495–496] Further literature.[497–498] Melting equilibrium with an eutectic at −33·5°; 33 mole per cent benzene.[405] Boiling and melting equilibria.[499]

N-methylaniline—Benzene
Vapour-liquid equilibria at 25, 35 and 50°.[498]

N,N′-Dimethylaniline—Benzene
Vapour-liquid equilibrium at 70°.[349] Solubility in benzene.[500]

Diphenylamine—Benzene
Solubility of diphenylamine in benzene from 0–40°.[501]

Systems of the diaminobenzenes with benzene
Melting equilibria of *ortho*-, *meta*- and *para*-phenylenediamine with benzene. Solid *meta*-phenylenediamine is at 53·8° in equilibrium with two liquids resp. containing 31·5 and 90 mole per cent benzene. Upper critical solution point at 69° and 68·5 mole per cent benzene.[502]

m-*Toluidine—Benzene*
Melting equilibria.[503] Eutectic at −54° and 21·2 mole per cent benzene.

Pyrrole—Benzene
Solid-liquid equilibria. Eutectic at −36·6° and 22·6 mole per cent benzene.[504]

Pyridine—Benzene
Vapour-liquid equilibrium at 1 atm.[505] Melting equilibria.[283, 421, 462, 506–507] Eutectic at −57°; 25·5 mole per cent benzene. Mixed crystals from 0–16 and 61–100 mole per cent benzene.

Indole—Benzene
Melting equilibria.[508] Eutectic at −9·5° and 72·0 mole per cent benzene.

TABLE IV.21

Solubility Range and/or Eutectic of Various Amines and Amine Salts with Benzene

Compound	*Temperature range,* °C	*Composition range weight % benzene*	*Eutectic temp.,* °C	*Eutectic comp., weight % benzene*	*Refs.*
Diethylamine	−54·5 to +5·4	0 – 100			283
Triethylamine	−117·7 to +5·5	0 – 100	−117·7		484
Tetrapentylammonium iodide	25				485
Tetraisopentylammonium iodide	25				485, 486
Dioctylamine (α form)	26·7 to −0·4	0 – 63·3	−0·4	63·3	487
Dioctylamine (β form, metastable)	14·6 to −1·5	0 – 58·4	−1·5	58·4	487
Dioctyldimethylammonium chloride					488
Trioctylamine			−38·5	8·9	489
Decylamine	16·1 to −5·6	0 – 68·3	−5·6	68·3	490
Dodecylamine	28·3 to 0·2	0 – 82·0	0·2	82·0	490
Dodecylammonium chloride	74·4 to 45·5	60 – 99·75			491
Dodecylammonium acetate	69 to 20	0 –> 99			492, 493
Dodecylmethylammonium chloride	78·3 to 20·9	70·1– 99·04			491
Dodecyldimethylammonium chloride	80 to 15	91·3– 99·8			491
Didodecylamine	51·8 to 10	0 – 96·8			487
Didodecyldimethylammonium chloride					488
Tridodecylamine			0·3	57·5	489
Tridecylammonium acetate ($C_{13}H_{27}NH_3OOCCH_3$)	52 to 26	30 – 98			493
Ditridecylamine	56·5 to 10	0 – 98·3			487
Tetradecylamine	38·2 to 3	0 – 89·2	3·0	89·2	490
Tetradecylammonium acetate ($C_{14}H_{29}NH_3OOCCH_3$)	60 to 35	33 – 98			493
Ditetradecylamine	60·6 to 10	0 – 99·1			487
Pentadecylammonium acetate	60 to 35	30 – 98			493
Dipentadecylamine	63·3 to 10	0 – 99·5			487
Hexadecylamine	46·8 to 4·6	0 – 93·9	4·6	93·9	490
Hexadecylammonium acetate	65 to 40	30 – 98			493
Heptadecylammonium acetate	67 to 40	30 – 98			493
Octadecylamine	53·1 to 5·2	0 – 97·8	5·2	97·8	490
Octadecylammonium acetate	85 to 20	0 –> 99			492, 493
Dioctadecylamine	72·3 to 10	0 –> 99·9			487
Trioctadecylamine	54·0 to 5·5	0 – 100	5·5	99·8	489

Isoquinoline—Benzene

Melting equilibria.[279] Eutectic at −20·0°, 62·8 mole per cent benzene.

3-Methylisoquinoline—Benzene

Solid-liquid equilibria.[509] Eutectic at −3·3°, 85·8 mole per cent benzene.

Carbazole—Benzene

Solubility of carbazole in benzene from 15·5 to 50°.[298]

4-Methylimidazole—Benzene

Melting equilibria.[510] Eutectic at 3·2° and 62·5 weight per cent benzene.

Nicotine—Benzene

Melting equilibria.[511]

1,3,5-Tridodecylhexahydro-sym-*triazine—Benzene*

1,3,5-Trioctadecylhexahydro-sym-*triazine—Benzene*

Solubility in benzene.[512]

Systems of the chloroanilines with benzene

Melting equilibria.[513] Eutectic *o*-chloroaniline-benzene at −27° and 44 mole per cent benzene. Eutectic *m*-chloroaniline-benzene at −30·6° and 37 mole per cent benzene. Eutectic *p*-chloroaniline-benzene at 1·5° and 94·7 mole per cent benzene.

Some systems of a nitrile with benzene

Vapour-liquid equilibria of acetonitrile[514–515] and of benzonitrile (at 70·0°)[349] with benzene.

Data of the eutectics of some nitriles with benzene are collected in Table IV.22.

Acetonitrile and benzene form a solid compound $C_2H_3N.2C_6H_6$, incongruently melting at −24·0°, the liquid containing about 40 mole per cent C_6H_6.[516–517]

TABLE IV.22

Eutectics of Some Nitriles with Benzene

Compound	*Eutectic temperature* °C	*Eutectic composition weight %*
Acetonitrile	−51·4;[484] −48·3;[516] −49·5[517]	9·1;[484] 17·0;[516] 15·0[517]
Propionitrile[484]	−96	8·0
Butyronitrile[484]	−117·8	8·3
Valeronitrile[484]	−102·8	8·5
Caprinitrile[518]	−25·9	33·3
Myristonitrile[518]	−7·6	61·0
Palmitonitrile[518]	−2·6	71·6
Stearonitrile[518]	+1·2	80·8

2-Aminoethanol—Benzene

Solid-liquid equilibria.[519] Eutectic at 5° and 10 mole per cent C_6H_6. Immiscibility in the liquid phase, the monoethanolamine-rich layer contains 10 mole per cent C_6H_6, the other one 99·9 mole per cent at 5°.

Systems of the aminophenols with benzene

Solid-liquid equilibria of *ortho-*, *meta-* and *para-*aminophenol with benzene.[520] At 110·6° solid *meta-*aminophenol coexists with two liquids, containing 87·7 and 38·2 mole per cent benzene resp. The upper critical solution point in this system is given as 122·3°; 67·7 mole per cent benzene.

Systems of the aminoacetanilides with benzene

Solid-liquid equilibria of *ortho-*, *meta-* and *para-*aminoacetanilide with benzene.[502] At 185° the solid *meta-*isomer coexists with two liquids containing resp. 81 and 36 mole per cent benzene. The upper critical solution point in this system is 266°, 70 mole per cent benzene. At 146·8° solid *para-*$C_8H_{10}N_2O$ coexists with two liquids containing 94 and 42 mole per cent benzene. The upper critical solution point in this system is 188°, 81 mole per cent benzene.

Ethyl urethane (ethyl carbamate)—Benzene

The eutectic temperature increases from 4·2° at 1 atm to 59·4° at 3920 atm, the eutectic composition changing from 3·9 to 80·0 mole per cent urethane.[281, 521]

Systems of ortho-, meta- *and* para-*aminobenzoic acid with benzene*

Solid-liquid equilibria.[522]

Formamide—Benzene

Melting equilibria.[484] Eutectic at 1·5°. At 5·2° immiscibility in the liquid phase, 5 and 95 mole per cent benzene.

N,N-Dimethylformamide—Benzene

Vapour-liquid equilibrium at 1 atm,[523] at 30–60°.[524]

Acetamide—Benzene

Immiscibility in the liquid phase. Upper critical solution temperature 143·60°.[525]

N-Methylacetamide—Benzene

Solid-liquid equilibria.[526]

Systems of some aliphatic amides and their N-phenyl- and N,N-diphenyl-derivatives with benzene

The solubilities of some amides and N-phenyl-amides in benzene expressed in weight per cent amide (*c.q.* derivative) at the indicated temperature are collected in Table IV.23.

*Systems of N-acetyl-*ortho- *and N-acetyl-*para-*toluidine with benzene*

Melting equilibria of *ortho-*[528] and *para-*acetotoluidide[529] with benzene.

Nitromethane—Benzene

Vapour-liquid equilibrium at 1 atm. Azeotrope at 0·843 mole fraction benzene.[244] Further literature.[530–531] Melting equilibria. Eutectic at −36·0°.[484]

TABLE IV.23

Solubility of Some Amides, N-phenyl- and N,N-diphenyl-amides in Benzene[527]

Temperature, °C	*10*	*30*	*50*	*60*	*70*	*80*	*80·1*
Compound				*Solubility, weight % amide*			
Capryl amide	0·5	0·6	3·9	12·1	34·2		60·5
Capramide	0·6	0·8	3·1	10·2	36·3		66·1
N-Phenylcapramide	9·2	40·1	73·9	87·8			
Metastable N-Phenylcapramide	12·0	52·4					
N,N-Diphenylcapramide	50·0	76·7					
Lauramide	0·6	1·0	6·3	21·0	45·4		70·4
N-Phenyllauramide	1·5	12·8	60·8	77·3	91·0		
Metastable N-Phenyllauramide		35·9					
N,N-Diphenyllauramide	35·9	64·2	90·2				
Myristamide	0·3	0·4	1·5	4·9	21·5	48·8	
Palmitamide	0·3	0·4	2·0	7·1	29·2	59·2	
N-Phenylpalmitamide	1·0	2·2	23·4	49·2	69·0		86·5
N,N-Diphenylpalmitamide	15·8	44·8	73·9	87·8			
Stearamide	0·3	0·4	1·6	4·9	20·0	47·1	
N-Phenylstearamide	0·7	1·5	12·4	35·9	59·2		84·7
N,N-Diphenylstearamide	9·9	39·4	69·7	83·8	97·2		

Tetranitromethane—Benzene
Melting equilibria.[532] Eutectic at −16·4°; 50 mole per cent benzene.

Nitroethane—Benzene
1-Nitropropane—Benzene
2-Nitropropane—Benzene
Total and partial vapour pressures at 25°.[531]

Systems of a nitrocompound of benzene, toluene, xylene or naphthalene with benzene
Data of solubilities at the indicated temperatures and eutectic data are collected in Table IV.24.

Boiling and melting equilibria in the system Nitrobenzene—Benzene;[499] vapour-liquid equilibria at 70°[351] and further literature about this system.[531] Further literature about the system *p*-Dinitrobenzene—Benzene.[354, 536] In the system 1,3,5-Trinitrobenzene—Benzene there is a compound $C_6H_3(NO_2)_3.2C_6H_6$, melting point 71·2°.

Systems of the nitrophenols with benzene
Melting equilibria. The system with *ortho*-nitrophenol has an eutectic at −5° and 83·5 mole per cent benzene.[544–546] The eutectic with *meta*-nitrophenol is near the melting point of benzene,[544–545] whereas the eutectic of the *para*-nitrophenol—benzene system is at 5·0° and 99·8 mole per cent benzene.[544–546]

TABLE IV.24

Solid-Liquid Equilibria of Some Nitrocompounds of Aromatic Hydrocarbons with Benzene

Compound	*Eutectic temperature,* °C	*Eutectic composition, mole % C_6H_6*	*Solubility in weight-% nitrocompound*	*At temperature,* °C
Nitrobenzene[169, 324, 499, 533]	−24	49		
o-Dinitrobenzene[534, 535]			2·8 –73·7	4·3–95·5
m-Dinitrobenzene	−1·0	90·75[533]		
	0·5	89·0[535]		
p-Dinitrobenzene[534, 535]			1·3 –18·4	4·8–80·9
1,3,5-Trinitrobenzene[535–538]	2	98*		
	70·8	57†		
1,2,4,6-Tetranitrobenzene[539]			2·24	23·0
p-Nitrotoluene[536]			35·9	15
2,4-Dinitrotoluene[534]	−0·5	89·0		
2,6-Dinitrotoluene[534]	0·1	91·3		
3,4-Dinitrotoluene[534]	−4·5	77		
2,4,6-Trinitrotoluene[540, 541]			11·5 –95·3	0 –75
2,4,6-Trinitroxylene[542]			0·92 –10·0‡	15 –80
1-Nitronaphthalene[543]			53·3 –68·1	18 –29·5
1,5-Dinitronaphthalene[543]			0·381– 1·479	16 –50
1,8-Dinitronaphthalene[543]			0·750– 2·754	16 –50
1,3,8-Trinitronaphthalene[543]			0·407– 0·987	15 –50
1,4,5-Trinitronaphthalene[543]			0·917– 6·874	15 –50

* Eutectic $(C_6H_6)s+(C_6H_3(NO_2)_3 \cdot 2C_6H_6)s+$liquid; melting point of the compound 71·2°.
† Eutectic $(C_6H_3(NO_2)_3 \cdot 2C_6H_6)s+(C_6H_3(NO_2)_3)s+$liquid.
‡ g per 100 ml solution.

Systems of the dinitrophenols and the symmetric trinitrophenol with benzene

Solubilities of benzene in 2,3-,[547] 2,4-,[546–548] 2,5-,[547] 2,6-,[547] 3,4-[547] and 3,5-dinitrophenol[549] are reported from their melting points to about 10 weight per cent dinitrophenol. For the system picric acid—benzene an eutectic is given at 4·2° and 98·7 mole per cent benzene.[406, 550–553]

2,4,6-Trinitrocresol—Benzene

At 17·0° 11·9 g and at 50° 35·9 g of $C_7H_5N_3O_7$ are soluble in 100 g benzene solution.[554]

Systems of some nitrophenyl ethers with benzene

The solubility of the compound in benzene at the indicated temperature is given in Table IV.25 in weight per cent of the ether.

TABLE IV.25

Solubilities of Some Nitrophenyl Ethers in Benzene

Temperature, °C	*15*	*20*	*23*	*25*	*29*	*50*
Compound			*Solubility*			
2,4-Dinitroanisole[555]	23·3					55·4
3,5-Dinitroanisole[554]		9·4				
2,4,6-Trinitroanisole[555]	48·7					85·7
2,3,5-Trinitroanisole[554]				15·3		
2,3,5,6-Tetranitroanisole[554]					1·53	
2,4-Dinitrophenetole[555]	39·0					67·7
2,4,6-Trinitrophenetole[555]	33·0					74·9
2,2',4,4',6-Pentanitrodiphenyl ether[556]			0·53			0·87

Systems of the nitrobenzaldehydes with benzene

Melting equilibria. The eutectics of *ortho-*, *meta-* and *para-*nitrobenzaldehyde with benzene are reported[557] respectively at −1·7°, 84·3 mole per cent benzene; −0·8°, 91·2 mole per cent benzene and 4·1°, 98·2 mole per cent benzene. In the system with the *meta-*compound a solid compound: $C_7H_5NO_3 . 2C_6H_6$ occurs,[313] incongruently melting at 7°, 76 mole per cent benzene.

Systems of the nitrobenzoic acids with benzene

Melting equilibria of *ortho-*, of *meta-* and of *para-*nitrobenzoic acid with benzene.[453, 558]

2,4,6-Trinitrobenzoic acid—Benzene

At 25° 0·3 g of $C_7H_3N_3O_8$ is soluble in 100 g benzene solution.[556]

rac.-2,4-Dinitro-2'-methyl diphenyl -6-carboxylic acid—Benzene

Melting equilibria.[559] Two compounds: $C_6H_6 . C_{14}H_{10}N_2O_6$, melting point 128° and $C_6H_6 . 2C_{14}H_{10}N_2O_6$, melting point 138°.

Dimethyl ester of 4,4',6,6'-tetranitrodiphenic acid—Benzene

Melting equilibria.[560] The compound: $C_6H_6 . 2C_{16}H_{10}N_4O_{12}$ melts incongruently at 156°, liquid composition 36 mole per cent benzene.

Systems of the nitroanilines with benzene

Melting equilibria of *ortho-*, of *meta-* and of *para-*nitroaniline with benzene.[513, 544, 561]

Systems of some nitro-substituted aromatic amines with benzene

The solubility of some nitro-substituted aromatic amines expressed in g/100 g benzene is given in Table IV.26.

TABLE IV.26

Solubilities of Some Nitro-Substituted Aromatic Amines

Compound	*Temp., °C*	*Solubility*	*Ref.*
2,4-Dinitro-N-methylaniline	17	0·32	539
2,4,6-Trinitroaniline	19·5	0·91	539
2,4,6-Trinitro-N-ethylaniline	24	51·8	556
2,4,6-Trinitro-N,N-diethylaniline	20	4·96	556
2,4-Dinitrodiphenylamine	15	2·118	536
	50	6·977	
2,2',4,4'-Tetranitrodiphenyl amine	15	0·320	536
	50	0·998	
Bis(2,4,6-Trinitrophenyl) amine	17	insoluble	536
	50	0·399	

o-, m- *and* p-*Nitroacetanilide—Benzene*
Melting equilibria.[513]

1,3,5-Trinitrohexahydro-s-triazine—Benzene
Solubility of 1,3,5-trinitrohexahydro-*s*-triazine (or: cyclo-trimethylenetrinitro-amine) in benzene at temperatures: 10–79·5°.[562–563]

2,4,6-Trinitrophenylnitramine—Benzene
At 3·9° 0·13 g $C_6H_3N_5O_8$ is soluble in 100 g benzene.[540, 564]

2,4,6-Trinitrophenyl-N-methylnitramine—Benzene
At 18° 4·8 g $C_7H_5N_5O_8$ is soluble in 100 g benzene solution.[539]

2,4,6-Trinitrophenyl-N-ethylnitramine—Benzene
At 25° 16·5 g and at 50° 38·6 g $C_8H_7N_5O_8$ is soluble in 100 g benzene solution.[556]

Pentaerythritol tetranitrate—Benzene
Solubility of $C_5H_8N_4O_{12}$ in benzene from 10–80·2°.[539, 565]

Systems of barbituric acid and derivatives with benzene
The solubility of barbituric acid, diethyl-, dipropyl- and N-methyl ethyl barbituric acid in benzene is given as: 7·1; 51·3; 182·0 and 178·7 mg/100 ml respectively.[566]

Quinine—Benzene
Solubility of quinine in benzene.[567]

Some systems of an 8-hydroxyquinolinate with benzene
Solubility of Cu, Al, Fe(III) and Zr 8-hydroxyquinolinate in benzene.[568]

syn- *and* anti-*Benzilmonoxime—Benzene*
Melting equilibria.[569] *Anti-* or β-benzilmonoxime gives a solid compound $C_6H_6.2C_{14}H_{11}NO_2$ incongruently melting at about 62°, 58 mole per cent C_6H_6.

Methyl ethyl ketoxime—Benzene
Vapour-liquid equilibria.[570]

4,4′-Dimethoxyazoxybenzene—Benzene
Melting equilibria.[571] Eutectic at 4·7°, 99 mole per cent benzene.

sym-*Diethydiphenylurea—Benzene*
At 0° 37·9 g, at 20° 50·4 g and at 50° 80·5 g $C_{17}H_{20}N_2O$ is soluble in 100 g benzene solution.[572]

Octanal semicarbazone—Benzene
At 20° 10·7 g of $C_9H_{19}N_3O$ is soluble in 100 cm^3 benzene solution.[573]

Systems of three octanone semicarbazones with benzene
The solubility of the semicarbazones of 2-octanone, 3-octanone and 4-octanone in benzene at 20° is 0·722; 6·28 and 23·0 weight per cent respectively.[573]

Systems of the chloro- and bromo-nitrobenzenes with benzene
Some data of the solid-liquid equilibria are collected in Table IV.27.

TABLE IV.27

Eutectics of Benzene with the Chloro- and Bromo-nitrobenzenes

Compound	*Eutectic temperature,* °C	*Eutectic composition mole % benzene*	*Refs.*
o-chloronitrobenzene	—	—	544
m-chloronitrobenzene	−5·7	81·9	544
p-chloronitrobenzene	−5	84	544, 574
o-bromonitrobenzene	−8	76	544
m-bromonitrobenzene	−4	84	544
p-bromonitrobenzene	3·5	97	544

1-Chloro-2,4-dinitrobenzene—Benzene
At 16° 158·43 g and at 31° 359·6 g $C_6H_3N_2O_4Cl$ is soluble in 100 g benzene.[574]

1-Chloro-2,4,6-trinitrobenzene—Benzene
Melting equilibria.[538, 574] The compound $2C_6H_6.C_6H_2N_3O_6Cl$ melts congruently at 39·0°. Eutectic $C_6H_6+2C_6H_6.C_6H_2N_3O_6Cl$: 2·5°, 94·5 mole per cent benzene. Eutectic $2C_6H_6.C_6H_2N_3O_6Cl+C_6H_2N_3O_6Cl$: 38°, 61 mole per cent benzene.

o-*Nitrobenzyl chloride—Benzene*
At 30° 75·2 g of $C_7H_6NO_2Cl$ are soluble in 100 g benzene solution.[575]

p-*Nitrobenzyl chloride—Benzene*
At 25·0° 37·4 g[576] and at 30° 42·6 g[575] of $C_7H_6NO_2Cl$ are soluble in 100 g benzene solution.

o-, m- *and* p-*Chloroacetanilide—Benzene*
Melting equilibria.[513]

Allyl mustard oil (allyl isothiocyanate)—Benzene
Melting equilibria.[484, 577] Eutectic at −108°, 8 mole per cent benzene.

N,N'-Allyl phenyl thiourea—Benzene
Melting equilibria.[355, 577–578] Eutectic at 5·42°, 99·6 mole per cent benzene.

Thiazole—Benzene
Solid-liquid equilibria. Eutectic at −48·5°, 30 mole per cent benzene.[579]

2-Undecylbenzothiazole—Benzene
Melting equilibria.[580] Eutectic at −9·3°, 42 weight per cent benzene.

2-Heptadecylbenzothiazole—Benzene
Melting equilibria.[580] Eutectic at 1·0°, 71·5 weight per cent benzene.

Bis(2,4,6-trinitrophenyl) sulphide—Benzene
At 23·0° 0·66 g $C_{12}H_4N_6SO_{12}$ is soluble in 100 g benzene.[581]

4·3·3. Phase Equilibria in Ternary and Multicomponent Systems

Hydrogen—Nitrogen—Benzene
Solubility of mixtures of hydrogen and nitrogen in benzene at 25° and pressures 100–500 atm.[582]

*Hydrogen—Hexane—*cyclo-*Hexane—Benzene*
Vapour-liquid equilibria.[583–584]

Oxygen—Nitrogen—Benzene
Solubility of air in benzene.[585–586]

Water—Hydrogen chloride—Benzene
Partition of HCl between water and benzene as a function of the HCl-pressure at 20°[587] and 25°.[588]

Water—Perchloric acid—Tributyl phosphate—Benzene
Liquid-liquid equilibria.[589]

Water—Iodine—Benzene
Liquid-liquid equilibria.[590]

Water—Sulphuric acid—Benzene
Solubility of over 75 weight per cent H_2SO_4 in benzene.[591–592]

Water—Ammonia—Benzene
Liquid-liquid equilibria.[593]

Water—Sodium chloride—Benzene
Liquid-liquid equilibria at temperatures from 20–300°.[594–595]

Water—Sodium chloride—Ethylene oxide—Benzene
Liquid-liquid equilibria.[596]

Water—Sodium chloride—Dioxane—Benzene
Solubility of NaCl in water—dioxane—benzene mixtures.[597]

Water—Potassium chloride—Benzene
Critical phenomena.[598]

Water—Potassium chloride—Methanol—Benzene
Mutual solubilities.[599]

Water—Potassium chloride—Acetone—Benzene
Solid-liquid equilibria from 16–80°.[600]

Water—Potassium iodide—Benzene
Critical phenomena.[598]

Water—Potassium sulphate—Methanol—Benzene
Mutual solubilities.[599]

Water—Potassium nitrate—Acetone—Benzene
Solid-liquid equilibria from 16–80°.[600]

Water—Ammonium sulphate—Ethanol—Benzene
Liquid-liquid equilibria.[601]

Water—Barium chloride—Methanol—Benzene
Mutual solubilities.[599]

Water—Silver perchlorate—Benzene
Mutual solubility at 25°.[602]

Water—Ethylene carbonate—Benzene
Liquid-liquid equilibria.[596]

Water—Hexane—Methanol—Benzene
Liquid-liquid equilibria at 20, 40 and 60°[603] and at 25°.[604]

Water—Hexane—Ethanol—Benzene
Vapour-liquid equilibria. Quaternary azeotrope.[605]

Water—Heptane—Benzene
Liquid-liquid equilibria;[606–607] in the critical region.[608]

Water—Heptane—Methanol—Benzene
Liquid-liquid equilibria at 20, 40 and 60°,[603] at 25°.[604]

Water—Heptane—Ethanol—Benzene
Vapour-liquid equilibria.[609]

Water—Heptane—Monomethylformamide—Benzene
Liquid-liquid equilibria.[606–607]

Water—cyclo-*Hexane—Benzene*
Vapour-liquid equilibrium at 1 atm and 68°. Two immiscible liquid phases.[610]

Water—cyclo-*Hexane—Methanol—Benzene*
Liquid-liquid equilibria at 20, 40 and 60°[603] and at 25°.[604]

Water—Methyl-cyclo-*hexane—Ethanol—Benzene*
Vapour-liquid equilibria. Quaternary azeotrope.[605]

Water—Carbon tetrachloride—Acetic acid—Benzene
Liquid-liquid equilibria at 25°.[611]

Water—Methanol—Benzene
Liquid-liquid equilibria.[612–615]

Water—Methanol—Ethanol—Benzene
Liquid-liquid equilibria.[616]

Water—Methanol—Glucose—Benzene
Mutual solubilities.[599]

Water—Methanol—Aniline—Benzene
Liquid-liquid equilibria.[617]

Water—Ethanol—Benzene
Vapour-liquid and liquid-liquid equilibria at 1 atm. Ternary azeotrope with vapour pressure maximum: t = 64·85°, mole fraction benzene = 0·539, mole fraction ethanol = 0·228. Two immiscible liquid phases.[618–620] Three-phase liquid-liquid-vapour equilibrium at 35, 45, 55 and 64°.[621] Further literature.[622–623]

Water—Ethanol—Ethyl isovalerate—Benzene
Liquid-liquid equilibria.[617]

Water—Ethanol—Anthranilic acid—Benzene
Solid-liquid equilibria.[624]

Water—1-Propanol—Benzene
Liquid-liquid equilibria from 20–70°.[625]

*Water—*iso-*Propyl alcohol—Benzene*
Liquid-liquid equilibria.[626] Vapour-liquid equilibria at 200–760 mm. Minimum azeotrope: 66·3°, 24·9 mole per cent H_2O, 56·4 mole per cent benzene[627–628] at atmospheric pressure. Further literature.[629]

*Water—*sec.-*Butylalcohol—Benzene*
Liquid-liquid equilibria at 30·0°. Vapour-liquid equilibria. Ternary azeotrope[630] (p = 665 mm, t = 65·5°, 6 weight per cent sec.butylalcohol, 86 weight per cent benzene). Critical liquid-vapour points.[631]

Water—Hexanol—Benzene
Vapour-liquid equilibria.[632]

Water—Phenol—Benzene
Liquid-liquid equilibria at 25°,[633–634] at 15–30°,[635] at 40, 60, 80 and 100°.[636]

Water—Glycol—Benzene
Liquid-liquid equilibria.[612, 637]

Water—Triethylene glycol dimethylether—Benzene
Mutual solubilities and plait point at 25°.[638]

Water—Ethylene oxide—Benzene
Liquid-liquid equilibria.[596]

Water—Acetone—Benzene
Liquid-liquid equilibria at 20, 40 and 60°.[639–640]

Water—Acetone—N-Acetylaniline—Benzene
Solubility of acetanilide in water-acetone-benzene mixtures.[641]

Water—Formic acid—Benzene
Vapour-liquid equilibria from 20–70°.[642]

Water—Acetic acid—Benzene
Liquid-liquid equilibria.[643]

Water—Propionic acid—Benzene
Liquid-liquid equilibria at 30, 40, 60 and 80°.[644]

Water—Butyric acid—Benzene
Liquid-liquid equilibria.[645]

Water—Acrylic acid—Benzene
Liquid-liquid equilibria.[646]

Water—Butyl acetate—Benzene
Liquid-liquid equilibria.[647]

Water—Naphthalene soaps—Benzene
Liquid-liquid equilibria.[648–649]

Water—Ethylene chlorohydrin—Benzene
Liquid-liquid equilibria.[650]

*Water—*cyclo-*Hexylamine—Benzene*
Liquid-liquid equilibria.[651–652]

Water—Pyridine bases—Benzene
Liquid-liquid equilibria in the systems water-benzene with pyridine, piperidine, 2-picoline, 3-picoline, 4-picoline, 2,4-lutidine, 2,5-lutidine, 2,6-lutidine, 3,5-lutidine and 2,4,6-collidine.[653–660]

Water—Monomethyl formamide—Benzene
Liquid-liquid equilibria.[606–607]

Water—Diethyl formamide—Benzene
Liquid-liquid equilibria at 25°.[661]

Water—Acetonitrile—Benzene
Liquid-liquid equilibria.[662]

Water—Caprolactam—Benzene
Solubility isotherms at 20°.[663–664]

Water—Dimethyl sulphoxide—Benzene
Liquid-liquid equilibria.[665–666]

Hydrogen chloride—Heptane—Benzene
Solubility of HCl gas at $-78{\cdot}51°$ in a mixture of 96·25 weight per cent heptane and 3·75 weight per cent benzene at pressures from 35–72·5 mm.[667]

Hydrogen chloride—Toluene—Benzene
Solubility of HCl gas at $-78{\cdot}51°$ in a mixture of 92·2 weight per cent toluene and 7·8 weight per cent benzene at pressures from 1·9–5·8 mm.[667–668]

Hydrogen chloride—Methanol—Benzene
Liquid-liquid equilibria.[669]

Hydrogen chloride—Methoxyethanol—Benzene
Liquid-liquid equilibria at 0°.[670]

Hydrogen bromide—Aluminum bromide—Benzene
Solubility of $AlBr_3$ at different partial pressures of HBr at $-78{\cdot}5$ and 0°.[671–672]

Hydrogen bromide—Methanol—Benzene
Solid-liquid equilibria at $-5°$.[669] Liquid-liquid equilibria at 0°.[670]

Hydrogen bromide—Methoxyethanol—Benzene
Liquid-liquid equilibria.[673]

Hydrogen bromide—Acetic acid—Benzene
Liquid-liquid equilibria at 20°.[670]

Sulphur—Ammonia—Benzene
Solid-liquid equilibria.[593]

Sulphur—Carbon disulphide—Benzene
Liquid-liquid and solid-liquid equilibria.[674]

Sulphur dioxide—90 per cent $SbCl_3$, 10 per cent H_2O—Benzene
Liquid-liquid equilibria.[675]

Sulphur dioxide—some Hydrocarbons—Benzene
Liquid-liquid equilibria of sulphur dioxide and benzene with *n*-butane at $-28{\cdot}9°$; with *n*-hexane at $-28{\cdot}9°$; with *n*-heptane at $-17{\cdot}8$, $-28{\cdot}9$ and $-37{\cdot}2°$; with *n*-decane at $-28{\cdot}9°$, with *cyclo*-hexane at $-28{\cdot}9°$[676] and with diesel oil.[675] Vapour-liquid equilibria of sulphur dioxide and benzene with *n*-pentane.[677–678]

Sulphur dioxide—some Hydroxycompounds—Benzene
Liquid-liquid equilibria of sulphur dioxide and benzene with glycol, with diethylene-glycol and with glycerol.[675]

Sulphur dioxide—Formic acid—Benzene
Liquid-liquid equilibria.[675]

Sulphur dioxide—Maleic acid—Benzene
Liquid-liquid and solid-liquid equilibria.[675]

Sulphur dioxide—Formamide—Benzene
Sulphur dioxide—Diethanolamine—Benzene
Liquid-liquid equilibria.[675]

Sulphuric acid—Methyl sulphate—Benzene
Sulphuric acid—Diisopropyl ether—Benzene
Oleum—Methyl sulphate—Benzene
Oleum—Ethyl sulphate—Benzene
Liquid-liquid equilibria.[674]

Sulphur hexafluoride—Hexane—Benzene
Liquid-liquid equilibria.[674]

Ammonia—Aqueous sodium hydroxide—Benzene
Ammonia—Aqueous potassium carbonate—Benzene
Liquid-liquid equilibria.[593]

Ammonia—Ammonium thiocyanate—Benzene
Liquid-liquid and solid-liquid equilibria.[593]

*Ammonia—*n-*Hexane—Benzene*
Liquid-liquid equilibria at 20°.[679]

Ammonia—Heptane—Benzene
Liquid-liquid equilibria.[593]

*Ammonia—*cyclo-*Hexane—Benzene*
Liquid-liquid equilibria at 0° and 20°,[679] at 25°.[593]

Ammonia—some Hydroxycompounds—Benzene
Liquid-liquid equilibria of ammonia and benzene with glycol,[593, 637] with glycerol[593] and with diethylene glycol.[637]

Ammonia—some Organic nitrogen compounds—Benzene
Liquid-liquid equilibria in the ternary systems of ammonia and benzene with formamide, ethanolamine, diethanolamine and urea;[593] solid-liquid equilibria in the ammonia—urea—benzene system.[593]

Antimony chloride—Hexane—Benzene
Solid-liquid equilibrium.[680]

Carbon monoxide—Various organic compounds—Benzene
Solubility of CO at 1 atm and 25° in mixtures of benzene with naphthalene, phenanthrene, ethanol, α-naphthol, β-naphthol, acetic acid, aniline or nitrobenzene.[681]

Carbon disulphide—Methyl sulphate—Benzene
Liquid-liquid equilibria.[674]

Carbon disulphide—Nitromethane—Benzene
Liquid-liquid equilibria.[682]
Lithium chloride—Methanol—Benzene
Solid-liquid and liquid-liquid equilibria.[669]

Lithium chloride—Ethanol—Benzene
Solid-liquid-vapour equilibria.[683]

Sodium hydrosulphate—Ethyl sulphate—Benzene
Solid-liquid and liquid-liquid equilibria.[674]

Sodium oleate—some Alcohols—Benzene
Solubility of Na oleate in mixtures of methanol and benzene at 25°; propylene glycol and benzene at 20°; *cyclo*hexanol and benzene at 25°.[684]

Potassium acetate—Ethanol—Benzene
Calcium nitrate—Ethanol—Benzene
Solid-liquid-vapour equilibria.[683, 685]

Copper(II) chloride—Methanol—Benzene
Solid-liquid equilibria at 30°.[686]

Some organic copper salts—Ethanol—Benzene
Solubility of copper methylbenzoate, copper isopropylbenzoate and copper tert.-butylbenzoate in mixtures of ethanol and benzene.[687]

Aluminum chloride—some Organic compounds—Benzene
Liquid-liquid equilibria in the ternary systems of $AlCl_3$ and benzene with pentane, heptane, acetone, ethyl acetate and 2,2′-dichlorodiethyl ether, and solid-liquid equilibria in the systems with acetone, ethyl acetate and 2,2′-dichlorodiethyl ether.[680]

Aluminum bromide—Diethyl ether—Benzene
Solid-liquid equilibria.[688] At 15·20 and 25° two compounds are formed $AlBr_3.(C_2H_5)_2O$ and $Al_2Br_6.(C_2H_5)_2O$.

Gallium chloride—Nitrobenzene—Benzene
Melting equilibria. A compound is formed: $GaCl_3.C_6H_5NO_2$.[689]

Methyl sulphate—some Organic compounds—Benzene
Liquid-liquid equilibria of methyl sulphate and benzene with pentane, with hexane, with heptane, with *cyclo*hexane and with methylsulphonic acid.[674]

Ethyl sulphate—some Organic compounds—Benzene
Liquid-liquid equilibria in the ternary systems of ethyl sulphate and benzene with heptane, diethyl phosphate, methylsulphonic acid and chlorosulphonic acid.[674]

Some Tetraalkylammonium halogenides—Methanol—Benzene
Solid-liquid equilibria in the ternary systems of tetramethylammonium chloride, tetraethylammonium bromide, tetraethylammonium iodide and tetrabutylammonium iodide with methanol and benzene.[690]

Tetraethylammonium iodide—1,2-Dichloroethane—Benzene
Solid-liquid equilibria.[691]

Ethylene carbonate—Diethylene glycol—Benzene
Liquid-liquid equilibria.[637]

Methyldichlorosilane—Methylphenyldichlorosilane—Benzene
Vapour pressure of liquid mixtures.[692]

Methane—Hexane—Benzene
Liquid-vapour equilibria.[693]

Methane—2,2,3-Trimethylbutane—Benzene
Vapour-liquid equilibria at 60° and 200–300 atm.[694]

Methane—Cyclohexane—Benzene
Vapour-liquid equilibria at 0–60° and 50–400 atm.[695]

Acetylene—Methanol—Benzene
Vapour-liquid equilibria.[696]

Hexane—Methylcyclopentane—Benzene
Vapour-liquid equilibria at 1 atm;[697] at 60°.[698]

Hexane—Methylcyclopentane—Cyclohexane—Toluene—Benzene
Vapour-liquid equilibria.[699]

Hexane—Methylcyclopentane—Ethanol—Benzene
Vapour-liquid equilibria at 1 atm.[697, 700]

Hexane—Cyclohexane—Benzene
Vapour-liquid equilibria.[701–702] Solid-liquid equilibria.[702]

Hexane—Naphthalene—Benzene
Solid-liquid equilibria.[703]

n-*Hexane—Decahydronaphthalene—Phenol solution (80 per cent)—Benzene*
Liquid-liquid equilibria at 20°.[704]

Hexane—Ethanol—Benzene
Vapour-liquid equilibria.[605, 705–706] Azeotrope.[605]

n-*Hexane—Phenol solution (70–90 per cent)—Benzene*
n-*Hexane—Phenol solution (70–90 per cent)—White oil—Benzene*
Liquid-liquid equilibria at 30°.[707]

Hexane—Glycol—Benzene
Liquid-liquid equilibria.[612, 637]

Hexane—Diglycol—Benzene
Liquid-liquid equilibria.[612, 708]

Hexane—Triglycol—Benzene
Liquid-liquid equilibria.[612]

Hexane—Ethylenediamine—Benzene
Vapour-liquid equilibria at 20°.[709]

Hexane—Dimethylformamide—Benzene
Liquid-liquid equilibria.[710]

Hexane—Dimethyl sulphoxide—Benzene
Liquid-liquid equilibria.[711]

Hexane—Sulpholane—Benzene
Liquid-liquid equilibria[712] at 20°, in the system hexane—tetrahydrothiophene-1,1-dioxide ('sulpholane')—benzene.

Heptane—Toluene—Diethylene glycol—Benzene
Liquid-liquid equilibria.[708]

Heptane—1,2-Dichloroethane—Benzene
Vapour-liquid equilibria.[713]

Heptane—Perfluorochloroheptane (∼10 per cent Cl)—Benzene
Liquid-liquid equilibria.[680]

Heptane—Methanol—Benzene
Liquid-liquid equilibria.[669]

Heptane—Ethanol—Benzene
Vapour-liquid equilibria.[714-716]

Heptane—Propanol-1—Benzene
Vapour-liquid equilibria at 75°.[717]

Heptane—Butanol-1—Benzene
Vapour-liquid equilibria.[718]

Heptane—Furfuryl alcohol—Benzene
Liquid-liquid equilibria.[673]

Heptane—Glycol—Benzene
Liquid-liquid equilibria.[637]

Heptane—Methoxyethanol—Benzene
Liquid-liquid equilibria.[673]

Heptane—Diethylene glycol (2,2'-Oxydiethanol)—Benzene
Liquid-liquid equilibria.[637, 708, 719]
Solid-liquid equilibria.[720]

Heptane—Benzoic acid—Benzene
Solid-liquid equilibria.[721]

Heptane—Aniline—Benzene
Liquid-liquid equilibria.[722-723]

Heptane—Monomethylformamide—Benzene
Liquid-liquid equilibria.[606-607]

Heptane—Acetonitrile—Benzene
Liquid-liquid equilibria[662, 724] and solid-liquid equilibria.[662]

Heptane—Ten different nitriles—Benzene
Liquid-liquid equilibria with hydrocarbon mixture of 0, 25, 50 and 70 per cent aromatics.[725]

Heptane—Nitromethane—Benzene
Liquid-liquid equilibria.[682]

Heptane—1-nitro-2-propanol—Benzene
Liquid-liquid equilibria.[726]

Heptane—Dimethyl sulphoxide—Benzene
Liquid-liquid equilibria.[727]

Heptane—2,2′-Dithiodipropionitrile—Benzene
Solid-liquid equilibria.[728]

2,3-*Dimethylpentane—Perfluorotributylamine—Benzene*
Vapour-liquid equilibria from 84·7° to 104·2° at 1 atm.[729]

Octane—Urea—Benzene
Solid-liquid equilibria.[730]

Octane—Dimethyl sulphoxide—Benzene
Liquid-liquid equilibria.[731]

Isoöctane—Cyclohexane—Furfural—Benzene
Isoöctane—Furfural—Benzene
Liquid-liquid equilibria at 25°.[732]

Methylcyclopentane—Cyclohexane—Benzene
Vapour-liquid equilibria.[733]

Methylcyclopentane—Ethanol—Benzene
Vapour-liquid equilibria.[734]

Cyclohexane—Toluene—Benzene
Vapour-liquid equilibria at 1 atm.[735]

Cyclohexane—Toluene—Isopropyl alcohol—Benzene
Liquid-vapour equilibria.[736]

Cyclohexane—Naphthalene—Benzene
Solid-liquid equilibria.[703]

Cyclohexane—Carbon tetrachloride—Benzene
Vapour-liquid equilibria.[737]

Cyclohexane—Trichloroethylene—Benzene
Vapour-liquid equilibria at 1 atm.[738]

Cyclohexane—Methanol—Benzene
Vapour-liquid equilibria at 53° and 1 atm.[610] Solid-liquid equilibria.[739] Further literature.[740–741]

Cyclohexane—Ethanol—Benzene
Vapour-liquid equilibria at 1 atm and 64 and 65°.[610] Ternary azeotrope 58·8 weight per cent cyclohexane, 10·8 weight per cent benzene, 65·05°.[741] Further literature.[733, 742–744]

Cyclohexane—1-Propanol—Benzene
Vapour-liquid equilibria at 1 atm and 79°.[610] Ternary azeotrope.[741, 745] Further literature.[746]

Cyclohexane—2-Propanol—Benzene
Vapour-liquid equilibrium at 70° and 1 atm.[610] Further literature.[747–751]

Cyclohexane—Isopropyl alcohol—Acetone—Benzene
Liquid-vapour equilibria.[752]

Cyclohexane—Isopropyl alcohol—Methyl ethyl ketone—Benzene
Liquid-vapour equilibria.[753]

Cyclohexane—Butyl alcohol—Benzene
Vapour-liquid equilibrium. Ternary azeotrope: 48 weight per cent cyclohexane, 48 weight per cent benzene, t = 77·42°.[741] Vapour-liquid equilibria, no ternary azeotrope.[754]

Cyclohexane—Isobutyl alcohol—Benzene
Vapour-liquid equilibria. Azeotrope 50 weight per cent cyclohexane, 42 weight per cent benzene, 76·73°.[755] Further literature.[756–757]

Cyclohexane—tert.-Butyl alcohol—Benzene
Vapour-liquid equilibria at 1 atm.[758]

Cyclohexane—Phenol—Benzene
Vapour-liquid equilibria at 1 atm and 92°.[610]

Cyclohexane—1,2-Ethanediol (Glycol)—Benzene
Vapour-liquid equilibrium at 1 atm and 76°.[610] Two immiscible liquids.

Cyclohexane—Methoxyethanol—Benzene
Vapour-liquid equilibria at 1 atm and temperatures from 73·0–95·5°. Azeotrope: t = 73·0°, mole fraction cyclohexane = 0·4992, mole fraction benzene = 0·4055.[610, 759]

Cyclohexane—2-Ethoxyethanol—Benzene
Vapour-liquid equilibria at 1 atm and 95°.[610]

Cyclohexane—2-(2-Ethoxyethoxy)ethanol ('Carbitol')—Benzene
Vapour-liquid equilibrium at 1 atm and 87°. At room temperature two immiscible liquids.[610]

Cyclohexane—Diethylene glycol—Benzene
Vapour-liquid equilibrium at 1 atm and 75°. Two immiscible liquid phases.[610]

Cyclohexane—Triethylene glycol—Benzene
Liquid-liquid equilibria.[637]

Cyclohexane—Furfurol—Benzene
Vapour-liquid equilibria at 1 atm and temperatures ranging from 79 to 108·7°.[760] At 20° two immiscible liquids.[610] Liquid-liquid equilibria at 25°.[732]

Cyclohexane—Acetone—Benzene
Vapour-liquid equilibrium at 1 atm and 55°.[610] The same at 1 atm and temperatures from 54.5–77·8°.[761]

Cyclohexane—2-Butanone—Benzene
Vapour-liquid equilibria at 1 atm and 72°;[661, 762] at 1, 8 and 12·7 atm.[763] Although all three binary systems form azeotropes, the ternary one does not. Solid-liquid and liquid-vapour equilibria.[764] Eutectic −97°, 61 mole per cent butanone, 26 mole per cent benzene.

Cyclohexane—Methyl isobutyl ketone—Benzene
Vapour-liquid equilibria at 1 atm.[765]

Cyclohexane—4-Hydroxy-4-methyl-2-pentanone—Benzene
Vapour-liquid equilibria at 82 and 92° and at 1 atm.[610]

Cyclohexane—Acetic acid—Benzene
Vapour-liquid equilibria at 1 atm.[640, 766] Ternary azeotrope.[766]

Cyclohexane—Methyl acetate—Benzene
Vapour-liquid equilibria.[767]

Cyclohexane—Ethyl acetate—Benzene
Vapour-liquid equilibria at 1 atm.[768–769]

Cyclohexane—1,4-Dioxane—Benzene
Vapour-liquid equilibria at 1 atm and 86 and 89°.[610]

Cyclohexane—Di(2-chloroethyl) ether—Benzene
Vapour-liquid equilibria at 1 atm and 105°.[610]

Cyclohexane—Ethylene diamine—Benzene
Liquid-liquid-vapour equilibrium at 20, 40 and 60°.[770–771]

Cyclohexane—Aniline—Benzene
Vapour-liquid equilibrium at 1 atm and 93°.[610] At room temperature two immiscible liquid phases occur. Further literature.[772–774]

Cyclohexane—Pyridine—Benzene
Vapour-liquid equilibrium at 1 atm and 93°.[610]

Cyclohexane—Formamide—Benzene
Vapour-liquid equilibrium at 1 atm and 76°. Two immiscible liquid phases.[610]

Cyclohexane—Acetonitrile—Benzene
Vapour-liquid equilibria at 64 and 65° and at 1 atm.[610] At 25° two immiscible liquid phases.[662]

Cyclohexane—Nitromethane—Benzene
Vapour-liquid equilibria at 1 atm and temperatures from 72–78°.[610, 775] Two immiscible liquid phases. Liquid-liquid equilibria.[776]

Cyclohexane—Nitrobenzene—Benzene
Vapour-liquid equilibria at 1 atm and 102°.[610]

Cyclohexane—2-Nitro-1-propanol—Benzene
Liquid-liquid equilibria.[726]

Methylcyclohexane—Ethanol—Benzene
Vapour-liquid equilibria. Azeotrope.[605]

Decahydronaphthalene—Acetic anhydride—Benzene
Liquid-liquid equilibria.[777]

*Decahydronaphthalene—*p-*Phenetidine—Benzene*
Liquid-liquid equilibria.[778]

*Toluene—*m-*Xylene—Benzene*
Solid-liquid equilibria.[779–780]

*Toluene—*m-*Xylene—Naphthalene—Benzene*
Solid-liquid equilibria.[781]

Toluene—Ethylbenzene—Benzene
Vapour-liquid equilibria.[782–783]

*Toluene—Ethylbenzene—*o-*Xylene—*m-*Xylene—*p-*Xylene—Naphthalene—Benzene*
Solid-liquid equilibria.[784]

Toluene—Naphthalene—Benzene
Solid-liquid equilibria.[780, 785]

Toluene—Anthracene—Benzene
Solid-liquid equilibria.[786]

Toluene—Methanol—Benzene
Critical temperatures and pressures.[787] Vapour-liquid equilibria.[788]

Toluene—Ethanol—Anthranilic acid—Benzene
Solid-liquid equilibria.[624]

Toluene—Diethylene glycol—Benzene
Liquid-liquid equilibria.[708]

Toluene—2,4,6-Trinitrophenol—Benzene
Solid-liquid equilibria.[789]

m-*Xylene—Naphthalene—Benzene*
Solid-liquid equilibria.[790]

Indene—Naphthalene—Benzene
Solid-liquid equilibria.[791]

Naphthalene—Triphenylmethane—Benzene
Solid-liquid equilibria.[792]

Naphthalene—Ethanol—Benzene
Solid-liquid equilibria.[703]

Naphthalene—2-Naphthol—Benzene
Solid-liquid equilibria.[793]

Naphthalene—Diphenylamine—Benzene
Solid-liquid equilibria.[793]

Naphthalene—Isoquinoline—Benzene
Solid-liquid equilibria.[794]

Naphthalene—2,4,6-Trinitrophenol—Benzene
Solid-liquid equilibria[795] at temperatures from 5–45°.

Phenanthrene—Anthracene—Carbazole—Benzene
Solid-liquid equilibria.[796]

Anthracene—Chloroform—Benzene
Solid-liquid equilibria.[786]

Anthracene—Methanol—Benzene
Solid-liquid equilibria.[789]

Anthracene—Ethanol—Benzene
Solid-liquid equilibria.[786]

Anthracene—Propanol—Benzene
Solid-liquid equilibria.[789]

Anthracene—Diethyl ether—Benzene
Solid-liquid equilibria.[789]

Phenanthrene—Carbazole—Benzene
Solid-liquid equilibria.[793]

Gasoline—Acetic acid—N-Methylaniline—Benzene
Gasoline—Acetic acid—N,N'-dimethylaniline—Benzene
Liquid-liquid equilibria.[797]

Lube oil—Nitromethane Benzene
Liquid-liquid equilibria.[682]

Polystyrene—Polyvinyl acetate—Benzene
Liquid-liquid equilibria.[798]

Rubber—Ethanol—Benzene
Rubber—Acetone—Benzene
Liquid-liquid equilibria[799] at 20, 40 and 60°.

Chloroform—Methanol—Benzene
Liquid-vapour equilibria.[623]

Chloroform—Methanol—Acetic acid—Benzene
Vapour-liquid equilibria.[800]

Chloroform—Acetone—Benzene
Vapour-liquid equilibria at 1 atm.[801–802] No ternary azeotrope.[803] Solid-liquid equilibria at 1·5, 2·5, 3·5 and 4·5°.[805] Further literature.[623, 804]

Chloroform—Anthraquinone—Benzene
Solid-liquid equilibria.[789]

Chloroform—Methyl acetate—Benzene
Vapour-liquid equilibria at 1 atm.[806]

Chloroform—Diantipyrylmethane dithiocyanate—Benzene
Liquid-liquid equilibria.[807]

Carbon tetrachloride—Methanol—Benzene
Vapour-liquid equilibria at 34·68 and 55·0° and various pressures.[808] Further literature.[809]

Carbon tetrachloride—Ethanol—Benzene
Vapour-liquid equilibria at 34·8, 50, 60 and 66° at various pressures.[810–811] Further literature.[809]

Carbon tetrachloride—1-Propanol—Benzene
Vapour-liquid equilibria.[809]

Carbon tetrachloride—2-Propanol—Benzene
Vapour-liquid equilibria at 1 atm.[812]

Carbon tetrachloride—Butanol—Benzene
Vapour-liquid equilibria.[809]

Carbon tetrachloride—Acetone—Benzene
Vapour-liquid equilibria.[813]

Carbon tetrachloride—Methyl ethyl ketone—Benzene
Vapour-liquid equilibria.[814–815]

Carbon tetrachloride—Acetic acid—Benzene
Vapour pressure and boiling temperature as function of liquid composition.[816]
Vapour-liquid equilibria at 20°.[817]

Carbon tetrachloride—Salicylic acid—Benzene
Solid-liquid equilibria.[721]

Carbon tetrachloride—Methyl acetate—Benzene
Vapour-liquid equilibria.[818]

Carbon tetrachloride—Butyl acetate—Benzene
Liquid-vapour equilibria.[819]

Carbon tetrachloride—2,4,6-Trinitrophenol—Benzene
Solid-liquid equilibria[789, 820] at 25°.

Bromoform—Formic acid—Benzene
Liquid-liquid equilibria at 25, 50 and 70° with one single heterogeneous region.[821]

Chlorobenzene—Acetone—Benzene
Vapour-liquid equilibria.[822]

Methanol—various Organic compounds—Benzene
Liquid-liquid equilibria in the ternary systems of methanol and benzene with: glycol,[637] resorcinol[669] and glycerol[669] at 25°, olive oil[823] at 20° and ethanolamine, diethanolamine, formamide and 2-hydroxypropionitrile at 25°.[669]

Solid-liquid equilibria in the systems of methanol and benzene with resorcinol[669] at 25°, fumaric acid,[824] *l*-malic acid,[824] *dl*-malic acid[824] and N-acetylaniline[825] at 20° and picric acid[669, 826] at 0, 12·5 and 25°.

Ethanol—various Organic compounds—Benzene
Solid-liquid equilibria in systems of ethanol and benzene with: fumaric acid,[824] *l*-malic acid,[824] *dl*-malic acid[824] and N-acetylaniline[825] at 20° and picric acid[789, 820, 826] at 0, 12, 12·5 and 25°.

Liquid-liquid equilibria in such systems with glycol[827] and with glycerol[828] at 25° and in the system with propylene glycol at 30°.[829]

Vapour compositions as function of liquid composition at 750 mm in the system with glycerol.[830] Liquid-vapour equilibria in the systems with 2-butanone[831] and with nitrobenzene.[832]

1-Propanol—Ethylene glycol—Benzene
Liquid-liquid equilibria at 25°.[833]

1-Propanol—N-Acetylaniline—Benzene
Solid-liquid equilibria[825] at 20°.

1-Propanol—2,4,6-Trinitrophenol—Benzene
Solid-liquid equilibria at 0 and 25°.[826]

Isopropyl alcohol—Methyl ethyl ketone—Benzene
Vapour-liquid equilibria at 50° and 760 mm.[834]

Isopropyl alcohol—2,4,6-Trinitrophenol—Benzene
1-Butanol—2,4,6-Trinitrophenol—Benzene
Solid-liquid equilibria[826] at 0 and 25°.

Phenol—Aniline—Benzene
Solid-liquid equilibria at 1·5, 2·5, 3·5 and 4·5°.[805]

Benzyl alcohol—Antipyrene—Benzene
Liquid-liquid equilibria[835] at temperatures between 50 and 150°.

2-Naphthol—2,4,6-Trinitrophenol—Benzene
Melting equilibria between 4 and 157°.[836] Two binary solid compounds are formed: $C_6H_6.C_6H_2(NO_2)_3OH$, melting at 84·3 and $C_{10}H_7OH.C_6H_2(NO_2)_3OH$ melting at 157·0°

Glycol—some Organic compounds—Benzene
Liquid-liquid equilibria in the ternary systems of glycol and benzene with ethylene oxide,[637] carbitol (diethylene glycol monoethyl ether),[637] acetone,[637] formamide,[637] acetonitrile[662] and nitromethane,[637] all at 25° and with acetone also at 27°.[827]

Glycerol—Acetic acid—Benzene
Liquid-liquid equilibria[837] at 25°.

Diethyl ether—2,4,6-Trinitrophenol—Benzene
Solid-liquid equilibria[820] at 12°.

Dimethoxymethane—Fumaric acid—Benzene
Dimethoxymethane—l-*Malic acid—Benzene*
Dimethoxymethane—dl-*Malic acid—Benzene*
Solubilities of these acids in mixtures of dimethoxymethane (formaldehyde dimethyl acetal or methylal) and benzene at 20°.[824]

Diethylene glycol—Acetonitrile—Benzene
Liquid-liquid equilibria[662] at 25°.

Diethylene glycol—Nitromethane—Benzene
Diethylene glycol—Thiophene—Benzene
Liquid-liquid equilibria[637] at 25°.

Triethylene glycol—2-Picoline—Benzene
Liquid-liquid equilibria at 20°.[656]

Acetone—some Carboxylic acids—Benzene
Solid-liquid equilibria in the ternary systems of acetone and benzene with fumaric acid, *l*-malic acid and *dl*-malic acid at 20°[824] and with benzoic acid[838] and salicylic acid[839] at 25°.

Acetone—Formamide—Benzene
Liquid-liquid equilibria.[840]

Acetone—N-Acetylaniline—Benzene
Solid-liquid equilibria[641] at 25°.[789, 838]

Acetone—2,4,6-Trinitrophenol—Benzene
Solid-liquid equilibria at 12°[820] and at 25°.[789]

Formic acid—Nitrobenzene—Benzene
Liquid-liquid equilibria.[682]

Acetic acid—Aniline—Benzene
Acetic acid—Pyridine—Benzene
Solid-liquid equilibria at 1·5, 2·5, 3·5 and 4·5°.[805]

Palmitic acid—Stearic acid—Benzene
Solid-liquid equilibria[841] between 5 and 70°.

Benzoic acid—Ethyl acetate—Benzene
Solid-liquid equilibria[838] at 25°.

Salicylic acid—Ethyl acetate—Benzene
Solid-liquid equilibria.[839]

1,4-Dioxane—Fumaric acid—Benzene
*1,4-Dioxane—*l-*Malic acid—Benzene*
*1,4-Dioxane—*dl-*Malic acid—Benzene*
Solid-liquid equilibria at 20°.[824]

Aniline—Allyl mustard oil—Benzene
Solid-liquid equilibria[842] in the system $C_6H_5NH_2$–C_3H_5NCS–C_6H_6.

2-Aminoethanol—Nitromethane—Benzene
Liquid-liquid equilibria at 25°.[682]

2-Aminoethanol—Thiophene—Benzene
Liquid-liquid equilibria at 25°.[674]

Mixed polyamides—Benzene
Solubility in benzene.[843]

*Nitrobenzene—*m-*Dinitrobenzene—Benzene*
Solid-liquid equilibria.[844] Eutectic at −31°, 46 mole per cent nitrobenzene, 49 mole per cent benzene. Peritectic at −19°, 25 mole per cent nitrobenzene and 60 mole per cent benzene.

REFERENCES

1 Baker, H. B., *J. Chem. Soc.*, 1912, **101**, 2339; 1922, **121**, 568; 1923, **123**, 1223; 1927, 2902; 1928, 1051.
2 Rossini, F. D., Mair, B. J. and Streiff, A. J., '*Hydrocarbons from Petroleum*' (New York, 1953).
3 Heilbron and Bunbury, *Dictionary of Organic Compounds.*
4 Spielmann, P. E., *The Constituents of Coal Tar*, 1924, p. 31.
5 Benzole Producers Ltd., Half-yearly Report of the Technical Department, September 1959.
6 Claxton, G. (ed.), '*Benzoles, Production and Uses*' (National Benzole and Allied Products Assoc., 1961), p. 593.
7 Lewis, J. S., *J. Inst. Petrol. Tech.*, 1925, **11**, 152 and *ibid.*, 1926, **12**, 32.
8 Glasgow, A. R., Murphy, E. T., Willingham, C. B. and Rossini, F. D., *J. Res. Nat. Bur. Standards*, 1946, **37**, 141.
9 Claxton, G. (ed.), '*Benzoles, Production and Uses*' (National Benzole and Allied Products Assoc., 1961), pp. 572 *et seq.*
10 British Standard Specification No. 135, Benzene. Revised edition 1972.
11 Claxton, G. (ed.), '*Benzoles, Production and Uses*' (National Benzole and Allied Products Assoc., 1961), p. 563.
12 Claxton, G., Rept. Joint Benzole Research Comm., 1929, p. 129.
13 Kurtz, S. S. and Sankin, A., '*Physical Chemistry of the Hydrocarbons*', Vol. 2 (Academic Press Inc., New York, 1953), p. 14.
14 Claxton, G., '*Physical and Azeotropic Data*' (National Benzole and Allied Products Association, 1958).
15 Moore, H., *J. Soc. Chem. Ind.*, 1917, **36**, 109, and *J. Inst. Pet. Tech.*, 1920, **6**, 186.
16 Zabetakis, M. G., Scott, G. S. and Jones, C. W., U S Bureau of Mines Report Investigations 4824, 1951.

[17] Briand, M., Dumanois, P. and Laffitte, P., *Compt. rend.*, 1933, **197**, 322.
[18] Rossini, F. D. *et al.*, '*Selected Values of Properties of Hydrocarbons*' (API Research Project No. 44, US Bureau of Standards, C. 467, 1947).
[19] Perry, J. H., '*Chemical Engineers' Handbook*', 4th Ed., 1963 (McGraw-Hill Book Co., USA).
[20] Meyer, J., *Z. physik. Chem.*, 1910, **72**, 225.
[21] Richards and Shipley, *J. Am. Chem. Soc.*, 1914, **36**, 1835.
[22] Horrocks, J. K. and McLaughlin, E., *Proc. Roy. Soc.*, 1963, A**273**, 259.
[23] Venart, J. E. S., *Journal of Chemical Engineering Data*, 1965, **10**, 239.
[24] Scheffy, W. J. and Johnson, E. F., *ibid.*, 1961, **6**, 245.
[25] Briggs, D. R. H., *Ind. Eng. Chem.*, 1957, **49**, 418.
[26] McAdams, W. H., '*Heat Transmission*', 2nd Ed., 1942 (McGraw-Hill Book Co., USA).
[27] Evans, E. B., *J. Inst. Petrol.*, 1938, **24**, 38, 321, 537.
[28] Hecks, J. R. and Orban, E., *J. Phys. Chem.*, 1956, **60**, 1025.
[29] Timmermans, J., '*Physico—Chemical Constants of Pure Organic Compounds*' (Elsevier Publishing Co., Inc., New York, 1950).
[30] Claxton, G. (ed.), '*Benzoles, Production and Uses*' (National Benzole and Allied Products Assoc., 1961), p. 653.
[31] Joris, G. G. and Taylor, H. S., *J. Inst. Phys.*, 1948, **16**, 45.
[32] Sidgwick, N. V. and Spurrell, W. J., *J. Chem. Soc.*, 1920, **177**, 1397.
[33] Shepherd, F. M. E., *J. Inst. Petrol. Tech.*, 1934, **20**, 294.
[34] Leibnitz, E. *et al.*, *J. Prakt. Chem.*, 1956, **4** (*4*), 105.
[35] Langmuir, I., *Phys. Rev.*, 1915, **6** (*2*), 79; *J. Am. Chem. Soc.*, 1915, **37**, 1139; 1916, **38**, 1145, 2221.
[36] Claxton, G. (ed.), '*Benzoles, Production and Uses*' (National Benzole and Allied Products Assoc., 1961), p. 313.
[37] International Labour Conference, Recommendations adopted at the 56th (1971) Session, HM Stationery Office, Cmnd 5003, June 1972.
[38] Organick, E. I. and Studhalter, W. R., *Chem. Eng. Prog.*, 1948, **44**, 847.
[39] Gornowski, E. J., *Ind. Eng. Chem.*, 1947, **39**, 1348.
[40] Canjar, L. N. and Manning, F. S., '*Thermodynamic Properties and Reduced Correlations for Gases*' (Gulf Publishing Co., Houston, Texas, 1967).
[41] Kobe, K. A. and Long, E. G., *Petrol. Refiner*, 1949, **28**, 145.
[42] Connolly, J. F. and Kandalic, G. A., *J. Chem. Eng. Data*, 1962, **7**, 137.
[43] Flynn, L. W. and Thodos, G., *ibid.*, 1961, **6**, 457.
[44] Lambert, J. D. *et al.*, *Proc. Roy. Soc.*, 1955, A**231**, 280.
[45] Clever, Battino, Sailor and Gross, *J. Phys. Chem.*, **61** (1957), 1078.
[46] Lannung, *J. Amer. Chem. Soc.*, **52** (1930), 68.
[47] Ramsay, Collie and Travers, *J. Chem. Soc. London, Trans.*, **67** (1895), 684.
[48] N. V. Philips Gloeilampenfabriek (Netherlands), FP 823,529 (1937).
[49] Körösy, *Trans. Faraday Soc.*, **33** (1937), 416.
[50] Ramstedt, *J. Phys. Radium*, **8** (1911), 253.
[51] Szeparowicz, *Sitz.-Ber. Akad. Wiss. Wien, Abt. IIa*, **129** (1920), 437.
[52] Schulze, *Z. Phys. Chem. Stoechiom. Verwandschaftslehre*, **95** (1920), 257.
[53] v. Hevesy, *Physik. Z.*, **12** (1911), 1214.
[54] Kling, *Chem. Ing. Tech.*, **38** (1966), 1080.
[55] Bradley and King, *J. Chem. Phys.*, **47** (1967), 1189.
[56] Jellenck and Wannow, *Z. Electrochem.*, **41** (1935), 346.
[57] Cook, Hanson and Aldes, *J. Chem. Phys.*, **26** (1957), 748.
[58] Thompson, Univ. Microfilms (Ann Arbor, Mich.), no. 64–8249.
[59] Thompson and Wayne, *A.I.Ch.E.J.*, **11** (1965), 457.
[60] Connolly, *Proc. Amer. Petrol. Inst.*; Sect. III, **45** (1965), 62.
[61] Schläpfer, Audykowski and Bukowiecki, *Schweiz. Arch. Angew. Wiss. Tech.*, **15** (1949), 299.
[62] Horiuti, *Sci. Papers Inst. Phys. Chem. Res. Tokyo*, **17** (1931), 125.
[63] McKeown and Hibbard, *Anal. Chem.*, **28** (1956), 1490.

[64] Fischer and Pfleiderer, *Z. Anorg. Allg. Chem.*, **124** (1922), 61.
[65] Morgan and Pyne, *J. Phys. Chem.*, **34** (1930), 2045.
[66] Umano, Kawasaki, Nakano and Hayano, *Tokyo Kogyo Shikensho Hokoku*, **53** (1958), 397.
[67] Roddy and Coleman, *Talanta*, **15** (1968), 1281.
[68] Arnold, Plank, Erickson and Pike, *Chem. Eng. Data Ser.*, **3** (1958), 253.
[69] Davis, *Ind. Chemist*, **38** (1962), 522.
[70] Guseva and Parnov, *Vestn. Mosk. Univ., Ser. II, Khim.*, **18**, no. 1 (1963), 76.
[71] McAuliffe, *Nature*, **200** (1963), 1092; *Am. Chem. Soc., Div. Petrol. Chem., Preprints*, **9** (1964), 275; *J. Phys. Chem.*, **70** (1966), 1267.
[72] Connolly, *J. Chem. Eng. Data*, **11** (1966), 13.
[73] Landolt-Börnstein, '*Zahlenwerte und Funktionen*' Vol. II 2a, (Heidelberg, 1960), p. 385; *ibid* Vol. II 2b, (Heidelberg, 1962), p. 3–395; Hertz, *Ber. Deut. Chem. Ges.*, **31** (1898), 2669; Hantzsch and Sebaldt, *Z. Phys. Chem. Stoechiom. Verwandschaftslehre*, **30** (1899), 295; Niini, *Suom. Kemistilehti*, A**11** (1938), 19.
[74] Uspenski, *Neft. Khoz.* (*Petrol. Ind. USSR*), **17** (1929), 713.
[75] Miyake, *Mem. Coll. Eng., Kyushu Imp. Univ.*, **6** (1931–2), 106.
[76] Jaeger, *Brennstoff-Chem.*, **4** (1923), 259.
[77] Umano and Hayano, *Kogyo Kagaku Zasshi*, **60** (1957), 1436.
[78] Kentaro Ito, *Sci. Papers Inst. Phys. Chem. Res. Tokyo*, **55** (1961), 176.
[79] Kudchadker and McKetta, *Hydrocarbon Process. Petrol. Refiner*, **41** (1962), 191.
[80] Thompson and Snyder, *J. Chem. Eng. Data*, **9** (1964), 516.
[81] Kentaro Ito, *Kôatsu Gazu Kyôkaishi*, **22** (1958), 518.
[82] Niini, *Ann. Acad. Sci. Fenn., Ser. A*, **55**, no. 8 (1940).
[83] Scheffer, *Versl. Kon. Akad. Wet. Amsterdam*, **23** (1914), 682.
[84] Kaplan, *A.I.Ch.E.J.*, **14** (1968), 821.
[85] Alwary and Schneider, *Ber. Bunsenges. Phys. Chem.*, **71** (1967), 633; Schneider, *Chem. Eng. Progr. Symp. Ser.*, **64** (1968), no. 88, 9.
[86] Broells, Peter and Schneider, *Ber. Bunsenges. Phys. Chem.*, **74** (1970), 682.
[87] Rebert and Kay, *A.I.Ch.E.J.*, **5** (1959), 285; Rebert and Hayworth, *ibid*, **13** (1967), 118.
[88] Klatt, *Z. Anorg. Allg. Chem.*, **234** (1937), 189.
[89] Simons, *J. Amer. Chem. Soc.*, **53** (1931), 83.
[90] Posin and Opykhtina, *Zh. Priklad. Khim.*, **20** (1947), 523.
[91] Bell, *J. Chem. Soc. London*, **1931**, 1371.
[92] Fairbrother and Balkin, *J. Chem. Soc. London*, **1931**, 1564.
[93] Knight and Hinshelwood, *J. Chem. Soc. London*, **1927**, 466.
[94] O'Brien, Kenny and Zuercher, *J. Amer. Chem. Soc.*, **61** (1939), 2405.
[95] O'Brien, *J. Amer. Chem. Soc.*, **63** (1941), 2709.
[96] Saylor, *J. Amer. Chem. Soc.*, **59** (1937), 1712.
[97] Cook, Lupien and Schneider, *Can. J. Chem.*, **34** (1956), 964.
[98] Stevovic, Tanaka and Milman, *Inorg. Nucl. Chem. Lett.*, **3** (1967), 415.
[99] Maass and Russell, *J. Amer. Chem. Soc.*, **40** (1918), 1561.
[100] Kapustinsky and Malzev, *Zh. Fiz. Khim.*, **14** (1940), 105.
[101] Arctowski, *CR Acad. Sci.*, **121** (1895), 123.
[102] Hildebrand and Jenks, *J. Amer. Chem. Soc.*, **42** (1920), 2180.
[103] Herrero, *An. Soc. Espan. Fis. Quim.*, **31** (1933), 416.
[104] Hildebrand and Jenks, *J. Amer. Chem. Soc.*, **43** (1921), 2173.
[105] Cossa, *Ber. Deut. Chem. Ges.*, **1** (1868), 138.
[106] Alexejew, *Wied. Ann.*, **28** (1886), 138.
[107] Larkin, Katz and Scott, *J. Phys. Chem.*, **71** (1967), 352.
[108] Haase, Rehse and Dallman, *Monatsh. Chem.*, **98** (1967), 922.
[109] Garelli and Falciola, *Atti Accad. Naz. Lincei*, **13** (1904), 110; *idem*, *Gazz. Chim. Ital.*, 34 II (1904), 1.
[110] Terres and Vollmer, *Petroleum* (*Berlin*), **31** (1935), no. 19, 1.
[111] Lloyd, *J. Phys. Chem.*, **22** (1918), 300.
[112] Seyer and Peck, *J. Amer. Chem. Soc.*, **52** (1930), 14.
[113] Smith, B. and Smith, G., *J. Chem. Soc. London*, **1965**, 5514.

[114] Gjaldbaek and Hildebrand, *J. Amer. Chem. Soc.*, **71** (1949), 3147.
[115] Just, *Z. Phys. Chem. Stoechiom. Verwandschaftslehre*, **37** (1901), 342.
[116] Parabella and Sagramora, *Ric. Sci.*, **29** (1959), 2605.
[117] Lehmann and Ruschitzky, *Chem. Tech. (Berlin)*, **18** (1966), 280.
[118] Stock and Somieski, *Ber. Deut. Chem. Ges.*, **54** (1921), 755.
[119] Garelli and Monath, *Atti Reale Accad. Sci. Torino*, **61** (1926), 4.
[120] Klemenc and Spitzer-Neumann, *Monatsh. Chem.*, **53** (1929), 413.
[121] Addison and Sheldon, *J. Chem. Soc. London*, **1956**, 1941.
[122] Christomanos, *Z. Anorg. Chem.*, **45** (1905), 132.
[123] Semenchenko and Shakhparonov, *Zh. Fiz. Khim.*, **22** (1948), 243.
[124] Puschin, *Glas. Hem. Drus. Kralj. Jugoslav.*, **7** (1936), 73.
[125] Breck, Harvey and Haendler, *J. Phys. Colloid Chem.*, **53** (1949), 906.
[126] Menshutkin, *Zh. Russ. Fiz.-Khim. Obshchest. Chast. Khim.*, **43** (1911), 393.
[127] Kelley and McCusker, *J. Amer. Chem. Soc.*, **65** (1943), 1307.
[128] Menshutkin, *Zh. Russ. Fiz.-Khim. Obshchest., Chast. Khim.*, **43** (1911), 393, 401 and 425. See also Toshiaki Kishikawa *et al.*, *Bull. Chem. Soc. Jap.*, **43** (1970), 1276.
[129] Dolezalek, *Z. Phys. Chem. Stoechiom. Verwandschaftslehre*, **64** (1908), 727.
[130] Gjaldbaek, *Acta. Chem. Scand.*, **6** (1952), 623.
[131] Skirrow, *Z. Phys. Chem. Stoechiom. Verwandschaftslehre*, **41** (1902), 139.
[132] Gjaldbaek, *Acta. Chem. Scand.*, **7** (1953), 537.
[133] Wan and Dodge, *Ind. Eng. Chem.*, **32** (1940), 95.
[134] Sander, *Z. Phys. Chem. Stoechiom. Verwandschaftslehre*, **78** (1911–12), 513.
[135] Baskerville and Cohen, *Ind. Eng. Chem.*, **13** (1921), 333.
[136] Kirejew, Kaplan and Vasneva, *Zh. Prikl. Khim.*, **6** *(68)* (1936), 799.
[137] Hirshberg, *Bull. Soc. Chim. Belg.*, **41** (1932), 163.
[138] Sameshima, *J. Amer. Chem. Soc.*, **40** (1918), 1503.
[139] Schmidt, *Z. Phys. Chem. Stoechiom. Verwandschaftslehre*, **121** (1926), 221.
[140] Carveth, *J. Phys. Chem.*, **3** (1899), 193.
[141] Levy, *Ind. Eng. Chem.*, **33** (1941), 928.
[142] Negisi, *Rev. Phys. Chem. Jap.*, **15** (1941), 98.
[143] Kir'yanova and Pinsker, *Khim. Prom.*, **43** (1967), 30.
[144] Lespieau, *CR Acad. Sci.*, **140** (1905), 855.
[145] Shakhparonov, Lel'chuk, Korchemskaya, Martynova, Baburina and Voronina, *Zh. Prikl. Khim.*, **33** (1960), 2699.
[146] Goates, Ott and Mangelson, *J. Chem. Eng. Data*, **9** (1964), 330.
[147] Nisol'son, Edel'shtein and Ivanov-Emin, *Zh. Neorg. Khim.*, **4** (1959), 954.
[148] Wheeler and Keating, *J. Phys. Chem.*, **58** (1954), 1171.
[149] Cranston and Livingstone, *J. Chem. Soc. London*, **1926**, 501.
[150] Smith and McBain, *J. Phys. Colloid Chem.*, **51** (1947), 1189.
[151] Palit, *J. Amer. Chem. Soc.*, **69** (1947), 3120.
[152] McBain and Eaton, *J. Amer. Chem. Soc.*, **131** (1928), 2166.
[153] Steams, Oppenheimer, Simon and Harkins, *J. Chem. Phys.*, **15** (1947), 496.
[154] Cole, *Philipp. J. Sci.*, **47** (1932), 351.
[155] Henstock, *J. Chem. Soc. London*, **1934**, 1340.
[156] Picon, *Bull. Soc. Chim. Fr. Mem.*, *(4)* **49** (1931), 399.
[157] Fairbrother, Nixon and Prophet, *J. Less.-Common Metals*, **9** (1965), 434.
[158] Belozerskii, Krichevskaya, Mel'tser, Vereshchagina, Kolobova and Anisimov, *Zh. Neorg. Khim.*, **12** (1967), 2481.
[159] Garwin and Hixson, *Ind. Eng. Chem.*, **41** (1949), 2299.
[160] Kutzelnigg, *Z. Anorg. Allg. Chem.*, **256** (1948), 46.
[161] Hill, *J. Amer. Chem. Soc.*, **44** (1922), 1163.
[162] Spurgeon, Thesis, Michigan 1941.
[163] Anders, *Z. Phys. Chem. Abt. A*, **164** (1933), 151.
[164] Linebarger, *Silliman's J. Sci.*, **49** *(3)* (1895), 52.
[165] Spencer and Voigt, *J. Phys. Chem.*, **72** (1968), 464.
[166] Von Laszcynski, *Ber. Deut. Chem. Ges.*, **27** (1894), 2285.

[167] Eley and King, *Trans. Faraday Soc.*, **47** (1951), 1 287.
[168] Fairbrother, Scott and Prophet, *J. Chem. Soc. London*, **1956**, 1164; Menshutkin, *Zh. Russ. Fiz.-Khim. Obshchest.*, **41** (1909), 1089.
[169] Fialkov and Fokina, *Ukr. Khim. Zh.*, **25** (1959), 288.
[170] Del Pino and Greenwood, *An. Real. Soc. Espan. Fis. Quim.*, Ser. B, **61** (1965), 963.
[171] Ensslin and Lessmann, *Z. Anorg. Chem.*, **254** (1947), 92.
[172] Sidgwick and Sutton, *J. Chem. Soc. London*, **1930**, 1461.
[173] Schulze and Hock, *Z. Phys. Chem. Stoechiom. Verwandschaftslehre*, **86** (1914), 445.
[174] Kapusstinski and Kessler, *Izv. Akad. Nauk. SSSR, Otd. Khim. Nauk.*, **1956**, 889; *Bull. Acad. Sci. USSR, Div. Chem. Sci.*, **1956**, 911.
[175] Dorfman and Hildebrand, *J. Amer. Chem. Soc.*, **49** (1927), 729.
[176] McDermott, *J. Amer. Chem. Soc.*, **33** (1911), 1963.
[177] Jacobson and Holmes, *J. Biol. Chem.*, **25** (1916), 29.
[178] Kharchenko, Mikhailov, Afanas'ev and Ponomareva, *Izv. Sib. Otd. Akad. Nauk. SSSR, Ser. Khim. Nauk.*, **1964** (3), 30.
[179] Bjerrum and Józefowicz, *Z. Phys. Chem.*, Abt. A, **159** (1932), 194.
[180] Strohmeier and Hümpfner, *Chem. Ber.*, **90** (1957), 2339.
[181] Strohmeier, *Chem. Ber.*, **88** (1955), 1218.
[182] Krause and Reissaus, *Ber. Deut. Chem. Ges.*, **54** (1921), 2063; **55B** (1922), 888.
[183] Elbishlawi and Spencer, *Ind. Eng. Chem.*, **43** (1951), 1811.
[184] Savvina and Velikovskii, *Gazov. Prom.*, **1957**, no. 2, 26; *Zh. Fiz. Khim.*, **30** (1956), 1596.
[185] Savvina, *Tr. Vses. Nauch.-Issled. Inst. Prir. Gazov.*, **1962**, 185.
[186] Savvina and Velikovskii, *Tr. Vses. Nauch.-Issled. Inst. Prir. Gazov.*, **1962**, 163.
[187] Stepanova, Vybornova and Velikovskii, *Gaz. Delo. Nauchn. Tekhn. Sb.*, **1965** (9), 3.
[188] McDaniel, *J. Phys. Chem.*, **15** (1911), 587.
[189] Metschl, *J. Phys. Chem.*, **28** (1924), 417.
[190] Kay and Nevens, *Chem. Eng. Progr. Symp. Ser.*, **48** (1952), no. 3, 108.
[191] Zhuse and Zhurba, *Izv. Akad. Nauk. SSSR, Otd. Khim. Nauk.*, **1960**, 364; *Bull. Acad. Sci. USSR, Div. Chem. Sci.*, **1960**, 335.
[192] Lyubetskii, *Zh. Prikl. Khim.*, **35** (1962), 141; *J. Appl. Chem. USSR*, **35** (1962), 125.
[193] Konobeev and Lyapin, *Khim. Prom.*, **43** (1967), 114.
[194] Bezemer, Thesis, Delft 1953; Miller, S. A., '*Ethylene and its Industrial Derivatives*' (Benn, London, 1969), p. 219.
[195] Zhurba and Zhuse, *Izv. Vyssh. Ucheb. Zaved., Neft. Gaz.*, **1960**, no. (1), 101.
[196] Zhuse, Zhurba and Esakov, *Izv. Akad. Nauk. SSSR, Otd. Khim. Nauk.*, **1960**, 361.
[197] Garelli, *Atti Reale Accad. Naz. Lincei, Rend* (6), **2** (1925), 120.
[198] McKinnis, *Ind. Eng. Chem.*, **47** (1955), 850.
[199] Lebeau, *Bull. Cl. Sci. Acad. Roy. Belg.*, **1908**, 304.
[200] Glanville, Sage, Hashmall and Othmer, *Ind. Eng. Chem.*, **32** (1940), 48.
[201] Mehra and Thodos, *A.I.Ch.E.J.*, **8** (1962), 604.
[202] Rozhnov, *Khim. Prom.*, **43** (1967), 288.
[203] Fried and Vitovec, *Collect. Czech. Chem. Commun.*, **25** (1960), 1642.
[204] Myers, *Ind. Eng. Chem.*, **47** (1955), 2215.
[205] Gotthard and Minea, *Rev. Chim. (Bucharest)*, **14** (1963), 520.
[206] Tongberg and Johnston, *Ind. Eng. Chem.*, **25** (1933), 733.
[207] Haldenwanger, *Chem. Ing. Tech.*, **23** (1951), 437.
[208] Kumarkrishna Rao, Swami and Narasinga Rao, *A.I.Ch.E.J.*, **3** (1957), 191.
[209] Rao and Satyanarayna, *Proc. Int. Symp. Distill.*, 1960, 99.
[210] Gotthard and Minea, *Rev. Chim. (Bucharest)*, **11** (1960), 642.
[211] Gotthard and Minea, *Rev. Chim. (Bucharest)*, **12** (1961), 489.
[212] Prabhu and Van Winkle, *J. Chem. Eng. Data*, **8** (1963), 210.
[213] Garber and Zelenevskaya, *Zh. Prikl. Khim.*, **36** (1963), 2306.
[214] Bizostowski, *Bull. Acad. Pol. Sci., Ser. Sci. Chim.*, **9** (7) (1961), 471.
[215] Smith and Robinson, *J. Chem. Eng. Data*, **15** (1970), 391.
[216] Jain, Gupta and Lark, *Indian J. Chem.*, **8** (1970), 815.
[217] Dojcansky, Heinrich and Surovy, *Chem. Zvesti*, **21** (1967), 713.

[218] Brown and Ewald, *Aust. J. Sci. Res.*, Ser. A **4** (1951), 198.
[219] Sieg, *Chem. Ing. Tech.*, **22** (1950), 322.
[220] Stage and Schultze, *Oel Kohle*, **40** (1944), 90.
[221] Barker, *Aust. J. Chem.*, **6** (1953), 207.
[222] Nielsen and Weber, *J. Chem. Eng. Data*, **4** (1959), 145.
[223] Bizostowski, *Rocz. Chem.*, **35** (1961), 291.
[224] Bizostowski, *Bull. Acad. Pol. Sci., Ser. Sci. Chim.*, **8** (1960), 291 and 297.
[225] Chew and Reneau, *Hydrocarbon Process Petrol. Refiner*, **45** (1966), 189.
[226] Housek and Hala, *Collect. Czech. Chem. Commun.*, **35** (1970), 1030.
[227] Jean Fu and Lu, *J. Appl. Chem.*, **16** (1966), 324.
[228] Loiseleur, Clechet and Merlin, *CR Acad. Sci.*, Ser. C **262** (1966), 1635.
[229] Loiseleur, Clechet and Merlin, *Bull. Soc. Chim. Fr.*, **1967**, 1057.
[230] Richards and Hargreaves, *Ind. Eng. Chem.*, **36** (1944), 805.
[231] Kyle and Tetlow, *J. Chem. Eng. Data*, **5** (1960), 275.
[232] Stephenson and Van Winkle, *J. Chem. Eng. Data*, **7** (1962), 510.
[233] Harrison and Berg, *Ind. Eng. Chem.*, **38** (1946), 117.
[234] Schnuse, Thesis, Braunschweig 1945.
[235] Elshayal and Lu, *J. Appl. Chem.*, **18** (1968), 277.
[236] Ralston, Hoerr and Crews, *J. Org. Chem.*, **9** (1944), 319.
[237] Kenny, *Chem. Eng. Sci.*, **6** (1957), 116.
[238] Hildebrand and Watcher, *J. Phys. Colloid Chem.*, **53** (1949), 886.
[239] Seyer, *J. Amer. Chem. Soc.*, **60** (1938), 827.
[240] Myers, *Ind. Eng. Chem.*, **48** (1956), 1104.
[241] Hermsen and Prausnitz, *Chem. Eng. Sci.*, **18** (1963), 485.
[242] Scatchard, Wood and Mochel, *J. Amer. Chem. Soc.*, **62** (1940), 712.
[243] Thornton and Garner, *J. Appl. Chem.*, 1st Suppl. Iss. (1951), 68.
[244] Weck and Hunt, *Ind. Eng. Chem.*, **46** (1954), 2521.
[245] Galska Krajewska, *Bull. Acad. Pol. Sci., Ser. Sci. Chim. Geol. Geogr.*, **6** (1958), 257.
[246] Susarev and Lyzlova, *Zh. Fiz. Khim.*, **36** (1962), 437.
[247] Boublik, *Collect. Czech. Chem. Commun.*, **28** (1963), 1771.
[248] Darmois, R. and Darmois, J., *J. Chem. Phys.*, **61** (1964), 830.
[249] Nataraj and Raja Rao, *Indian J. Technol.*, **5** (1967), 212.
[250] Dojcansky, Heinrich and Surovy, *Chem. Zvesti*, **22** (1968), 514.
[251] Diaz-Pena and Rodrequez-Cheda, *An. Quim.*, **66** (1970), 721.
[252] Storonkin and Morachevskii, *Zh. Fiz. Khim.*, **31** (1957), 42.
[253] Kravchenko and Eremenko, *Zh. Prikl. Khim.*, **23** (1950), 613; *J. Applied Chem. USSR*, **23** (1950), 647.
[254] Rastogi and Nigam, *Trans. Faraday Soc.*, **55** (1959), 2005.
[255] Sponer and Spangler, *Proc. Int. Symp. Mol. Struct. Spectros*, 1962 (B 305), 4.
[256] Donald and Ridgway, *J. Appl. Chem.*, **8** (1958), 403.
[257] Griswold and Ludwig, *Ind. Eng. Chem.*, **35** (1943), 117.
[258] Kurnakov and Jefremov, *Zh. Russ. Fiz.-Khim. Obshchest.*, **44** (1912), 1992.
[259] v. Huhn, *Forsch. Ingenieurw.*, **2** (1931), 109 and 129.
[260] Todd, *Ind. Eng. Chem.*, **32** (1940), 287.
[261] Rollet, Elkaïm, Tolédano and Sénèz, *CR Acad. Sci.*, **242** (1956), 2560.
[262] Joung, *J. Chem. Soc. London Trans.*, **83** (1903), 68.
[263] Schuberth, *J. Prakt. Chem.*, **6** (1958), 129.
[264] Heertjes, *Chem. Process Eng.*, **41** (1960), 385.
[265] Omoto and Ezaki, *Kagaku Kogaku*, **30** (1966), 709.
[266] Kesselman, Hollenbach, Mijers and Humphrey, *J. Chem. Eng. Data*, **13** (1968), 34.
[267] Kravchenko, *Zh. Fiz. Khim.*, **13** (1939), 133.
[268] Mitsukuri and Nakatsuchi, *J. Soc. Chem. Ind. Jap.*, **29** (1926), 26.
[269] Robert, *Bull. Soc. Chim. Fr.* (4) **25** (1919), 463.
[270] Tarasenkov, *Zh. Prikl. Khim.*, **3** (1930), 153.
[271] Wettig, Thesis, Vienna 1949.
[272] Kravchenko, *Zh. Prikl. Khim.*, **22** (1949), 724.

[273] Kravchenko, *Zh. Fiz. Khim.*, **13** (1939), 989.
[274] Nakatsuchi, *Sci. Rep. Tohoku Imp. Univ.*, **15** (1926), 53.
[275] Nakatsuchi, *J. Soc. Chem. Ind. Jap.*, **29** (1926), 29.
[276] Paterno and Ampola, *Gazz. Chim. Ital.*, **27** (1897), I, 481.
[277] Paterno and Montemartini, *Gazz. Chim. Ital.*, **24** (1894), II, 208.
[278] Gasztych, *Zesz. Nauk. Politech. Slask. Chem.*, **13** (1963), 57.
[279] Kravchenko and Pastukhova, *Zh. Prikl. Khim.*, **25** (1952), 313.
[280] Schläpfer and Flachs, *Helv. Chim. Acta*, **10** (1927), 381.
[281] Deffet, *Bull. Soc. Chim. Belg.*, **45** (1936), 213.
[282] Hulett, *Z. Phys. Chem. Stoechiom. Verwandschaftslehre*, **28** (1899), 629.
[283] Pickering, *J. Chem. Soc. London Trans.*, **63** (1893), 998.
[284] Washbum and Read, *Proc. Nat. Acad. Sci. USA*, **1** (1915), 191.
[285] Cheng, *Chem. Eng. Sci.*, **13** (1963), 715.
[286] Smyth and Lewis, *J. Amer. Chem. Soc.*, **62** (1940), 949.
[287] Baxendale, Enüstün and Stem, *Phil. Trans. Roy. Soc. London*, A **243** (1950–1), 169.
[288] Everett and Penney, *Proc. Roy. Soc. London*, Ser. A **212** (1954), 164.
[289] Warner, Scheib and Svirbely, *J. Chem. Phys.*, **2** (1934), 590.
[290] Szabados, *KFKI Kozlem*, **15** (1967), 283.
[291] Everett and Swinton, *Trans. Faraday Soc.*, **59** (1963), 2476.
[292] Jannelli and Vitagliano, *Gazz. Chim. Ital.*, **91** (1961), 772.
[293] Mortimer, *J. Amer. Chem. Soc.*, **45** (1923), 633.
[294] Jannelli, Lamanna and Sciacovelli, *Gazz. Chim. Ital.*, **93** (1963), 1231.
[295] Jannelli, Della Monica, M. and Della Monica, A., *Gazz. Chim. Ital.*, **93** (1963), 953.
[296] Della Monica, A., Della Monica, M. and Jannelli, *Gazz. Chim. Ital.*, **94** (1964), 1369.
[297] Versmann, *Jahresber. Chem.*, **1874**, 423.
[298] Clark, *Ind. Eng. Chem.*, **11** (1919), 204.
[299] Tyrer, *J. Chem. Soc. London Trans.*, **97** (1910), 1778.
[300] Hildebrand, Ellefson and Beebe, *J. Amer. Chem. Soc.*, **39** (1917), 2301.
[301] Henstock, *J. Chem. Soc. London Trans.*, **121** (1922), 2124.
[302] Kravchenko, *Ukr. Khim. Zh.*, **18** (1952), 459.
[303] Kravchenko and Pastukhova, *Zh. Obshch. Khim.*, **29** (1959), 27.
[304] Kuriloff, *Z. Phys. Chem. Stoechiom. Verwandschaftslehre*, **23** (1897), 547.
[305] Mamedaliev and Musakhanly, *Zh. Prikl. Khim.*, **13** (1940), 735.
[306] Lumbuhob, *Zh. Fiz. Khim.*, **14** (1940), 782.
[307] De Minjer, Thesis, Delft 1939.
[308] Reinders and De Minjer, *Rec. Trav. Chim. Pays-Bas*, **59** (1940), 369.
[309] Tyrer, *J. Chem. Soc. London Trans.*, **99** (1911), 1633.
[310] Tyrer, *J. Chem. Soc. London Trans.*, **101** (1912), 81.
[311] Wroczinski and Guye, *J. Chim. Phys.*, **8** (1910), 189; Wyatt, *Trans. Faraday Soc.*, **24** (1928), 429.
[312] Ampola and Manuelli, *Gazz. Chim. Ital.*, **25** (1895), II, 91.
[313] Schmidlin and Lang, *Ber. Deut. Chem. Ges.*, **45** (1912), 899.
[314] Scatchard, Wood and Mochel, *J. Amer. Chem. Soc.*, **61** (1939), 3206.
[315] Rosanoff and Easly, *J. Amer. Chem. Soc.*, **31** (1909), 1953.
[316] Ocon and Espantoso, *An. Real. Soc. Espan. Fis. Quim.*, Ser. B **54** (1958), 401.
[317] Fowler and Lim, *J. Appl. Chem.*, **6**, Pt. 2 (1956), 74.
[318] Smit and Ruyter, *Rec. Trav. Chim. Pays-Bas*, **79** (1960), 1244.
[319] Haughton, *Brit. Chem. Eng.*, **10** (1965), 237.
[320] Campbell, Kartzmark and Chatterjee, *Can. J. Chem.*, **44** (1966), 1183.
[321] Markuzin, Gomerova and Lesteva, *Zh. Prikl. Khim.*, **39** (1966), 1878.
[322] Campbell and Musbally, *Can. J. Chem.*, **48** (1970), 3173.
[323] Baud, *Ann. Chim. Phys.* (*8*) **29** (1913), 124; Ott, Goates and Budge, *J. Phys. Chem.*, **66** (1962), 1387.
[324] Linard, *Bull. Soc. Chim. Belg.*, **34** (1925), 363.
[325] Wilhelm and Battino, *J. Chem. Thermodyn.*, **3**, no. 3 (1971), 379.
[326] Von Zawidzki, *Z. Phys. Chem. Stoechiom. Verwandschaftslehre*, **35** (1900), 129.

[327] Sieh, R. H. W., 'Het evenwicht vloeistof-damp in het stelsel benzeen-1,2-dichloorethaan' (The equilibrium liquid-vapour in the system benzene-1,2-dichloroethane), Thesis, Delft 1964, pp. 52–64.
[328] Smith and Matheson, *J. Res. Nat. Bur. Stand.*, **20** (1938), 641.
[329] Haldenwanger, *Chem. Ing. Tech.*, **23** (1951), 437.
[330] Zelvenskii, Shalygin and Titov, *Khim. Tekhnol. Topl. Masel*, **7**, no. 4 (1962), 5; Bushmakin, Lyslova and Molodenko, *Zh. Prikl. Khim.*, **26** (1953), 1258; Ebert and Tschamler, *Monatsh. Chem.*, **80** (1949), 473; Lumbuhob, *Zh. Fiz. Khim.*, **14** (1940), 562.
[331] Tschamler, *Monatsh. Chem.*, **79** (1948), 500.
[332] Baud, *Ann. Chim. Phys.* (*8*) **27** (1912), 89; Baud and Gay, *Bull. Soc. Chim. Fr.* (*4*) **9** (1911), 114.
[333] Sumarokov and Shefer, *Zh. Obshch. Khim.*, **26** (1956), 3247.
[334] Wyatt, *Trans. Faraday Soc.*, **25** (1929), 43.
[335] Liebermann and Kohler, *Monatsh. Chem.*, **99** (1968), 2514.
[336] Neckel and Volk, *Monatsh. Chem.*, **89** (1958), 754.
[337] Kirschbaum, Fischer and Wolf, *Chem. Ing. Tech.*, **34** (1962), 423.
[338] Podder, *Z. Phys. Chem.* (*Frankfurt am Main*), **35** (1962), 267.
[339] Subba Rao and Venkata Rao, *Indian J. Technol.*, **1** (*4*) (1963), 143.
[340] Lazár, *Chem. Zvesti*, **9** (1955), 227.
[341] Hildebrand, Fisher and Benesi, *J. Amer. Chem. Soc.*, **72** (1950), 4348.
[342] Hoerr and Harwood, *J. Org. Chem.*, **16** (1951), 771.
[343] Hildebrand and Cochran, *J. Amer. Chem. Soc.*, **71** (1949), 22.
[344] Scott, *J. Amer. Chem. Soc.*, **70** (1948), 4090.
[345] Banks and Musgrave, *J. Chem. Soc. London*, **1956**, 4682.
[346] Duncan and Swinton, *J. Phys. Chem.*, **70** (1966), 2417.
[347] Duncan and Swinton, *Trans. Faraday Soc.*, **62** (1966), 1082; Patrick and Prosser, *Nature*, **187** (1960), 1021.
[348] Gaw and Swinton, *Nature*, **212** (1966), 283; *Trans. Faraday Soc.*, **64** (1968), 2023.
[349] Martin and Collie, *J. Chem. Soc. London*, **1932** Part II, 2658.
[350] Schay, Varsanyi and Billes, *Period. Polytech. Chem. Eng.*, **1** (1957), 131; Serwinsky and Michalovski, *Zesz. Nauk. Politech. Lodz. Chem.*, no. **16** (1965), 45.
[351] Martin and George, *J. Chem. Soc. London*, **1933** Part II, 1413.
[352] Sumarokov and Shefer, *Zh. Obshch. Khim.*, **26** (1956), 3247.
[353] McGlashan and Wingrove, *Trans. Faraday Soc.*, **52** (1956), 470.
[354] Schröder, *Z. Phys. Chem. Stoechiom. Verwandschaftslehre*, **11** (1893), 449.
[355] Shishokin, *Z. Anorg. Allg. Chem.*, **181** (1929), 137.
[356] Kohler, *Monatsh. Chem.*, **91** (1960), 1113.
[357] Paternó, *Gazz. Chim. Ital.*, **26** (1896) II, 1.
[358] McLaughlin and Mesker, *J. Chem. Soc.*, A., **1966**, 1106.
[359] Gunther, *J. Amer. Chem. Soc.*, **67** (1945), 189.
[360] Fritzweiler and Dietrich, *Beiheft Z. Ver. deut. Chem.* No.**4** (1933); *Angew. Chem.*, **45** (39) (1932), 605 and **46** (17) 1933), 241.
[361] Kumarkrishna Rao, Sarma, Swami and Narasinga Rao, *J. Sci. Ind. Res.*, Sect. B **16** (1957), 4.
[362] Krichevskii, Khazanova and Linshits, *Zh. Fiz. Khim.*, **31** (1957), 2711.
[363] Tallmadge, Schroeder, Edminster and Canjer, *Chem. Eng. Progr.*, *Symp. Ser.* No. 10, Vol. 50 (1954), 137.
[364] Haywood, *J. Amer. Chem. Soc.*, **21** (1899), 994.
[365] Lee, Siang Chieh, *J. Phys. Chem.*, **35** (*2*) (1931), 3558.
[366] Wehe and Coates, *A.I.Ch.E.J.*, **1** (1955), 241.
[367] Ocon and Espantoso, *An. Real. Soc. Espan. Fis. Quim.*, Ser. B **54** (1958), 421.
[368] Soday and Bennett, *J. Chem. Educ.*, **7** (1930), 1336.
[369] Scatchard, Wood and Mochel, *J. Amer. Chem. Soc.*, **68** (1946), 1957.
[370] Perrakis, *J. Chim. Phys.*, **22** (1925), 280.
[371] Wieth, *Phys. Z.*, **30** (1930), 126.
[372] McCracken and Smith, *A.I.Ch.E.J.*, **2** (1956), 498.

[373] Skaates and Kay, *Chem. Eng. Sci.*, **19** (1964), 431.
[374] Schreinemakers, *Z. Phys. Chem. Stoechiom. Verwandschaftslehre*, **47** (1904), 445.
[375] Brown and Smith, *Aust. J. Chem.*, **7** (1954), 264.
[376] Szapiro, *Zesz. Nauk. Politech. Lodz. Chem.*, no. **7** (1958), 3; McCracken, Storvick and Smith, *J. Chem. Eng. Data*, **5** (1960), 130.
[377] Gardek, Univ. Microfilms (Ann Arbor, Mich.), no. 64–9159.
[378] Ellis, *Trans. Inst. Chem. Eng.*, **30** (1952), no. 1, 58.
[379] Landwehr, Yerazunis and Steinhauser, *Chem. Eng. Data Ser.* **3** (1958), 231.
[380] Giona and Passino, *Rass. Chim.*, **11**, no. 2 (1959), 9.
[381] Zharov and Morachevskii, *Zh. Prikl. Khim.*, **36** (1963), 2397.
[382] Vega and Alvarez Gonzalez, *An. Real. Soc. Espan. Fis. Quim.*, Ser. B **61** (1965), 831.
[383] Ratkovics, *Acta Chim. (Budapest)*, **49** (1966), 85.
[384] Zharov and Buevick, *Zh. Prikl. Khim.*, **41** (1968), 2032.
[385] Viala, *Bull. Soc. Chim. Fr.* (*4*) **15** (1914), 5.
[386] Washburn, Hnizda and Vold, *J. Amer. Chem. Soc.*, **53** (1931), 3237.
[387] Kumarkrishna Rao, Swami and Narasinga Rao, *J. Sci. Ind. Res.*, Sect. B **16** (1957), 233.
[388] Brown and Smith, *Austr. J. Chem.*, **12** (1959), 407.
[389] Olsen and Washburn, *J. Amer. Chem. Soc.*, **57** (1935), 303.
[390] Isamu Nagata, *J. Chem. Eng. Data*, **10** (1965), 106.
[391] Udovenko and Mazanko, *Zh. Fiz. Khim.*, **41** (1967), 1615.
[392] Rabinovich, Nikolaev, Gochaliev and Tret'yakova, *Dokl. Akad. Nauk. SSSR*, **110**, 241.
[393] Allen, Lingo and Felsing, *J. Phys. Chem.*, **43** (1939), 425.
[394] Mann, Shemilt and Waldichuck, *J. Chem. Eng. Data*, **8** (1963), 502.
[395] Yerazunis and Plyushchev, *Zh. Neorg. Khim.*, **9** (1964), 2433.
[396] Vijayaraghavan, Deshpande and Kuloor, *Indian J. Technol.*, **3** (1965), 267.
[397] Pahlavouni, *Bull. Soc. Chim. Belg.*, **36** (1927), 533.
[398] Reddy and Venkata Rao, *Indian J. Technol.*, **3** (1965), 45.
[399] Fajans, *Bull. Acad. Pol. Sci., Ser. Sci. Chim.*, **12** (1964), 877.
[400] Javier Quintanilla, *Riv. quím. ing. quím. Monterrey*, **2** (1956), 23.
[401] Jayarama Rao, Dakshinamurty and Venkata Rao, *J. Appl. Chem.*, **12** (1962), 274.
[402] Hoerr, Harwood and Ralston, *J. Org. Chem.*, **9** (1944), 267.
[403] Skaates, Univ. Microfilms (Ann Arbor, Mich.), no. 62–2161; Diss. Abstr., **22** (1962), 3962.
[404] Piatti, *Z. Phys. Chem.*, Abt. A **152** (1931), 36.
[405] Kremann and Borjanovics, *Monatsh. Chem.*, **37** (1916), 59.
[406] Kuriloff, *Z. Phys. Chem. Stoechiom. Verwandschaftslehre*, **23** (1897), 673.
[407] Leibnitz, Könnecke and Niese, *J. Prakt. Chem.* (*4*) **4** (1957), 278.
[408] Walker, Collett and Lazzell, *J. Phys. Chem.*, **35** (1931), 3259.
[409] Rothmund, *Z. Phys. Chem.*, **26** (1898), 433.
[410] Francis, *Ind. Eng. Chem.*, **36** (1944), 764, 1096.
[411] Ruehelman and Haines, *J. Gas Chromatogr.*, **5** (1967), 290.
[412] Ebert and Tschamler, *Monatsh. Chem.*, **80** (1949), 473.
[413] Yamamura, *Bull. Chem. Soc. Jap.*, **1** (1926), 183.
[414] Ramachandran, *Indian J. Technol.*, **8** (1970), 323.
[415] Paternó, *Gazz. Chim. Ital.*, **26** (1896) II, 9.
[416] Johnson and Francis, *Ind. Eng. Chem.*, **46** (1954), 1662.
[417] Skinner, *Ind. Eng. Chem.*, **47** (1955), 222.
[418] Ismailov, Shakhverdieva and Ramazanzade, *Izv. Vyssh. Ucheb. Zaved. Neft Gaz*, **11** (1968), 43.
[419] Puschin and Rikovsky, *Z. Phys. Chem.*, Abt. A **142** (1929), 211.
[420] Othmer, *Ind. Eng. Chem.*, **35** (1943), 614.
[421] Hatcher and Skorrow, *J. Amer. Chem. Soc.*, **39** (1917), 1939.
[422] Tsakalatos and Guye, *J. Chim. Phys.*, **8** (1910), 340.
[423] Paterno, *Gazz. Chim. Ital.*, **25** (1895), I, 411.
[424] Sidgwick and Allott, *J. Chem. Soc. London Trans.*, **123** (1923), 2819.
[425] Brown and Smith, *Austr. J. Chem.*, **10** (1957), 423.

[426] Steinhauser and White, *Ind. Eng. Chem.*, **41** (1949), 2912; Lutugina, Molodenko and Dement'ev, *Vestn. Leningrad. Univ. Fiz. Khim.*, **22** [no. 4] (1967), 107.
[427] Kumarkrishna Rao, Swami and Narasinga Rao, *J. Sci. Ind. Res.*, Sect. B **16** (1957), 195.
[428] Dakshinamurty, Jaya Rama Rao, Raghavacharya and Venkata Rao, *J. Sci. Ind. Res.*, Sect. B **16** (1957), 340.
[429] Hoerr, Reck, Corcoran and Harwood, *J. Phys. Chem.*, **59** (1955), 457.
[430] Garland, Hoerr, Pool and Ralston, *J. Org. Chem.*, **8** (1943), 344.
[431] Pariaud, *Bull. Soc. Chim. Fr.*, **1950**, 103.
[432] Kirejev, Kaplan and Vasneva, *Zh. Fiz. Khim.*, **5** (1934), 739.
[433] Loskit, *Z. Phys. Chem. Stoechiom. Verwandschaftslehre*, **134** (1928), 156.
[434] Ewins, *J. Chem. Soc. London Trans.*, **105** (1914), 350.
[435] Gordon and Reid, *J. Phys. Chem.*, **26** (1922), 773.
[436] Wrewsky, Held and Shchukarev, *Z. Phys. Chem. Stoechiom. Verwandschaftslehre*, **133** (1928), 377.
[437] Wrewsky, Savaritzkii and Scharloff, *Z. Phys. Chem. Stoechiom. Verwandschaftslehre*, **112** (1924), 97.
[438] Othmer, *Ind. Eng. Chem.*, Anal. Ed., **4** (1932), 232.
[439] Beckmann and Liesche, *Z. Phys. Chem. Stoechiom. Verwandschaftslehre*, **92** (1918), 421.
[440] Marek, *Collect. Czech. Chem. Commun.*, **20** (1955), 1490.
[441] Volpicelli and Zizza, *Chim. Ind. (Milan)*, **45** (1963), 1502.
[442] Werner, *J. Prakt. Chem.*, **29** (1965), 26.
[443] Haughton, *Brit. Chem. Eng.*, **12** (1967), 1102.
[444] Markuzin, *Zh. Fiz. Khim.*, **42** (1968), 1415.
[445] Tan, Krieger and Miller, *J. Amer. Chem. Soc.*, **74** (1952), 6181.
[446] Roloff, *Z. Phys. Chem. Stoechiom. Verwandschaftslehre*, **17** (1895), 325.
[447] Molochnikov, Markova and Kogan, *Zh. Prikl. Khim.*, **40** (1967), 2083; *J. Appl. Chem. USSR*, **40** (1967), 2011.
[448] Ralston and Hoerr, *J. Org. Chem.*, **7** (1942), 546.
[449] Powney and Addison, *Trans. Faraday Soc.*, **34** (1938), 625.
[450] Ralston and Hoerr, *J. Org. Chem.*, **10** (1945), 170.
[451] Hoerr and Harwood, *J. Phys. Chem.*, **56** (1952), 1068.
[452] Chipman, *J. Amer. Chem. Soc.*, **46** (1924), 2445.
[453] Sidgwick and Ewbank, *J. Chem. Soc. London Trans.*, **119** (1921), 979.
[454] Sidgwick and Clayton, *J. Chem. Soc. London Trans.*, **121** (1922), 2263.
[455] Davies and Griffiths, *Trans. Faraday Soc.*, **49** (1953), 1405.
[456] Carr and Krop Koller, *J. Chem. Eng. Data*, **7** (1962), 26.
[457] Isamu Nagata, *Kanazawa Daigaku Kogakubu Kiyo*, **3** (1964), 197.
[458] Bushmakin, Baldangiin and Molodenko, *Zh. Prikl. Khim.*, **38** (1965), 1417; *J. Appl. Chem. USSR*, **38** (1965), 1392.
[459] Sedgwick, Hoerr and Harwood, *J. Org. Chem.*, **17** (1952), 327.
[460] Loskit, *Z. Phys. Chem. Stoechiom. Verwandschaftslehre*, **134** (1928), 135.
[461] Tonn and Schoch, *Ind. Eng. Chem.*, **38** (1946), 413.
[462] Kravchenko and Eremenko, *Zh. Prikl. Khim.*, **23** (1950), 613; *J. Appl. Chem. USSR*, **23** (1950), 647 (Eng. transl.).
[463] Gropşianu, Z., Kyri J. and Gropşianu, R., *Acad. Rep. Populare Romine, Baza cercetari stiint. Timisoara, Studii cercetati stiint., Ser. stiinet chim.*, **4** (1957), no. 3–4, 73–85; *Chem. Abstr.*, **53** (1959), 19501e.
[464] Gropşiana and Muraressu, *Acad. Rep. Populare Romine, Baza cercetari stiint. Timisoara, Studii cercetari stiint., Ser. stiinte chim.*, **3** (1956), no. 1–2, 81–6; *Chem. Abstr.*, **51** (1957), 16028a.
[465] Serebrennaya and Byk, *Zh. Prikl. Khim.*, **39** (1966), 1441; *J. Appl. Chem. USSR*, **39** (1966), 1350.
[466] Snijder and Gilbert, *Ind. Eng. Chem.*, **34** (1942), 1519.
[467] Murto, Kivinen, Strandman and Virtalaine, *Suom. Kemistilehti*, B **41** (1968), 375.
[468] Myers, Smith, Katz and Scott, *J. Phys. Chem.*, **70** (1966), 3341.
[469] Sidgwick and Turner, *J. Chem. Soc. London Trans.*, **121** (1922), 2256.

[470] Tschamler, *Monatsh. Chem.*, **79** (1948), 162.
[471] Auméras and Minangoy, *Bull. Soc. Chim. Fr.*, **1950**, 326.
[472] Kendall and Booge, *J. Amer. Chem. Soc.*, **38** (1916), 1712.
[473] Menshutkin, *Zh. Russ. Fiz.-Khim. Obshch.*, **45** (1913), 1701; *Chem. Zbl.*, **85** (1914), I, 463.
[474] Kenttämaa, Lindberg and Nissema, *Suom. Kemistilehti*, B **33** (1960), 189.
[475] Lindberg, Kenttämaa and Nissema, *Suom. Kemistilehti*, B **34**, No. 9 (1961), 98.
[476] Beckmann, *Z. Phys. Chem. Stoechiom. Verwandschaftslehre*, **22** (1897), 609.
[477] Fawcet and Rasmussen, *J. Amer. Chem. Soc.*, **67** (1945), 1705.
[478] Kravchenko, *Zh. Prikl. Khim.*, **23** (1950), 288.
[479] Kitaigirodskii, *Dokl. Akad. Nauk. SSSR*, **113** (1957), 604.
[480] Magnesco and Ode, *Scientia*, **31** (1964), 5.
[481] Sapunov, *Koks Khim.*, **1967**, no. 8, 28.
[482] Spryskov, *Zh. Obshch. Khim.*, **22** (1952), 2035.
[483] Halban, *Z. Phys. Chem. Stoechiom. Verwandschaftslehre*, **84** (1913), 129.
[484] Joukovsky, *Bull. Soc. Chim. Belg.*, **43** (1934), 397.
[485] Vernon and Masterson, *J. Amer. Chem. Soc.*, **64** (1942), 2822.
[486] Vernon, Luder and Giella, *J. Amer. Chem. Soc.*, **63** (1941), 862.
[487] Hoerr, Harwood and Ralston, *J. Org. Chem.*, **9** (1944), 201.
[488] Ralston, Reck, Harwood and Du Brow, *J. Org. Chem.*, **13** (1948), 186.
[489] Ralston, Hoerr and Du Brow, *J. Org. Chem.*, **9** (1944), 259.
[490] Ralston, Hoerr, Pool and Harwood, *J. Org. Chem.*, **9** (1944), 102.
[491] Broome and Harwood, *J. Amer. Chem. Soc.*, **72** (1950), 3257.
[492] Hoerr and Harwood, *Ind. Eng. Chem.*, **44** (1952), 2923.
[493] Harwood, Ralston and Selby, *J. Amer. Chem. Soc.*, **63** (1941), 1916.
[494] Hosseini and Schneider, *Z. Phys. Chem. (Frankfurt am Main)*, **36** (1963), 137.
[495] Kortüm, Freier and Woerner, *Chem. Ing. Tech.*, **25** (1953), 125.
[496] Pannetier, Abello and Mignotte, *Bull. Soc. Chim. Fr.*, **1964**, 2964.
[497] Deshpanda and Pandya, *Trans. Faraday Soc.*, **63** (1967), 380.
[498] Pannetier and Abello, *Ann. Chim. (Paris)*, **11** (no. 1–2) (1966), 7.
[499] Falqui and Putzolu, *Rend. Semin. Fac. Sci. Univ. Cagliari*, **27** (1957), 74.
[500] Ampola and Rimatori, *Gazz. Chim. Ital.*, **27** (1897), I, 51.
[501] Desvergnes, *Ann. Chim. Anal. Chim. Appl.* [2] **10** (1928), 253.
[502] Sidgwick and Neill, *J. Chem. Soc. London Trans.*, **123** (1923), 2813.
[503] Dessart, *Bull. Soc. Chim. Belg.*, **35** (1926), 9.
[504] Jannelli and Orsini, *Rend. Accad. Sci. Fis. Mat., Naples* [4] **26** (1959), 246.
[505] Vriens and Medcalf, *Ind. Eng. Chem.*, **45** (1953), 1098.
[506] Giordano Orsini and Jannelli, *Gazz. Chim. Ital.*, **87** (1957), 159.
[507] Cioffi and Messer, *J. Chem. Eng. Data*, **11** (1966), 532.
[508] Kravchenko and Pastukhova, *Zh. Fiz. Khim.*, **26** (1952), 1191.
[509] Kravchenko and Pastukhova, *Dokl. Akad. Nauk. SSSR*, **130** (1960), 771.
[510] Hückel, Datow and Simmersbach, *Z. Phys. Chem.*, A**186** (1940), 137.
[511] Wroczynski and Guye, *J. Chim. Phys.*, **8** (1910), 214.
[512] Hoerr, Rapkin, Brake, Warner and Harwood, *J. Amer. Chem. Soc.*, **78** (1956), 4667.
[513] Sidgwick and Rubie, *J. Chem. Soc. London Trans.*, **119** (1921), 1013.
[514] Brown and Smith, *Aust. J. Chem.*, **8** (1955), 62.
[515] Mato and Sanchez, *An. Real. Soc. Espan. Fis. Quim.*, Ser. B **63** (1967), 1.
[516] Goates, Ott and Budge, *J. Phys. Chem.*, **65** (1961), 2162.
[517] Chistozvonova, Dugacheva and Anikin, *Zh. Fiz. Khim.*, **41** (1967), 82; *Russ, J. Phys. Chem.*, **41** (1967), 42.
[518] Hoerr, Binkerd, Pool and Ralston, *J. Org. Chem.*, **9** (1944), 68.
[519] Zemlyanov and Bergman, *Zh. Obshch. Khim.*, **34** (1964), 3536; *J. Gen. Chem. USSR* **34** (1964), 3584.
[520] Sidgwick and Callow, *J. Chem. Soc. London Trans.*, **125** (1924), 522.
[521] Puschin, *Z. Phys. Chem. Stoechiom. Verwandschaftslehre*, **118** (1925), 447; Puschin and Masarovich, *Zh. Russ. Fiz.-Khim. Obshchest.*, **46** (1914), 1366.

[522] Lazzell and Johnston, *J. Phys. Chem.*, **32** (1928), 1331.
[523] Delzenne, *Chem. Eng. Sci.*, **2** (1953), 220.
[524] Quitzsch, *J. Prakt. Chem.*, **28** (1965), 69.
[525] Poppe, *Bull. Soc. Chim. Belg.*, **44** (1935), 640.
[526] Bonner, Woolsely and Philp, *J. Chem. Eng. Data*, **13** (1968), 218.
[527] Ralston, Hoerr and Pool, *J. Org. Chem.*, **8** (1943), 473.
[528] Hall, Collett and Lazzell, *J. Phys. Chem.*, **37** (1933), 1087.
[529] Pollock, Collett and Lazzell, *J. Phys. Chem.*, **50** (1946), 23.
[530] Brown and Smith, *Aust. J. Chem.*, **8** (1955), 501.
[531] Saunders and Spaull, *Z. Phys. Chem. Frankfurt am Main*, **28** (1961), 332.
[532] Ioffe and Lilitsch, *Zh. Obshch. Khim.*, **24** (1954), 81; *Chem. Abstr.*, **48** (1954), 7412g.
[533] Ravich and Bogush, *Zh. Obshch. Khim.*, **31** (1961), 716; *J. Gen. Chem. USSR*, **31** (1961), 656.
[534] Kremann, *Monatsh. Chem.*, **29** (1908), 863.
[535] Lobry de Bruyn, *Rec. Trav. Chim. Pays-Bas*, **13** (1894), 116.
[536] Desvergnes, *Moniteur scient. Quesneville* [5] **15** (1925), 149.
[537] Desvergnes, *Chim. Ind. Paris*, **25** (1931), 3.
[538] Hammick, Hills and Howard, *J. Chem. Soc. London*, **1932**, 1530.
[539] Desvergnes, *Revue Chim. industrielle*, **40** (1931), 34.
[540] Desvergnes, *Moniteur scient. Quesneville* [5] **14** (1924), 121.
[541] Taylor and Rinkenbach, *J. Amer. Chem. Soc.*, **45** (1923), 44.
[542] Desvergnes, *Ann. Chim. Anal. Chim. Appl.* [2] **2** (1920), 278.
[543] Desvergnes, *Moniteur scient. Quesneville* [5] **16** (1926), 73.
[544] Bogoyavlenski, Bogolyubov and Vinogradov, *Schrift. Dorpater Naturforscherges.*, **15** (1906), 216; *Chem. Zbl.*, **78** (1907), I, 1738.
[545] Carrick, *J. Phys. Chem.*, **25** (1920–1), 628.
[546] Desvergnes, *Revue Chim. industrielle*, **36** (1927), 194.
[547] Sidgwick and Aldous, *J. Chem. Soc. London Trans.*, **119** (1921), 1001.
[548] Desvergnes, *Chim. Ind. Paris*, **26** (1931), 71.
[549] Sidgwick and Taylor, *J. Chem. Soc. London Trans.*, **121** (1922), 1853.
[550] Desvergnes, *Rev. chim. industrielle*, **36** (1927), 224.
[551] Findlay, *J. Chem. Soc. London Trans.*, **81** (1902), 1217.
[552] Mindowich, *Zh. Fiz. Khim.*, **30** (1956), 1082.
[553] Piatti, *Z. Angew. Chem.*, **44** (1931), 519.
[554] Desvergnes, *Rev. chim. industrielle*, **38** (1929), 66.
[555] Desvergnes, *Moniteur scient. Quesneville* [5] **14** (1924), 249.
[556] Desvergnes, *Moniteur scient. Quesneville* [5] **16** (1926), 201.
[557] Sidgwick and Dash, *J. Chem. Soc. London Trans.*, **121** (1922), 2586.
[558] Collett and Lazzell, *J. Phys. Chem.*, **34** (1930), 1838.
[559] Hammick and Williams, *J. Chem. Soc. London*, **1935**, 1856.
[560] Hammick and Sixsmith, *J. Chem. Soc. London*, **1939**, 972.
[561] Collett and Johnston, *J. Phys. Chem.*, **30** (1926), 70.
[562] Urbanski and Kwiatkowski, *Rocz. Chem.*, **13** (1933), 585.
[563] Desvergnes, *Chim. Ind. Paris*, **28** (1932), 1038.
[564] Taylor and Rinkenbach, *J. Amer. Chem. Soc.*, **45** (1923), 1218.
[565] Urbanski and Kwiatkowski, *Rocz. Chem.*, **13** (1933), 740.
[566] Leikola, Snellman and Kauppila, *Farmaseuttinen Aikakauslehti*, **66** (1957), 41.
[567] Van Iterson-Rotgans, *Z. Phys. Chem. Stoechiom. Verwandschaftslehre*, **87** (1914), 305.
[568] Zharovskii and Ryzhenko, *Zh. Anal. Khim.*, **22** (1967), 1142.
[569] Taylor and Marks, *J. Chem. Soc. London*, **1930**, 2302.
[570] Quitzsch, Wunderlich and Geiseler, *J. Prakt. Chem.*, **30** (1965), 119.
[571] Bogojawlenski and Winogradow, *Schrift. Dorpater Naturforscherges*, **15** (1906), 230; *Chem. Zbl.*, **78** (1907), I, 1741.
[572] Desvergnes, *Ann. Chim. Anal. Chim. Appl.* [2] **10** (1928), 226.
[573] Asinger, Geiseler and Laue, *Chem. Ber.*, **90** (1957), 485.
[574] Desvergnes, *Moniteur scient. Quesneville* [5] **15** (1925), 73.

[575] McCombie, Scarborough and Smith, *J. Chem. Soc. London*, **1927**, 802.
[576] Von Halban, *Z. Phys. Chem. Stoechiom. Verwandschaftslehre*, **84** (1913), 144.
[577] Kurnakov and Plaksina-Shishokina, *Izv. Inst. Fiz-Khim. Anal., Akad. Nauk. SSSR*, **5** (1931), 29.
[578] Shishokin, *Izv. Inst. Fiz.-Khim. Anal., Akad. Nauk. SSSR*, **4** (1928–30), 195; Kurnakov and Plaksina, *ibid*, **4** (1928–30), 317.
[579] Meyer, Bourrelly and Metzger, *CR Acad. Sci.*, Ser. C **267**, no. 2 (1968), 114.
[580] Du Brow, Hoerr and Harwood, *J. Amer. Chem. Soc.*, **74** (1952), 6241.
[581] Desvergnes, *Rev. chim. industrielle*, **38** (1929), 265; *Chem. Zbl.*, **101** (1930), I, 820.
[582] Krichevsky and Jefremova, *Zh. Fiz. Khim.*, **22** (1948), 11.
[583] Brainard, Univ. Microfilms (Ann Arbor, Mich. USA), No. 65–5273; *Dissertation Abstr.*, **25** (1965), 7126.
[584] Brainard and Brymer Williams, *A.I.Ch.E.J.*, **13** (1967), 60.
[585] Dolezalek, *Z. Phys. Chem.*, **71** (1910), 191.
[586] Robinet, *CR Acad. Sci.*, **58** (1864), 608.
[587] Knight and Hinshelwood, *J. Chem. Soc. London*, **1927**, 466.
[588] Wynne-Jones, *J. Chem. Soc. London*, **1930**, 1064.
[589] Siekierski and Gwozdz, *Nukleonika*, **5** (1960), 205.
[590] Veksler, *Khim. Prom.*, **1960**, 483.
[591] Olsson, *Ark. Kemi.*, **15** (1960), 259.
[592] Cerfontain and Telder, *Rec. Trav. Chim. Pays-Bas*, **84** (1965), 545.
[593] Francis, *J. Chem. Eng. Data*, **10** (1965), 327.
[594] Umano and Hayano, *Kogyo Kagaku Zasshi*, **60** (1957), 1936.
[595] Umano, Kawasaki, Nakano and Hayano, *Tokyo Kogyo Shikensho Hokoku*, **53** (1958), 397.
[596] Francis, *Chem. Eng. Sci.*, **22** (1967), 627.
[597] Barkan and Beketova, *Uch. Zap. Beloruss. Gos. Univ. Im V.I. Lenina Ser. Khim.* (1956) No. 29–222; *Chem. Abstr.*, **54** (1960), 6260g.
[598] Schneider, *Ber. Bunsenges. Phys. Chem.*, **73** (1969), 294.
[599] Barkan, *Uch. Zap. Beloruss. Gos. Univ. Im. V.I. Lenina Ser. Khim.*, **1955**, no. 24, 102.
[600] Barkan and Demyanchick, *Tr. Khim. Fak. Beloruss. Gos. Univ.*, **1960**, 97.
[601] Radyshevskaya, Nikurashina and Mertslin, *Zh. Obshch. Khim.*, **32** (1962), 673; *J. Gen. Chem. USSR*, **32** (1962), 673.
[602] Hill, *J. Amer. Chem. Soc.*, **44** (1922), 1163.
[603] Leibnitz, Könnecke and Lippert, *J. Prakt. Chem.* (*4*) **3** (1956), 311.
[604] Atsushi Ikari and Isao Arakawa, *Kogyo Kagaku Zasshi*, **68** (1965), 257.
[605] Swietoslawski, Zieborak and Galska-Krajewska, *Bull. Acad. Pol. Sci., Ser. Sci., Chim. Géol. Géogr.*, **7** (1959), 43; *Chem. Abstr.*, **54** (1960), 16068g.
[606] Steib, *J. Prakt. Chem.*, **28** (1965), 252.
[607] Steib, *J. Prakt. Chem.*, **30** (1965), 39.
[608] O'Grady, *J. Chem. Eng. Data*, **12** (1967), 17.
[609] Hollo, *Period. Polytech.*, **2** (1958), 113.
[610] Updike, Langdon and Keyer, *Trans. Amer. Inst. Chem. Eng.*, **41** (1945), 717.
[611] Prince and Hunter, *Chem. Eng. Sci.*, **6** (1957), 245.
[612] Leibnitz, Könnecke and Niese, *J. Prakt. Chem.* (*4*) **4** (1957), 278.
[613] Udovenko and Mazanko, *Zh. Fiz. Khim.*, **37** (1963), 2324.
[614] Mertslin, Kamaevskaya and Nikurashina, *Zh. Fiz. Khim.*, **40** (1966), 2539; *Russ. J. Phys. Chem.*, **40** (1966), 1359.
[615] Ivekovic and Gertner, *Croat. Chem. Acta*, **39** (1967), 103.
[616] Kamaevskaya and Mertslin, *Zh. Fiz. Khim.*, **44** (1970), 2564; *Russ. J. Phys. Chem.*, **44** (1970), 1453.
[617] Frolov, *Zh. Fiz. Khim.*, **39** (1965), 2877.
[618] Barbaudy, Thesis, Paris 1925.
[619] Norman, *Trans. Inst. Chem. Eng.*, **23** (1945), 66.
[620] Mertslin, Nikurashina and Kamaerskaya, *Zh. Fiz. Khim.*, **35** (1961), 2628; *Russ. J. Phys. Chem.*, **35** (1961), 1299.

[621] Morachevskii and Belonsov, *Vestn. Leningrad Univ.*, **13**, No. 4, *Ser. Fiz. Khim.*, No. 1 (1958), 117.
[622] Giona and Roberto, *Rass. Chim.*, **11**, No. 2 (1959), 9.
[623] Vogelpohr, *Chem. Ing. Tech.*, **36** (1965), 907.
[624] Volkov, *Uch. Zap. Perm. (Gos.) Univ.*, **13** (1959), no. 3, 73.
[625] Udovenko and Mazanko, *Zh. Fiz. Khim.*, **37** (1963), 1151; *Russ. J. Phys. Chem.*, **37** (1963), 610.
[626] Udovenko and Mazanko, *Zh. Fiz. Khim.*, **38** (1964), 2984; *Russ. J. Phys. Chem.*, **38** (1964), 1626.
[627] Masahiro, Yorigana and Yoshimura, *Hiroshima Daigaku Kogakubu Kenkyo Hokoku*, **13** (1965), 41.
[628] Morachevskii and Legochkina, *Zh. Prikl. Khim.*, **38** (1965), 1789; *J. Appl. Chem. USSR*, **38** (1965), 1751.
[629] Udovenko and Mazanko, *Zh. Fiz. Khim.*, **41** (1967), 395; *Russ. J. Phys. Chem.*, **41** (1967), 197.
[630] Davis and Evans, *J. Chem. Eng. Data*, **5** (1960), 401.
[631] Zawisza, *Bull. Acad. Pol. Sci., Ser. Sci. Chim. Géol. Géogr.*, **7** (1959), 259.
[632] Kogan and Tolstova, *Zh. Fiz. Khim.*, **33** (1959), 276.
[633] Mitsuho Hirata and Shigefumi Fujita, *Kagaku Kikai*, **21** (1957), 201.
[634] Papafil and Beldia, *Analele Stiint. Univ. 'A.I. Cuza'*, Iasi Sect. I, **9**, No. 1 (1963), 183.
[635] Shiji Nagata and Waratu Eguchi, *Mem. Fac. Eng. Kyoto Univ.*, **19** (1957), 102.
[636] Tagliavini and Arich, *Ric. Sci.*, **28** (1958), 2557.
[637] Francis, *J. Chem. Eng. Data*, **10** (1965), 260.
[638] Murphy and McConell, *J. Chem. Eng. Data*, **5** (1960), 143.
[639] Arich and Tagliavini, *Ric. Sci.*, **28** (1958), 1620.
[640] Möbius, *J. Prakt. Chem.* (4) **2** (1955), 95.
[641] Volkov, *Uch. Zap. Perm. (Gos.) Univ.*, **9** (1955), no. 4, 171; *Chem. Abstr.*, **53** (1959), 13737d.
[642] Udovenko and Aleksandrova, *Zh. Fiz. Khim.*, **37** (1963), 52.
[643] Christian, Affsprung and Taylor, *J. Phys. Chem.*, **67** (1963), 187.
[644] Biancani and De Filippo, *Gazz. Chim. Ital.*, **88** (1958), 1202; Krishnamurty, Jayarama Rao and Venkata Rao, *Trans. Indian Inst. Chem. Eng.*, **6** (1953–4), 161; *Chem. Abstr.*, **50** (1956), 3051i.
[645] Krishnamurty and Venkata Rao, *Trans. Indian Inst. Chem. Eng.*, **8** (1955–6), 52.
[646] Ababi and Mihaila, *Stud. Univ. Babes-Bolyai, Ser. Chem.*, **8** (1963), 429.
[647] Udovenko, Efimova and Chaikhorskii, *Radiokhimiya*, **2** (1960), 675.
[648] Ismailov and Mamedova, *Tr. Azerb. Gos. Univ. Ser. Khim.*, **1959**, 87.
[649] Ismailov, Mamedova and Alekperova, *Uch. Zap. Azerb. Gos. Univ., Fiz.-Mat. Khim. Ser.*, **1959**, no. 3, 77; *Uch. Zap. Azerb. Gos. Univ., Fiz.-Mat. Khim. Ser.*, **1959**, no. 4, 73; *Chem. Abstr.*, **55** (1961), 27863h and **56** (1962), 10984g.
[650] Ababi and Popa, *Analele Stiint. Univ. 'A.I. Cuza'*, Iasi Sect. I, **8**, No. 1 (1962), 233.
[651] Omote and Nakamura, *Kogyo Kagaku Zasshi*, **70** (1967), 1599.
[652] Koperska, M. and Cieslak, M., *Przem. Chem.*, **47** (4) (1968), 201; *Chem. Abstr.*, **69** (1968), 5641.
[653] Mato, Hernandy and Bueno, *Stud. Chem. Univ. Salamanca*, **1968**, no. 3–51.
[654] Krzewski and Smutek, *Collect. Czech. Chem. Commun.*, **31** (1966), 515.
[655] Mertslin and Nikurashina, *Zh. Obshch. Khim.*, **32** (1962), 3122; *J. Gen. Chem., USSR*, **32** (1962), 3071.
[656] Charles and Morton, *J. Appl. Chem.*, **7** (1957), 39.
[657] Bylicki and Jankun-Pinska, *Bull. Acad. Pol. Sci. Ser. Sci. Chim.*, **12** (1964), 837 and 843.
[658] Jankun-Pinska, *Bull. Acad. Pol. Sci. Ser. Sci. Chim.*, **13** (1965), 121 and 125.
[659] Porada, *Bull. Acad. Pol. Sci. Ser. Sci. Chim.*, **14** (1966), 695 and 699.
[660] Porada, *Bull. Acad. Pol. Sci. Ser. Sci. Chim.*, **15** (1967), 205.
[661] Vasenko, *Nauch. Zap. L'vov Politekh Inst.*, No. **62** (1957), 325.
[662] Francis, *J. Chem. Eng. Data*, **10** (1965), 145.
[663] Morachevskii and Sabinin, *Zh. Prikl. Khim.*, **33** (1960), 1775.

[664] Tettamanti, Nogrady and Sawinsky, *Periodica Polytech.*, **4** (1960), 201.
[665] Soszewinski and Wolski, *Przem. Chem.*, **44** (1965), 189.
[666] Soszewinski and Wolski, *Przem. Chem.*, **46** (1967), 678.
[667] Brown and Brady, *J. Amer. Chem. Soc.*, **74** (1952), 3570.
[668] Brown and Brady, *J. Amer. Chem. Soc.*, **71** (1949), 3573.
[669] Francis, *J. Chem. Eng. Data*, **11** (1966), 96.
[670] Francis, *J. Phys. Chem.*, **62** (1958), 579.
[671] Brown and Wallace, *J. Amer. Chem. Soc.*, **75** (1953), 6268.
[672] Eley and King, *J. Chem. Soc. London* **1952**, 2517 and 4972.
[673] Francis, *J. Chem. Eng. Data*, **12** (1967), 269.
[674] Francis, *J. Chem. Eng. Data*, **11** (1966), 557.
[675] Francis, *J. Chem. Eng. Data*, **10** (1965), 45.
[676] Satterfield, Powell, Oster and Noyen, *Ind. Eng. Chem.*, **47** (1955), 1458.
[677] Bowden, Univ. Microfilms (Ann Arbor, Mich. USA), No. 65–8576.
[678] Bowden, Staton and Smith, *J. Chem. Eng. Data*, **11** (1966), 296.
[679] Kiyoharu Ishida, *Bull. Chem. Soc. Jap.*, **30** (1957), 612.
[680] Francis, *J. Chem. Eng. Data*, **12** (1967), 380.
[681] Skirrow, *Z. Phys. Chem.*, **41** (1902), 139.
[682] Francis, *J. Chem. Eng. Data*, **11** (1966), 234.
[683] Alvarez, Gonzalez and Vegahea, *An. Real. Soc. Espan. Fis. Quim.*, Ser. B **63** (1967), 749.
[684] Moore, Wilson and Hutchinson, *Biochem. J.*, **4** (1909), 347.
[685] Alvarez, Gonzalez and Vegahea, *An. Quim.*, **64** (1968), 89.
[686] Dorn, Campbell, Cochran, Hall, Scheller and Stuart, *J. Chem. Eng. Data*, **9** (1964), 28.
[687] Crawford and Stewart, *J. Chem. Soc. London*, **1955**, 288.
[688] Gorenbein and Pivnutel, *Zh. Obshch. Khim.*, **27** (1957), 20; *J. Gen. Chem. USSR*, **27** (1957), 19.
[689] Fialkov and Fokina, *Ukr. Khim. Zh.*, **25** (1959), 288.
[690] Goldberg and Vernon, *J. Amer. Chem. Soc.*, **73** (1951), 2845.
[691] Vernon and Sheard, *J. Amer. Chem. Soc.*, **70** (1948), 2035.
[692] Shakhparonov, Balamutova, Lel'chuk, Mikheev, Shutova, Glushkova and Martynova, *Zh. Fiz. Khim.*, **34** (1960), 1734.
[693] Stepanova, Vybornova and Velikovskii, *Gaz. Delo Nauchn.-Tekhn. Sb.*, **1965** (*10*), 9; *Chem. Abstr.*, **64** (1966[1]), 10455f.
[694] Savvina and Velikovskii, *Tr. Vses. Nauchn. Issled. Inst. Prir. Gazov.*, **1962**, 185.
[695] Stepanova, Vybornova and Velikovskii, *Gaz. Delo Nauchn.-Tekhn. Sb.*, **1967** (*1*), 24; *Chem. Abstr.*, **67** (1967[11]), 68067.
[696] Shleinikov, *Zh. Vses. Khim. Obshchest.* (im D. J. Mendeleeva), **11** (1966), 116.
[697] Belknap and Weber, *J. Chem. Eng. Data*, **6** (1961), 485.
[698] Schuberth and Leibnitz, *J. Prakt. Chem.*, **27** (1965), 276.
[699] Weatherford and Van Winkle, *J. Chem. Eng. Data*, **15** (1970), 386.
[700] Kaes and Weber, *J. Chem. Eng. Data*, **7** (1962), 344.
[701] Butler and Ridzway, *J. Appl. Chem.*, **17** (1967), 191.
[702] Ridzway and Butler, *J. Chem. Eng. Data*, **12** (1967), 509.
[703] Yoshida Takatoshi, Yamaaa Ikuho and Machii Hirooki, *Kagaku Kogaku*, **32** (1968), 1091.
[704] Toshiro Takahashi and Tamechika Yamamoto, *Kogyo Kagaku Zasshi*, **59** (1956), 639.
[705] Waldo and Weber, *J. Chem. Eng. Data*, **8** (1963), 349.
[706] Kuan Shih Yuan, Lu, Ho and Keshpande (Ho and Lu), *J. Chem. Eng. Data*, **8** (1963), 549 (553).
[707] Toshiro Takahashi and Tamechika Yamamoto, *Kogyo Kagaku Zasshi*, **59** (1956), 634.
[708] Ustraikh, *Tr. Vses. Nauch.-Issled. Inst. Neftekhim Protsessov* **1962**, no. 5, 101.
[709] Markuzin and Plekhotkin, *Zh. Fiz. Khim.*, **36** (1962), 1973.
[710] Prokopets and Zeliznyi, *Izv. Vyssh. Ucheb. Zaved. Neft. Gaz.*, **5** (1962), no. 7, 51.
[711] Waksmundzki, Stelmack and Wolski, *Przem. Chem.*, **43** (1964), 194.
[712] Avdei Pokorskii and Yablochkina, *Tr. Vses. Nauch.-Tekh. Soveshch. Protsessy Zhidk. Ekstr. Khemosorbtsii 2nd*, **1964**, 329.

[713] Lutugina, Shandalova, Korobchenkova and Polyakov, *Zh. Fiz. Khim.*, **44** (1970), 2110.
[714] Wagner and Weber, *Chem. Eng. Data Ser.* **3** (1958), 220.
[715] Nielsen and Weber, *J. Chem. Eng. Data*, **4** (1959), 145.
[716] Oakeson and Weber, *J. Chem. Eng. Data*, **5** (1960), 279.
[717] Jean Fu and Lu, *J. Chem. Eng. Data*, **13** (1968), 6.
[718] Vyayaraghavan, Deshpande and Kulon, *J. Indian Inst. Sci.*, **48** (1966), 138.
[719] Grigoriu(m) and Precup, *Petrol Gaze*, **18** (1967), 198.
[720] Johnson and Francis, *Ind. Eng. Chem.*, **46** (1954), 1662.
[721] Herz and Levi, *Kolloid Z.*, **50** (1930), 21.
[722] Mandik, *Ropa Uhlie*, **8** (1966), 143.
[723] Mandik, *Collection Czech. Chem. Commun.*, **31** (1966), 3869.
[724] Werner and Schuberth, *J. Prakt. Chem.*, **31** (1966), 225.
[725] Ripa, De Angelis and Berti, *Riv. Combust.*, **10** (1956), 803.
[726] Diyarov, Kozlov and Bureeva, *Neftepererab, Neftekhim Moscov*, **1966** (9), 30.
[727] Waksmundzki, Stelmach and Wolski, *Przem. Chem.*, **43** (1964), 445.
[728] Skinner, *Ind. Eng. Chem.*, **47** (1955), 222.
[729] Kyle and Tetlow, *J. Chem. Eng. Data*, **5** (1960), 275.
[730] Calderbank and Nikolov, *J. Phys. Chem.*, **60** (1956), 1.
[731] Waksmundzki, Wolski, Stelmach and Jurkiewicz, *Przem. Chem.*, **43** (1964), 548.
[732] Henty, McManamey and Prince, *J. Appl. Chem.*, **14** (1964), 148.
[733] Susarev and Gorbunov, *Zh. Fiz. Khim.*, **40** (1966), 68.
[734] Sinor and Weber, *J. Chem. Eng. Data*, **5** (1960), 243.
[735] Subba Rao and Venkata Rao, *Can. J. Chem. Eng.*, **42** (1964), 266.
[736] Vitman and Zharov, *Zh. Fiz. Khim.*, **44** (1970), 2916.
[737] Lutugina and Kovalichev, *Vestn. Leningrad Univ.*, 21 (10), *Ser. Fiz. Khim.*, no. 2 (1966), 91.
[738] Jayarama Rao, Dakshinamurty and Venkata Rao, *J. Sci. Ind. Res.*, Sect. B **20** (1961), 218.
[739] Jones and Amstell, *J. Chem. Soc. London*, **1930**, 1316.
[740] Morachevskii and Komarova, *Vestn. Leningrad Univ.*, 12, no. 4, *Ser. Fiz. Khim.*, no. 1 (1957), 118.
[741] Zieborak and Galska-Krajewska, *Bull. Acad. Pol. Sci., Ser. Sci., Chim. Géol. Géogr.*, **6** (1958), 763.
[742] Morachevskii and Zharov, *Zh. Prikl. Khim.*, **36** (1963), 2776.
[743] Deshpande and Lu, *J. Appl. Chem. London*, **15** (1965), 136.
[744] KuanShih-Juan Wu, *Bull. Inst. Chem. Acad. Sinica*, no. **13** (1967), 31.
[745] Haase, *Z. Phys. Chem., Leipzig*, **195** (1950), 362.
[746] Morachevskii and Ch'eng, *Zh. Fiz. Khim.*, **35** (1961), 2335.
[747] Storonkin and Morachevskii, *Zh. Fiz. Khim.*, **30** (1956), 1297.
[748] Storonkin, Morachevskii and Kudryavtseva, *Zh. Fiz. Khim.*, **31** (1957), 375.
[749] Szapiro, *Zesz. Nauk. Politech. Lodz. Chem.*, no. **6** (1957), 3.
[750] Zarubin, *Zh. Fiz. Khim.*, **33** (1959), 147.
[751] Isama Nagata, *Can. J. Chem. Eng.*, **42** (1964), 82.
[752] Vitman and Zharov, *Vestn. Leningrad. Univ. Fiz. Khim.*, **1970**, no. 2–107.
[753] Lutugina and Kolbina, *Zh. Prikl. Khim.*, **41** (1968), 2766.
[754] Subba Rao and Venkata Rao, *J. Appl. Chem.*, **18** (1968), 61.
[755] Galska-Krajewska, *Bull. Acad. Pol. Sci., Ser. Sci., Chim. Géol. Géogr.*, **6** (1958), 257.
[756] Kudryavtseva and Susarev, *Zh. Prikl. Khim.*, **39** (1966), 629.
[757] Nataraj and Raja Rao, *Trans. Indian. Inst. Chem. Eng.*, **1968** (April), 95; Publ. in *Indian Chem. Eng.*, **10**(2).
[758] Reddy and Venkata Rao, *Indian J. Technol.*, **5** (1967), 66.
[759] Thornton and Garner, *J. Appl. Chem.*, **1**, Suppl. Issue no. 1 (1951), 68.
[760] Thornton and Garner, *J. Appl. Chem.*, **1**, Suppl. Issue no. 1 (1951), 61.
[761] Kurmanadha Rao, Krishnamurty and Venkata Rao, *Rec. Trav. Chim. Pays-Bas*, **76** (1957), 769.
[762] Dakshinamurty and Venkata Rao, *J. Appl. Chem.*, **7** (1957), 654.

763 Swami, D. R., KumarKrishna Rao, V. N. and Narasinga Rao, M., *J. Sci. Ind. Res.*, **15B** (1956), 550.

764 Donald and Ridgway, *J. Appl. Chem.*, **8** (1958), 408.

765 Dakshinamurty, Jayarama Rao, G., Acharya and Venkata Rao, C., *Chem. Eng. Sci.*, **9** (1958), 69.

766 Baradarajan and Satyanarayana, *J. Chem. Eng. Data*, **13** (1968), 148.

767 Isamu Nagata, *J. Chem. Eng. Data*, **7** (1962), 461.

768 Kwang-Chu Chao, Univ. Microfilms (Ann. Arbor, Mich., USA), Publ. no. 19076; *Dissertation Abstr.*, **16** (1956), 2114.

769 Chao and Hougen, *Chem. Eng. Sci.*, **7** (1958), 246.

770 Komarova and Kogan, *Zh. Prikl. Khim.*, **37** (1964), 1570.

771 Komarova and Kogan, *Zh. Prikl. Khim.*, **38** (1965), 2690.

772 Jost, *Z. Naturforsch.*, **1** (1946), 576.

773 Chiraranjan Podder, *Z. Phys. Chem. (Frankfurt am Main)*, **39** (1963), 79.

774 Murit, Sriram and Narasinshamurty, *Brit. Chem. Eng.*, **1967**, no. 12, 1882.

775 Weck and Hunt, *Ind. Eng. Chem.*, **46** (1954), 2521.

776 Suri, Rao and Ramakrishna, *Indian J. Chem.*, **4** (1966), 194.

777 Angelescu and Ganea, *Rev. Chim. Acad. Repub. Pop. Roum.*, **2** (1957), 41.

778 Angelescu and Ciplea, *Rev. Chim. Acad. Repub. Pop. Roum.*, **2** (1957), 35.

779 Kravchenko, *Zh. Fiz. Khim.*, **13** (1939), 989; *Chem. Abstr.*, **34** (1940), 6513^8.

780 Kravchenko, *Zh. Fiz. Khim.*, **13** (1939), 1523; *Chem. Abstr.*, **35** (1941), 375^4.

781 Kravchenko, *Acta Physicochim. URSS*, **20** (1945), 567; *Chem. Abstr.*, **40** (1946), 4283^8.

782 Joseph-Natan, Manjarrez and Bravo, *Rev. Soc. Quim. Mex.*, **12** (1968), 9.

783 Velazquez, Manjarrez and Joseph-Natan, *Rev. Soc. Quim. Mex.*, **12** (1968), 169A.

784 Kravchenko, *Zh. Fiz. Khim.*, **24** (1950), 1033; *Chem. Abstr.*, **45** (1951), 2763i.

785 Kravchenko, *Zh. Fiz. Khim.*, **13** (1939), 133; *Chem. Abstr.*, **34** (1940), 315^2.

786 Disselkamp, *Z. Phys. Chem.*, **123** (1926), 99.

787 Nozdrev and Grechkin, *Zh. Fiz. Khim.*, **38** (1964), 2663; *Russ. J. Phys. Chem.*, **38** (1964), 1444.

788 Burke and Williams, *Trans. Kentucky Acad. Sci.*, **27** (1966), 29; *Chem. Abstr.*, **69** (1968), 46411.

789 Mahieu, *Bull. Soc. Chim. Belg.*, **45** (1936), 667.

790 Kravchenko, *Zh. Fiz. Khim.*, **14** (1940), 248; *Chem. Abstr.*, **36** (1942), 3724^9.

791 Kravchenko, *Zh. Fiz. Khim.*, **26** (1952), 1284.

792 Khaishbashev, *Izv. Sekt. Fiz.-Khim. Anal., Inst. Obshch. Neorg. Khim. Akad. Nauk. SSSR*, **14** (1941), 227; *Chem. Abstr.*, **40** (1946), 2725^2.

793 Bruni, *Gazz. Chim. Ital.*, **28** (1898), II, 508.

794 Kravchenko and Passtuchova, *Zh. Fiz. Khim.*, **27** (1953), 822.

795 Piatti, *Z. Angew. Chem.*, **44** (1931), 519.

796 Bednov, Borkova, Novikov and Privalov, *Khim. Prod. Koksovaniya Uglei Vostoka SSSR, Poluchenie, Obrabotka, Ispol'z., Metody Analizov, Vost. Nauchn.-Issled. Uglekhim. Inst. Sb. Statei*, **2** (1964), 126; *Chem. Abstr.*, **62** (1965), 8889f.

797 Ponomarev, *Uch. Zap. Molotov Gos. Univ.*, **3** (1939), no. 4, 65–76; *Khim. Referat Zh.*, **4**, no. 6 (1941), 22; *Chem. Abstr.*, **37** (1943), 5903^7.

798 Lipatov, *Kolloid. Zh.*, **22** (1960), 639; *Chem. Abstr.*, **56** (1962), 985b.

799 Caspari, *J. Chem. Soc. London*, **107** (1915), 162.

800 Shishkin Kotsyuba and El'tsova, *Ukr. Khim. Zh.*, **30** (1964), 137.

801 De Minjer, 'Damp-vloeistofevenwichten . . .', Thesis, Delft 1939.

802 Reinders and de Minjer, *Rec. Trav. Chim. Pays-Bas*, **59** (1940), 369.

803 Ewell and Welch, *Ind. Eng. Chem.*, **37** (1945), 1224.

804 Campbell, Kartzmark and Chatterjee, *Can. J. Chem.*, **44** (1966), 1183.

805 Cherkashin, *Visn. L'viv. Derzh. Univ. Ser. Khim.*, **1963** (*6*), 20.

806 Isamu Nagata, *J. Chem. Eng. Data*, **7** (1962), 360.

807 Mochalov, Zhivopistsev, Petrov and Yakovleva, *Zh. Obshch. Khim.*, **37** (1967), 25; *J. Gen. Chem. USSR*, **37** (1967), 21.

808 Scatchard and Ticknor, *J. Amer. Chem. Soc.*, **74** (1952), 3724.

[809] Hirata and Hirose, *Mem. Fac. Technol. Tokyo Metrop. Univ.*, **11** (1961), 876.
[810] Schreinemakers, *Z. Phys. Chem.*, **47** (1904), 445.
[811] Schreinemakers, *Z. Phys. Chem.*, **48** (1904), 257.
[812] Isamu Nagata, *J. Chem. Eng. Data*, **10** (1965), 106.
[813] Subba Rao and Venkata Rao, *J. Chem. Eng. Data*, **11** (1966), 158.
[814] Lutugina, Molodenko and Dement'ev, *Vestn. Leningrad Univ.*, **1967**, 22 (22), *Fiz Khim.*, no. 4, 107.
[815] Lutugina, Molodenko and Pantyushova, *Zh. Prikl. Khim.*, **41** (1968), 2018.
[816] McMillan and McDonald, *J. Phys. Chem.*, **49** (1945), 10.
[817] Markuzin and Lazeeva, *Vestn. Leningrad Univ. Fiz. Khim.*, **1970**, no. 3, 117; *ibid*, **1970**, no. 4, 108.
[818] Isamu Nagata, *J. Chem. Eng. Data*, **15** (1970), 213.
[819] Lutugina, Molodenko and Orliebskaya, *Zh. Prikl. Khim.*, **39** (1966), 1774; *J. Appl. Chem. USSR*, **39** (1966), 1657.
[820] Angelescu and Dumitrescu, *Z. Phys. Chem. Stoechiom. Verwandschaftslchre*, **132** (1928), 217.
[821] Avenarius and Tarassenkow, *Zh. Obshch. Khim.*, **16** (1946), 1777; *Chem. Abstr.*, **41** (1947), 5372g.
[822] Free and Hutchison, *J. Chem. Eng. Data*, **4** (1959), 193 and 306.
[823] Rigamonti and Vacirca, *Ann. Chim. (Rome)*, **48** (1958), 478.
[824] Descamps, *Bull. Soc. Chim. Belg.*, **49** (1940), 91.
[825] Gregg-Wilson and Wright, *J. Chem. Soc. (London)*, **1928**, 3111.
[826] Duff and Bills, *J. Chem. Soc. (London)*, **1932**, 881.
[827] Trimble and Frazer, *Ind. Eng. Chem.*, **21** (1929), 1063.
[828] McDonald, *J. Am. Chem. Soc.*, **62** (1940), 3183.
[829] Degalcesan and Laddha, *J. Appl. Chem.*, **12** (1962), 111.
[830] McDonald, *J. Phys. Chem.*, **45** (1941), 706.
[831] Zharov and Buevich, *Zh. Prikl. Khim.*, **41** (1968), 2032; *J. Appl. Chem. USSR*, **41**, (1968), 1911.
[832] Kogan, *Khim. Prom. (Moscov)*, **1957**, 356.
[833] Shukla, *A.I.Ch.E. J.*, **14** (1968), 666.
[834] Zharov, Malegina and Morachevskii, *Zh. Prikl. Khim.*, **38** (1965), 2132; *J. Appl. Chem. USSR*, **38** (1965), 2089.
[835] Krupatkin, *Zh. Obshch. Khim.*, **27** (1957), 567.
[836] Kuriloff, *Z. Phys. Chem. Stoechiom. Verwandschaftslchre*, **24** (1897), 441.
[837] Siggia and Hanna, *Anal. Chem.*, **21** (1949), 1086.
[838] Marden and Dover, *J. Amer. Chem. Soc.*, **38** (1916), 1235.
[839] Marden and Dover, *J. Amer. Chem. Soc.*, **39** (1917), 1.
[840] Blank, *Zh. Fiz. Khim.*, **42** (1968), 3056; *Russ. J. Phys. Chem.*, **42** (1968), 1628.
[841] Ralston and Hoerr, *J. Org. Chem.*, **10** (1945), 170.
[842] Kurnakov and Plaksina-Shishokina, *Izv. Inst. Fiz.-Khim. Anal., Akad. Nauk. SSSR*, **5** (1931), 29.
[843] Frunze and Korshak, *Izv. Akad. Nauk. SSSR, Otd. Khim. Nauk.*, **1958**, 344.
[844] Ravich and Bogush, *Zh. Obshch. Khim.*, **31** (1961), 716; *J. Gen. Chem. USSR*, **31** (1961), 656.

. The Editor regrets that references 45 onwards differ from the standard normally followed in these books in that the volume is given before the date. The authors in this section have followed 'Access' published by the American Chemical Society and in view of the some 800 references the editor decided not to transpose them in view of the risk of typographical errors and the resultant nullification of the value of this section.
Names of periodicals are abbreviated as given in 'Access' and/or 'the Chemical Abstracts'.

CHAPTER 5

ANALYSIS AND HANDLING

By K. H. V. French

5.1. SPECIFICATIONS

SPECIFICATIONS for benzene in the 1970's are mainly directed towards the definition of its quality for use in chemical synthesis. Due to its toxic nature, benzene should only be used as solvent under strictly controlled and totally enclosed conditions. This situation is recognised in most countries and specifications for benzene as a solvent will not be considered.

The purity of benzene marketed for synthesis purposes is usually greater than 99·5 per cent and all but a few parts per million of the impurities are toluene and hydrocarbons with boiling points close to that of benzene. Specifications for benzene, therefore, include clauses for four purposes:

(1) Identification;
(2) Indication of purity;
(3) Limitation of minor impurities;
(4) Limitation of contaminants.

Benzene is normally identified by its density and its boiling point; its purity is primarily indicated by its crystallising point, and also by its distillation range; specialised tests have been developed for the limitation of impurities and contaminants.

This grouping of specification clauses is illustrated by reference to the (draft) British Standard Specification for Benzene (1973) (Table V.1).

TABLE V.1

Grouping of Specification Clauses

Identification	Density at 20°C	0·878–0·881
	Distillation range to include	80·1°C at 760 mm
Purity	Distillation range 1 to 96%	0·6°C max.
	Crystallising point	5·35°C min.
Limitation of Impurities	Total sulphur	2·0 p.p.m. max.
	Free sulphur	0·5 p.p.m. max.
	Hydrogen sulphide	negative*
	Mercaptans	negative*
	Neutrality	neutral*
Limitation of contaminants	Acid wash test	0·1 max.
	Residue on evaporation	5 mg/100 ml max.

* By specified qualitative tests.

In Table V.2 are a number of typical national specifications for synthesis grade benzene. It is customary to include clauses to ensure that the product is free from undissolved water and is colourless.

TABLE V.2

National Specifications for Benzene

		UK *BS 135:1973* *Benzene*	*USA* *ASTM D 2359* *Refined benzene* *535*	*Germany* *DIN 51 633* *Reinbenzol*
Appearance		—	Clear liquid	
Colour		20 max. (ASTM D1209)	20 max. (ASTM D1209)	water white
Specific gravity		—	0·8820–0·8860 at 15·56/15·56°C (ASTM D891)	—
Density	g/ml	0·878–0·881 at 20°C	—	min. 0·883 at 15°C (DIN 51 757)
Refractive index	min. at 20°C	—	—	1·5003
Distillation		*0·6°C range 1–96% including 80·1°C	1·0°C range drop to dry including 80·1°C	0·6°C range drop to dry (DIN 51 761)
Crystallising point	min. °C	5·35	5·35 (ASTM D 852)	5·20 (DIN 51 798)
Acid wash test	max.	0·2†	1·0† (ASTM D 848)	0·2 (DIN 51 762)
Bromine index	max. g/100 ml	—	—	0·05
Total sulphur	max. p.p.m. by mass	2·0	—	30
Thiophen	max. p.p.m.	—	1·0 (ASTM D 1685)	—
Free sulphur	max. p.p.m. by mass	0·5	—	—
Corrosive sulphur		—	—	
Copper corrosion		—	none (ASTM D 849)	—
Hydrogen sulphide		Free	Free	Free
Mercaptans (Doctor test)		Negative	—	Negative
Neutrality		Neutral	Free of SO_2 (ASTM D 847)	—
Non-aromatics	max. % wt/wt		0·15 (ASTM D 2360)	—

* It has recently been decided to omit the distillation clause from the BS 135 on the grounds that the other clauses define the benzene adequately.

† See Section 5.2. and Table V.3.

Customers may add extra clauses to their specifications to limit impurities to which their particular process is sensitive. Some examples of these additional clauses are given below:

Toluene 50 mg/kg maximum. Larger quantities of toluene in benzene cause difficulties in the purification of certain intermediates produced from benzene, e.g. nitrobenzene, aniline and chlorinated diphenyls.

n-*Heptane*—400 mg/kg maximum. The absence of straight chain paraffins is important in benzene that is to be used for the production of caprolactam by the cyclohexane route. *n*-Heptane for example gives a ketone on oxidation which in turn yields a volatile base that contaminates the final product.

Total aliphatics—1000 mg/kg or 3000 mg/kg maximum depending on the application. The aliphatics may be non-reactive and build up in the recycle, e.g. nitration and alkylation. In the production of maleic anhydride they are oxidised and the evolution of heat may be detrimental to the process.

Lead—0·02 mg/kg maximum. Lead poisons many catalysts. Its occasional inclusion in a specification stems from a fear of contamination in road or rail vehicles which have carried gasoline containing lead alkyls.

Chlorine—2·0 mg/kg maximum. Organically combined chlorine has been reported as occurring in trace quantities in benzene, but contamination is the most likely explanation. The limitation of chlorine content is to avoid damage to catalysts.

5.2. CHEMICAL METHODS OF ANALYSIS

5.2.1. Total Sulphur

The severe limitation of total sulphur content that appears in all specifications for synthesis grade benzene reflects the serious poisoning effect traces of sulphur have on catalysts used for the conversion of benzene to other products.

The usual methods for the determination of total sulphur involve either combustion and determination of the sulphur dioxide formed or reduction and determination of the hydrogen sulphide formed.

When combustion methods are used for materials with only trace sulphur contents, the laboratory atmosphere must be excluded from the combustion chamber and the oxygen and any carrier gases must be freed from all sulphur compounds. Three methods meeting these requirements will be considered in some detail.

Throughout this section sulphur contents are expressed in parts per million by mass (p.p.m. or mg/kg) of sulphur no matter in what combined form it may be present.

5.2.1.1. Wickbold method Since this method was first described in 1957,[1] it has been studied in a number of countries and has appeared as an Institute of Petroleum method in the UK,[2] as a German standard method[3] and is being considered as an ISO standard method by committees, ISO/TC 28 (Petroleum Products) and ISO/TC 47 (Chemistry). The differences between these descriptions of the method are minor.

The general arrangement of the equipment may be seen from Fig. 5.1; details of the burner are given in Fig. 5.2. The burner is constructed in quartz or stainless steel. The combustion tube with its condenser is of quartz. The hydrogen burns in the primary oxygen; the sample is drawn up into the centre tube of the burner and is atomised by the secondary oxygen into the oxyhydrogen flame where the sample is completely burnt. The sulphur dioxide formed is absorbed by the solution in the tower. A blank test must be run without the sample to correct for any sulphur in the oxygen, hydrogen and reagents.

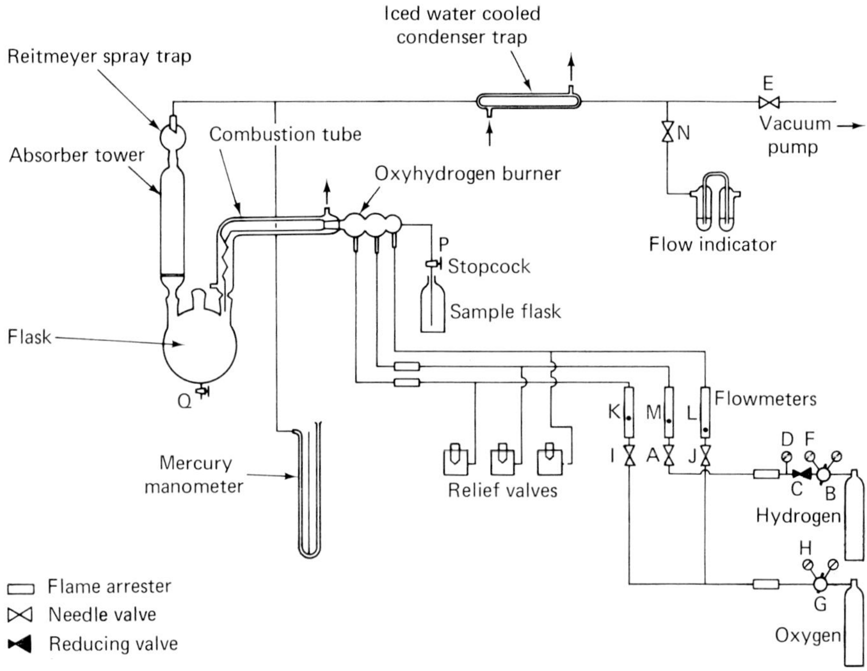

FIG. 5.1 *Wickbold Rapid Combustion Apparatus for Liquid Samples Other than LPG*

The absorbing solution is dilute hydrogen peroxide (1·5 or 3 per cent). The sulphuric acid formed may be determined in a number of ways:

A visual indicator method may be used in which the absorbing solution is titrated with 0·01N barium perchlorate solution using Thorin as the indicator (Thorin is the sodium salt of 2(2-hydroxy-3, 6 disulpho-1-naphthyl azo) benzene arsenic acid). The end point is indicated by a change from yellow to pink. Some operators find the end point is enhanced by the addition of methylene blue when the change becomes greenish-yellow to purple. Lighting conditions are also an important factor in making the end point clear.

The difficulties associated with the visual end point may be eliminated by using a nephelometric titration procedure. The absorbing solution is transferred to the cell of a spectrophotometer and titrated with 0·005N barium chloride solution. The end point is determined graphically from plots of the galvanometer reading against volume of titrant added. A suitable wavelength for these measurements is 415 nm. Full details of these finishing procedures have been published.[2, 3] The repeatability and reproducibility standard deviations at 1 p.p.m. level are 0·35 and 0·7 p.p.m. respectively.

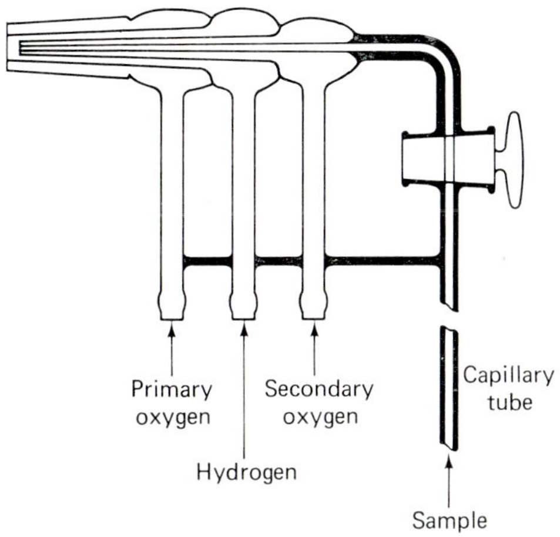

FIG. 5.2 *Wickbold Burner*

5.2.1.2. Martin and Floret method The original work on this method, which is also known as the 'Rhône Poulenc method', was published by Martin and Floret in 1958.[4] The Total Sulphur Working Group of the International Conference of Benzole Producers made an extensive study of the application of the method to benzene. A detailed description of their version of the method together with the results of a co-operative test programme were published in 1964.[5]

The apparatus is constructed in glass as opposed to quartz or stainless steel which means the cost is lower than that for the Wickbold unit. The general layout is given in Fig. 5.3. The sample is fed dropwise from the burette into the inner bulb of the burner where it evaporates into the stream of carbon dioxide. The outer bulb of the burner is heated to ensure complete evaporation of the sample. Oxygen is passed through the annular space between the two bulbs to the burner outlet where it forms an envelope round the carbon dioxide/sample stream and ensures complete combustion. The absorber is in the form of a condenser with flattened bulbs. The gases being scrubbed bubble through the absorption solution which is retained in these bulbs.

Unlike the Wickbold method, the sample is evaporated (not atomised) which means there is a possibility of involatile sulphur remaining in the evaporator bulb. It is not possible to drive off and burn any residual material at the end of the combustion because the flame extinguishes as soon as the last of the sample is evaporated. If part of the carbon dioxide is replaced by hydrogen a flame can be maintained without the sample, but the intense heat produced creates problems in an all-glass apparatus.

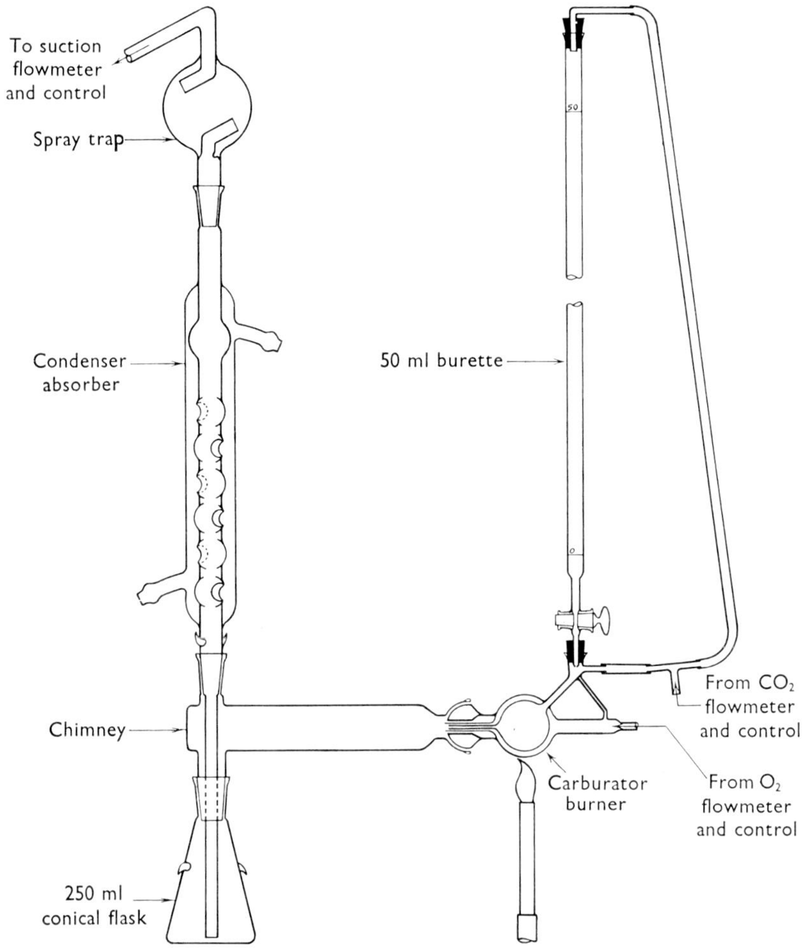

FIG. 5.3 *Rhône-Poulenc Apparatus for the Determination of Sulphur*

It is unusual for benzene to contain involatile sulphur and thus the method can be satisfactory.

As in the Wickbold method, the absorbing solution is dilute hydrogen peroxide. The Thorin indicator or nephelometric methods of determining the sulphate are applicable, but a conductometric procedure is preferred particularly at sulphur levels below 2 p.p.m.

The conductivity cell should have a cell constant of about 0·5 cm^{-1}; two platinum electrodes 10 mm × 10 mm and 10 mm apart are satisfactory. Methanol is added to the absorbing solution prior to titration with 0·01N barium acetate solution. The end point is determined by means of a high sensitivity conductometer.

The co-operative tests on samples containing 1 to 4 p.p.m. total sulphur showed the repeatability standard deviation to be 0·12 p.p.m. and the reproducibility standard deviation to be 0·35 p.p.m.

5.2.1.3. Microcoulometric method This method, which is sometimes called the 'modified Dohrmann Method', is described in detail by Killer and Underhill.[6] The method is applicable to all light hydrocarbons, it is rapid and is particularly valuable when a large number of samples are to be tested.

By means of a microsyringe, 1 to 2 µl of sample is injected through a septum into the end of a specially designed quartz combustion tube (Fig. 5.4). The sample is evaporated into a stream of nitrogen in the first section of the tube which is maintained at 400–500°C and is then carried by the nitrogen to a wider section of the tube where it mixes with a stream of oxygen. The temperature in this zone is 900°C and complete oxidation of the sample is obtained.

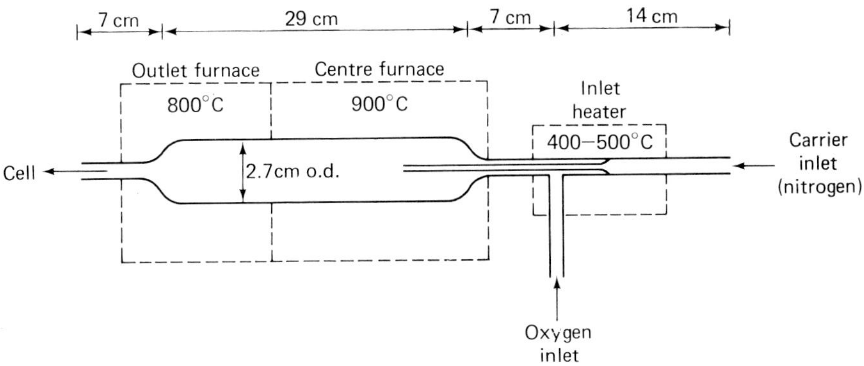

FIG. 5.4 *Modified Combustion Tube for Total Sulphur Determination*

The products of combustion pass to a microcoulometer cell (e.g. a Dohrmann T-300-P iodine cell) containing a dilute acetic acid/potassium iodide electrolyte. The titration of the sulphur dioxide is carried out with electro-chemically generated iodine. The output from the cell passes to a recorder fitted with an integrator. By suitable calibration, the integrator reading can provide a direct measure of the total sulphur content of the sample.

The repeatability standard deviation at the 1 p.p.m. level is 0·5 p.p.m.

5.2.1.4. Granatelli reduction method The publication of this method is 1959[7] aroused considerable interest, because it provided a simple, rapid method that was applicable to samples containing only a few p.p.m. of sulphur. More recently the method has been published as a standard STPTC procedure.[8]

The apparatus is shown in Fig. 5.5. Raney nickel is treated with sodium hydroxide solution at 75–80°C in flask A and the nickel residue is washed with water and then with isopropanol. The apparatus is then assembled and a slow stream of nitrogen is passed through it. Isopropanol and a measured volume of the sample are run into the flask and the contents boiled for 40 minutes. Sulphur compounds in the sample are reduced and retained as nickel sulphide.

Hydrochloric acid is now added slowly from the dropping funnel B. Hydrogen and hydrogen sulphide are evolved and are swept through the condenser D to the absorber E.

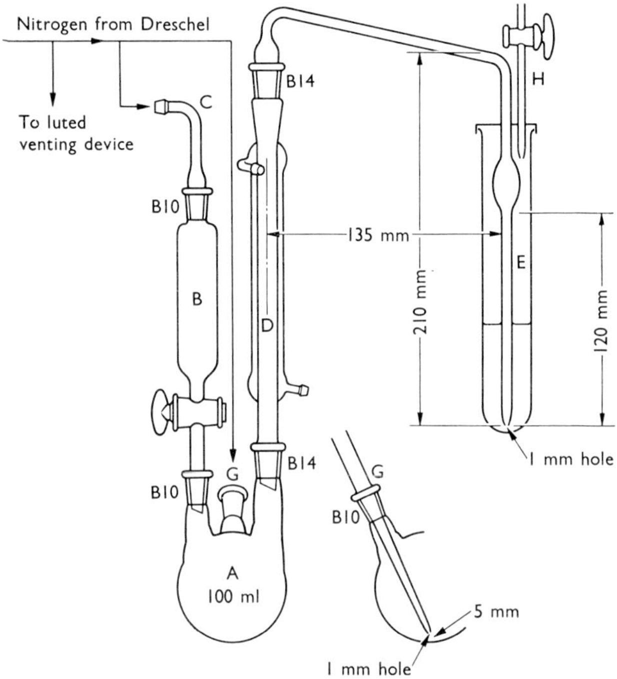

FIG. 5.5 *Reduction Apparatus*

In the STPTC method the absorbing solution is a mixture of equal parts of 1N sodium hydroxide solution and acetone. To this dithizone indicator is added and as the hydrogen sulphide is absorbed, the solution is titrated with mercuric acetate solution. The colour change is from yellow to pink.

The precision of the method has been established by co-operative tests by a STPTC working group. The repeatability standard deviations at 1 and 10 p.p.m. were found to be 0·06 and 0·2 p.p.m. respectively. The corresponding reproducibility standard deviations were 0·1 and 0·6 p.p.m.

Attempts have been made to improve the finishing procedure and to extend the applicability to lower levels than 1 p.p.m. Pitt and Rupprecht describe a colorimetric finish in which the hydrogen sulphide, after liberation from the nickel, is converted to methylene blue. The intensity of the colour produced is measured spectrophotometrically.[9] The procedure is probably more sensitive than the titration procedure, but it is slower and some workers have found it rather erratic.

Fensom *et al.* describe a modification of the procedure which allows the method to be extended downwards to 0·01 p.p.m.[10] The Raney nickel is suspended in a water/isopropanol mixture during the reaction with the sample. The sulphided nickel is separated on a filter before being treated with hydrochloric acid. The liberated hydrogen sulphide is titrated with mercuric acetate using dithizone as an indicator. The titration is carried out in the cell of a spectrophotometer and the optical density at 550 nm is measured after each addition of titrant. The end-point is obtained from a plot of the readings.

The Granatelli method and the later developments of that method have many advantages over combustion methods. It is the most precise method, the only method by which one operator can carry out several (up to six) tests simultaneously and it uses standard laboratory apparatus. Unfortunately, it has some restrictions. It cannot be used with samples containing appreciable quantities of unsaturated hydrocarbons and some oxygenated sulphur compounds are not reduced under the conditions of the test.

Synthesis grade benzene would not normally contain sufficient unsaturated hydrocarbons to interfere and it might be expected to be free from oxygenated sulphur compounds. However, sulpholane is used widely for the extraction of aromatics from petroleum fractions and sulphones can be formed in acid treatment of hydrocarbons. These materials are high-boiling and should not be present in the finished benzene, but the fact that they may be present on the plant means contamination is just possible.

It is for this reason that Raney nickel reduction methods are used for routine purposes, but a combustion method is preferred for referee purposes.

5.2.2. Individual Sulphur Compounds

In addition to methods for the direct determination of total sulphur, procedures are available for the determination of individual sulphur compounds that may be present. For specification purposes these tests are rarely of interest, but they can be valuable for plant control, acceptance testing and in other circumstances where speed is of paramount importance.

5.2.2.1. Thiophen Thiophen is conveniently determined by colorimetric procedures based on the colours formed when that compound reacts with alloxan or isatin in sulphuric acid.[11, 12] The alloxan method is the simplest procedure: the benzene sample (2 ml) is shaken for 30 seconds with a solution of 0·01 per cent wt/vol alloxan in 90·5±0·5 per cent wt/wt sulphuric acid. After settling for 30 minutes, the purple colour produced in the acid layer is compared with the colours produced by standard solutions of thiophen in benzene or with glass colour standards.[13] The range of applicability of the test is 2–10 p.p.m. sulphur as thiophen.

The isatin method may be performed in a similar manner, but the sample is shaken with a mixture of 0·05 per cent wt/vol isatin in chloroform and 0·01 per cent wt/vol ferric sulphate in 92 per cent wt/wt sulphuric acid.[13] The range of applicability can be modified by adjustments of the relative proportions of reagents and sample; and the precision and ease of operation can be improved by using a photoelectric absorptiometer to measure the optical density of the acid layer.[13, 14, 15] An appropriate wavelength is 570 nm and the instrument will require calibration by use of standard solutions of thiophen in benzene. When applied in this way, the isatin method is suitable for thiophen sulphur in the range 0·1 to 15 p.p.m.

In either the alloxan or isatin method, thiophen-free benzene and pure thiophen are required for calibration purposes. Thiophen-free benzene may be prepared from commercially pure benzene by agitating the benzene with three separate 5 per cent portions of 98 per cent wt/wt sulphuric acid, followed by a water wash, a wash with dilute sodium hydroxide solution and a further water wash. The treated benzene is then dried and distilled.

Thiophen of adequate purity is not usually available commercially. Thiophen as received from the laboratory supplier may be purified by cooling it in a bath of methanol and solid carbon dioxide until a thick slurry is obtained. The thiophen crystals are filtered off through a coarse sintered glass filter, melted and recrystallised a further twice. The product is then washed with dilute silver nitrate solution and finally with water. After filtration, the thiophen is suitable for preparing standards in thiophen-free benzene.

Because the sample under test is shaken with sulphuric acid, both the isatin and the alloxan methods become inapplicable if there are present significant quantities of impurities (e.g. certain olefins) that produce colourations with sulphuric acid.

Samples of synthesis benzene normally pass a strict acid wash test clause and do not present any problem in these thiophen methods.

5.2.2.2. Carbon disulphide The most effective methods for the determination of low concentrations of carbon disulphide in benzene are colorimetric procedures based on the interaction of carbon disulphide, diethylamine and a copper salt to form the yellow coloured copper diethylthiocarbamate.[16] Similar methods have been described in which other bases have been used in place of diethylamine.

In the standard STPTC procedure based on this reaction, diethylamine (1 ml), cupric acetate solution (1 ml of a 0·03 per cent solution in alcohol) and a suitable volume of sample are mixed.[17] The yellow colour which rapidly develops is compared with the colours produced by standard concentrations of carbon disulphide. Alternatively, a photoelectric absorptiometer may be used to measure the intensity of the colour and the concentration of carbon disulphide determined from a calibration curve. A suitable wavelength for these measurements is 430 nm.

The method is applicable to concentrations above 0·1 p.p.m. and can be used at concentrations as high as 1000 p.p.m. The repeatability and reproducibility standard deviations are 3 per cent and 10 per cent of the concentration respectively.

5.2.2.3. Free sulphur Free or elementary sulphur can rarely be detected in synthesis benzene, but specifications invariably contain a clause to ensure its absence. Occasionally a quantitative or semi-quantitative test for free sulphur is included, but more often some form of copper or silver strip test or a mercury test is used.

A quantitative test which can be made specific to free sulphur and is applicable to the range 0·5 to 50 p.p.m. depends upon the reaction between sulphur and potassium cyanide to form potassium thiocyanate and the blood-red colour of ferric thiocyanate.[18] The sample is first shaken with potassium chloride/mercuric chloride solution to remove any sulphur compounds that might interfere. Solutions of potassium cyanide and ferric chloride in aqueous acetone are mixed with a known volume of the sample and the absorbance of the resulting solution is measured at 465 nm in a photoelectric absorptiometer. For the lowest concentrations a long cell (4 or 10 cm) is required.

When a sample containing traces of free sulphur is shaken with mercury, a black sulphide is formed which appears as tarnish on the mercury surface or as a suspension in the sample. The absence of these effects is indicative of the absence of free sulphur.

This procedure has been made semi-quantitative.[19] A volume of sample, which may be as large as 100 ml for the lowest concentrations, is shaken with 2 ml of mercury. The sample is decanted as shown in Fig. 5.6 through a filter holder (Fig. 5.7) so that the precipitate is transferred evenly on to the strip of filter paper clamped in the holder. The circular stain that is produced on the filter paper is compared with similar stains produced from standard solutions of free sulphur in benzene.

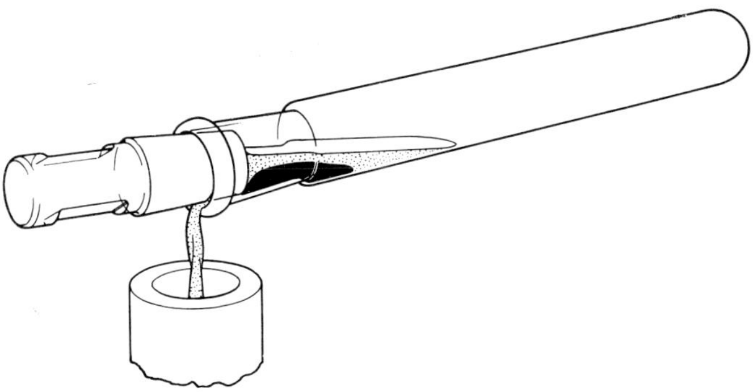

FIG. 5.6 *Corrosive Sulphur (Mercury Test)—Decanting the Sample*

Printed standards have been prepared which make the use of standard solutions unnecessary.[20] The method will determine free sulphur in the range 0·2 to 20 p.p.m.

In the ASTM specification, a copper corrosion clause is included. The method[21] requires 200 ml of sample to be refluxed with a polished strip for 30 minutes. For the sample to be regarded as non-corrosive, the strip shall show no signs of iridescence or discoloration of the surface. Corrosion of the strip by benzene samples can usually be taken to indicate the presence of free sulphur, but other impurities can cause changes in the appearance of the strip.

Silver strip tests are performed in much the same way, but the interpretation in terms of sulphur present is unreliable.

5.2.2.4. Mercaptans, hydrogen sulphide and sulphur dioxide These sulphur compounds rarely occur in commercially pure benzene and qualitative tests are usual to confirm their absence. The doctor test[22] serves for the detection of both mercaptans and hydrogen sulphide. The sample is shaken with sodium plumbite solution. If a black precipitate is formed immediately hydrogen sulphide is present. If there is

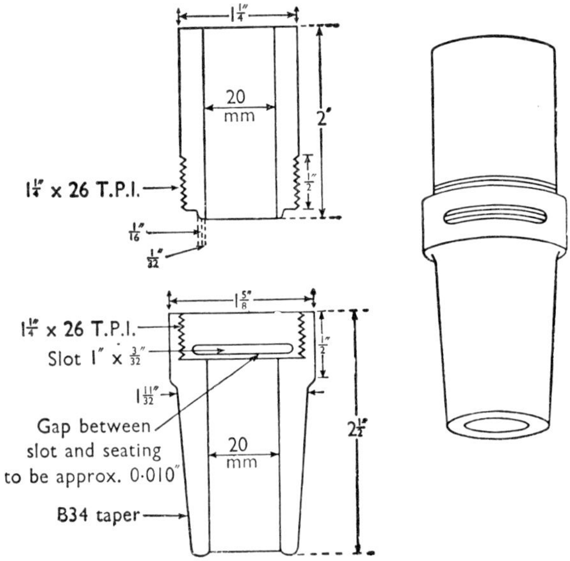

FIG. 5.7 *Corrosive Sulphur (Mercury Test)—Nylon Filter-holder*

no black precipitate, flowers of sulphur are added and the tube reshaken. A brown or black precipitate indicates the presence of mercaptans. The sensitivity of the test is such that it will detect 2 p.p.m. of sulphur as mercaptans and 10 p.p.m. of sulphur as hydrogen sulphide. It follows that the total sulphur clause in specifications for benzene is usually more restrictive than the doctor test and the latter is not essential in specifying the product. If a quantitative determination of traces of mercaptans is required, a number of methods are available. An amperometric method described by Kolthoff and Harris[23] involves a silver nitrate titration and is the basis of the ASTM method D1323.[25] It is applicable to samples containing more than 3 p.p.m. S as mercaptans. A colorimetric titration method is somewhat more sensitive.[24] The sample, in ammoniacal isopropylalcohol, is titrated with silver nitrate solution and dithizone is used as the indicator. The end-point may be visual (pale orange to deep red) or by means of a spectrophotometer (615 nm). The method is applicable to a lower level of 0·2 p.p.m. S as mercaptans. A potentiometric method that is slightly more sensitive is described by Hammerich and Gondermann.[26]

The presence of sulphur dioxide in a sample is usually detected by its acidity. In the ASTM method the sample is considered satisfactory if less than 2 drops of 0·1N sodium hydroxide are required to produce a pink colour in 100 ml of neutral distilled water which has been shaken with 100 ml of sample, the indicator being phenolphthalein.[27]

In a British method, 50 ml of sample is shaken with 10 ml of distilled water to which 0·3 ml of NBA mixed indicator has been added. A change from grey-green to yellow indicates acidity. The indicator is a mixture of 0·04 g bromocresol green dissolved in 10 ml of ethanol and 0·2 g of alizarin red S dissolved in distilled water.

The solutions are mixed and diluted to 100 ml. Either hydrochloric acid or sodium hydroxide (both 0·01N) are added to the indicator solution until 0·3 ml of the treated indicator solution added to 10 ml of distilled water shows the neutral shade. The advantage of this method is the speed with which it can be applied. It will also detect basic materials in the sample—the indicator then changes to blue or purple.[28]

5.2.3. Acid Wash Test

In the acid wash test the sample is shaken with concentrated sulphuric acid and the colour produced in the acid layer is measured. The original purpose of the test was to indicate whether or not the sample had been adequately refined by treatment with sulphuric acid. Although the bulk of benzene now produced is refined by methods other than acid refining, acid wash test clauses remain in specifications.

In the U K and Germany the standard procedure involves shaking together equal volumes of sample and 95 per cent wt/wt sulphuric acid and comparing the colour produced in the acid layer with standard solutions of potassium dichromate in sulphuric acid. The result is described in terms of the number of grams of potassium dichromate per litre of the standard solution, i.e. an acid wash test of 0·2 matches 0·2 g $K_2Cr_2O_7$ per litre.[29, 30]

The ASTM procedure is slightly different in that one volume of sample is shaken with three volumes of 96 per cent wt/wt sulphuric acid and the colour produced is compared with a range of reference colour standard solutions. These solutions contain various proportions of cobalt chloride, ferric chloride and potassium dichromate.[31]

An exact comparison of the British and ASTM acid wash tests is not possible due to the difference in procedure, but the approximate relationship shown in Table V.3 holds for benzene and most similar products.

TABLE V.3

Acid Wash Tests
(Relationship Between ASTM D848 and STPTC RLB 10–67).

ASTM	*STPTC*
1	0·1 –
2	0·1
3	0·1 +
4	0·2 –
5	0·2 +
6	0·4 –
7	0·5 –
8	0·5 +
9	0·6 –
10	0·6

The acid wash test is some measure of unsaturated compounds present in the sample, but it is far from quantitative. Some olefines produce little colour with sulphuric acid under the conditions of the test whereas traces of others produce intense colorations. There is some evidence that traces of nitrogen compounds also produce colours in the acid wash test.

In modern refining plant, a clay treater is often included as the final stage in the benzene refining. The acid wash test of the product rises as the clay treater becomes exhausted and has proved a useful test for monitoring these units.

The bromine index[32] is a better measure of olefin content, but is rarely applied to benzene.

5.2.4. Organically-combined Chlorine

There is little evidence that organically-combined chlorine compounds are normal impurities in commercial benzene, but they may be introduced by contamination. As a few specifications include a clause limiting the chlorine content, a reliable method of test is required. Most combustion methods for the determination of trace quantities of sulphur can be adapted to allow for the determination of similar quantities of combined chlorine. In the combustion all the chlorine compounds are converted to hydrogen chloride. After absorption, the chloride is determined by titration with silver nitrate.

A quicker and easier method is based on the observation by Liggett that all chlorine compounds react with a solution of sodium diphenyl in ethylene glycol dimethyl ether.[33] As the method does not appear to be published it is described in some detail:

Twenty millilitres of sample are pipetted into a 250 ml beaker and warmed on a hot plate until hot to the touch, but not boiling. The beaker is removed from the hot plate and 10 ml of sodium diphenyl reagent (available commercially) is added. The mixture is swirled and warmed on the hot plate for one minute. Two millilitres of methanol are then added to destroy the excess sodium diphenyl. When the effervescence has stopped, 10 ml of 5 vol hydrogen peroxide is added. The beaker is covered with a watch-glass and the mixture boiled for 20 minutes. The beaker is removed from the hot plate and 30 ml xylene, 30 ml methanol, 1 ml 50 per cent sulphuric acid and 50 ml acetone are added.

The mixture is cooled and titrated potentiometrically with 0·0025N silver nitrate solution. During the titration the mixture is stirred by means of a magnetic stirrer. A silver/glass electrode system and a sensitive pH meter has been found satisfactory.

Providing great care is taken over the quality of reagents, and contamination is avoided, the method may be applied to chlorine levels down to 0·1 p.p.m.

5.2.5. Lead

Lead is a very unusual impurity in benzene, but clauses restricting this element (to 0·02 p.p.m. for example) have been noted in some specifications. Lead in benzene may be determined by treatment with a solution of bromine in chloroform followed by extraction with dilute nitric acid. The lead is then converted to lead dithizonate which is estimated colorimetrically either visually or by measuring the absorption at 525 nm.[34, 35]

5.3. PHYSICAL METHODS OF ANALYSIS

The principal physical methods which are used in the analysis of benzene are the determination of density, refractive index, crystallising point and distillation characteristics. A consideration of the effect of impurities on these properties and the precision with which they may be determined leads to the conclusion that the crystallising point is the important index of the quality of benzene. The other properties may prove valuable in identifying the product as benzene or detecting gross deviations from specification.

Analyses based on gas liquid chromatography are important in the determination of specific hydrocarbon impurities.

5.3.1. Crystallising Point

Pure benzene has a crystallising point of +5·533°C.[36] The convenient temperature level makes the measurement of the crystallising point of benzene samples a simple means of assessing their purity. Current specifications are mostly in the range 5·2–5·4°C which means the essential measurement is a crystallising point depression in the range 0·1–0·3°C and the precision of the method must be compatible with these depression values. As benzene is hygroscopic, any attempt to determine the crystallising point of a dry sample will be complicated by its tendency to take up water from the atmosphere. Crystallising points are, therefore, determined on samples saturated with water and a correction (0·09°C) is applied for the depression caused by the dissolved water.

A general description of the procedure for the determination of the crystallising point of benzene is as follows:

The sample under test is saturated with water and a quantity of the wet sample is transferred to the tube of the crystallising apparatus and cooled slowly to a temperature slightly below its crystallising point. Crystallisation is induced; the temperature rises and the maximum is noted. The crystallising point of the dry sample is calculated from this value by appropriate application of the correction for thermometer error and the correction due to the presence of water.

Differences occur in details of procedure published by the various standardising bodies. In the ASTM procedure about 8 ml of sample is used in the test.[37] The sample is placed in a small test tube within a larger one which is in an ice bath. A thermometer and stirrer are in the benzene. The sample is stirred continuously during the cooling and crystallising stages. Duplicate determinations by this procedure should not differ by more than 0·02°C.

The British (STPTC) method utilises 80 ml of saturated sample and the test is performed in the apparatus shown in Fig. 5.8.[38] The sample is stirred with the thermometer during the cooling period and for 5 seconds after crystallisation. It is then kept stationary until the maximum temperature is reached. The precision of this procedure is superior to that of the ASTM procedure—duplicate tests should not differ by more than 0·01°C (95 per cent probability).

The STPTC procedure includes a number of refinements aimed at improving the accuracy of the method. The thermometer is graduated in 0·02°C and is tapped during use to ensure consistent readings. The thermometer is calibrated by determining the crystallising point of a certified benzene sample and the thermometer correction is calculated from this result. In the UK, benzene with a certified crystallising point is obtainable from The National Physical Laboratory, Teddington.

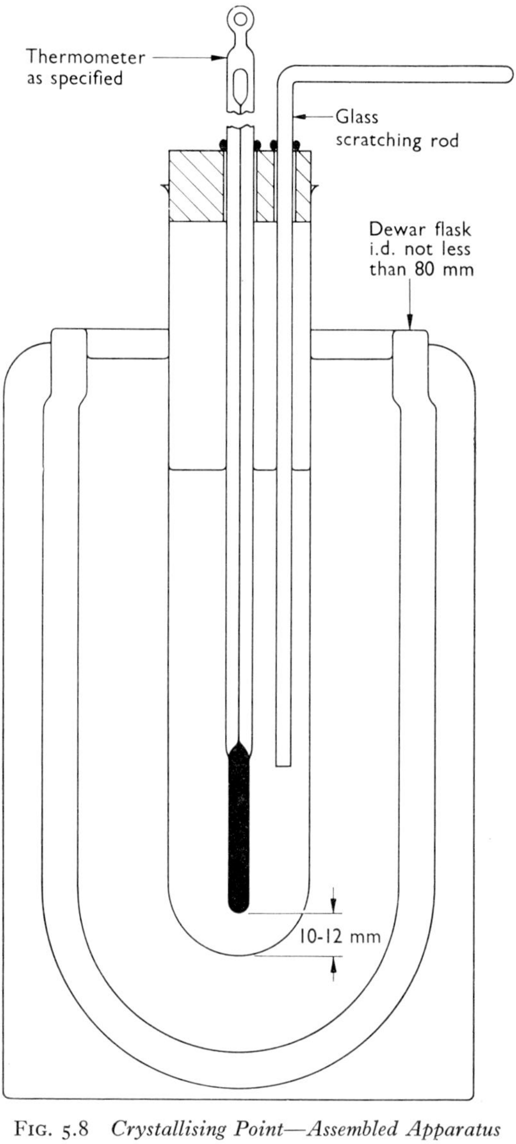

FIG. 5.8 *Crystallising Point—Assembled Apparatus*

The crystallising point of benzene is widely used for control of benzene production units. Where large numbers of tests are required an automatic or semi-automatic procedure for the determination of crystallising point can eliminate subjective errors and save operator time. These methods require, a cooling system, a device for promoting crystallisation and a means of sensing and recording the temperature changes. A unit described by Hickman utilises a thermopile for sensing temperature[39] whereas that of Girling uses a thermistor.[40] The repeatability is somewhat better than the STPTC method.

The effect of various impurities on the crystallising point of benzene may be calculated from the cryoscopic constant for benzene. The depression of the crystallising point of benzene produced by some of the commoner impurities is given in Table V.4. The value for thiophen is an experimental value which is lower than the theoretical value due to the non-ideal nature of the system. All values are applicable to dilute solutions only.

TABLE V.4

Effect of Impurities on Crystallising Point of Benzene

Impurity	*Depression of c.p. per 1% wt/wt impurity* (°C)
Heptanes	0·51
Methylcyclohexane	0·52
Toluene	0·56
Hexanes	0·60
Cyclohexane	0·61
Thiophen	0·39
Carbon disulphide	0·67

5.3.2. The Distillation Test

The role of the distillation test as an indication of the purity of benzene is now of little value because of the high purity of benzene produced today. In spite of this, specifications continue to include a distillation range clause and some consideration will be given to the significance of the various test methods and ranges.

The temperatures at the points chosen for the extremes of the ranges to be used in distillation tests for the assessment of the quality of benzene should be sensitive to high—or low—boiling impurities and they should be precise. The precision of the temperature readings will depend to a considerable extent on the rate of temperature rise at the point chosen. Generally speaking, the wider the range selected, the greater the sensitivity, but the poorer the precision. The best choice must thus be a compromise.

A series of tests following the BS 135 procedure[41] were performed on a sample of benzene alone and to which known quantities of toluene had been added. The temperature ranges for drop to dry, 5–95 ml, 5–97 ml and 1–96 ml, are shown in Table V.5 from which it may be seen that the change in range for a given quantity of impurity is much greater for the drop to dry range than for the others.

TABLE V.5

Effect of Toluene on Distillation Range of Benzene

Range °C	*99·60% wt/wt benzene*	*+0·15% wt/wt toluene*	*+0·31% wt/wt toluene*
Drop-dry	0·50	1·10	2·30
5–95 ml	0·09	0·33	0·54
5–97 ml	0·11	0·71	1·35
1–96 ml	0·30	0·69	0·78

The absolute rate of temperature rise at these points will depend upon the nature of impurities, but rates of rise at the drop and dry points usually prove to be 3–4 times those at the 1 per cent and 96 per cent points respectively. Typical rates are 0·025°C/sec (i.e. 0·3°C/ml) at the 96 per cent point and 0·075°C/sec (i.e. 0·9°C/ml) at the dry point. The quantity of water dissolved in benzene is difficult to control because of the hygroscopic nature of the hydrocarbon. The drop point is very sensitive to changes in water content whereas the 1 per cent point and later points are not affected.

For these reasons the precision of measurements at the drop and dry points is poor.

Table V.6 shows the distillation ranges for eleven commercial samples of benzene, the purity of these samples and the freezing points from which the purities were calculated. As would be expected the correlation between purity and distillation range is poor; the presence of an equi-molar quantity of toluene or cyclohexane would have about the same effect on the freezing point (and purity), but would have profoundly different effects on the distillation range.

TABLE V.6

Relationship Between Purity and Distillation Range (Commercial Benzene Samples)

Freezing point °C	*Purity* mol %	*Drop-dry* °C	*5–95 ml* °C	*5–97 ml* °C	*1–96 ml* °C
4·60	98·56	1·60	0·33	0·72	0·78
4·97	99·13	0·92	0·19	0·28	0·56
4·98	99·15	0·90	0·24	0·33	0·58
5·20	99·49	1·08	0·27	0·48	0·61
5·23	99·54	1·07	0·27	0·48	0·56
5·33	99·69	0·57	0·10	0·16	0·34
5·37	99·67	0·45	0·12	0·24	0·25
5·38	99·76	0·71	0·14	0·16	0·40
5·38	99·76	0·50	0·18	0·29	0·31
5·40	99·80	0·75	0·14	0·17	0·41
5·40	99·80	0·80	0·15	0·30	0·46

In the United States and Germany the drop to dry range with its greater sensitivity, but poorer precision appears to be preferred. In the United Kingdom, France and Belgium the 1–96 per cent range is used. At best either of these ranges can control (rather coarsely) the toluene content of benzene, but neither can be taken as a measure of purity.

5.3.3. Refractive Index and Density

The refractive indices and densities of benzene and some of the more common hydrocarbon impurities in benzene are as follows:[36]

	n_D^{20}	D at 20°C
Benzene	1·50112	0·87901
Toluene	1·49693	0·86694
Cyclohexane	1·42623	0·77855
Methylcyclohexane	1·42305	0·76936
n-Hexane	1·37486	0·65937
n-Heptane	1·38764	0·68376

The wide range of these values makes it apparent that these properties give little information about the purity of a sample, unless it is 100 per cent benzene. Their main value is thus one of general identification.

None of these impurities is likely to be present in commercial pure benzene in quantities in excess of 0·1 per cent wt/wt.

This quantity of *n*-hexane would produce changes in refractive index and density of benzene of 0·0001 and 0·0002 respectively. The corresponding values for cyclohexane are 0·00007 and 0·0001. The values for toluene are less than one-tenth of these.

It follows that, if these impurities are to be detected by refractive index or density, high precision in the measurements is essential. Having obtained the required precision, the results cannot be interpreted until the composition is known.

Measurements of density and refractive index are thus of little value.

A refractometer of the Abbé type can measure refractive index to 0·0001. Immersion instruments are available that facilitate measurement to 0·00002, but careful control of temperature then becomes essential.

The most suitable methods for the determination of density of benzene are by hydrometer[42] or pycnometer.[43, 44] The precision of the methods is not seriously different providing the hydrometer has been certified by a competent body. Repeatability and reproducibility standard deviations are of the order 0·0001 and 0·00015 respectively.

The main use for density determination is for the interconversion of mass and volume and this precision is more than adequate.

5.3.4. Gas Liquid Chromatography

This well-known technique is particularly suitable for the analysis of volatile mixtures and is, therefore, widely used for the determination of hydrocarbon impurities in benzene. In principle, this technique depends upon the injection of the sample into a suitable carrier gas. The sample vaporises into the gas and is carried through a tube filled with a finely divided inert solid (solid support) which supports a non-volatile liquid (stationary phase). The sample thus comes into a series of partition equilibria with respect to the liquid. The components of the sample having low partition coefficients with respect to the liquid are preferentially removed from the liquid phase by the pure gas following the vapour along the tube. As a result intermittent local concentrations of pure components are eluted from the tube in order of their increasing partition coefficients with respect to the liquid. Various detectors are in use to detect and determine the compounds as they leave the column. Highly reliable commercial instruments are available that either record a chromatogram from which the composition of the sample may be calculated or integrate the chromatogram and print out the analysis results.

Standardised methods for the analysis of benzene by this technique may be divided into procedures for the determination of toluene and those for the determination of the non-aromatic constituents of benzene. A method for toluene in benzene that is applicable to toluene concentrations 0·01–0·1 per cent wt/wt uses dilaurylphthalate as stationary phase.[45] Both IP[46] and ASTM[47] methods are available for the determination of non-aromatic hydrocarbons.

5.4. TOXICITY OF BENZENE

5.4.1. Introduction

Benzene can enter the body via the lungs and through the skin. It has both acute and chronic toxic action.

The acute effects resulting from exposure to high concentrations are quickly apparent and profound. They are non-specific and typical of fat soluble, volatile solvents. Inhalation of high concentrations of benzene vapour will produce drowsiness, headache, giddiness, unconsciousness and, if prolonged, death. A concentration of 20,000 p.p.m. is fatal after 5–10 minutes; a concentration of 2,500 p.p.m. causes serious disturbance on exposure for 60 minutes. In a case of acute benzene poisoning immediate removal of the subject from the toxic atmosphere is essential. Artificial respiration, oxygen and respiratory stimulants should then be given.

The chronic effects, which result from prolonged exposure to lower concentrations, are insidious in nature and constitute an even greater hazard. The remainder of this section is, therefore, concerned with this aspect of the toxicity of benzene.

It has long been known that exposure to relatively low concentrations of benzene in the atmosphere can cause chronic poisoning. Early symptoms are varied, vague and not specific to benzene exposure. As the poisoning progresses, there may be damage to the bone marrow that may lead to severe or fatal blood changes. In the last few years the hazard of chronic benzene poisoning has been causing greater concern due to reports of chromosome aberrations associated with exposure to benzene[48] and to the suggestion that benzene might be an initiator of the leukaemia process.[49, 50] Leukaemia would not be expected to develop until there is contact with a second substance known as the promoter. No proposals for a promoter for this process are given, but latent periods of fifteen and twelve years after cessation of exposure to benzene are reported.

The International Labour Office arranged a Meeting of Experts on the Safe Use of Benzene and Solvents Containing Benzene in Geneva in May 1967. Delegates from twelve countries were present together with representatives from the International Labour Office, World Health Organisation, Council of Europe and European Economic Community. It was agreed that benzene poisoning presented a major problem of occupational hygiene. Aspects discussed included determination of acceptable levels of exposure, safety measures, medical control and health education. The meeting favoured the replacement of benzene by less toxic products whenever possible, the strict limitation of the amount of benzene in solvents and, in particular, those that may contain it as an impurity, and a reduction of the tolerable limit in the working atmosphere. It was recognised that the last point had little practical significance until adequate analytical methods were developed.

The International Labour Office has published 'Benzene: Uses, Toxic Effects, Substitutes'. This volume includes the Report of the 1967 Meeting of experts together with papers on many aspects of the hazards associated with benzene. It includes an extensive survey of Laws and Regulations concerning the use of benzene in industry.[51]

Under Section 63(1) of the 1961 United Kingdom Factories Act all practicable measures are to be taken to protect employees against the inhalation of any injurious fume. The Ministry of Labour recommend a maximum concentration of benzene in the atmosphere of 25 p.p.m. vol/vol (i.e. 80 mg/m^3).[52] Any new health recommendations that are published will undoubtedly take note of recent views on the toxicity of benzene and new limitations to the conditions under which benzene may be used safely must be anticipated.

If the proposed New Safety, Health and Welfare Legislation (UK) became law and included some of the items in the 'First Consultative Document',[53] benzene would be within the broad definition of a 'Toxic Material' as it is a substance, which, if breathed, can present a long term risk to health. The provisions of the Act would require that, except where the process is enclosed, operations on 'toxic materials' must be performed under an efficient exhaust draught system. The Factory Inspector would be empowered to require periodic atmospheric testing on sites where a 'toxic material' was being handled. It should be noted that the scope of the proposed legislation would include any premises, whether in the open air or not, in which people were employed.

Although improvements in working conditions that reduce risk of poisoning are most desirable, the following points help to keep the hazard associated with exposure to benzene in perspective.

(1) Elkins states that a total of 140 fatal cases of benzene poisoning have been recorded in the literature up to 1959. He also notes that fatal poisonings are not common with concentrations below 200 p.p.m.[54]

(2) In the United Kingdom chronic benzene poisoning is compulsorily notifiable to the Factory Department and Browning reports that, in eighteen years to 1965, eleven such cases were notified.[55] Between 1965 and 1972 a further fourteen cases were notified and three of these were fatal.[56]

(3) Cases of poisoning by benzene are usually associated with its use in solvents for rubber and paints [56, 57, 58] In the UK there have been few cases associated with

production, refining, transport or utilisation of benzene in bulk for chemical synthesis. The good record in this field was spoiled in 1969 when six non-fatal cases were notified. These men were all working on maintenance of the same solvent plant at a refinery. The periodic haematological examination undertaken by the employer showed abnormalities that required the cases to be notified.[59]

5.4.2. Benzene in Air

The maximum allowable concentration (MAC) or threshold limit value (TLV) for an air-borne substance refers to the concentration of that substance in air to which workers may be exposed day after day without adverse effect. Until 1939 a value of 100 p.p.m. was accepted for benzene by the American Standards Association (ASA) although some authorities as long ago as 1926 stated that there was a substantial hazard below this concentration.[60]

In 1938 and 1946 studies were made in leather-cloth factories where workers were exposed to concentrations of benzene between 40 and 70 p.p.m.[61] When one worker died of aplastic anaemia typical of benzene poisoning, the remaining fifty were examined and thirteen were found to have blood abnormalities. In 1951 ASA dropped the limit to 35 p.p.m. and 1957 to 25 p.p.m. British recommendations have followed the American pattern and may be expected to continue to do so. The recommended TLV for benzene for 1970 published by the American Conference of Governmental Industrial Hygienists is a ceiling value of 25 p.p.m., but it is possible that a time-weighted average (TWA) value of, perhaps, 10 p.p.m. may be introduced at some future date. A TWA value of 10 p.p.m. means that excursions up to 20 p.p.m. are permissable for periods not exceeding 15 minutes providing each excursion above 10 p.p.m. is compensated by a comparable excursion below the limit. There are considerable practical difficulties in monitoring TWA values without installation of continuous recording equipment.

From a critical study of the literature Truhaut concluded that there was some evidence that concentrations of 25 p.p.m. could be injurious to animal or human subjects. A safety factor was necessary and he, therefore, proposed a TLV of 5 p.p.m.[62] At the Geneva meeting of experts some participants favoured 20 p.p.m. for the ceiling value and 5 p.p.m. for the average concentration.[51]

The MAC values given in Table V.7 are for a number of countries some of which endeavour to control exposure to benzene by recommendations or regulations limiting the benzene concentration in working atmospheres.[63] Most European countries, however, do not use this criterion but rely on labelling solvents and preparations to indicate their benzene content. They then restrict the purposes for which the various classes of material may be used.

The EEC countries are considering the following labelling system:

Label	*Classification*	*Benzene content*
Skull	Poisonous	Above 1 per cent
St Andrew's Cross	Harmful	0·25 to 1 per cent
No label	—	Less than 0·25 per cent

A St Andrew's Cross may also be required for materials containing toluene and/or xylene. The limiting concentrations of these hydrocarbons have not yet been fixed.

TABLE V.7

MAC Values for Benzene in Air

Country	*p.p.m. vol/vol*	*mg/m³*
Bolivia	100	320
Bulgaria	17	50
Chile	35	110
Czechoslovakia	17	50
Finland	25	80
France	31	100
Germany (Federal Republic of)	25	80
Hungary	6	20
Japan	25	80
Malagasy Republic	31	100
Morocco	31	100
Poland	31	100
Rumania	17	50
Spain		
Men	70	220
Women	35	110
USSR	6	20
United States*		
Florida	25	80
Georgia	25	80
Maine	35	110
Uruguay	310	1000

* Under the Walsh-Healey Public Contract Act, applicable to those who supply the federal Government, the figure for benzene is 25 p.p.m. (80 mg/m^3). This limit is also recognised by most states of the United States.

In the United Kingdom, the British Rubber and Resin Adhesive Manufacturers' Association (BRRAMA) specify the labelling of hazardous solvents. These include benzene, petroleum, naphtha containing more than 2 per cent benzene and other solvents containing more than 0·5 per cent benzene. A green label indicates that the adhesive contains no hazardous solvent. A purple label is used when a hazardous solvent is present. The label bears the word 'Warning', the name of the hazardous solvent and the proportion in the solution. Suggestions are made on the type of ventilation that should be used with green and purple labelled materials. The Paintmakers Association has recommended to its members that no products should contain more than 1 per cent of benzene. A safe interpretation of this recommendation is that toluene used for paintmaking should not contain more than 1 per cent of benzene. A standard method of determining benzene in toluene has been published by STPTC.[64]

In the United Kingdom in 1971 a Statutory Instrument 'The Benzene Solvents (Limitation of Use) Regulations' is in draft form. These regulations will prohibit the use as a solvent of any liquid containing more than 1 per cent by volume of benzene.

5.4.3. Metabolism of Benzene

From experiments on volunteers, Teisinger reported that about 46 per cent of benzene inhaled was retained in the body and rapidly absorbed into the blood. Twelve per cent of that absorbed was subsequently eliminated via the lungs. In the urine about 0·1 per cent of that absorbed was eliminated as benzene, 29 per cent as phenol, 3 per cent as pyrocatechol and 1 per cent as hydroquinone.[65] From animal experiments it appears that about 80 per cent of the benzene absorbed is eliminated in two or three days. The remainder is eliminated slowly.

Experiments described by Sherwood and Carter show the rate of excretion of phenol in urine from a human subject exposed to 25 p.p.m. of benzene for $4\frac{1}{2}$ hours. The half-life of the phenol in the urine appears to be in the range 4–8 hours.[66]

The hypothesis that the chronic toxic action of benzene is at least in part due to the phenolic compounds produced from it is receiving considerable support. There are reports from autopsies of significant amounts of benzene in the bone marrow.[67, 68] If it is oxidised there to phenols, the products could accumulate to yield a concentration sufficient to produce cytotoxic effects. This mechanism would account for the differences in the toxic effects of benzene and phenol, the latter being converted in the blood to water-soluble alkali salts that are rapidly eliminated in the urine. It is also possible that unchanged benzene sensitises the bone marrow to the antimitotic action of the phenols.[62] Differences in the ability of individuals to eliminate the phenols or conjugate them into less toxic excretory compounds could account for the variation in their sensitivity to benzene.[55]

5.4.4. Assessment of Exposure—Atmosphere Tests

To monitor benzene concentrations in air usefully it is essential to have analytical methods capable of operating accurately and repeatably in the range of 1–20 p.p.m. The current method published by HM Factory Inspectorate[69] is inapplicable to concentrations below 12 p.p.m. Commercially available detection tubes such as those marketed by Dräger Normalair Ltd., are suitable for concentrations above 15 p.p.m. Various organisations are working on the development of methods applicable to lower concentrations but few reliable methods appear to have been published.

A serious disadvantage of normal benzene-in-air tests, even if low levels can be accurately measured, is that the result gives a value for benzene in the atmosphere over a short time interval, but little indication of the benzene actually inhaled by a worker. It is well known that the concentration of impurities in the air varies with small changes in location. One measure to overcome this deficiency in air testing is the use of a personal sampler procedure described by Sherwood and Carter.[66] The worker carries in a pocket or on a strap a unit that samples the air near the nose level throughout the working period. A battery-operated pump draws air at a known rate through a silica gel tube and the benzene adsorbed is subsequently determined by gas chromatographic analysis. An intrinsically safe version of the sampler is available.

5.4.5. Assessment of Exposure—Breath Tests

Benzene can be detected in exhaled breath for quite considerable periods after exposure to benzene has ceased. Hunter has described procedures by which as little as 0·1 p.p.m. of benzene can be detected.[70, 71] The technique involves the direct analysis of the breath on a gas chromatograph. Sherwood and Carter have increased the usefulness of the technique by perfecting a method of breath collection in tubes that can be returned to the laboratory for analysis.[66] In the $4\frac{1}{2}$-hour exposure tests already mentioned, they were able to detect benzene in the exhaled breath more than 50 hours after exposure had occurred. They favour this method for monitoring benzene intake.

5.4.6. Assessment of Exposure—Urine Tests

Examination of urine for excreted metabolites of the benzene to which the individual may have been exposed provides a measure of the benzene absorbed by that individual, gives a value averaged over several hours and provides a means of detecting exposure before any damage to bone marrow occurs.

For many years the urine sulphate ratio test has been used for this purpose.[54] The interpretation of the test is based on the premise that benzene is partly metabolised to organic derivatives conjugated with sulphate radicals. The inorganic sulphate is determined on the urine as received and the total sulphate is determined after hydrolysis of the urine with hydrochloric acid. A rapid gravimetric method for this procedure has been described by Gerarde.[72] The interpretation is as follows:

Ratio inorganic/total sulphate	*Exposure*
Above 80 per cent	Negligible
70–80 per cent	Mild
30–70 per cent	Dangerous
Below 30 per cent	Extremely hazardous

According to Hunter,[70] this method has the disadvantage that it neglects what may be the most important factor in the toxicity of benzene, namely free or unconjugated phenol. A considerable number of methods for determining phenol in urine have been published. These have been reviewed by Docter and Zielhuis.[73] These and other workers[74, 75] show correlations between phenol in the urine and benzene in the air inhaled. Docter and Zielhuis drew attention to the differences in the quantitative relationships appearing in the literature. However, they conclude that for groups of workers a quantitative relationship does exist, but for individuals further work is required under clearly described conditions.

With the present state of knowledge, there are problems in the interpretation of phenol-in-urine results. If, however, the total phenols found in a 24-hour urine sample exceeds 100 mg, then the subjects' working conditions should be investigated.[51] Benzene in air tests by means of a personal sampler technique would be an appropriate method.[66]

5.4.7. Assessment of Exposure—Blood Tests

Browning has summarised the value of regular blood tests as follows:

'The outstanding characteristic of chronic benzene poisoning is disturbance of the blood picture, which is often in its initial stages entirely dissociated from any complaint of ill-health which might arouse suspicion. It is for this reason that routine blood examinations are especially valuable in detecting signs of absorption which, though not necessarily indicative of true poisoning, might be followed by severe injury to the bone marrow if exposure is continued.

'Air estimations, urinary sulphate ratio and phenol tests are all valuable in assessing and controlling the amount of exposure, and clinical examinations may reveal mild symptoms, mainly subjective, but the one criterion which should lead to removal from exposure is a deterioration in the blood picture from that shown by a pre-employment examination'.[55]

She stresses that the criterion of abnormality should be based on *changes* observed during periodical blood examinations and tabulates the deviations from normal that suggest chronic benzene poisoning.

Blood tests should include red and white cell counts and a differential count of leucocytes.

5·5. STORAGE AND TRANSPORT

5·5·1. Introduction

In the storage and transport of benzene the characteristics which require special attention are the hydrocarbon's flammability, its toxicity and its high crystallising point. These three aspects are considered briefly in this section. For details it is essential that the specific references and regulations applicable to the country and area are consulted.

5·5·2. Flammability

Petroleum-spirit' as defined in Section 23 of the Petroleum (Consolidation) Act, 1928[76] includes hydrocarbon oils having a flashpoint below 73°F (22·8°C) and benzene, therefore, comes within the provisions of this Act. In the United Kingdom a licence is required for the storage of 'petroleum spirit' and hence for benzene. These licences are issued by local authorities and are subject to annual renewal. The practical and legal aspects of storage of petroleum spirit have been considered in detail by Hughes.[77] The Institute of Petroleum publish a 'Model Code of Safe Practice in the Petroleum Industry'[78] and much of the material included is applicable to the storage of benzene.

The volatility of benzene and the flammable atmosphere which may be produced requires that all electrical equipment conform to the regulations appropriate to the area (see ref. (78) Part 1).

It is well established that petroleum and benzene vapours may be ignited by a discharge of static electricity which can accumulate on equipment and in the hydrocarbon whilst it passes through pipes. Detailed information on this subject is given in a monograph by Klinkenberg and van Der Minne[79] and by Hughes.[77] The safety precautions in any storage installation must include measures which will disperse static electrical charges as quickly as possible.

The conveyance of benzene by road in United Kingdom is governed by the Petroleum-Spirit (Conveyance by Road) Regulations 1957 and an amendment dated

1966.[80] The regulations restrict the volume of petroleum spirit carried to 6,600 gallons and the vehicle must be of approved design.

There are no statutory regulations relating to the conveyance of petroleum spirit by rail in the United Kingdom, but conditions are specified by the clearing house of British Rail.

5.5.3. Toxicity

The toxic nature of benzene both as an acute and as a chronic poison has been discussed in Section 4 of this chapter. In all stages of storage and transport attention must be given to the toxic properties. Personnel must avoid breathing benzene vapour even at low concentrations and there must be no contact of liquid benzene with the skin. Benzene poisoning is a serious industrial disease and, in the United Kingdom is notifiable under section 82 of the Factories Act 1961 and is a compensatable disease under the National Insurance (Industrial Injuries) Act, 1946.

The maximum allowable concentration of benzene in a working atmosphere is 25 p.p.m. vol/vol (i.e. 80 mg/M^3). As this concentration is below the level that most people can detect by odour, it follows that, if benzene can be smelt, a hazardous concentration is present. If no odour of benzene can be detected, it does not necessarily follow that the atmosphere is safe.

Whenever possible, operations involving benzene should be carried out in totally enclosed vessels which are vented to the atmosphere away from the working area. Personnel engaged in operations involving benzene (e.g. loading and unloading of road and rail vehicles) should always be up-wind of the source of benzene vapour.

When benzene is open to the atmosphere within an enclosure or building (e.g. in drum filling), it should be in an area fitted with a fume hood and efficient ventilation.

If exposure to benzene vapour is unavoidable, a suitable respirator must be worn. Canister-type respirators (gas masks) will only give protection against low concentrations of benzene, and this protection is only for a short time period. For example the appropriate canister conforming to BS 2091 will give protection against a maximum concentration of 1 per cent by volume of benzene for 30 minutes.[81] Fresh-air pressure tube or self-contained types of breathing apparatus are generally preferable to the respirator.

Benzene must not be allowed to come into contact with the skin or clothes. PVC-coated gloves provide reasonable protection for the hands, but some benzene may be absorbed and evaporate from the gloves after use. It is, therefore, important that they should be stored in a ventilated location away from personnel. The all-too-common practice of carrying the gloves in the cab of the road vehicle is to be deplored.

Any clothing or boots wet with benzene must be changed and allowed to dry in a safe area. Great care must be taken in the disposal of any rags wet with benzene. Both flammability and toxicity hazards must be considered.

The operations in which exposure to benzene has been found to occur include:

(a) Uncoupling hoses which have not been properly drained.

(b) Manual dipping of road vehicles.

(c) Sampling tanks.

Leaking glands on meters and pumps are also a major source of danger. The highest possible standards of maintenance are essential on all equipment handling benzene.

5.5.4. High Crystallising Point

Since commercially pure benzene freezes at about 5·35°C, provision must be made for heating the equipment handling this product in climates where lower temperatures may occur. Lines and valves must be provided with steam, hot water or electrical tracing. Storage tanks, road and rail tanks must be fitted with heating coils.

Leakage of benzene in cold weather can prove a particularly serious hazard. Patches of frozen benzene or benzene icicles may not be recognised as benzene and can, therefore, be particularly dangerous from both toxic and flammability points of view.

5.5.5. Labelling

When benzene is conveyed by road, the Inflammable Substances (Conveyance by Road) (Labelling) Regulations 1968 apply[82] and the vehicle must display the prescribed notice at the front and rear. The notice is a red square 20 × 20 cm bearing a flame symbol and the words 'INFLAMMABLE LIQUID' in black.

Vehicles should also carry a 'Transport Emergency Card' such as that issued by the European Council of Chemical Manufacturers Federation. The purpose of this card is to provide the essential information for dealing with an accident. A reproduction of the Benzene card is provided in Fig. 5.9. A similar card is produced in the USA by the Manufacturing Chemists' Association.

Drums containing benzene should bear a warning label which in the United Kingdom conforms to 'Marking Containers of Hazardous Chemicals'.[83] The Benzene label is reproduced in Fig. 5.10. In countries of the European Common Market, benzene containers must be marked with both flame and skull symbols to indicate flammability and toxicity respectively.[84]

TRANSPORT EMERGENCY CARD (Road)

CEFIC TEC(R)-7
May 1971 Rev. 1
Class IIIa ADR
Marg. 2301, 1°a

Cargo **BENZENE**

Colourless liquid with perceptible odour
Immiscible with water
Lighter than water

Nature of Hazard

Highly inflammable (flashpoint below 21°C)
Volatile
The vapour is invisible, heavier than air and spreads along ground
Can form explosive mixture with air particularly in empty uncleaned receptacles
Heating will cause pressure rise with risk of bursting and subsequent explosion
The substance poisons by absorption through skin and inhalation
The vapour has effects resembling alcoholic intoxication

Protective Devices

Suitable respiratory protective device
Goggles giving complete protection to eyes
Plastic or synthetic rubber gloves
Eyewash bottle with clean water

EMERGENCY ACTION — Notify police and fire brigade immediately

- Stop the engine
- No naked lights. No smoking
- Mark roads and warn other road users
- Keep public away from danger area
- Use explosionproof electrical equipment
- Keep upwind

Spillage

- Shut off leaks if without risk
- Contain leaking liquid with sand or earth
- Prevent liquid entering sewers, basements and workpits, vapour may create explosive atmosphere
- Warn inhabitants—explosion hazard
- If substance has entered a water course or sewer or contaminated soil or vegetation, advise police

Fire

- Keep containers cool by spraying with water if exposed to fire
- Extinguish with dry chemical, foam, halones or waterspray
- Do not use water jet

First aid

- If the substance has got into the eyes, immediately wash out with plenty of water for several minutes
- Remove contaminated clothing immediately and wash affected skin with soap and water
- Seek medical treatment when anyone has symptoms apparently due to inhalation

Additional information provided by manufacturer or sender

TELEPHONE

Prepared by CEFIC (CONSEIL EUROPEEN DES FEDERATIONS DE L'INDUSTRIE CHIMIQUE, EUROPEAN COUNCIL OF CHEMICAL MANUFACTURERS' FEDERATIONS) Zurich, from the best knowledge available; no responsibility is accepted that the information is sufficient or correct in all cases

Obtainable from NORPRINT LIMITED, BOSTON, LINCOLNSHIRE

Acknowledgment is made to V.N.C.I. and E.V.O. of the Netherlands for their help in the preparation of this card

Applies only during road transport English

FIG. 5.9

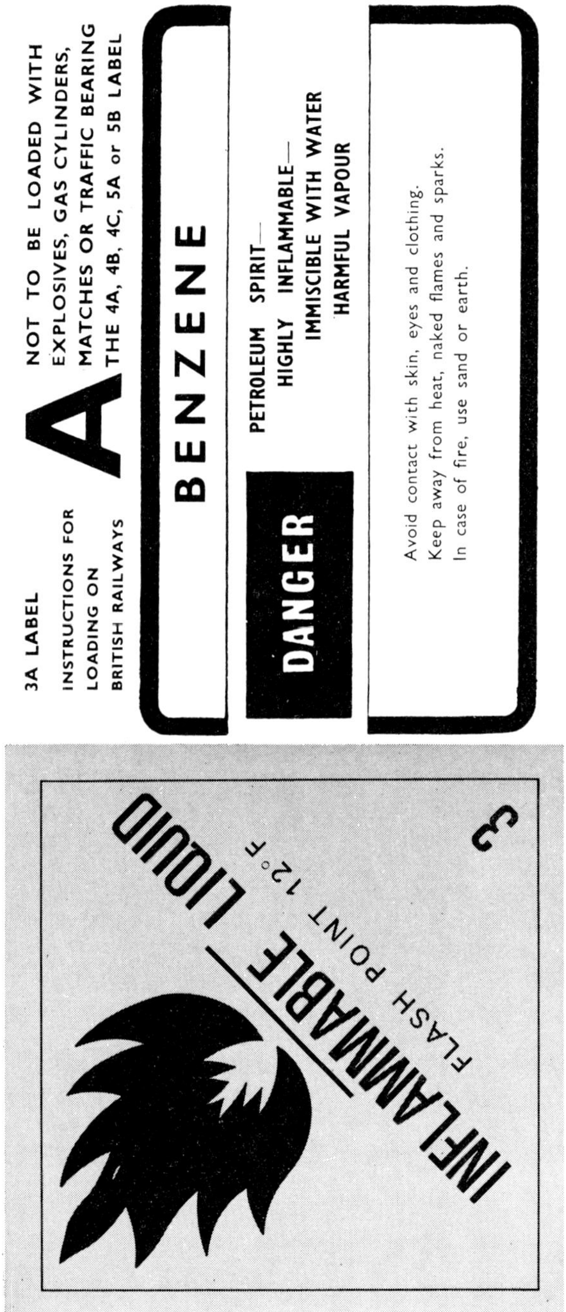
3A LABEL
INSTRUCTIONS FOR LOADING ON BRITISH RAILWAYS
A
NOT TO BE LOADED WITH EXPLOSIVES, GAS CYLINDERS, MATCHES OR TRAFFIC BEARING THE 4A, 4B, 4C, 5A or 5B LABEL
BENZENE
DANGER
PETROLEUM SPIRIT—
HIGHLY INFLAMMABLE—
IMMISCIBLE WITH WATER
HARMFUL VAPOUR
Avoid contact with skin, eyes and clothing.
Keep away from heat, naked flames and sparks.
In case of fire, use sand or earth.
INFLAMMABLE LIQUID
FLASH POINT 12° F
3

Fig. 5.10

REFERENCES

[1] Wickbold, R., *Angew. Chem.*, 1957, **69**, 530.
[2] '*IP Standards for Petroleum and Its Products*', Method IP 243/70T (Institute of Petroleum, London, 1971).
[3] Deutsche Normen, DIN 51 409.
[4] Martin, F. and Floret, A., *Chim. Anal.*, 1958, **40**, 120.
[5] French, K. H. V., *Anal. Chem.*, 1964, **36**, 339.
[6] Killer, F. C. A. and Underhill, K. E., *Analyst*, 1970, **95**, 505.
[7] Granatelli, L., *Anal. Chem.*, 1959, **31**, 434.
[8] '*Standard Methods for Testing Tar and Its Products*', 6th Edition, Method No. RLB 30–67, STPTC, Gomersal, Cleckheaton, Yorkshire, 1967.
[9] Pitt, E. H. H. and Rupprecht, W. E., *Fuel*, 1964, **43**, 417.
[10] Fensom, A. *et al.*, *Analyst*, 1971, **96**, 186.
[11] French, K. H. V., *J. Soc. Chem. Ind.*, 1946, **65**, 15.
[12] French, K. H. V., *J. App. Chem.*, 1952, **2**, 664.
[13] STPTC Method No. RLB 28–67, *loc. cit.*
[14] ASTM Method D 1685, *ASTM Standards*, Part 20, American Society for Testing and Materials, Philadelphia, USA.
[15] Barnett, H. A. *et al.*, *Anal. Chem.*, 1960, **32**, 842.
[16] Dick, T. A., *J. Soc. Chem. Ind.*, 1947, **66**, 253.
[17] STPTC, Method No. RLB 17–67, *loc. cit.*
[18] STPTC, Method No. RLB 29–67, based on Bartlett, J. K. and Skoog, D. A., *Anal. Chem.*, 1954, **26**, 1008, 1955, **27**, 369, *loc. cit.*
[19] STPTC, Methods No. RLB 25–67 and 29–67, *loc. cit.*
[20] Benzole Producers Ltd., 132–5 Sloane Street, London, SW1X 9BB.
[21] ASTM Method D 849, *ASTM Standards*, Part 20.
[22] IP Method, IP 30/56, *loc. cit.*
[23] Kolthoff, I. M. and Harris, W. E., *Ind. Eng. Chem., Anal. Ed.*, 1946, **18**, 161.
[24] Kunkel, R. K. *et al.*, *Anal. Chem.*, 1959, **31**, 1098.
[25] ASTM Method D 1323, *ASTM Standards*, Part 17, *loc. cit.*
[26] Hammerich, T. and Gondermann, H., *Erdöl u. Kohle*, 1964, **17**, 20.
[27] ASTM Method, D 847, *ASTM Standards*, Part 20, *loc. cit.*
[28] STPTC, Method RLB 5–67, *loc. cit.*
[29] STPTC, Method RLB 10–67, *loc. cit.*
[30] Deutsche Normen, DIN 51 762.
[31] ASTM Method D 848, *ASTM Standards*, Part 20, *loc. cit.*
[32] ASTM Method D 1491, *ASTM Standards*, Part 20, *loc. cit.*
[33] Liggett, L. M., *Anal. Chem.*, 1954, **26**, 748.
[34] Griffing, M. E. and Rozek, A., *Anal. Chem.*, 1957, **29**, 191.
[35] Analytical Methods Committee, *Analyst*, 1959, **84**, 127.
[36] Rossini, F. D. *et al.*, '*Hydrocarbons from Petroleum*' (Reinhold Publishing Corp., New York, 1953); Glasgow, A. R. *et al.*, *J. Res. Nat. Bur. Standards*, 1946, **37**, 141.
[37] ASTM Method D 852, *ASTM Standards*, Part 20 *loc. cit.*
[38] STPTC Method RLB 24–67; BS 135, *loc. cit.*
[39] Hickman, J. B., *J. Chem. Ed.*, 1948, **25**, 163.
[40] Girling, G. W. *et al.*, *Benzole Digest*, 1969, **3**, 17.
[41] *Specifications for Benzene*, BS 135: 1963 British Standards Institution, London.
[42] ASTM Method, D 1298, *ASTM Standards*, Part 18; *IP Standards for Petroleum and Its Products*, Method 160/68.
[43] STPTC Method GP1, *loc. cit.*
[44] ASTM Method, D 941, *ASTM Standards*, Part 17; IP Method 189/69, *loc. cit.*
[45] STPTC Method RLB 31–67, *loc. cit.*
[46] IP Method 282/72T, *loc. cit.*
[47] ASTM Method, D 2360, *ASTM Standards*, Part 20, *loc. cit.*
[48] Tough, I. M. and Court Brown, W. M., *Lancet*, 1965, **1**, 684.

[49] Devowin, R. L., *J. Am. Med. Assoc.*, 1963, **185**, 748.
[50] Vigliani, E. C. and Saita, G., *New. Engl. J. Med.*, 1964, **271**, 872.
[51] 'Benzene: Uses, Toxic Effects, Substitutes', *Occupational Safety and Health Series No.* **12**, International Labour Office, Geneva, 1968.
[52] *Dust and Fumes in Factory Atmosphere, Safety, Health and Welfare Booklet, New Series No. 8*, Ministry of Labour, 1966.
[53] *New Safety, Health and Welfare Legislation*—First Consultative Document, Ministry of Labour, December 1967.
[54] Elkins, H. B., '*The Chemistry of Industrial Toxicology* (John Wiley, New York, 1959).
[55] Browning, E., '*Toxicity and Metabolism of Industrial Solvents* (Elsevier, London, 1965), p. 26.
[56] *Annual Reports of HM Chief Inspector of Factories*, HM Stationery Office, London.
[57] Vigliani, E. C. and Saita, G., *New England J. of Med.*, 1964, **271**, 872.
[58] Hernberg, S. *et al.*, *Brit. J. industr. Med.*, 1966, **23**, 204.
[59] *Annual Report of HM Chief Inspector of Factories*, HM Stationery Office, London, 1969, 61.
[60] National Safety Council, *Final Report of the Committee of Benzol*, National Bureau of Casualty and Surety Underwriters, 1926.
[61] Hardyland, H. L. and Elkins, H. B., *J. Ind. Hyg. Toxicol.*, 1948, **30**, 196.
[62] Truhaut, R., *Arch. Maladies Prof. Med. Travail et Securité Sociale*, 1968, **29**(*1–2*), 5.
[63] *Protection against Hazards arising from Benzene*, Reports VI(1) and (2), Agenda for the International Labour Conference, Geneva, 1971, International Labour Office, Geneva, 1970
[64] STPTC Method RLB 32–67 *loc. cit.*
[65] Teisinger, J. *et al.*, *Pracovni Lekarstvi*, 1952, **4**, 175.
[66] Sherwood, R. J. and Carter, F. W. G., *Ann. Occup. Hyg.*, 1970, **13**, 125.
[67] Duvoir, M. *et al.*, *Arch. Maladies Prof. Med. Travail et Securité Sociale*, 1946, **7**, 230.
[68] Perrauld, M. *et al.*, *ibid*, 1944, **5**, 239.
[69] *Methods for the Detection of Toxic Substances in Air*, Booklet No. 4, HM Factory Inspectorate, 1972.
[70] Hunter, C. G., *Ann. Occup. Hyg.*, 1966, **9**, 191.
[71] Hunter, C. G., *Proc. R. Soc. Med.*, 1968, **61**, 913.
[72] Gerarde, H. W., *Am. Ind. Hyg. Assoc. J.*, 1960, **21**, 511.
[73] Docter, H. J. and Zielhuis, R. L., *Ann. Occup. Hyg.*, 1967, **10**, 317.
[74] Van Haaften, A. B. and Sie, S. T., *Am. Ind. Hyg. Assoc. J.*, 1965, **26**, 52.
[75] Walkley, J. E. *et al.*, *ibid*, 1961, **22**, 362.
[76] Petroleum (Consolidation) Act, 1928, HM Stationery Office, London, WC2.
[77] Hughes, J. R., '*Storage and Handling of Petroleum Liquids: Practice and Law*' (Griffin, London, 1967).
[78] '*Model Code of Safe Practice in the Petroleum Industry* (Institute of Petroleum, London), Part 1 Electrical Safety Code, 1966; Part 2 Marketing Safety Code, 1965; Part 3 Refining Safety Code, 1965.
[79] Klinkenberg, A. and van der Minne, J. L., '*Electrostatics in the Petroleum Industry*' (Elsevier, Amsterdam, 1958).
[80] *The Petroleum-Spirit (Conveyance by Road) Regulations 1957 and (Amendment) 1966.* Statutory Instruments 1957, No. 191 and 1966, No. 1190.
[81] British Standard 2091: 1969.
[82] *The Inflammable Substances (Conveyance by Road) (Labelling) Regulations* 1968. Statutory Instrument 1968, No. 928.
[83] *Marking Containers of Hazardous Chemicals* (Chemical Industries Association Ltd., London, 1970).
[84] J. Officiel des Communautés Européenes, 16 August 1967, No. 196/81.

CHAPTER 6

CYCLOHEXANE AND ITS DERIVATIVES

by F. Eschard, A. Forge, Dang Vu Quang, J. Alagy, C. Busson, Gordon D. Allen and J-C. Marion

6.1. CYCLOHEXANE

by F. Eschard, A. Forge and Dang Vu Quang

CYCLOHEXANE is one of the most important intermediate products in petrochemistry since it forms the second step in the chain of products leading from benzene to the various monomers used in the manufacture of Nylon 6 and Nylon 6/6. Although other routes exist, those which pass through cyclohexane are current practice and have been for many years the most important. This is largely due to the simplicity of the method of producing cyclohexane on an industrial scale by the hydrogenation of benzene in a high degree of purity and to the high degree of efficiency reached by the oxidation reactions leading directly to a mixture of cyclohexanol and cyclohexanone.

The oxidation reactions will be described in a subsequent section of this chapter. In the present section we shall describe the properties, production and other uses of cyclohexane.

6.1.1. Physical and Chemical Properties of Cyclohexane

Cyclohexane is present in very small quantities in all crude petroleum, an average content being 0·2 per cent. A heavy paraffinic crude may contain as little as 0·05 per cent, a light crude up to 0·5 per cent. Fractions prepared in the refining process usually contain about 5 per cent cyclohexane while separation of the C6 cut can give products containing 20 per cent or more; this quantity can vary widely according to the naphthenic character of the crude. Cyclohexane is always accompanied by its isomer, methylcyclopentane, and benzene; the former is present in a quantity between one and one and a half times the cyclohexane, while the benzene varies from one half to the same as the quantity of cyclohexane present.

The principal physical properties of cyclohexane are shown in Table VI.1. Its vapour pressure as a function of temperature has been plotted in Fig. 6.1. More details of the physical properties can be obtained from specialist publications.[1,2]

The boiling point at normal pressure is almost identical with that of benzene (80·1 °C), about 10° higher than that of methylcyclopentane, and mid-way between those of the normal- and iso-C6 and C7 paraffins. The lowering of freezing point constant is shown in Table VI.1 in view of its practical importance. Industrial specifications for cyclohexane always incorporate a crystalllising point clause with a minimum value, the total quantity of impurities can be estimated by observing this property. For the same quantity of impurities the lowering of the freezing point for cyclohexane is four times as great as that for benzene. Fig. 6.2 shows how the crystallising point varies with the content of impurities for cyclohexane and benzene, the molecular weight of the impurities being taken as 84.

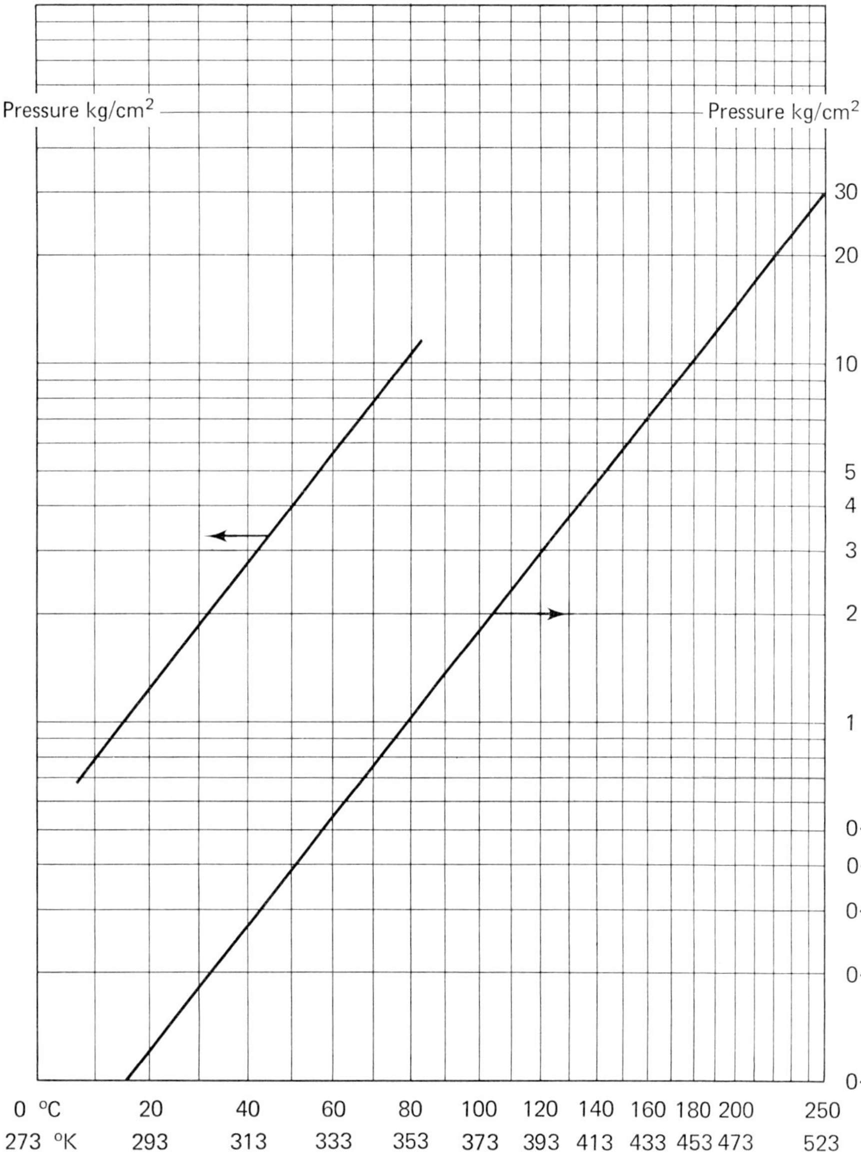

FIG. 6.1

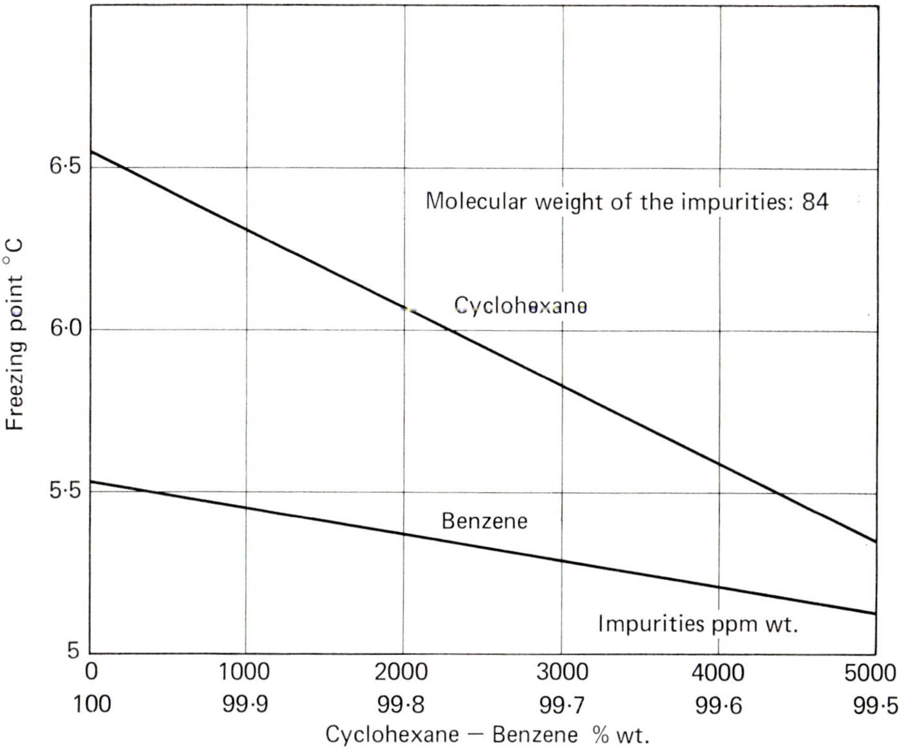

FIG. 6.2

TABLE VI.I

Physical Properties of Cyclohexane

C_6H_{12}

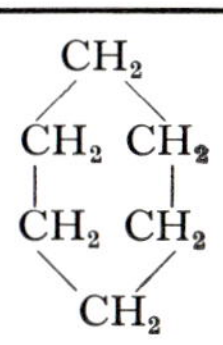

Molecular weight		84·16
Density at 20°C gm/cm³		0·7791
Refractive index $n_D^{19,5}$		1·4226
Freezing (crystallising) point °C		6·55
Cryoscopic constant °C/M*		20·2
Vapour pressure at 20°C in millibars		106
Boiling point at 1014 millibars (760 mm) in °C		80·8
Flammability limits in air, % vol.	lower	1·26
	upper	7·75
Solubility of water at 20°C %wt.		0·0107

*M = no. of moles of the solute dissolved in 1000 gm of solvent.

The chemical properties of cyclohexane are generally those of a saturated hydrocarbon with a ring structure and in particular those that might be expected from a compound with six CH_2 groups in the molecule.

As do all saturated hydrocarbons, cyclohexane has a low degreee of reactivity. It only reacts at temperatures above 150°C with very reactive compounds or at a lower temperature with compounds of which the reactivity has been increased in some way e.g. by the action of light.

Thus the oxidation of cyclohexane to form cyclohexanol and cyclohexanone by molecular oxygen is carried out between 155° and 165°C. Similarly, nitration by nitric acid needs the same order of temperature. On the other hand nitrosation by the reaction with nitrosyl chloride can be carried out at room temperature using a photochemical technique. These three reactions have resulted in industrial applications which will be dealt with in more detail in later paragraphs.

Cyclohexane also reacts at room temperature with chlorine under the influence of ultra-violet light. Contrary to the nitrosation reaction the free radical chain developed is very long, the yield being as much as 20,000 at 20°C. This means that monochlorocyclohexane can be obtained with a very low consumption of energy.

Thus a pilot plant producing 150 kg per day was found to consume 16 kWh of energy gross per ton of product produced, i.e. including such factors:yield of light from the lamps, ratio of light consumption, and the actual characteristics of the reaction.[3]

Reactions involving heterogeneous catalysis can bring about the dehydrogenation of cyclohexane to benzene, its isomerisation to methylcyclopentane or, by hydrocracking, the production of hexane by ring-opening, the two latter reactions, whatever the catalyst used, are much slower than the first-mentioned.

In practice these reactions are important as they take place in the catalytic reforming process and convert the cyclohexane and the methylcyclopentane present in the fractions prepared from crude petroleum into benzene. This benzene which can be separated from the other hydrocarbons by solvent extraction, or, more recently, by extractive distillation is in a very pure state (99.9 per cent). By hydrogenation cyclohexane can be obtained again and in a similar state of purity. Although this route may seem complicated it is in fact the principal method of obtaining pure cyclohexane.

Handling cyclohexane presents the same flammability risks as are shown by other saturated hydrocarbons of similar volatility. It is however not nearly so dangerous from the toxic angle as benzene, the latter being ten times as poisonous as is cyclohexane.

6.1.2. Methods of Production of Cyclohexane

Practically all cyclohexane is manufactured by the hydrogenation of pure benzene, the fractionation of cuts rich in cyclohexane being the only other method to be exploited industrially.

The reason for the production from benzene lies in the very high degree of purity demanded by the users particularly those carrying out oxidation processes. The benzene content of the cyclohexane must be of an exceedingly low order as if it accumulates it inhibits the oxidation reaction. Likewise the presence of other saturated hydrocarbons leads to the formation of undesirable oxidation products, the presence of which complicates the purification stages in the production of the cyclohexanol-cyclohexanone mixture.

6.1.2.1. Distillation processes In spite of their limitations it is worth while mentioning these briefly. Many companies have operated or at least studied this technique.

Phillips who are the only company still operating this process have increased its efficiency by coupling the fractionation proper with a stage in which methylcyclopentane is isomerised to cyclohexane; this improves the yield of cyclohexane and also its purity which can reach 99 per cent. The production capacity of this process at the Phillips plant in the USA has never been disclosed. It was indicated in 1969 that it was less than 20 per cent of the total production in the USA which was then of the order of 150–200,000 tons per annum.

The Humble process is based on an extractive distillation using phenol. Starting from a fraction containing 76 per cent cyclohexane the latter is obtained with a yield of 60 per cent and a degree of purity above 99 per cent.

Recently the German company URBK has developed a new extractive distillation process based on phenoxyethanol.[4] The main column has 180 plates and enables cyclohexane to be produced with a purity above 99·9 per cent with a recovery of 97 per cent. This process too is coupled with a hydrogenation process in which the raw material is the liquid fraction from a steam-cracking process containing 3–5 per cent of cyclohexane but 80 per cent of benzene. Compared with the traditional treatment of such a fraction this process has the advantage of making use of the whole of the benzene and the cyclohexane. Its industrial use has not been reported.

6.1.2.2. Hydrogenation of benzene

6.1.2.2.1. Introduction The various industrial processes using this reaction differ from each other in the techniques used to solve problems posed by the thermochemical and kinetic properties of the reaction and to make use of the various grades of benzene and hydrogen available industrially.

Benzene contains very small quantities of foreign hydrocarbons amongst which is cyclohexane. As we shall see later these hydrocarbons are recovered unchanged and have therefore no effect on the hydrogenation.

It is different with sulphur; nevertheless this problem has to-day lost its importance; ten years ago coal tar benzene contained appreciable amounts of sulphur, 2–5 ppm or exceptionally 10 ppm, now it is rare to meet samples with more than 2 ppm, very frequently the sulphur content is only 0·5 ppm.

The quality of the hydrogen varies according to its source, e.g. hydrogen from synthesis gas, gas from the catalytic reforming of naphthas, gas from steam cracking and even gas from dealkylation of toluene units. Table VI.2 shows some typical compositions. Both the hydrogen content and the quantity and type of impurities vary.

TABLE VI.2

Examples of the Composition of Hydrogen-rich Gas Streams

	Gas from steam reforming (*1*)	*Gas from steam cracking* (*1*)	*Gas from catalytic reforming*	*Gas from hydro-dealkylation*
H_2% by vol	96	76·5	78·4	57
$CO+CO_2$ ppm vol	10	5	—	—
CH_4 % vol	4	23·2	7·9	28·8
C_2s % vol	—	0·3	5·4	13·5
C_3s % vol	—	—	4·9	0·13
C_4s % vol	—	—	2·4	0·02
C_5s % vol	—	—	1·0	0·01
Toluene	—	—	—	0·18
H_2S ppm wt.	1	—	5	—

(*1*) after methanation

6.1.2.2.2. Thermodynamics of the reaction The hydrogenation of benzene is an exothermic reaction which takes place with contraction of volume. The reaction is written:

$$C_6H_6+3H_2 = C_6H_{12} \quad \Delta H = -51{\cdot}2 \text{ Kcal/mole at } 200°C$$

The reaction is therefore favoured by a low temperature and by a high partial pressure of hydrogen.

Fig. 6.3 shows the correlation between pressure and temperature for a fixed residual content of benzene. It can be seen that to keep within the prescribed specifications for cyclohexane (less than 100 ppm of benzene) a temperature of 235°C must not be exceeded during the course of the reaction whatever the partial pressure of hydrogen.

6.1.2.2.3. Kinetic data The catalytic hydrogenation of benzene to cyclohexane is one of the longest-known reactions in organic chemistry. However the quantitative kinetic data were only determined about twenty years ago.[5]

The principal characteristics of the reaction are the following:

—the reaction can be carried out either in a mixed liquid-vapour phase or completely in the vapour phase,

—catalysts which can be used for this reaction comprise the precious metals such as platinum, paladium, ruthenium, and nickel,

—the speed of the reaction remains constant in terms of the rate of transformation of the benzene (zero order reaction),

—the purity of the cyclohexane rather than the yield is influenced by two secondary reactions: isomerisation to methylcyclopentane and cracking.

The isomerisation equilibrium at 200°C corresponds to a conversion of 68 per cent of the cyclohexane to methylcyclopentane and this increases with temperature to 83 per cent towards 300°C. Catalysts must therefore be chosen which do not effect this isomerisation; with nickel catalysts methylcyclopentane makes its appearance from about 250°C.

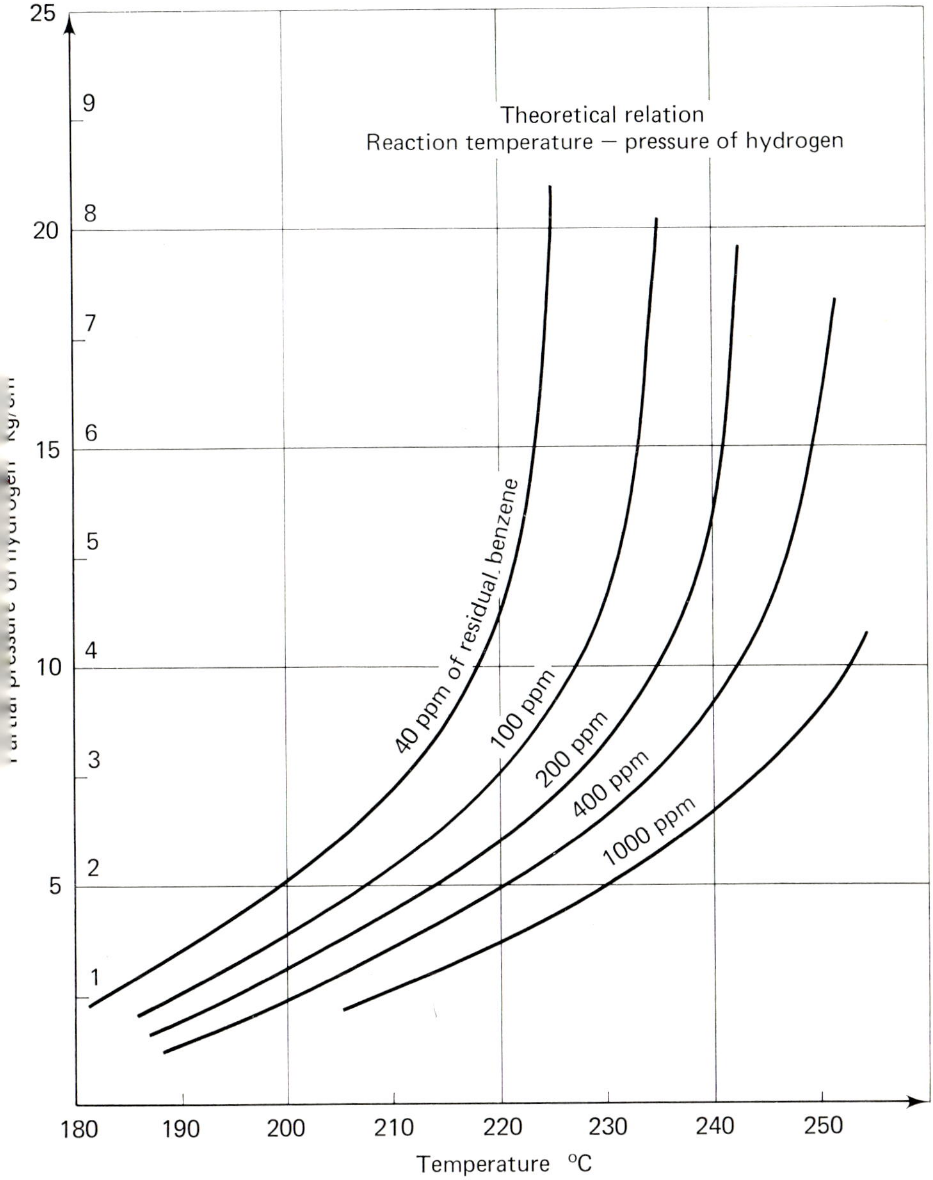

Fig. 6.3

6.1.2.2.4. Catalyst poisons Sensitivity to poisons depends on the nature of the metal used as catalyst.

Nickel, the most generally used metal, is irreversibly poisoned by sulphur, mercury (which can be introduced with hydrogen produced electrolytically) and reversibly poisoned by carbon monoxide and carbon dioxide. Precious metals show the best resistance to catalyst poisons; these when poisoned by sulphur can be regenerated by a stream of pure hydrogen at a sufficiently high temperature.

It is therefore necessary to remove at least a part of these poisons which, as has been said above, can be carried by the hydrogen stream or in the case of sulphur by the benzene.

Purification of the gas is very simple; if it contains sulphur in the form of hydrogen sulphide a wash with soda removes it quickly and efficiently; carbon monoxide and carbon dioxide can best be removed by methanation.

The low concentrations of sulphur in industrial benzene no longer justify either the use of precious metals or the special arrangements, which could be quite complex, previously used for protecting the hydrogenation catalysts.

6.1.2.2.5. Processes The economic importance of cyclohexane has resulted in the development of a large number of industrial processes, some of these processes have been put into operation and exclusively used by the manufacturing companies themselves.

Processes differ in the following ways:

—the nature of the catalyst

—the operating conditions under which the reaction takes place on the catalyst, e.g. in the liquid or vapour phase,

—the design of the reactor, i.e. whether the reaction takes place on a fixed bed or on a moving bed,

—the method used for dissipating the heat formed due to the exothermicity of the reaction.

We have already underlined the great importance of this last point since, taking into account thermodynamic considerations, the reaction temperature must not be allowed to exceed a limit, the value of which will depend on the amount of benzene present. Now in an adiabatic system each 1 per cent of benzene converted raises the temperature of the benzene-hydrogen mixture by 7–8°C. Moreover the latent heat of evaporation of cyclohexane is some 4·42 Kcal/mole at 200°C, 11·6 moles of liquid cyclohexane must therefore be vaporised to absorb the heat of the reaction at constant temperature.

In a fixed bed the problem can be resolved in either of two ways:

—several reactors may be installed in series and the temperature lowered between each reactor either by a quench or by cooling through heat exchangers. This method is difficult and cumbersome.

—a combined reactor-heat exchanger is used. The necessity for filling each reactor tube with catalyst as well as the greater number of tubes needed makes this method both costly and inconvenient to operate.

In the IFP process described later a moving bed-liquid phase method is used.

6.1.2.2.6. Industrial vapour-phase processes

6.1.2.2.6.1. Bexane process (DSM Stamicarbon) (Fig. 6.4) Benzene mixed with make-up and recirculated hydrogen is vaporised by heat exchange with the reaction products. The mixture is led into a multitubular reactor filled with a catalyst based on a noble metal.[6]

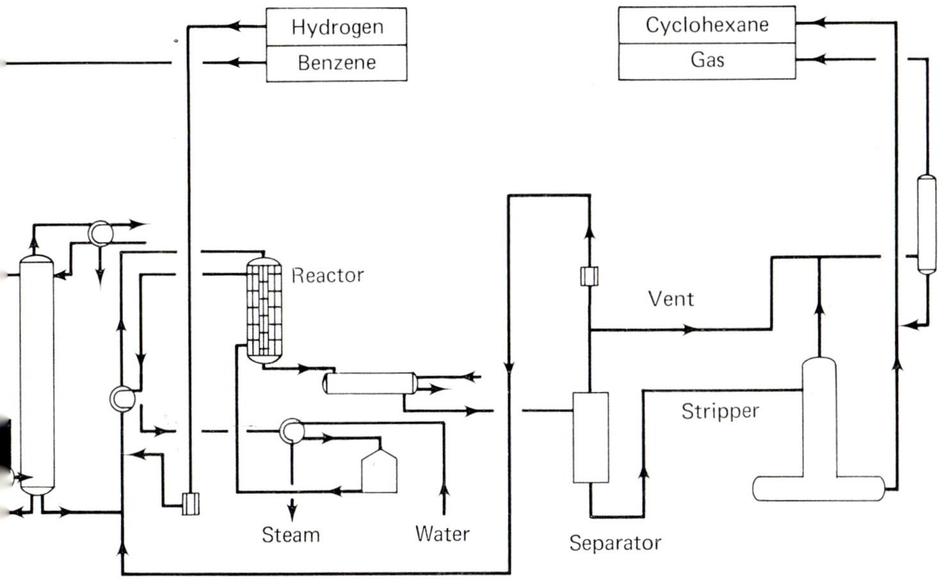

FIG. 6.4

The reaction develops heat which is removed across the surface of the tubes in the form of low-pressure steam. The area of the surface is so calculated that the maximum temperature does not exceed 370°C and that the temperature at the exit from the tubes is brought back to 220°C.

Cooling at the conclusion of the reaction is essential to re-establish equilibrium conditions favourable to the complete conversion of the benzene. The operating pressure is about 35 atmospheres and the purity of the product of the order of 99·9 per cent. The catalyst used has been particularly developed to avoid isomerisation reactions and cracking of the cyclohexane formed. The purity is adequate for cyclohexane to be used for the production of nylon.

Considerable recycling of the gases is necessary as is shown by a not inconsiderable consumption of electrical energy.

A Bexane unit of 60,000 t/a is on stream at Geleen in Holland. Four other units, each of capacity 25,000 tons, are in operation in USSR.

6.1.2.2.6.2. Process HA-84. This process results from collaboration between Engelhard Industries, who developed the catalyst, and Sinclair Research Inc., who were responsible for the actual industrial set up. The process is licensed by the Chem Systems Co.

The catalyst is based on a noble metal and the whole process is similar in principle to the Bexane process. It is however different from the latter in an additional benzene evaporator and a lower ratio of recycle gas.

The residual benzene content is estimated at 5 ppm and some 83 per cent of the total quantity of hydrogen is usefully consumed.

A plant with a capacity of 45,000 t/a belonging to Atlantic-Richfield is on stream at Watson in California using this process.[7]

6.1.2.2.6.3. The Hydrar Process (UOP). The Hydrar process was the first industrial process for the hydrogenation of benzene to be made available.

Twelve plants are at present in operation and recently a new one with a capacity of 3956 barrels per day was announced in the Caprolactam complex belonging to the Empresa Colombiana de Petroleos.

Originally the Hydrar process was based on a platinum catalyst promoted with lithium salts. Since then a nickel-based catalyst has been developed.

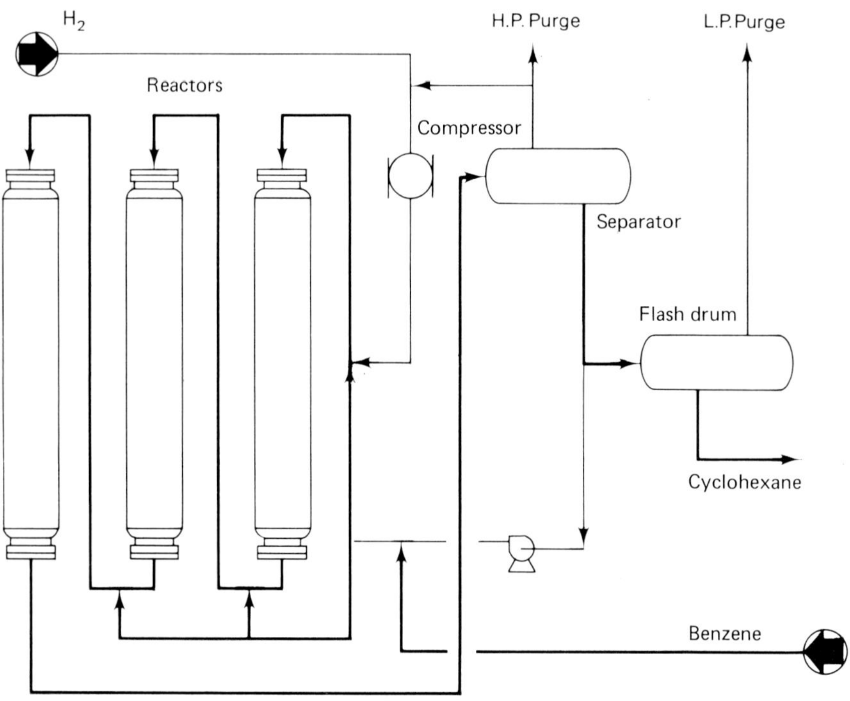

FIG. 6.5

Impurities such as H_2S, CO, and CO_2 act as poisons and must be removed. On the other hand sulphur compounds present in the benzene could be tolerated up to a limit of 300 ppm when platinum was used as the catalyst.

Contrary to the Bexane and HA-84 processes the Hydrar process makes use of adiabatic reactors, the heat of the reaction being removed by cooling between each reactor. Fig. 6.5 shows the lay-out of this process in principle. Various different arrangements of the reactors, the heat exchangers and cooling of the cyclohexane are possible.

The temperature of the catalyst bed varies from about 200-350°C. However in view of thermodynamic considerations the exit temperature of the last reactor is kept below 275°C.

Pressure in the reactor is kept at about 30 atmospheres while the hydrogen/hydrocarbon molar ratio is maintained at 2 to 1 by recycling. Purging of the gas is carried out in such a way that a minimum of 30 per cent hydrogen is retained whatever the circumstances.

6.1.2.2.6.4. The Houdry Process. This process is similar in principle to the Hydrar process.

The reactors are adiabatic, benzene is injected in stages while the ratio of hydrogen to hydrocarbon is the same (2 to 1). The temperatures are nevertheless lower and residual contents of 10 ppm have been announced. On the other hand the hydrogen-hydrocarbon ratio both of liquid and gas seems to be higher.

A plant with a capacity of 100,000 t/a is in operation at the Milford Haven site of Gulf Oil Eastern in Wales. Another unit has been in operation in Spain since 1970.[8]

6.1.2.2.7. Liquid phase processes

6.1.2.2.7.1. The I.F.P. Process (Fig. 6.6).[9] In the I.F.P. process the benzene and hydrogen are injected directly without preheating into a reactor containing a suspension of catalyst in liquid cyclohexane. The heat evolved in the reaction is removed by heat exchange between the catalyst suspension and the water under pressure. To carry this out the catalyst suspension is passed through a circulating pump at a high speed across an exterior heat exchanger and then returned to the reactor. This operation plays an important part in keeping the catalyst in suspension. The expansion of the cooling water produces low-pressure steam.

The reaction therefore takes place in a reactor which is vigorously agitated; the concentration of benzene is consequently very low but this is no disadvantage as the reaction rate remains constant up to a very high conversion ratio.

The cyclohexane produced is removed from the reactor as a vapour in the current of inert gases and the small excess of hydrogen which is essential to complete the conversion of the benzene. The mixture of gases then passes into a finishing reactor containing catalyst on a fixed bed where the last traces of benzene can, if necessary, be hydrogenated but with a rise in temperature of only a few degrees.

The hydrogenation is thus carried out at a constant temperature under perfectly controlled conditions, avoiding altogether any formation of by-products and yielding a product with a virtually nil-benzene content.

Since cyclohexane is not recycled, a very small supply of residual hydrogen is adequate to produce a partial pressure of hydrogen sufficient to produce complete conversion; it is not necessary to employ a recycling technique to make use of a low grade of hydrogen, e.g. 55 per cent. With such a gas 87 per cent of the hydrogen can be usefully consumed by suitable adjustment of the operating conditions.

Where the hydrogen feedstock is sufficiently pure, e.g. 90 per cent of hydrogen, the residual gas is recirculated and this enables 98 per cent of the hydrogen to be usefully consumed.

In the most usually met-with operations the purity of the hydrogen used is intermediate, e.g. 75 per cent, in such cases a once-through operation is sufficient at a pressure of 25–30 atms to attain the optimum useful consumption of hydrogen.

The catalyst used in the liquid phase reactor is Raney nickel. While the earlier units incorporated a preparation section, nowadays a non-pyrophoric type of catalyst is used which can be handled in air in the form of a dry powder; it is mixed directly into suspension in the cyclohexane. An arrangement allows for the catalyst to be 'made up' without interrupting the operation of the unit; this affords a very simple method of compensating for the loss of activity of the catalyst brought about by any sulphur in the benzene.

Benzene containing as much as 5 to 10 ppm of sulphur can thus be easily hydrogenated without any appreciable reduction in the overall speed of the reaction with the sole disadvantage that the catalyst cost will be increased.

One of the important problems which has been solved during the development of this process has been how to handle a suspension of powered catalyst in the cyclohexane. Efficient equipment has been developed both for the circulating pump as well as for all control points to enable the solid-liquid mixture to be properly handled.

At the present time there are ten plants where the I.F.P. process is being used, four of these have capacities of 100,000 t/a. Plants with higher capacities could be erected without difficulty.

6.1.2.2.7.2. The Arosat Process. This process, which was developed by Lummus, is a liquid phase process using a fixed-bed nickel catalyst. It would seem that the the greater part of the heat developed in the process is removed by the evaporation of the cyclohexane, the vapour of which being recondensed with the production of low-pressure steam. All the operations take place in a single reactor.

As the ratio of hydrogen to cyclohexane vapour must remain constant in order to give suitable partial pressure of hydrogen, this process involves a considerable degree of recycling. As a result of this the consumption of electricity is much closer to that of vapour phase processes than other processes using a liquid phase.

The residual benzene content will be less than 20 ppm.

The process data has been established on the results from a pilot plant with a capacity of 1 kg/hour. No large-scale plant has been announced using this process.

6.1.2.2.7.3. The British Petroleum Process.[11] Like the Arosat the BP process uses a fixed bed and the heat is removed by the evaporation of the cyclohexane.

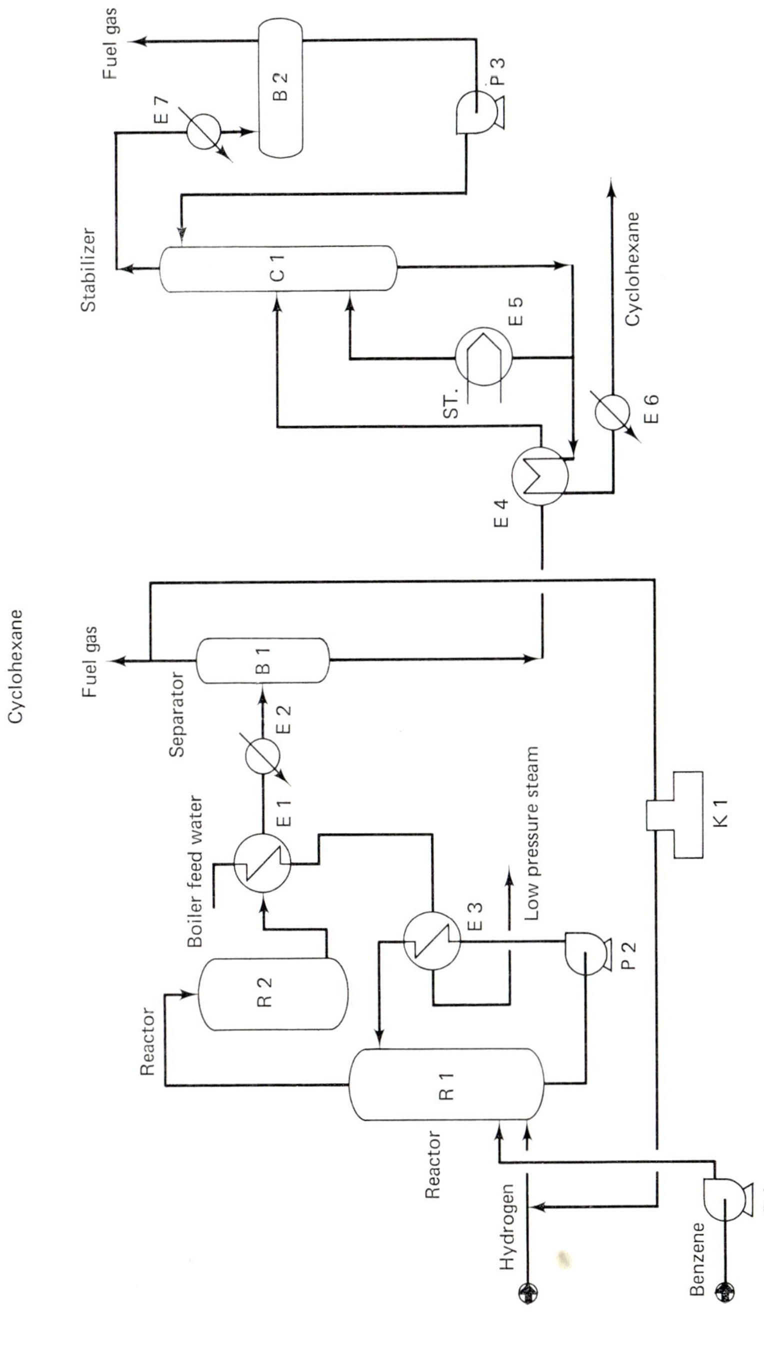
Cyclohexane
Fuel gas
B 2
E 7
P 3
Stabilizer
C 1
Cyclohexane
E 5
ST.
E 6
E 4
Fuel gas
Separator
B 1
E 2
E 1
Boiler feed water
Low pressure steam
E 3
P 2
K 1
Reactor
R 2
Reactor
R 1
Hydrogen
Benzene
P 1

FIG. 6.6

To avoid any substantial amount of recycling the partial pressure of hydrogen is maintained at a comparatively low value and the conversion limited to 95 per cent in the first reactor.

The product thus obtained is then treated in a second reactor where in the presence of fresh hydrogen and in the absence of cyclohexane vapour the reaction can be completed thus leaving a residual benzene content of less than 10 p.p.m.

The catalyst used is one based on nickel. Safeguards based on absorbent materials are included in the reactor scheme to protect the catalyst from being poisoned by sulphur.

The catalyst shows excellent selectivity, isomerisation to methylcyclopentane being less than 50 p.p.m.

The process has been developed from pilot runs of 5000 hours duration. There does not appear to have been any development on an industrial scale.

6.1.3. Quality of Product: Starting Materials Consumed

The purity of the cyclohexane obtained depends mainly on the quality of the benzene used as starting material combined to some extent also with the effectiveness of the process used for removing the last traces of the benzene and for avoiding the production of by-products. The purest cyclohexane results from benzene produced by the dealkylation of toluene. Table VI.3 shows for example the composition of a bulk-produced sample made by this route (case 1) and that of the corresponding cyclohexane produced. Comparative compositions for an average benzene and the cyclohexane produced are shown in the same table (case 2).

TABLE VI.3

Quality of Industrial Cyclohexane

	Case 1		*Case 2*	
	Benzene (hydrodealkylation)	*Cyclohexane*	*Benzene average*	*Cyclohexane*
Melting point	5·52°C	6·52°C	5·47°C	6·38°C
Saybolt colour	30+	30+	30+	30+
Cyclopentane	—	—	40 ppm	30 ppm
n-Hexane	<1 ppm	4 ppm	40 ppm	30 ppm
Methyl pentanes	5 ppm	—	20 ppm	10 ppm
Methyl hexanes	—	—	55 ppm	45 ppm
n-Heptane	—	—	100 ppm	85 ppm
Methylcyclopentane	2 ppm	3 ppm	300 ppm	310 ppm
Cyclohexane	32 ppm	>999900 ppm	150 ppm	999300 ppm
Methylcyclohexane	1 ppm	30 ppm	50 ppm	140 ppm
Benzene	>999900 ppm	<1 ppm	999000 ppm	<1 ppm
Toluene	16 ppm	<1 ppm	100 ppm	<1 ppm
Other compounds	—	—	<50 ppm	<50 ppm

Stoichiometrically the chemical reaction needs 800 m³ at S.T.P. of hydrogen and 0·93 tons of benzene to give 1 ton of cyclohexane, in other words 1·08 tons of cyclohexane are produced from 1 ton of benzene or 1·22 m³ of cyclohexane from 1 m³ of benzene.

The only losses from an industrial unit result from the purge of gas in the high pressure separator; normally it pays to minimise this by cooling the gas to be purged down to about 10°C; the purity of the hydrogen used has little effect on the loss of cyclohexane. It is realistic to reckon on a production of 1·05 tons of cyclohexane from 1 ton of benzene but a figure above 1·06 tons is frequently obtained.

The actual consumption of hydrogen depends on the purity of the gas stream used as well as on the actual process, particularly whether the gas stream is recirculated or not. Taking into account the losses at the high pressure separation stage a consumption per ton of cyclohexane of between 820 m³ for a stream rich in hydrogen to 900 m³ for one of lower quality, both figures being given at S.T.P.

6.1.4. Uses of Cyclohexane

The greater part of the cyclohexane produced in the world is used for the production of Nylons 6 and 66, other applications amounting to no more than 10 per cent of the total. The main chemical uses of cyclohexane are shown diagramatically in Fig. 6.7 which the thickness of the arrows indicates qualitatively the relative importance of the various production sequencies. The processes for the oxidation of cyclohexane to a mixture of cyclohexanol and cyclohexanone are described in a later part of this chapter, this including also the dehydrogenation of this mixture to all cyclohexanone. Other processes for the production of nylon precursors from cyclohexane are described in 6.3.

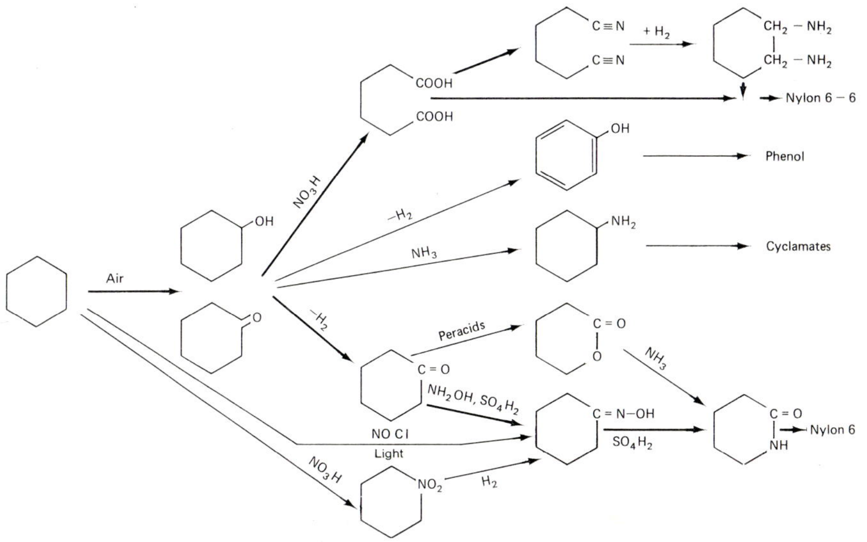

FIG. 6.7

Among other cyclohexane[13] derivatives cyclohexylamine is of importance, this can also be produced starting from cyclohexanol-cyclohexanone mixtures (Fig. 6.7). Thus using the Leonard process, in which the starting material is reacted with ammonia in the presence of catalysts, a mixture of mono-, di, and tri-amines is produced. Cyclohexylamine, which can also be obtained by the hydrogenation of aniline, is used for the manufacture of cyclamates, the capacity for which in the USA was a little over 10,000 t/a but this product is no longer used in foodstuffs in USA as a result of FDA regulations.

We have already drawn attention to the use of cyclohexane as a solvent. This outlet absorbs about 10 per cent of the total consumption.

6.1.5. Economic Data

As nearly 90 per cent of the cyclohexane produced is used for the synthesis of nylon 6 and nylon 66, world production has tended to follow the development and the fluctuations in the demand for these two materials.

However a major change has taken place during the last ten years in that instead of being confined almost exclusively to the USA production of cyclohexane has spread to all parts of the world and in particular Western Europe, Japan, Central America, and the Eastern block.

Generally cyclohexane is produced by petrochemical companies attached to large groups of refineries from which cheap benzene and hydrogen are readily available. The cyclohexane is then sold to the producers of fibres.

In some large complexes the integrated operations start from crude naphtha and cyclohexane is produced and consumed in the same group of plants.

The much rarer case where the producer of monomer for fibres himself starts from benzene and hydrogenates it seems to have been adopted to a greater extent in Japan than in any other countries.

6.1.5.1. Cost of Production of Cyclohexane

Economic data relating to various processes for the production of cyclohexane are shown in Table VI.4.[13] The figures in this table come from too diverse sources to form a basis for a strict comparison between them; some indications of the characteristics of different processes can be obtained particularly under the heads of investment and consumption of utilities.

Calculation of the production cost of cyclohexane is shown in Table VI.5, set out for a unit with a capacity of 100,000 t/a. In the case under examination it is assumed that a gas stream containing 75 per cent of hydrogen is used and without recycling.

From this it can be seen that the cost of the hydrogen makes up half the cost of the conversion of benzene to cyclohexane. This proportion is even higher when the hydrogen used is synthesis gas produced by the steam reforming of gas or naphtha, the cost of which is 800 to 1000 F.Fr. per ton. In spite of the reduced consumption to 0·074 tons per ton the cost of hydrogen works out at F.Fr.74 per ton of cyclohexane or about 75 per cent of the cost of the conversion, which is practically doubled.

TABLE VI.4

Processes for the Production of Cyclohexane: Economic Data

	Hydrar	*IFP*	*Houdry*	*H.A.84*	*Arosat*	*BP*	*Bexane*
Capacity t/a	25,000	100,000	60,000	50,000	100,000	160,000	60,000
Capital cost of plant F.Fr. $\times 10^6$	3·5	3.4 (*2*)	2·6 (*2*)	4·4	6.2 (*1*)	6·5 (*2*)	4.0 (*2*)
Initial charge of catalyst, F.Fr. $\times 10^6$	(*3*)	0·15	(*3*)	(*3*)	(*3*)	(*3*)	(*3*)
Consumption per ton of cyclohexane produced							
Raw materials:							
Benzene (t)	0·93	0·930	0·93	0·935	0·935	—	0·930
Hydrogen gas m³ at S.T.P.	943 at 95%	1350 at 75%	935 at 100%	1136 at 85%	910 at 90%	—	935 at 100%
Catalysts & Chemicals, F.Fr.	3·2	3	1·7	1	—	1·6	1·6
Utilities:							
Steam consumed (tons)	0·35	0·185	0·20	—	—	—	0·08
Steam produced (tons)	—	0·95	—	0·51	0·58	—	0·63
Electricity kWh	125	12	72	46·5	44	28	20
Cooling water m³	12·5	4	4·5	23·5	17	50	30
Heating water m³	12·8	0·95	—	—	—	—	0·67
Fuel 10^6 Kcal	0·25	—	—	—	0·85	0·02	—
Labour (operatives per shift)	2	1	1	1	1	1	1

Notes:

(*1*) Complete plants including any pretreatment, additional compressors and, if necessary, the equipment for preparing the catalyst is included.

(*2*) Plants limited solely to the hydrogenation of the benzene.

(*3*) Included in capital cost of plant.

TABLE VI.5

Cost of Production of Cyclohexane by Hydrogenation of Benzene Capacity 100·000 t/a of Cyclohexane, year 1972

Capital		*French Francs*
Cost of the actual plant unit		$4{\cdot}0 \times 10^6$
Engineering and installation costs, licences, etc.		$5{\cdot}15 \times 10^6$
Initial charge of catalyst		$0{\cdot}15 \times 10^6$
Working capital (one month's operating costs)		$3{\cdot}3 \times 10^6$
Operating costs in Francs/ton		
Variable charges		
Benzene, 0·93 ton at 400 Fr/ton	372.0	
Hydrogen, 0·082 ton at 300 Fr/ton (75%)	24.6	
Utilities including credit for steam	−5.7	
Catalyst cost	3.0	
Total		393.9
Fixed charges		
Labour, 1 operator per shift	3.0	
Interest 7% on capital cost	6.6	
Maintenance 4% on capital cost	3.7	
Overheads 3% on capital cost	2.8	
Depreciation 12% on capital cost	11.5	
Interest on working capital 9%	3.0	
Total		30.6
Cost of production of 1 ton of cyclohexane		424.5
Cost of converting 0·93 ton of benzene to 1 ton of cyclohexane		52.5

It follows therefore that the availability of cheap hydrogen is a decisive element in the planning of a cyclohexane production unit. This condition will be fulfilled in a petrochemical complex containing steam cracking units or catalytic reformers for the production of aromatics and also close to a refinery with an excess of hydrogen.

6.1.5.2. Production and Consumption of Cyclohexane

The relation between the production of nylon 6 and nylon 66 and the consumption of cyclohexane varies considerably according to the economic set up locally for it depends on the chain of products; thus a part of nylon output continues to be made starting from phenol. Table VI.6 shows the percentages of nylon made from different starting materials and how these percentages will change over the years in different areas of the world.[14] The part played by cyclohexane in Japan stands out while in other parts of the world the proportion while at present less, is increasing.

The growth in the production of polyamide synthetic fibres has rather marked time in the last few years and this has led to some slowing down of the world consumption of cyclohexane.

Table VI.7 gives the production of nylon 6 and nylon 66 as well as the corresponding consumption of cyclohexane in 1971 and estimates for 1975.

TABLE VI.6

Relative Production of Nylon 6 and Nylon 66 and the Proportion of Cyclohexane in the raw materials

	Repartition between Nylon 6 and Nylon 66			
	1970		*1975 (estimate)*	
	Nylon 6 %	*Nylon 66* %	*Nylon 6* %	*Nylon 66* %
USA	34·7	65·3	34·7	65·3
Wesen Europe	43·0	57·0	42·0	58·0
Japan	80·0	20·0	90·0	10·0

	Percentage produced from Cyclohexane					
	1970			*1975 (estimate)*		
	Nylon 6 %	*Nylon 66* %	*Total* %	*Nylon 6* %	*Nylon 66* %	*Total* %
USA	51·0	95·0	79·7	59·0	96·0	83·1
Western Europe	64·0	60·0	61·7	64·0	70·0	67·6
Japan	93·0	100·0	93·0	94·0	100·0	95·0

TABLE VI.7

Consumption of Nylon and Cyclohexane ·000 t/a

	1971	*1975 (estimate)*
USA		
Nylon 6	240	305
Cyclohexane for Nylon 6	150	230
Nylon 66	320	400
Cyclohexane for Nylon 66	465	580
Cyclohexane for Nylon 6 and 66	615	810
Total cyclohexane produced	800	1100
Western Europe		
Nylon 6	285	415
Cyclohexane for Nylon 6	230	335
Nylon 66	380	550
Cyclohexane for Nylon 66	350	595
Total cyclohexane for nylon	580	930
Japan		
Nylon 6	350	510
Cyclohexane for Nylon 6	230	335
Nylon 66	6	60
Cyclohexane for Nylon 66	10	100
Total cyclohexane for nylon	450	750

6.1.5.3. Capacity for Cyclohexane Production

The capacity for cyclohexane production installed in various parts of the world in 1972 in thousands of tons is:

USA	1000
Central America	260
Western Europe	1100
Japan	>440

Except in Japan production capacity clearly exceeds consumption: in the USA many plants have been closed in recent years while in Europe new units have been brought on stream.

6.2 THE OXIDATION OF CYCLOHEXANE

by J. Alagy and C. Busson

6.2.1. Historical

Oxidation of cyclohexane is one of the most important reactions in petrochemistry. It forms one of the stages in the industrial process leading from petrochemical raw materials to polyamide fibres.

Work on the application of this reaction started during the 1930s. The cyclohexane used was obtained by direct fractional distillation from petroleum cuts rich in naphthenes. The purity of these fractions never exceeded 85 per cent and it was oxidised in a single stage process to adipic acid. Adipic acid obtained in this way was relatively impure and needed a costly purification step. The increasing demand for adipic acid resulting from the rapid growth of polyamide fibres caused producers to start using phenol as the raw material. The latter gave cyclohexanol by hydrogenation and this was either oxidised by nitric acid to adipic acid or else dehydrogenated to cyclohexanone which was then converted to caprolactam.

Appearance on the market of very high purity cyclohexane obtained by the hydrogenation of benzene has led to a spectacular development of oxidation processes and these now make up the principal methods of producing cyclohexanol and cyclohexanone.

6.2.2. The Chemical Reaction Involved

Large scale application of the oxidation of cyclohexane has given rise to much research in both fundamental and applied fields and this has resulted in applications for numerous patents. Understanding of the reaction has increased in depth and the factors influencing it have been localised. This has led to a progressive improvement in the performance of the process on an industrial scale.

The complex mechanism of the reaction has been studied in detail.[15,16,17,18] A free radical chain mechanism takes place thus:

Initiator+R	→	R'	chain initiation	
$R'+O_2$	→	RO'_2	chain propagation	(1)
RO'_2+RH	→	$ROOH+R'$		(2)
$R'+R'$	→	R—R		(3)
$R'+RO'_2$	→	ROOR	chain termination	(4)
$RO'_2+RO'_2$	→	Inert products		(5)

The reaction is initiated by certain intermediates capable of forming the radical R'. The radicals formed in this way take up molecular oxygen to form the peroxide radicals RO'_2 which react with the hydrocarbon raw material to form the hydroperoxide and regenerate the radical R' thus allowing the chain reaction to continue. Chain termination, except under very low partial pressures of oxygen takes place mainly through reaction (5).

Cyclohexyl hydroperoxide, being relatively unstable under the reaction conditions, decomposes into free radicals. These bring about degenerate chain branching thus: Monomolecular decomposition:

$$ROOH \quad — \quad RO'+OH$$

Bimolecular decomposition

$$2ROOH \quad — \quad RO'+RO'_2+H_2O$$

Interaction of cyclohexane and hydroperoxide

$$RH+ROOH \quad — \quad RO'+R'+H_2O$$

The radicals thus formed initiate the reaction. Further, the hydroperoxide formed can react with the free radicals present in the reaction medium. The cyclohexanol and cyclohexanone thus formed are oxidised in their turn according to similar mechanisms. Gathering together the various reactions gives rise to an extremely complex scheme, a simplified representation of which has been given by Emanuel *et al.*[18]

This scheme can be verified experimentally to a certain degree. Thus:

cyclohexyl hydroperoxide gives on thermal decomposition cyclohexanol and cyclohexanone with good selectivity.[20,21]

cyclohexanol is oxidised to cyclohexanone with a molar selectivity of the order of 95 per cent.[18,21,22]

cyclohexanone is oxidised as a first step to the keto hydroperoxide and then by a series of complex reactions the final products are produced.[17,18]

By neglecting the various hydroperoxides, the low concentration of which reaches a stationary state the reaction as a whole can be looked at as a co-oxidation of the three principal starting materials: cyclohexane cyclohexanol and cyclohexanone. The reaction constants of the rupture of the hydroperoxides of all three compounds are of the same order. We can therefore understand that in the whole set up chain interaction does not affect the special inhibition or entrainment phenomena.

To a first approximation, therefore, we can consider that we are dealing with a complex system in which parallel first order reactions are taking place.

Thus the system can be expressed:

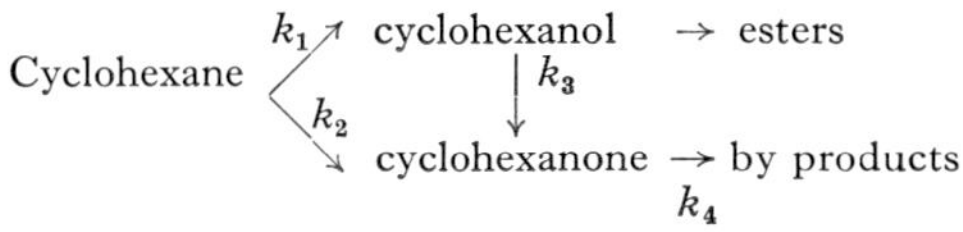

The determination of the ratio of the reaction constants has been carried out.[19]

$$\frac{k_1}{k_2} = 3{\cdot}7 \quad \frac{k_3}{k_4} = 1{\cdot}4 \quad \frac{k_3}{k_1} = 24 \quad \frac{k_4}{k_2} = 66$$

These ratios are but little affected by temperature.[19]

The value of the constants show:

(a) that cyclohexane is much less readily oxidised than the intermediate products it is wished to produce.
(b) that most of the cyclohexanone is formed by passing through the cyclohexanol stage.
(c) that the by-products are mainly formed from cyclohexanone as starting material.

6.2.3. Industrial Processes

The information collected is being utilised more and more for the improvement of industrial processes.

Thus the earlier processes carried out the oxidation of cyclohexane to adipic acid in a single stage. At first sight this seemed obvious but the selectivity obtained was not sufficient and finally the two stage process has been preferred. The latter incorporates the air-oxidation of cyclohexane to a mixture of cyclohexanol and cyclohexanone in the first stage which, after purification, can be oxidised by nitric acid to adipic acid.

It should however be noted that workers in the field have not given up hope of developing a one-stage process and work in this field appears to be just on the point of meeting with success.

Much work has been devoted to the air-oxidation of the cyclohexanol-cyclohexanone mixture to adipic acid, however the selectivity of such reactions always appears slightly inferior to that obtained by nitric acid and they have not been used to any extent.

6.2.3.1. Oxidation of Cyclohexane to Cyclohexanol-Cyclohexanone

Industrial processes have been directed to obtaining a cyclohexanol-cyclohexanone mixture with the maximum degree of selectivity. The various parameters of the reaction now fall within the following limits:

Temperature	155–156 C
Residence time	15–30 mins
Pressure	10–15 atm

6.2.3.1.1. Effect of the conversion factor Since the intermediate products are much more readily oxidised than cyclohexane itself there is a marked influence of the rate of conversion on the selectivity (Fig. 6.8). The choice of the conversion rate is not easy as it involves the complete optimisation of the process (chemical selectivity, size of equipment, consumption of utilities, etc.). Further, owing to the low conversion rate, the material balance is very poor, both on pilot plant and on the industrial scale. Nevertheless with increasing capacity of the industrial units it is noticeable that the conversion ratio normally used falls, e.g. from 10 to 4 per cent per pass. Actually it is usually in the region of 4–5 per cent. At this conversion the molar selectivity obtained is in the range of 76–78 per cent. Often units are designed to work at rather higher, e.g. up to 6–7 per cent conversion with selectivities of the order of 70–72 per cent. There is therefore a considerable degree of flexibility.

6.2.3.1.2. Effect of the number of reaction stages In Fig. 6.8 curve (a) shows the selectivity as a function of the conversion factor for discontinuous operation. This curve shows also the maximum selectivity that can be obtained from a tilted reactor. Curve (b) shows the results that can be obtained with a thoroughly agitated one stage reactor. For the highest possible values in curve (a) a calculation shows that several reactors, thoroughly agitated, must be used in series (Fig. 6.9). The number of reactors used is a case of balancing the additional cost resulting from the increase in the number of reactors (at constant reaction volume) against the resulting gain in selectivity. Generally three reactors are employed, the installation of a fourth is only justified in plants with capacities exceeding 50,000 t/a.

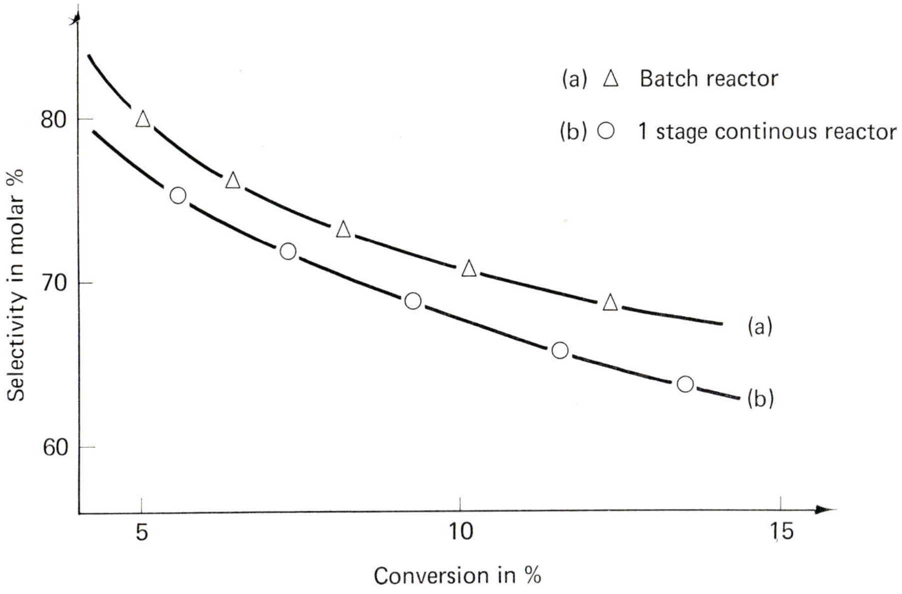

FIG. 6.8

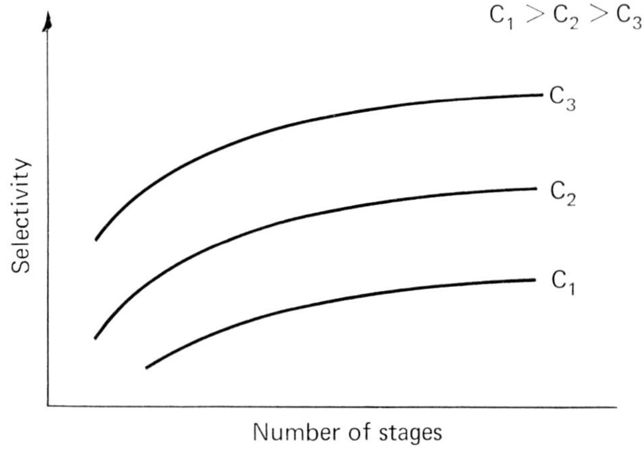

FIG. 6.9

6.2.3.1.3. Effect of catalysts Metallic salts, soluble in hydrocarbons, usually salts of cobalt (naphthenate, oleate, etc.) have been used. Their action consists in decomposing the hydroperoxides present in the medium by an oxidation-reduction mechanism.

$$M^{2+} + ROOH \rightarrow M^{3+} + RO' + OH^-$$
$$M^{3+} + ROOH \rightarrow M^{2+} + ROO' + H^+$$

These catalysts, in addition to their effect in initiating and terminating reaction chains enable a limited control of the composition of the reaction products to be carried out. Thus in a noncatalytic auto-oxidation the ratio of cyclohexanone to cyclohexanol is about 2, in a cobalt-catalysed oxidation it falls to about 0·6. The reduction in the concentration of cyclohexanone in the medium produces a favourable effect on the selectivity. Most industrial processes use soluble salts of cobalt in the proportion of about 100 p.p.m.

More elaborate attempts have been made to reduce the concentration of cyclohexanone in the reaction product.[24] These are based on the oxidation of cyclohexane at a very low conversion followed by hydrogenation over a fixed-bed catalyst which reduces both the hydroperoxide and the cyclohexanone to cyclohexanol, the two reactions being carried out either in two reactors in a cyclic manner or alternatively in a series of reactors. In spite of claims for improvements in the selectivity it does not appear that this process has up till now been used industrially.

Recent work has been directed towards achieving maximum selectivity in cyclohexyl hydroperoxide production. To achieve this it is necessary to operate at very low conversions in the complete absence of catalysts and further it is necessary to use steel reactors, enamel-coated to avoid the peroxides decomposing on the surface of the stainless steel reactors normally used. In this way the hydroperoxides accumulate, being much less easily oxidised than the cyclohexanol or the cyclohexanone and a gain in selectivity thus results,[18] provided that the hydroperoxides can be decomposed, after concentration, in a second stage with the aid of a selective catalyst such as chromium octoate.[25]

An elegant method using boric acid has been proposed,[26,27] it enables a selective decomposition of cyclohexyl hydroperoxide into cyclohexanol to be carried out followed by protection of the latter in the form of its boric esters which are resistant both to oxidation and to heat. The catalytic effect of boric acid has been studied in the case of the oxidation of cyclo-dodecane and a hypothesis put forward that the hydroperoxide-boric ester complex decomposes into the boric ester and the corresponding alcohol.[28]

To carry out the operating conditions will be decided on the basic data provided, partly from the oxidation reaction we have just discussed and partly from the esterification of the alcohol with boric acid.

It is known that boric esters are particularly easily hydrolysable.[29,30] For this reason to obtain them it is necessary to orientate the esterification equilibrium by the maximum possible removal of water, e.g. by azeotropic methods.[29]

To increase the speed of esterification the factors requiring attention are the temperature and the concentration of boric acid; the stationary concentration of free alcohol, other things being equal, being fixed by the concentration of water in the reaction medium.

The speed of esterification is however greater the higher the temperature; this is however limited by the necessity of the selectivity of the oxidation reaction remaining high.

The stationary concentration of free alcohol must be fixed as a function of two opposing parameters: the need for speed of esterification being such as to give a viable process while at the same time limiting as far as possible resulting oxidation of ketones and by-products.

It appears from fundamental data that the esterification of cyclohexanol by boric acid is the slowest reaction and this limits the overall velocity of the process.

The industrial processes actually used are:

(1) Minor variations on the oxidation process catalysed by cobalt salts. These are based on traditional processes and continue to operate covering the greater part of the cyclohexane oxidised.
(2) Processes using boric acid, these are normally confined to new plants.

6.2.3.2. Oxidation of cyclohexane catalysed by soluble cobalt salts

A mixture of make-up and recycled cyclohexane containing 100 p.p.m. of cobalt naphthenate is injected into the first oxidation reactor together with a quantity of air corresponding to the conversion ratio desired (Fig. 6.10). The liquid effluent from the first reactor passes to the second and then to the third reactor, each time with the appropriate quantity of air, which is thoroughly distributed throughout the mass of the liquid. The operating pressure is near to 10 atm. The mean temperature of the reactors is in the region of 150-160°C. Frequently the temperatures fall between the first and third reactor. The gaseous effluent containing nitrogen, carbon dioxide, water, formic acid, oxygen and cyclohexane vapour is cooled in a heat exchanger, often in a column with direct contact. The cyclohexane and water are condensed,

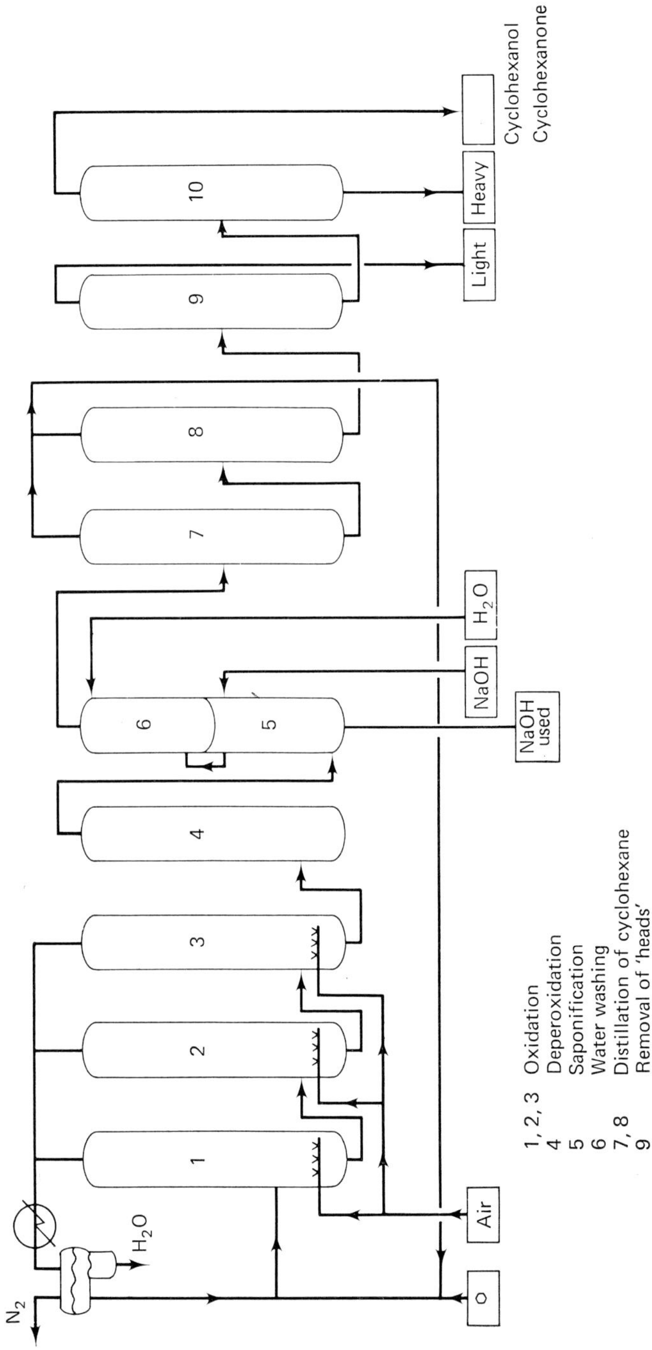
Cyclohexanol
Cyclohexanone
Heavy
Light
H_2O
NaOH
NaOH used
Air
N_2
H_2O
1, 2, 3 Oxidation
4 Deperoxidation
5 Saponification
6 Water washing
7, 8 Distillation of cyclohexane
9 Removal of 'heads'

FIG. 6.10

separated by gravity, the water is treated to recover the small quantities of cyclohexanol-cyclohexanone it contains while the cyclohexane is recycled. The inert gases, after having been washed with a high-boiling product to recover residual cyclohexane, are rejected to atmosphere. The liquid effluent from the third reactor has a very complex composition. The main products are cyclohexanol and cyclohexanone. In spite of the presence of catalyst appreciable traces of the peroxides persist; mono-acids and di-acids from C_1 to C_6 as well as their esters are present. Peroxy compounds are decomposed by adding either a soluble catalyst or by means of a heterogeneous catalyst in a fixed bed. The reaction takes place in the absence of air at a temperature of the order of 100°C. Saponification followed by washing enables the greater part of the lower boiling esters as well as all the acids to be removed. Unreacted cyclohexane is purified by a preliminary distillation at atmospheric pressure followed by a distillation under vacuo and is then recycled. Low boiling impurities are removed in a column dealing with the 'heads' then the purified cyclohexanol-cyclohexanone mixture is recovered at the head of the 'tail' column. The sections on saponification and purification will be treated in more detail in the course of the description of the oxidation process using boric acid since in this sphere the problems raised in the two types of reaction are essentially similar.

6.2.3.2.1. Value of adipic acid Adipic acid is the chief by-product of this reaction, it is formed both in the free state and as an ester of cyclohexanol. For companies manufacturing only caprolactam the viability of such an operation is less obvious except for companies producing a very large tonnage or deliberately producing a wide range of compounds. Thus the choice of the optimum conditions for the process depends to some extent on the value placed on the adipic acid.

6.2.3.2.2. Reactor problems It has already been mentioned that the actual technique used in bringing about gas-liquid contact can have a profound influence on the selectivity.[31] This can be explained qualitatively as follows:

The oxidisability of cyclohexane is much less than that of cyclohexanol and cyclohexanone, further, cyclohexyl hydroperoxide which is much more stable to oxidation than cyclohexane cannot have any effect as a buffer product in the mechanism of the contact phenomenon since this product undergoes thermal decomposition under the conditions of the reaction.[32] As a result, in a system of consecutive reactions of this type, a high selectivity to intermediate products can be obtained by limiting the amount of conversion. However oxidation takes place at the liquid-gas interface and each bubble can be looked at as a microreactor. In such conditions, to avoid very high orders of local conversion, the speed of transfer of oxygen to the gas-liquid interface must be less than the speed of transfer of the oxidised products within the mass of liquid. As a result of this, the optimum conditions have to be determined from parameters the effects of which are contradictory, e.g. the size and length of life of the bubbles, the degree of agitation of the liquid, the ratio of the two phases, etc. The design of the reactor can thus only follow from tests on pilot units combined with hydrodynamic tests on a rough model to obtain data for carrying out the extrapolation on to an industrial scale.

Another very important problem arising from this process is the soiling of the reactor brought about by the appearance of a second brownish, very viscous phase, insoluble in cyclohexane and containing essentially adipic acid and other oxygenated products. This extremely sticky phase shows a tendency to deposit on the reactor and to stop up the pipes resulting in it being necessary to shut down the reactor and clean it down with alkali and water. These shut-downs can prove very frequent if the operating conditions are not carefully chosen. One suggestion that has been put forward is to interpose a water wash between each oxidation reactor. This sticky phase can be eliminated by these means.[33,34] Such a technique is particularly valuable where adipic acid is a desired product. It is also possible to reduce the production of the sticky phase mentioned above by a close control of the operating conditions and by restricting the degree of conversion. The design of the reactor thus becomes of paramount importance. Particularly good results have been obtained on the industrial scale using a reactor in the form of a column specially designed for this reaction.[35]

6.2.3.3. Oxidation of Cyclohexane in Presence of Boric Acid

In the oxidation of cyclohexane in the presence of boric acid the operating conditions must be so defined to allow a selective oxidation of cyclohexane combined with a selective decomposition of the hydroperoxide and the formation of boric esters sufficiently rapidly to capture the cyclohexanol as and when formed. The speed of the oxidation must therefore be limited by restricting the supply of oxygen to the system.

Some significant results obtained in batch experiments are shown in Table VI.8.

TABLE VI.8

No. of test	1	2	3	4	5	6
Temperature in °C	165	165	165	165	165	165
Pressure in atm	12	12	12	12	12	12
% vol oxygen in oxidising gas	4	10	4	4	4	4
Reaction time in min.	150	150	90	150	150	150
Quantity of boric acid introduced in g/kg of cyclohexane starting material	70	70	80	80	30	70
Cyclohexanol in mols/kg of cyclohexane starting material	1·25	1·26	1·20	1·27	1·17	1·25
Cyclohexane in mols/kg of cyclohexane starting material	0·11	0·14	0·10	0·12	0·20	0·11
Acids ,in equivs/kg of cyclohexane starting material	0·14	0·24	0·30	0·14	0·28	0·14
Ratio of cyclohexan-ol -one produced to cyclohexane reacted in mole %	90·2	86·1	83·7	90·3	85·6	90·2
% of cyclohexane reacted	12·0	12·9	12·6	12·3	13·6	12·0

The industrial process is shown in Fig. 6.11.

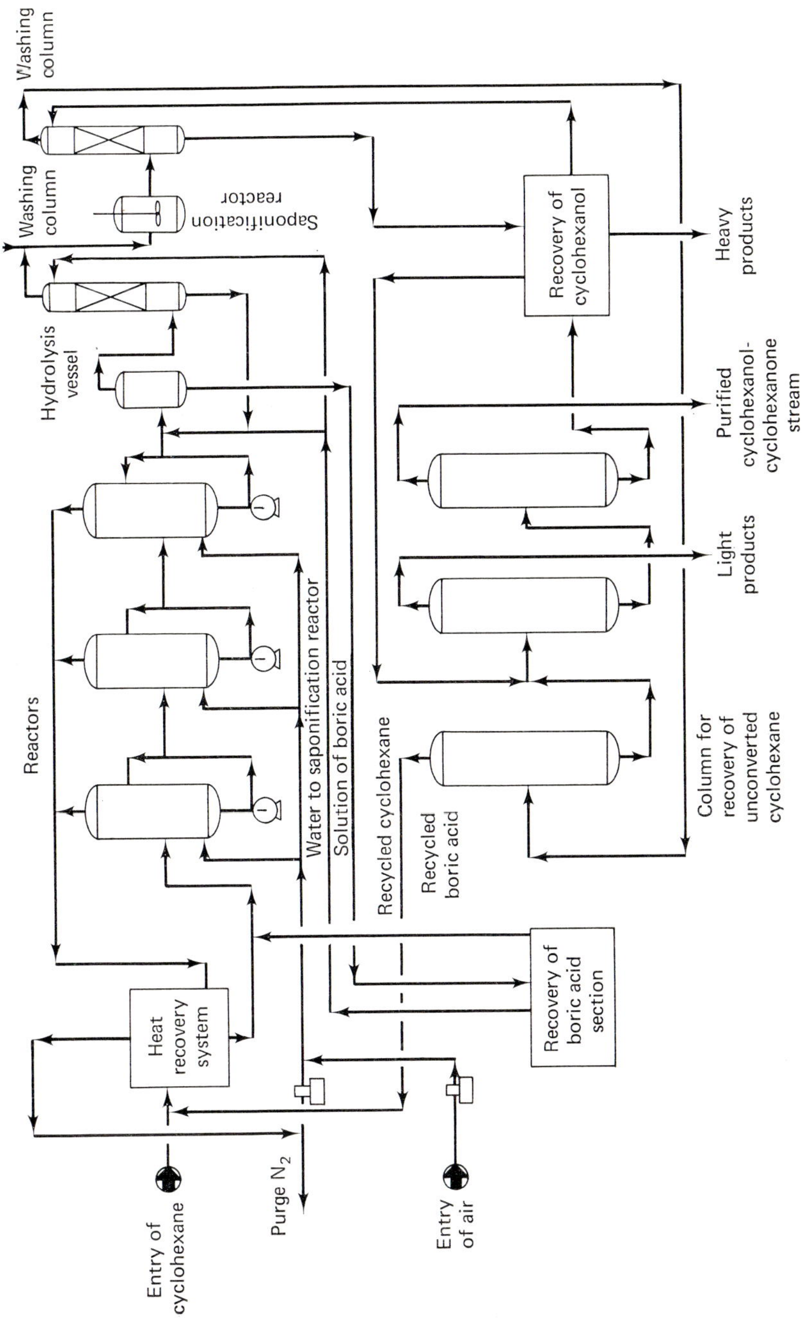

Washing column
Washing column
Saponification reactor
Hydrolysis vessel
Reactors
Water to saponification reactor
Solution of boric acid
Recycled cyclohexane
Recycled boric acid
Heat recovery system
Recovery of boric acid section
Entry of cyclohexane
Purge N_2
Entry of air
Recovery of cyclohexanol
Heavy products
Purified cyclohexanol-cyclohexanone stream
Light products
Column for recovery of unconverted cyclohexane

FIG. 6.11

Fresh cyclohexane mixed with recycle material is injected into the first of the three reactors placed in series. The necessary air, diluted by a part of the recycled nitrogen saturated with cyclohexane is injected at the base of the oxidation reactors. The inert gases leaving the reactor and containing an appreciable quantity of cyclohexane vapour are passed into heat exchanger in which the cyclohexane together with the water are condensed, separated by decantation and the cyclohexane recycled. The necessary quantity of inert gas needed to dilute the air to an oxygen content of 8–10 per cent in the oxidising gas is recycled so that the quantity of inert gas, corresponding to the nitrogen in the air necessary for the reaction, is purged from the system after cooling down.

The mean operating conditions in the three reactors are:

Temperature 165°C.
Pressure 12 atm.
Residence time 3 hours in the three reactors together.

The metaboric acid in suspension in the cyclohexane is injected into the first reactor. It represents about 10 per cent by weight of the fresh cyclohexane injected for a conversion ratio of 10 per cent.

The liquid effluent containing the boric acid esters of cyclohexanol is treated with water, the boric esters are very rapidly hydrolysed into cyclohexanol and orthoboric acid, the latter remaining in aqueous solution.

After separation the aqueous phase is sent to the boric acid recovery unit.

6.2.3.3.1. Treatment of the organic phase After washing with water to eliminate the traces of boric acid remaining, the reaction products under optimum reaction conditions are shown in Table VI.9.[36]

Products with boiling points above that of cyclohexanol amounted in toto to 3·2 per cent by wt. of the products of the oxidation and consisted of—

Cyclohexylidene-cyclohexanol-2
Cyclohexane diol
Cyclohexane dione
Lactones, dicyclohexyl and ethyl cyclohexane, etc.

An examination of the composition of the organic phase shows the great diversity of the compounds obtained and hence the complexity of the oxidation reactions.

Most of the degradation reactions have been worked out by Berezin[18] and our results are in general agreement with his.

Thus the presence of formic acid and formaldehyde and, in particular, carbon monoxide and carbon dioxide show that some actual breakdown of large molecules by oxidation takes place. The presence of mono acids is the result of a shortage of oxygen. The appreciable quantity of valeric and butyric acids (20 per cent molar of the total quantity of acids) explains the quantity of carbon dioxide and the formic acid produced.

The phase is treated with alkali to remove the residual free acids and to saponify the esters of the organic acids. This treatment is essential to destroy volatile esters such as cyclohexyl formate and acetate which are not separable by fractionation and hence liable to contaminate the final cyclohexanol-cyclohexane mixture. On the

TABLE VI.9

	Moles
Cyclohexanol	116·10
Cyclohexanone	13·50
Peroxides	2·08
Formic acid	1·64
Acetic acid	0·48
Propionic acid	0·05
Butyric acid	1·00
Valeric acid	1·06
Caproic acid	0·72
Succinic acid	0·55
Glutaric acid	0·11
Adipic acid	1·95
Hydroxycaproic acid	traces
Cyclohexyl formate	4·41
Cyclohexyl acetate	traces
Cyclohexyl propionate	traces
Cyclohexyl butyrate	0·09
Cyclohexyl caproate	0·31
Cyclohexyl succinate	traces
Cyclohexyl glutarate	traces
Cyclohexyl acid adipate	0·57
Cyclohexyl neutral adipate	1·22
n-Pentanol	0·19
Butanol	traces
Formaldehyde	0·17
Carbon monoxide	12·56
Carbon dioxide	2·72
Butane	0·67
Pentane	0·10

other hand too drastic saponification results in a loss of cyclohexanone by condensation. About 80 per cent of the unreacted cyclohexane is removed by fractionation at atmospheric pressure and then the remainder in a column under vacuo to prevent the loss of cyclohexanone by condensation at the foot of the column where it is essential to keep the temperature below 140°C.

The crude cyclohexanol-cyclohexanone is finally purified by the removal of a complex mixture of compounds more volatile than cyclohexanone, the chief of which are *n*-pentanol and aldehydes. In this final distillation it is usual to accept a slight loss of cyclohexanone in the front fraction to enable a column of optimum size to be used.

The efficiency of the purification steps has a major influence on the purity of the caprolactam end product.

The product from the base of the column is therefore distilled, still under vacuo, and at a temperature such that the base of the column does not exceed 140°C, the fronts then being specification cyclohexanol-cyclohexanone. The product from the base of the second column contains the heavy esters remaining after the preliminary saponification and are submitted to more drastic alkali treatment to recover residual cyclohexanol.

If alcohols such as pentanol had been allowed to remain in the purified product the corresponding aldehydes and ketones would be formed in the dehydrogenation step and these react in the subsequent stages of the process forming impurities which appear in the caprolactam end product.

Analysis of the cyclohexanol-cyclohexanone product shows the presence of impurities to the extent as given in Table VI.10.

TABLE VI.10

Impurities Present in Cyclohexanol-Cyclohexanone Mixture as Shown by Vapour Phase Chromatography

	% wt
Cyclohexyl formate	0·01
n-Pentanol	0·03
Cyclopentanol	0·04
Cyclohexyl propionate	0·01
Other impurities	0·13

The relative proportions of the main ingredients are:

Cyclohexanone	13·8%
Cyclohexanol	86·2%

A chemical estimation of acids and esters present showed that acids were not detectable and esters were present to the extent of 0·18 equivalents per kilogram.

The purity of the mixture controls the quality of the caprolactam end product. Two main types of impurities can be distinguished:

(a) Cyclohexyl esters which tend to inhibit the dehydrogenation stage.

(b) Aldehydes and ketones other than cyclohexanone, also the alcohols they can produce by dehydrogenation.

All aldehydes and ketones react to form oximes and undergo the Beckmann transformation giving amides which take part in the polycondensation reaction and bring about rupture in the polyamide chains.

By oxidation of cyclohexane in presence of boric acid a cyclohexanol-cyclohexanone mixture is obtained, the purification of which does not give rise to any special problems and enables high purity caprolactam to be produced.

6.2.3.3.2. Treatment of the aqueous phase The aqueous solution containing the orthoboric acid is treated to recover and recycle the orthoboric acid. It has an extremely complex composition since it is in equilibrium with the corresponding organic phase and hence its composition is controlled by the distribution coefficients of the various constituents such as cyclohexanol, cyclohexanone and mono- and di-acids with one to six carbon atoms. The solution is first cooled to 50°C in a crystalliser under vacuo. The crystals of orthoboric acid are centrifuged out, then converted to metaboric acid in a rotating oven. The metaboric acid obtained in this way is recycled to the oxidation reactor.

The greater part of the mother liquor is recycled to the reaction vessels in which the boric acids were hydrolysed, the remainder being purged. The actual proportion is determined by the degree of purity needed to prevent the formation of coloured residues arising from thermal decomposition of organic impurities when the orthoboric acid is reconverted to metaboric acid.

The purged liquor is crystallised separately by a forced evaporation and the boric acid thus obtained returned to the first crystallisation units.

The process using boric acid (in spite of the higher capital costs arising for the need for boric acid recovery units) can nevertheless reduce the cost of the cyclohexanol-cyclohexanone mixture appreciably due to the increased selectivity of the reaction.

Information relating to the consumption of utilities and chemicals is given in Table VI.11.

TABLE VI.11

	Consumption per ton of the cyclohexanol-cyclohexanone mixture
Cyclohexane	925 Kg
Boric acid	5 Kg
Caustic soda 100%	33 Kg
Steam at 25 atm	1·4 tons
Steam at 10 atm	4·0 tons
Electricity	300 Kwh
Brine (5 °C)	60,000 Kcal
Cooling water	280 tons
Process water	2 tons

6.2.4. Cyclohexanol

6.2.4.1. Production of Cyclohexanol

Cyclohexanol is manufactured, as we have just seen, by the oxidation of cyclohexane. It can also be produced by the hydrogenation of phenol but this route is less used than the former one. Hydrogenation of phenol is carried out at an elevated temperature and pressure in the presence of the standard hydrogenation catalysts such as copper, nickel or a mixture of the two or their oxides. Actually for reasons explained below, the reverse reaction, the production of phenol from cyclohexanol is of interest.

6.2.4.2. Physical Properties.

Cyclohexanol is a mobile colourless liquid. Its physical constants are given in Table VI.12.[37]

TABLE VI.12

Melting point	25·15°C
Boiling point	161·1°C
Density $F^{20°}/_{4°}$	0·9493
Refractive index $n_D^{25°}$	1·4648
Viscosity, cP at 25°C	4.6
Flash point (open cup)	67·2°C
Vapour pressure at 25°C	1·1 mm Hg

6.2.4.3. *Uses*

The principal use of cyclohexanol is as an intermediate in the production of adipic acid and caprolactam.

Cyclohexanol is also used as a stabiliser for soap, as a solvent for lacquers and varnishes and in the production of plasticisers such as dicylohexyl phthalate.

Finally cyclohexanol is a potential raw material for phenol.

To propose the production of phenol from a cyclohexanol-cyclohexanone mixture would seem somewhat paradoxical since in 1970 25 per cent of phenol in the USA and 40 per cent in Europe was hydrogenated to cyclohexanol and then reduced to cyclohexanone with a view to producing adipic acid and caprolactam.[63] This percentage is however falling each year.

The inverse reaction, cyclohexanol to phenol, has been suggested as a novel possibility due to the improved selectivity in the oxidation of cyclohexane, up to 90 per cent having been obtained using processes involving boric acid. But the important dehydrogenation step must now be considered in detail.

To obtain pure phenol it is necessary to obtain in a single pass a very high conversion since phenol forms an azeotrope with cyclohexanol, rich in phenol, the composition of which varying according to the pressure (75% phenol under a pressure of 0·12 atm).

Reactions leading to the dehydration of cyclohexanol into light products must be avoided, this can be done using nickel or platinum catalysts on an inert support. The operation is carried out at atmospheric pressure in tubular reactors heated in such a way as to maintain the temperature of the catalyst at about 400°C. In this way the reaction is virtually complete with a selectivity of 98 per cent.

Coupling this reaction with the selective oxidation of cyclohexane the quantity of benzene necessary for the production of one ton of phenol is:

Cyclohexanol-cyclohexanone mixture	1·085
Cyclohexane	1·005
Benzene	0·957

which corresponds to an overall selectivity of benzene to phenol of 87 per cent.

This is not very different from that obtainable from other routes for the production of phenol from benzene: 0·93 to 1·0. The relative merits of the different routes must therefore be decided on other economic criteria.

Overall the route via cumene possesses the most advantages, others will only be used when the coproduction of acetone with the phenol presents problems.

The cyclohexane-cyclohexanol-phenol process however shows to advantage over other processes starting from benzene: sulphonation, chlorination (Dow) or oxy-chlorination (Hooker-Raschig), for the technology employed is relatively simple and does not give rise to corrosion problems. A more interesting replacement route on the other hand is that developed by Dow using toluene as a starting material (see Chapter VIII). The cyclohexane route to phenol will show to advantage where it is possible to integrate a dehydrogenation unit into a complex incorporating major capacities for the hydrogenation of benzene and the oxidation of cyclohexane as part of the production of caprolactam.

A plant using the Scientific Design process had been put in operation by Monsanto in Australia but is understood no longer to be used.

6.2.5. Cyclohexanone

6.2.5.1. Production of Cyclohexanone

Cyclohexanone is generally manufactured by the dehydrogenation of cyclohexanol. The overall reaction is:

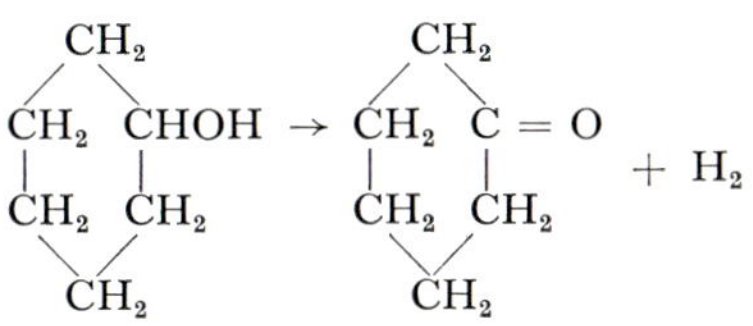

The reaction is endothermic (H = 13 Kcal) and is reversible. The value of the equilibrium constants and the per cent converted when operating with pure cyclohexanol is shown as a function of the temperature in Table VI.13.[38]

TABLE VI.13

Temperature in C°	*Equilibrium constant*	*Conversion* %
220	0·3	0·480
250	0·75	0·654
275	1·5	0·775
350	8·5	0·945
375	16·5	0·971

The desired reaction is by no means the only one, there are parallel reactions leading to cyclohexane, dehydrogenation to benzene and cyclohexane, dehydrogenation of the nucleus giving phenol and condensation of cyclohexanone leading to the formation of heavy products cyclohexenyl-cyclohexanone.

The dehydrogenation is based on a vapour phase reaction in the presence of a solid catalyst. In the past red oxide of iron has been the most commonly used catalyst. But while red oxide of iron is a good catalyst for the dehydrogenation of pure cyclohexanol this does not apply if the mixture from the oxidation of cyclohexane is the starting material.[39]

Manufacturers of cyclohexanone now almost always use this mixture and new catalysts based on zinc have been developed.

Examples:

Scientific Design Co.	Zinc oxide[39]
Inventa AG	Mixture of zinc and calcium oxides and carbonates.[40]
BASF	Zinc oxide.[41]
Institut français du Pétrole	A mixture of oxides of zinc, calcium and potassium, deposited on silica.[42]

As is shown in Table VI.14 these catalysts are very selective as well as giving a good conversion ratio.

TABLE VI.14

Catalyst	*Scientific Design*	*Inventa AG*	*BASF*	*Inst fr. du Pétrole*	*van Hansdieter Hofmann*
Temperature in °C	375–400	425–435	350–370	380–425	360–420
Pressure in atm	1	1		1	
Velocity*	2–4	1	1	2–4	0·5–0·8
Conversion %	—	80	83	80	65–75
Molar selectivity	—	98	98	98	98

* Velocity is expressed in terms of litres of raw material converted per litre of catalyst per hour.

The scheme of the process is shown in Fig. 6.12.[44]

The cyclohexanol-cyclohexanone mixture resulting from combining the product from the oxidation of cyclohexane with recycled cyclohexanol containing the dehydrogenation by-products is vaporised. The non-volatile portion consisting of cyclohexanol, cyclohexanone and the greater part of the by-products is rejected or sent back to the oxidation unit, actually to the purification stage before recovery of the alcohol and the ketone.

The cyclohexanol and cyclohexanone vapours are passed to the hydrogenation reactors, heat being conveyed by the circulation of molten salts. The product passing out is cooled and condensed. Gas containing mainly hydrogen is removed. The liquid phase is distilled in a primary column allowing water and light hydrocarbons to be removed at the head of the column. The residue is distilled in a second column which separates the pure cyclohexanone from the remaining cyclohexanol and the by-products.

6.2.5.2. Properties of Cyclohexanone

6.2.5.2.1. Physical properties Cyclohexanone is a colourless, slightly volatile liquid. It has a characteristic ketone odour. It is slightly soluble in water (9 %wt at 20°C) but miscible with numerous organic solvents (ethyl alcohol, ethyl ether, benzene, acetone, etc.).

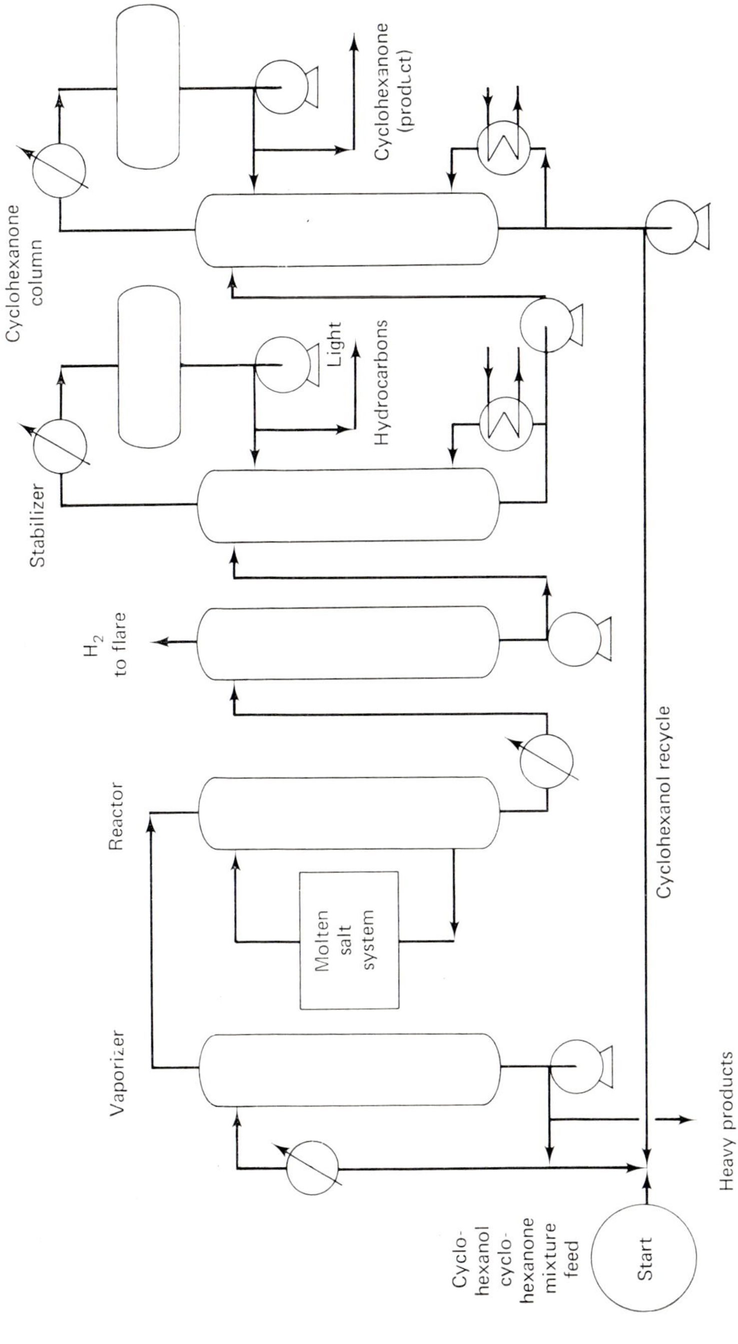
Cyclohexanone column
Cyclohexanone (product)
Stabilizer
Light Hydrocarbons
H_2 to flare
Reactor
Molten salt system
Vaporizer
Cyclohexanol recycle
Heavy products
Cyclo-hexanol cyclo-hexanone mixture feed
Start

Fig. 6.12

Cyclohexanone is an excellent solvent for cellulose acetate and nitrate, vinyl resins, latex, fats, oils, etc.

The physical constants are given in Table VI.15.

TABLE VI.15

Source	*Kirk Othmer*[37]	*Allied Chemical and Dye Corp*[45]
Melting point	—47°C	—
Boiling point	156·7°C	155·5°C
Density $^{20}_{4}$	0·9478	0·9478
Refractive index n_D^{20}	1·4507	
n_D^{25}		1·4489
Viscosity in cP at 20°C	2·2	2·2
Flash point (open cup) °C	54	51
Vapour pressure at 20°C in mm Hg	3·95	see Fig. 6.13
Composition and temperature diagram		see Fig. 6.14

6.2.5.2.2. *Physiological properties*[46] Cyclohexanone seems to have only slight toxic properties. However repeated or prolonged contact with the skin can produce dermatitis as a result of the fat-solvent properties.

Some irritation of the retina has been noted while cyclohexanone can also act on the lining of the upper respiratory tract. At high concentrations cyclohexanone has a narcotic effect. The maximum allowable concentration of the vapour in air was fixed in 1961 at 50 pts per million or 200 mg/m³ by the American Conference of Industrial Hygiene.

6.2.5.2.3. *Chemical properties* Cyclohexanone possesses chemical properties similar to all compounds with the carbonyl group. Thus it reacts with hydroxylamine, phenyl hydrazine, Grignard reagents, hydrogen cyanide, sodium bisulphite, etc. to produce addition compounds. It also reacts with other acids and alkalis to give condensation compounds.

6.2.5.2.4. *Specifications* A commercial product must contain at least 99.7 per cent of cyclohexanone, not more than 0·05 per cent of water, a density at 25/25°C between 0·9440 and 0·9460, a refractive index at 25°C of between 1·4460 and 1·4490. The distillation range between 5 and 95 per cent distilling must not exceed 1·5°C.[37]

When cyclohexanone is used as the starting material for the production of caprolactam, in order to produce a fibre of satisfactory quality, the product must conform to the following specifications:

Cyclohexanone (minimum)	99·8%
Cyclohexanol (maximum)	0·02%
n-Pentanol (maximum)	0·05%
Cyclohexyl formate and similar esters (maximum)	0·08%
Acetic acid and similar acids (maximum)	0·01%
Apha colour index	10

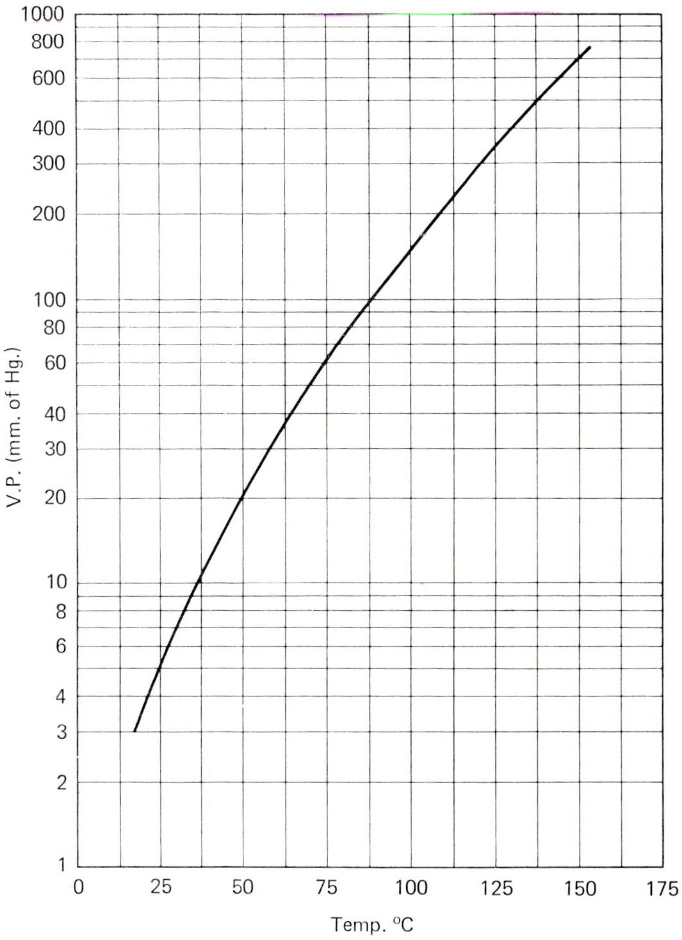

FIG. 6.13

6.2.5.2.5. *Uses* Thanks to its great solvent power cyclohexanone is used in industries dealing with lacquers, varnishes, leather, rubber, printing ink and insecticides, etc.

But its principal use is in the manufacture of caprolactam where it is the starting material, the monomer of Nylon 6, either by the cyclohexanone-oxime route (the one most frequently used) or by the caprolactone route (Union Carbide). These will be dealt with in detail in later sections.

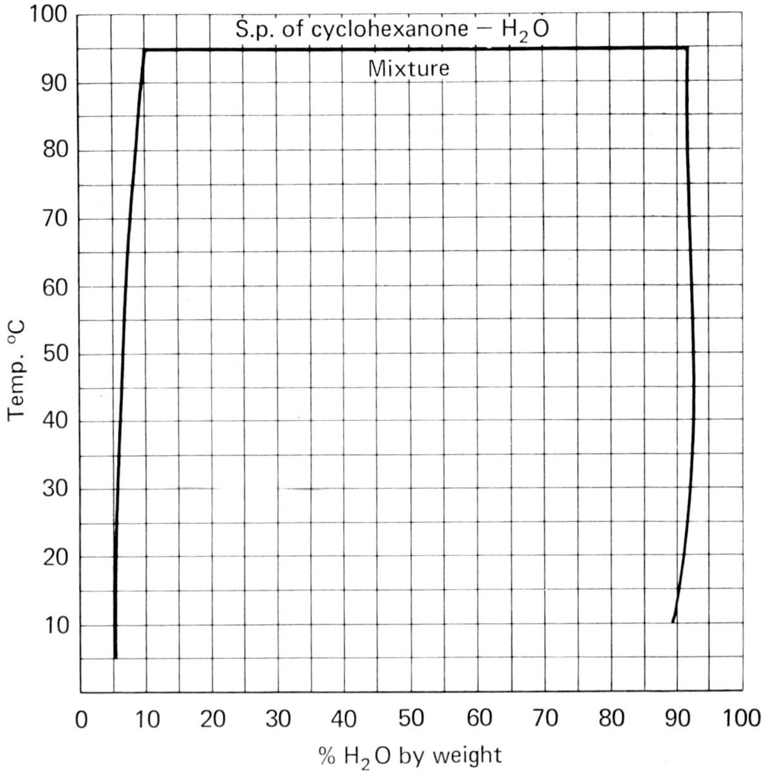

FIG. 6.14

6.3. POLYAMIDE FIBRES AND THEIR PRECURSORS

by Gordon D. Allen and J-C. Marion

Polyamides, frequently referred to in some countries by the generic term 'nylon', are condensation products that contain recurring amide groups as integral parts of the main polymer chains. Condensation takes place between bifunctional monomers such as amino acids, diamines and dibasic acids, although occasionally polyamides are formed by addition polymerisation especially where the monomers are cyclic lactams. Typical structural formulae of linear polyamides may be represented as:

$$H_2NRNH(COR'CONHRNH)_nCOR'COOH$$

or

$$H_2NRCO(NHRCO)_nNHRCOOH$$

depending on methods of formation, where R and R′ represent chains between bifunctional groups in the reactants and n represents the degree of polymerisation or number of recurring groups in the polymer chain.[47]

Although considerable research and development has been carried out with various combinations of amino acids, diamines-dibasic acids, only two polyamides—polyamide 66 (formed by condensation of hexamethylene diamine and adipic acid) and polyamide 6 (polycaprolactam) are produced on a large scale commercially—with world-wide output around 2 million t/a of which around 55 per cent is polyamide 66:

$$[\text{—NH(CH}_2)_6\text{—NH—CO(CH}_2)_4\text{CO—}]_n \quad \text{polyamide 66}$$

$$[\text{—NH(CH}_2)_5\text{CO—}]_n \quad \text{polyamide 6}$$

From the fibre consumption viewpoint, polyamide 66 and polyamide 6 are very similar. The production routes to their monomer and subsequently polymerisation and fibre-forming are very different. They are, therefore, treated separately. The production of cyclohexanol-cyclohexanone mixture, known as KA Oil, has previously been described, here the subsequent steps are described below. But here it is noted that fibre-grade monomers have to have a very high purity in order to yield commercially acceptable fibres.

6.3.1. Polyamide 66

In the manufacture of nylon 66, first commercialised in 1938 by E. I. Du Pont de Nemours and Company in the USA, two monomers, adipic acid and hexamethylene diamine, are required to produce the intermediate polyamide salt (hexamethylene diamine adipate) which undergoes the condensation polymerisation to polyamide 66.

The production of adipic acid and hexamethylene diamine is, in general, made by a few giant chemical enterprises which frequently operate an integrated production system ranging from raw material to fibrous product. Polyamide 66 technology has been particularly exploited in the UK (ICI) and France (Rhône-Poulenc) amongst European nations and patent coverage did not expire in some areas until the 1960's.

The number of monomer processes operational is limited, but over the years, producers have introduced process improvements which are not disclosed but are known to have an important impact on economics of production.

6.3.1.1. Adipic Acid

Adipic acid is produced commercially by oxidation of KA Oil (cyclohexanol-cyclohexanone mixture) using:

(1) nitric acid

(2) air

The former is by far the more important in practice, as air oxidation routes—associated with Du Pont and Scientific Design, do not attain the high purity level essential for fibre-grade adipic acid. From the economic viewpoint, the combination of cyclohexane/air catalytic oxidation to KA Oil followed by nitric acid oxidation of the KA Oil to adipic acid, apparently offers the more attractive economics because of the better overall yields, in spite of slightly higher initial capital investment.

6.3.1.1.1. Nitric acid and oxidation of KA Oil: The oxidation of KA Oil to adipic acid, a flow sheet for which is shown in Fig. 6.15, is carried out in a two stage reactor

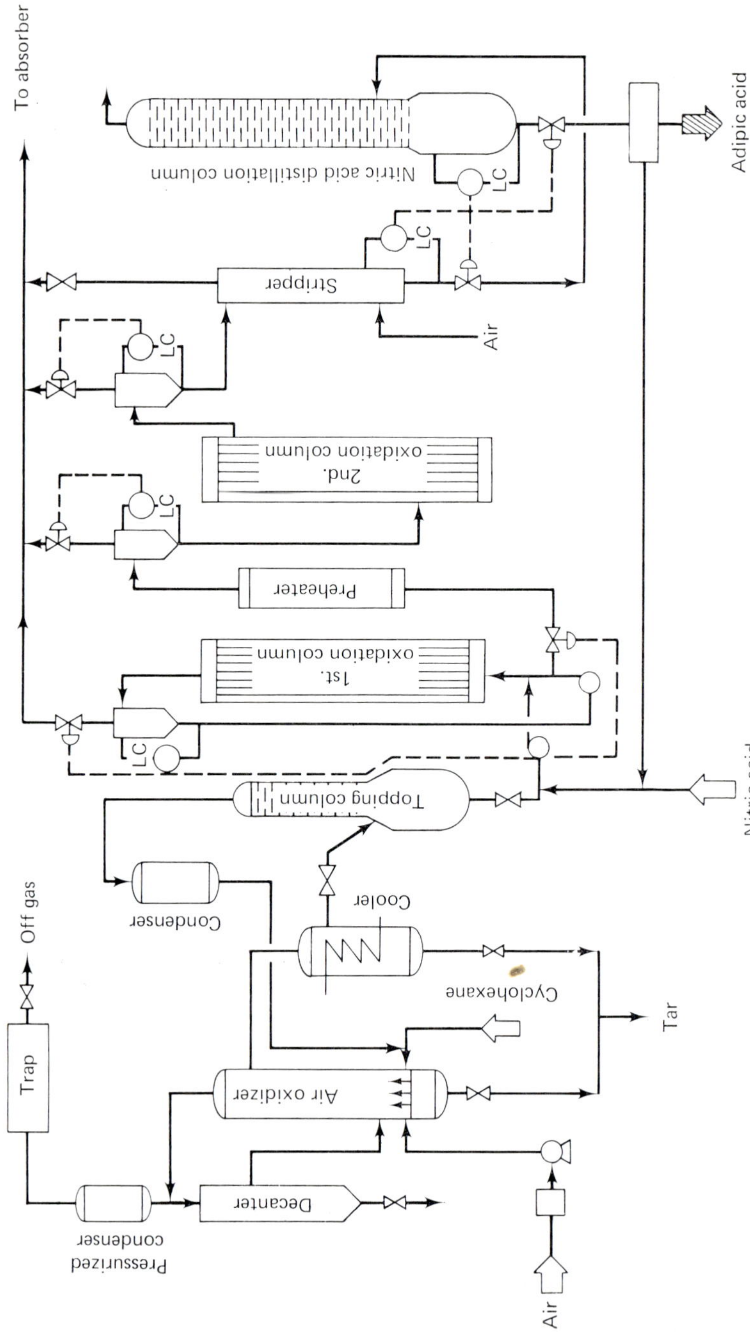

FIG. 6.15 *Flow sheet for adipic acid manufacturing process (two-stage process for air oxidation and nitric acid oxidation)*

system.[48,49] Nitric acid at 60 wt per cent concentration is fed into the first stage reactor in ratios of 3:1 or higher to KA Oil, commonly along with a catalyst of copper and vanadium compounds.

The copper, usually in the form of cupric nitrate serves to retard the formation of by-product acids formed from the cyclohexanol or other intermediates, while the vanadium normally in the form of ammonium vanadate retards the formation of by-products from the 'dione' intermediate. In the reaction which is carried out for about 45 min. at between 65–70°C and a pressure of about 30 p.s.i.g. 90 per cent of the oil is converted. The reaction continues for a further 45 min. to completion in a second stage reactor where the temperature range is between 90-110°C.

Crude product from the reaction system is sent to the crystallisation section where it is subject to two stages of crystallisation. The first stage gives about 95 per cent purity after centrifuging. It is then redissolved in water and recrystallised to produce fibre grade adipic acid. The economics of the process are closely tied up with the usage of nitric acid. If the ratio of the nitric acid to KA Oil utilised in the reaction is increased, higher yields of adipic acid are obtained. However this involves additional operating costs and emphasises the need for efficient recovery techniques.

6.3.1.1.2. Air oxidation of KA Oil The alternative air oxidation of KA Oil process, carried out in glacial acetic acid solvent, is seldom used because the overall time for the two stage reaction is four times longer than that of the nitric acid process.

In the process KA Oil, catalyst (usually copper and manganese acetates) and the acetic acid solvent are fed to the first stage reactor which operates at 160–175°C and 100 p.s.i.g. Air is then sparged into the bottom of the reactor. The exothermic heat of reaction, completion of which takes about 3 hours, is removed by internal cooling coils. An overhead condenser recovers the solvent and oil carried over with the reactor off-gas.

The effluent from the first stage is fed to the second stage reactor for a similar period of treatment at 80°C and 100 p.s.i.g. The adipic acid product is then purified as before by two-stage crystallisation, while the solvent is recovered, and after treatment recycled.

Other techniques have been investigated to produce adipic acid including the oxidation of such products as cyclohexene, chlorocyclohexane, caprolactone and cyclohexylamine as well as reacting acetylene with formaldehyde via the Reppe process, tetrahydrofuran and carbon monoxide, acetylene and acetic acid and the dimerisation of acrylic ester, but none are likely in the forseeable future to replace the existing KA Oil oxidation process.

6.3.1.2. Hexamethylene diamine

Hexamethylene diamine is the other monomer required for polyamide 66. It is obtained industrially by several different processing techniques. Fig. 6.16 gives a schematic diagram of the manufacturing processes in current use:

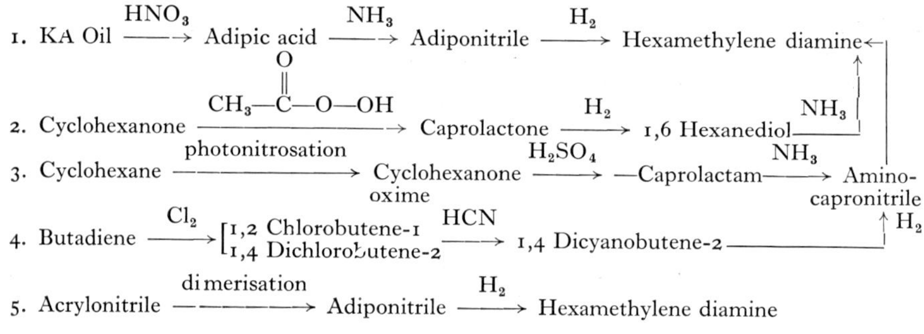

KA Oil is Cyclohexanol-Cyclohexanone mixture

FIG. 6.16 *Hexamethylene diamine manufacturing processes in current use*

The latter two processes are included in this section owing to their commercial importance and for completeness in describing polyamide 66 precursor production. The KA Oil/adipic acid route is the most widely used; together with the butadiene route, the two processes account for most of the World's production of hexamethylene diamine.

6.3.1.2.1. KA Oil/adipic acid process: The oxidation of KA Oil to adipic acid is described in Section 6.3.1.1. The main two stages in the subsequent processing are the conversions of:
(1) adipic acid to adiponitrile
(2) adiponitrile to hexamethylene diamine
which are carried out consecutively without the separation of adiponitrile.

6.3.1.2.1.1. *Adiponitrile from Adipic Acid.* The process for producing adiponitrile (Fig. 6.17) is based on the vapour phase reaction of adipic acid with ammonia in the presence of a dehydration catalyst:[50]

$$HOOC(CH_2)_4COOH + 2NH_3 \rightarrow NC(CH_2)_4CN + 4H_2O$$

Solid adipic acid is melted and kept in liquid form at 205°C under an ammonia atmosphere. Anhydrous ammonia is vaporised at 20°C and 120 p.s.i.g. in an evaporator and then superheated to 280°C prior to entering a specially designed multitubular reactor packed with catalyst, usually boric acid or boron phosphate and phosphoric acid supported with silica. The liquid adipic acid is then heated to 280°C and fed to the bottom of the reactor where atomisation and vaporisation takes place. The vaporised acid is mixed with ammonia in a molar ratio of 1:10 where a strongly exothermic reaction takes place at around 280°C. The vapours leaving the reactor are cooled down to 230°C and separated into a vapour and a liquid phase. The vapour phase, containing mostly ammonia and water with some adiponitrile and unreacted acid, is condensed leaving the non-condensed ammonia to be compressed, and recycled back to the reactor.

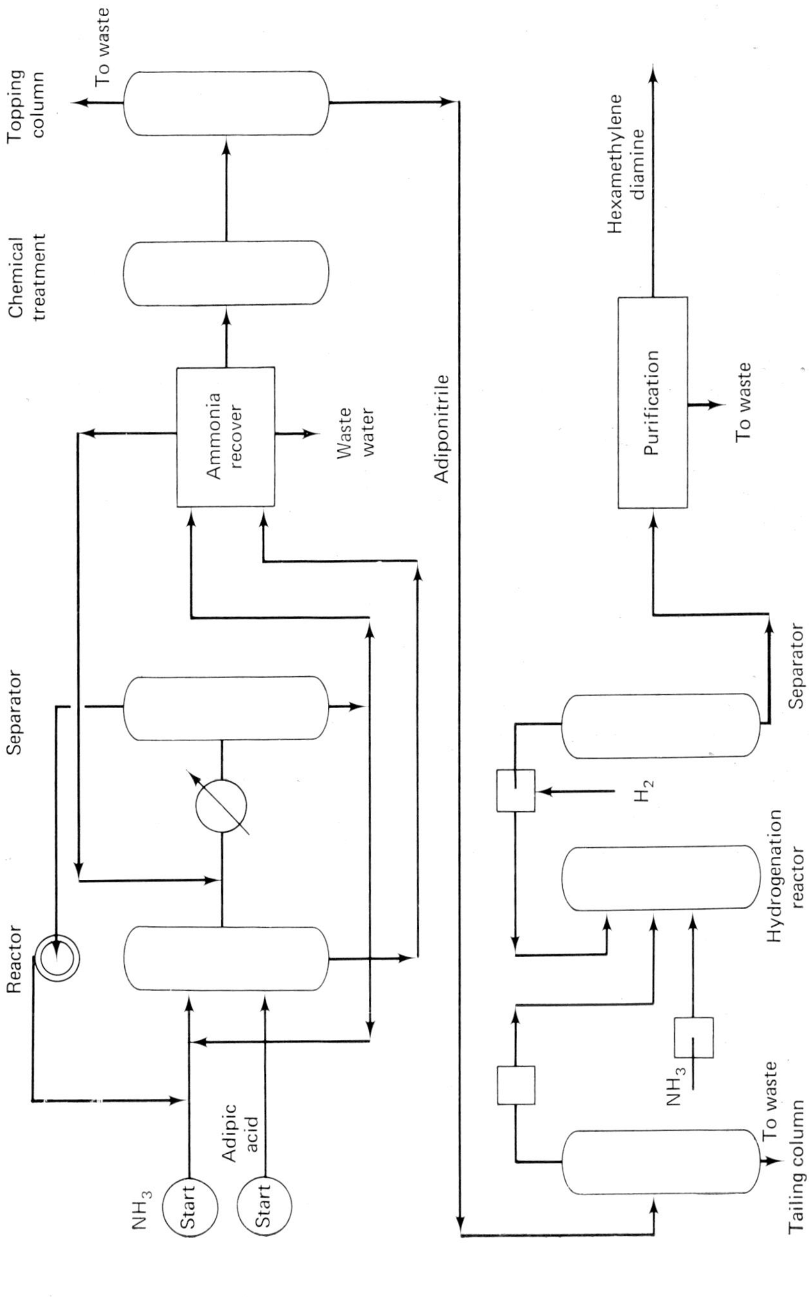

FIG. 6.17 *Simplified flow chart for the production of Hexamethylene Diamine. Two stage process: adiponitrile from adipic acid hexamethylene diamine from adiponitrile*

The liquid phase, which contains adipic acid, partially converted acid, adiponitrile and by-products of the reaction, is fed to a second reactor to complete the reaction at 280°C. The vapours from this reactor are distilled to give crude adiponitrile which then undergoes dehydration to remove water and some of the heavy by-products, and vacuum distillation to remove some of the light component impurities. Further distillation and chemical treatment gives pure adiponitrile suitable for dehydrogenation to hexamethylene diamine.

The main problem with the process is that the yields obtained can be relatively low—around 80 per cent because of adipic acid loss by decomposition during vaporisation, but this is being improved by the development of instant vaporisation techniques and the use of fixed bed and fluidised bed reactors, so that yields in excess of 90 per cent are possible. Techniques have also been developed to allow the reaction to take place in the liquid phase.

6.3.1.2.1.2. *Hexamethylene Diamine from Adiponitrile.* Almost all hexamethylene diamine is currently produced commercially by the hydrogenation of adiponitrile:[49]

$$NC(CH_2)_4CN + 4H_2 \rightarrow H_2N(CH_2)_6NH_2$$

This hydrogenation (Fig. 6.17) is carried out in the liquid phase by contacting liquid adiponitrile, toluene and ammonia with a gaseous mixture of hydrogen, ammonia, and minor amounts of adiponitrile and toluene in a fixed bed reactor containing cobalt oxide catalyst. It is essential that all the feedstocks are of a very high purity to prevent poisoning of the catalyst, common in hydrogenation processes. The reaction requires high pressures in the range of 200 to 700 atmospheres and temperatures between 100–250°C. Ammonia is used as a solvent for the adiponitrile, sometimes in conjunction with an aromatic solvent to reduce harmful temperature gradients in the reactor. The reaction must be carefully controlled, but is characterised by its high selectivity and yields.

Crude hexamethylene diamine is removed and undergoes refining first by overhead azeotropic distillation with water and then several vacuum distillations to give the high purity of product required for the manufacture of polyamide 66. Failure to produce highly purified hexamethylene diamine can mean low molecular weight polyamide 66 with poor dyeability, colour, and physico-chemical properties.

The reactor used need not always be a fixed bed variety but can be of the powdered catalyst suspension type. However then both the equipment and operations tend to become complex. Other techniques tried include reducing the pressure to around 20 atmospheres and using different catalysts including nickel, copper, tungsten, zinc, and iron.

6.3.1.2.2. Caprolactone process: the main starting material for this process is cyclohexanone, which is obtained from cyclohexane by air oxidation in the liquid phase to give KA Oil. Cyclohexanol is separated out and undergoes dehydrogenation to convert it also into cyclohexanone.

Cyclohexanone is converted to caprolactone, which is the cyclic ester of hydroxycaproic acid and quite resistant to polymerisation, by a continuous oxidation process with peracetic acid,[51] that in addition produces acetic acid as a co-product:

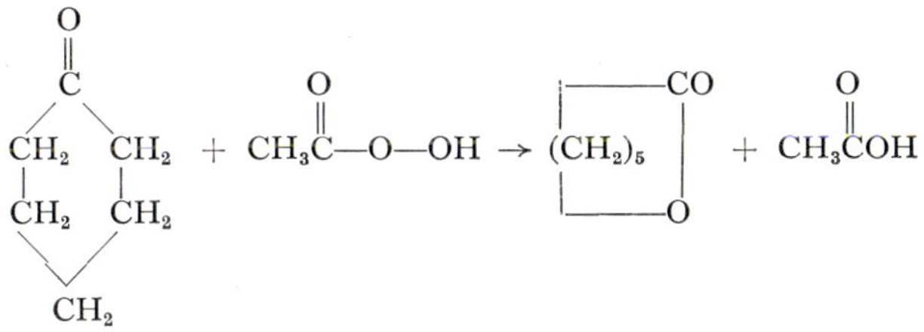

The peracetic acid required for the reaction is produced by the liquid phase oxidation of acetaldehyde in an inert solvent (ethyl acetate or acetone) with oxygen containing 2 wt% of ozone as promoter. The intermediate product formed is acetaldehyde monoperacetate, which is then thermally cracked by vapour phase pyrrolysis to form peracetic acid. This is recovered in an ethyl acetate solvent, while the acetaldehyde is recycled to undergo further oxidation:

$$CH_3\overset{\overset{\displaystyle O}{\|}}{C}H + O_2 \rightarrow CH_3\overset{\overset{\displaystyle O}{\|}}{C}\text{—}OOH$$

The reaction which takes place in a continuous liquid phase reactor tank at between 10–20°C must be carefully controlled, especially with regard to the concentration of acetaldehyde monoperacetate because it can be highly explosive. The reactor effluent is partially vaporised under pressure to remove excess acetic acid, before being fed continuously through a tubular reactor. This is externally steam-heated to 100°C resulting in the complete and instantaneous vaporisation of the solution and simultaneous thermal breakdown of acetaldehyde monoperacetate. The vapour is then passed to a fractionating column to separate out the peracetic acid. Purification of the peracetic acid is completed by redistillation.

Peracetic acid can be produced by an alternative process using a reactor and azeotropic column, which involves the liquid phase oxidation of acetic acid with 50 per cent hydrogen peroxide in the presence of sulphuric acid and an entraining agent. The reaction takes place at 40°C and atmospheric pressure or below and requires a total of about 2 hours to complete:

$$CH_3\text{—}\overset{\overset{\displaystyle O}{\|}}{C}\text{—}OH + H_2O_2 \xrightleftharpoons{\text{acid catalyst}} CH_3\text{—}\overset{\overset{\displaystyle O}{\|}}{C}\text{—}O\text{—}OH + H_2O$$

The reaction proceeds to an equilibrium mixture of peracetic acid in an excess of acetic acid and aqueous hydrogen peroxide. For the process to be economical, complete utilisation of the high priced hydrogen peroxide is essential. This requires large excesses of acetic acid (10 moles acetic acid per mole of hydrogen peroxide) and the use of entraining solvents such as ethyl, *n*-propyl, or isopropyl acetates to remove the water formed and thus alter the shift of the equilibria.

The liquid phase oxidation to caprolactone involves a two step reaction of 25 per cent peracetic acid solution with cyclohexanone in acetic acid and ethyl acetate. In the initial reactor, the reaction time is 4 hours at around 40°C and 90 mm of mercury. The heat of reaction is removed by reflux of the ethyl acetate solvent, while the reaction product is then fed to the second vessel, where after a further ½ hour the oxidation is complete.

Caprolactone and by-products then pass to a stripper column, where the cyclohexanone, acetic acid and ethyl acetate are separated out and purified for re-use. The lactone passes into a film evaporator, where the product is vaporised at 50 mm of mercury. The lactone is pumped through a product cooler and finally into a tank ready for the hydrogenation plant.

The hydrogenation reaction, which converts caprolactone continuously into 1,6 hexanediol, is carried out in a fixed-bed type reactor with the catalyst (copper chromite/barium oxide activated) submerged by liquid feed. Hydrogen in the ratio 10:1 to the caprolactone is charged into the bottom of the reactor and bubbles up through the liquid phase catalyst bed mixture. The operating temperature is 260–275°C and the pressure ranges from 4000–4800 p.s.i.g.

The liquid reactor product flows to a separation unit where it is cooled by heat exchange. The product goes to flash drums to assist in the recovery of the hydrogen and finally to a series of columns, where pure 1,6 hexanediol is extracted and unreacted products returned for recycle:

$$\underbrace{CH_2(CH_2)_4C=O}_{\text{—O—}} + 2H_2 \xrightarrow[\text{barium oxide activated}]{\text{copper chromite}} HOCH_2\text{—}(CH_2)_4\text{—}CH_2OH$$

The final stage of the process is the ammonolysis of 1,6 hexanediol to hexamethylene diamine, whereby 1,6 hexanediol is reacted with ammonia and a small amount of hydrogen at high temperatures and pressures:

$$HOCH_2(CH_2)_4CH_2OH + 2NH_3 \xrightarrow[\text{catalyst}]{\text{Raney}} H_2NCH_2(CH_2)_4CH_2NH_2 + H_2O$$

Hydrogen gas is supplied to avoid the possibility of dehydrogenation of hexamethylene diamine to adiponitrile. The catalysts suitable for the ammonolysis include Raney cobalt, Raney nickel, reduced copper, Raney copper and nickel on Kieselguhr but Raney nickel is normally preferred.

6.3.1.2.3. Caprolactam process: This process was developed by the Toyo Rayon Company, "Toray," in Japan and is being operated on a small scale. According to patents taken out by the company, caprolactam and ammonia are contacted in the gaseous phase at about 350°C to form aminocapronitrile in the presence of a phosphate catalyst (e.g. magnesium, aluminium, calcium, barium, or zinc). The yield is said to be quantitative. Aminocapronitrile is hydrogenated to give hexamethylene diamine by a process similar to that of the hydrogenation of adiponitrile already described.

The feasibility of the process for large scale operation is unproven, but it could be economical on a small scale as one of the methods to effectively utilise off-grade lactam at integrated polyamide 6 producers.

Toray, which developed the PNC process for caprolactam production (treated in polyamide 6 section), is also able as a result of this process to obtain hexamethylene diamine as a by-product of photonitrosation.

Although this section is basically limited to the production of hexamethylene diamine from cyclohexane and/or KA Oil, it would not be complete without a brief description of two important commercial processes starting from other raw materials.

6.3.1.2.4. Butadiene process: This process was developed by Du Pont and is being used by them for much of their hexamethylene diamine production.

The first step of the process[52] is the vapour phase chlorination of butadiene, which is carried out at between 200–250°C and pressures slightly above atmospheric. The molar ratio is 4:1 butadiene to chlorine, but can be higher at 6·5:1 to reduce by-product formation. The products of the reaction are 1,4 dichlorobutene-2 and 1,2 dichlorobutene-1 which are formed in almost equal ratios:

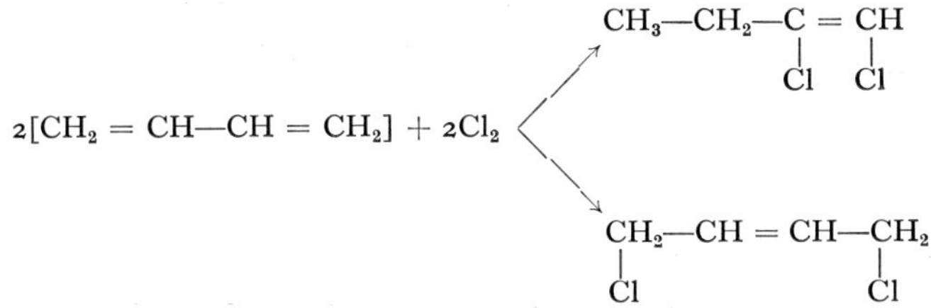

1,2 chlorobutene-1

1,4 dichloro butene-2

The dichlorobutenes mixture and hydrogen cyanide are reacted with each other in an organic solvent at 100–150°C in the presence of a hydrogen chloride acceptor such as calcium carbonate to obtain 1,4 dicyanobutene-2 at a yield of about 95 per cent 1,4 dicyanobutene-2 is then isomerised at 50–100°C in an inert solvent (benzene) with alkali hydroxide (sodium hydroxide) or amine as the catalyst. The 1,4 dicyclobutene isomerisate is then hydrogenated at normal pressure in the gaseous phase at about 250°C on a palladium catalyst; which results in adiponitrile being obtained almost quantitatively:

$$NCCH_2CH = CHCH_2CN + H_2 \rightarrow NCCH_2CH_2CH_2CH_2CN$$

The conversion of the adiponitrile to hexamethylene diamine follows the same steps as those already described for the adipic acid process.

The advantages of the process are a high conversion rate at each reaction stage and high yield. Profitability depends on the use of low price butadiene and the availability of large quantities of relatively cheap chlorine and hydrocyanic acid.

6.3.1.2.5. Acrylonitrile dimerisation process: This process was first commercialised by the Monsanto Company in the USA. Subsequently, Asahi Chemical Company in Japan now also claim to have developed a unique process for acrylonitrile dimerisation. Various other companies have also carried out investigations in this field particularly acrylonitrile producers throughout the world.

In the Monsanto process,[53] acrylonitrile is dissolved in an electrolyte at a concentration of more than 10 per cent and introduced into the cathodic cell. The electrolyte is an aqueous solution of tetraethyl ammonium *p*-toluene sulphate and the cathode is lead. An ion exchange membrane is placed between the cathodic compartment and the anodic cell, which contains dilute sulphuric acid. With the pH in the catholyte maintained at 7–9·5, acrylonitrile is dimerised through electrolyte reduction to form adiponitrile. A yield of more than 90 per cent is claimed.

The bis-cyanoethyl ether impurity formed in this process is decomposed with the aid of such agents as magnesium carbonate, alumina, chromium trioxide, sodium permanganate in order to facilitate refining, while the other impurities which include propionitrile, ethylene cyanhydrin, and 2 cyanoethyl adiponitrile can be separated out by distillation:

$$2CH_2CHCN + H_2 \rightarrow NC(CH_2)_4CN$$

The conversion of adiponitrile to hexamethylene diamine follows the same steps as for adipic acid.

The main advantage of the process is relatively cheap acrylonitrile raw material, but this is balanced by heavy power consumption and increased equipment costs.

6.3.1.3. Polymerisation of Polyamide 66

The first stage in the polymerisation of polyamide 66[54] is the formation, isolation and purification of hexamethylene diamine adipate (AH salt). This is carried out by mixing stoichiometric proportions of hexamethylene diamine and adipic acid in water, where it is 47 per cent soluble at 18°C; to ensure the balance of reactants for the preparation of high molecular weight polymer. This solution is purified and concentrated as necessary for the next stage of the process. If the AH salt is to be isolated as white crystals melting around 190°C, extraction is carried out with benzene. Subsequent conversion back to an aqueous solution involves dissolving the crystals in demineralised water and removing the impurities by filtration.

Actual commercial polymerisation processes vary from company to company with both batch and continuous techniques available. Normally an aqueous 47–60 per cent solution of AH salts is charged into an autoclave under an inert atmosphere, usually nitrogen, together with dosing reagents such as 0·6– 1·2 mole per cent of acetic acid stabiliser to control viscosity to between 2·2 and 2·8, delustrants, pigments, and heat and light stabilisers. The pressure in the reactor is normally about 250 p.s.i.g. but pressures as high as 400 p.s.i.g. have been employed. The pressure is kept constant during the polycondensation to polyamide 66 by allowing the steam produced to escape into the atmosphere. This is necessary to prevent the polymer from freezing out before it reaches a degree of polymerisation (the number of amide links in the average polymer chain—in this case 100) suitable for fibre spinning.

Towards the end of the reaction, the pressure is gradually lowered to atmospheric and the temperature held between 250 and 300°C for about 1 hour to complete the reaction. If further polymerisation is required to give a high viscosity product, the melt is subjected to vacuum treatment. The polymer melt is then removed, cooled, dried, and chipped ready for storage.

In continuous processes fibres can, with process modifications, be spun directly from the melt.

6.3.1.4. Spinning and Drawing[55]

Polyamide 66 is melt spun. The ribbon of polyamide is broken into chips. These are fed through a hopper into a spinning vessel. After this process, they fall on to an electrically heated grid which has a mesh too small to pass the chips until they have melted. The molten polyamide passes into the pool where the level of molten polyamide must be kept practically constant.

The melt (fusion temperature 288°C) is metered by a pump through a filter to the orifices of the spinnerette. The melt solidifies immediately on passing through the spinnerette and the so-formed filaments pass through a cooling chamber in which a cold air current is swept across them. The yarn passes through a steam chamber to moisten before winding.

The filaments as they are first spun are not very strong and very dull. They are, however, cold drawn by stretching by about 400 per cent, giving brighter filaments with better tenacity and elongation-at-break, and wound onto the primary spinning package.

Polyamide yarn passes around guides between a pair of input feed nip rollers determining its critical speed and passes over a deflector and then two or three times around the output feed roller; thence via a guide on to the take-up bobbin, twist being inserted into the yarn during take-up. Staple is obtained by cutting filaments, spinning the short lengths and winding on to a former.

6.3.2. Polyamide 6

Polyamide 6 is the homopolymer of caprolactam (2-oxohexamethylenimine). Its development stems from original work associated with Paul Schlack of IG Farbenindustrie in Germany during the late 1930's. Prior to World War II, it had attained commercial status and after hostilities, polyamide 6 was produced in other countries, including Italy, Japan, and the USSR. Today, almost all entrants to polyamide fibre production including small units in developing nations—manufacture polyamide 6 in preference to the more technologically complex 66-fibre.

6.3.2.1. Caprolactam

There are a number of production routes to caprolactam, associated with such companies as DSM, Inventa, Scientific Design, etc. The mixture of cyclohexanol and cyclohexanone, KA Oil, is now one of the most important intermediates in the manufacture of caprolactam required for polyamide 6 fibre production.

Traditionally, phenol has been used as key raw material for caprolactam. It is, however, more expensive than the cyclohexane derived route. Even such companies as Allied Chemical Corporation, DSM, and BASF, which possess proprietary technology and/or in-house production facilities are switching away from phenol. Below we discuss the commercial processes:

(a) via KA Oil (notably oxime processes)
(b) cyclohexane (photosynthesis: Toray)

(c) Toluene (notably SNIA)
other process are mentioned in the literature. The aim of current research is two-fold:
—lower caprolactam costs
—elimination of by-product ammonium sulphate
and progress commercially in the various techniques is outlined below: (Also see Fig. 6.18 which gives a schematic diagram of the manufacturing processes in current use.)

1. KA Oil (cyclohexanone) $\xrightarrow{NH_2OH}$ cyclohexanone oxime $\xrightarrow{H_2SO_4}$ caprolactam

2. KA Oil (cyclohexanone) $\xrightarrow{CH_2CO+HNO_3}$ nitrocyclohexanone $\xrightarrow{H_2O}$ nitrocaproic acid $\xrightarrow{H_2}$ amino caproic acid $\xrightarrow{\text{cyclisation}}$ caprolactam

3. KA Oil (cyclohexanone) $\xrightarrow{CH_3-C(=O)-O-OH}$ caprolactone $\xrightarrow{NH_3}$ caprolactam

4. Cyclohexane $\xrightarrow{NOCl\ +\ h\nu}$ cyclohexanone oxime $\xrightarrow{H_2SO_4}$ caprolactam

5. Toluene $\xrightarrow{O_2}$ benzoic acid $\xrightarrow{H_2}$ cyclohexane carboxylic acid $\xrightarrow{NOHSO_4}$ caprolactam

FIG. 6.18 *Caprolactam manufacturing processes in current use*

6.3.2.1.1. Caprolactam from KA Oil There are three main methods of this transformation:
(1) oxime processes
(2) Techni-Chem
(3) Union Carbide
The oxime process in its various proprietary versions is traditional and by far the most current.[56] In the above cited three methods, KA Oil is treated with caustic soda to remove unwanted acid and ester impurities. The cyclohexanone is then separated from the mixture by distillation and the cyclohexanol fraction is converted to cyclohexanone by vapour phase dehydrogenation normally over zinc or copper-based catalysts.

6.3.2.1.1.1. Oxime Processes. Cyclohexanone is converted to the oxime which readily forms caprolactam via the Beckmann rearrangement:

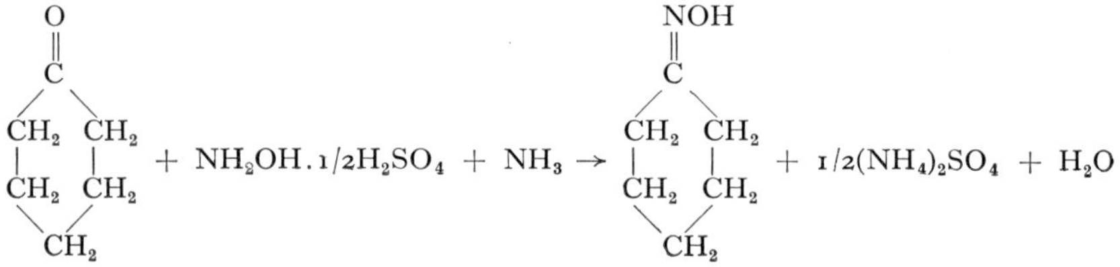

This involves condensing high purity cyclohexanone with hydroxylamine sulphate normally in a reactor system operating at temperatures between 80–110°C. Sulphuric acid liberated in the reaction is neutralised with 25 per cent aqueous ammonia to approximately pH 7. When recirculation through the oxime reactors is complete, a two-phase product stream is withdrawn. Separation of the organic phase containing the oxime from the heavier ammonium sulphate layer is carried out by decantation. The ammonium sulphate recovered is crystallised and used mainly as a fertiliser.

The oxime, in the presence of oleum as a catalyst, undergoes conversion to the lactam according to the Beckmann rearrangement. This reaction combined with neutralisation of oleum may be represented by the following equation:

$$(CH_2)_5C = NOH + H_2SO_4 + 2NH_4OH \rightarrow (CH_2)_5\begin{cases} C = O \\ | \\ NH \end{cases} + (NH_4)_2SO_4$$

Reaction conditions The reaction heat is removed in the reactant cooler through which the reactant is continuously recirculated. The mixture of caprolactam and sulphuric acid is fed to a vapour disengaging unit where the acid is neutralised with a 13 per cent aqueous ammonia solution and recycled back to the degassing vessel to prevent a violent rise in temperature. This neutralisation method prevents deterioration of the caprolactam.

The caprolactam solution and ammonium sulphate solution obtained by neutralisation are cooled and separated in the crude lactam vessel. Each component product then undergoes further processing and purification before being collected and stored.

The reaction requires about 2 kg of sulphuric acid per kg of oxime and consequently on neutralisation produces large quantities of, often unwanted, ammonium sulphate, the selling price for which can greatly affect the overall economics of the whole process. However, yields of the lactam are almost quantitative and as the oxime produced does not contain any unreacted cyclohexanone, the quality of the caprolactam produced is high which is an essential requirement if it is to be used for fibre production.

An alternative route to the oxime which avoids both the oxidation of the cyclohexane and the need for hydroxylamine is by the nitration of the cyclohexane.

Nitrogen can be introduced into the cyclohexane molecule by means of the NO_2 group. The reaction is carried out by reacting cyclohexane with nitric acid either in the liquid phase at temperatures approaching 180°C and pressures of about 30 atm or in the gaseous phase at 300°C and atmospheric pressure.

In both cases there is only a low conversion of cyclohexane while in the liquid phase there is simultaneous production of adipic acid.

By catalytic reduction of the nitrocyclohexane by hydrogen the oxime of cyclohexanone is produced directly which as in other methods can be converted to caprolactam by the Beckmann transformation.

This process has been applied industrially in the USA by Du Pont in a plant with a capacity of the order of 25,000 t/a. However it appears that the plant has not been working for some years.

By-product ammonium sulphate To prevent the formation of ammonium sulphate as a result of the Beckmann rearrangement, agents other than sulphuric acid have been tried. Patent literature indicates that phosphoric acid, boric acid, and silica alumina have been tried but as yet none of these agents are utilised commercially.

Recently, DSM developed a commercial caprolactam process, which totally avoids the formation of by-product ammonium sulphate. The process, known as the sulphuric acid recirculation process, is a further development of DSM's earlier hydroxylamine phosphate oxime (HPO) process[57] for cyclohexanone oxime, the intermediate for caprolactam.

In the HPO process hydroxylamine is produced by bringing hydrogen gas into contact with a circulating inorganic liquid containing nitrite ions, phosphoric acid, as a buffer and a noble metal catalyst. The hydroxylamine produced is reacted with pure cyclohexanone in the pressure of toluene as a solvent. This reduced the quantity of ammonium sulphate produced to 60% of that produced in the conventional process. In the new process ammonium sulphate production is reduced to zero.

Characteristics of this process are:

(a) Beckmann rearrangement of cyclohexanone oxime using oleum to give a sulphuric acid/caprolactam mixture. (The oxime is produced by DSM's HPO process, in which no ammonium sulphate by-product is formed.)

(b) The sulphuric acid-caprolactam mixture is neutralised to ammonium bisulphate and the caprolactam separated from this by solvent extraction.

(c) Oleum recovery from ammonium bisulphate. This is achieved by pyrolysis of the ammonium bisulphate at 1000°C utilising natural gas burners to give sulphur dioxide, water vapour, and nitrogen. Sulphur dioxide is then converted into oleum by a conventional process, and the oleum recycled to the Beckmann rearrangement step.

Thus the ammonia used for neutralisation of the sulphuric acid in the new process is minimised because neutralisation is stopped at the acid sulphate stage, to provide a more easily pyrolysed material. In the pyrolysis, the ammonium moiety of the acid salt is destroyed, giving nitrogen and water only, and the sulphur dioxide which is produced can be converted into the desired oleum by a conventional route. By destruction of the ammonia in this way, production of by-product ammonium sulphate is completely avoided.

The new DSM caprolactam process has not so far been licensed but has obvious applications in locations where there is low fertiliser demand for ammonium sulphate or where there are limitations on discharge of nitrogen-containing effluent to river waters.

Purification Initial treatment of the crude lactam involves removal of the organic reaction phase, extraction with a solvent such as benzene, re-extraction with water and then steam distillation.

Final purification of the lactam is carried out by crystallisation where the crude caprolactam proceeds in one direction with the recycle mother liquor passing counter-currently to the stream of crystals moving forward to be purified. From the crystalliser section, a slurry stream is withdrawn and charged to a filter where the crystals are separated and washed with the mother liquor from the succeeding crystallisation stage. Impurities present remain with the mother liquor, leaving a purified crystalline product. Caprolactam crystals are melted, flash-dried to yield a flaked, high quality fibre-grade monomer.

6.3.2.1.1.2. The Techni-Chem Process. This process[56] has been shown to be technically feasible and economically attractive at pilot plant level but no large scale commercial plant has yet been built. Its main advantage is that it produces no unwanted by-products.

In this process, cyclohexanone produced by oxidation of cyclohexane undergoes first, acylation, with the aid of an acetic anhydride medium to form cyclohexenyl acetate and then nitration to nitrocyclohexanone.

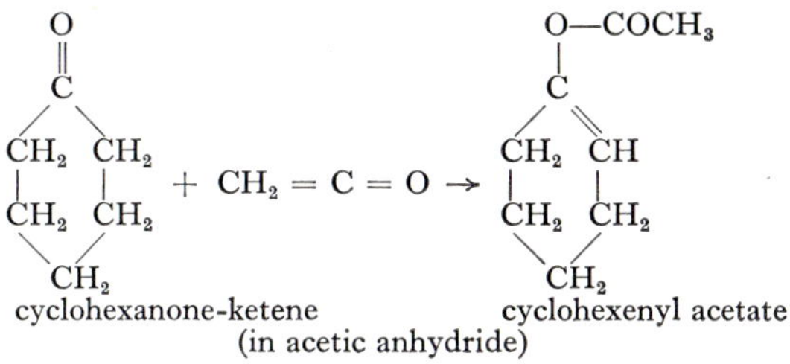

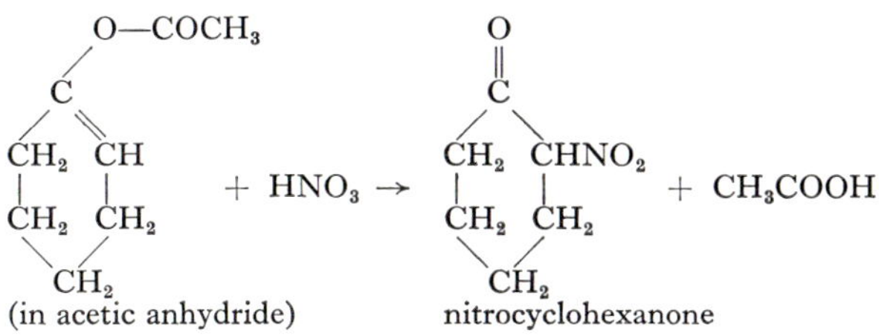

The nitration step is really the key to this process because by proceeding via cyclohexenyl acetate in an acetic anhydride medium, selective mono-nitration is achieved effectively eliminating problems inherently encountered in nitrations. There is also no necessity for appreciable quantities and high concentrations of nitric acid in the reactions.

Cleavage of the ring is by hydrolysis in an aqueous base, and then the resulting solution undergoes a hydrogenation reaction in the presence of a catalyst where at moderate temperature and pressure reduction is accomplished:

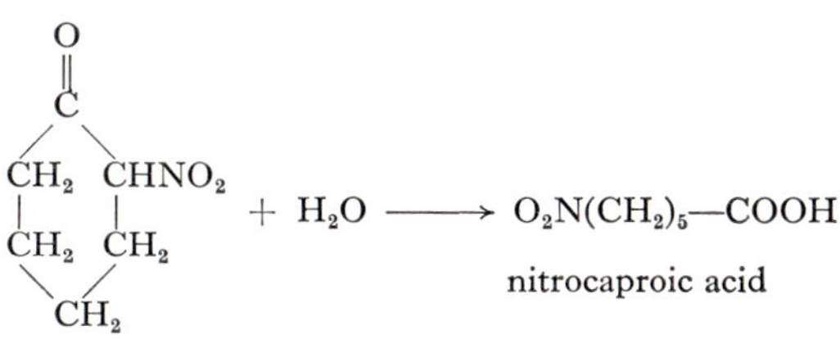

$$O_2N(CH_2)_5\text{—}COOH \xrightarrow{+H_2} H_2N(CH_2)_5\text{—}COOH$$

aminocaproic acid

Aminocaproic acid is then thermally cyclised to caprolactam in aqueous solution, any unreacted acid being separated for recycling in the aqueous solution to regenerate lactam thus enabling high overall yields to be achieved.

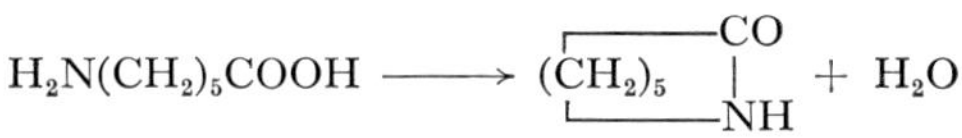

Crude caprolactam is readily extracted from the reaction mixture by means of a suitable organic solvent. which is recovered and recycled from the purification section.

This process, although going through cyclohexanone, avoids the conventional oximation and Beckmann rearrangement reactions, thereby eliminating the formation of by-product ammonium sulphate. The by-product acetic acid produced can be recycled in its entirety to regenerate ketene and as such the overall economics appear attractive.

6.3.2.1.1.3. Union Carbide Process. Another process which avoids the production of unwanted ammonium sulphate was operated by Union Carbide at a plant in the USA. In this process,[58] cyclohexanone is oxidised to caprolactone with peracetic acid. This intermediate is then reacted with an aqueous solution of ammonia at 100°C and 220 atmospheres to form caprolactam which can be extracted and purified:

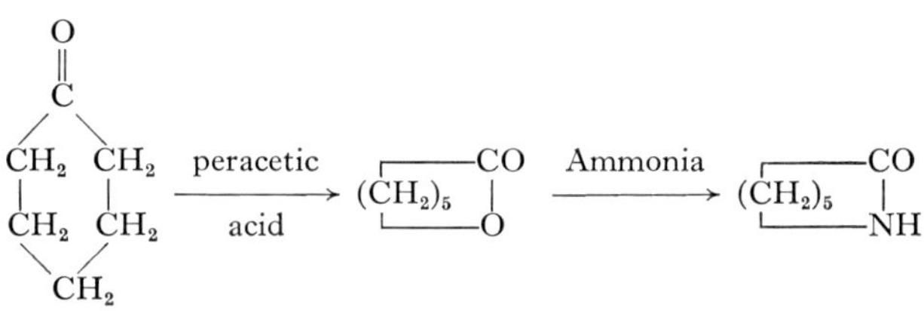

This process had the disadvantage of a relatively low conversion rate and the fact that an almost equal amount of acetic acid is produced as by-product from the production of caprolactone.

Various other modifications and process developments steps have been developed by companies such as Degussa of West Germany, Teijin of Japan, and BP Chemicals in the UK, but it is expected that those going through the oxime step will continue to be the most favoured commercially for some years.

6.3.2.1.2. Caprolactam via Photonitrosation-Toray process The need to find shorter and cheaper production routes to caprolactam has led to the development of several processes which either have fewer process stages, utilise different raw materials, or do not produce unwanted by-products.

In the first category, one of the more novel and attractive economically processes was developed by the Toyo Rayon Company (Toray) in Japan. Research on photonitrosation[56,59,62] to caprolactam was started in 1950, the process commercialised 1962/3 and recently expanded to 150,000 t/a.

The principal innovation of the process is a photochemical reaction brought about by using actinic light from mercury vapour lamps and nitrosyl chloride, which converts cyclohexane directly into cyclohexanone oxime hydrochloride at over 80 per cent yield.

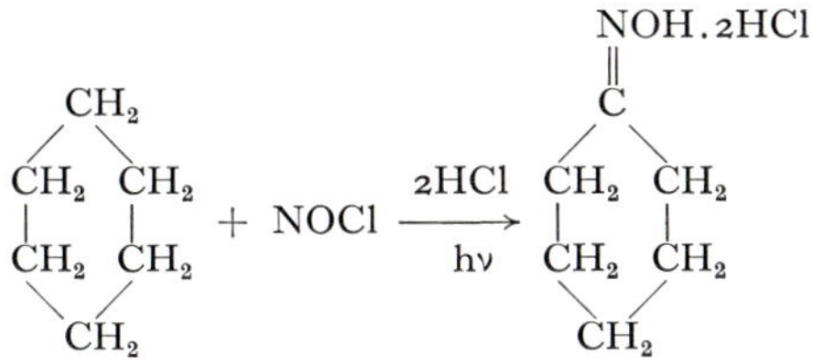

The chemistry of this photochemical reaction can be expressed as follows:

(a) Nitrosyl chloride dissociates into a chlorine radical and a nitric oxide radical in the presence of actinic light:

$$NOCl \xrightarrow{\text{light}} NO^{\cdot} + Cl^{\cdot}$$

(b) The chlorine radical removes a hydrogen atom from cyclohexane molecule to produce a hydrogen chloride molecule and cyclohexyl radical:

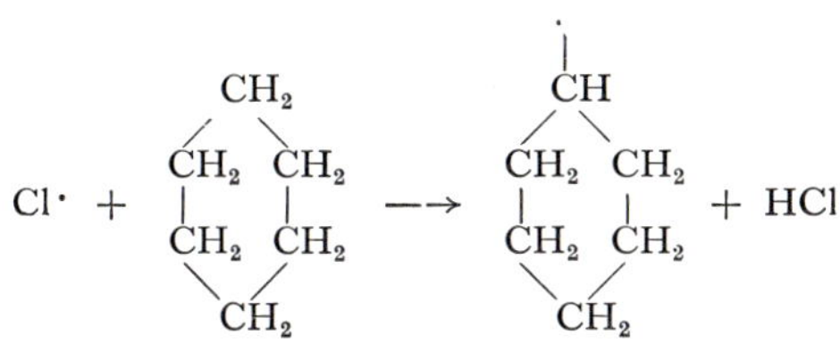

(c) The cyclohexyl radical combines with the nitric oxide radical to form nitroso-cyclohexane, which is chemically unstable under the reaction conditions and is either dimerised to bis-nitrosocyclohexane or in the presence of hydrogen chloride is easily isomerised to the oxime hydrochloride.

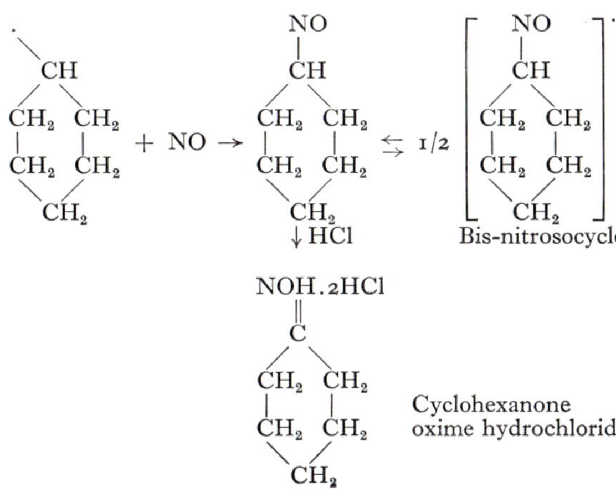

Bis-nitrosocyclohexane

Cyclohexanone oxime hydrochloride

Cyclohexanone oxime hydrochloride thus derived is then converted directly to caprolactam with the evolution of hydrogen chloride gas by the Beckmann rearrangement in the presence of oleum.

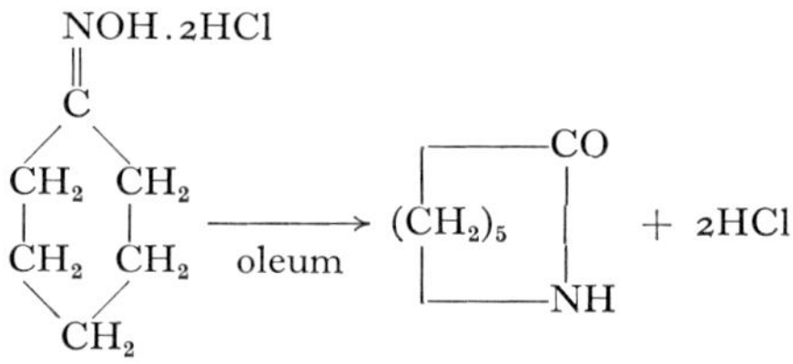

In the production process (flowsheet Fig. 6.19) the first step in the reaction is the manufacture of nitrosyl chloride, obtained by the following reactions:

(a) Firstly catalytic oxygenation of ammonia with air at 700°C and atmospheric pressure using platinium-rhodium gauzes as catalyst:

$$2NH_3+3O_2 \rightarrow N_2O_3+3H_2O$$

(b) The nitrogen oxides are cooled and then concentrated sulphuric acid is reacted with the nitrogen trioxide in an absorption column to give a sulphuric acid solution of nitrosylsulphuric acid:

$$2H_2SO_4+N_2O_3 \rightarrow 2NHOSO_4+H_2O$$

(c) This mixture then passes to a further absorption column where nitrosylchloride is generated by reacting the nitrosylsulphuric acid with preheated hydrogen chloride:

$$HNOSO_4+HCl \rightarrow NOCl+H_2SO_4$$

The mixture of nitrosylchloride and hydrogen chloride (which is recycled) is collected from the top of the column while the sulphuric acid leaving the bottom is concentrated for reuse in the production of further nitrosylsulphuric acid.

The photonitrosation reaction takes place in a horizontal cylindrical reactor with a flat lid through which the mercury lamps are inserted. The nitrosylchloride mixture is reacted with high purity cyclohexane in the presence of the actinic light to give cyclohexane oxime hydrochloride. This is separated out from the base of the reactor, then subjected to treatment with oleum to produce a sulphuric acid solution of caprolactam and hydrogen chloride by the Beckmann rearrangement. As for other caprolactam processes, the sulphuric acid solution is neutralised with ammonia to produce an aqueous ammonium sulphate solution and crude caprolactam, which is purified and flaked ready for use in fibre production.

A very important factor in the industrial application of this reaction lies in the consumption of light, i.e. electrical energy, since the reaction yield is very low. The fact that nitrosyl chloride absorbs light with a wave length below 6000 Å enables the light emitted by the lamps down to 3600 Å to be used; below this wave length the production of tarry by-products is noted. With very powerful lamps (20kW) becoming available the yield of photochemical energy becoming available reaches 30 per cent of the power of the lamp. Overall an output of oxime slightly above 0·3 Kg per kWh can be achieved.[12,62]

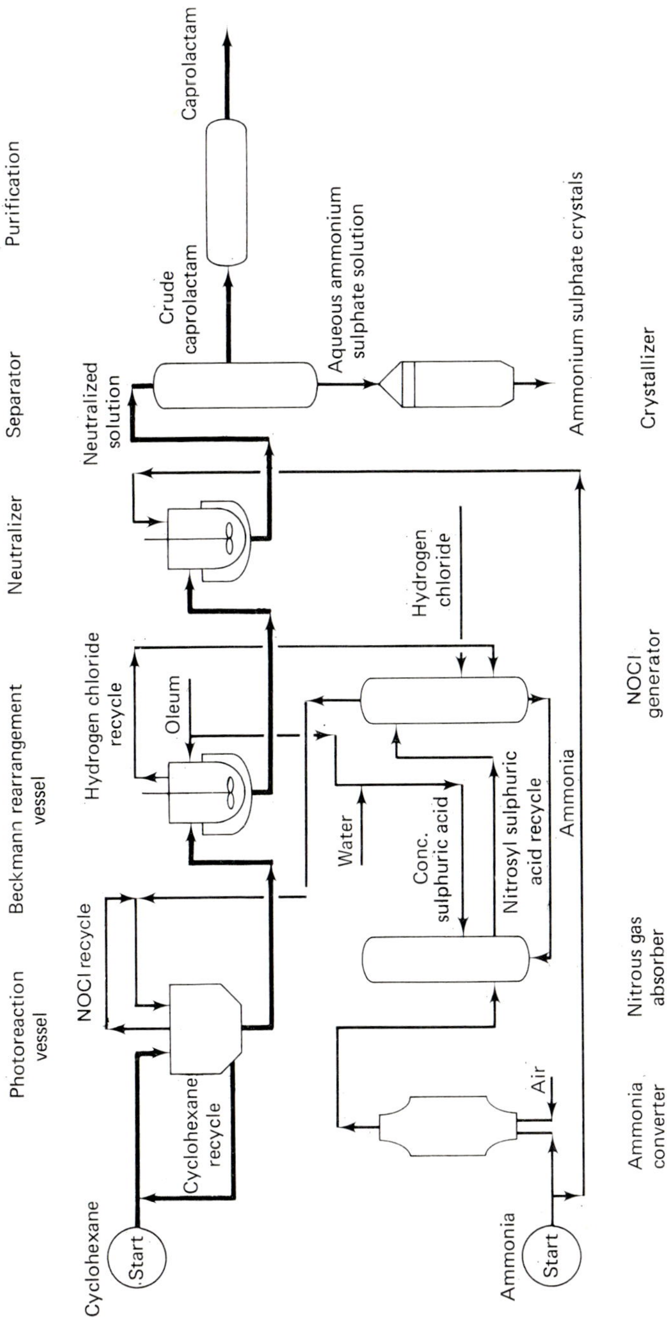

FIG. 6.19 *Toyo Rayon Photochemical Process Flowsheet*[207]

This process was regarded some years ago with much interest because of the highly simplified flow and the low ammonium sulphate production (2·3 Kg per Kg of caprolactam). In comparison to the traditional process, all intermediate steps between starting material and oxime have been eliminated, while by-product production has been reduced to 50 per cent.

6.3.2.1.3. SNIA Viscosa process Currently an important industrial process which starts from a raw material other than cyclohexane is that developed by SNIA Viscosa[60] of Italy during the early 1960s. Here relatively cheap toluene is employed which unlike cyclohexane is not derived directly from benzene. In this process caprolactam is produced via benzoic acid which subsequently undergoes hydrogenation and then nitrosation to produce caprolactam.

Toluene is first converted to benzoic acid by oxidation with air. Reaction is carried out in liquid phase with a cobalt salt as catalyst, at 160–170°C and 8–10 atm pressure:

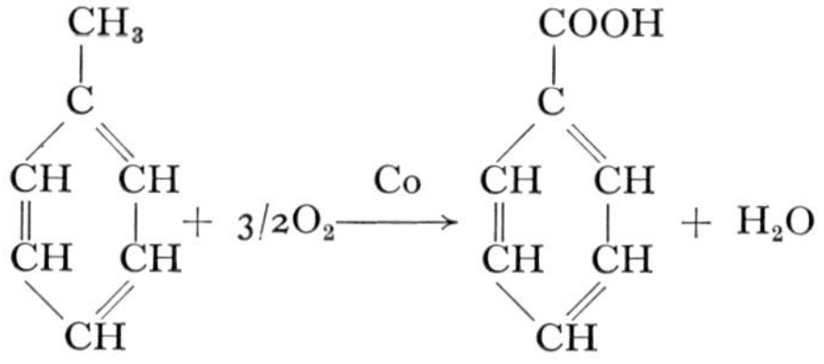

A single continuous reactor is employed. Conversion is close to 30 per cent and reaction yield is about 92 per cent of theoretical.

The next stage is the hydrogenation of benzoic acid to cyclohexane carboxylic acid. This is performed in a series of stirred reactors, in liquid phase, in presence of palladium supported in active carbon. Temperature is about 170°C and pressure 10–17 atm. Hydrogen or ammonia-synthesis gas is employed as reducing agent. The reaction is carried out with 99 per cent conversion and an almost quantitative yield:

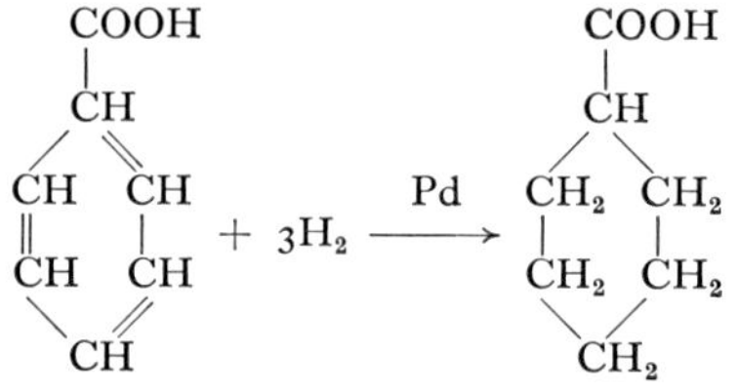

The nitrosating agent of SNIA reaction, nitrosyl sulphuric acid, is used in sulphuric solution at 73 per cent. It is prepared from ammonia, by catalytic oxidation with air on a Pt--Rh net, followed by absorption of N_2O_3 gases in oleum:

$$2NH_3 + 3O_2 \rightarrow NO.NO_2 + 3H_2O$$
$$NO.NO_2 + H_2S_2O_7 \rightarrow 2NOHSO_4$$

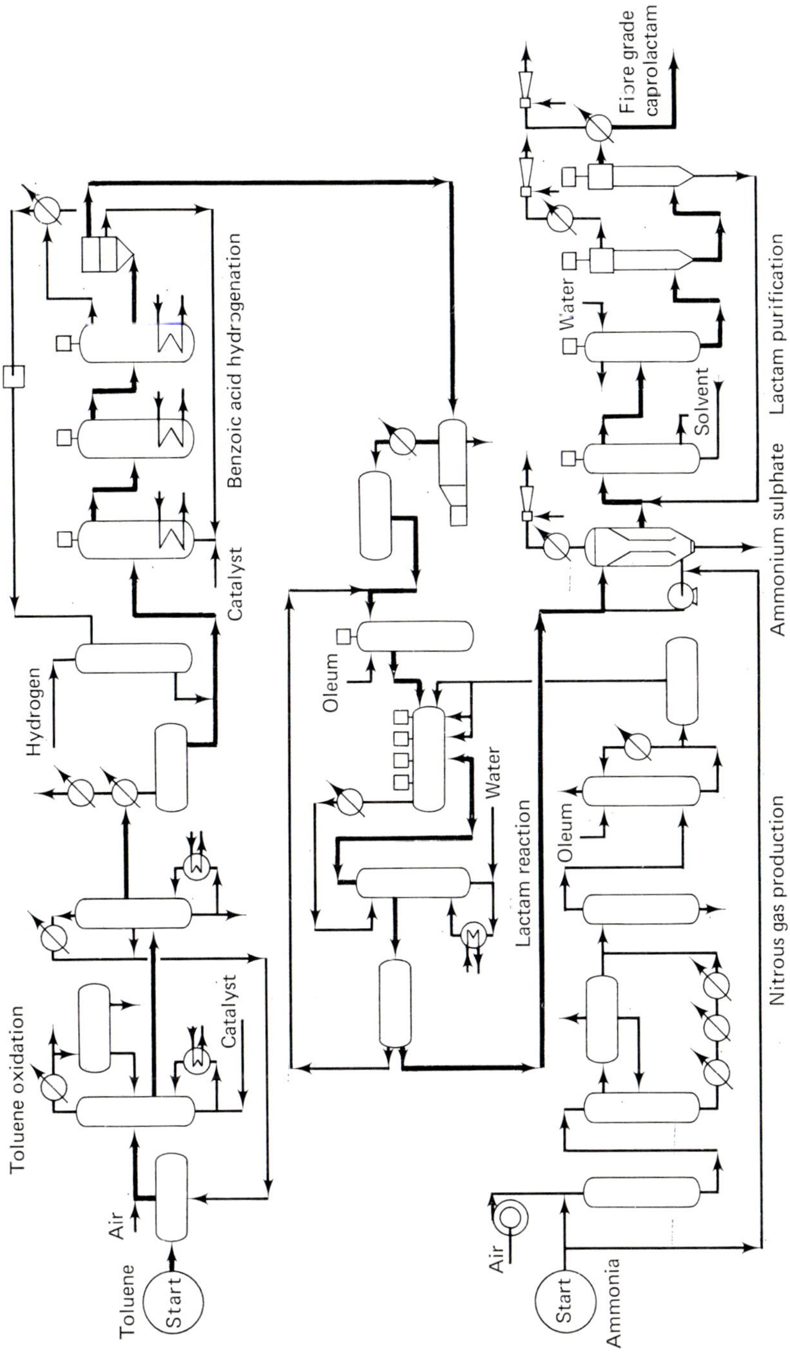

FIG. 6.20 *SNIA Viscosa Process Flowsheet*[207]

Both cyclohexane carboxylic acid and nitrosyl sulphuric acid solution are sent to the caprolactam synthesis units. The organic acid is premixed with oleum and fed to the nitrosation reactor. This is a multi-stage full mixed unit. To each stage $NOHSO_4$ solution is fed in the proper proportion to get complete conversion of the nitrosating agent while cyclohexane carboxylic acid overall conversion is kept at 50 per cent. The main reaction is as follows:

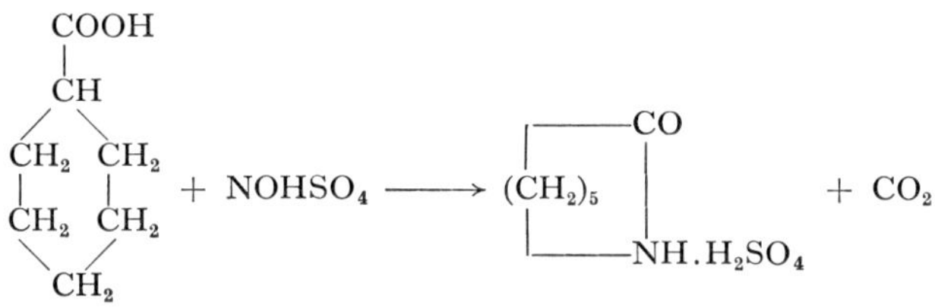

The reaction mass is hydrolysed with water at low temperatures. Cyclohexane effluent in liquid phase from the reactor extracts unconverted cyclohexane carboxylic acid. Its last traces are recovered from the sulphuric solution by counter current extraction with distilled cyclohexane as solvent.

Neutralisation followed by the extraction and purification of caprolactam employ techniques similar to those described previously for other processes. Ammonium sulphate is produced as a by-product by the quantity is only half that produced by the traditional processes.

6.3.3. Polymerisation to Polyamide 6

Originally polymerisation was carried out using a conventional batch autoclave process where caprolactam was heated to about 260°C often in the presence of some 10 per cent water with controlled emission of steam. The polymer was extracted, then extruded, quenched with water and cut into chips. The plants were usually small, which allowed adaptability; important if a wide range of fibres is required. However, currently almost all new plants in developed nations are larger, and for improved economics and greater product uniformity, operate continuous polymerisation techniques. Spinning and drawing are often undertaken directly as filaments emerge from the spinnerettes.

Representative of the various continuous polymerisation processes developed for converting caprolactam into polyamide 6 filament or chip are those offered by Vickers-Zimmer of West Germany, having built plants throughout the world using specially designed reactors.

In a process,[61] which can be represented by the simplified flow diagram in Fig. 6.21, solid caprolactam in powder or flake form is initially melted—(1) the melt is then fed batch-wise into the mixer (2), where the catalyst and the chain terminator to control and maintain viscosity to between 2·5 and 3·2 are added. The monomer mixture is drained through a filter into an intermediate vessel, (3) and thence to a dosing and mixing system, (4) where a delustrant and other chemicals may be added. The addition of a delustrant to reduce transparency, increase whiteness, and prevent undesirable gloss in finished fabrics is normally accomplished by dispersing 0·3–2·0 per cent of finely divided titanium dioxide, which has been treated in a colloid mill and then in a settler. Colouring pigments, optical brighteners and heat and light stabilisers in finely divided form may also be added.

In the polymerisation stage, the melt passes through the reactor (5) in which it descends slowly at atmospheric pressure through four successive heating zones where water acts as a catalyst, prior to vacuum treatment of the melt. In this carefully controlled polymerisation reaction, which is slightly exothermic, formation of aminocaproic acid takes place followed by polycondensation of this acid to give polyamide 6.

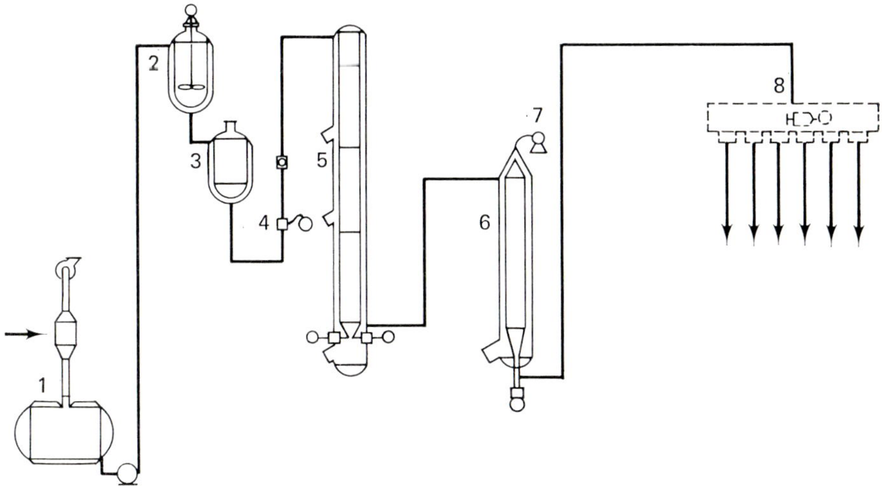

FIG. 6.21 *Flow diagram for polyamide 6 polymerisation (Vickers-Zimmer)*

From the reactor, the melt is transferred continuously to the vacuum system (6) and (7) for the removal of low molecular weight components by distillation. The product coming from the vacuum stage enters the spinning plant through a distribution system (8) and is spun directly, thus eliminating the conventional chip production and handling stages.

An alternative process in which continuous polymerisation is carried out at elevated pressure follows a similar sequence of operations, but in this case polymerisation under pressure is followed by equalisation in an after-polymerisation stage at atmospheric pressure before the melt continues to the vacuum and later processing stages.

If chip is required, the melt leaving the reactors is fed through a proportioning pump which meters the polyamide 6 melt to a spinning head from which it emerges spagehetti-like from the spinnerettes. After cooling in a large vat, cutting takes place to give uniform length chip, which falls into an extractor from where the lower molecular weight components are washed out with demineralised water in a continuous and fully automatic procedure. The product passes through a separator and ultimately to a drumdrier where after vacuum drying the product is sent to storage or a processing plant.

6.3.4. Spinning and Drawing[55]

As previously noted in modern continuous polymerisation polyamide 6 units, spinning and drawing equipment are fed directly from the spinnerettes. In the case of non-continuous units—usually smaller and/or of older design—polycaprolactam chip is washed and dried and then melted to a clean liquid which enters the spinnerettes.

The method used for polyamide 66 of passing the yarn as it is spun through a steam chamber is not suitable for polyamide 6 because there is such a relatively high residual concentration of caprolactam that the filaments would become sticky if steamed and liable to adhere either to the walls of the spinning equipment and/or to each other.

Next the yarn is stretched to about five times its original length to orient the constituent linear molecules and thereby to make the filaments strong, supple, and unshrinkable. The yarn is washed again with water to remove any low molecular weight polymer, dried, and cone wound. If staple fibre is required instead of continuous filament, then instead of being cone wound the yarn is crimped, cut, and baled.

6.3.5. The Role of Polyamide Fibres

World output of polyamide 66+6 is close to 2 million tons, and polyamides represent nearly 40 per cent production of synthetic fibres. After some forty years of commercialisation, they are to be regarded as mature synthetics with an annual growth rate of some 8 per cent, i.e. rather lower than that of total synthetics and of polyesters in particular. Polyamide 66 still is the more important, though new producers normally make the 6-fibre, as caprolactam is readily available in commodity amounts. Moreover, while there are relatively few sources of 66-technology, many polyamide 6 plants have been built world-wide, notably by chemical plant contractors. Finally, polyamide 6 technology is less complex than the polymerisation of adipic acid and hexamethylene diamine to polyamide 66.

TABLE VI.16

Estimated Geographical Distribution of Polyamide-6 and -66 Fibres (1972)[64]

	Polyamide 6		*Polyamide 66*	
	'000 tons	%	*'000 tons*	%
West Europe	264	40	400	60
USA	178	25	534	75
Japan	307	99	2	1

But polyamide 66 and 6 find similar and very diversified commercial usage in apparel, domestic textiles and in industrial applications. In principle, polyamide 66 is preferred for the more exacting industrial usages (e.g. belting, tyre cord), while polyamide 6 is more common in fine denier, hosiery-type apparel. It must, however, be emphasised that the two fibres are essentially interchangeable and end-use pattern of each type varies between countries, often reflecting the installation of production units of the one type before the other polyamide was actively marketed in a given area.

REFERENCES

[1] Gallant, R. W., *Physical Properties of Hydrocarbons* (Gulf Publishing Company, Houston USA, 1970), Vol. II, p. 154.

[2] Kilpatrick *et al.*, *J. Research Natl. Bur. Standards*, 1947, **39**, 523.

[3] Pauw, F. de, Alexandre, M. and Limido, G. E., *XXVII⁰ Congrès de Chimie Industrielle Bruxelles*, 1954, Vol. II, p. 168.

[4] Nettesheim, G., *Chemie-Ingenieur Technick*, 1969, **41**, 1260–65.

[5] De Ruyter, E. and Jungers, J. C., *Bull Soc. Chim. Belges*, 1949, **58**, 210.

[6] Boonstra H. J. and Zwieterning, P., *Chem. and Ind.*, December 1966, p. 2039.

[7] Field, S. and Dalson, M. H., *Hydrocarbon Processing*, 1967, **45**, 169.

[8] Craig, R. C., *Chemical Engineering*, 1970, **77**, 108.

[9] Dufau, A. F., Eschard, F., Haddad, A. C. and Thonon, C., *Revue de l'Institut Francais du Pétrole*, Vol. XIX, No. 9, September, 1964.

[10] Reily, J. W. and Sze, M. C., *Chemical Engineering Progress*, 1967, **62**, p. 73.

[11] Peet, W. A. and Winsor, J., *Oil and Gas Journal*, 6 March 1967, p. 107.

[12] Deschamps, A., *Thesis*, Paris, 1966.

[13] Leprince, P. *et al.*, *Procédés de Pétrochimie* (Paris 1973) Technip Eds., Chap. III.

[14] Leprince, P., *Hydrocarbon Processing*, July 1971, p. 75–83.

[15] Berezin, I. V., Dzantiev, B. G., Kazanskaya, N. F., Sinochkina, L. I. and Emanuel, N. M., *Zh. Fiz. Khim.*, 1957, 3.

[16] Berezin, I. V., Denisov, E. T. and Emanuel, N. M., '*Symposium on chemical kinetics catalysis and reactivity* (Moscow Academy of Science of USSR, 1955).

[17] Berezin, I. V. and Kazanskaya, N. F., *Bull. Acad. Sci. USSR*, 1959, 126.

[18] Berezin, I. V., Denisov, E. T. and Emanuel, N. M., '*The oxidation of cyclohexane* (Moskva-Pergamon Press, 1966).

[19] Seree de Roch, I., Private communication (Institut Français du Pétrole, 1965).

[20] Brown, N., Hartig, M. J., Roedel, M. J., Anderson, A. W. and Schweitzer, C. E., *J. Amer. Chem. Soc.*, 1955, 77.

[21] Parlant, C., Seree de Roch, I. and Balaceanu, J. C., *Bull. Soc. Chim. de France*, 1963, 2452.

[22] Pritzkow, W., *Chem. Ber.*, 1956.

[23] Alagy, J., Clement, G. and Balaceanu, J. C., *Bull. Soc. Chim. de France*, 1959, 1325.

[24] Société des Usines Chimiques Rhone-Poulenc, FP 1,327,137 (1963).

[25] Société des Usines Chimiques Rhone-Poulenc, FP 1,580,206 (1969).

[26] Bashkirov, A. N., Kamzolkin, V. V., Sokova, K. M. and Andreyeva, T. P., *Sixth World Petroleum Congress*, 1963, section IV, paper 38.

[27] Bashkirov, A. N. and Kamzolkin, V. V., *Fifth World Petroleum Congress*, 1959, section IV, paper 15.

[28] Broich, F. and Graseman, H., Private communication (Chemische Werke Huls, 4370 Marl).

[29] Pascal, P., '*Nouveau traité de chimie minérale*' (Masson, Paris, 1961), Vol. VI.

[30] Steinberg, H., *Organoboron chemistry* (John Wiley, New York, 1964), Vol. I.

[31] Steerman, J. W. M., Kaarsemaker, S. and Hoftyzer, P. J., *Second European Symposium on Chemical reaction engineering*, November, 1963.

[32] Burguieu, J. P. and Van Landeghem, H. (in the press).
[33] BASF, GP 1,114,801 (October 1957).
[34] E. I. Dupont de Nemours and Co., USP 2,825,742 (1957).
[35] Institut Français du Pétrole, FP 1,527,718 (1968).
[36] Alagy, J. and colleague, *Seventh World Petroleum Congress PD* **18**, Mexico, 1967.
[37] Kirk Othmer, '*Encyclopedia of Chemical Technology*', (Interscience Publishers), second edition, Vol. 6, p. 683.
[38] Cubberley, A. H. and Mueller, M. B. *JACS*, 1947, **69**, 1535.
[39] Scientific Design Company Inc., BP 909, 227 (1960).
[40] Inventa AG, BP 947,112 (1964).
[41] BASF, FP 1,384,217 (1964).
[42] Institut Français du Pétrole, FP 1,543,933 (1968).
[43] Von Handsdieter Hofmann, *Chemiker Zeitung*, 1966, **90**, 595.
[44] Publication de l'Institut Français du Pétrole.
[45] National Aniline Division, Allied Chemical and Dye Corporation, Technical Bulletin I-19.
[46] Patty, F. A., '*Industrial Hygiene and Technology*' (Interscience Publishers, 1963), Vol. II, p. 1765.
[47] '*Encyclopedia of Polymer Science and Technology*' (Interscience Publishers ,1969), Vol. 10, pp. 348–460.
[48] Special Supplement to *Chemical Engineering Science* (1954), Vol. 3.
[49] *Hydrocarbon Processing*, November 1971.
[50] *Japan Quarterly* **5–6**, October 1969.
[51] *J. Am. Chem. Soc.*1958, **80**, 4079–82.
[52] *C.E.E.R.*, May 1972, Vol. 4, No. 5, pp. 47–60.
[53] *Nitrogen*, No. 40, March/April 1966, p. 14.
[54] *C. & E.N.*, 28 June 1965.
[55] Moncrieff, R. W., '*Man-Made Fibres*', (London, 1963), Chap. 19.
[56] *E.C.N*, Caprolactam Process Survey, 2 May 1969.
[57] *Hydrocarbon Processing*, November 1972, pp. 92–95.
[58] USP 3,025,306; USP 3,064,008; USP 3,000,880.
[59] *Chemical Engineering*, 25 March 1968.
[60] *Hydrocarbon Processing*, November 1970, pp. 137–145.
[61] Technical Literature and General Sources.
[62] Hulme, P. and Turner, P. E., *Chem. and Proc. Eng.*, November 1967, p. 96.
[63] Leprince, P. *et al.*, '*Procédés de Pétrochimie*' (Paris 1971), Technip Editions, Chap. IX.
[64] Meeks, J. E. and Winter, C. L., Paper presented to Royal Aeronautical Soc. (London), December 1972.

SOURCES

A vast number of articles exists in the literature on one or another phase in the production of polyamide 66 and 6 and their precursors respectively, notably articles in technical magazines, reference books and the trade brochures of individual process owners and/or licensees (e.g. CJB, DSM, Inventa, SNIA, Vickers-Zimmers, etc.). A review of such literature was made and the above were selected as the more important or significant sources. It will be noted that many sources listed give references to other secondary sources.

CHAPTER 7

ETHYLBENZENE, STYRENE AND THE STYRENE POLYMERS

by Ralph Landau, Rex E. Lidov, J. Habeshaw and A. N. Roper

7.1. INTRODUCTION by Ralph Landau and Rex E. Lidov

THERE is little commercial interest in ethylbenzene except as a raw material for the manufacture of styrene. There are of course some of its derivatives which find their way to the market place; acetophenone, for example, is one. Similarly, a number of its derivatives are used in perfumery, though the total volume involved is not large, and other, still smaller, markets absorb some more. Nonetheless, the total volume of such compounds, derived from ethylbenzene, in commerce would hardly warrant a place for it in a book like this devoted to benzene derivatives.

Perhaps, surprisingly, much the same thing can be said of styrene itself. It finds its major importance as the monomer which leads to its homopolymers and its many co- and ter-polymers. By itself it would be, if not uninteresting, without great significance. Again, some of its compounds do indeed enter commerce. Its oxide, for example, is employed in minor quantities as an inhibitor for substances which can generate hydrogen chloride on storage and in which that hydrogen chloride acts as auto-catalyst for more such generation. It is a source for other compounds which occasionally find use. But overall we consider it because of its polymers.

The polymers are another matter entirely. They were probably the first really high molecular weight substances to be characterised and to find use. The simple homopolymer is still one of the most important thermoforming plastics in every-day trade. Modified in substantially minor fashion it gives rise to so-called high impact molding compounds and to what are called 'heat resisting' compounds.

The co-polymers extend the usefulness of styrene monomer to an even greater extent. With butadiene it forms elastomers, which find an enormous market in tyres and other 'rubber' articles; with acrylonitrile and butadiene, it leads to the ABS series of high molecular weight ter-polymers, ranging from almost rigid plastics to highly flexible substances—which find a place in a whole line of almost unbreakable products.

With all that, the industry can be characterised only as one which is extremely mature. One does not look to it for surprises. Changes when they occur, are in general evolutionary rather than revolutionary and manufacturing processes, particularly for ethylbenzene and styrene have changed very little in the last decade or two. It was in 1952, that Ray H. Boundy and Raymond F. Boyer of the Dow Chemical Company obtained a copyright on what might be regarded even today as the definitive monograph in the field.[1] This is not to say of course that there have not been important developments since that time which must be recounted; there have.

It is, however, worth saying that the Boundy and Boyer volume consists of some thirteen hundred pages in which both the page and the type size is similar to that in the book you now hold. It is clear, under these circumstances, that this chapter must represent a severe selection from available material; in no sense is it a detailed patent review or even the whole story of its subject matter.

In what follows, an attempt will be made to discuss: the manufacture of ethylbenzene needed for the production of styrene, the manufacture of styrene from ethylbenzene and the polymerisation of styrene in terms of the major polymer species derived from the latter. The discussion of manufacturing processes will be directed, hopefully, to an understanding of common factors, rather than of historical differences, which have evolved while the selection of polymers for discussion will be based esentially on the overall significance of the species on the current market.

7.2. THE MANUFACTURE OF ETHYLBENZENE

by Ralph Landau and Rex E. Lidov

7.2.1. Sources of Ethylbenzene

Almost all of the ethylbenzene used for the manufacture of styrene is obtained by the alkylation of benzene with ethylene. There is, however, one other source which must be briefly considered; aromatic streams, generally from petroleum sources, which contain ethylbenzene. These streams provide roughly 9 per cent of the ethylbenzene converted to styrene.[1a] While there is some tendency to speak of this material as 'Natural' ethylbenzene, this is not, in fact, accurate. In general the ethylbenzene in such streams is produced as a concomitant to the manufacture of xylene by catalytic hydroforming or platforming.

In general, ethylbenzene occurs in refinery xylene fractions to the extent of about 15 per cent, an amount which approximates the thermo-dynamic equilibrium composition for the C_8- aromatic hydrocarbons at 1000°K.[2] Since ethylbenzene boils 2·1°C below *p*-xylene, it is separable by fractional distillation. Unfortunately, it is not easily separable; obtaining ethylbenzene satisfactory for use in a styrene dehydrogenation plant requires super-fractionation using columns having well in excess of 100 theoretical plates which must, in addition, be provided with high reflux ratios.

The fractionation was first commercialised by Cosden Oil and Chemical Company.[3] The Cosden separation process used three columns in series, each with 130 actual plates operating with a reflux ratio of 87:1. In this connection, it is worth noting that Cosden had a particularly favourable xylene stream for fractionation since it contained 28 per cent of ethylbenzene.

Despite the difficulties, there are about seven commercial plants in the United States which accomplish this separation using streams with lower concentrations of ethylbenzene (20 per cent is more common) than did Cosden.

Still, as must be obvious, the separation by fractionation is neither easy nor inexpensive and there is some evidence that the percentage of total ethylbenzene so obtained will decrease with time.

7.2.2. Ethylbenzene by the Ethylation of Benzene

The remaining 90 to 91 per cent of the ethylbenzene used for the manufacture of styrene is derived by direct synthesis; manufacture is based on the reaction of ethylene with benzene in the presence of a catalyst. The equation is given in (1):

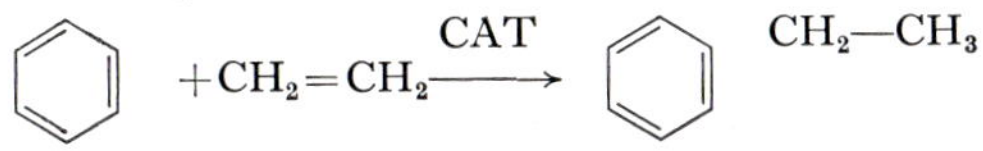

(1)

Essentially two kinds of processes are used. In the first, the indicated reaction is carried out in the vapour phase and at elevated temperatures and pressures; the catalyst is solid phase and clearly heterogeneous. In the second, the reaction is liquid phase, temperatures and pressures are moderate (100°C and ambient to 100 lb p.s.i.g.) and, while the catalyst is present as an apparently heterogeneous second liquid phase, the probability is that the catalytic material functioning in the system is in solution so that the reaction is homogeneous. The second process utilizes a classical Friedel-Crafts reaction and it is, by far, the process most used both in the United States and in Europe.

The catalyst for the vapour-phase reaction first used was a so-called 'solid phosphoric acid' (H_3PO_4/P_2O_5 on kieselguhr) but, for reasons which will be later discussed, this gave way to a boron tri-fluoride modified anhydrous γ-alumina.[4] Other catalysts have been used in the vapour-phase process. Moreover, it is reported that there are active molecular sieves which can be used as catalysts and that improved results are thereby possible. However, plant data obtained using a molecular sieve catalyst have not yet been published. One of the most interesting aspects of the seemingly widely different benzene alkylation processes is that all can be regarded as proceeding by essentially the same mechanism. Clearly, there are differences in detail between the liquid-phase processes and the vapour-phase processes but the major reaction pathways must be regarded as essentially identical.

7.2.2.1. Reaction mechanism The vapour-phase process has been used with phosphoric acid on kieselguhr, with boron tri-fluoride on alumina, with a more or less classical silica-alumina of the type used in cracking catalysts and with various other combinations of metallic phosphates and hydrogen fluoride or with boron tri-fluoride.

The Friedel-Crafts reaction can employ, although not with equal success, the usual range of Friedel-Craft catalysts: aluminium chloride, ferric chloride, stannic chloride, zinc chloride and the like. Aluminium chloride is, by a wide margin, the most useful, and most usually used, catalyst.

The Friedel-Crafts reaction ordinarily requires that an activator be added; the usual activator is either hydrogen chloride, or some substance which can lead to hydrogen chloride such, for example, as ethyl chloride.

It is immediately obvious that all of the catalysts, for either the vapour-phase or the liquid-phase processes, are acids, either ordinary protonic acids or generalised Lewis acids. It is, accordingly, reasonable to assume that the mechanism is common, at least in its major aspects, in all of these cases. On the other hand, it would appear unlikely that all of the steps which might be written for the solution reaction at about 100°C with aluminium chloride can also be assumed to take place, in the same discreet fashion, for the reaction carried out at 400° to 600°C in vapour phase over a silica-alumina catalyst. All of the steps involved, or possibly involved, are best illustrated for the Friedel-Crafts reaction; there is strong evidence suggesting that each discreet step illustrated in the equations which follow does, in fact, occur.

The initial reaction step is indicated in equation (2) and possible pathways in (2a) and (2b):

$$HCl + AlCl_3 + CH_2{=}CH_2 \rightarrow CH_3{-}\underset{\substack{| \\ AlCl_4^-}}{CH_2^+} \qquad (2)$$

$$HCl + AlCl_3 \rightarrow H^+AlCl_4^- \qquad (2a)$$

$$H^+AlCl_4^- + CH_2{=}CH_2 \rightarrow CH_3{-}\underset{\substack{| \\ AlCl}}{CH_2^+} \qquad (2b)$$

While stable salts of the acid $H^+AlCl_4^-$ are known, there is no evidence for the existence of the free acid. Accordingly, while equations (2a) and (2b) are useful as an aid to a conceptionalisation of the reaction represented by equation (2), they are no more than that. The actual formation of the sigma-bonded ethyl aluminium tetrachloride must be assumed to result from a concerted reaction of all three reactants.

The fact that the same ethyl aluminium tetrachloride is readily formed from ethyl chloride and aluminium chloride explains why either hydrogen chloride or ethyl chloride can function as an activator for the reaction:

$$CH_3CH_2Cl + AlCl_3 \rightarrow CH_3\underset{\substack{| \\ AlCl_4^-}}{CH_2^+} \qquad (3)$$

The next event in the reaction, illustrated in equation 4,

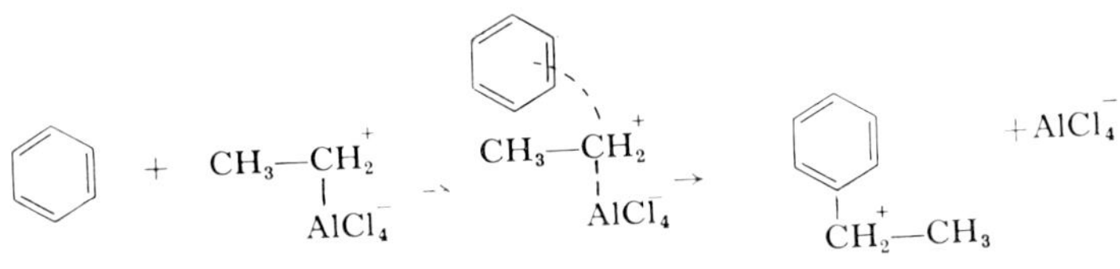

(4)

a. b.

is the formation of a π-bonded phenylethyl carbonium aluminum tetrachloride. As indicated in brackets, the reaction is shown as proceeding via displacement of the aluminium tetrachloride by the π-cloud of the aromatic ring. As an alternative in this step one might write the dissociation of the aluminium compound into an ethyl carbonium ion and the aluminium tetrachloride ion and the subsequent direct attack by the carbonium ion on the aromatic ring. It seems quite clear in the case of a compound like isobutylene, which can form the highly stable *t*-butyl carbonium ion, that the reaction does occur via the carbonium ion. On the other hand, it is equally clear that in the case of methyl chloride, the reaction proceeds by way of a displacement of the type here indicated.

The situation for the ethyl compound is somewhat less clear and the available evidence is ambiguous. It is worth noting, however, that in the ordinary Friedel-Craft ethylation the amount of sec.-butylbenzene, formed a customary by-product in ethylation processes, is well under 1 per cent. The butylbenzene obtained in an ethylation reaction arises from the dimerisation of ethylene to butene and the subsequent ethylation of the benzene with that butene. Given 'free' ethyl carbonium ion in the solution one would anticipate rather larger amounts of the butybenzene than is, in fact, found; under the circumstances, the assumption of an S_N-2 displacement reaction is reasonable. The fact that the initial aromatic compound formed is a π-complex, has been unambiguously demonstrated; in some cases it has been shown that the formation of the π-complex is the rate determining the step for the reaction.[5]

After formation, the π-complex re-arranges to the more strongly bonded sigma-complex, as shown in the first part of equation (5):

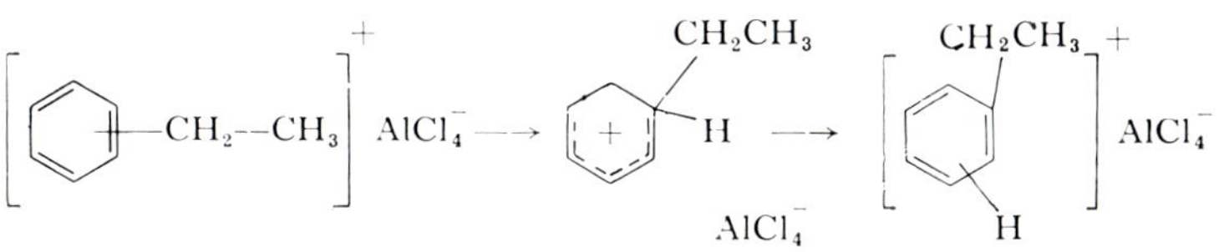

(5)

The intermediate sigma-complex must regain aromatic stability and does so by a second re-arrangement which gives rise to a hydrogen bonded π-complex. Breaking of the hydrogen π-bond leads finally to the newly formed ethylbenzene and provides active catalyst for initiation of a new cycle.

Equation (6) illustrates the final step:

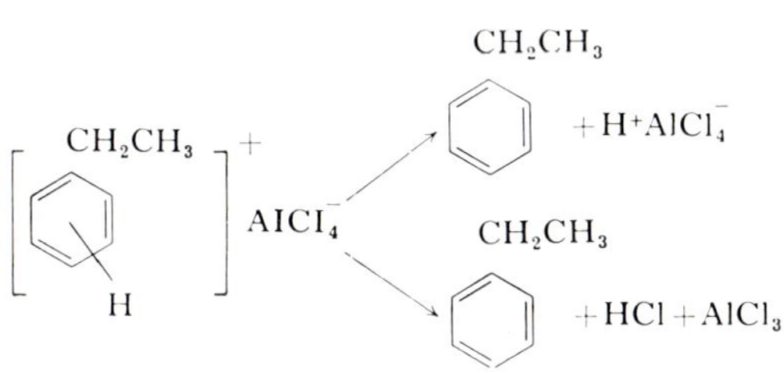

(6a)

(6b)

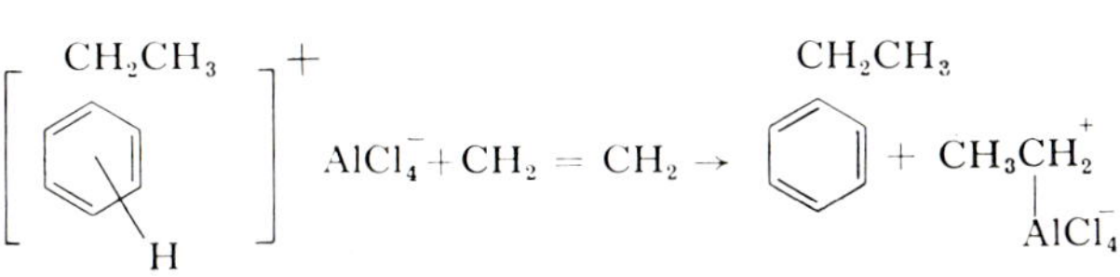

(6c)

Here again three routes can be postulated; (6a) since it leads to an acid not known to exist undoubtedly does not occur. More generally, it is assumed the reaction path is the one shown in (6b) which provides for the regeneration of the activating hydrogen chloride and aluminium chloride. However, so far as we know, there are no data which eliminate the path shown in (6c); the fact that at 100° ethylene is rapidly absorbed at the surface of the benzene solution in the reactor does suggest that it may well be a direct reaction of ethylene with the π-bonded hydrogen aluminium tetrachloride is the course that the reaction takes.

It is worthy of mention that for the aprotic Lewis catalyst, aluminium chloride, the reaction requires the addition of proton as the first step; the 'activator', hydrogen chloride, is needed to provide this proton. Alternatively, in the context of the reaction, ethyl chloride can be substituted. But, what is here important is that generalisation of the mechanism to catalysts which can themselves readily supply proton is immediately straightforward and need not here be laboured. The only reaction system for which the generalisation may not be immediately evident is that using silica-alumina as the catalyst.

The difficulty of extending the mechanism to reactions using the cracking catalysts as catalysts for ethylation is more apparent than real. Benesi demonstrated, as early as 1957, that the silica-alumina cracking catalysts have sites of very strong acidity present on their surfaces.[6] Benesi was the first to use classical acid indicators (although in a rather non-classical way) to demonstrate the presence of sites of varying acid strengths ranging from weak to highly acidic on the silica-alumina surface. Even earlier, workers with these materials had demonstrated that the catalytic activity of these substances is increased with small amounts of water and is lost if the catalysts are completely dehydrated.[7] It can be reasonably assumed that coordination of water at appropriate sites on the silica-alumina surfaces provides proton in the same way that the coordination of hydrogen chloride with aluminum chloride does. Moreover, using spectroscopic methods, Evans was able to show that double bonds are protinated over a clay catalyst.[8] Overall, then, it takes no particular stretching of assumptions to conclude that the general mechanism for the vapour-phase alkylation over a silica-alumina catalyst is, in major respects, similar to that already shown for aluminium chloride coordinated with hydrogen chloride.

To say that, overall, mechanisms are common for a number of reactions is not, of course, to say that they are identical. Indeed, it is necessary to postulate significant differences between the alkylation carried out over aluminium chloride and that carried out over silica-alumina. That this is so, is suggested by important aspects of the overall reactions.

A facet of the alkylation reaction which has not yet been discussed is the fact that when it is carried out to any useful conversion both dialkyl benzene and higher polyalkyl benzenes are always produced together with the desired ethylbenzene. These polyalkylbenzenes (including the dialkylbenzenes), when re-cycled with benzene over some catalysts, e.g., aluminium chloride activated with hydrogen chloride, are converted, in a reaction referred to as transalkylation, to monoalkylbenzene. The fact that this is possible is, of course, of very great economic importance. While transalkylation proceeds quite normally with most of the catalysts and in most of the processes currently in use, it does not occur over silica alumina. Obviously then, the mechanism of the reaction over silica alumina must somehow differ from the one earlier described. It seems probable that over catalysts of silica-alumina type

π-bonded alkyl intermediates are not involved in the alkylation. The reason that π-complexes of benzene with the alkyl do not occur can be assumed to be that the benzene π-electrons are not available for the purpose.

The earlier statement that the first step over a silica-alumina catalyst is the donation of a proton from a strongly acidic site on the catalyst implies that the new carbonium ion thus formed is chemisorbed onto the catalyst surface. If this 'bound' carbonium ion is to be able to react with a benzene molecule, the latter, too, will have to be held on the surface of the catalyst. Such adsorption undoubtedly involves the π-electrons of the aromatic ring and they are thus unavailable for bonding with the ethyl carbonium ion. The carbonium ion, under these circumstances, directly attaches to one of the carbon atoms of the ring and, at the temperatures involved, the hydrogen atom already there departs as a proton, almost, if not, simultaneously and thus regenerates the original highly acid site on the catalyst surface. If this, indeed, is the reaction course, it is not difficult to see why transalkylation does not take place; attack by proton leading to the formation of a much more energetic carbonium ion does not ordinarily occur.

Protonation of the π-electron cloud of benzene does, however, occur. With this as the initiating reaction, it is easy, in contrast, to write a reaction series for the liquid phase Friedel-Crafts system similar to that already shown in which, in a reversal of the alkylation reaction, ethyl carbonium ion alumina tetrachloride becomes available in the system and subsequently reacts with a benzene molecule to form ethylbenzene. The overall effect of such a series is to provide one molecule of polyalkyl benzene in which the number of alkyl groups has been reduced by one and a molecule of ethylbenzene. Provided only that the ratio of unalkylated benzene to polyalkyl benzene is adequately high, the polyalkyl benzene is converted, to whatever degree may be desired, to mono-ethylbenzene. The adverse economic effects of di- and polyalkyl benzene formation can thus be substantially eliminated.

7.2.4. Comparison of Various Ethylation Processes

Before proceeding to an examination of a typical plant used for the manufacture of ethylbenzene it is desirable briefly to examine the processes which are used for such manufacture. Processes can be examined both from the standpoint of their relative yields and in terms of the operating advantages or disadvantages which each possesses. It seems at once evident that processes which have survived and are still in use must have yields which are quite similar: for those cases, comparison of yields is not likely to prove highly significant. It is thus more instructive to examine the yield of one of the current processes as against those obtained from processes now no longer in use. On the other hand, the relative operating advantage/disadvantages of the processes currently employed are of greater interest than are the same factors for processes which have not survived.

7.2.4.1. Yield comparisons Table VII.1 provides a comparison of three processes only the first of which is still in use.

TABLE VII.1[9]

Product Distribution for Various Ethylation Processes: benzene to ethylene in 10:1 mol ratio

	Product distribution ratios			*sec-butyl benzene as*
	EB	*di EB*	*poly EB*	*% of EB*
Catalyst				
Aluminium Chloride	35	0·99	0·8	0·03
Silica-alumina	10	0·95	0·3	0·5
Phosphoric acid-Kieselguhr	38	0·70	0·3	0·79

Of the processes for which data are given in the table, the one using aluminium chloride as a catalyst is liquid phase; it is the process most extensively used today. The two processes utilising the other two catalysts shown are both vapour phase and no longer employed. Each of the vapour-phase catalysts produces far more sec.-butyl benzene than is obtained in the process using aluminium chloride (16 times and 26 times as much). While one half per cent of butylbenzene per 100 lb of ethylbenzene may not seem very much, it represents a complete waste of many, many tons of raw material a year. This is one of the factors which led to abandonment of the processes using silica-alumina and phosphoric acid on Kieselguhr.

There is, in addition, another factor. For neither of the two vapour-phase processes is recycling of the diethylbenzene and polyethylbenzene possible since over neither of these two catalysts does transalkylation occur. By contract, substantially all of the diethylbenzene and most of the polyethylbenzene produced by the process using aluminium chloride can be converted to monoethylbenzene simply by recycling these heavier materials to the reaction vessel. Not only is ethylbenzene produced but, since the process involved is transalkylation, an additional weight of ethylbenzene is obtained over and above that of the diethyl and polyethylbenzene converted. Thus, for example, 1 lb of diethylbenzene leads to almost 2 lb of ethylbenzene and, in like fashion, 1 lb of polyethylbenzene gives almost 'poly-pounds' of monoethylbenzene. The yield structure of the aluminium chloride process will be given more fully after the process itself is described in more detail in a typical embodiment.

7.2.4.2. Comparison of operating advantages/disadvantages Only two processes are currently used to an extent which warrants consideration. Of these, the most widely used are liquid-phase processes employing aluminium chloride as the catalyst; less extensively employed is the Alcar process licensed by UOP (Universal Oil Products Co.). The latter is a vapour-phase process using a boron trifluoride modified anhydrous gamma-alumina as the catalyst with continuous addition of BF_3 with the feed to the alkylator.[10] The situation with respect to the catalyst is not wholly clear since at least one informant suggests that the catalyst (as of about 9 January 1967) is silica gel impregnated with phosphoric acid with a continuous feed of boron trifluoride.[11] In any event, the published data seemed to indicate that the process uses a reaction zone operating at about 290°C and 900 p.s.i.; transalkylation, accomplished in a separate reactor, is possible in this system. Transalkylation is carried out at about 200°C and 500 p.s.i. The temperatures in the transalkylation reactor seem to be low enough and the pressures high enough to suggest that there may be a condensed benzene phase on the catalyst; it may thus be that the process, as conducted in the transalkylator, uses a liquid or a pseudo-liquid system which makes transalkylation possible.

Among the advantages of the vapour-phase system are that it is much less corrosive, overall, than is the aluminium chloride process; accordingly, it does not require lined reactors or special alloys. It is a system which was designed to operate on refinery gases. On the other hand, it requires very high recycle rates since it operates with very low conversion; approximately 10 units of benzene are recycled for every unit of ethyl benzene produced. Moreover, the water level in the feed must not exceed about 0·01 parts per million which is at least an order of magnitude lower than that requirement for the aluminium chloride reaction. The low conversion leads to high capital costs and high utility charges.

The aluminium chloride process can boast a technology which is extremely well known and a plant which runs extremely smoothly with minimum personnel. Moreover, although the process is corrosive and requires a brick-lined reactor, maintenance is extremely low, provided that the lined reactor design is good and that the lining bricks were properly installed. The major disadvantage of the aluminium chloride process is that the reaction is completely killed by slugs of water and, once this happens, it may require as long as three or four days to bring the system on stream again. The usual practice is to continue to pump aluminium chloride and ethyl chloride through the system until all the water has reacted and this requires that a substantial inventory of ethyl chloride be maintained. Moreover, in contrast with the Alcar system which requires a separate reactor for transalkylation, in the aluminium chloride process the transalkylation reaction proceeds in the alkylator so that alkylation and transalkylation proceed simultaneously.

7.2.5. Preliminary Considerations: Aluminium Chloride Alkylation Process Plant

Consideration of the actual flows in an aluminium chloride process-benzene ethylation plant will be simplified if we consider first feed purity requirements and, briefly, required materials of construction.

7.2.5.1. Feed purity requirements The aluminium chloride-ethylation plant is not particularly demanding with respect to the quality of the raw materials charged to it. Except for water, the effect of which tends to be 'disastrous', as has already been noted, most impurities in raw materials are tolerated with, at worst, some minor decrease in yield or catalyst life or both. In this instance, a decrease in catalyst life means that more fresh aluminium chloride must be fed than would otherwise be required in order to maintain adequate activity in the reactor. The single factor which must be closely watched is that no substances are adventitiously introduced which will subsequently prove to be inseparable from the styrene which is the ultimate goal.

Overall, *sulphur* in the feed is undesirable; if present as *thiophen* it may adversely affect the catalyst and, in some cases, it may lead to the formation of heavy tars in reboilers. *Paraffins* in the benzene are undesirable to the extent that, on recycle, they build up in the system and ultimately lead to dilution of the benzene. *Toluene* in the benzene is ethylated as is benzene; its presence can therefore lead to subsequent distillation difficulties not only in the separation of ethylbenzene but ultimately in the purification of the styrene. *Xylenes* must of course be avoided. The alkylation plant has nowhere nearly enough distillation capacity to separate xylenes from ethylbenzene or, ultimately, completely from styrene. Note, however, that for the benzene impurities which have been mentioned, the problem which they present is, for the most part that of subsequent product purification rather than any interference in the ethylation reaction as such. Nonetheless, they are to be avoided.

The ethylene feed may contain almost any desired amount of *paraffinic* gases provided only that the overall plant has been designed with adequate capacity and the means to eliminate them from the system.

The situation is quite different with respect to *olefinic* impurities in the ethylene. These, will react as readily as does ethylene itself leading to alkyl benzenes corresponding to whatever olefinic impurities were present. This not only uses up benzene, but it can significantly increase the purification load in the plant. Olefinic impurities in the ethylene are thus undesirable. *Acetylene*, which is sometimes found in some ethylene sources, is also highly undesirable. In general, it leads to divinylbenzene and tars or both. Here again, benzene is wasted and distillation problems are rapidly multiplied.

One other significant 'feed' stream to an ethylation plant is the recycle stream of diethylbenzene and polyethylbenzenes. In general, there is some tar, or very heavy ends, formed in an aluminium chloride ethylation plant and these would normally recycle with the polyethylbenzenes. These heavy end materials tend both to deactivate the catalyst, and to dilute the catalyst phase, adversely affecting the reaction equilibrium. Thus, in order to insure a wholly satisfactory recycle stream, it is necessary to top the polyethylbenzenes from the heavy tars with which they are associated.

In general, commercial aluminium chloride is wholly satisfactory for the process.

7.2.5.2. Materials of Construction In order to indicate what special materials of construction are required in an aluminium chloride ethylation plant, it will first be necessary briefly to describe the flow of the reaction product as it leaves the alkylator and the kind of treatment this reaction product is ordinarily given. It has already been noted that the alkylation reactor system is highly corrosive; this corrosiveness is inherent in the mixture in the alkylator and so the stream which leaves that vessel is equally corrosive.

The stream flows from the alkylator through a pipe and pump to a cooler and from the cooler it flows to a hydrolyser. The function of the hydrolyser is to decompose the catalyst contained in the stream; not infrequently, contact between the water used for hydrolysis and the stream is obtained in a pump after which the water-hydrocarbon mixture flows to a separator. A highly acidic water stream is discharged from the separator and the hydrocarbon stream is treated with a caustic solution, again, frequently, in a pump, in order to eliminate all traces of acid. After this step the hydrocarbon mixture is essentially noncorrosive and can be handled in carbon steel.

For the most part, alkylators are brick-lined, carbon steel vessels; they can also be glass-lined. In general, the brick-lining properly installed, has proved to be very highly serviceable and somewhat less expensive than the corresponding glass-lined equipment. In a number of plants in the United States, a furan cement is placed between the brick and the shell; the cement lining has to be relatively thick in order to fill the space between shell and brick. The brick itself is secured and sealed with an acid proof silicate cement.

Bottom and top heads must also, of course, be lined. In some cases the lining for these is much the same as for the shell itself but less expensive lining consisting of multi-layered glass cloth secured and impregnated with furan cement has been used.

It is necessary to be sure that exit and entry points for the vessel which, in general, consist of nozzles going through the protective brickwork, are properly designed. In the first place the design must ensure that there is no leakage of corrosive materials around them back to the vessel shell and, in the second, the design must be such that the lining is not dislodged by expansion and contraction of the nozzles. The nozzles themselves are usually of Hastelloy and thus require no protection.

For the same reason the nozzles into the vessel are made of Hastelloy, so that they will resist corrosion, the ethylene inlet sparger is also frequently of Hastelloy.

The reaction is, of course, exothermic. Cooling is ordinarily provided by means of a side stream passing from the reactor through a cooler and back to the reactor. Piping to and from the cooler from the reactor are either Hastelloy or glass-lined steel pipe. The cooler itself is either of Karbate or of an impervious graphite.

The product stream withdrawn from the vessel passes via a pipe, again either Hastelloy or glass-lined steel, to a heat exchanger (cooler) of Karbate or impervious graphite and from the cooler via pipe, again Hastelloy or glass-lined, to a pump. Water and the hydrocarbon streams enter the pump through a tee at its inlet and the pump provides the necessary contacting of the aqueous and hydrocarbon phases. In some cases, it has been found desirable to place an expanded, packed pipe section at the inlet to the pump so that the turbulence of the steam is reduced and the temperature attained on first mixing the hydrocarbon and water streams is also brought down.

The stream passes from the pump, again through pipes of the type previously mentioned, to a separator vessel. This vessel is, once again, a carbon steel shell, brick-lined as is the alkylator itself. A highly acidic aqueous stream containing the aluminium chloride catalyst and hydrochloric acid is withdrawn from the bottom of the separator.

From the top of the separator the now water washed hydrocarbon stream is again sent through corrosion resistant piping to a second Karbate pump wherein it is contacted with a caustic stream in order to free it completely from any traces of acid. The pump mix, which can now be handled in carbon steel goes to a second separator.

The balance of the plant requires no special corrosion resistant materials.

7.2.6. The Aluminium Chloride Alkylation Process Plant

A schematic diagram for an aluminium chloride process alkylation plant is shown in Fig. 7.1.; note that the schematic includes both the alkylation portion of an overall styrene plant as well as the dehydrogenation section. We will here be concerned with only the first portion of the overall plant shown in the diagram.

Benzene, ethylchloride and aluminium chloride (via a hopper emptied by means of a screw feeding device) are introduced into the alkylator. Over a period of time a red-brown liquid separates. This is the catalyst layer required before actual alkylation will start. Catalyst formation is relatively rapid and, in any event, is a requirement only at the start-up of the plant. Thereafter, ethylene, benzene and ethyl chloride, the promoter, are added continuously, catalyst formation also takes place continuously and a product stream is continuously removed.

It should be noted that the incoming benzene stream contains not only fresh benzene but recycle benzene and recycle diethylbenzene and polyethylbenzene both of which are returned to the reactor for conversion to ethylbenzene by means of a transalkylation reaction.

The alkylation reaction is ordinarily carried out at about atmospheric pressure and the boiling point of the reactants, a little higher than 80°C. Under these conditions the reaction is sufficiently rapid so that the ethylene introduced is completely absorbed and reacted and the difficulty of operating a highly corrosive system under super-atmospheric pressure is avoided. But, while this is the case in many or most plants, it is not always the case.

Some designers choose to carry out the reaction at pressures as high as 125 p.s.i.g. and at temperatures as high as 130°C.[12] While the references cited are relatively old there are modern engineering organisations which still prefer to design for super-atmospheric pressures in the alkylator. The reasons for this are quite clear. Operating at pressures above normal and, consequently, at elevated temperatures, not only is the rate of reaction enhanced, but, in the heat exchange required for cooling the alkylator, steam, which can be used for power, is generated. It is also probable that at the higher temperatures the transalkylation more nearly attains its equilibrium values, thus reducing the load of di- and ethylbenzenes recycled through the plant.

It is worth observing that except for the skill of the process and plant designer, little can be done by the manufacturer of styrene to make his styrene less expensive than that of another producer. In a method as old as this all of the basic reaction facts are not only well known but all are in the public domain. It is only the skill of the designer in ringing every possible advantage from the plant as a whole which can beneficially affect styrene costs. Since a styrene plant requires fairly large amounts of distillation and refrigeration (for cooling, transfer and sometimes storage of the final product) and is thus a heavy energy consumer, any design feature which recovers energy is highly useful.

Thus, the use of superatmospheric pressures in the alkylator remains attractive and in the actual case, with a properly applied brick-lining, presents far fewer difficulties than might, *a priori*, have been expected.

It is still necessary to indicate the usual concentrations of the various feeds in the reactor. These, in general, are based on the equilibria in the system which establish

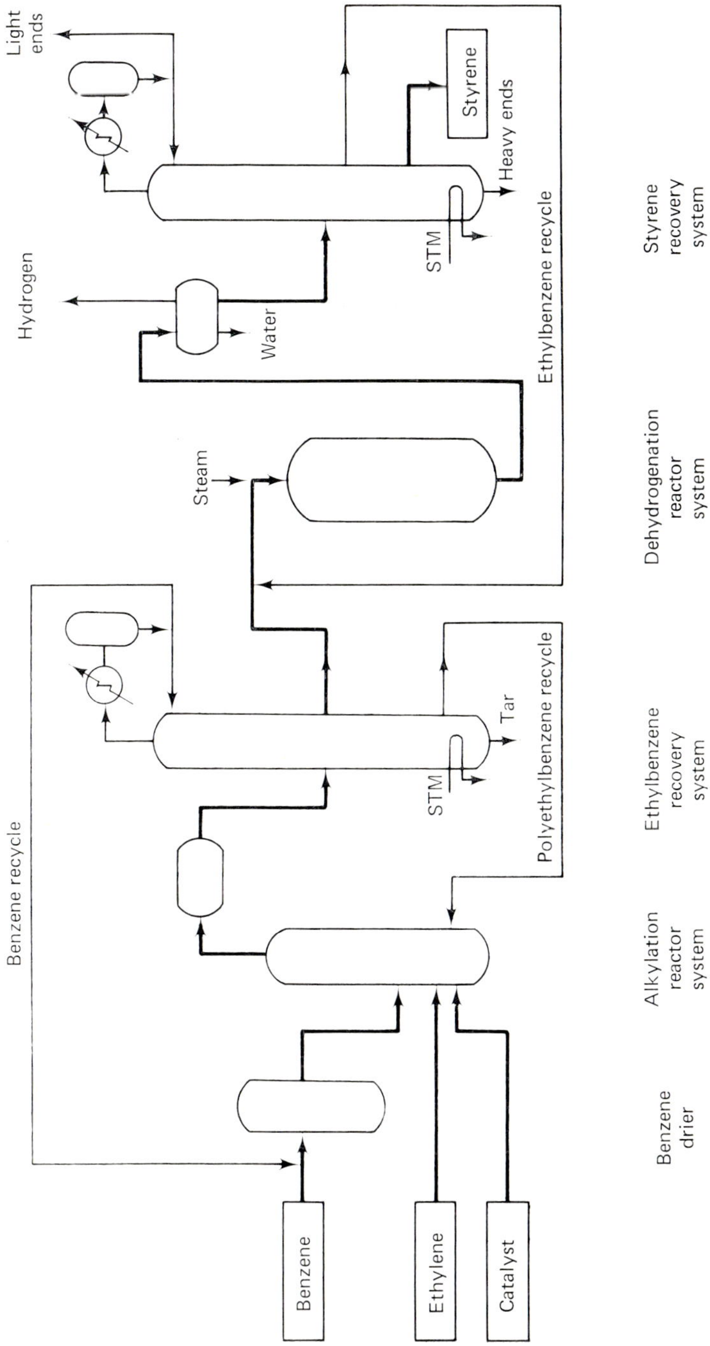
Benzene
Ethylene
Catalyst
Benzene recycle
Polyethylbenzene recycle
STM
Tar
Steam
Ethylbenzene recycle
Hydrogen
Water
STM
Heavy ends
Styrene
Light ends
Benzene drier
Alkylation reactor system
Ethylbenzene recovery system
Dehydrogenation reactor system
Styrene recovery system

FIG. 7.1

the ratio of benzene, ethylbenzene and polyethylbenzene in the solution as a function of the ratio between ethyl groups and benzene rings in the mixture. The pertinent data were obtained from earlier operations in a Dow Chemical Company plant and were given by Boundy and Boyer in the book previously cited. A reproduction of their data in the form of a graph is given in a chapter by J. W. Donaldson in Miller's *Ethylene and its Industrial Derivatives*.[13]

According to the data graphed by Boundy and Boyer, ethylbenzene tends to approach a maximum in the reaction, beyond which it rises only very slowly, when the ratio of ethyl groups to benzene rings is 0·6. At this point the polyethylbenzenes (including the diethylbenzene) quantity is relatively low but it rises relatively rapidly beyond the 0·6 ratio. Accordingly, plants are generally operated at the 0·6 ratio point.

There is some evidence which indicates that the transalkylation reaction at 80°C is somewhat less rapid than needed to attain the Boundy and Boyer equilibrium; the reaction mixture thus contains not 10 per cent of polyethylbenzene, as these writers indicate, but rather a concentration of the order of 17 per cent. There are also data available which indicate that in an alkylator operating under pressure, and consequently at higher temperature, the polyethylbenzene content of the system is not only as low as that indicated by Boundy and Boyer but perhaps a very little bit lower.

In general, then, feeds are adjusted so that the composition in the reactor, exclusive of catalyst layer material, tars, and the like, is ethylbenzene 40 per cent, benzene, approximately 50 per cent, and polyethylbenzene 10 per cent, more or less depending on transalkylation equilibria in the specific system.

Reference to the diagram will show that the benzene and recycle benzene pass through a drier before entering the alkylator. In terms of what already has been said about the adverse effects of water on the alkylation reaction, the reason for the drier is obvious. In point of fact, the incoming fresh benzene is ordinarily sufficiently dry to make this step unnecessary. However the recycle benzene, since it is recycled from aqueous washing steps, contains more water than is permissable for the reaction system and it must be dried.

Drying is almost always by azeotropic distillation; in such a dehydration system only a very small amount of the total benzene need be vaporised and carried overhead. Accordingly, the cost of the dehydration step is small and the inclusion of all the incoming benzene adds an additional measure of safety against the possibility of introducing unwanted water into the reactor.

Referring once again to Fig. 7.1, the product from the alkylation reactor system is seen to pass through the ethylbenzene recovery system by way of what looks like a simple holding tank in the line. That holding tank is simply a diagrammatic representation of a number of essential steps which must be accomplished before ethylbenzene can be recovered from the products stream. In most plants, the actual flow includes a cooling step, two washing steps and a settling step.

Product stream from the alkylator passes through a cooler. If the plant is one in which the alkylator operates at approximately atmospheric pressure, cooling is by water and the heat contained in the stream is largely lost. Conceivably, of course, some plants may operate using the recycle benzene as coolant; in such cases some of the heat in the product stream would be recovered. In plants operating at sufficient super-atmospheric pressure so that the product stream temperature is above 100°C, the heat exchanger can, in part, take the form of a steam generator, a system which provides more useful power conservation. In any event, as the result of the cooling, a highly viscous catalyst phase separates out.

While, as a result of the cooling, a second catalyst phase comes out of solution, no attempt is made in modern plants to isolate and recycle the separated catalyst material. There are a number of reasons for this. In the first place, the material is so viscous that its separation and recycle are extremely difficult. Of course, if the material separated represented significant economic values, a solution to the difficulty thus presented would by now have been sought and found; the situation however is otherwise. Overall, the reaction uses very little catalyst so that its separation for re-use has almost no economic significance. Moreover, the aluminium chloride in the separated catalyst phase is so contaminated with heavy tar that its activity is very low and thus its recycle provides little benefit. Consequently, product stream and the undissolved catalyst phase which it contains is passed directly to a hydrolyser in which the stream is brought into contact with water.

In most plants, the hydrolyser takes the form of a pump which provides excellent and rapid contacting between the incoming water and the incoming product streams. The hydrolysis results in decomposition of the aluminium chloride catalyst complex. A solution of the aluminium chloride, and such hydrogen chloride as is contained in the complex, makes up the water phase while hydrocarbon-soluble substances contained in the catalyst complex are dissolved in the organic product stream.

The mixture from the pump is sent to a settling tank wherein the phases are separated and the aqueous phase is removed for discard. In order to insure substantially complete separation of aqueous and organic phases, the time allowed in the separator may run up to six or more hours.

From the aqueous separator the organic phase again is sent to the inlet of a pump where it is combined with an aqueous caustic stream; the pump again acts as a contactor for the two phases. This aqueous caustic stream insures complete removal of acids from the organic phase. Once again the mixture from this pump goes to a separator from which the caustic stream is recycled (with additional makeup as needed and with some discard into the aqueous drain stream from the first separator). Here again some six to ten hours storage in the separator is provided for the organic stream and, in some plants, at least, the organic phase from the first caustic separator goes to a second holding tank, again with a storage time of six to ten hours in order to insure that the final product stream leaving the settlers is free of caustic.

These then are the steps contained in the 'Black Box' represented in the diagram by the horizontal holding drum between the alkylator reactor system and the ethylbenzene recovery system.

(It might in passing be worth noting the reason for a two step hydrolysis and caustic wash rather than a single step process, which, on its face, appears possible by using a caustic solution in the first contacting step. It would, *a priori*, appear that such a single stage process would obviate the use of one karbate pump and of the first brick-lined separator at the cost of only a small amount of caustic solution. Unfortunately, such a single step hydrolysis cannot be used because the alkali precipitates an aluminium oxyhydroxide or an aluminium hydroxide. The presence of this finely divided solid phase makes separation of the two liquid phases extremely difficult or virtually impossible. Thus, the two stage process using hydrolysis followed by an alkali wash is a necessity.)

Returning once again to the diagram, it is seen that the reaction product, following the elimination of catalyst, is passed to the ethylbenzene recovery system. The ethylbenzene recovery system, represented diagramatically by a single fractionating tower, ordinarily consists of three fractionating towers. In the first tower benzene is removed for recycle; in the second tower, ethylbenzene is recovered as an overhead product and in the third tower, polyethylbenzenes, including diethylbenzene, are separated from tars and recovered for recycle.

Here, too, the way the towers in such a system are utilised depends, to a large extent, on the skill and the design philosophy of the process designer. It depends, too, on the sources of heat available for the reboilers of the various towers.

In presumably every plant, the first tower is normally operated at atmospheric pressure. Obviously, it is desirable that little or no ethylbenzene be recycled with the benzene. A more significant decision, one affecting the entire operation is where to cut between ethylbenzene and benzene since some materials boiling between the two are always present; all of these might be returned to the reactor with the benzene in order to provide maximum means for separating pure ethylbenzene in the following column. If this is done, one must face the problem of an increasing buildup of impurities in the alkylator, leading ultimately to an undesirable dilution of the reaction occurring therein. Of course, if the recycled material contains toluene, this will ultimately reach an equilibrium value in the reactor inasmuch as toluene is ethylated. The ethyl toluene is a 'heavy' in the benzene column and is necessarily removed with the ethylbenzene into the second column.

Ethyltoluene will also move with the bottoms stream in the second, ethylbenzene, column. Some thought must however be given to the possibility not only that it will complicate the separation of the ethylbenzene but also that if some of it moves with the ethylbenzene into the dehydrogenator the resulting methyl styrene will adversely affect the styrene purification.

Another possibility that the designer can consider at this stage is the use of a pasteurising section in the ethylbenzene column. In this case, the cut in the first column can be made rather sharply for benzene, relying on the pasteuring section of the second column to remove undesired light ends, including toluene, from the ethylbenzene overhead product of the second column.

The second, or ethylbenzene, column is, in most plants, operated at reduced pressure (*ca.* 200 mm). The exact conditions depend, in part, on the sources of heat for the column reboiler and, in part, on the designers estimate of how he can achieve the separations which are essential in that tower. It is of particular importance that the ethylbenzene produced as a final product be free of diethylbenzene. The presence of diethylbenzene in ethylbenzene leads to the formation of divinylbenzene during the course of the dehydrogenation to form styrene. Divinylbenzene in styrene results in the formation of gels when the styrene is polymerised and since, for many purposes, the presence of gel is highly undesirable, it is necessary to insure that the ethylbenzene dehydrogenated to make styrene be free of diethylbenzene.

The bottoms stream from the ethylbenzene is pumped to a third column in which diethylbenzene and the lighter boiling polyethylbenzenes are, as earlier indicated, separated from the heavy polyethylbenzenes and tar. The former materials are returned to the alkylator for conversion to ethylbenzene by means of the transalkylation reaction. The latter stream, consisting of the heavy polyethylbenzenes and tar are either discarded or, in some cases, added to the heavy ends resulting from styrene purification to render the latter more fluid. In almost all cases, the third column is operated at reduced pressures (*ca.* 40 mm).

At this point, we have a reasonably complete description of the typical ethylation process plant which uses aluminium chloride as a catalyst and this, as already noted, represents the vast majority of plants now in operation. There is, however, a minor addition important in some plants, which should be discussed.

It has been assumed, for the plant described, that the feed ethylene is of high purity; this, indeed, is generally the case. However, it is quite possible to use an impure ethylene feed in an alkylation plant such as has just been described provided only that the impurities are non reactive. In the practical case, this means that the impurities will be paraffinic gases.

When an impure feed ethylene is employed, it is necessary to make provision for eliminating the inert gaseous impurities from the system. Ordinarily, this is accomplished by venting the gaseous impurities from the alkylation reaction.

The vented gas will, of course, carry benzene with it and it may also carry some hydrogen chloride as well. The benzene is recovered from the gas stream by passing the vent gas through an absorber. The absorbent used is ordinarily recycle polyethylbenzene which makes isolation of the recovered benzene unnecessary; the benzene simply passes back to the reactor with the recycle polyethylbenzene. If removal of hydrogen chloride is also necessary, that is accomplished in a single water or dilute alkali washing column before the gas is finally discharged from the plant, generally to be used as fuel.

7.2.7. Yield and Consumption Data for the Aluminium Chloride Catalysed Alkylation Reaction

It is rather difficult to give precise data with respect either to yield or raw material consumption for the aluminum chloride-catalysed ethylation of benzene. For one thing, yields based on either ethylene or benzene are above 95 per cent; this has been the case since Boundy and Boyer indicated yield data in their 1952 book. At that time they reported that the yield of ethylbenzene, based on benzene, was 95·5 per cent and that the yield, based on ethylene was 96·8 per cent. More recently, Donaldson suggests that a yield, on benzene, of 96·3 per cent and, on ethylene, of 95·6 per cent may perhaps be typical. Either set of figures is respectable and we might be justified in assuming that the differences hardly warrant much thought. However, the fact that over the years, if the data are to be taken at face value, yield on benzene has risen about 1 per cent while that on ethylene has dropped by an equal amount suggests either that the data are not real or, more likely, that at this point precise data are unavailable to anyone but the manufacturers.[14] It is altogether certain however, that *conversion* of ethylene is complete in modern plants.

Donaldson also provides utilisation data for the remaining chemicals consumed in the process. He indicates that aluminum chloride is required in the amount of 10 kg per 100 kg of ethylbenzene produced. On the same basis, the promotor, calculated as hydrogen chloride, is needed to the extent of 5·5 kg/1000 kg of ethylbenzene and 100 per cent caustic soda is used to the extent of 10 kg/100 kg of ethylbenzene. These figures are quite clearly of the right order and can be assumed to be sufficiently accurate for our present purposes.

Donaldson also gives utilities figures. While the figures themselves do not, on examination, seem unlikely, there are some difficulties in obtaining such figures which deserve mention. In the first place, not only the actual values for each of the utilities mentioned may well change, depending on the ingenuity of the designer of the plant in conserving utilities, but, the relationships between them may change by virtue of which of several choices he has exercised at various points in the plant. Steam, power and fuel gas are all, in effect, energy sources and frequently interchangeable depending upon what is available and least expensive in a particular plant. It thus seems less likely that typical utility consumption figures can be cited with confidence although figures for *a plant* or *a given type of plant* might be given. There is another difficulty; in general, ethylbenzene plants and ethylbenzene dehydrogenation plants are tied together. It is precisely this tie-in that gives the resourceful designer increased possibilities for reducing the overall utilities consumption load. While consumption figures for the overall plant are, without question, known quite accurately there may, in a number of instances, be some real difficulty in separating utilisations for the two sections of the overall plant. Still, having said all of this, it seems wise to add that Donaldson's utilization figures provide a reason able sense of proportion about utilities requirements in an ethylbenzene plant.

TABLE VII.2

Sales Specifications for Ethylbenzene

Product	*Allied*	*Carbide*	*Koppers*	*Dow*
Purity (typical)—wt%	98	—	—	—
Dist. range—max., °C	1·8	1·8	1·2	—
including, °C	136·2	136·2	136·2	—
Dist. range, °C	—	—	—	134–137
5–95%within, °C	—	—	—	1
Colour, Co-Pt	15	15	15	—
Sp. grv., $\frac{15·5°}{15·5°}$	0·868–0·872		0·870–0·872	—
Sp. grv., 20°/20°	—	0·8676–0·8684	—	—
Free acid	None	N.1	Neutral	—
Benzene, % max.	—	0·3	0·1	1·0
Diethylbenzene, % max.	—	0·17	0·6	0·04
Minimum purity, wt%	98·0	—[1]	00.0	—
S compounds	free from H_2S and SO_2	—	—	—
Cu corrosion	free from iridescence	—	—	—

[1] Given as 99+ mol% in 1970.

7.2.8. Ethylbenzene Sales Specification

There are sales specification for ethylbenzene; once again it is difficult to know precisely what they mean. By far the largest amount of ethylbenzene is made for the ethylbenzene dehydrogenation plant in which it is to be converted into styrene. It is, for the most part, neither sold nor bought. If the situation were otherwise, it is doubtful that a boiling range specification would be the major criterion of purity for a commodity for which purity is so important a gauge of suitability.

In general, ethylbenzene plants are not sold on the basis that the ethylbenzene they produce will meet this or that specification, but more realistically on the basis that the ethylbenzene produced will lead to a satisfactory styrene in the dehydrogenation plant for which it is intended. In short, the real specification for ethylbenzene is: 'This ethylbenzene, when dehydrogenated in the plant for which it is intended, will be converted into styrene which meets all applicable specifications.' Table VII.2 lists sales specifications of a number of American ethylbenzene producers as of the middle of the 1960s.

7.3. THE MANUFACTURE OF STYRENE MONOMER

by J. Habeshaw

7.3.1. General

Practically all styrene monomer now produced is made from ethylbenzene, and this hydrocarbon will certainly continue to be by far the most important styrene precursor. Ethylbenzene can be made from ethylene and benzene (at present the source of most of the ethylbenzene used), or may be co-produced with other aromatics (particularly *p*-xylene) from aromatic extracts or reformates made in refineries. Whatever the source of the ethylbenzene, conversion to styrene can follow the main routes:

(1) Chlorination followed by dehydrochlorination.
(2) Oxidation followed by hydrogenation of the acetophenone so formed to carbinol, which is then dehydrated to styrene.
(3) Oxidation of ethylbenzene to the hydroperoxide, which is then reacted with an olefin (propylene) to make the olefin oxide, with reduction of the hydroperoxide to carbinol, which is then dehydrated to styrene. This process coproduces styrene and propylene oxide.
(4) Direct dehydrogenation of ethylbenzene to styrene.

The last of these is by far the most important and also the best known of the methods, but all are discussed in more detail in following sections.

Several sources of styrene alternative to ethylbenzene have been described in the literature. Large ring oligomers of diolefins, particularly by-product oligomers of butadiene such as cyclo-octadiene formed during the production of Nylon-12 precursors can be converted to styrene,[15] as can vinylcyclohexane[16,17,21] using fairly conventional dehydrogenation catalysts. It is unlikely that these butadiene based routes can ever provide more than a very minor source of styrene.

Toluene based processes could be more interesting, but data so far disclosed on the reaction of toluene and methanol over ion-exchanged X-type Zeolites[18] and on the pyrolysis of toluene/alkane mixtures[22] does not seem attractive enough to be competitive with the ethylbenzene based routes. Direct reaction of ethylene with benzene to styrene without intermediate formation of ethylbenzene using organometallic catalysts has been described,[19,20] but this method is not yet close to practical application. Nor is the pyrolysis of diarylaliphatic hydrocarbons[23] such as 1·2-diphenylbutane (producible from toluene and isopropenylbenzene) likely to be used on any significant scale in the foreseeable future. Of known routes to styrene not based on ethylbenzene the only one likely to find application soon on a commercially significant scale is the extraction from appropriate fractions of the pyrolysis gasoline made in large modern ethylene plants as a substantial co-product when cracking liquid feedstocks. The amount of styrene so producible is of course dependent on the scale of ethylene manufacture. Despite this limitation economical extraction processes could well find application in Japan and Western Europe, where there is a very large installed capacity for ethylene production from naphtha or heavier feedstocks.

7.3.2. Styrene Production from Ethylbenzene

The separation of ethylbenzene and styrene is difficult, and styrene polymerises thermally very readily. This separation was a major problem in the early days of styrene manufacture, and for a time made the chlorination-dehydrochlorination and oxidation to acetophenone more attractive. However once the design of fractionation systems operating under vacuum and development of polymerisation inhibitors for addition during the fractionation steps made the separation possible commercially, both these methods fell into disuse. However the cost of the separation was still considerable, and much work to increase the per-pass conversion in the dehydrogenation step has been done. The conversion must be increased without significant loss in efficiency of use of the ethylbenzene, since raw material costs represent a substantial part of the styrene cost.[24] In the direct dehydrogenation considerable heat is absorbed during reaction, and the rate of undesirable side reactions relative to styrene formation increases very rapidly with rise in temperature, although the equilibrium proportions of styrene increase with rise in temperature. Changes in the reaction conditions and process design must not increase the impurities left in the styrene monomer, nor introduce new and difficultly separable impurities, since the purity of the styrene product is very critical in polymer formation, which is by far the largest outlet for styrene. These factors tend to make drastic changes in the technology slow of acceptance in industry. Methods under study will be considered later after reviewing the present state of the direct dehydrogenation process.

7.3.2.1. Direct Dehydrogenation of Ethylbenzene to Styrene

The general form of the process is shown diagrammatically in Fig. 7.2, showing the essential steps in the dehydrogenation of ethylbenzene in the presence of steam. The main features of the process are largely dictated by the thermodynamics of the dehydrogenation reaction, and by the difficulty of separating ethylbenzene and styrene without causing excessive polymer formation. Ethylbenzene is preheated either alone or in the presence of part of the steam fed to the process, and is then joined by the remainder of the steam superheated to a level which will give the required inlet temperature to the reactors when blended with the ethylbenzene feed. After reaction the product is rapidly cooled and passed into the fractionation system, where pure styrene and recycle ethylbenzene are separated from the by-products (mainly benzene and toluene). Thermodynamic equilibrium limits the conversion at any given temperature, and there is a large heat input needed for the dehydrogenation reaction. The variation of equilibrium constant and heat of reaction with reaction temperature is shown in Table VII.3.

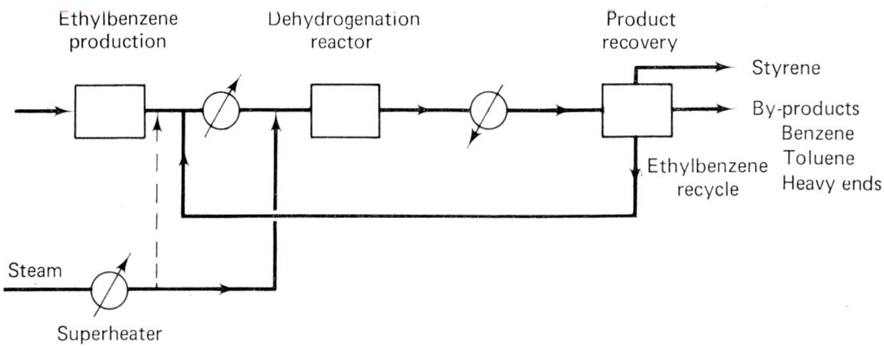

FIG. 7.2

TABLE VII.3

Temperature, °C	427	527	627	727
Heat of reaction, H Kcal/mole	29·55	29·71	29·82	29·89
Log 10 K_p	−2·482	−1·327	−0·4252	0·8022
K_p, atm	$3{\cdot}30 \times 10^{-3}$	$4{\cdot}71 \times 10^{-2}$	$3{\cdot}75 \times 10^{-1}$	2·00
Temperature, °C	827	927	1027	
Heat of reaction, H Kcal/mole	29·92	29·91	29·87	
Log 10 K_p	0·896	1·387	1·8090	
K_p, atm	7·87	$2{\cdot}44 \times 10$	$6{\cdot}44 \times 10$	

(Quoted by Reference 17: see also References 13, 26, 27, 28)

Styrene production is clearly favoured by rising temperature over the whole range of practicable operating temperatures, and reduced pressure is also obviously favourable. The heat input needed will, in adiabatic reactors, lead to large falls in temperature over the reaction path, unless very large amounts of diluent are used. This, besides the need to use steam to reduce side reactions and maintain catalyst activity, would dictate the use of large proportions of steam diluent at least in adiabatic reactors. The upper limit of operating temperature is fixed by the need to keep down unwanted side reactions, which increase relative to styrene formation as reaction temperature is increased. American plants use largely adiabatic reactors, and the reaction heat is supplied largely as sensible heat of the steam diluent. German plants use multitubular reactors with the catalyst in tubes and flue gas in the shell supplying heat to the essentially isothermal reactors. The isothermal system can give a higher per-pass conversion for a given reaction efficiency and requires less steam, but is a more costly system to construct. The adiabatic reactors cannot make as close an approach to the optimum temperature profile as the isothermal systems, and hence give a lower per-pass conversion for a given reaction efficiency and need more steam. An optimum temperature profile will adjust temperatures along the reaction path so as to obtain the best economic balance between high conversion to styrene and lowest possible formation of unwanted by-products. The factors involved have been discussed in some detail in recent reviews. Miller, Hornibrook and Donaldson have considered mainly the American (Dow) type of process,[29,30,31] whilst Ohlinger and Stadelmann have dealt mainly with the German (BASF) process. Some further details on the more important factors updating the information in these reviews is given below.

7.3.2.1.1. Catalysts for direct dehydrogenation—Generally iron oxide based catalysts promoted and stabilised in various ways have continued to be mainly used in adiabatic reactors, whilst the isothermal systems use mainly zinc-based catalysts. Both types of catalyst are self-regenerative, with lives of one to two years. The main objective of continued catalyst development has been to increase the per-pass conversion without loss of reaction efficiency, particularly with the iron oxide based systems. The alkali-promoted chromia-stabilised iron oxide catalysts are said to be improved by the addition of zinc or copper oxides.[33] Addition of nickel and cobalt oxides is claimed to allow conversion of carbon oxides to methane in the reactor, so supplying some of the heat of reaction and pushing the equilibrium further towards the styrene side.[34] Although earlier work[36] indicated a poisoning effect of chloride on the catalysts, and certainly the presence of large amounts of chloride in the feed causes difficulties through formation of chloride deposits,[31] addition of small amounts of chloride to the catalyst is said to reduce toluene formation.[37] The use of carbon[37] and alumina[38,39] as components of the iron oxide catalyst has been described, and addition of magnesium oxide is claimed to increase both the activity and mechanical strength of the catalyst.[40] Initial reduction of the catalyst is claimed as beneficial,[41,42] and initial heating to temperatures around 600°C to improve activity and selectivity of the catalysts.[39, 43] A number of modifications in the detailed preparative methods for the two main catalyst types have been described,[44,45,46,47] and the incorporation of Group VIII metals claimed. The two main types of catalyst, albeit with improvements in composition and preparative techniques, seem likely to continue in use for the foreseeable future.

7.3.2.1.2. Conditions of the dehydrogenation reaction—The adiabatic (Dow) type of reactor system is shown in Fig. 7.3, adapted from Reference 31. With the very large plants now essential for economically competitive styrene production, heat economy receives much emphasis. In particular recovery and reuse of the heat of condensation of the large amount of steam in the effluent from the reactors is necessary. The

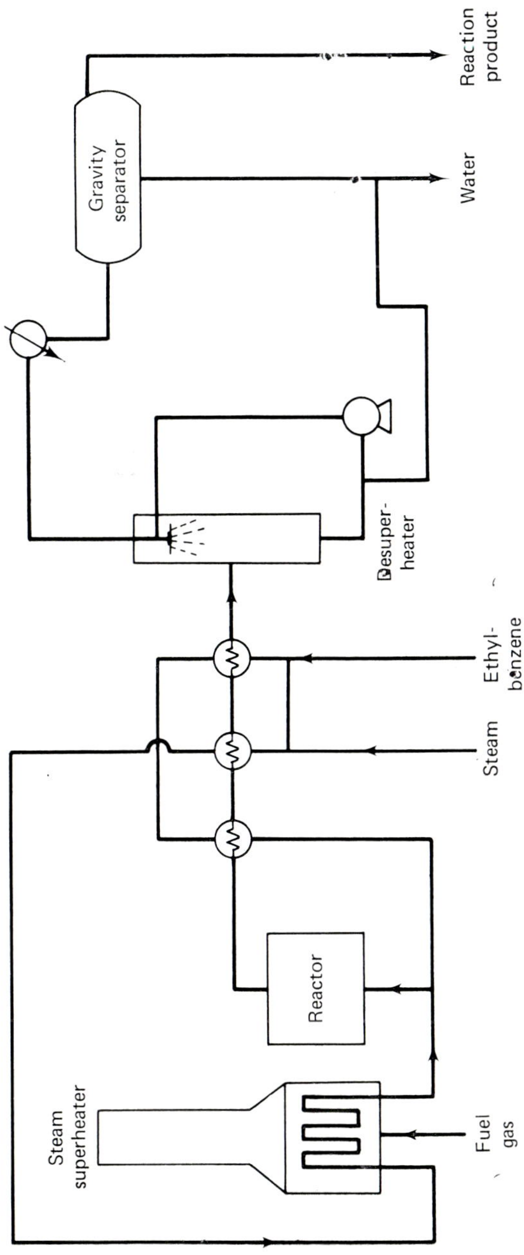
Reaction
product
Gravity
separator
Water
Desuper-
heater
Ethyl-
benzene
Steam
Reactor
Steam
superheater
Fuel
gas

FIG. 7.3

heat of the reactor effluent is used in indirect exchangers (see Fig. 7.3) to preheat separately the steam and ethylbenzene feeds, some steam being fed into the first ethylbenzene preheater to reduce inefficient thermal cracking reactions. The steam feed is further superheated in a gas fired heater, commonly of the 'Petro-Chem' type, to around 720°C. The superheated steam and ethylbenzene feeds are then combined as close to the catalyst bed inlet as possible, again to minimise thermal cracking reactions. In adiabatic reactors the temperature falls from about 630°C at the inlet to around 530°C at the exit. After heat exchange with feed and steam (Fig. 7.3) the reactor effluent is cooled to its condensation temperature by direct exchange, followed by condensation in conventional exchangers. The water phase from the condensate is stripped of organic compounds and passed from the system, or may be recycled. The organic portion passes forward to the separation system. Usual pressure ranges in the reaction system are from about 2 atm gauge at the superheater to about 0·1 atm gauge at the condenser exit.

In the isothermal reaction system, steam and ethylbenzene are preheated together and fed into the reactor tubes. Reaction heat is supplied by flue gases on the shell side of the catalyst packed-tubes, raising the temperature from about 530°C at the inlet to about 630°C at the exit. As indicated above, the 'isothermal' system gives somewhat higher per-pass conversions for a given reaction efficiency than the adiabatic reactors, and requires rather less steam. However, the overall ultimate yield of styrene obtained is about the same in both the systems, the products of reaction are similar, and subsequent work up by vacuum fractionation is essentially similar.

A good deal of work to improve the per-pass conversion in adiabatic reactors without loss in efficiency of reaction or heat economy has been done. The commonest method proposed is to carry out the reaction in more than one catalyst zone, with intermediate heat supply between zones. Patents to Halcon describe reaction in two zones, with the reaction mix reheated with steam in fixed heaters between zones.[50] UOP[51] use combustion gases from part of the products to reheat the reaction mix between zones. French workers also propose[54] re-heating the product with steam before feeding to a second reaction zone. Monsanto[55] use high-selectivity catalyst in the first reaction zone, followed by high activity material (e.g. the Shell iron oxide based catalyst) in a second zone. UOP claim that addition of preheated toluene before the second zone of multi-zone reactors improves the reaction efficiency,[52] and have also described means of minimising the temperature drop across the catalyst bed.[53] A published mathematical analysis of the reaction system[56] confirms that the two-reactor system will give better efficiency of styrene production than the single zone adiabatic reactor.[56] There is also a published analysis of the temperature profiles and conversion levels throughout the reaction path in tubular reactors.[57] Advantage has also been claimed for the use in a first heated reactor of the promoted iron catalyst, followed by reaction over the same catalyst in a subsequent adiabatic reaction zone.

The use of nickel-free steels to avoid catalysis of cracking reactions has increased the difficulties of reactor fabrication, but patents to Badger[59] and Union Carbide[60] indicate that at least some nickel containing steels can be used in this service; the former still provide nickel-free inlet nozzles to the reactors. A high production rate per unit of reactor volume is claimed for a radial flow reactor[61] where the feed is passed radially through the catalyst bed.

Advances in the large scale reactor systems have continued as a number of relatively small improvements, with as a result appreciable improvements in heat economy, and probably (although precise figures are difficult to derive from diverse published data) an increase in per-pass conversion of around 10 per cent without significant loss in overall efficiency of conversion.

7.3.2.1.3. Recovery of styrene from direct dehydrogenation products—The two-column system required in earlier styrene plants for the separation of ethylbenzene and styrene is being replaced by a single-column system in new plants. This shortened recovery system is shown diagrammatically in Fig. 7.4 (based on Reference 29). In this system styrene and tar are left as residue in the first large column, the other products (benzene, toluene and ethylbenzene mainly) being taken overhead. The styrene is separated from the tar and polymer in a second column (see Fig. 7.4), any styrene left in the tarry residue being recovered in a stripping column and returned to the styrene tower. Tar and polymer are discarded from the base of this stripper. The distillate from the first column is passed to the ethylbenzene tower, which takes benzene and toluene overhead, leaving ethylbenzene as residue, which is recycled to the dehydrogenation reaction. The benzene and toluene by-products are separated in the final tower.

The substantial advance involved was made possible by the development of low pressure-drop and low hold-up column trays [62,63,64,65] which enabled the temperature and residence time of styrene in the column to be kept low enough to avoid excessive polymerisation rates in the single column. Inhibition of polymerisation, mainly with sulphur in the American process and with hydroquinone in the German process continues to be essential in the styrene recovery system.[31]

Improvements have also been made in the heat economy of the process, particularly in recovering and using the large heat of condensation of the steam present in the reactor effluent from, more especially, the adiabatic reactors. Numerous patents, (see for example references 66, 67, 68, 69) describe detailed means of accomplishing improved heat economy. Recovery of all the hydrocarbon from the aqueous effluent from the process is becoming more important with increasing emphasis on the environmental needs. This is usually achieved by extraction with ethylbenzene feed followed by steam stripping to take out the last traces of hydrocarbon.[66,70] The use of thin film evaporators to recover further styrene and aromatic values from the fractionator residues is described[71] and the addition of cut-back oils to supply some further heat and stripping action to these residues has been patented.[72] The similar addition of fuel oil to the residues to give a more efficiently combustible material is also claimed.[73] The use of hydrodesulphurisation and hydrodealkylation to generate benzene and ethylbenzene from the residues from the system has also been patented.[74]

Although the use of extraction with silver nitrate[75] and silver salt solutions[76] such as fluoroborate for selective removal of styrene from reaction mixtures has been described, no basically novel methods of recovering styrene monomer from the dehydrogenation reaction products have come into actual use during the last few years.

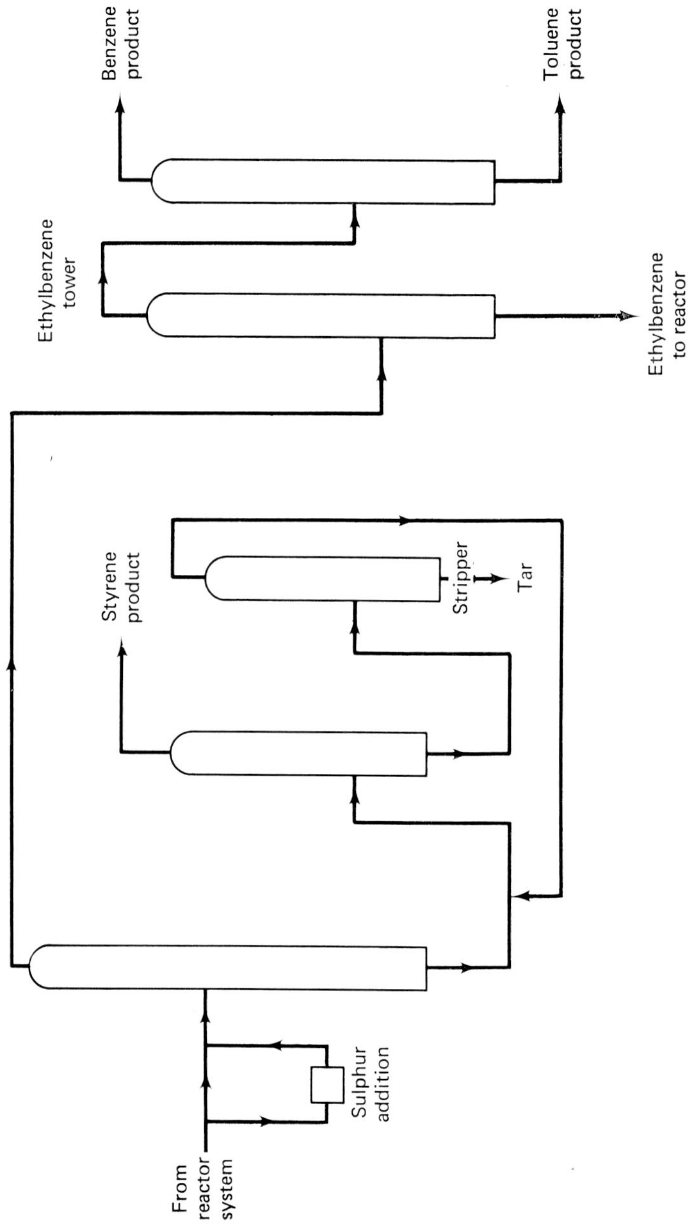
Benzene product
Toluene product
Ethylbenzene tower
Ethylbenzene to reactor
Styrene product
Stripper
Tar
From reactor system
Sulphur addition

FIG. 7.4

7.3.2.1.4. General position of the direct dehydrogenation process—This is now a mature and fully developed process, further advances being likely only in matters of detail, although these may of course have considerable value in the very large plants which are now essential. As indicated above improvements in per-pass conversion and heat economy have been achieved, and it seems doubtful whether any great further economies of scale are to be had. Other possible methods of improving the economics of converting ethylbenzene to styrene have therefore been examined, and these are described in the next section.

7.3.2.2. Ethylbenzene dehydrogenation using hydrogen acceptors

One obvious way of shifting the equilibrium between ethylbenzene and styrene is to react the hydrogen present in the reaction mix with some acceptor. This of course involves loss of the hydrogen by-product from the process, but may eliminate the very large heat requirement of the styrene formation reaction and simplify the reaction system. From the published information the main difficulty at present is to obtain reaction efficiencies comparable with those of the direct reaction while still increasing the per-pass conversion sufficiently to make full scale development interesting.

7.3.2.2.1. Oxidative dehydrogenation Oxygen is the most obvious hydrogen acceptor and a considerable patent literature now exists on this: this parallels the development of oxidative dehydrogenation processes for butadiene production. The majority of the patents are concerned with catalyst for the ethylbenzene to styrene reaction, and the more important of these are summarised in Table VII.4.

TABLE VII.4

Catalysts for Ethylbenzene Oxidative Dehydrogenation

Catalyst and conditions	*References*
Iodine or HI at over 400°C	77
Halogens at 400–800°C	78
Bromine at 300–400°C	79
Fluidised quartz and iodine at 400–700C	80
Molybdates and oxides of K, Gp V, VI or VIII metals at 480–500°C	81
CdO on carrier (silica)	82
GeO_2	83
Fischer-Tropsch catalysts	84
Mixed oxides of W and Cr, Nb with K	85
Fluidised Cr_2O_3 catalysts on carriers with alkali metal promoters	86

A study of two-component catalysts on carriers by Balandin[87] showed that cobalt-molybdenum on magnesia was the best catalyst tested, giving maximum styrene yields of around 55 per cent. High selectivities were given by ammonium molybdate catalysts[88] but supported bismuth-molybdenum and cobalt-molybdenum catalysts gave only low conversions in Italian tests. The use of fluidised catalyst systems is favoured in German patents.[80,86] No major problems in purifying styrene made by oxidative dehydrogenation have been reported so far, but from the patents and literature there is no firm evidence that any of the catalysts reported so far are likely to lead to a commercially viable styrene process.

7.3.2.2.2. Use of hydrogen acceptors other than oxygen The use of elemental sulphur and sulphur dioxide as hydrogen acceptors has been proposed, and effective catalysts for the reactions concerned described. The more interesting of these are listed in Table VII.5 below.

TABLE VII.5

Catalysts for Ethylbenzene Dehydrogenation with Sulphur or Sulphur Dioxide as Hydrogen Acceptor

Hydrogen acceptor	*Catalysts and conditions*	*References*
Sulphur dioxide	Metal phosphates: $BiFePO_4$; $CuNiPO_4$, etc.	89
	Spinel or Perovskite catalysts, e.g., $CrFeZnO_4$	91
	Magnesia	92
	Iron oxide-active carbon	93
	Charcoal	90
Sulphur	NiS/MoS_2	94
	Alumina	95

The possible process using sulphur dioxide has been briefly described in the literature.[96] The main overall reaction is:

$$C_6H_5.C_2H_5+\tfrac{1}{3}SO_2 \rightarrow C_6H_5.CH = CH_2+\tfrac{1}{3}H_2S+\tfrac{2}{3}H_2O$$

Benzene, toluene and benzothiophen are main by-products, with small amounts of elemental sulphur, carbon monoxide, carbon dioxide and other organic sulphur compounds being formed. No detailed description of a practical scheme for the recovery of styrene from the reaction products made in this reaction has so far been published.

Other sulphur compounds have been suggested as hydrogen acceptors, including hydrogen sulphide and thiourea[97] and carbon oxysulphide,[98] and the use of nitrobenzene as acceptor is said[99] to give styrene and aniline simultaneously.

7.3.2.2.3. Direct oxidation of ethylbenzene and co-production of styrene and propylene oxide The old manufacturing method of Union Carbide involving oxidation to acetophenone, hydrogenation of this to the carbinol and catalytic dehydrogenation of the carbinol to styrene is no longer a significant commercial source of styrene (see Reference 100 for details of this method).

However, ethylbenzene can be efficiently oxidised to the hydroperoxide by methods analogous to the cumene oxidation used in the cumene process for phenol and acetone. This hydroperoxide will react with olefins to form the epoxide corresponding to the olefin, the hydro-peroxide being reduced to the corresponding alcohol. The overall reactions with ethylbenzene as hydrocarbon and propylene as the olefin are:

$$C_6H_5.CH_2.CH_3 \xrightarrow{O_2} C_6H_5CH(OOH).CH_3$$

$$C_6H_5.CH(COOH) + CH_2 = CH.CH_3 \longrightarrow C_6H_5.CH(OH).CH_3 + \underset{\diagdown \; O \; \diagup}{CH_2—CH_2—CH_3}$$

$$C_6O_5CH(OH)CH_3 \xrightarrow[—H_2O]{} \underset{\textit{Styrene}}{C_6H_5.CH.CH_2} \qquad \qquad \textit{Propylene oxide}$$

General descriptions of this process have been given[101,102] from these and the patents to Halcon International[104] and Scientific Design[103] the main features are as follows. The formation of ethylbenzene peroxide and the dehydration of phenylethyl alcohol to styrene, which were known before, have been improved, but the most important innovation step is the efficient reaction of the olefin with the hydroperoxide to form the alcohol and the epoxide. Ethylbenzene is oxidised at around 130°C and at a few atmospheres pressure to form the hydroperoxide as a dilute solution in ethylbenzene. This is then concentrated by distillation to form an approximately 35 per cent solution of the hydroperoxide in ethylbenzene. This concentrated solution is then reacted in the presence of catalysts with propylene at about 110°C. From the patents the naphthenates of tungsten and molybdenum appear to be the most successful of the catalysts disclosed. The crude product is then fractionated to remove unreacted propylene, and the stabilised residue hydrogenated at about 40°C over a nickel catalyst to reduce any unconverted peroxide to alcohol. Fractionation of the hydrogenated product then gives propylene oxide product, ethylbenzene for recycle and phenyl ethyl alcohol. This last is then dehydrated over, for example titania, to give high yields of styrene.

Little technical detail has been disclosed on the process, and its technical and economic position compared with other styrene processes will not be clear until commercial plant operation has continued for some time. A full scale plant using the technology is said to be in operation in Spain,[105] so the long term prospects of the process should become clearer during the next few years. About 2·5 tons of styrene are made for each ton of propylene oxide produced: in view of the very much larger markets for styrene than for propylene oxide this ratio of products may not in the long run prove such a disadvantage as might appear at first sight.

7.3.2.2.4. Other methods of converting ethylbenzene to styrene Chlorination of ethylbenzene followed by dehydrochlorination to styrene was an early method used to circumvent the difficulty of separating ethylbenzene from styrene, which was a major stumbling block to the commercial production of styrene in those days.[25] However this method left nuclear chlorinated compounds in the styrene after separation, and the process degraded chlorine to hydrogen chloride with adverse effects on the economics. With the solution of the problems in separating styrene from the dehydrogenation product this process therefore passed out of use, despite the high yield and relatively simple styrene recovery system needed (see Reference 25 for a review of this early technology).

Uncatalysed thermal dehydrogenation gives about a 30 per cent yield of styrene from ethylbenzene,[25,106] and obviously this method cannot compete with the overall yields in excess of 90 per cent obtained by the catalytic reaction, quite apart from the formidable separation problems involved in recovering pure styrene from the complex mixture formed. However, pyrolysis as a source of styrene is coming into the industrial picture with the technical solution of the problems of separating styrene from the appropriate fractions of the liquid cracked products made in large modern ethylene plants using liquid feedstocks (see below).

Other ways of forming styrene from ethylbenzene have been disclosed, for example reaction with bromotrichloromethane[107] and hydrogen exchange with ethylene to give ethane and styrene.[108] None of these has as yet come into serious consideration as the basis of a commercial process.

7.3.3. Styrene Production from Pyrolysis Products

The presence of styrene in the liquid fractions from the steam cracking of naphtha and other liquid feedstocks has long been known (see Reference 109). The absolute amount produced in the earlier low severity cracking plants, and the styrene concentration in the liquid fractions were too low to be attractive commercially, although the amount formed relative to ethylene was not drastically lower than in the modern medium and high severity units. The absolute amount rose with the enormous increase in size of ethylene plants, and the concentration in the separated C_8 fractions increased with the use of more severe cracking conditions, so that styrene extraction has become a much more attractive proposition. The amount of styrene made in ethylene plants is around 5 per cent of the ethylene produced (dependent to some extent on the feedstock and conditions used), so that about 20,000 T/A of styrene is now available in the liquid products from a typical modern ethylene plant. The C_8 (xylene) fractions of the liquid product have fairly high styrene contents and ethylbenzene is also present, so that after hydrogenation the product has a relatively high ethylbenzene content compared with similar fractions from catalytic reformates. Such aromatic fractions are therefore a less good source of xylenes (especially *p*-xylene) than the reformates (see Table VII.6 below), which reduces their value and helps to make them a more economic source of styrene, their value being broadly that of a gasoline component. The approximate proportions of the constituents present in the pyrolysis products, compared with analogous aromatic extracts from reformates, with the boiling points of the main components are shown in Table VII.6.

TABLE VII.6

Main Constituents of Cracked Gasoline C_8 Fractions

Component	*Boiling point °C/760 mm*	*Amount present wt% (approx.)*	*Amount present in typical cat. reformate wt% (approx.)*
C8's non-aromatics	135–146	2	—
Ethylbenzene	136·2	9	18
p-xylene	138·4	12	19
m-xylene	139·1	26	42
o-xylene	144·4	14	21
Styrene	145·2	30–35	—
Other aromatics	—	2	—

Toray Industries now claim to have overcome the technical problems in separating pure styrene from the C_8 fraction of cracked gasoline, and offer a process in which styrene is separated by extractive distillation. Brief general descriptions of the process have been published,[110,111,112,113] the main features being as follows.

The C_8 fraction is separated from the liquid cracked products by conventional fractionation, to give a liquid of composition typically as shown in Table VII.6 above. The styrene is then removed by fractionation in the presence of a solvent (not disclosed), the styrene-rich solvent then being separated, the styrene removed and the solvent recycled. Development of a novel chemical treatment for the C_8 fraction before the extractive distillation was necessary before acceptably pure styrene could be obtained from the extractive distillation. The distillation operates at a rather higher temperature than the vacuum fractionation used in recovering styrene from dehydrogenation products, and a new inhibitor to prevent polymerisation of the styrene had to be found before uneconomic limitations on the operating conditions of the extractive distillation column could be removed. The xylene fraction taken overhead from the extractive distillation is of roughly the same composition as the catalytic reformates (see Table VII.6 above), and is suitable for feeding to xylene recovery units. The process thus upgrades in value the residual aromatic fraction (at least so long as there is a xylenes unit available to accept the material) as well as recovering the styrene present. The process is claimed to give high quality styrene and to show lower costs than the conventional direct dehydrogenation process.[111] The extractive distillation solvent and novel inhibitor actually used have not been disclosed. Patents to Toray Industries list some fourteen solvents said to be effective, including 2-pyrrolidone, N-methyl-2-pyrrolidone, pyridine, morpholine, ethylene diamine, cyclohexylamine, propylene carbonate, phenol and aniline. The choice made depends on ability to keep the extractive distillation temperature low enough to allow of economical column design at reasonable cost, and give easy separation of solvent from the styrene-rich extract.

7.3.4. General Position of Styrene Monomer Manufacturing Processes

There is no real doubt that the direct catalytic dehydrogenation of ethylbenzene, which at present provides much the greatest part of the styrene of commerce, will continue to do so for many years to come.[114,115] With rising costs of ethylene and benzene, increasing amounts of the ethylbenzene needed may well be derived from petroleum refinery aromatic streams, notably reformates, by methods similar to those in use in the Cosden fractionation process.[116] Despite the difficulty of the separation of ethylbenzene from the close boiling xylene isomers (see Table VII.6 above) the economics of this method are likely to improve relative to the synthetic production of ethylbenzene from benzene and ethylene.

The only other styrene manufacturing processes to have commercial potential in the short and medium term are the co-production of styrene and propylene oxide from ethylbenzene and propylene, and the extraction from pyrolysis gasoline described above. The expansion of the former will be limited by the propylene oxide market, and of the latter by the scale of ethylene production from liquid feedstocks. They are not therefore likely to provide a replacement for the present dehydrogenation route. The extent to which their use grows will depend on their technical performance as well as on the market factors, and this will not be clear until they have been in full-scale use for some time.

7.4. POLYMERISATION OF STYRENE

by A. N. Roper

7.4.1. Introduction

Although styrene and its polymer were first reported in 1839[117] it was not until early this century that the true nature of polymerisation reactions began to be understood. At first the styrene polymerisation reaction was thought to be one of oxidation.[118] Then Ostromislensky[119] proposed his concept of compound reactions in 1911. It was, however, Staudinger in 1920[120] who first put forward the long chain structure for the polymer, although the broad outlines of the reaction mechanism were not clearly defined until much later.[121] He suggested the name 'polystyrene' in place of the term 'metastyrene' previously used.

Commercial production of polystyrene was begun by I.G. Farbenindustrie in Germany in 1933, and by Dow Chemical Co. in the USA in 1937. During World War II production capacity for styrene monomer was vastly increased to supply the synthetic rubber plants. The resulting ready availability of monomer after the war led to a very rapid expansion in the production and usage of styrene polymers and copolymers which is still continuing.

7.4.2. Polymerisation Mechanisms

The polymerisation reaction can be represented simply as:

$$n\,\begin{matrix} CH = CH_2 \\ | \\ C_6H_5 \end{matrix} \rightarrow -\left[\begin{matrix} -CH-CH_2- \\ | \\ C_6H_5 \end{matrix}\right]_n-$$

and there are two chemically dissimilar methods of promoting this reaction. Polymerisation can be initiated either by free radicals, which may be thermally or chemically generated, or by an ionic reaction.

Free radical initiation is by far the most important method industrially, although anionic initiation is now assuming increasing importance in, for example, the production of block copolymers of styrene which possess unique properties.

Polymerisation is accompanied by a number of observable changes. The polymer is completely soluble in the monomer and thus the viscosity of the polymerising mass rises rapidly. The initially mobile liquid forms an increasingly sticky mass and finally a hard, clear, water white solid. There is considerable evolution of heat, the heat of polymerisation being $17{\cdot}4 \pm 0{\cdot}2$ Kcal/mole.[122] Specific gravity increases from 0·905 to 1·045 and there is thus a contraction in volume.

7.4.2.1. Free radical polymerisation Styrene polymerises spontaneously on heating in vacuum or an atmosphere from which oxygen has been excluded. Very few other vinyl monomers possess this property.

The nature of the spontaneous initiation reaction has been the subject of considerable study. At first the initiation reaction was thought to be second order and formation of a biradical by combination of two molecules of styrene was proposed by Flory to account for this.[123] This would, however, tend to give cyclic polymers of low molecular weight. Mayo[124] found that the overall reaction rate was 5/2 order in styrene, requiring a third order initiation reaction, and proposed that three styrene molecules combined in one step to produce two monoradicals. Study of the dimerisation of styrene by a number of workers produced a scheme which accounts both for the chemical initiation reaction and the dimeric products viz:

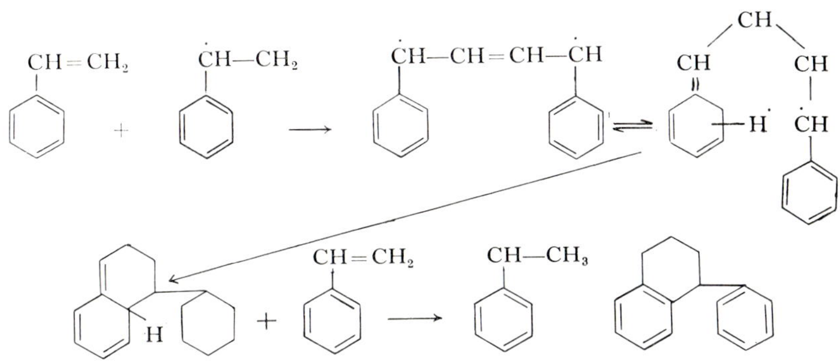

This mechanism is now generally accepted.

The initial rate of thermal polymerisation is a function solely of temperature (see Fig. 7.5),[125] and below 90°C the rates obtained are too low to be useful commercially. The molecular weight of the polymer produced is, moreover, inversely proportional to temperature and below 90°C very high molecular weight polymers are obtained which are not easily processed. Suitable rates and molecular weights are obtained in the 120–160°C range, but to complete polymerisation it is necessary to raise the temperature to 180–200°C and this produces some less desirable low molecular weight polymer. By initiating polymerisation with a compound which liberates free radicals on heating, it is possible to obtain high rates and complete polymerisation at lower temperatures. Organic peroxides and azo compounds are the free radical catalysts usually employed, benzoyl peroxide, which has a useful operating temperature range of 60–90°C, being the most frequently used. It decomposes by a first order reaction and the initial rate of polymerisation is proportional to the square root of the catalyst concentration. Benzoyl peroxide is susceptible to radical attack however and when this is by a growing polystyryl radical, chain transfer occurs and the molecular weight is reduced.

$$-\left[-\underset{C_6H_5}{\underset{|}{CH}}-CH_2-\right]_n- + C_6H_5-\underset{\underset{O}{\|}}{C}-O-O-\underset{\underset{O}{\|}}{C}-C_6H_5 \rightarrow -\left[-\underset{C_6H_5}{\underset{|}{CH}}-CH_2-\right]_n-O-\overset{\overset{O}{\|}}{\underset{\underset{C_6H_5}{|}}{C}}$$

$$+C_6H_5\underset{\underset{O}{\|}}{C}\cdot$$

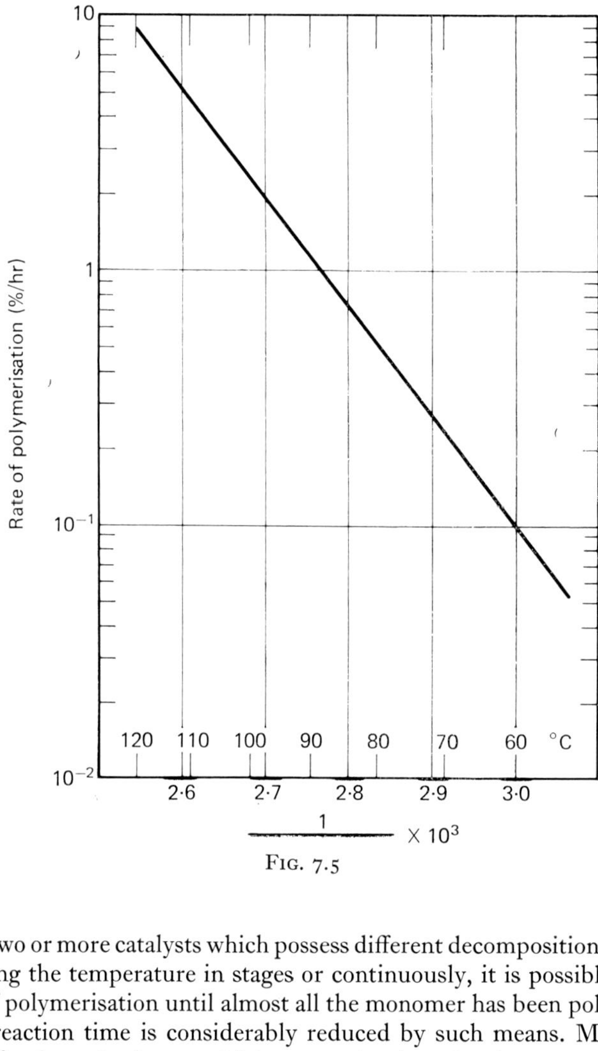

FIG. 7.5

By using two or more catalysts which possess different decomposition temperatures and increasing the temperature in stages or continuously, it is possible to maintain high rates of polymerisation until almost all the monomer has been polymerised and the overall reaction time is considerably reduced by such means. Maintenance of high-rates of polymerisation to a high conversion is assisted to some extent by the 'gel' effect; the retardation, due to the rapidly increasing viscosity of the system, of the termination process which requires that two active polymer radicals collide together.[126]

Free radical polymerisation of styrene can also be carried out in a non-homogeneous system where the styrene is dispersed in water. When an oil soluble catalyst is used and the styrene is maintained in suspension by continuous agitation and the use of a stabiliser such as a finely divided powder or a protective colloid, the reaction kinetics are identical to those of the bulk homogeneous process. Polystyrene particles or beads of 0·1–2·0 mm diameter are obtained. When, however, reaction is initiated by using a water soluble peroxide. and a stable emulsion is formed by using a soap or other surfactant as stabiliser, the kinetics are quite different. Polymerisation now takes place in the soap micelles in which styrene is solubilised and to which it diffuses from the styrene droplets[127] and an emulsion of submicronic polymer particles is obtained. High rates of polymerisation and high molecular weight polymers are obtained simultaneously in contrast to the situation with bulk or suspension polymerisation.

7.4.2.2. Ionic polymerisation Ionic polymerisation proceeds through a species which acquires either a partial negative or positive charge, the active sites in the two cases being a carbanion and a carbonium ion respectively.

Generation of the active centre in anionic polymerisation is followed by rapid addition of the monomer (propagation) but there is no spontaneous termination reaction. Using purified reagents the resulting polystyryl anions remain active even when all monomer has been consumed. The term 'living' polymer was suggested by Szwarc[128] for such polymeric anions.

A much higher concentration of active centres can be maintained than in free radical initiation and this permits very high polymerisation rates. As initiation is very fast compared to propagation virtually all the polymer chains are growing simultaneously and the molecular weight distribution is thus very narrow. Since each polymer chain is generated by one molecule of initiator, the molecular weight of the polymer can be calculated from the initiator concentration. Organometallic compounds such as butyl lithium[129] and alkali metals can act as anionic initiators. The exact mechanism of initiation with butyl lithium is dependent on the polarity of the solvent and other factors. For example, in the presence of an ether it occurs by ionisation of the butyl lithium followed by solvation of the ion pair and direct attack on the styrene double bond by the anion. Another anionic initiator, the sodium naphthalene complex however, initiates by an electron transfer process viz:

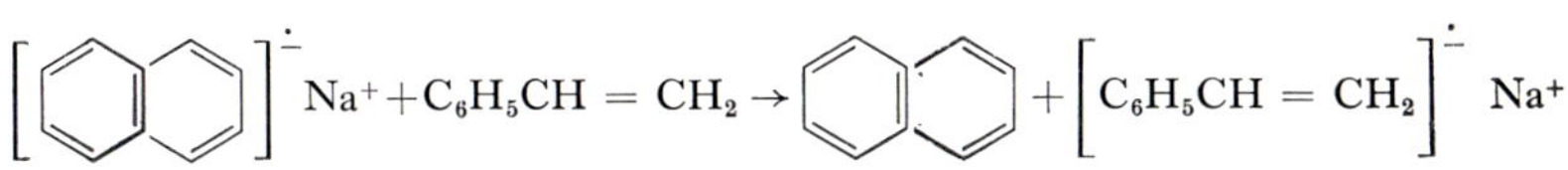

$$2\left[C_6H_5CH = CH_2\right]^{\dot{-}} Na^+ \rightarrow \overset{+}{Na}\overset{-}{C}H(C_6H_5)CH_2CH_2\overset{-}{C}H(C_6H_5)\overset{+}{Na} \rightarrow \text{Polymerisation}$$

Because the polymeric anion remains active after all monomer has been consumed it is possible to use it to initiate polymerisation of either more styrene or a second monomer. It is this ability to produce true block copolymers of styrene which has led to the present commercial exploitation of the anionic polymerisation process.

Although cationic polymerisation has similarities with anionic polymerisation there is normally a termination step. Cationic catalysts may be proton-donating acids such as sulphuric, hydrochloric or perchloric, or compounds generating carbonium ions by Friedel Crafts reaction, for example $AlCl_3$, $TiCl_4$ or $SnCl_4$. Trace quantities of water are required as co-catalysts. Polymerisation rates are again very high but the products are mainly of low molecular weight and little commercial importance.

One further type of ionic polymerisation is that in which initiation is by a Ziegler-Natta type of catalyst and which produces stereoregular or isotactic polystyrene.[130,131] These catalysts are usually inhomogeneous, being combinations of a transition metal halide, for example $TiCl_3$, and an aluminium alkyl. Although isotactic polystyrene has interesting properties, notably a much higher softening point than the normal (atactic) polymer, in practice it is difficult to realise these properties. They are only obtained when the polymer has crystallised and this can only be achieved by careful annealing. Amorphous isotactic polystyrene is little different from the normal atactic form. Isotactic polystyrene remains an interesting laboratory product without commercial importance.

7.4.3. Industrial Polymerisation Processes

Three fundamental factors control the design of plants and processes used for the large scale polymerisation of styrene. These are:

(a) The kinetic limitations of the polymerisation process, whether it is initiated thermally or by catalysis.

(b) The exothermic nature of the reaction.

(c) The need for cleanliness and absence of contamination to produce maximum clarity of product.

Of these three the problem of temperature control resulting from the heat of polymerisation and the increasing viscosity of the polymerising mass presents most difficulties on a plant scale. If styrene is allowed to polymerise adiabatically the temperature will rise to about 350°C, high pressures will be produced and the molecular weight will be unacceptably low. The problem has been overcome by skilful plant design and by modification to the process. Methods by which satisfactory heat transfer through a mass of viscous polymer have been achieved will be discussed in the section on bulk or mass polymerisation. To avoid the problem, suspension, emulsion and solution processes have been developed and will be discussed separately.

7.4.3.1. Bulk Polymerisation Satisfactory temperature control can only be achieved in bulk polymerisation by providing sufficient cooling surface area to remove the heat produced. The problem of heat removal is compounded by the rapidly increasing viscosity and poor thermal conductivity of the polymerising mass. The maximum distance between heat exchange surfaces must be carefully controlled. Thus the temperature at the centre of a tube containing styrene monomer and having a wall temperature of 100°, will rise excessively if the tube diameter exceeds 2 in. The mass can be agitated to improve heat transfer until the viscosity becomes too great. This is usually when about 40 per cent of the styrene has polymerised.

In the first commercial process operated by I.G. in Germany[132] the styrene monomer was polymerised initially in a stirred jacketted reactor to 30–40 per cent conversion. It was then pumped to the top of a tower which was maintained liquid-full by feeding alternately from two such batch reactors or 'prepolymerisers'. The temperatures in the tower were controlled by means of jackets and internal coils in 5 zones ranging from 100–110°C at the top to 180–200°C at the bottom. The polymer melt was continuously removed from the base of the tower by an extruder and the extrudate cooled and granulated. Residence time in the tower was of the order of 28 hours. In a later modification of the process control at the top of the tower was improved by allowing some styrene monomer to boil off and returning the condensed monomer to the process.[133]

Batchwise mass polymerisation processes have been developed in which individual containers are filled either with styrene monomer or 30–40 per cent prepolymer and then heated over a fixed temperature cycle. One such process dates back in its origins to 1938[134] and has been widely used. In this process the 30–40 per cent prepolymer is pumped into a modified plate and frame filter press. The heat control medium, steam or hot oil, passes through the plates, and the frame depth used is small enough to give good temperature control. When the press has been filled the temperature is gradually raised until the polymerisation is completed. After cooling down the press is opened and the polymer obtained as large blocks. These have to be crushed before the polystyrene can be used and this is one disadvantage of the process. Control is however good and if desired catalysts can be used to obtain complete polymerisation at lower temperatures than are possible in the tower type process.

The original tower process gave a polymer which still contained a small amount of styrene. Vacuum stripping was introduced as a means of removing this residual monomer. In one process 40 per cent prepolymer is made continuously in a stirred vessel with a minimum of top to bottom mixing and pumped into a slow moving heated screw conveyor.[135] A temperature gradient along the conveyor raises the conversion to near completion. The hot polymer is then fed into the top of a vacuum chamber through a die producing fine strands, volatiles being removed as the strands fall down the chamber. The molten polymer is extruded from the base, cooled and granulated.

A separate stirred prepolymeriser is avoided in a Dow process using three towers in series, each of which is equipped with a slow moving agitator and internal grids of heating or cooling pipes.[136] Each tower has three temperature zones and the overall temperature rise is from 95° to 225°. To reduce the viscosity and assist heat transfer, up to 25 per cent of ethyl benzene is added to the styrene monomer feed. From the final tower the polymer melt is pumped to a falling strand vacuum devolatiliser where residual styrene and the ethyl benzene are removed.

In another process developed by Dow no agitation is required.[137] The styrene monomer is polymerised to about 95 per cent conversion in a horizontal tank fitted with closely spaced pipes filled with Dowtherm heat exchange fluid, using a heating cycle lasting about 30 hours. The tank contents are then heated to 240°C and pumped into a holding tank where a vacuum is applied to remove volatiles and the molten polymer is fed to an extruder. By using two primary tanks alternately continuous production of finished polymer is obtained from the holding tank.

Unavoidably in all these processes involving high temperature finishing, low molecular weight polymer is produced in the later stages, broadening the molecular weight distribution. To obtain more uniform high molecular weight polymer a process was developed by I G Farbenindustrie in which the 40 per cent prepolymer was fed to a two roll mixing mill in a vacuum chamber, instead of to a tower. The unreacted monomer was recovered and recycled.[138] In a similar process developed by Union Carbide the agitated prepolymerisation is taken much nearer to completion, vacuum being applied to the reactor to boil the styrene at reduced temperature.[139] Devolatilisation is done by a two roll mill in a vacuum chamber. The rolls are helically threaded to form essentially a twin screw extruder, simplifying recovery of the product in handleable form.

Since all these continuous or semi-continuous bulk processes can be operated in totally enclosed systems, elimination of oxygen, which causes discolouration of the polymer, is comparatively easy. Thus polystyrene of excellent colour and clarity can be produced readily, and probably at least half of the world production is still made using one or other of these processes as described or modified.

7.4.3.2. Suspension Polymerisation If styrene is dispersed in a non-solvent medium as small droplets (0·1–1·0 mm dia.) and polymerised, the heat of polymerisation is easily removed from the continuous phase since this does not increase in viscosity. Water is almost universally used as the suspension medium and this limits the temperatures which can be used for polymerisation. Hence, oil soluble initiators, usually organic peroxides, are employed to obtain realistic rates of polymerisation.

The initial droplet size obtained depends on the intensity of agitation, since there is an equilibrium between the two processes of droplet breakup and coalescence. As polymerisation proceeds and the droplet viscosity rises breakup becomes more difficult and the drop size increases. Ultimately there is a point at which shear forces are inadequate to break up the coalescing drops and complete failure of the suspension occurs. To prevent this happening it is necessary to use a suspension stabiliser and two main classes of compounds are employed for this purpose:

(a) Colloids and water soluble macromolecules such as starch, gelatine, polyvinylalcohol, hydroxyethyl cellulose, polyvinylpyrollidone and polyacrylate salts.
(b) Water insoluble inorganic compounds such as zinc oxide, aluminium hydroxide, barium sulphate, calcium and magnesium carbonates or phosphates or more complex natural materials such as talc or bentonites. To perform satisfactorily these must all be very finely divided.

The water soluble compounds act by increasing the viscosity of the aqueous phase and making coalescence between colliding droplets more difficult. The insoluble powders are wetted by the aqueous phase and coalescence is prevented by particles lying between colliding droplets.

The suspension process was first developed in the 1930s for the polymerisation of acrylate esters, but the first application to styrene polymerisation was not until 1946,[140] tricalcium phosphate being the stabiliser used. Since this could be completely removed subsequently by treatment with acid, production of clear, colourless polystyrene by a suspension route became feasible. The effectiveness of calcium phosphates was soon found to depend on the particle size, submicronic particles being preferred,[141] and the ratio of CaO to P_2O_5 in the phosphate, a non stoichiometric compound of ratio greater than 1·35 to 1 (w/w) being the most effective.[142]

The efficiency of inorganic powders such as calcium phosphate can also be improved by adding very small amounts of an 'extender', for example an anionic surfactant.[143,144,145]

Suspension polymerisation is normally carried out as a batchwise operation using a jacketed reactor fitted with an agitator and baffle. Deionised water and styrene are charged to the reactor at a ratio between 3:1 and 4:6 (v:v). The initiator, most commonly benzoyl peroxide, is added, the styrene dispersed by agitation and the suspension stabiliser then added. The quantity of stabiliser required is of the order of 0·2–1·0%wt on styrene. The reactor is then heated through a time/temperature cycle starting at 80–90°C and finishing at 130–140°C. In order to speed up polymerisation in the later stages when the main initiator will have been virtually all consumed, it is common practice to add with the benzoyl peroxide a minor amount of a second initiator having a higher decomposition temperature, for example tert. butyl perbenzoate or di-tert. butyl peroxide.[146] This procedure is effective in reducing the residual styrene monomer to a very low level ($<0·5$ per cent). The total heating cycle is 8–12 hours and when completed the reactor is cooled down and the slurry of polystyrene beads discharged to a buffer vessel where the phosphate stabiliser may be destroyed with acid. The slurry is then pumped to a dewatering centrifuge where the beads are washed free of stabiliser residues, and discharged through a flash or similar continuous drier. The dry beads are either extruded directly to give moulding granules, or if a coloured product is required, dry blended with dyes or pigments before extrusion.

Methods of operating suspension processes continuously have been proposed but the problems of maintaining a stable suspension through a cascade of reactors or in a tower without any back mixing have so far prevented industrial exploitation.

The main advantages of the suspension process lie in the ease with which polymers of different molecular weight may be produced simply by adjusting the catalyst concentration and heating cycle and in the ability to obtain polymer of very low residual monomer content without a devolatilisation stage. It is, however, more difficult to obtain a perfectly clear waterwhite polymer than is the case with bulk polymerisation.

7.4.3.3. Emulsion Polymerisation As with the suspension process the heat in an emulsion polymerisation is removed rapidly via the aqueous phase. A water soluble initiator is used, however, in this case and in contrast with suspension polymerisation where the polymer particle size is controlled by agitation and stabiliser concentration, the particle in emulsion polymerisation grows from a soap micelle[127] and agitation has no effect on size. Stabilisation of the emulsion is by means of a soap or synthetic surfactant, used often in combination with buffers and protective colloids. Normally very high molecular weight polystyrene is produced at high rates of polymerisation. Changing the temperature has little effect on molecular weight, but this can be reduced by the use of a chain transfer agent such as a mercaptan.

It is difficult to remove all traces of the emulsifiers used when recovering the polymer after coagulation of the emulsion. Thus the colour of polystyrene obtained by this route is inferior to that of bulk or suspension polymer, and it is not a significant source of homopolystyrene for moulding purposes. The emulsion polymerisation route is, however, used extensively in making styrene copolymers such as ABS, and polystyrene emulsions are used as such for certain applications.

7.4.3.4 Solution Polymerisation Bulk polymerisation in which the mass viscosity is reduced by addition of a solvent, for example the Dow process described above in which ethyl benzene is used, might perhaps be classified as a solution polymerisation. The term is here, however, used to refer to processes in which the solvent is the major component. The only such processes involving the polymerisation of styrene which are of industrial importance are those using anionic initiation.

The rates of polymerisation obtained with anionic initiation are such that complete polymerisation can be achieved in minutes rather than the hours associated with the processes already discussed. This is a consequence of the virtually simultaneous initiation of all the polymer chains. Such an explosive reaction can only be controlled by carrying it out in dilute solution. Anionic initiators are also very sensitive to impurities such as air and water and these and other contaminants must be eliminated. Industrial application of the process thus presents considerable problems. However the manufacture by an anionic process of block copolymers of styrene and butadiene or isoprene, the so called 'thermoplastic rubbers', is well established, indicating that these problems can be solved.[147]

Patents have been published, describing anionically initiated polymerisation processes for the manufacture of polystyrene[148] but it is not known that any such processes are being used commercially.

7.4.4. Styrene Polymers

7.4.4.1. Unmodified polystyrene (General purpose polystyrenes) The styrene homopolymers marketed today are high molecular weight (Mw = 200–300,000) crystal clear thermoplastics, hard rigid and free from odour and taste. Moulded and extruded products of low unit cost can be obtained due to their combination of ease of fabrication, thermal stability, low specific gravity and low cost. Virtually any colour can be obtained by blending with dyes and pigments and scrap is readily reprocessed.

The first commercial products were essentially pure polystyrene containing a small amount of residual styrene. The presence of styrene monomer was, however, found to reduce the heat distortion temperature of the polymer and cause crazing in moulded articles. Hence, devolatilisation stages were introduced into the manufacturing processes to remove unpolymerised monomer. By this means the heat distortion temperature was raised, but the melt flow and hence mouldability were reduced. Two possible ways of improving melt flow are either to reduce the molecular weight or to incorporate lubricants or plasticisers in the polymer. Molecular weight can only be reduced to a certain extent before the polymer becomes too brittle, whilst lubricants improve flow at the expense of heat distortion temperature. Both methods are however used. The most common lubricants are highly refined hydrocarbon oils (white oils) or esters such as butyl stearate used in concentrations up to 5–6 %wt. These do not adversely affect the colour or clarity of the product.

As a result of these developments there is now available a group of styrene homopolymers, often referred to as general purpose grades. These range from the

TABLE VII.7

Properties of Unmodified Polystyrenes

Property	*Method of test*	*Units*	*Medium flow*	*High flow*	*Heat resistant*	*High mol wt*	*Heat resistant high mol wt*
Physical properties							
Specific gravity	ASTM D792–64T	—	1·05–1·07	1·05–1·06	1·05–1·07	1·05–1·07	1·05–1·07
Viscosity	BS 2782–4048	Centipoise	1·70–2·00	1·70–2·00	1·90–2·10	2·70–3·10	2·70–3·10
Monomer content	BS 1493:67 AppE	%	0·2–0·5	0·2–0·5	<0·1	0·2–0·5	<0·1
Refractive index	ASTM D542–50	$n^{25}D$	1·59	1·59	1·59	1·59	1·59
Light transmission	ASTM D1003–61	%	88–90	88–90	88–90	88–90	88–90
Thermal properties							
Specific heat	—	kJ/kg/K	0·134	0·134	0·134	0·134	0·134
Coefficient of linear expansion	ASTM D696–44	mm/mm/K	$6–8 \times 10^{-5}$	$6–8 \times 10^{-5}$	$6–8 \times 10^{-5}$	$6–8 \times 10^{-5}$	$6–8 \times 10^{-5}$
Heat distortion temp.	ASTM D648–56						
at 66 p.s.i.		°C	87–91	80–83	96–98	85–87	96–98
at 264 p.s.i.		°C	82–85	75–78	93–95	80–82	93–95
Softening point	BS2782:102C	°C	89–92	82–85	100–102	87–89	100–102
Mechanical properties							
Tensile strength	BS2782:301J	MN/m^2	35–41	35–41	45–52	45–52	52–55
Elongation	BS2782:301J	%	1·0–2·5	1·0–2·5	1·0–2·5	1·0–2·5	1·0–2·5
Modulus intension	ASTM D638–64T	MN/m^2	$3·1–3·8 \times 10^3$	$3·1–3·8 \times 10^3$	$3·4–4·1 \times 10^3$	$3·1–3·8 \times 10^3$	$3·4–4·1 \times 10^3$
Flexural strength	ASTM D790–63	MN/m^2	62–76	62–76	76–96	69–83	76–96
Deflection	ASTM D790–63	mm	3·2–7·6	3·2–7·6	3·2–7·6	3.2–7·6	3·2–7·6
Compression strength	ASTM D695–63T	MN/m^2	96–103	96–103	96–103	96–103	96–103
Izod impact strength (notched)	BS2782:308A	Joules/6·4 mm notch	0·08–0·12	0·08–0·12	0·08–0·12	0·08–0·12	0·08–0·12
Rockwell hardness	ASTM D785–62	M Scale	70–80	70–80	70–80	70–80	70–80
Electrical properties							
Power factor	BS2782:207	—	$1–5 \times 10^{-4}$	$1–5 \times 10^{-4}$	$1–5 \times 10^{-4}$	$1–5 \times 10^{-4}$	$1–5 \times 10^{-4}$
Dielectric constant	BS2782:207 —	—	2·4–2·6	2·4–2·6	2·4–2·6	2·4–2·6	2·4–2·6
Volume resistivity	BS2782:202	ohm cm	$> 10^{13}$	$> 10^{13}$	$> 10^{13}$	$> 10^{13}$	$> 10^{13}$

heat resistant high molecular weight polymers of low flow and high softening point, through polymers of increasing flow to the high flow or easy processing grades of low softening point. Typical properties of representative materials in this range are shown in Table VII.7. Polymers for special applications are made by compounding one or other of these materials with additives. For example U.V. absorbers and antioxidants are incorporated to give a light stable polymer, and antistatic additives to give polymers less susceptible to dust pick-up. The accumulation of static electrical charges which causes dust pick up is, in most applications, an undesired side-effect of the excellent electrical properties of polystyrene.

These homopolymers are manufactured by both bulk and suspension processes, the latter being more flexible when changes in molecular weight are required.

7.4.4.2. Rubber modified polystyrene (Impact polystyrenes) A basic deficiency of unmodified polystyrene is its poor resistance to impact. One method of improving the impact strength is to incorporate a minor amount of rubber in the polymer and this can be done in one of three ways.

(a) Mechanical blending of solid polystyrene and rubber.
(b) Blending of rubber or polystyrene emulsions.[149]
(c) Polymerisation of a solution of rubber in styrene monomer.[150,151]

The first method is inefficient in its usage of rubber; moreover to achieve a reasonable improvement in impact strength the amount of rubber used is such that other properties, for example tensile strength, mouldability and surface finish, are adversely affected. A mechanical blending process is thus no longer employed to make toughened or high impact polystyrenes. Emulsion blending is also inefficient in rubber usage but a better dispersion of the rubber in the polystyrene and hence better moulding properties than with mechanical blends are obtained. Although not used for impact polystyrene production, modified latex blending processes are used extensively for the manufacture of ABS type copolymers.

The third method gives the best products provided one essential step is not omitted. The solution of rubber in styrene (styrene-butadiene or cis-polybutadiene rubber) must be agitated during the initial part of the polymerisation, that is to say up to at least 25 per cent conversion. Subsequently polymerisation can be completed without agitation either by one of the bulk processes already described[151] or by dispersion of the prepolymer in water and suspension polymerisation.[152,153] The reasons why agitation is essential become apparent when a detailed examination is made of the changes which occur during this initial period.[154] As polystyrene starts to form in the solution there is a phase separation due to the mutual incompatability of the two polymers, rubber and polystyrene. A dispersion of fine droplets of polystyrene dissolved in styrene begins to form in the rubber/styrene solution. As polymerisation proceeds the volume of this disperse phase increases until it exceeds that of the continuous phase. Inversion then occurs and the rubber/styrene solution becomes the disperse phase. Some of the polymerising styrene has by this stage become grafted onto the butadiene molecules and this styrene grafted rubber now acts as a stabiliser for the rubber/styrene droplets. If the system has not been agitated complete inversion is not achieved and the final polymer will still contain a more or less continuous network of grafted and crosslinked rubber which has little or no reinforcing action. When the inversion has occurred satisfactorily appreciable amounts of the polystyrene/styrene are occluded in the new disperse phase droplets of rubber/styrene. It is important to retain this material since it increases the volume of the rubber particles in the final product, and hence their effectiveness. Continuation of agitation for too long after inversion has been completed leads to excessive loss of this occluded material. In the final stages of polymerisation the rubber is cross-linked and becomes insoluble. Excessive heating at this stage will over cross-link the rubber and reduce its effectiveness. A good impact polystyrene will contain a dispersion of rubber particles 1–10μm in diameter which contain appreciable inclusions of polystyrene. Thus starting with 5 per cent of rubber it is possible to obtain a final polymer containing up to 40 per cent by volume of disperse phase.

The manner in which the rubber particles improve impact strength is by absorbing the energy of impact and preventing crack propagation. To do this effectively they must retain rubberiness, i.e., they must not have been over cross-linked, they must not be too small and they must be present in sufficient numbers, i.e., the volume fraction must be high. Careful control of conditions at all stages of polymerisation is essential to satisfy these requirements.

As already noted, after carrying out the agitated prepolymerisation stage the polymerisation can be finished by any one of a number of different processes. By using more rubber products of increasing impact strength are obtained. Melt flow and softening point are adjusted by inclusion of lubricants such as a white oil or butyl stearate. These properties are often tailored to a particular application, but the properties of some typical products are given in Table VII.8.

7.4.4.3. Expanded Polystyrene Expanded or foam polystyrene is a rigid cellular form of the polymer in which the individual cells are nonintercommunicating and are of the order of 50–150μm in diameter. It can be made at densities down to less than 16 kg/m^3.

The first commercial production of an expanded polystyrene, Styrofoam, was by Dow in the USA and took place in 1943. This product was made by dissolving normally gaseous organic compounds such as methyl chloride in polystyrene under pressure and at temperatures between 100° and 200°C. The resultant doughy mass is extruded by gas pressure through a shaping die where the gas vapourises and expands the soft polymer to a closed cell structure.[155,156] Cell size is controlled by the use of additives such as ammonia[157] or metal stearates.[158] The product is obtained as a rough log or billet of any desired length but with height and width limited by the size of the shaping die. Blocks, sheets and sections are cut from the log by sawing or hot wire cutting. A disadvantage of the process is the high transport charges incurred in transferring such low density material between manufacturer and user. To overcome this deficiency processes have been developed to make an expandable polystyrene, i.e., a polymer of normal density which can be expanded by the user. The first such product was made by BASF in Germany, the process involving a batchwise bulk polymerisation of styrene in which a low boiling paraffin such as pentane or hexane had been dissolved.[159] The polymer block was crushed to yield granules which would expand when heated in boiling water or steam. These granules

TABLE VII.8

Properties of Rubber Modified Polystyrenes

Property	Method of test	Units	Impact heat resistant	High impact	High impact medium flow	High impact easy flow
Physical properties						
Specific gravity	ASTM D792–64T	—	1·05–1·07	1·05–1·07	1·05–1·07	1·05–1·07
Monomer content	BS1493:62 AppE	%	0·10	0·10–0·20	0·10–0·20	0·10–0·20
Thermal properties						
Specific heat	—	kJ/kg/K	0·134	0·134	0·134	0·134
Coefficient of linear expansion	ASTM D696–44	mm/mm/K	$6–8 \times 10^{-5}$	$6–8 \times 10^{-5}$	$6–8 \times 10^{-5}$	$6–8 \times 10^{-5}$
Softening point	BS2782:102C	°C	96–100	84–86	79–81	72–75
Mechanical properties						
Tensile strength	BS2782:301	MN/m^2	34·5	21–28	21–28	17–27
Elongation	BS2782:301	%	10–15	25–33	27–33	30–40
Modulus in tension	ASTM D638–64T	MN/m^2	$2·3–3·1 \times 10^3$	$1·9–2·4 \times 10^3$	$1·9–2·4 \times 10^3$	$1·9–2·4 \times 10^3$
Flexural strength	ASTM D790–63	MN/m^2	48–62	35–41	31–38	28–41
Deflection	ASTM D790–63	mm	12–20	33–41	38–48	41–53
Izod impact strenth (notched)	BS2782:306A	Joules/6·4 mm notch	0·40	0·50	0·50	0·50
Rockwell hardness	ASTM D785–62	M scale	40–60	40–60	40–60	40–60
Electrical properties						
Power factor	BS2782:207	—	$5–15 \times 10^{-4}$	$5–51 \times 10^{-4}$	$5–51 \times 10^{-4}$	$5–15 \times 10^{-4}$
Dielectric constant	BS2782:207	—	2·5–2·7	2·5–2·7	2·5–2·7	2·5–2·7
Volume resistivity	BS2782:202	ohm cm	10^{13}	10^{13}	10^{13}	10^{13}

could be stored for long periods, provided they were kept in a sealed container, and expanded and moulded when required.

The most significant advance, however, took place when expandable polystyrene beads were successfully produced by using a suspension polymerisation process. The expanding agent, normally *iso*- or *n*-pentane or a mixture of the two, can be added to the monomer before polymerisation is begun,[160] or during the polymerisation[161] or completely polymerised beads can be impregnated in a separate heating cycle.[162] To obtain good retention of the expanding agent and achieve low density products, polystyrene beads of larger size (0·5–1·5 mm diameter) than are normally produced for general purpose or impact grades are required and stabiliser systems have been specially developed to produce these large beads.[173]

The correct technique must be used in moulding expandable beads if satisfactory results are to be achieved. It is necessary to heat the beads just above the softening point of the polymer and boiling water or more usually low pressure steam is used for this purpose. Moulds with perforated walls through which the steam can be blown directly are employed. The increase in volume occurring when beads are expanded to a density of 16 kg/m^3 is more than sixty times, and this degree of expansion cannot be achieved by direct expansion of the beads in one step. A technique has been developed, however, in which the beads are first allowed to expand freely in a steam chamber or 'pre-expander'. The free flowing 'pre-expanded' beads are then allowed to stand in air for several hours. As the beads cool down after the pre-expansion step a partial vacuum is created in the cells and air slowly diffuses in to restore the pressure in the cells to atmospheric. The mould is completely filled with these 'matured' pre-expanded beads, closed and steam heated and then cooled according to a preset cycle. During the moulding step the air expands as well as residual expanding agent and provides additional pressure to fuse together the individual beads into a coherent block. It is possible to obtain mouldings of even lower density by carrying out a second pre-expansion and maturing step before finally moulding the beads.

Foamed polystyrene sheets up to 2·3 mm in thickness which are biaxially oriented are made by extruding expandable beads using the bubble blowing extrusion process by which polyethylene film is normally produced. In order to produce fine cells in the sheet, which give it a good texture and strength, it is necessary to add nucleating agents, e.g., a mixture of citric acid and sodium bicarbonate, to the beads before extrusion.[163] An alternative method of making this thin expanded sheet is a reversion to the direct foaming process. An unmodified heat resistant grade of polystyrene is extruded and the hydrocarbon expanding agent pumped into the extruder barrel.[164]

One serious drawback to more extensive use of expanded polystyrene in certain applications is its flammability, and grades which are self extinguishing are therefore manufactured. These are made by incorporating a fire retardant chemical, particularly an organic bromine compound in the beads, at some stage during manufacture. Tris 2, 3-dibromopropyl phosphate[165] and hexabromocyclododecane[166] are compounds which have been used. The amount of bromo-compound required can be significantly reduced, say from 5 to 0·5 per cent, if a free radical generator such as an organic peroxide is also present.[167,168]

For a number of reasons expanded polystyrene has achieved a unique position amongst rigid thermoplastic foams, the low cost of the polymer and expanding agent, satisfactory shelf life and ease of moulding with inexpensive equipment all being important factors.

The properties of some typical expanded polystyrenes are given in Table VII.9.

TABLE VII.9

Properties of Expanded Polystyrene

Property	*Test*	*Units*	*Extruded block*	*Block moulded from beads*	*Block moulded from beads*
Density	—	kg/m^3	30·0	16·0	24·0
Thermal properties					
Thermal conductivity	BS3837	W/m/K	1·8	1·7	1·7
Max. service temp.		°C	80	80	80
Coefficient of linear expansion		mm/mm/K	$7·0 \times 10^{-5}$	$7·2 \times 10^{-5}$	$7·2 \times 10^{-5}$
Mechanical properties					
Tensile strength	BS3837	MN/m^2	0·48	0·22	0·24
Flexural strength	BS3837	MN/m^2	0·48	0·22	0·31
Compression strength	BS3837	MN/m^2	0·24	0·08	0·21
Electrical properties					
Power factor (1 MHz)	BS2067	—	—	$1·0 \times 10^{-3}$	$1·0 \times 10^{-3}$
Dielectric constant (1 MHz)	BS2067	—	1·05	1·02	1·04
Absorption					
Water vapour permeability (38°C 90% RH)	BS3837	$kg/m^2/s/50$ mm	—	6×10^{-7}	6×10^{-7}

7.4.5. Styrene Copolymers

Only those copolymers of commercial importance which contain a major proportion of styrene will be considered. The rubbery diene copolymers and the acrylonitrile-butadiene-styrene (ABS) family of polymers are thus excluded from discussion, as well as the very extensive use of styrene in unsaturated polyester resin laminates.

7.4.5.1. Styrene-acrylonitrile copolymers (SAN) Styrene and acrylonitrile copolymerise readily in bulk, emulsion or solution. The reactivity ratios of the two monomers are such, however, that a homogeneous copolymer is only obtained by using a monomer mixture containing about 39 per cent molar of acrylonitrile (approx. 25%wt). With monomer mixtures of less than 15 per cent or greater than 40%wt acrylonitrile, homopolystyrene or homopolyacrylonitrile respectively are formed as polymerisation nears completion. The resultant polymers are inhomogeneous and not transparent.

Copolymers produced commercially contain usually 25–35%wt acrylonitrile and during polymerisation precautions are taken to minimise any changes in the composition of the copolymer. The copolymerisation reaction is highly exothermic, being faster than styrene homopolymerisation and also producing much higher molecular weight polymer. To obtain a copolymer of low enough molecular weight to be readily mouldable, the copolymerisation must either be carried out at high temperature, which increases the temperature control problem, or chain transfer agents must be employed. Emulsion polymerisation has been used[169] but it is difficult to remove completely the emulsifiers and obtain a clear product, and the bulk method is preferred. In one process the monomer mixture is recirculated by means of a pump through a coil maintained at constant temperature (130–180°C), fresh monomer mixture being added and partially polymerised copolymer solution removed continuously.[170] Modifiers such as benzene or toluene are added to help in temperature control and these are removed, together with the unreacted monomers, in a devolatilisation step and recycled to the monomer feed stream.

Although copolymerisation can be carried out in suspension the process is complicated by the fact that acrylonitrile is appreciably soluble in water. As the concentration of acrylonitrile monomer remaining in the dispersed droplets decreases by polymerisation, more acrylonitrile diffuses into them from that dissolved in the water, changing the composition in the beads. Production of a homogeneous copolymer is thus impossible.

Styrene-acrylonitrile copolymer has marked advantages over polystyrene in certain areas. In addition to being slightly tougher it possesses much greater resistance to solvents such as non-aromatic and chlorinated hydrocarbons. It has much higher resistance to stress particularly in environments such as milk or butter, which are stress-cracking agents for polystyrene. As more acrylonitrile is incorporated in the copolymer these properties improve, but the unpigmented material becomes increasingly yellow. The yellow tint is usually masked by incorporation of a blue dye. Typical properties of the copolymer are given in Table VII.10.

TABLE VII.10

Properties of Styrene Acrylonitrile Copolymer

Property	*Method of test*	*Units*	*Typical values*
Physical properties			
Specific gravity	ASTM D792–64T	—	1·06–1·08
Refractive index	ASTM D542–50	N_D	1·577–1·565
Thermal properties			
Specific heat	—	kJ/kg/K	0·08
Coeff. of linear expansion	ASTM D696–44	mm/mm/K	$6{\cdot}7–6{\cdot}9 \times 10^{-5}$
Heat distortion temp. at 264 p.s.i.	ASTM D648–56	°C	99–104
Mechanical properties			
Tensile strength	ASTM D638–64T	MN/m^2	57–76
Elongation	ASTM D638–64T	%	2·1–3·7
Modulus in tension	ASTM D638–64T	MN/m^2	$3{\cdot}6–3{\cdot}7 \times 10^{-3}$
Impact strength	BS2782.308A	J/6·4 mm notch	0·10–0·15
Rockwell hardness	ASTM D785–62	M scale	80–82
Electrical properties			
Dielectric constant	ASTM D150–59T	—	2·75–3·0
Dissipation factor	ASTM D150–59T	—	0·011–0·012

7.4.5.2. Styrene-divinylbenzene copolymer Copolymerisation of styrene with divinylbenzene yields a cross-linked copolymer which is insoluble and infusible. This type of product has found considerable use as the base material for the production of ion-exchange resins by sulphonation. The degree of cross-linking and, therefore, the amount by which the sulphonated product swells in water, depends on the amount of divinylbenzene used and normally this is about 8 per cent. The copolymer is usually required in bead form and hence is made by a suspension polymerisation process.[171] Suspension stabilisers used are of the types already described but considerable expertise is involved in obtaining a stable suspension. The polymerisation reaction is faster than that of styrene alone and, once a three dimensional network has been established by crosslinking, the bead size is fixed. Beads of between 0·2 and 1·0 mm diameter are normally required for this application.

7.4.6. Uses of Styrene Polymers and Copolymers

In Tables VII.11 and VII.12 the consumption of styrene polymers for the last five years in the U K and the U S A respectively, is broken down into various fields of application. Separate figures for styrene-acrylonitrile copolymer are not available for the U K but the total consumption is low.

Packaging, housewares, toys and consumer durables appear as the major outlets in both lists. Differences are apparent in some of the smaller applications, which reflect different attitudes in the two countries. For example there is appreciable usage of impact polystyrene for drain pipes in the U S A but none in the U K, and expanded polystyrene vending cups have a much larger share of the market in the U S A.

TABLE VII.11

UK Usage of Styrene Polymers
(1000 tons)

	1968	*1969*	*1970*	*1971*	*1972*
Polystyrene and Impact polystyrene					
Packaging	} 38·4	} 56·9	46·3	43·7	49·8
Vending cups			24·0	15·2	18·3
Sheet (other than packaging)	7·5	5·7	7·1	5·6	6·1
Consumer durables	6·2	9·1	9·1	9·1	11·2
Household	11·9	13·4	13·4	18·6	20·3
Toys and novelties	11·5	12·4	13·9	12·2	13·2
Shoe heels	3·6	3·6	6·1	5·1	6·6
Light fittings	2·6	3·1	3·6	3·6	3·9
Bathroom accessories	3·6	4·1	2·8	4·6	5·6
Flower pots				5·1	5·8
Building			4·1	3·0	3·6
Furniture			1·5	1·5	2·0
Combs, brushes and trinkets	1·3	1·5	1·3		
Miscellaneous	1·8	12·2	3·5	2·7	3·0
Total	88·4	122·0	136·7	130·0	149·4
Expanded polystyrene					
Refrigeration and cold stores	1·6	} 5·7	} 5·5	} 5·7	} 6·3
Building	4·2				
Retail (DIY and tiles)	4·0	4·6	5·3	5·1	5·1
Packing (excl. sheet)	2·4	3·7	3·9	4·4	4·9
Miscellaneous (incl. boats, chairs and cups)	0·5	0·8	1·0	1·0	1·0
Total	12·7	14·8	15·7	16·2	17·3

TABLE VII.12

USA Usage of Styrene Polymers
(1000 tons)

	1968	*1969*	*1970*	*1971*	*1972*
Polystyrene and impact polystyrene					
Consumer durables	45·3	62·1	60·8	108·6	120·5
Disposable servicewear				5·4	7·2
Drainpipe	18·2	22·7	31·8	38·6	43·2
Furniture	23·6	34·0	38·6	47·3	56·5
Electrical/Electronics	25·9	36·3	38·6	39·1	47·4
Housewares	81·6	97·5	113·5	125·0	159·8
Lighting fittings	18·2	19·0	22·7	17·3	19·0
Monofilament	2·3	2·3	2·3	2·3	2·3
Packaging	182·0	230·0	245·0	272·6	336·2
Personal care products	15·0	18·2	20·0	21·8	23·2
Plumbing and bath fixtures				2·7	3·0
Siding and panels				4·1	4·3
Signs				2·6	2·7
Toys	95·5	109·0	125·0	131·8	142·2
Wallcoverings				2·3	2·3
Export				12·7	17·0
Miscellaneous	135·0	162·0	149·0	169·7	209·0
Totals	642·6	793·0	847·2	1003·9	1196·0
Styrene acrylonitrile copolymer					
Total	16·8	18·2	20·0	33·0	43·0
Expanded polystyrene					
Expanded board	20·4	21·8	22·3	21·2	21·5
From expandable beads					
Board and billet	19·5	22·3	24·5	27·3	35·4
Cups	19·5	21·4	23·2	27·5	31·7
Other mouldings	15·0	16·4	17·3	21·0	25·2
Packaging	18·2	20·4	22·7	29·0	38·1
From crystal extruded as sheet					
Disposables				5·4	7·0
Packaging	} 25·0	} 38·6	} 56·7	63·6	88·3
Other				1·4	1·8
Total	117·6	140·9	166·7	196·5	249·0

7.4.6.1. Methods of Fabrication Unmodified and impact polystyrenes can readily be processed by all the methods commonly used in the plastics industry. Articles can be produced by injection or blow moulding, sheet extrusion and vacuum forming or profile extrusion. Expandable polystyrene beads require specialised techniques of moulding, as already described, but they can also be extruded to thin sheet and then vacuum formed. The ease of fabrication and generally desirable properties of styrene polymers have led to a very great diversity of applications. Major areas of use are in packaging, the production of consumer durables and housewares and in toy manufacture. Examples of the more important applications are discussed below.

7.4.6.2. Packaging applications Packaging is undoubtedly the largest area of application of styrene polymers. Containers of all shapes and sizes for the packaging of foodstuffs, pharmaceutical and cosmetic products and domestic chemicals such as bleaches or scouring powder, are injection or blow moulded from both general purpose and impact polystyrenes. Specially formulated high flow anti-static grades of impact polystyrene are used extensively in the high speed injection or blow moulding of cream and yoghurt containers. An alternative method of making this type of container is by extrusion of a flat sheet and vacuum forming on a continuous basis. After removal of the formed containers the remaining scrap sheet is granulated and recycled. Cups for vending machines, food and produce trays and egg containers are also produced by vacuum forming from sheet. Biaxially oriented extruded crystal polystyrene film is used for packing fresh produce. Injection moulded screw caps for bottles and jars, either plastic or glass, are extensively used. A particular application for impact polystyrene is in the production of cotton reels and bobbins.

The use of expanded polystyrene in packaging is increasing very rapidly. This is particularly the case with moulded packs which are designed to fit exactly the articles being packed. Because of the low pressures involved, only low cost easily produced moulds are required. This method of protection for goods in transit is used not only for very fragile objects such as glass or chinaware but also for scientific instruments, household appliances, items of industrial equipment and numerous other products. Expanded beads or granules are also used as loose-fill packing in certain applications. Thin sheet is extruded either from expandable beads or from crystal polystyrene with direct injection of pentane into the extruder, and vacuum formed to make such things as fruit packing spacers, produce trays and egg containers. A special small size expandable bead is supplied for the high speed moulding of vending cups, which have the advantage over toughened polystyrene cups of providing insulation for hot drinks.

Where resistance to stress corrosion is important in food packaging, for example with butter or vegetable oils, styrene-acrylonitrile copolymer may be used instead of polystyrene.

7.4.6.3. Applications in consumer durables Both unmodified and impact polysyrenes are used widely and to varying degrees in such articles as refrigerators, radio and television sets, vacuum cleaners, washing machines, cameras, record players, tape cassettes, etc.

In refrigerators the use of impact polystyrene liners vacuum formed from thick sheet has declined in the face of competition from ABS, but other interior fittings are often polystyrene. A recent development has been the introduction of 'structural foam' radio, record player and loudspeaker cabinets. Structural foam is made by injection moulding impact polystyrene in which is incorporated a chemical expanding agent. The polymer is injection moulded and expands in the mould to a density of 700–800 kg/m^3. The surface is then either laminated with a wood grain finish foil or lacquered to simulate wood, the result being so close to timber in appearance as to defy any but the closest examination.

Vacuum cleaner and food mixer/blender housings are injection moulded in impact dolystyrene as are less expensive camera bodies. Crystal polystyrene is used for tape cassettes and tape reels. For applications where extra toughness and chemical resistance as well as transparency is required, such as washing machine doors and food blender goblets, styrene acrylonitrile copolymer is preferred.

7.4.6.4. Applications in housewares and furniture Kitchen and bathroom equipment and accessories are a large outlet for both the unmodified and impact grades. Containers for storing cakes, bread and other foods, kitchen scales and bathroom cabinets are a few examples. In the home generally such things as shirt and dress hangers, brush handles, combs and many small items are made in polystyrene. Polystyrene picnic tableware is lightweight and inexpensive. Where better resistance to impact is required, as for example in drinking tumblers, styrene acrylonitrile copolymer is used.

Two important applications in the manufacture of furniture use expanded polystyrene and structural foam respectively. Chair shells, particularly easy chairs, are moulded in expandable polystyrene at a density of about 20–25 kg/m^3. Often the shell is moulded round a sub-frame to which the legs can be attached. The shell is then padded and upholstered to give a lightweight inexpensive easy chair. Legs for chairs, tables or cabinets are moulded in structural foam impact polystyrene, often complete with elaborately 'carved' detail, and then lacquered to simulate wood. Drawer fronts and cupboard doors can similarly be produced to make reproduction antique furniture.

7.4.6.5. Building and construction applications The mechanical properties of unmodified and impact polystyrenes exclude them from widespread use as building components. Impact polystyrene sheet is, however, employed as shuttering when casting concrete, to produce smooth or decorative surface finishes. Toilet cisterns moulded in impact polystyrene and polystyrene wall tiles are other minor applications.

Expanded polystyrene is used extensively in a number of different ways. Expanded boards of density 16–32 kg/m^3, either cut from large moulded blocks or made by the direct extrusion process, are used as insulation both in ordinary buildings and in cold stores. The self-extinguishing grade is preferred for this application. The expanded board can be put beneath concrete floors and into the wall cavity during construction of normal buildings. The walls, ceiling and floors of cold stores are, however, lined with expanded polystyrene slabs which are then plastered over and tiled. A technique has been developed of continuously heat welding together strips of expanded polystyrene to form self-supporting domes. Laminated panels sandwiching expanded polystyrene between skins of aluminium, plasterboard or plywood are used for the construction of partition walls or as external wall cladding. In a further application, lightweight concrete is obtained by mixing freely expanded beads in the concrete before casting. Expanded polystyrene ceiling tiles are cut from blocks or directly moulded, and are used to give decorative ceiling effects. A thin veneer of expanded polystyrene (1–2 mm thick), cut continuously from a block, is used beneath wallpaper to provide wall insulation.

7.4.6.6. Electrical applications Although polystyrene has excellent electrical properties and many small components are moulded in polystyrene the total quantity used for electrical purposes is not great. A major use in this field is for light fittings. Specially formulated light stabilised and opalescent grades are used for the extrusion or injection moulding of light diffusers, particularly for use in fluorescent tube lighting systems.

7.4.6.7. Miscellaneous applications The majority of toys marketed today contain at least one part moulded in polystyrene. Whilst individually the quantities may not be great the overall consumption in the toy market is very high.

Generally speaking there is not a wide usage of polystyrenes in the transport field. Some minor components and trim parts are, however, made in polystyrene or styrene-acrylonitrile copolymer and the latter is also used for lightweight battery cases.

The heels for women's shoes are moulded in a specially formulated impact polystyrene which resists splitting on nailing and can be solvent lacquered. The thick soles for the fashionable platform soled shoes are moulded in foamed impact polystyrene to reduce weight.

Synthetic paper having very good dimensional stability is made by the surface treatment or coating of extruded impact polystyrene film.

7.4.7. Production and Consumption of Styrene Polymers

The widespread use of styrene polymers throughout the world is reflected in the number of countries possessing manufacturing facilities. Table VII.13 gives an idea of the production capacity available in 1967. Since then polystyrene plants have been started up in additional countries and the capacities in those countries listed considerabily enlarged. Table VII.14 gives an indication of how capacity has increased in the USA since 1969 by over 40 per cent to a figure approaching the world capacity in 1967. This rate of expansion has, moreover, been exceeded both in Britain (Table VII.15) and in the rest of Western Europe (Table VII.16).

The United States remains, however, the largest producer and consumer, with a total consumption of polystyrene (general purpose, impact and expanded) and styrene-acrylonitrile copolymer for 1972 of almost 1·5 million tons (Table VII.12). The West European consumption (including the UK) of styrene polymers was about 1·2 million tons (Table VII.17), but this figure probably included some non-moulding grades of polymer. The individual UK production and consumption figures are quoted in Table VII.18. Japan is the third largest producer after Western Germany and recent figures for Japanese production are given in Table VII.19.

The total world consumption, therefore, of moulding and extrusion grade styrene polymers and copolymers exceeds 3·0 million tons annually and is continuing to increase.

TABLE VII.13

World Production Capacity for Styrene Polymers (1967)
(1000 tons)

N. America	United States	800	836
	Canada	20	
	Mexico	16	
S. America	Argentina	20	30
	Brazil	5	
	Colombia	3	
	Chile	2	
Asia	Japan	250	253
	India	3	
Australasia	Australia	4	4
Europe	West Germany	175	586
	Britain	115	
	France	85	
	Italy	80	
	Poland	75	
	Spain	12	
	Sweden	10	
	USSR	10	
	Greece	6	
	Rumania	6	
	Yugoslavia	6	
	Netherlands	5	
	Norway	1	
			1709

TABLE VII.14

US Production Capacity for General Purpose and Rubber Modified Polystyrenes

(1000 tons)

Manufcturer	*End 1969*	*End 1970*	*End 1971*	*End 1972*
Amoco Chemicals	77	77	77	77
American Electric	—	18	18	18
A & E Plastic	—	9	9	9
BASF/Wynadotte	36	36	36	36
Cosden Oil & Chemical	66	90	90	90
Dart Industries	60	60	60	60
Dow Chemical	320	320	320	410
Foster Grant	86	86	86	160
Hammond Plastic	11	11	11	11
Howard Industries	7	7	7	7
Monsanto	170	170	170	340
Richardson Co.	23	23	23	23
Shell Chemical	36	36	36	36
Sinclair-Koppers	136	136	136	136
Solar Chemical	27	27	27	27
Southern Petrochemicals	18	18	18	45
Union Carbide	77	77	77	77
USS Chemicals	—	—	90	90
	1150	1267	1357	1652

TABLE VII.15

UK Production Capacity for Polystyrene Manufacture

(1000 tons)

Manufacturer	*End 1967*	*End 1968*	*End 1970*	*End 1971*	*End 1972*
BP Plastics	7	18	30	30	30
Distrene (Dow)	26	40	70	70	70
Kaylis	6	6	6	6	6
Monsanto	21	25	25	25	25
Shell Chemicals UK	25	30	40	60	90
Sterling	30	34	45	45	114*
Total	115	153	216	236	335

* Planned

TABLE VII.16

West European Production Capacity for Polystyrene Manufacture (1000 tons)

Country	*1967*	*1972*
West Germany	175	650
France	85	205
Italy	80	185
Netherlands	5	100
Spain	12	55
Sweden	10	12
Greece	6	11
Total	373	1218

TABLE VII.17

West European Consumption of Styrene and Related Polymers[172] (1000 tons)

	1968	*1969*	*1970*	*1971*	*1972*
West Germany	245	275	300	350	425
Great Britain	118	145	168	169	191
France	102	145	171	145	238
Italy	127	178	193	220	259
Belgium	26	33	38	47	54
Netherlands	20	28	31	47	51
Total	638	804	901	1028	1218

TABLE VII.18

UK Production and Consumption of Styrene Polymers (1000 tons)

	1968	*1969*	*1970*	*1971*	*1972*
Polystyrene and impact polystyrene					
Production	101·6	134·2	142·3	142·8	167·7
Exports	30·5	20·3	22·9	22·9	37·6
Imports	22·4	6·1	16·8	10·2	19·3
Consumption	93·5	121·9	134·2	130·1	149·4
Expandable polystyrene					
Production	17·8	23·4	25·4	27·2	28·9
Exports	7·6	11·6	12·6	14·2	16·8
Imports	2·5	3·0	2·7	2·8	4·6
Consumption	12·7	14·7	15·6	16·2	17·3

TABLE VII.19

Japanese Production of Styrene Polymers
(1000 tons)

	1967	*1968*	*1969*	*1970*	*1971*
General purpose and impact polystyrenes	176	235	316	404	409
Expandable/expanded polystyrene	39	54	73	89	90
Styrene-acrylonitrile copolymer	20	25	33	37	42
	235	314	422	530	541

REFERENCES

[1] Boundy and Boyer, '*Styrene, its Polymers, Copolymers and Derivatives*' (Reinhold Publishing Corp., 1952).
[1a] '*Chemical Economics Handbook*' (Stanford Research Institute, Menlo Park, California), Section Benzene 618, 5023M, March 1962.
[2] Kirk-Othmer, '*Encyclopaedia of Chemical Technology*', **15**, 189 (1st edn).
[3] Krausse and Randall (Cosden Oil & Chemical Co.), GP 1,418, 688 (1960).
[4] Bloch (Universal Oil Products), USP 3,183,233 (1965).
[5] Olah, Kuhn and Floor J. *Am. Chem. Soc.*, 1962, **84**, 1688.
[6] Benesi, *J. Phys. Chem.*, 1957, **61**, 970.
[7a] Hansford, *Ind. Eng. Chem.*, 1947, **39**, B49.
[7b] Hansford, *et al.*, *Joint Symposium of the Use of Isotopes in Petroleum Chemistry*, 118th Meeting, Am. Chem. Soc., Chicago, (Sept. 1950).
[7c] Hinden, Mills and Oglad, *J. Am. Chem. Soc.*, 1951, **73**, 278.
[7d] Tamalo, *Faraday Soc. Discussion*, 1950, **8**, 270.
[8] Evans, *Faraday Soc. Discussion*, 1950, **8**, 302.
[9] Kutz, *et al.*, *J. Organic Chem.*, 1951, **16**, 699.
[10a] Grote and Gerard, *Chem. Eng. Prog.*, 1960, **56** (1), 60.
[10b] Cosner, *Eur. Chem. News*, Large Plant Supplement, Sept. 30, 1969, p. 29.
[10c] Fenske (Universal Oil Products), USP 3,200,163 (1965).
[10b] Bloch (Universal Oil Products), USP 3,183,233 (1965).
[10e] Universal Oil Products, BP 905,051 (1962).
[11] Private Communication.
[12a] Sherwood, *Industrial Chem.*, 1954, **30**, 25.
[12b] *Sanders, et al.*, *Ind. Eng. Chem.*, 1953, **45**, 2.
[13] '*Ethylene and its Industrial Derivatives*', ed. Miller (Ernest Benn, Ltd., London, 1968), Fig. 11.1, p. 876.
[14] *Ibid.*, p. 283.
[15] Shell Oil Company, USP 3,305,595.
[16] R. B. Valitov, *et al.*, USSR 227,307; 279,614.
[17] Sosnieva, I. E. *et al.*, *Vest Mosk. Khim.*, 1970, **11**(6) 785: C.A. **74**, 87174.
[18] Siderenko, Yu. N. *et al.*, USSR 188,958.
[19] Temkin, O. N. *et al.*, *Kinet. Katal*, 1970, **11** (6), 1592–3.
[20] Sumitomo Chem. Co. Ltd., German Offen. 2,043,549.
[21] Maruzen Oil Co., USP 3,502,736.
[22] Mobil Oil Co., USP 3,396,206.

[23] Dow Chemical Co., USP 3,441,625.
[24] *European Chemical News*, 13 October 1972, pp. 10–11, 14.
[25] Ghosh, J. C. *et al.*, *Current Science*, 1945, **14**, 269.
[26] Mitchell, J. E. *Amer. Inst. Chem. Engrs.*, 1946, **42**, 293–308.
[27] Webb, G. A. and Corson B. B., *Ind. Eng. Chem.*, 1947, **39**, 1153–56.
[28] Pitzer, K. S. *et al.*, *J. Amer. Chem. Soc.*, 1944, **66**, 2120–1.
[29] Miller, S. A. and Donaldson J. W., *Chemical and Process Engrg.*, 1967, **48**, 37–42, 47.
[30] Hornibrook J. N., *Chemistry and Industry*, 19 May 1962, pp. 827–878.
[31] Donaldson J. W., *Ethylene*, ed. S. A. Miller (Ernest Benn Ltd.), chapter 11.
[32] Ohlinger H. and Stadelmann, S., *Chim. Ing Tech.*, 1965, **37**, 361.
[33] Monsanto Company, USP 3,387,053.
[34] Foster Grant Co. and Chemetron Corp., USP 3,424,808.
[35] Monsanto Co., USP 3,306,942, cf. Monsanto Chemical Co., USP 3,306,947.
[36] Kearby, K. K., *Ind. Eng. Chem.*, 1950, **42** (2), 295.
[37] Halcon International Inc., Belg.P 659,161.
[38] Phillips Petroleum Co., USP 3,304,693.
[39] Dow Chemical Company, GP 1,247,295.
[40] Kirov, S. M., USSR 182,697, see C.A., 1967, **66**, 10489.
[41] Garifzyanov, G. G. *et al.*, USSR 335, 226, C.A., 1972, **77**, 62488.
[42] Asahi Y. and Kusano, T., *Nippon Kagaku Kaishi*, 1972 (3), 518–22.
[43] Monsanto Company, USP 3,409,688.
[44] Shell Int. Res. Maats. N.V., Dutch P. 66,05,005.
[45] Standard Oil Co. of Indiana, USP 3,335,197.
[46] Kubica *et al.*, BP 1,232,966.
[47] Farbwerke Hoechst., BP 1,238,468.
[48] UOP Company, USP 3,682,838: 3,379,786.
[49] Foster Grant Co., USP 3,437,703.
[50] Halcon International Inc., Belg.P 664,144: USP 3,330,878: 3,294,856; Scientific Design Co. Inc., Belg. P. 618,932: 632,163.
[51] UOP Co., South African Appln. 67, 05637.
[52] UOP Co., USP 3,409,689.
[53] UOP Co., FP 2,093,068; USP 3,402,212; USP 3,542,889.
[54] Houilleres du Bassin de Lorraine, FP 1,540,367–8.
[55] Monsanto Chemical Co., USP, 3,223,743, cf. Csomontanyi, G. *et al.*, Romanian P. 52,748: C.A., 1972, **77**, 5998.
[56] Steel, J. G. P. and Cameron, H., *Can. J. Eng.*, 1969, **7** (2), 183–7.
[57] Roessler H., and Wenske, R., *Chem. Tech.* (*Berlin*), 1967, **19**, 471–4.
[58] Bad.An.Soda.Fab. A-G, GP 1,290,130.
[59] Badger Co. Ltd., South African Appln. 69, 05016.
[60] Union Carbide Corpn., USP 3,326,996; Union Carbide Corpn., South African Applns. 68, 01350, 67, 92232.
[61] UOP Co., USP 3,515,763.
[62] Billet, R., Conrad, S. and Grubb, C. M., *Proc. Int. Symp. Distn.*, 1969, **5**, 111–126.
[63] Frank, J. C. *et al.*, *Chem. Engrg. Prog.*, 1969, **65** (2), 79–86.
[64] Coulter, K. E., *Petrol.Refin.*, 1952, **31** (8), 95; (10), 145; (11), 56; (12), 137.
[65] Esso Res. and Engrg. Co., USP 3,309,295.
[66] UOP Co., USP 3,515,764–3,515,767.
[67] Houilleres du Bassin de Lorraine, FP 1,540,369.
[68] Halcon International, USP 3,256,355; FP 1,356,231; French Addn. 87,288.
[69] BASF Dutch P 66, 13630.
[70] Badger Co. Inc., BP 1,097,619.
[71] UOP Co., USP 3,575,647.
[72] UOP Co., USP 3,408,263–3,408,266.
[73] Klays Bergt *et al.*, East German P 57,109.
[74] Union Carbide Corpn., BP 1,280,848.
[75] Metallgeselschaft A-G, FP 1,409,563; GP 1,205,523; USP 3,427,362.

[76] Monsanto Company, USP 3,427,362; 3,513,764.
[77] Eastman Kodak Co., Belg.P 667,996.
[78] Petro-Tex Chemical Corpn., USP 3,247,273.
[79] Jap. Carbide Ind. Ltd., Jap. P 71, 10850.
[80] Bad. An. Soda Fab. A-G., Dutch P 66, 08919; GP 1,260,460; German Offen. 1,806,080.
[81] Bogdanova, O. G. *et al.*, USSRP 189,873, see C.A. **67**, 108417; **75** 65866.
[82] Firestone Tire and Rubber Co., USP 3,511,887.
[83] Marathon oil Co., USP 3,410,920.
[84] Marathon Oil Co., USP 3,406,219.
[85] Princeton Chemical Research, USP 3,371,125; FP 1,500,632.
[86] Bad. An. Soda Fab. A-G., BP 1,148,108.
[87] Balandin, A. A. *et al.*, *Neftekhimiya*, 1967, **7** (4), 544–7.
[88] Russo G. and Maselli, S., *Chim. Ind.* (*Milan*), 1969, **51** (4), 343–50.
[89] Shell Oil Co., USP 3,209,155.
[90] Soc. Nationale des Petroles d'Aquitaine, FP 1,601,793; French Addn. 2,067,531; GP 1,963,697.
[91] Gulf Res. and Dev. Co., USP 3,644,549.
[92] Esso Res. and Engrg. Co., GP 1,959,814; FP 2,025,041; 2,077,564.
[93] Standard Oil Co., Ohio, USP 3,652,698.
[94] Schumann, C. and Seymour, USP 3,344,201.
[95] Esso Res. and Engrg. Co., German Offen. 1,806,255.
[96] Schaffel, G. S., *Oil Gas J.*, 1972, **70** (18), 114–15.
[97] Monsanto Company, BP 1,089,239; FP 1,508,337.
[98] Mobil Oil Corpn., USP 3,399,243.
[99] Kozlov, N. S. *et al.*, USSRP 281,455.
[100] Sanders, H. J., Keag, H. F. and McCullough, H. S., *Ind. Eng. Chem.*, 1953, **45** (1), 2.
[101] *Chemical Week*, 1966, **99** (5), 19; (5) 49; (21) 115; 1967, **100** (4), 60.
[102] Stobaugh, R. B., Calarco, V. A., Morris, R. A. and Stroud, L. W., *Hydrocarb. Process.*, January 1973, pp. 99–108.
[103] Scientific Design Co. Inc., Belg.P 641,452; 644,090.
[104] Halcon International Inc., Dutch P 65, 15037; 66, 08484; USP 3,442,963.
[105] *European Chemical News*, 27 July 1973, p. 6.
[106] Naugatuck Chemical Co., USP 1,552,075.
[107] UOP Co. Co., USP 3,271,469.
[108] Malinovskii, S. and Krzywicki, A., *Przemysl. Chim.*, 1966, **45** (9), 489–92.
[109] Kraemer, G., *Ber.*, 1890, **23**, 3269.
[110] Sato, Masakai, *Hydrocarb. Process*, May 1973, pp. 141–44.
[111] *Petroleum and Petrochemical International*, 1972, **12** (12), 64–68.
[112] Yamanaka, Yoshio, *et al.*, *Chem. Econ. Eng. Dev.*, 1972, **4** (4), 27–30; 26.
[113] Toray Industries Inc., German Offen., 2,006,863; Jap.P 72, 16403; 72, 16426.
[114] Herrin, H. R. and Gough, M. W., *E.C.N.*, 13 October 1972, pp. 10–11, 14.
[115] *Chem. Eng. News*, 4 June 1973.
[116] Anderson, E. V. *et al.*, *Ind. Eng. Chem.*, 1960, **7**, 52.
[117] Simon, E., *Ann.*, 1839, **31**, 271.
[118] Stobbe H., and Posnjake, G., *Ann. Chem.*, 1909, **371**, 259.
[119] Ostromislensky, I., *J. Russ. Phys. Chem. Soc.*, 1911, **44**, 204.
[120] Staudinger, H., *Ber.*, 1920, **53**, 1073.
[121] Staudinger, H. and Frost, W., *Ber.*, 1935, **68**, 2351.
[122] Dainton, F. S., Ivin, K. J., and Walmsley, D. A. G., *Trans. Far. Soc.*, 1960, **56**, 1784.
[123] Flory, P. J., *J.A.C.S.*, 1937, **59**, 241.
[124] Mayo, F. R., *J.A.C.S.*, 1953, **75**, 6133.
[125] Haward, R. N. and Simpson, W., *Trans. Far. Soc.*, 1951, **47**, 212.
[126] Elly, J. Haward, R. N. and Simpson, W., *J. Appl. Chem.*, 1951, **1**, 347.
[127] Harkins, W. D., *J.A.C.S.*, 1947, **69**, 1428.
[128] Szwarc, M., Levy, M. and Milkovich, R., *J.A.C.S.*, 1956, **78**, 2656.
[129] Bywater, S. and Worsfold, D. J., *J. Polym. Sci.*, 1962, **58**, 571.

[130] Natta, G. and Corradini, P., *Makr. Chem.*, 1955, **16**, 77.
[131] Kern, R. J., Hurst, H. G. and Richard, W. R., *J. Polym. Sci.*, 1960, **45**, 195.
[132] Wulff, C. and Dorrer E., GP 634,278 (1936) to IG Farbenindustrie AG; *CIOS Rep.* XXIX, -62, 1945, 129.
[133] BASF, GP 1,112,631 (1961).
[134] Röhm, O., GP 656,469 (1938).
[135] Stober, K. E. and Amos, J. L. (Dow Chemical Co.), USP 2,530,409 (1950).
[136] Dow Chemical Co., USP 2,727,884 (1955).
[137] '*Encyclopaedia of Polymer Science and Technology*' (Interscience), Vol. 13, p. 133.
[138] Boundy, R. H. and Boyer R. F., '*Styrene*', 1952, p. 907.
[139] Union Carbide Corpn., USP 2,496,653.
[140] Hohenstein, W. P., USP 2,524,627 (1946).
[141] Grim, J. M. (Koppers Co. Inc.), USP 2,594,913 (1947).
[142] Haward, R. N. and Hohenstein W. P. (Petrocarbon Limited), BP 659,771 (1949).
[143] Grim, J. M. (Koppers Co. Inc.), USP 2,673,194 (1951).
[144] Hohenstein, W. P., Haward, R. N. and Elly, J. (Petrocarbon Limited), BP 690,877 (1950).
[145] Maclay, W. N., *J. Appl. Polym. Sci.*, 1971, **15**, 867.
[146] Koppers Co. Inc., BP 691,908 (1953).
[147] Holden, G. and Milkovich, R. (Shell Development Co.), USP 3,231,635 (1962).
[148] Dow Chemical Co., USP 3,530,105 (1967).
[149] Seymour, R. B. (Monsanto Co.), USP 2,574,439.
[150] Ostromislensky, I. (Naugatuck Chemical Co.), USP 1,613,673 (1928).
[151] Dow Chemical Co., USP 2,694,692.
[152] Styrene Products Ltd., BP 854,238 (1959).
[153] Monsanto Co., USP 2,862,906; 2,652,907 and 2,886,553.
[154] Molau, G. E. and Keskkula, H., *J. Polym. Sci.* A1, 1966, **4**, 1595.
[155] Dow Chemical Co., USP 2,450,436.
[156] Dow Chemical Co., USP 2,515,250 (1950).
[157] Dow Chemical Co., USP 2,576,911.
[158] Dow Chemical Co., USP 2,577,743.
[159] BASF, BP 695,826.
[160] BASF, BP 715,100.
[161] BASF, BP 1,024,195 (1962).
[162] BASF, BP 754,513 (1956).
[163] Koppers Co. Inc., USP 3,089,599 (1960).
[164] Sun Chemical Co., BP 949,655 (1961).
[165] BSAF, BP 825,611 (1956).
[166] BASF, GP 1,244,396 (1965).
[167] Dow Chemical Co., USP 3,058,926 (1959).
[168] Eichhorn, J., *J. Appl. Polym. Sci.*, 1964, **8**, 2497.
[169] Bakelite Ltd., BP 590,247 (1947).
[170] Dow Chemical Co., USP 2,739,142 (1956).
[171] Haward, R. N. (Petrocarbon Limited), BP 707,647.
[172] *Europlastics*, 1973, **46** (6), 160.
[173] Roper A. N. and Barber, E. G. (Shell Int Res. Maats) BP 1,056,470 (1965).

CHAPTER 8

PHENOL AND ITS DERIVATIVES

By W. T. Reichle, F. M. Konrad and J. R. Brooks

8.1. HISTORICAL by W. T. Reichle

IMPURE phenol was first isolated from coal tar by Runge about 1834. Others were able to obtain a pure product.[1] The first synthesis of phenol seems to have been by way of a reaction of aniline hydrochloride with silver nitrite.[2]

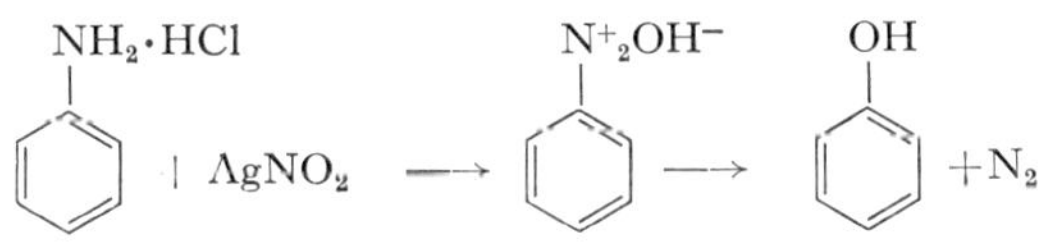

The first technically important phenol synthesis was the sulphonation of benzene followed by fusion with caustic. This was developed into an economic process around the turn of the century. Prior to that and until about World War I, coal tar supplied virtually all the phenol. The discovery that trinitrophenol (picric acid) was a very potent explosive of considerable military value produced a huge upsurge in phenol requirements particularly during the Great War. Most phenol was then made via the sulphonation synthesis.

Modern chemistry has placed an enormous demand on synthetic phenol production (about 1·7 thousand million lb were used in 1970). Virtually all new plant capacity now uses the cumene process because of its cheap raw materials, low investment and operating costs, and high efficiency.

8.2. TOXICOLOGY by W. T. Reichle

Phenol has very important antiseptic properties. Lister in 1867[3] discovered the bactericidal properties of phenol. Lister's phenol spray was very popular, it greatly aided surgery and the disinfection of wounds, etc. Prolonged exposure of patients and physician alike to phenol—by inhalation and absorption—caused chronic or subacute phenol poisoning. The quantities which passed through man were sometimes appreciable; several grams of phenol were excreted in the urine each day.[4]

Prolonged exposure to phenol (vapour or by skin absorption) can cause damage to lungs, liver, kidneys, heart, and the genitourinary tract. In animals, the 30–60 p.p.m. range of phenol vapour in air seems to cause harm. The recommended maximum air concentration for 8 hr. exposure is 5 p.p.m. (19 mg/ft^3); in animals the LD_{50} appears to be about 100–600 mg/kg (oral). In man the oral ingestion of phenol causes intense burning in the throat and abdominal pains. Muscular weakness, unconsciousness, weak pulse, characteristic phenol odour in the breath, and poor respiration are general symptoms. Death is due to respiratory failure. Skin contact causes 'burns' of varying severity—depending upon the length of exposure and the phenol concentration and temperature.

Phenol spilled on the skin should be washed off with generous quantities of warm water *at once*. In case of oral poisoning, vomiting should be induced (castor oil, or vegetable oil are useful) or gastric lavage with milk should be employed. Circulatory stimulants should be used as a symptomatic treatment.

Certain phenols have extraordinary powers to inhibit the growth of bacteria. Hexachlorophene is used in soaps as a very effective bactericide.

Deichmann and Keplinger[5] and Sax[6] have described the toxicology of phenol in some detail.

8.3. PHYSICAL PROPERTIES OF PURE PHENOL by W. T. Reichle

Table VIII.1 gives some of the physical properties of pure phenol; Table VIII.2 details the specifications of commercial USP phenol. While the latter is 99 per cent phenol, small amounts of impurities are always present. These are principally water and otherwise depend upon the synthesis route. For example, phenol made by the Raschig process contains small amounts of chlorophenols; cumene phenol will have acetophenone and alpha-methylstyrene as impurities (0·5 to 0·001 per cent range). Citric acid (to 50 p.p.m.) is frequently added to commercial phenol to act as antioxidant-stabiliser. Phenol which has been exposed to the air, particularly above its melting point, is frequently pink.

TABLE VIII.1

Physical Properties of Pure Phenol[7]

Melting point (°C)	40·9
Boiling point (°C, 760 mm Hg)	181·8
(100 mm Hg)	120·7
(30 mm Hg)	92·78
(10 mm Hg)	70·86
Heat of evaporation, b.p. (cal/g)	117·0
Critical temperature (°C)	419·2
Critical pressure (atm)	60·5
Heat of combustion (crystalline phenol to liquid water at 15°C, kcal/mole)	732
Density, 41° (g/ml)	1·0576
Refractive index N_D, 41°	1·54178
Solubility in water (wt %, 25°C)	8·45
Flash point (°C)	About 80

TABLE VIII.2

Specifications for Commercial Phenol, USP[8]

	Specification Limits
Distillation, 760 mm	90% shall distill within a 2°C range which shall include 182°C.
Phenol	99·0% by weight, minimum.
Freezing point	40·6°C, minimum.
pH of a 5% aqueous solution	neutral or slightly acidic to litmus paper.
Solubility in water	a solution of one part phenol and 15 parts of distilled water shall be clear and colourless with no turbidity.
Water content	0·1% by weight, maximum.
Nonvolatile matter	0·05% by weight, maximum.
Colour (molten) at time of shipment	20 platinum-cobalt, maximum.
	Shipping Data
Containers	55-gal ICC 17E iron drum with polyethylene gasket—475 lb.
Flash point, Cleveland open cup	185°F
Label required by Bureau of Explosives	Poison B.
Average weight per US gallon	8·76 lb at 50°C.

8.4. CHEMISTRY by W. T. Reichle

The phenol molecule is a highly reactive benzene derivative.[9, 10] It oxidizes readily in air at 50°C where it gradually turns pink and eventually brown due to the formation of quinones and polymers. The phenol molecule is also highly reactive towards electrophiles but quite unreactive with nucleophiles. This is due to the high electron density of the aromatic pi-system; the non-bonding oxygen-p-electrons delocalize charge into the ring, thereby raising the electron density considerably:

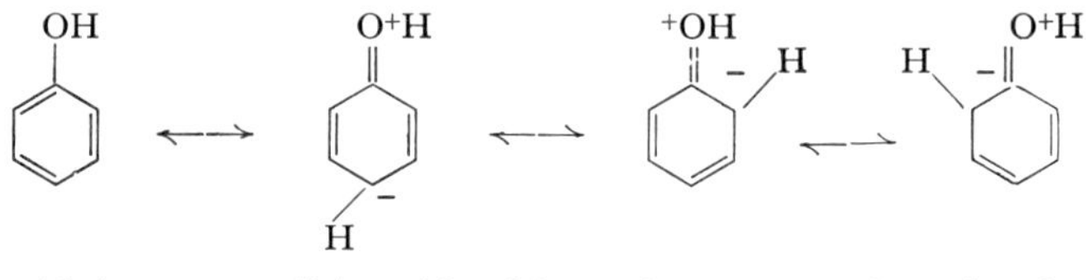

The phenoxide ion accomplishes this without charge separation; thereby producing an even more reactive ion:

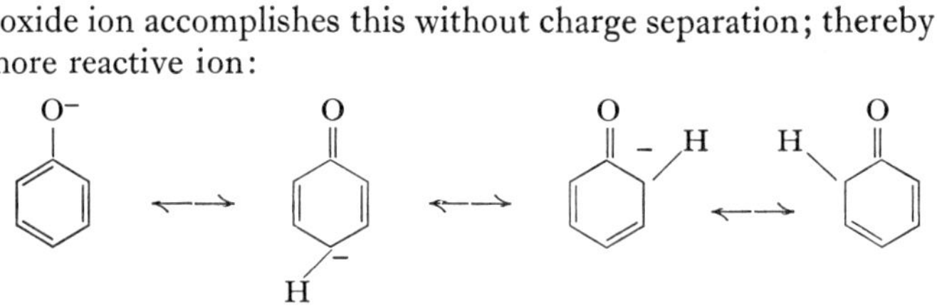

This aromatic ring is therefore highly activated towards electrophilic reagents; the hydroxyl group is overall rate-enhancing, relative to benzene, and is a strong ortho-para director.

TABLE VIII.3

Rates of Halogenation of Benzene Derivatives[11]

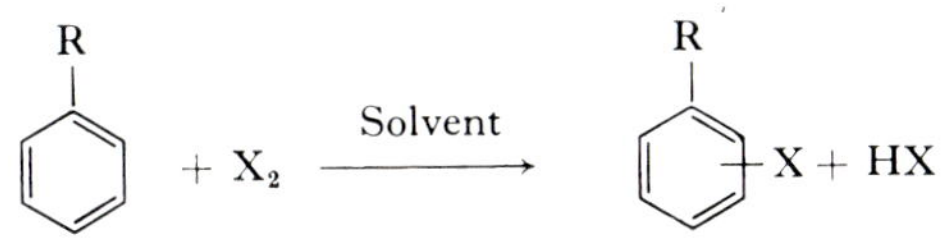

Substituent—R	*Relative rate*
$N(CH_3)_2$	5×10^{18}
OH	$1 \cdot 1 \times 10^{11}$
OCH_3	$1 \cdot 2 \times 10^{9}$
CH_3	$3 \cdot 4 \times 10^{2}$
H	(1·0)
Cl	0·11
COOH	$2 \cdot 6 \times 10^{-4}$
NO_2	$1 \cdot 8 \times 10^{-6}$

From the data in Table VIII.3 it can be seen that phenol is more reactive towards halogen (chlorine, bromine) than any benzene derivative except N,N′-dimethyl aniline. Phenol is 10^{11} times as reactive as benzene and 10^{17} times as reactive as nitrobenzene. This is an enormous range of reactivity. Thus the aromatic ring system of phenol undergoes a large number of reactions with even weak electrophiles with ease. Nitration leads, eventually, to the 2,4,6-trinitro derivative (picric acid),

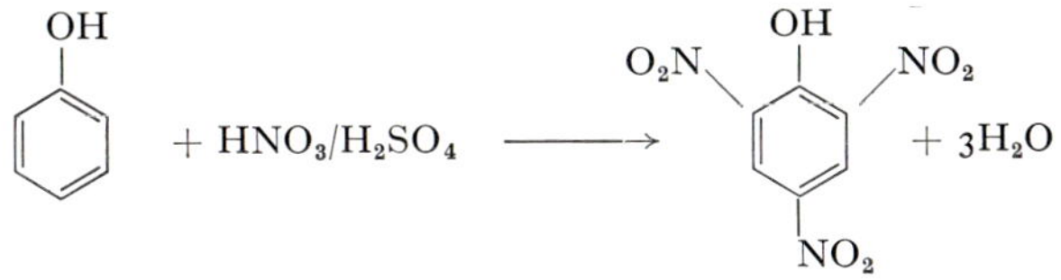

while halogenation with bromine or chlorine in polar solvents (water, acetic acid) leads rapidly to the 2,4,6-trihalophenol.

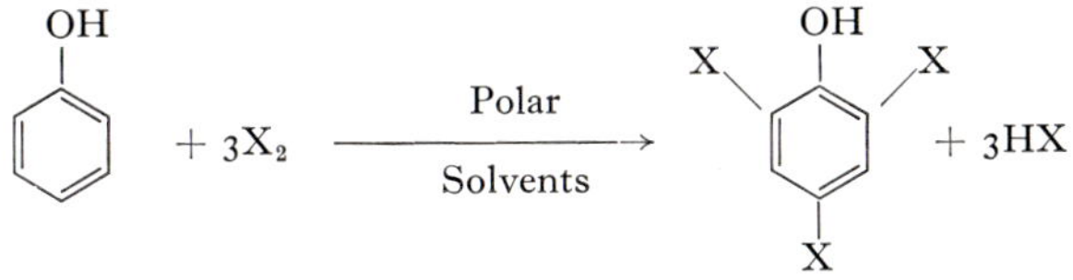

In non-polar solvents this reaction can lead to the *o* or *p*-mono or 2,4-di-substituted phenols. Bromination can be used to determine phenol quantitatively.[12]

The phenol molecule is highly reactive with the carbonyl groups of aldehydes or ketones. Reaction with formaldelyde, either acid or base catalyzed, leads to various phenol-formaldehyde polymers.[13, 14]

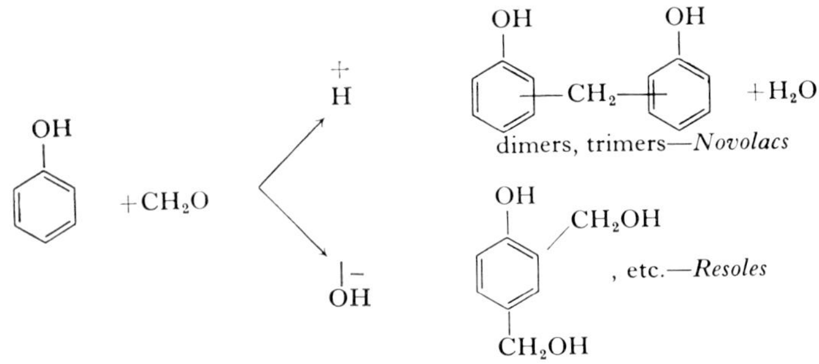

About half of all phenol produced is used in these phenol-formaldelyde condensates. They are discussed at length in a later section. Similarly the acid catalyzed reaction of phenol and acetone leads to bisphenol-A:

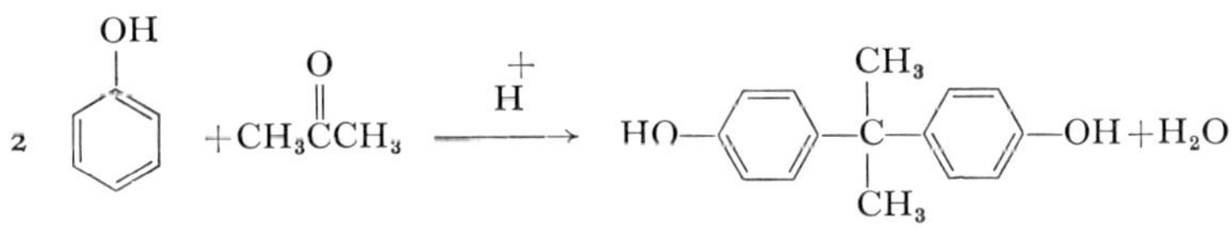

The nature and products of these commercially very important reactions are also discussed in a later section.

Phenol reacts readily with sulphuric acid or sulphur trioxide to yield the sulphonic acid (ortho and para), and isomeric sulphones:

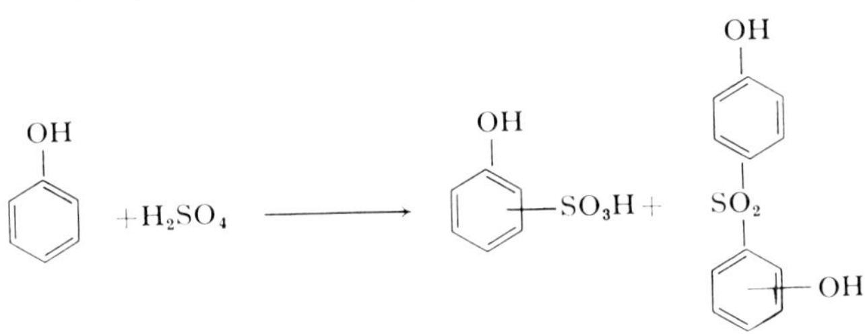

The diazonium ion couples with phenol to form an azo linkage,

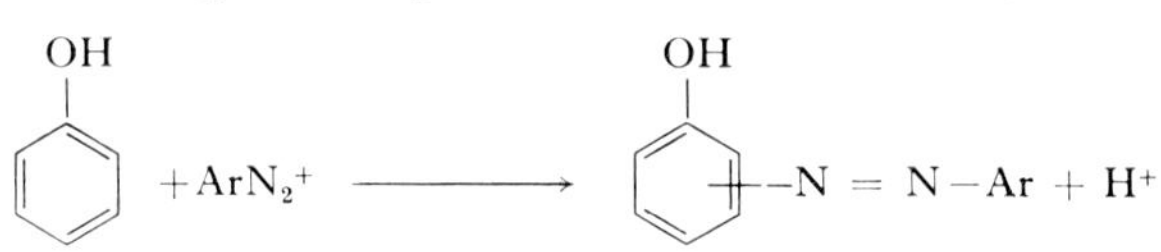

This reaction is the basis of the very important class of azo dyestuffs.[15]

Phenol is readily alkylated, under acid catalysis, with alcohols, ethers, or olefins to yield substituted phenols:

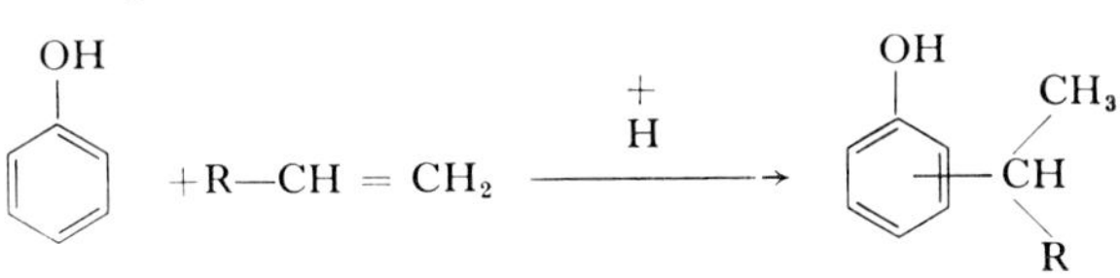

These are also discussed separately (p. 382).

The aromatic ring can be hydrogenated to cyclohexanol,

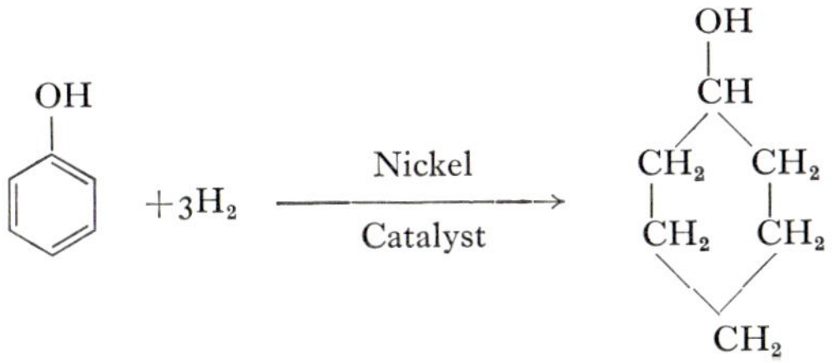

A very interesting reaction is the Bucherer reaction[16] in which, under forcing conditions, the hydroxyl group is exchanged against an amine group to yield aniline:

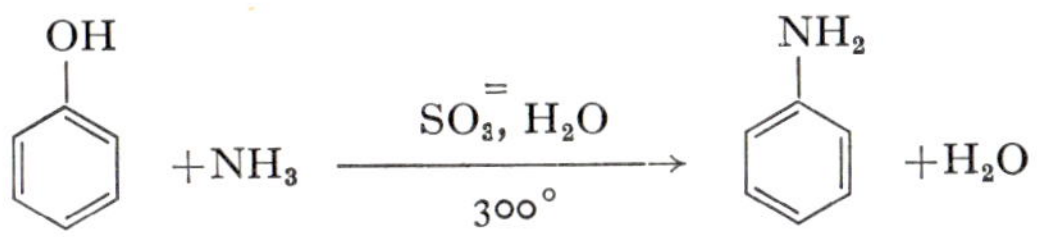

This reaction probably proceeds via the keto form of phenol with subsequent ligand exchange:

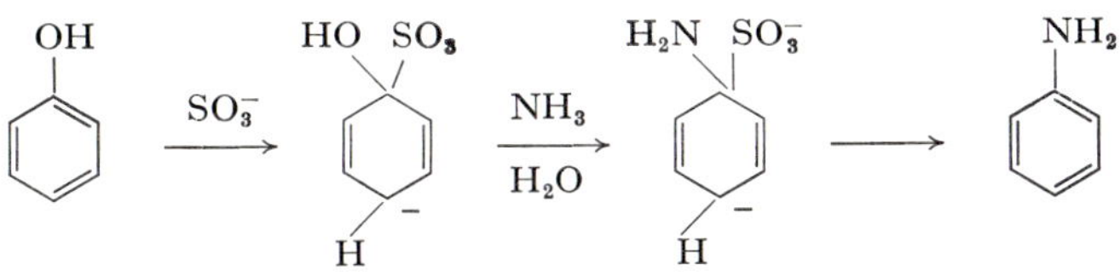

In the event of substantial quantities of aniline being needed, and nitric acid not being readily available, this method would probably be an acceptable commercial synthesis.[17]

The hydroxyl proton of the phenol molecule is much more acidic than the corresponding proton in a normal alcohol (pK_a phenol $\simeq 10$; cyclohexanol $\simeq 18$). Phenol is a weaker acid than carbonic acid, therefore, carbon dioxide will liberate phenol from aqueous sodium phenoxide. The phenolic hydroxyl will form esters with ease,

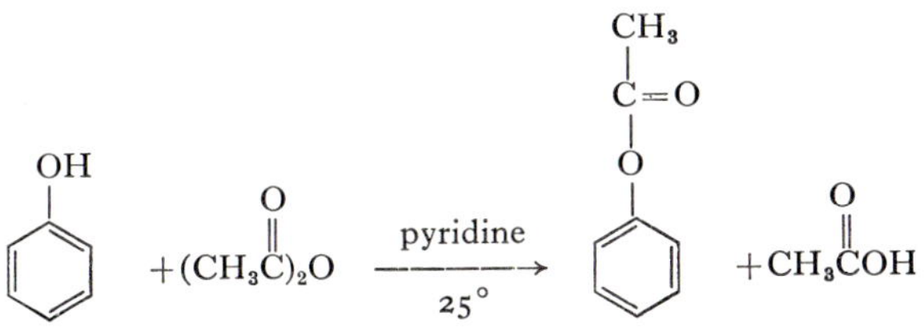

and react with diazomethane to yield the methyl ether,

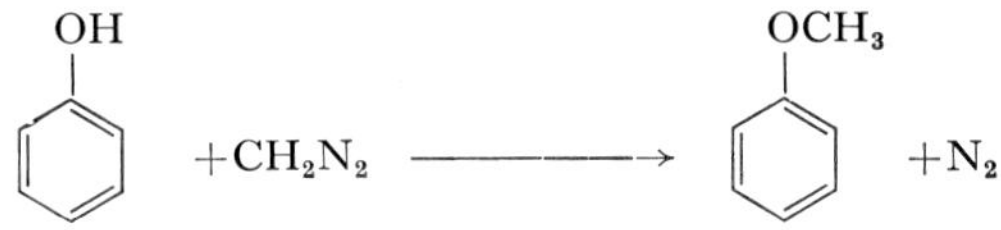

Alternatively, the combination of aqueous caustic-dimethyl sulphate also yields anisole. Reaction of the phenol anion with alkyl halides gives rise to two products, one is due to oxygen alkylation, the other carbon alkylation:

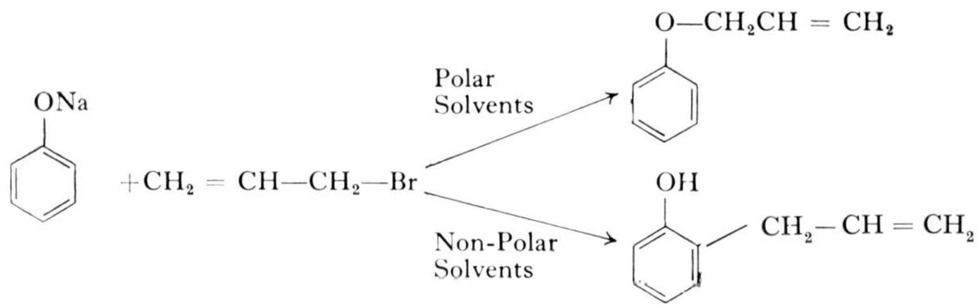

Phenol does *not* react with halogen acids, thionyl chloride or phosphorus halides to give halobenzenes.

Phenoxide ion undergoes two unexpected reactions. One is the Kolbe-Schmidt reaction[18] with carbon dioxide to give either salicylic or *para*-hydroxy benzoic acid:

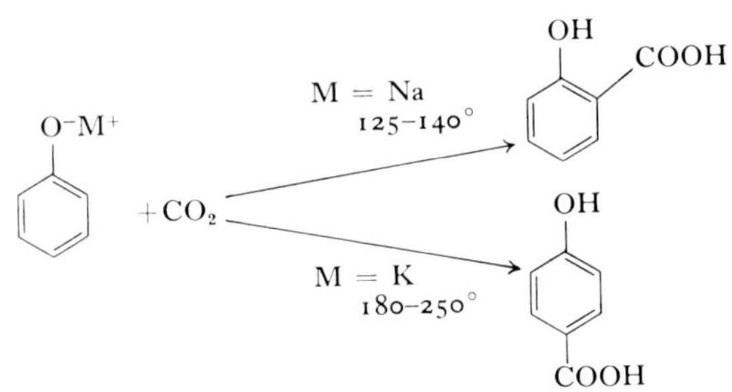

The other is a reaction with chloroform (Reimer-Tiemann reaction) to give salicylaldehyde:[19]

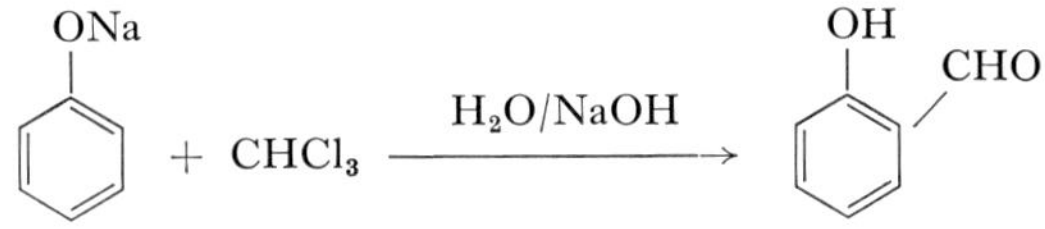

The phenol molecule is very reactive towards oxidants. Most often intractible polymers are formed which defy analysis. In all likelihood the phenol molecule is first oxidized to ortho or para quinones or hydroquinones. These then add phenolic hydroxyl to form polymers:[20]

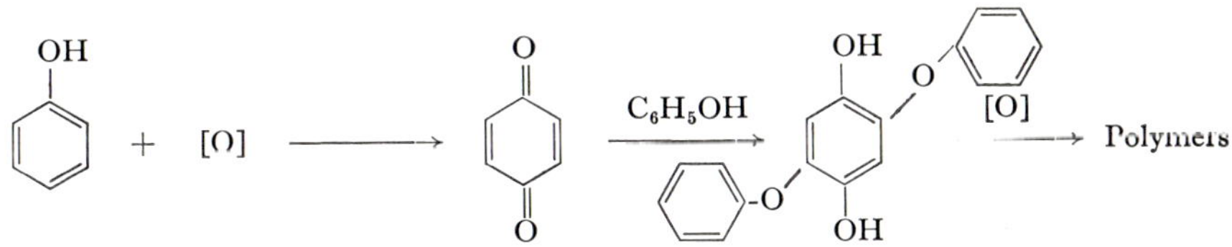

Some oxidizing agents, under carefully controlled conditions, lead to isolable products, although not necessarily in good yield:[21]

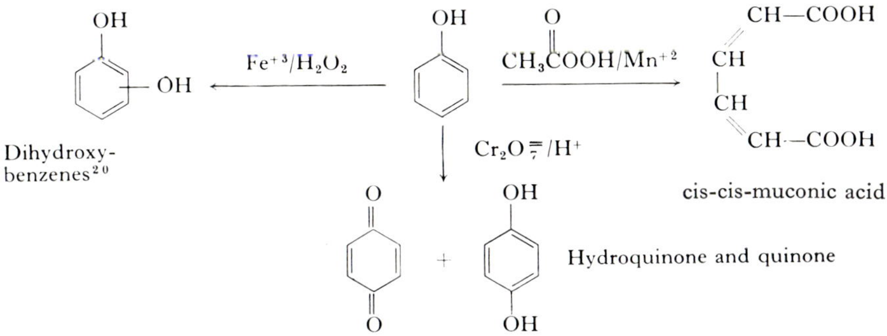

Homolytic attack on the phenol molecule takes place by oxygen-proton abstraction;

This aryloxy radical then forms polymers or addition compounds. Frequently, strongly hindered phenols (2,6-di-tert.-butylphenol, etc.) are excellent oxidation inhibitors (p. 392). Hydroquinone is used extensively and effectively as a polymerisation inhibitor for vinyl monomers which polymerise by homolytic paths (styrene, methyl methacrylate, etc.).

8.5. ECONOMICS by W. T. Reichle

The economic statistics cited here have been gathered from a variety of sources. Notable amongst them are the US Tariff Commission, other US Department of Commerce reports and projections, publications such as *Chemical Week*, *Chemical Engineering News* and the *Chemical Economics Handbook* published by the Stanford Research Institute, as well as data available within the Union Carbide Corporation. We are grateful to Mr James L. Brannon for much of this information.

Prior to World War I phenol was used primarily in pharmaceuticals and the synthetic dye industry. About 10 million lb were produced annually in the United States, all from coal tar. The advent of the Great War required large quantities of phenol for the high explosive picric acid (2,4,6-trinitrophenol). At the end of the war

United States consumption had reached 72 million lb per year.[22] This necessitated a large synthetic phenol capacity; virtually all was produced via the benzene sulphonation-caustic fusion route. World War II resulted in doubling of the United States phenol supply (from 92 million lb per year in 1940 to 188 million lb per year in 1945). In 1970 the United States phenol capacity was about 2 thousand million lb per year and worldwide capacity about 5·6 thousand million lb. This chemical has now become a very cheap (7–8 cents lb), tonnage commodity. Single production units as large as 400 million lb per year are being built in order to take advantage of economy of scale. The price of phenol is now only about twice the cost of the principal raw material—cumene—so that very little leeway exists for labour, operating costs, overhead, or capital charges. The profit margin on phenol is distressingly small.

Table VIII.4 gives the worldwide demand—capacity for phenol in 1970. Europe and the United States produced more phenol than the rest of the world combined. The Eastern Block (Russia and allies) is slightly behind the Far East (principally Japan); the remainder of the world's capacity/demand is virtually insignificant by comparison. China is reputed to purchase 'substantial' quantities of phenol from Japan.

TABLE VIII.4

Worldwide Demand-Capacity for Phenol—1970 (in millions of lb)

	Demand	*Capacity*
European Economic Community	1,301	2,055
European Free Trade Association	296	274
Eastern Block (Russia, etc.)	509	529
Far East (Japan, etc.)	533	587
Western Hemisphere (non-U S)	97	104
Other	80	71
United States	1,682	2,015
Worldwide	4,498	5,635

Tables VIII.5 and 6 give the phenol producers in the United States and Europe as well as the process used and the nominal plant capacities for 1970. The United States has about eighteen large or medium-sized phenol plants which use all known processes (principally cumene at present). Europe has, similarly, principally cumene capacity, with the older plants using largely the sulphonation method. Only one chlorobenzene hydrolysis and one toluene-oxidation plant exist, both in Germany. Throughout the remainder of the world there are an assortment of phenol plants all of relatively small capacity. For example, Argentina has a Raschig phenol unit which produces about 10–12 million lb per year. Australia and India may well be the next locations for large (100 million lb per year) synthetic phenol plants. Both countries have a fairly high and growing demand which is currently being covered by importation or small local plants which cannot match the efficiency of a large unit.

TABLE VIII.5

United States Phenol Producers—1970

Company (location/process)	*Capacity (million lb per year)*	*Remarks*
Dow		
Midland, Mich.—Chlorobenzene hydrolysis	220	To be phased out gradually
Kalama, Wash.—Toluene	45	Sold in 1970*
Oyster Creek, Tex.—Cumene	2–400	Due in 1972–3
Monsanto		
Sangett, Ill.—Sulphonation	120	To be shut down 1970
Choc. Bayou, La.—Cumene	140	To 400 million lb in late 1970
Allied		
Frankford, Pa.—Cumene	430	
Chevron		
Richmond, Calif.—Cumene	55	To be phased out (?)
Hooker		
Tonawanda, NY—Raschig	60	Shut down late 1970
South Shore, Ky.—Raschig	70	,, ,, 1971
Reichhold		
Tuscaloosa, Ala.—Sulphonation	150 (max.)	
Union Carbide		
Marietta, Ohio—Raschig	110	To be shut down 1971
Bound Brook, NJ—Cumene	150	
Ponce, Puerto Rico—Cumene	(300)	Scheduled 1972–3
Shell		
Houston, Tex.—Cumene	50	
Skelly		
El Dorado, Kans.—Cumene	50	
Clark		
Blue Island, Ill.—Cumene	75	Probably additional 10 million lb in 1971
Hercules		
Gibbstown, NJ—Cumene	55	Could be phased out 1972
US Steel		
Haverhill, Ohio—Cumene	200	
Georgia Pacific		
Rebecca, La.—Cumene	(75–150)	Due 1971
Various		
Natural phenol—coal tar, petroleum	60	

* Now owned by Kalama Chem. Co., a corporation organised by former Dow employees.

TABLE VIII.6

European and Japanese Phenol Producers—1970

Company	*Process*	*Capacity (million lb per year)*	*Remarks*
Phenol Chemie	Cumene	570	Germany
Scholven Chemie	Cumene	132	Germany
Bayer	Chlorobenzene hydrolysis	66	Germany
Dow	Toluene	132	Germany
ICI	Cumene	186	UK
British Petroleum	Cumene	150	UK
Rhone Petroleum	Cumene	220	France
Dutch State Mines	Cumene	187	Dutch
SIR	Cumene	352	Italy
SIFA	Cumene	165	Italy
Monte Edison	Cumene	143	Italy
USSR	Cumene	185	Italy
USSR	Sulphonation	205	Italy
Mitsui Petro Chemicals	Cumene	304	Japan
Mitsui Toatsu	Sulphonation	206	Japan

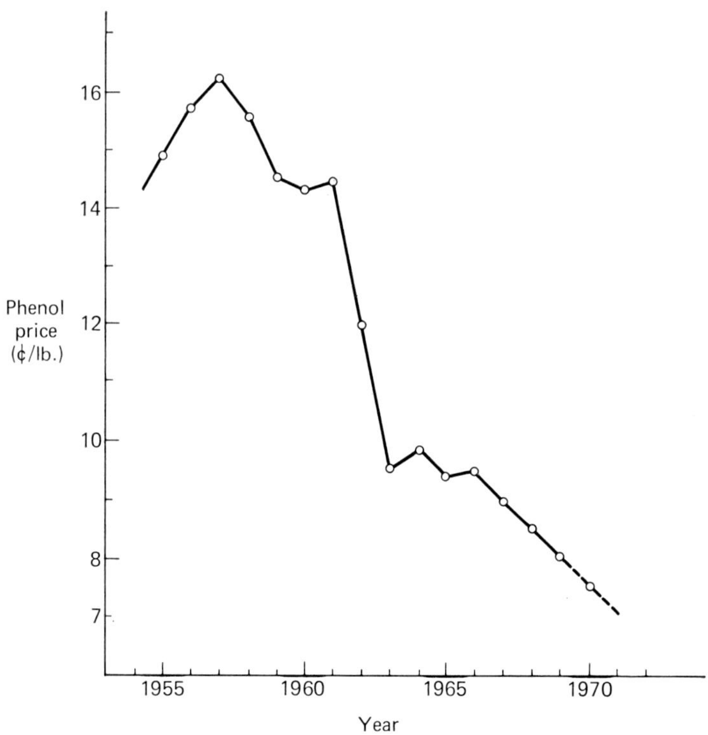

FIG. 8.1 *Phenol Prices*

Table VIII.7 illustrates the phenol and acetone price developments since 1955. In 1962–3 large cumene units came on stream, several were owned by oil companies which had no internal consumption and had to sell all the phenol they produced. The result was a 30–40 per cent price drop (about 14·3 ct/lb in 1960 to 9·5 ct/lb in 1963) which was very difficult to accept by high cost producers. Fig. 8.1 illustrates this more clearly. The future will witness further price declines although such a sharp drop as in 1960–2 will probably not be seen (baring price wars). The continued decline in phenol prices requires ever increasing plant sizes. Thus a good rule of thumb is that doubling plant capacity will increase the capital requirements by only 60 per cent. Hence, a truly large unit can have significantly lower depreciation charges than a smaller unit. Also, the operating and overhead costs of a 100 million lb plant and one twice or four times that size are probably not very different. This is another incentive for these very large units. In the next several years phenol prices are expected to decline to the 6–7 ct/lb range, but this is about the limit.

TABLE VIII.7

United States Phenol Prices
(Average Selling Price (ct/lb))

Year	*Phenol*	*Acetone*
1955	14·8	6·2
6	15·7	6·6
7	16·2	7·2
8	15·5	7·3
9	14·5	7·2
1960	14·3	7·0
1	14·4	6·2
2	12·0	5·6
3	9·5	5·0
4	9·7	4·8
1965	9·4	4·6
6	9·5	4·8
7	8·9	4·7
8	8·5	4·5
9	8·0 (est.)	—
1970	7·5 (est.)	—*

* Spot prices have been as low as 3ct/lb; presently a number of cumene–phenol producers have a considerable surplus of acetone.

Over the decades five large-scale, synthetic phenol processes have been devised (Table VIII.8). The most recent one employs cumene as starting material. The first plants which utilized this process were built in the 50's and since then virtually all new phenol capacity has been by this route*. In the United States in 1955—19 per

* The recent (1967–70) addition of 50 million lb per year of sulphonation phenol capacity to the Reichhold complex in Tuscaloosa, Alabama, is largely due to the favourable disposal of sodium sulphite by-product to the neighbouring pulp mills. Reichhold also has a large in-house consumption of phenol.

cent was cumene phenol; in 1974 about 84 per cent of all phenol will be made via the cumene route. Similarly, European cumene phenol in 1955 was 44 million lb, in 1970 about 675 million lb. Fig. 8.2 illustrates the demise of all synthetic phenol processes save the one based on cumene. The more favourable operating characteristics of the cumene route and lower capital costs, due to the production of two chemicals (phenol and acetone) on one plant site, made this change possible.

TABLE VIII.8

United States Phenol Capacity By Process Type (Quantities in million lb)

Year	*1955 (%)*	*1960 (%)*	*1965 (%)*	*1970 (%)*	*1971 (%)*
Cumene	120 (19·3)	235 (27·0)	615 (44·9)	1,195 (59·9)	2,255 (83·8)
Sulphonation	138 (22·2)	205 (23·6)	210 (15·3)	245 (12·1)	150 (5·6)
Chlorobenzene hydrolysis	210 (33·9)	220 (25·3)	220 (16·1)	220 (10·9)	110 (4·1)
Raschig	128 (20·6)	170 (19·5)	235 (17·2)	240 (11·9)	70 (2·6)
Toluene	0 (0)	0 (0)	40 (2·9)	45 (2·2)	45 (1·7)
Coal tar	25 (4·0)	40 (4·6)	50 (3·6)	60 (3·0)	60 (2·2)
Total	621	870	1,370	2,005	2,690

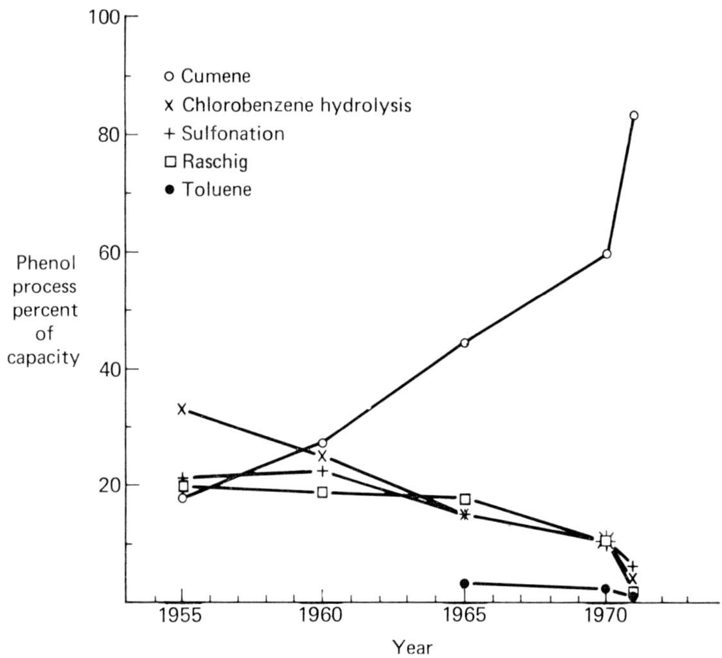

FIG. 8.2 *Phenol Process: Percent of Capacity*

This huge growth of cumene phenol capacity is in large part also due to the fact that the acetone consumption is growing at about the same rate as the phenol consumption is growing.[23] Table VIII.9 illustrates this point. Acetone consumption has grown at an 8–9 per cent per year rate (1955–70) while the phenol growth has been about 10 per cent per year (except in 1968–70 when there was little growth). Strangely enough, this large acetone capacity has not made a profound impact on acetone prices (Table VIII.7). Even during the 1962 phenol price break, acetone prices held up reasonably well.

TABLE VIII.9

Phenol and Acetone Growth Rates

Year	*Acetone production (million lb per year)*	*Growth* (% per year)*	*Phenol production (million lb per year)*	*Growth* (% per year)*
1945	350	7·4	205	
1950	480	2·5	320	11·0
1955	540	8·5	510	11·9
1960	770	9·0	770	10·2
1965	1,120	9·0	1,220	11·7
1970	1,620		1,550	5·4†

* Five year average.
† Since 1968 there has been little growth in phenol consumption.

TABLE VIII.10

United States Phenol Consumption

End use	*1969 (million lb)*	*1970 (million lb)*	*1970 (%)*
Phenolic resins	762	832	49·6
Bisphenol	153	168	10·0
Caprolactam	240	246	14·6
Adipic acid	50	50	2·9
Plasticisers	29	31	1·8
Lube oil refining	24	24	1·4
Alkyl phenols	91	92	5·5
Pesticides	71	74	4·4
Salicylic acid	39	41	2·4
Miscellaneous	105	109	6·5
Inventory	134	—	—
Export	15	15	0·9
	1,713	1,682	

In the United States three principal uses account for 75 per cent of all phenol consumption (Table VIII.10); a single one for about one-half. The phenolic resins, mostly phenol–formaldehyde condensates, have grown at a respectable pace in recent years largely because of the widely variable product properties, ease of manufacture, acceptable and essentially patent-free know-how, and low and stable prices of the products (15–30 ct/lb in bulk). Bisphenol A is also widely employed in epoxy resins and various polymers (polycarbonates, polysulphone).[24] There is only one consumer of phenol for caprolactam production, Allied Chemical Corp. in USA since the cyclohexane oxidation technique is more attractive; all new caprolactam plants use cyclohexane as starting material. In Europe a very large quantity of phenol is used by the older caprolactam plants.

The remaining phenol uses are smaller. Plasticisers (phosphate esters), alkyl phenols (detergents and oxidation inhibitors), and pesticides (notably the 2,4-D herbicides and pentachlorophenol wood preservative) absorb relatively small quantities of phenol but result in speciality products which have a unique economic-technical importance.

8.6. THE MANUFACTURE OF PHENOL by W. T. Reichle

There are six basic sources of phenol. Five are synthetic routes to this chemical, the other yields 'natural' phenol from coal tar or petroleum distillates. The latter is a very small and static source (about 4 per cent of the total). Three synthetic processes (chlorobenzene-caustic hydrolysis, Raschig process, and the sulphonation-caustic fusion method) are for various reasons no longer economic and no new plants using this technology are being planned.[25] The toluene oxidation method has not been practiced widely; only one small plant is operating in the United States and no plans for new ones are known. All new phenol capacity built in recent years, or is currently being planned, uses the cumene oxidation technology.[26, 27]

8.6.1. Preparation of Phenol by Sulphonation Route

The sulphonation process is the oldest source of synthetic phenol. This process is in essence based on the observation by Faraday, in the early part of the nineteenth century, that aromatic sulphonic acids when fused with caustic, yield phenols. There exist two first-hand reports on the practical aspects of this synthesis scheme. The first is the detailed government report on German phenol-sulphonation practice[28] and the other is a description of the Monsanto Company's, Sauget, Ill., phenol plant.[29] A fairly exhaustive study of the synthesis of cresol via toluene sulphonation and caustic fusion has also been published.[30]

The chemistry of this phenol synthesis is as follows:

(1) The sulphonation of benzene with sulphuric acid.

This step is a very important one. The product—benzene sulphonic acid—should contain as little sulphuric acid as possible, preferably no sulphone and no disulphonated benzene. Various processes[31–35] exist which accomplish this. Both the Leverkusen phenol process[28] as well as the Monsanto plant[29] employed excess sulphuric acid (100 per cent) in this sulphonation step. Presumably, the by-product sodium sulphate, which results from the neutralization step was a pure and saleable product. The German plant used 2 mole sulphuric acid per mole of benzene; hence, for every pound of benzene sulphonic acid produced, about 0·9 lb of sodium sulphate results as by-product.

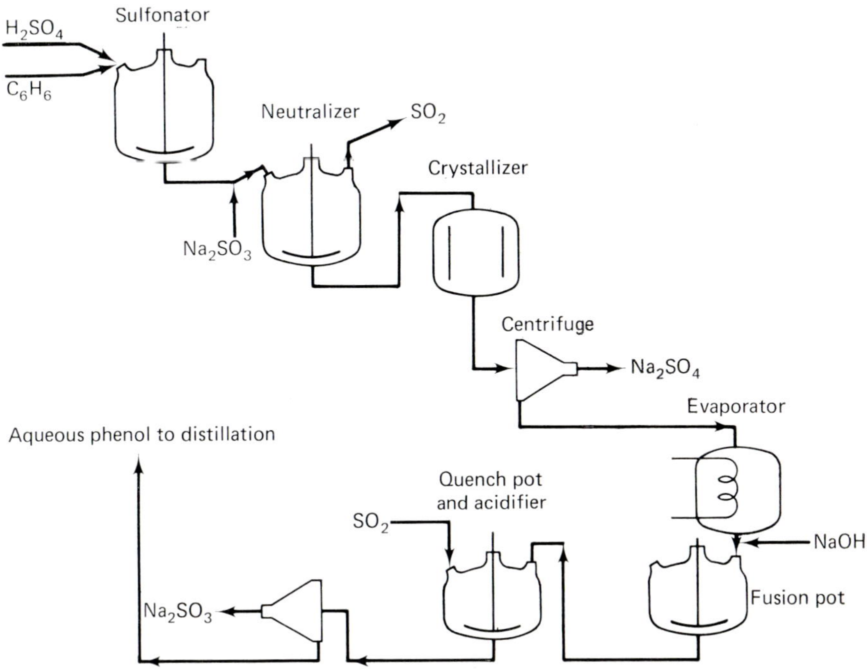

FIG. 8.3 *Monsanto Process*[29]

In areas where this sodium sulphate cannot be disposed of by sale or dumping, methods have to be found by way of which the sulphuric acid can be completely converted to benzene sulphonic acid. This can be accomplished by the Tyrer process[35] or similar schemes[36] in which excess benzene vapours remove the reaction product—water—from the sulphuric-sulphonic acid mixture and thereby push the equilibrium to completion. This can be accomplished readily in a counter-current distillation column of acid-proof ceramic. Such equipment imposes a much higher investment burden on the plant.

(2) Following sulphonation, the gross product is neutralised with caustic soda or sodium sulphite:

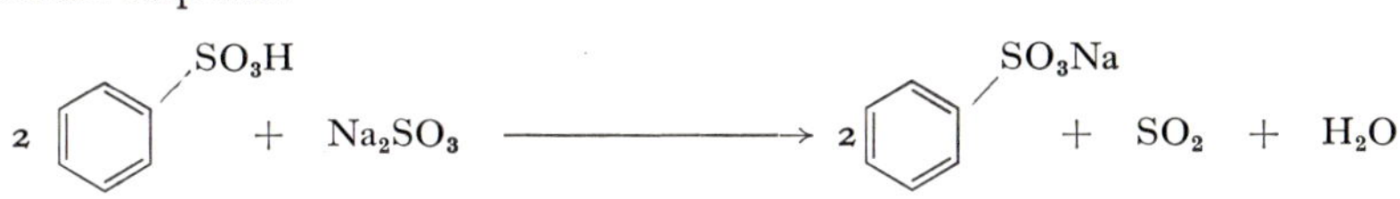

$$H_2SO_4 + Na_2SO_3 \longrightarrow Na_2SO_4 + SO_2 + H_2O$$

In the Monsanto plant[29] the acid mixture is neutralised with sodium sulphite and caustic soda to yield the sodium sulphate and sulphur dioxide. The latter is re-used to spring phenol from the sodium phenolate liquors or to be reoxidised to sulphuric acid. The resulting mixture of sodium sulphate and sodium benzene sulphonate is separated by crystallisation and the less soluble sodium sulphate dried and sold. The aqueous sodium benzene sulphonate solution is evaporated and the resulting solid dried to a low moisture level. Alternatively, German[28] practice used first sodium sulphite and completed the neutralisation with calcium carbonate.

(3) Caustic fusion is then employed to convert the sodium benzene sulphonate into phenol.

The solid sodium benzene sulphonate is added slowly to essential anhydrous, molten caustic at 300°C in a stirred and heated steel vessel. The reaction appears to be very rapid.

Heating to 330–340° completes the reaction. The molten mass is slowly dumped into boiling water. The heat of this reaction is removed as steam. The resulting solution is cooled and the sodium sulphite, which crystallises from the melt, is filtered off and re-used in the acid neutralisation step. Alternatively the sodium sulphite may be sold to the pulp and paper industries.

(4) The phenol is sprung from the sodium hydroxide-sodium phenolate solution. This can be accomplished with carbon dioxide, the German practice,[28] or with sulphur dioxide, Monsanto.[29] The by-product sodium sulphite or sodium carbonate is re-used in the neutralisation step following the sulphonation reaction.

This aqueous phenol solution is then subjected to distillation to remove water and to recover a substantially dry product which has a 1° boiling range and a freezing point of 40·5° (about 0·07 per cent water).

The stoichiometry of the reactions (net) indicates the consumption of at least 1 mole of sulphuric acid (probably almost 2 in practice), 1 mole of benzene, and 4 moles of sodium hydroxide to give theoretically 1 mole phenol, in practice 75–80 per cent, and 1 mole each of sodium sulphite and sodium sulphate. If these inorganic salts can be sold profitably then this synthesis scheme is, on the surface, a viable one. On the other hand, this sulphonation-caustic fusion process suffers from the disadvantage of producing corrosion, the problem of handling simultaneously gases, liquids and solids and also has a high pollution rating (aqueous phenol, sodium sulphite–sodium sulphate, sulphur dioxide, etc.).

8.6.2. The Raschig Process

The basic features of this process were developed at the works of Dr. F. Raschig G M B H, Ludwigshaven, Germany in 1929–35 by Dr. Walter Prahl.[37] This process was licensed to the Durez Company and a commercial plant (30 million lb/year) began operations in 1940. This unit was shut down recently (1970). In Germany, a fairly large pilot plant was operated until about 1942 and then shut down, largely because of uneconomical operating experience and a number of difficult technical problems. This Ludwigshaven plant was totally destroyed by bombing in 1944.

Durez, now a subsidiary of the Occidental Petroleum Company, operates an additional Raschig phenol plant in South Shore, Kentucky (60 million lb/year) and a smaller plant in Argentina (10–12 million lb/year). The Union Carbide Corporation has a large Raschig unit (110 million lb/year) in Marietta, Ohio which will be shut down in 1971.

It is unlikely that further Raschig phenol plants will be constructed, principally because of the relatively high capital investment required and the high utility load (steam and gas or oil). Additionally, corrosion is a serious problem and results in substantial and recurring maintenance charges. Oddly enough, corrosion is not severe in the areas where superheated hydrochloric acid is handled, but much corrosion is experienced in areas which handle acid below its dew point. Currently teflon-lined steel pipe (or glass-ceramic piping) is adequate. Relative to the cumene process this Raschig phenol synthesis—attractive as it appears on paper—is no longer competitive.

This process involves two basic chemical operations. In the first one ('First Stage') benzene is chlorinated with air and hydrochloric acid in the presence of a copper-on-alumina catalyst at 275°:

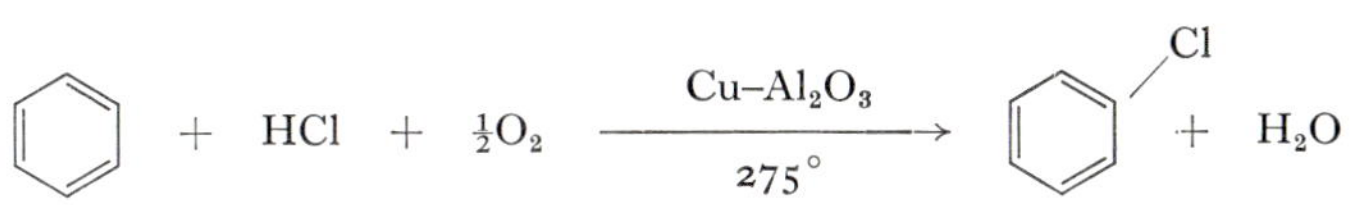

Basically, this scheme is the Deacon process[38] for the oxidation of hydrogen chloride to chlorine. Benzene intercepts the chlorine in this 'First Stage' to produce chlorobenzene. In practice a large excess of benzene is used in order to suppress the formation of polychlorinated benzenes. Even so, about 5–8 per cent dichlorobenzenes are produced. All hydrogen chloride is used up. The *p*-dichlorobenzene represents a saleable item after separation from the other di- and polychlorinated aromatics. These polychlorinated benzenes represent a net loss of chlorine values from the process. Consequently make-up chlorine or hydrogen chloride have to be employed.

The resulting gas mixture (water, benzene, chlorobenzenes, etc.) is condensed and fractionated. The excess benzene is recycled, the water discarded, the polychlorinated residues worked up for *p*-dichlorobenzene while the principal product—chlorobenzene is then used in the 'Second Stage' where it is hydrolysed to phenol. This involves contacting a mixture of chlorobenzene vapour and steam which is heated to 450–500°C, with a copper promoted calcium phosphate apatite catalyst[39] to cause hydrolysis to phenol and hydrochloric acid to occur:

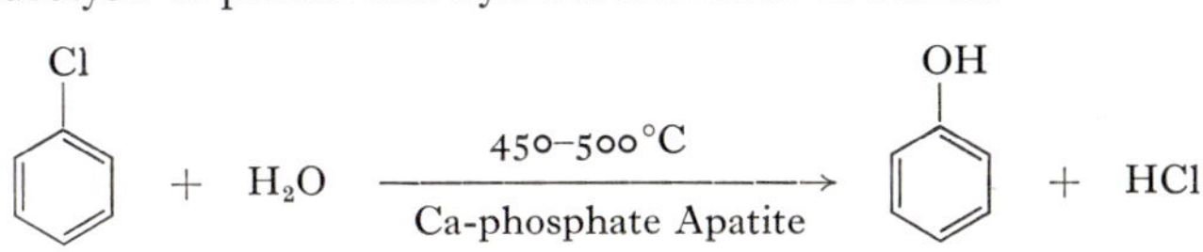

The yield per pass is seldom in excess of 12 per cent; hence a great deal of recycling and a heavy utility load is required for this step. Additionally, a number of severe

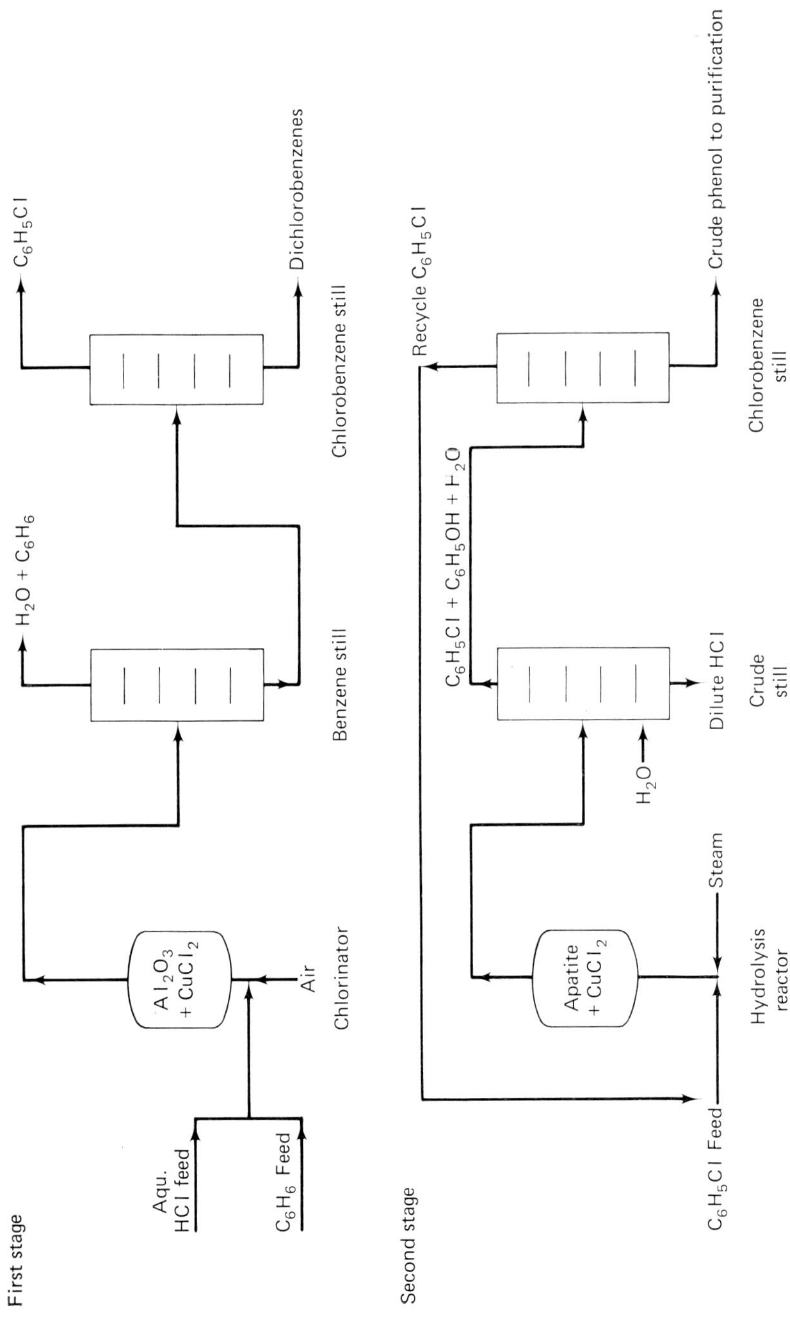

FIG. 8.4 *Flowsheet of the Raschig Phenol Process*

operational problems exist. The catalyst is frequently of variable quality and decreases in activity with use. This is compensated by an increase in the reaction temperature to the limit of 500–525 °C. The copper co-catalyst concentration has to be maintained at the 0·5–1 per cent level by frequent additions of metal powder or a copper compound. The reaction itself produces a carbon coating on the catalyst surface which serves to deactivate the catalyst. The carbon is periodically (every 4–8 hr) burned off with air.

The high operating temperatures, the carbon formation, the variable catalyst quality and life, and the high heat requirements make this hydrolysis step a difficult operation. Despite this, the over-all conversion of chlorobenzene to phenol is high, 90–93 per cent. With the exception of a small quantity of tars and benzene, virtually no by-products are obtained.

The nature[39] of this hydrolysis step is an interesting one. The first step appears to be a ligand exchange between the chlorobenzene molecule and the apatite:

This results in the chloro-apatite and phenol. In the second part of this reaction, the steam hydrolyses the chloro-apatite to the hydroxyl apatite.

$$2H_2O + Ca_{10}Cl_2(PO_4)_6 \longrightarrow Ca_{10}(OH)_2(PO_4)_6 + 2HCl$$

The copper co-catalyst functions by holding the chlorobenzene on the catalyst surface, so that ligand exchange can take place more readily. The displacement reaction itself does not seem to involve a benzyne[40, 41] intermediate but appears to be a straightforward nucleophilic displacement on an aromatic ring:

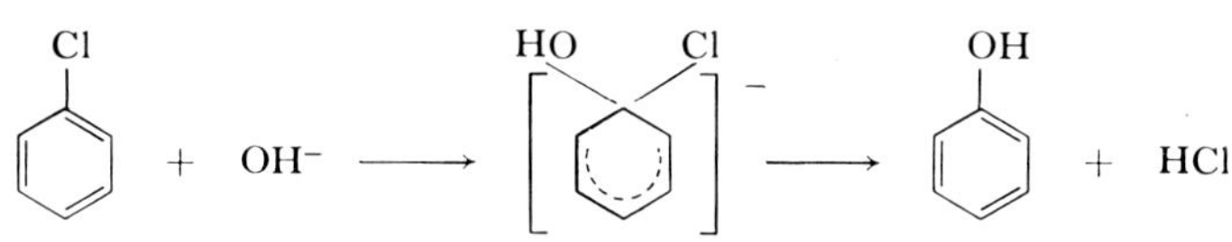

The entering hydroxyl group is bound to the same carbon atom which formerly held the chlorine atom.

The exit gases from the hydrolysis reactors are cooled, condensed and separated into their components. The aqueous hydrochloric acid and small quantities of benzene are recycled back to the first stage. Excess chlorobenzene is recycled. The phenol is subject to various purification steps (distillation) and sent to storage.

Overall, the efficiency of conversion of benzene to phenol is between 90–93 per cent and is one of the highest of the commercial phenol syntheses. Neglecting chlorine make-up and polychlorinated benzene losses, the net reaction is an indirect oxidation of benzene to phenol using air:

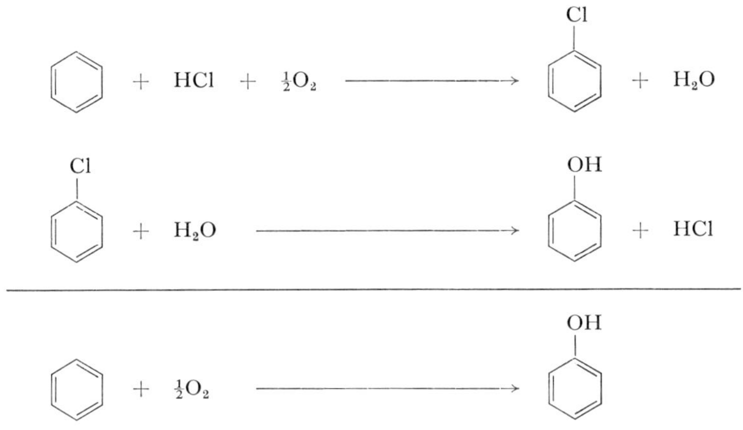

8.6.3. Chlorobenzene Hydrolysis with Caustic Soda[27]

The hydrolysis of chlorobenzene with aqueous caustic is on paper one of the simplest phenol syntheses, yet in fact it represents severe operating problems due to the extreme reaction conditions[42, 43] which are employed. The chlorobenzene can

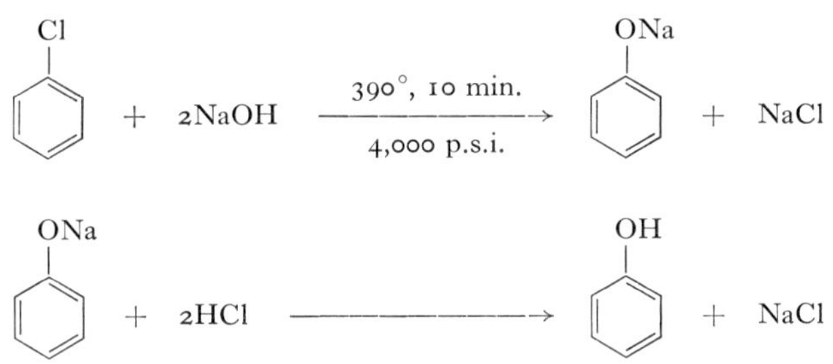

be made via a Raschig First Stage unit (where there is excess hydrochloric acid available) or by way of the usual chlorination of benzene. The hydrochloric acid by-product from the chlorination is used to acidify the sodium phenate obtained later in the hydrolysis step.

In the late 1920's process development work was carried out by the Dow Chemical Company and the I. G. Farben consortium in Germany which led to the construction of substantial synthetic phenol capacity by this route.

In practice the chlorobenzene is mixed with about 2·2 mole of aqueous sodium hydroxide (per mole of chlorobenzene) as well as the recycled diphenylether. This mixture is preheated and injected into long, tubular reactors made of nickel-lined steel.* The reaction takes place at 390°C, 4,000 p.s.i. and requires a residence time of about 10 minutes.

The conversion of chlorobenzene to phenol is about 85 per cent and to diphenyl ether about 10 per cent. Since there is no substantial market for this chemical (other

* Other materials of construction do not appear to be useful, chemical grade titanium might be servicable under these conditions.

than as a heat transfer fluid—Dowtherm—and some incidental chemical uses) most of the ether is recycled into the process. The discovery that the diphenyl ether is cleaved by aqueous caustic is a very important one and this helped to make this process an economic reality.

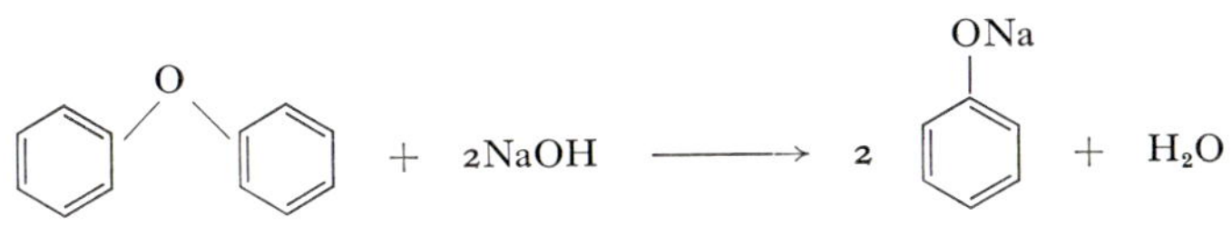

The crude reaction mixture—sodium phenate, sodium chloride, excess caustic water, diphenyl ether, phenyl phenols and a small amount of unreacted chlorobenzene, is extracted with an organic solvent (isopropyl ether or benzene) to remove the diphenyl ether, and separated by distillation. The aqueous liquors are acidified with hydrochloric acid and the phenol is separated by distillation. The phenyl phenol is also recovered; it is a valuable chemical intermediate for certain coating resins.

The large Dow plant in Midland, Michigan recycles the aqueous sodium chloride to electrolytic cells where chlorine and sodium hydroxide are made from it. These are then reused in the phenol plant. In this way substantially all by-products (sodium chloride and diphenyl ether, principally) are reclaimed and recycled into the process. The net products are phenol, phenyl phenols and, if desired, diphenylether. The overall phenol yield is about 90 per cent (based on chlorobenzene with recycled diphenyl ether).

The chemistry of this caustic hydrolysis process is quite involved and can result in numerous by-products. Unlike the Raschig chlorobenzene hydrolysis (over calcium apatite with the copper co-catalyst and in the presence of steam) this reaction seems to proceed via a benzyne mechanism.[40, 41] The first step is probably an abstraction by the hydroxyl group of a proton ortho to the chlorine group,

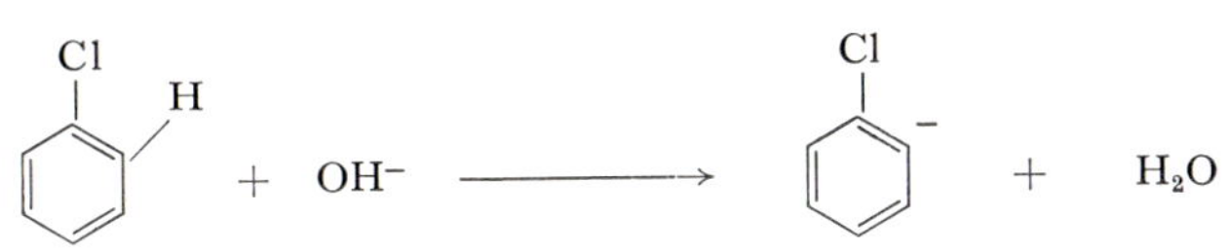

and, simultaneously, or shortly afterwards, the chlorine comes off. This results in a benzyne molecule, water and chloride ion,

This benzyne is a highly reactive, short-lived intermediate. Its electrophilic nature

causes it to react with any nucleophile in the neighbourhood, water and phenol are the principal ones:[44–46]

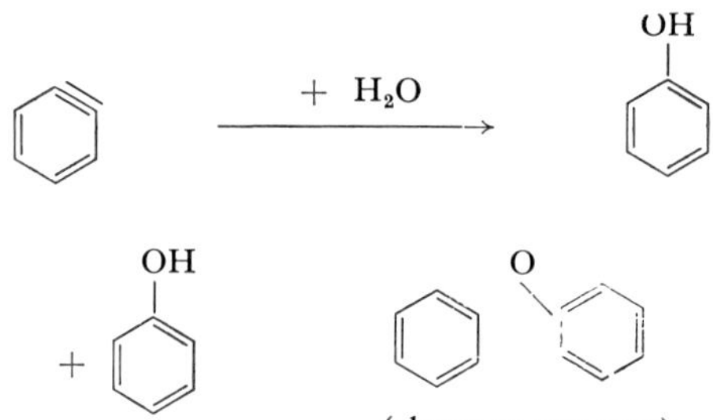

(about 10 per cent)

The phenyl phenols are probably the reaction products of the benzyne intermediate with various phenoxide resonance structures:

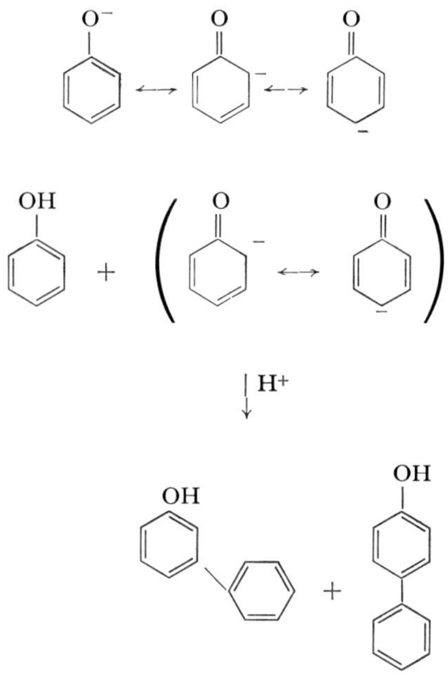

3·5 per cent   1·5 per cent

Other by-products also result from similar reactions (3–5 per cent polymeric residues).

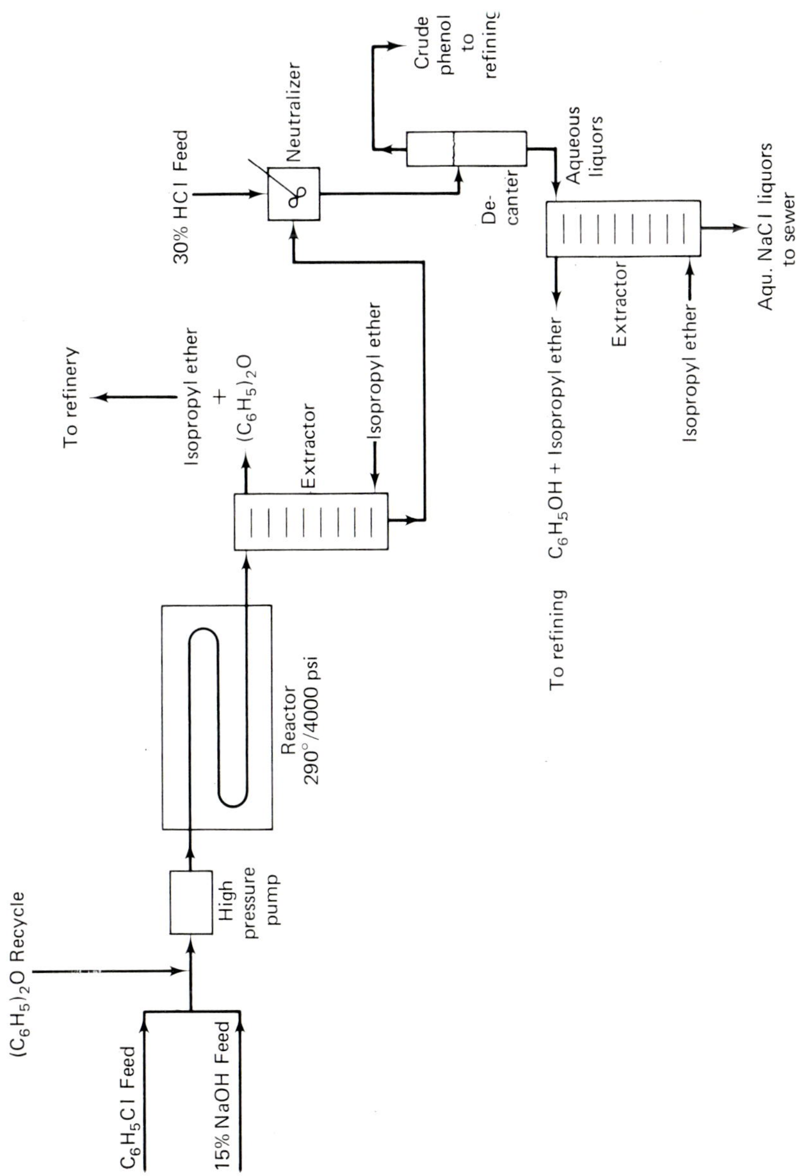

FIG. 8.5 *Flowsheet for Chlorobenzene—Caustic Hydrolysis Phenol Process*

8.6.4. Toluene Oxidation Process

This process[26, 27] was developed principally by the Dow Chemical Company as well as a number of other concerns. Only two plants which use this process have been built, both by the Dow Chemical Company. One is situated in Kalama, Washington[47] and the other is located in Germany (132 million lb/year capacity). The Kalama unit is relatively small (40–50 million lb/year) and has now been sold to former Dow employees.

This phenol synthesis employs toluene as a starting material and phenol is formed after two chemical steps. The first involves an oxidation of toluene to benzoic acid,

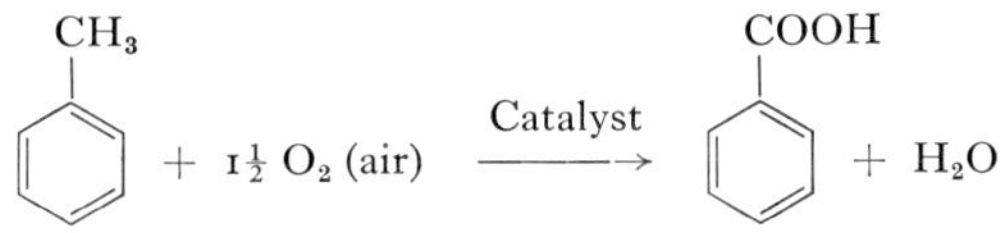

and the second a further oxidation to carbon dioxide and phenol,

The molecular weight of toluene is 92, compared with 78 for benzene (which is the starting material for all other phenol synthesis) places this synthesis at a decided disadvantage since the extra weight, the methyl group, is lost completely as carbon dioxide and water. Thus this process makes sense only when a mole of toluene has the same cost as a mole of benzene; e.g. the price of 1 lb toluene should be $78/92 = 0{\cdot}85$ times the price of 1 lb benzene. Currently this ratio is about 1·07. Also the two chemical steps are not quantitative as the above two equations might suggest. Numerous side reactions can subtract from the desired products, benzoic acid and phenol, and considerable care has to be exercised in the large scale execution of this phenol synthesis in order to maintain a reasonably high efficiency.

The first step is an air oxidation of toluene to benzoic acid in which cobalt bromide is used as a catalyst. This catalyst is unique in its activity. Both cobalt and bromide ion have to be present in order to maximise the reaction rate and the specificity to benzoic acid. The reaction appears to be free-radical in nature and involve benzylhydroperoxide, bromine radicals and a Co^{+2}/Co^{+3} complex.[48] It is carried out at 140–200°C, under oxygen or air pressure (100–400 p.s.i.) and in the presence of a solvent. The solvent may be excess product, i.e. benzoic acid, or toluene, water, etc. The heat of the reaction may be removed by employing a large recycle stream of product(s). The cobalt bromide–water combination can lead to corrosion unless proper materials of construction are chosen. The reaction appears to be 90 per cent specific; the by-products are methyl biphenyls, biphenyl, benzyl alcohol, benzaldehyde, and possibly the corresponding esters, etc. Purification of such a complex mixture is difficult—it can be accomplished by distillation or by a crystallisation of the benzoic acid from water.

The next step in this synthesis is, from a chemical standpoint, a very interesting one. Benzoic acid is reacted, in the presence of cupric ion, air and steam, to yield phenol and carbon dioxide. The actual chemical steps appear to be the following.[49] The benzoic acid first reacts with the cupric ion to form a salt,

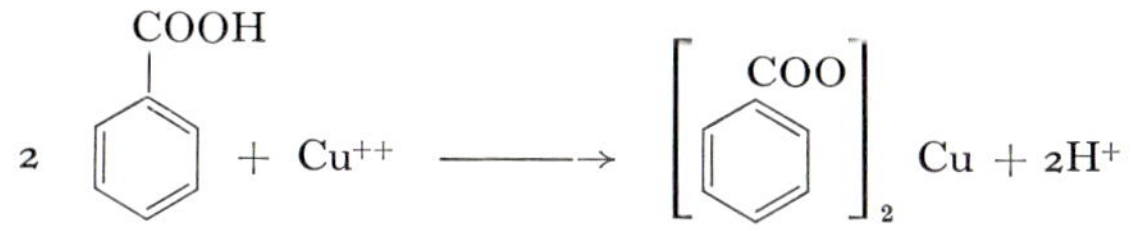

This decomposes to phenyl benzoate, carbon dioxide, and cuprous benzoate via a cyclic transition state which shows the incoming phenol-oxygen to appear in the positions ortho to the departing carboxylate group. Indeed, *p*-toluic acid yields *m*-cresol and 1-naphthoic acid exclusively 2-naphthol.[49]

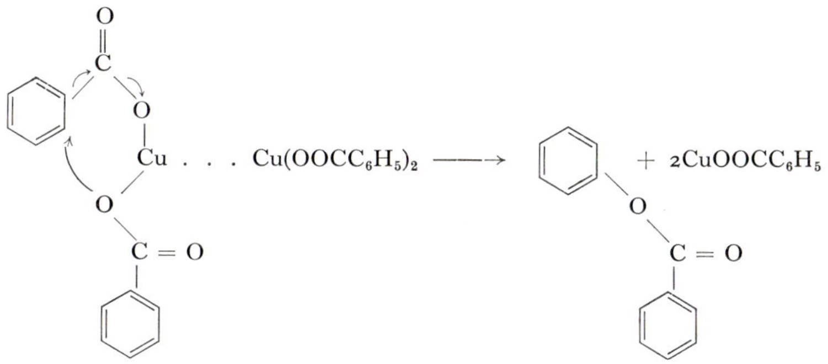

Subsequently the steam, which is admitted simultaneously with the air and benzoic acid, hydrolyses the ester to phenol and benzoic acid:

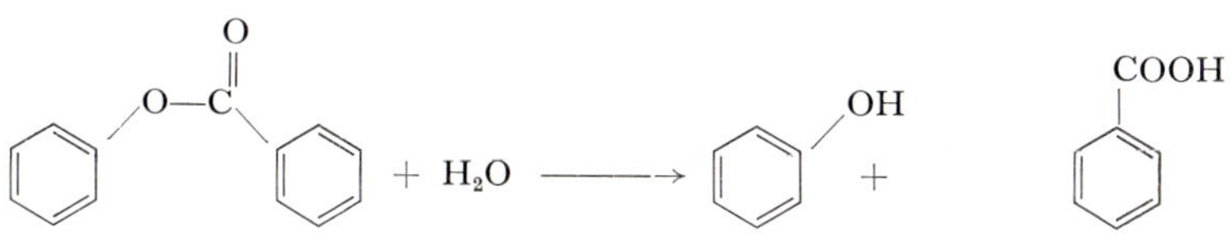

The air simultaneously oxidises the Cu^{+} back to Cu^{++}, and the benzoic acid in turn reacts to form the cupric benzoate, etc.

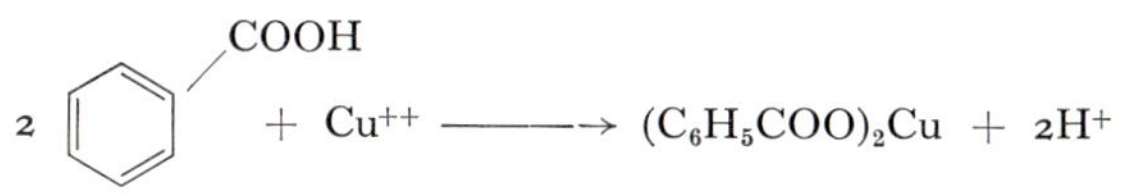

The decomposition of the cupric benzoate appears to be a thermal reaction which takes place between 220 and 280°C. Magnesium benzoate is reported to improve the yield, although this may also lead to by-products.[50]

In practice this entire cycle of reactions is carried out in one step by blowing air and steam through a molten mixture of benzoic acid and 1–5 per cent cupric ion; both air and steam are kept in considerable excess (to 2·5 and 50–70 times, respectively). The phenol so produced should be in contact with the reacting mass as briefly as possible, since phenol, particularly in the presence of cupric ion, oxidises rapidly to polymeric materials; these appear as 'tars' in this reaction. The specificity of the reaction is upwards of 90 per cent if the reaction variables are optimised with respect to product yield. By-products are tars, phenyl benzoate, biphenyl, biphenyl ether, etc. Materials of construction can be Hastelloy-C, titanium, or Hastelloy-B. The reaction specificity appears to a certain extent to depend upon the nature of the reactor walls.

Purification of the product is accomplished by a series of distillations. Water is removed with toluene and the phenol separated from the high boilers and tars by distillation. Final phenol purification is an additional distillation.

In view of the complexity of the reactions, the need for corrosion resistant materials of construction for both chemical steps, the irrevocable loss of the methyl group (as carbon dioxide and water), and a difficult purification cycle for both product and intermediate, it is not surprising that only two plants have been built and no further ones appear in the offing.

8.6.5. Cumene Oxidation

The cumene (isopropylbenzene) route to phenol[26, 27] is at present the favoured one. Virtually all new phenol capacity employs this synthesis. There are several reasons for this popularity. The process is basically non-corrosive, non-polluting, has few by-products, employs no exotic techniques or high pressure and yields two readily saleable chemicals phenol and acetone together with a small amount of by-product alpha-methyl styrene. These are obtained in a high degree of purity and at very reasonable costs. Also, the capital investment is relatively low, particularly since two major chemicals are produced on one plant site.

A potential drawback of this phenol synthesis is of course the co-production of acetone (0·6 lb per lb phenol) if the sale price of this co-product is less than its production cost. Happily, phenol and acetone have enjoyed about the same growth rates in the past two decades so that a really serious surplus has not developed.

There are two similar versions of this cumene-phenol process available for licensing. One is by the Allied Chemical Company and the other by a Hercules-Imperial Chemicals Industries consortium. Allied built its own plants, the Hercules-ICI plants are built by the Kellogg, Lummus, or Stone and Webster Corporations. There are marginal differences between these designs. After a careful study the Union Carbide Corporation in 1967 decided to use the Allied process. There are probably two or possibly three other versions available; a French one, a Russian design and a Japanese method. To what extent they infringe upon the US–English patents of Allied or Hercules-ICI is unknown.

8.6.5.1. Production of cumene The first step in the reaction is the production of cumene which is monoisopropyl benzene. This is relatively straightforward. Benzene and propylene are reacted in the vapour phase with an acid catalyst, e.g. phosphoric acid supported on pumice. A propane-propylene mixture can be used as the propane does not react and can readily be stripped from the reaction products.

Excess benzene is employed to suppress the formation of di-isopropyl benzenes. The unreacted benzene can readily be removed from the cumene by fractionation.

Mild steel equipment is satisfactory, the reaction taking place at about 250°C and 350 lb/sq. in. pressure.

Some 76·5 tons of benzene and 41·2 tons of propylene are converted to 100 tons of cumene, the yield being 94 per cent on the propylene. Small quantities of di-isopropyl benzenes and higher alkylated benzenes are produced, known as cumene bottoms and can be separated from the cumene by fractionation. It is possible to recover some propylene and benzene values from the bottoms.

The main reaction can be simply expressed by the equation:

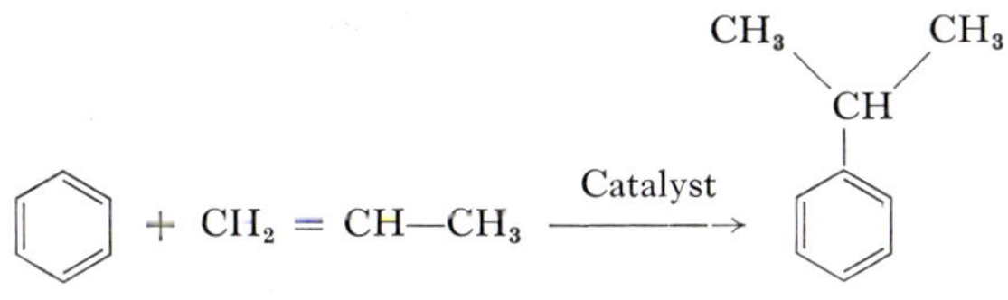

About two thousand million pounds of cumene were produced in the USA in 1970 at an approximate contract price of 3·8–4 cents per pound.

8.6.5.2. Peroxidation of Cumene The next stage in the production of phenol is the conversion of cumene to a hydroperoxide, the highly reactive alpha hydrogen of cumene being very sensitive to oxidation reactions.

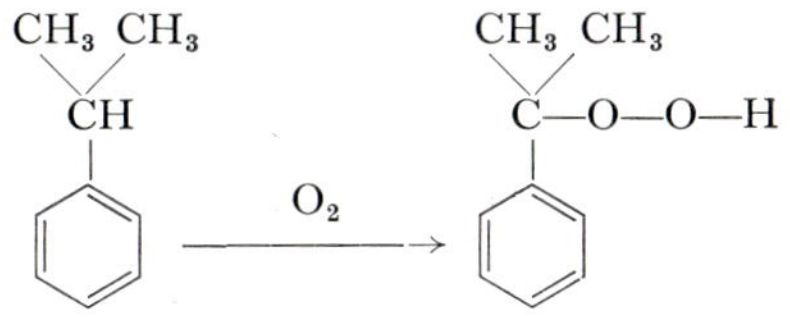

This reaction takes place at an acceptable rate in the 90—130°C temperature range. In practice the reaction is not pushed beyond 30 per cent conversion since otherwise undesirable side reations and premature peroxide decomposition leads to low cumene efficiencies. Catalysts and decomposition inhibitors may be added to the oxidising mass. These are usually mild bases (calcium carbonate, sodium carbonate, etc.) which inhibit acid catalysed decomposition of the hydroperoxide. Air, or air variously enriched in oxygen, can be employed. The former involves handling large inert gas volumes with possible cumene losses in the exit streams. This necessitates large hydrocarbon absorption units in order to trap these polluting off-gases. Air enriched in oxygen presents a greater safety problem and also an added cost factor but higher reaction rates can be obtained. The air or oxygen pressure is normally between 20 and 250 psi.

The oxidation is carried out in three to five reactors in series; the first reactor with fresh cumene feed is operated at the highest temperature (110°C), and the last reactor with the highest cumene hydroperoxide concentration has the lowest reaction temperature (90°C). Agitation can be supplied in a variety of ways; the chemical efficiency to hydroperoxide is between 85 and 95 per cent. The major by-product is dimethyl phenylcarbinol. Following the last reactor is a stripper column in which excess cumene is removed under vacuum to a concentration of about 85 per cent hydroperoxide in the stripper bottoms.

8.6.5.3. Decomposition of the Hydroperoxide The final step involves the decomposition of this hydroperoxide to phenol and acetone:

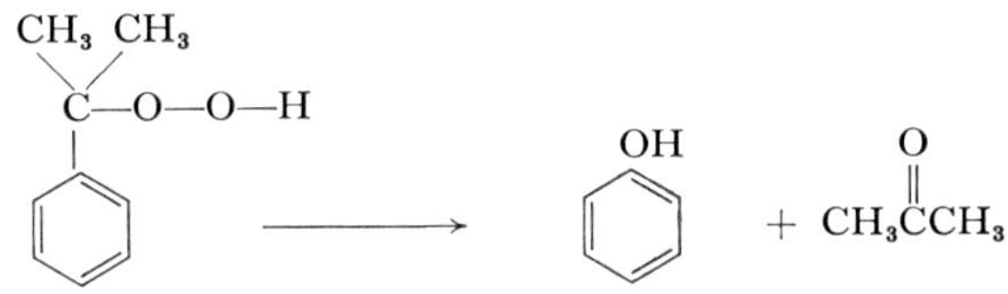

This reaction is catalysed by acids and involves, as the first step, the loss of water from the protonated hydroperoxide:

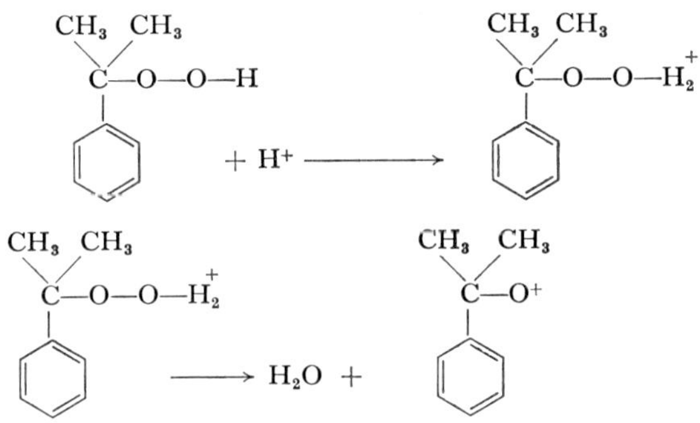

This loss of a water molecule is probably simultaneously accompanied by the migration of the phenyl group to the electro-positive oxygen atom[51]:

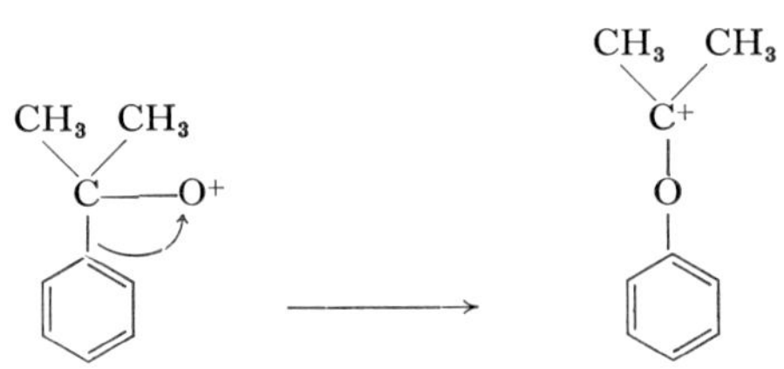

The resulting intermediate then reacts with water to generate phenol, acetone, and the proton acid:

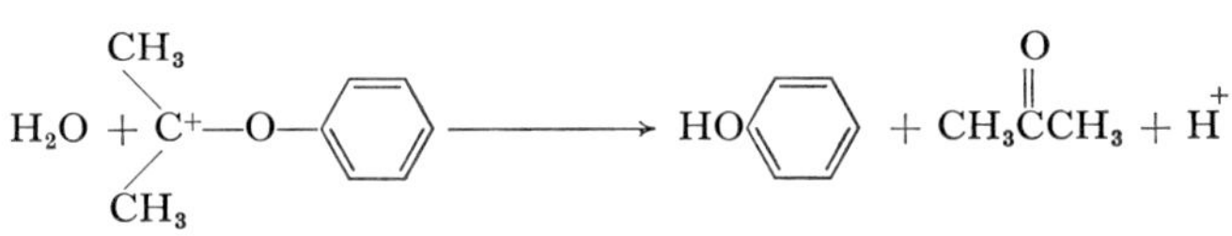

Various complications can arise at this point. The acetone and phenol, in the presence of acids, could react to form bisphenol-type products. The acetone itself could dimerise, etc. The hydroperoxide could decompose to the carbinol and dehydrate to form alpha-methyl styrene, which in turn can polymerise to low molecular weight oligomers.

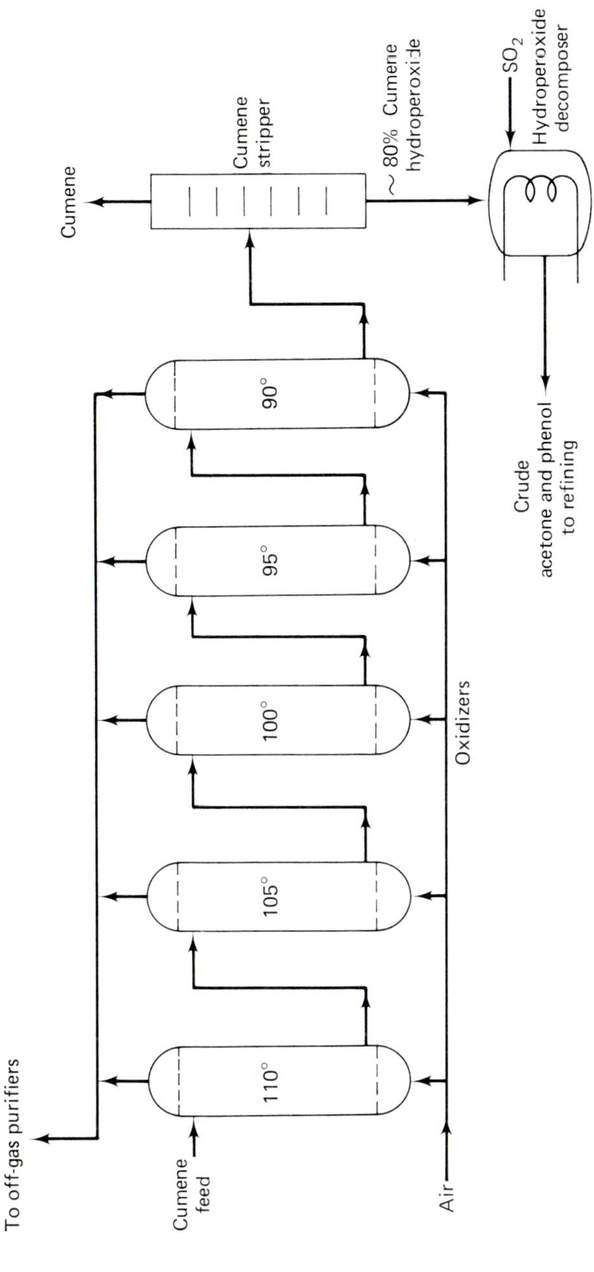

FIG. 8.6 *Flowsheet of Cumene—Phenol Oxidiser Section*

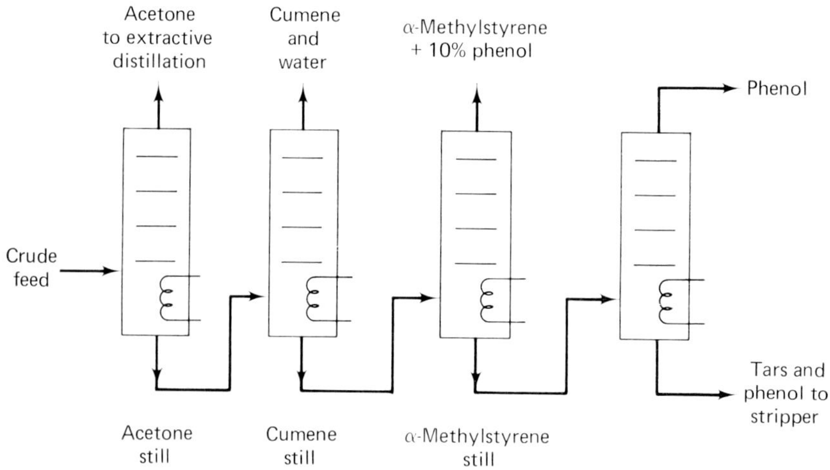

FIG. 8.7 *Acetone—Phenol Purification*

In the plant a wide variety of catalysts and reaction schemes are employed to carry out this second reaction under suitable conditions. It is a very rapid reaction, very exothermic and can readily lead to by-products. Strong mineral acids or acid ion-exchange resins are employed but they are a source of corrosion. Very small quantities of sulphur dioxide function quite efficiently. The decomposition reaction is carried out in the 60–80°C range and a large volume of cooled product is normally used to dilute the incoming hydroperoxide and serve as reaction solvent-moderator. Good agitation and heat transfer are very important. Simultaneously with this hydroperoxide decomposition, any dimethyl phenyl carbinol which is formed or present, dehydrates to alpha methyl styrene:

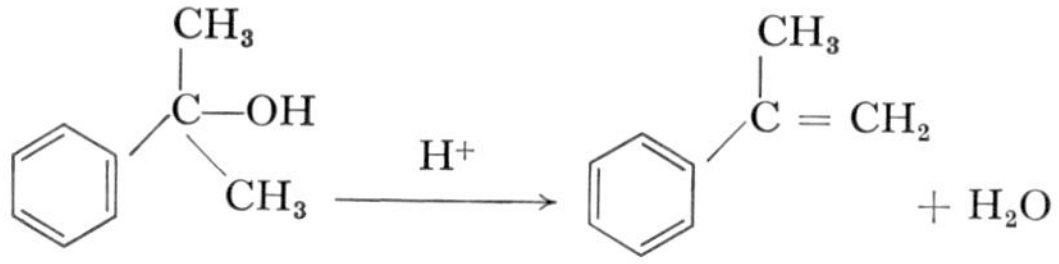

Purification of the crude reaction mixture is effected principally by distillation. First, the acetone is flashed from the cumene-phenol mixture and further purified by one or two distillations. The purification of the phenol is a complex matter and numerous schemes have been advanced to resolve this problem. Three to five distillations involving the separation of cumene (phenol-free for recycle), the alpha methyl styrene, some acetophenone, and various tars are currently employed. Side streams are further treated (by extraction) so that virtually all phenol or by-product values are recovered.

This cumene to phenol synthesis has a 92–3 per cent efficiency to phenol, 91–3 per cent to acetone while about 2 per cent acetophenone and 6–8 per cent alpha-methyl styrene are formed as by-products. Both of these latter chemicals can also be sold.

8.6.6. Cyclohexane Oxidation

This phenol synthesis scheme, although it looks very attractive on paper, has not yet found practical application. A very considerable amount of research and development effort has gone into defining the operating conditions and economic variables of this process; the conclusions seem to be that this method is not (yet?) competitive with the cumene synthesis. This cyclohexane oxidation process could become attractive if acetone supply would be in considerable excess of useage or no substantial in-house consumption of acetone existed. At present there appear to be no operating cyclohexane-phenol plants.

The chemistry of this synthesis involves three steps. The first is hydrogenation of benzene to cyclohexane. But unless surplus hydrogen is available on the plant site, it is usually advantageous to purchase cyclohexane. The difference in the contract price of benzene and cyclohexane is very small. The reaction is as follows:

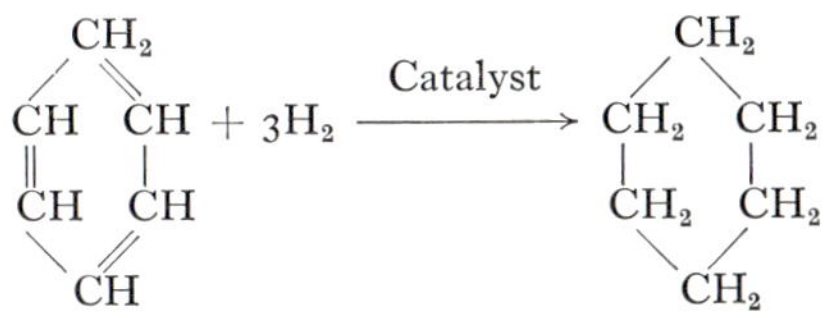

Normally, supported nickel-palladium or Raney-nickel catalysts are used. The temperatures are 200–250° and the hydrogen pressure is about 4–800 psi. The benzene must be scrupulously free of sulphur compounds (thiophene) as otherwise the catalyst will be poisoned rapidly. Alternatively, sulphided nickel or palladium catalysts can be used but the operating conditions are far more rigorous (about 450°C, 4,500 psi). The conversion to cyclohexane is quantitative.

The subsequent reaction step is an air oxidation of cyclohexane to, primarily, a mixture of cyclohexanone—cyclohexanol,

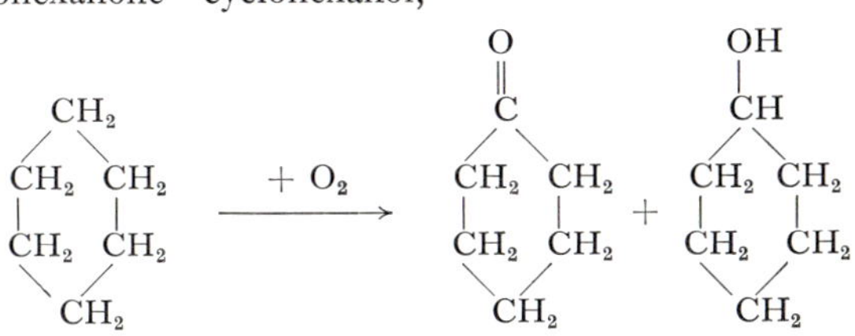

By-products are adipic, glutaric, and succinic acids as well as their esters and also carbon dioxide. Too rigorous oxidation increases by-products to unacceptable levels, while low levels of oxidation have to be paid for by excessive utility costs and investments in separation facilities.

The patent literature gives a wide variety of gas-liquid contacting devices, catalysts (metal salts), inhibitors (basic acid, etc.) which all, in one way or another increase the conversion to the ol/one mixture and suppress by-product formation. Generally, operating conditions are 150–200°C, variously enriched air (to 50 per cent O_2) and frequently elevated pressures (to 200 psi). The conversion per pass should be kept to the 4–6 per cent level so that a selectivity of about 80 per cent is not lost.

The purification of the product(s) of this step is cumbersome. The excess cyclohexane is stripped off and the acid by-products are extracted with caustic. This also hydrolyses the esters. Extraction separates the aqueous from the cyclohexane soluble products. The acids are liberated with mineral acid and can be further purified or burned.

The distilled ol/one stream is dehydrogenerated in the last step. This reaction proceeds:

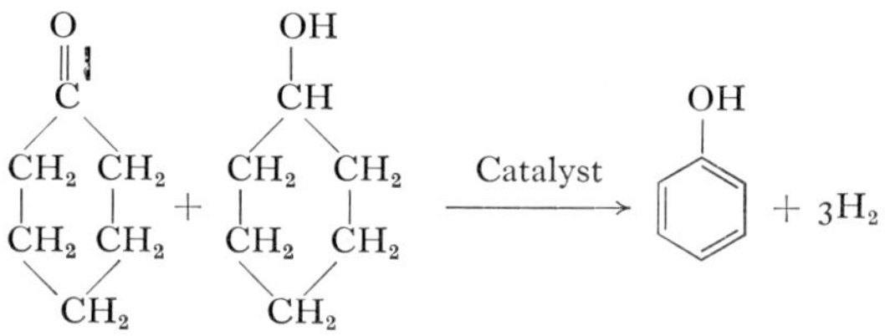

under forcing conditions (300–380°C, atmospheric or low pressure). At the low temperature range the conversion per pass is less than 80 per cent and 90±5 per cent in the higher temperature range. Conversely, at the lower temperature little or no benzene or cyclohexane are made; at the higher temperature the efficiency to phenol drops. The phenol product is then purified via a number of distillation steps.

Corrosion does not appear to be a problem in this process, nor are the operating conditions particularly rigorous. The major difficulty appears to be the cyclohexane oxidation step itself, the subsequent separation step and the by-product disposal.

(The production of cyclohexane and its oxidation is dealt with in more detail in Chapter 6.)

8.7 HALOPHENOLS by W. T. Reichle

Phenol undergoes halogenation with unusual ease at high or low temperatures, with or without solvent. In the case of chlorination, the products are mono to penta chlorophenols. These substances, or derivatives, have important biological effects. The 2,4-dichloro- and 2,4,5-trichlor-phenoxy acetic acids are amongst the best selective herbicides known; pentachlorophenol is the standard wood preservative. Other halophenol derivatives function as miticides, leather preservatives and bactericides.[52]

Only about 4·5 per cent of phenol produced goes into halophenols—but the products are of very considerable economic significance and utility.

8.7.1. Economics

The two principal uses of chlorinated phenols are in the area of herbicides and wood preservatives. In the United States in 1969 72 million lb of phenol was consumed in this area, 20 million lb was used for the production of pentachlorophenol, 32 million lb for 2,4-dichloro-phenoxy acetic acid and another 20 million lb for miscellaneous uses, much of it for the 2,4,5-trichloro-phenoxy acetic acid. Most of the pentachlorophenol (70 per cent) was used as wood preservative, 15 per cent was converted to the sodium salt and 15 per cent in miscellaneous uses. The Vietnam war has, due to the large utilisation of herbicides, boosted consumption of the 2,4-dichloro- and 2,4,5-trichloro-phenoxy- acetic acids significantly. In fact, between 1968 and late 1970 very little of the 2,4,5-trichloro-phenoxy acetic acid or derivatives could be purchased in the United States; only the 2,4-D acid derivatives were freely available to the civilian sector of the economy.

Recently, the 2,4,5-trichloro phenoxy acetic acid has been implicated as a possible cause of birth defects but at the time of writing (middle of 1971) the US Environmental Protection Agency has decided not to suspend the use of this herbicide.[53] Such doubts, as well as a termination of the Vietnam war will slow the growth of this chemical considerably. Table VIII.11 shows the production of the principal chlorinated phenols. For pentachlorophenol, the growth rate is very small 3–4 per cent/ year) while the 2,4-D and 2,4,5-T growth rates have been somewhat larger.

TABLE VIII.11

United States Production of Chlorinated Phenols (million lb.)

Year	C_6Cl_5OH	2,4,5—$C_6H_2Cl_3OH$*	2,4—$C_6H_3Cl_2OH$†
1968	48·6	28·1	—
7	44·2	25·3	77·1
6	43·2	17·9	68·2
5	40·0	13·0	63·3
4	36·9	13·7	53·7
3	33·9	12·4	46·3
2	38·8	11·8	43·0
1	—	10·9	43·4
1960	39·3	10·0	36·2
1956	31·4	6·7	28·8

* Prices $0·80—1·25/lb range.
† Prices $0·32—0·40/lb range.

8.7.2. Preparation

Two substantially different routes are employed in the synthesis of these chlorinated phenols. The first is the halogenation of phenol, or alkyl phenols, under varying conditions to produce the desired product(s). The alternate method involves hydrolysis of appropriate polyhalobenzenes.

2,4-Dichlorophenol is prepared by chlorination of phenol in a polar solvent (acetic acid or water) at low temperatures:

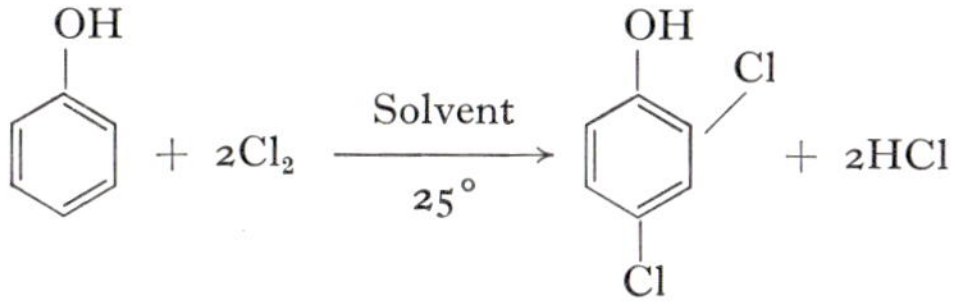

When this reaction is carried out in a non-polar solvent or with excess chlorine in a polar solvent, then the symmetric trichlorophenol results:

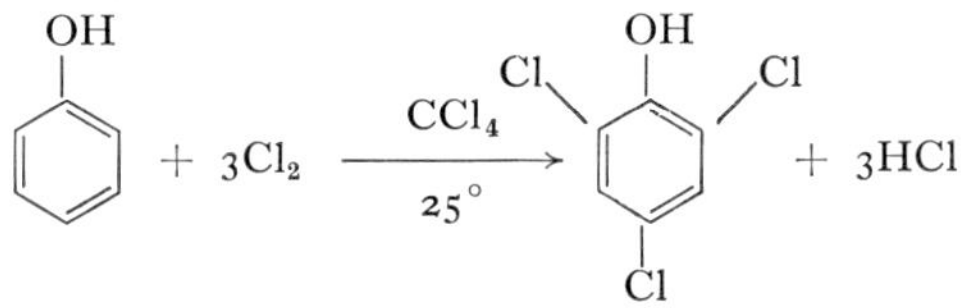

This reaction also proceeds with unusual ease with bromine and a method for derivatising and quantitatively assessing phenol is based on it.[12] This 2,4,6-tribromophenol is a stable compound with a sharp melting point (92·5–93·5°).

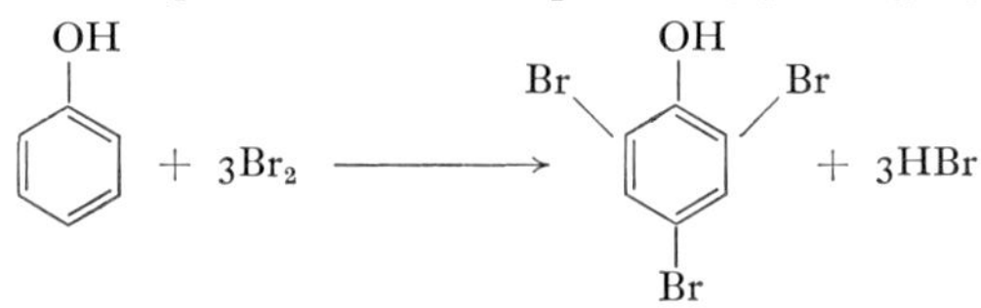

Pentachlorophenol is made similarly by exhaustively chlorinating phenol in the molten state at elevated temperatures and in the presence of a Friedel-Crafts catalysts ($FeCl_3$, $AlCl_3$, about 1 per cent).

The alternate hydrolytic method can also be used to prepare pentachlorophenol. When hexachlorobenzene is heated at 130–140° with sodium hydroxide, the sodium salt of pentachlorophenol results in good yield[54]:

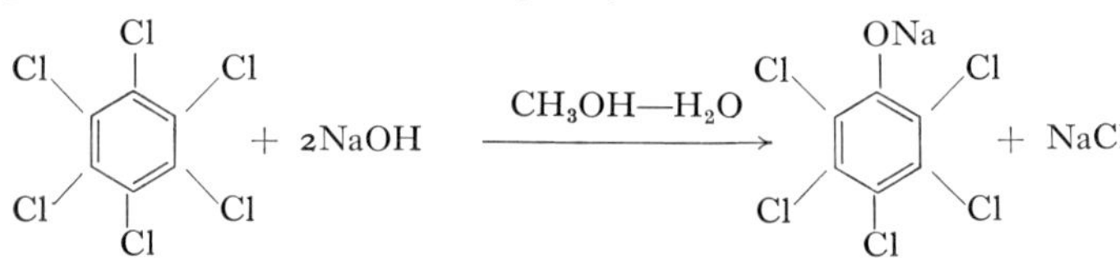

The 2,4,5-trichlorophenol has to be prepared by the hydrolytic route since chlorination of phenol does not yield this compound in acceptable quantity. To prepare 2,4,5-trichlorophenol, p-dichlorobenzene is chlorinated to the symmetrical tetrachlorobenzene,

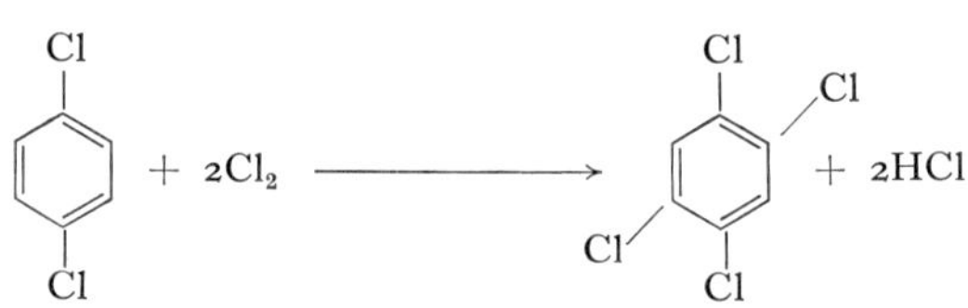

and this in turn is hydrolysed with aqueous caustic to yield the phenol:

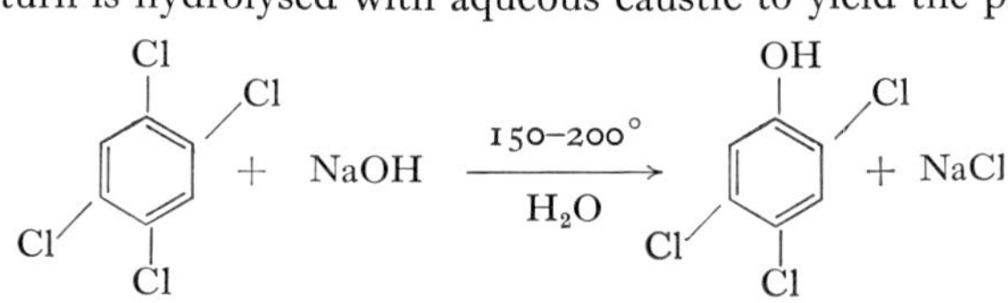

Only the desired isomer can result from this synthesis.

m-Chlorophenol must be prepared by indirect methods[55]. First nitrobenzene is chlorinated,

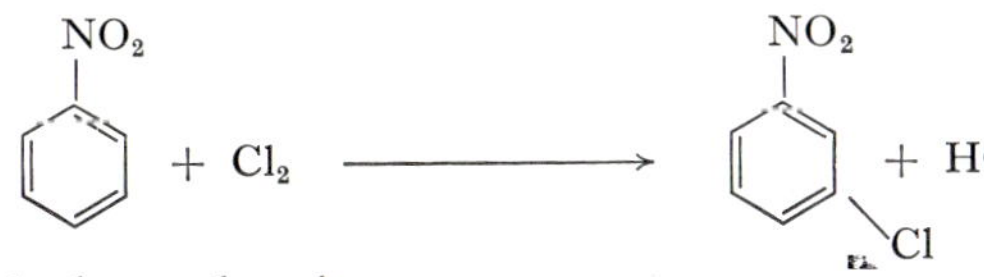

the nitro group is then reduced,

Iron or zinc and acid are the usual methods, although catalytic techniques could also be employed. The resulting amine is diazotised,

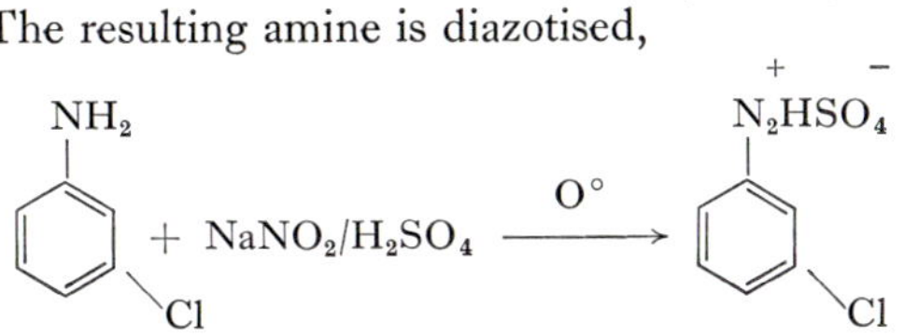

and the diazonium compound, on heating in dilute, aqueous sulphuric acid, decomposes to the *m*-chlorophenol,

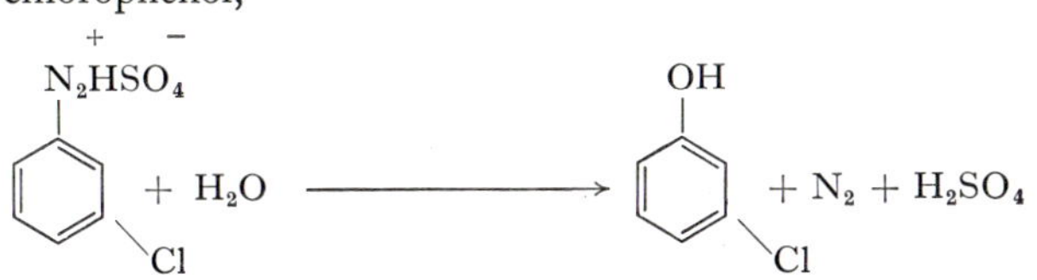

This is an involved and expensive synthesis and consequently the product enjoys a high price (about $5/lb) and has no appreciable consumption.

8.7.3.

TABLE VIII.12

Properties of Halophenols

Compound	*b.p.* (°C)	*m.p.* (°C)
Phenol	181·8	40·9
o-chlorophenol	175–6	8·7
m-chlorophenol	219	40–1
p-chlorophenol	215–7	32·8
2,4-dichlorophenol	210–1	43–4
2,4,5-trichlorophenol	245–6	68
Pentachlorophenol	309–10	190
o-bromophenol	194–5	5·6
m-bromophenol	236–7	32–3
p-bromophenol	238	63·5
2,4,6-tribromophenol	—	92·5–3·5
p-iodophenol	dec.	93–4
2,4,5-triodophenol	dec.	157–8

8.7.4. Physiological Effects

Mono to polychlorophenols have physiological characteristics[56,57] which are similar to those of the parent phenol molecule. On contact with the skin they cause dermatitis, the severity of which depends on the length of contact time. Oral ingestion in the rat resulted in convulsions and death with an apparent LD_{50} of about 0·6 g/Kg (in olive oil). This indicates that these compounds are relatively non-toxic.

Pentachlorophenol or the sodium salt, as vapours or dust, can cause severe eye irritations (at above 1 mg/m^3; below this level no effects are noted). They can also be absorbed through the skin. Deaths from skin absorption, injection or chronic exposure have been reported. Thorough washing of exposed skin and gastric lavage in case of ingestion can prevent serious injuries, provided treatment is carried out rapidly. There appears to be no specific antidote for chronic exposure. The LD_{50} is in the 50–200 mg/Kg range for the rat or rabbit (oral, subcutaneous). This does not therefore appear to be a very toxic substance.

8.7.5. Chemistry

Chlorophenols are stronger acids than phenol itself: K_a (25°) $= 1 \times 10^{-10}$ for phenol; $6{\cdot}6 \times 10^{-10}$ for p-chlorophenol and $1{\cdot}2 \times 10^{-5}$ for pentachlorophenol. Therefore, phenol can be separated from the chlorophenols by reacting the latter with aqueous sodium carbonate with which phenol does not react. These chlorinated phenols form salts with certain amines.

The replacement of a hydrogen atom in the phenol molecule by chlorine does not cause a marked loss in reactivity. Electrophilic substitution reactions, such as nitration, sulphonation, alkylation, Friedel-Crafts catalysed halogenations, etc., all proceed readily but at a somewhat slower rate than with the unsubstituted parent molecule.

The chlorine groups can be displaced by entering nucleophiles such as alkoxide, aryloxide or hydroxide and such reactions are accelerated by copper catalysts.[58, 59] The products are phenol-ethers or diphenols:

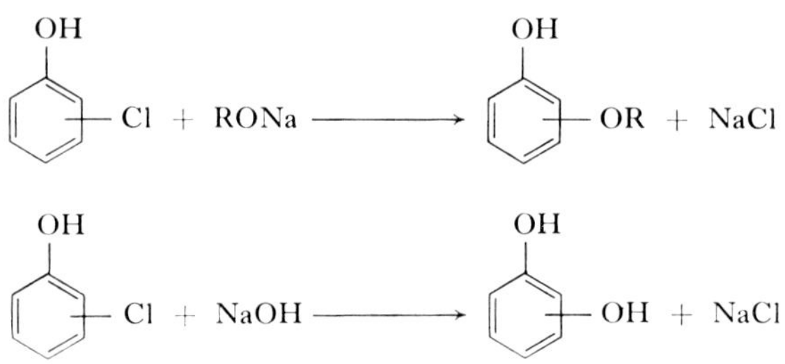

The phenolic hydroxyl also functions as a nucleophile in the anionic form. Thus 2,4-dichlorophenol sodium salt reacts with mono chloroacetic acid to yield the 2,4-dichloro-phenoxyacetic acid:

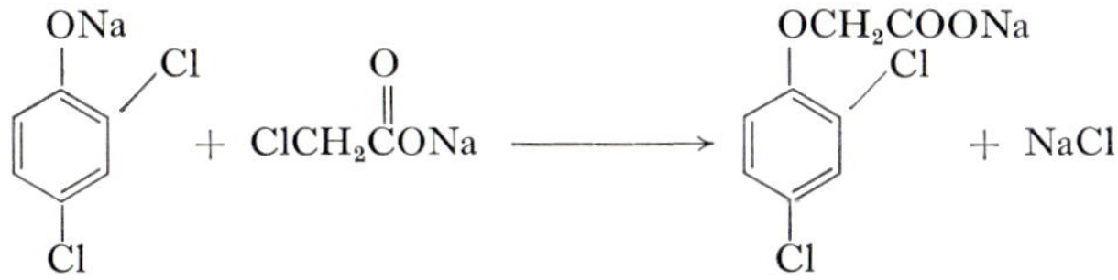

The chlorine group of the chloroacetic acid is a very reactive one due to the adjacent carboxyl group. The reaction proceeds readily and in high yield. The 2,4,5-trichlorophenoxyacetic acid is made similarly. Further reactions of these acids yield the desired esters or ethanolamine adducts which are potent selective herbicides.

These phenols also react to form phosphate esters,

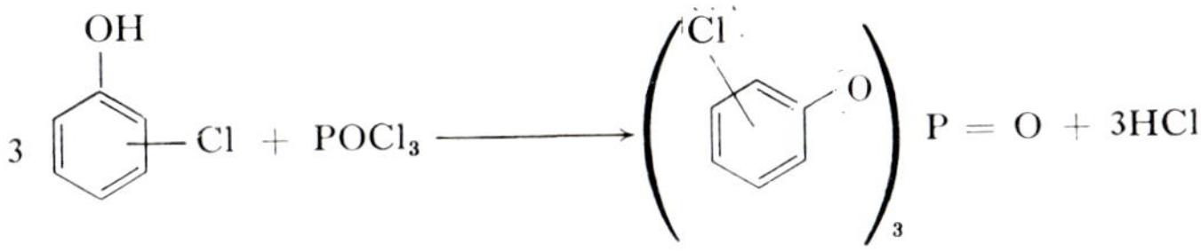

which are useful as plasticisers.

The chlorophenols also react with formaldehyde to yield chlorinated phenol-formaldehyde resins. These have somewhat greater flame retardancy than the unchlorinated resins. Reaction with ortho or para substituted chlorophenols is slow but the addition of small quantities of high meta cresol produces a reasonably reactive resin.[60]

8.7.6. Uses

The herbicide market is dominated by derivatives of 2,4-dichloro and 2,4,5-trichloro-phenoxyacetic acids.[61, 62] These acids are used as such, as the sodium salt or as the butyl, etc. esters or as the ethanolamine adducts. They are usually available as formulations, e.g. as concentrated emulsifiable oil solutions or dispersions, etc, which are then diluted with water for spray–application. Concentrated formulations are used for aerial spraying. The 2,4-D derivatives are relatively selective herbicides for broad-leafed weeds although in high concentration they prove to be lethal also to grasses and woody plants. The 2,4,5-T derivatives are far less specific and are consequently used in clearing operations—maintaining roadways and right-of-way. Frequently, applications of these herbicides includes the use of arsenicals to give particularly long lasting action.

Solutions of pentachlorophenol in hydrocarbons are used to impregnate wood in order to make it resistant to bacterial and fungal decay. Next to the crude coal tar creosote mixtures or chlorinated mixtures of these, pentachlorophenol has proved to be the standard preservative in the field. It has long and effective action, relatively low toxicity towards warm blooded animals, and is not overly phytotoxic.

Phenol and variously substituted phenols (alkyl, chloro, nitro, etc.) have very pronounced and broad antiseptic properties.[63, 64] The activity of the phenol is dependent upon the substituents attached to it, the mode of application and the particular bacterium in question. The effectiveness of the phenol is quantified via its phenol coefficient. The higher this coefficient the more effective the antibacterial action of the phenol. Several compounds in commercial use are *p*-chlorobenzylphenol, *o*-phenylphenol, *p*-tert.-amylphenol, etc. These are variously used as topical antiseptics. A chlorinated member of this series is hexachlorophene:

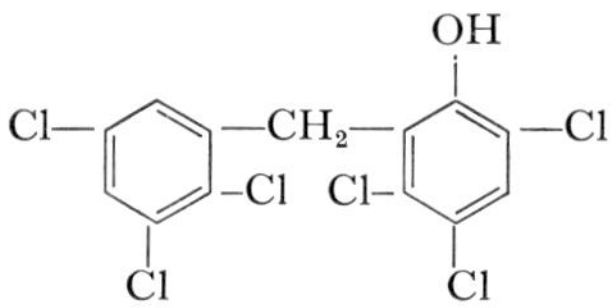

This is used extensively in soaps because of its very pronounced effect on bacterial skin flora. The poundage of phenol consumed in this area is, of course, very small.

8.8. ALKYLPHENOLS by W. T. Reichle

The total market for alkylphenols[65] is relatively small (less than 200 million lb in 1970 in the United States) in comparison with the total phenol market. The dollar volume though is quite high. Many of these alkylphenols function uniquely and in sometimes homeopathic quantities to upgrade specific product properties in a striking manner. For example, the addition of 0·005 per cent BHA (tert.-butyl-p-hydroxyanisole isomer mixture) to certain food stuffs (cereals, potato chips, roasted nuts, and fats in general) with or without co-antioxidants, increases the shelf life of these products from days to weeks. Needless to say, such alkyl phenols command a fairly good price (Food Grade BHA $4·75/lb) even though they are relatively easy to manufacture. From an economic standpoint, these phenol derivatives are the opposite of the parent compound phenol. They are made in relatively small volume, they have a high value due to their unique characteristics, they require a good deal of technology for proper application and for these reasons command a fairly high price.

There are four major areas which utilise alkyl phenols:

(1) As monomers in resin polymer applications (alkylphenol-formaldehyde polymers, bisphenols)
(2) Antioxidants
(3) Non-ionic surfactants
(4) Phosphate esters for plasticisers or lube-oil and gasoline additives.

The various aspects of the polymer applications are discussed elsewhere in this chapter. The following is a discussion of the economic aspects of the antioxidant, surfactant and phosphate ester derivatives of these alkylphenols.

The use of the cresols, the methylphenols, are not dealt with fully here as they will be covered in a subsequent book in more detail.

8.8.1. Preparative Methods

The very high reactivity of the phenol molecule towards electrophiles permits alkylation with olefins and alcohols using relatively mild acid catalysts.[66] These catalysts are normally sulphuric or phosphoric acid or the BF_3-phenol adduct. Strong acid ion exchange resins such as the sulphonated polystyrenes, acidic clays, silica-alumina catalysts, toluene sulphonic acid, etc., are all useful in this reaction. Alcohols and olefins react with these catalysts to form carbonium ions,

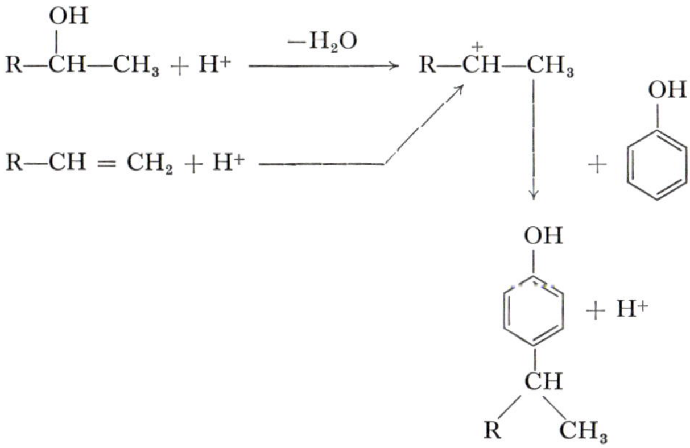

which in turn react with the phenol molecule to form ortho or para alkylphenols.

Di or trialkylation is readily carried out. For example, para-tert. butyl phenol is made on a large scale from *iso*-butylene and phenol using a sulphonated polystyrene catalyst:

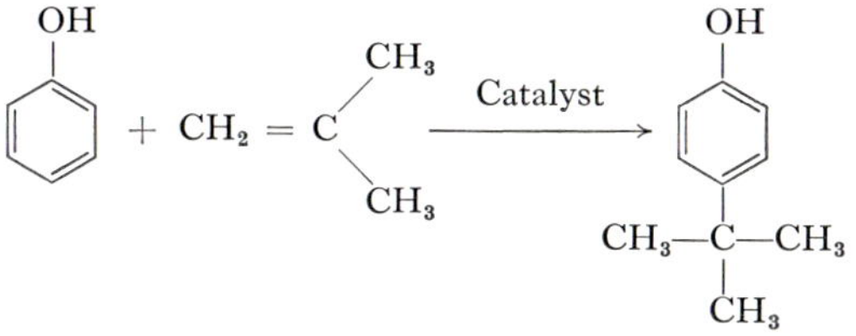

The relative amount of the ortho or para contents depends upon nature of the catalyst. In order to moderate the reaction the olefin can be diluted with inert gas (butane or *iso*-butane, etc.). Cresols react similarly with *iso*butylene to give mono- and di-tert. butyl cresols.[67, 68]

Dealkylation proceeds with ease in case of tert-butyl substituents. Thus, 2,2′6,6′-tetra (tert.-butyl) biphenol loses all alkyl groups on refluxing in phenol with 5 per cent aqueous sulphuric acid within about one hour.[69]

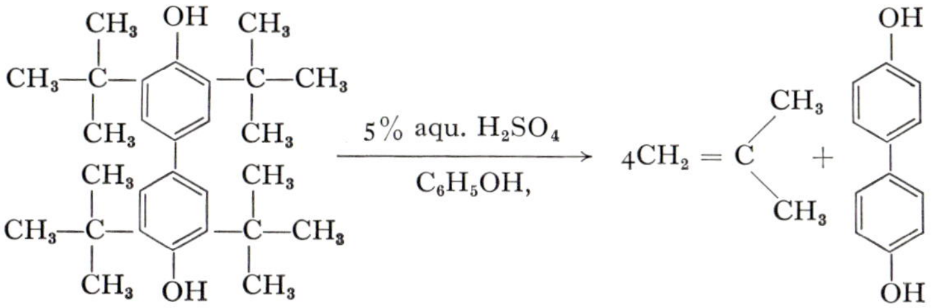

On heating alone the tert.-butyl groups are lost from the biphenol molecule in the 260–320° temperature range.

A surprisingly simple method for preparing ortho alkylphenols has been discovered in recent years. When a metal salt (Ca, Mg, Zn, preferably Al) of the phenol is used in catalytic quantities to catalyze the reaction of *iso*butylene and phenol (at about 150°C) principally ortho alkylated products are obtained[70]:

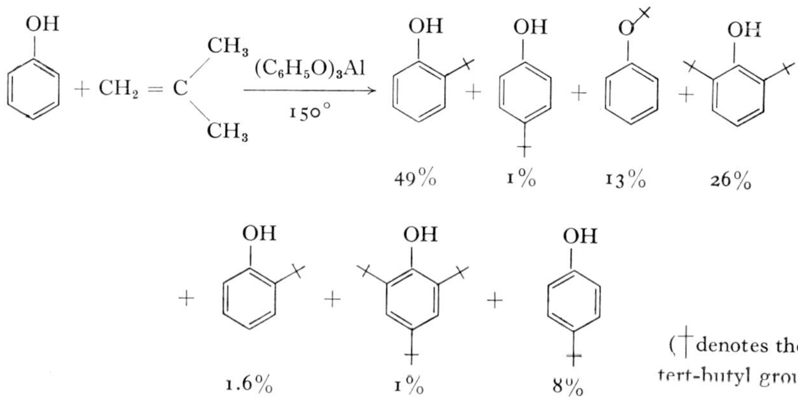

(† denotes the tert-butyl group)

In this reaction the tert.-butylphenyl ether is formed first:

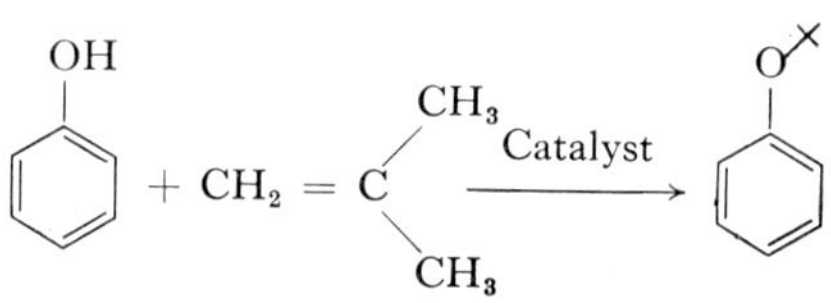

This compound, in the presence of the metal catalyst, rearranges to the ortho tert.-butylphenol, probably via an intramolecular reaction involving a cyclic transition state. The formation of meta or para tert.-butyl phenols results from side reactions. This technique for ortho alkylation is at present of considerable commercial importance since the products, such as 2,6-di-tert.-butylphenol, are important and widely employed as antioxidants.

TABLE VIII.13

Properties of Alkylphenols

RC_6H_4OH	*b.p.* (°C)	*m.p.* (°C)
o-CH_3	190·8	30·0
m-CH_3	202·2	11·5
p-CH_3	202·1	34·7
2,4-$(CH_3)_2$	210	27·0
2,5-$(CH_3)_2$	210	74·5
2,6-$(CH_3)_2$	212	—
2,6-Di-tert. butyl	253	37·0

Another alkylphenol is 2,6-xylenol. This substance is prepared from phenol or ortho-cresol and methanol, or dimethyl-ether in the presence of an alumina catalyst[71]:

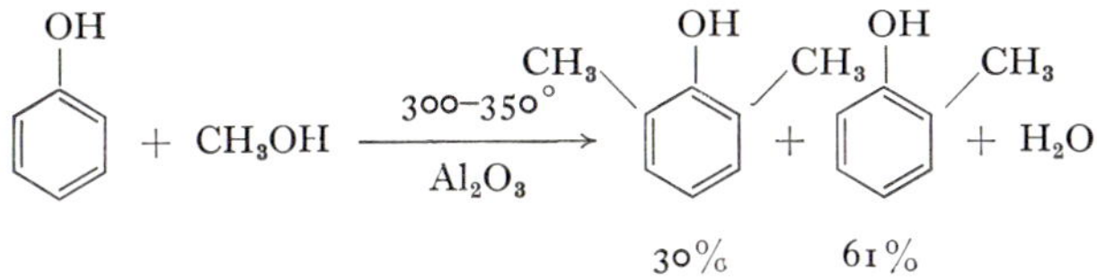

8.8.2. Chemistry of Alkylphenols

Like the parent molecule phenol, these alkylphenols are highly reactive towards electrophiles and oxidising agents but they are substantially unreactive towards nucleophiles. The reactivity of the aromatic ring towards electrophiles and oxidants is enhanced, beyond the reactivity of phenol itself, due to the electron-donating effect of the alkyl groups. Also, relative to phenol, the hydroxyl proton is slightly less acidic (pKa phenol 10·0, para cresol 10·17).

Alkylation or dealkylation of these alkylphenols has already been discussed above. The ease of dealkylation is tert. alkyl > sec. alkyl > prim. alkyl; in fact the dealkylation of cresols has to be carried out under drastic conditions.

These alkylphenols are very responsive towards halogenation, nitration and sulphonation. None of these reactions leads to products of large economic useage. Reactions with aldehydes or ketones leads to 'bisphenols':

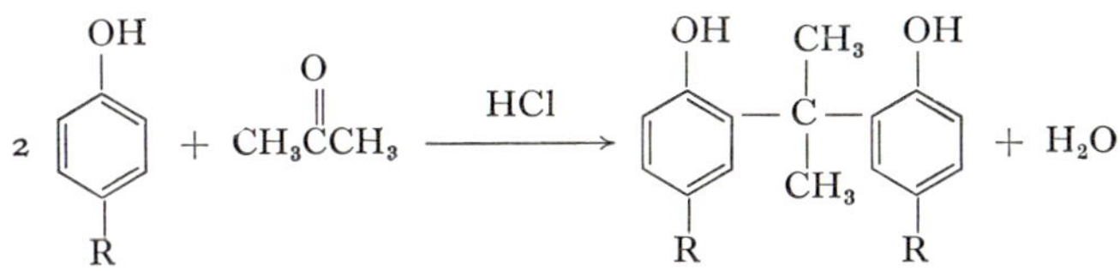

These products are discussed at length elsewhere in this chapter. Under other experimental conditions these alkylphenols will react with aldehydes to give methylol substituted phenols (base catalysed),

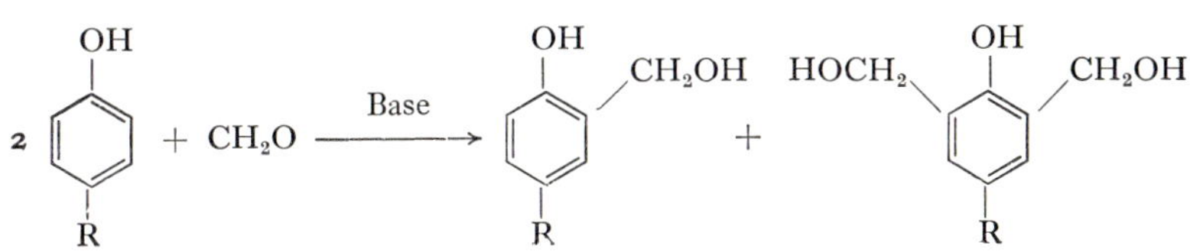

while under acid conditions bis and tris phenols are formed:—

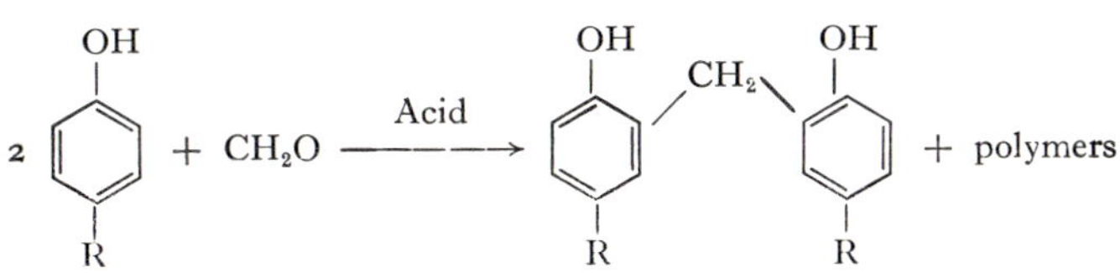

Sulphur dichloride (SCl_2) reacts with these phenols to give coupled products:

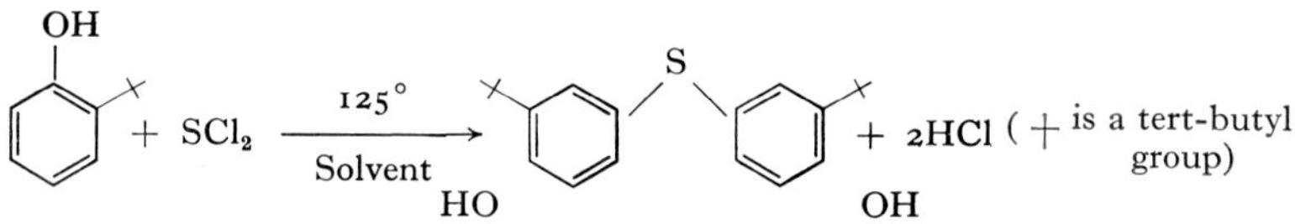

The reaction proceeds with ease at low temperatures to give products which have use as antioxidants and monomers.

The phenol hydroxyl group can also participate in a variety of reactions. From an economic standpoint the reaction with ethylene oxide is the most important one,

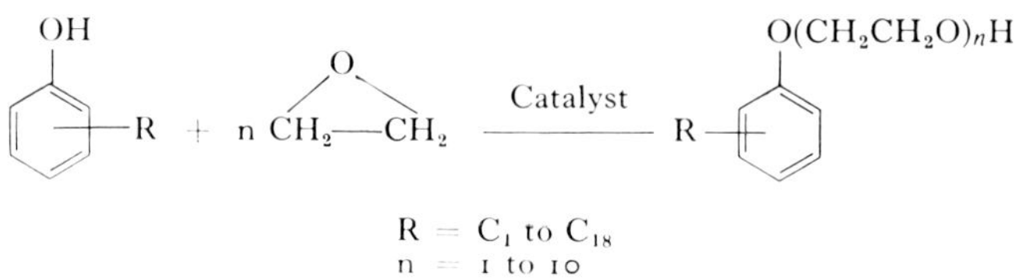

since it leads to the important non-ionic surfactants.

The oxidation of these phenols is an exceptionally interesting and important subject. Phenols readily lose a hydrogen atom, by radical abstraction, to form a 'phenoxy' radical. This can occur with a wide number of oxidising agents.

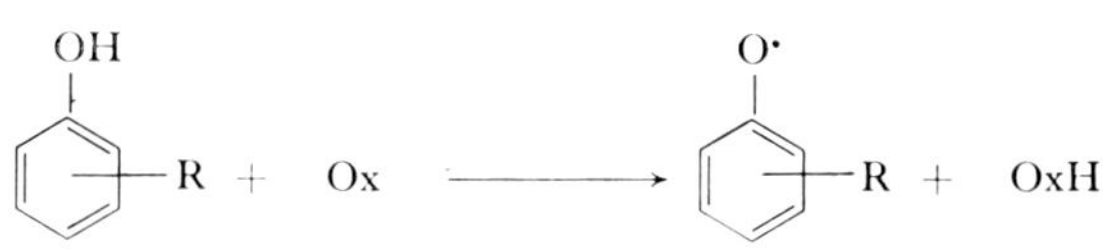

This radical exists as a resonance hybrid,

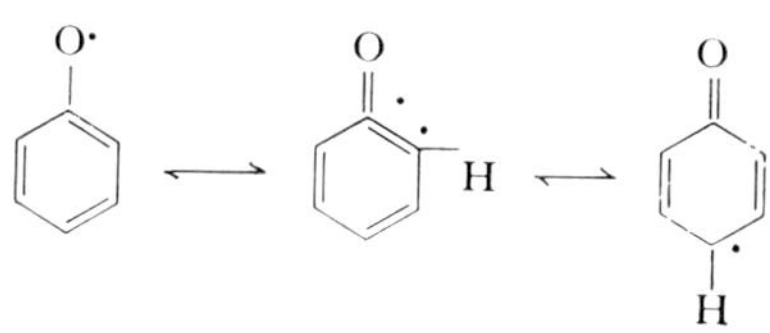

and readily reacts further, with itself or other suitable reagents. The nature of the product(s) depends strongly on the starting phenol, the oxidising agent and on reaction conditions.

For example, when sodium phenoxide is reacted with cupric salts unidentified polymers and cuprous compounds result.[59]

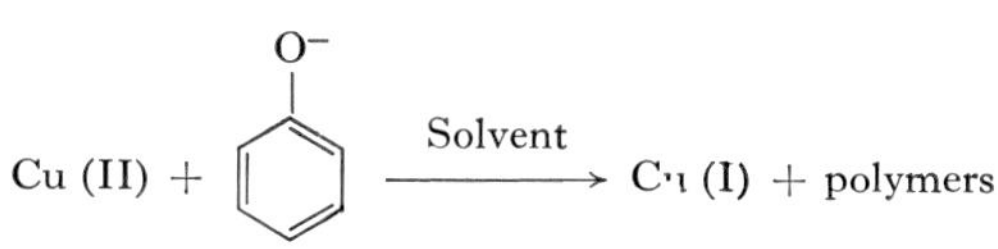

On the other hand, when 2,6-dimethylphenol is oxidised with air in the presence of pyridine (solvent) and copper salts the commercially important polyphenylene oxide (PPO) polymers result[72, 73]:

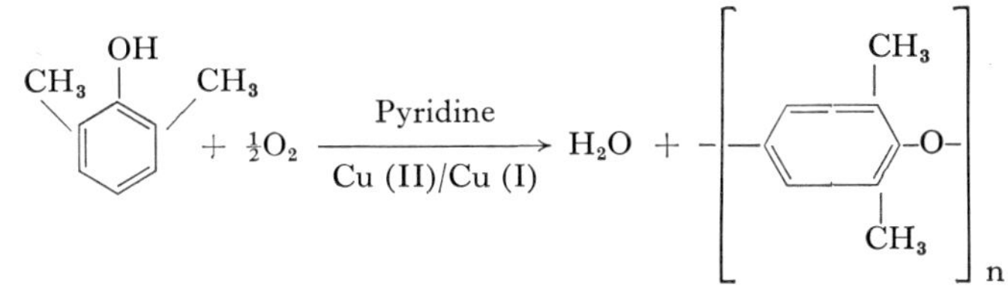

Under slightly different reaction conditions tail-to-tail coupling takes place to yield bisphenols or quinones instead of polymers.[72, 73]

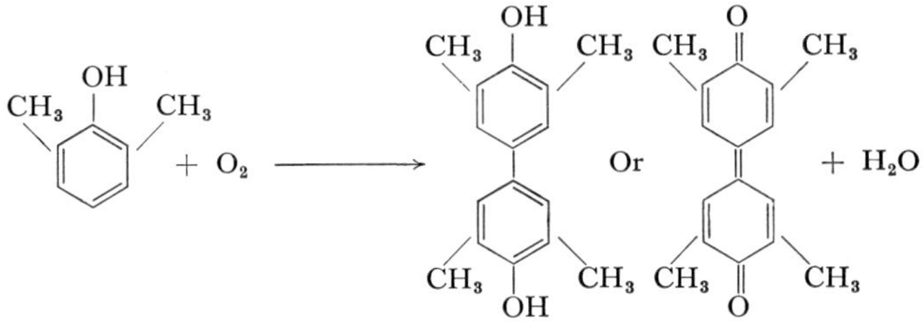

When the 2,6-alkyl groups are tert.-butyl (instead of methyl) then only tail-to-tail coupling takes place to yield biphenol/quinone products.[72]

The tri-tert. butyl phenol, on reaction with oxygen[74]:

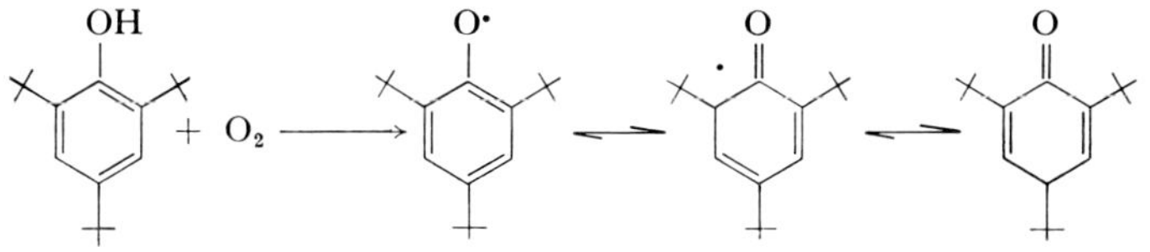

yields the phenoxy radical and this couples with more oxygen to yield a peroxide:

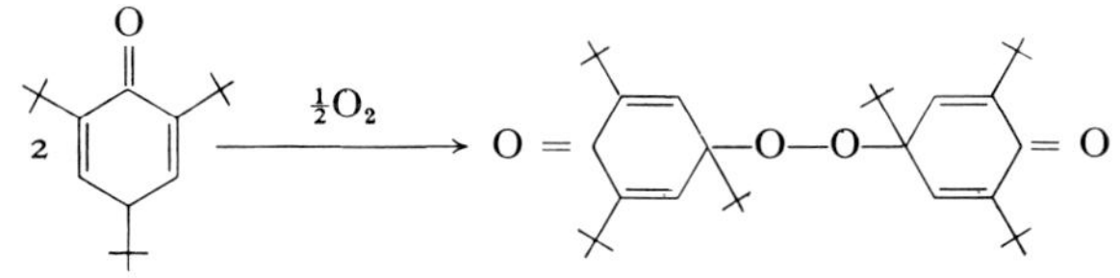

8.8.3. Physiological Aspects

Alkylphenols have about the same physiological activity as the parent molecule, phenol. Cresols,[75] and similar simple alkylphenols, should be treated with the same respect due to phenol itself: contact with the skin produces dermatitis or more severe burns if the phenol is not removed immediately. Ingestion should be followed by immediate gastric lavage; the alkylphenol vapours are however probably no more harmful than those of phenol itself.

One group of alkylphenols stand out in their physiological activity: the active principal of poison ivy Rhus (Toxicodenron radicans) and other members of Anacardiaceae family are long chain alkyl catechols which cause severe contact dermatitis in sensitised humans. A key compound is 3-(*n*-pentadecyl) catechol,[76]

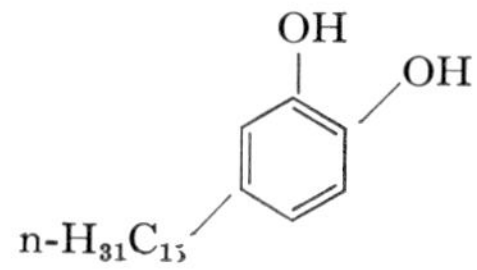

along with three analogs having unsaturation in the alkyl side chain. When the alkyl chain is small, e.g. methyl, tert.-butyl, etc., then there is little activity. The mechanism of this poison ivy dermatitis involves first a sensitisation of the individual followed by a more or less severe allergic response.

Many alkylphenols are used in skin-contact or for foodstuff preservation. The former have pronounced antibacterial behaviour (hexachlorophene) and the latter are excellent antioxidants for fats and oils (vide infra). Long term, high level feeding studies with dogs and rats show that these compounds are relatively innocuous.[77]

8.8.4. Uses

The following will discuss the economic impact (uses) which these alkyl phenols have had. The utilisation of these compounds in the alkylphenol—formaldehyde resins, the bisphenols, and polyphenylene oxide areas is disccussed elsewhere.

TABLE VIII.14

Non Ionic Surfactant Production in the United States Alkyl phenol ethylene oxide adduct* (Millon lb. per year)

Year	*Alkyl-C_9H_{19}*	*Alkyl-$C_{12}H_{25}$*
1961	—	29·4
1963	150	—
4	—	52·5
5	129	—
7	112	13·6
8	132·6	16·3

* 100% basis.

8.8.4.1. Alkylphenol-ethylene oxide surfactant adducts The reaction of ethylene oxide with alkyl phenols:

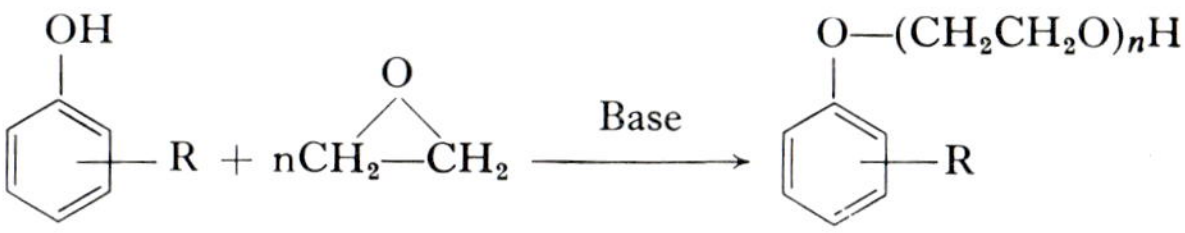

yields adducts which have a long polar end (ethylene oxide chain) and a long non-polar tail (alkyl groups C_6–C_{18}) linked together via a phenol residue.[78] These substances have found varied uses as non-ionic surfactants. Table VIII.14 gives the total phenol useage in this area as well as the ethylene-oxide-alkylphenol adduct production of compounds having C_9 and C_{12} alkyl chains.

Normal soaps and detergents have ionic hydrophilic groups (e.g.-

$$CO_2^-Na^+,\ -SO_3^-Na^+,\ -PO_3^-Na^+)$$

attached to a hydrophobic tail. These alkylphenol-ethylene oxide adducts are non-ionic surfactants due to the very polar but non-ionic nature of the polyethylene oxide chain. They are soluble in both polar and non-polar media and can function effectively in both. These surfactants are also very stable in both strongly alkaline and strongly acidic systems, they are stable to elevated temperatures and water hardness does not affect their surface activity.[79, 80]

The relative and absolute sizes of R and n determines the hydrophobic-hydrophilic properties of the products. The degree of branching of R determines biodegradability; linear alkyl chains are most readily degraded. In recent years there has been a sharp shift to all-linear alkylphenols in this area. Table VIII.15 gives a breakdown of the nature and uses of various of these products.

Some of these alkylphenol-ethylene-oxide adducts can be further reacted to form sulphonated surfactants. The free hydroxyl group of the ethylene oxide chain reacts to form the half ester:

$$\sim CH_2CH_2OH + SO_3 \longrightarrow \sim CH_2CH_2O\text{—}\overset{\overset{\large O}{\|}}{\underset{\underset{\large O}{\|}}{S}}\text{—}OH$$

which is then neutralised with caustic to yield the salt:

$$\sim CH_2CH_2O\overset{\overset{\large O}{\|}}{\underset{\underset{\large O}{\|}}{S}}\text{—}OH + NaOH \longrightarrow \sim CH_2CH_2O\text{—}\overset{\overset{\large O}{\|}}{\underset{\underset{\large O}{\|}}{S}}\text{—}ONa + H_2O$$

These now function very much like the normal sulphonated or sulfated surfactants. About 9·6 million lb of these sulphonated adducts were produced in 1966 (price about 23c/lb).

TABLE VIII.15

Properties and Uses of Alkylphenol Ethylene Oxide Adducts

Wt % ethylene oxide	*n*	*Uses*
20–40	1–3	Defoamers; intermediates for sulphation; petroleum oil dispersants.
40–60	3–7	Hydrocarbon soluble detergents; emulsifiers for insecticides and herbicides; intermediates for sulphation.
60–70*	7–10	Textile detergents; leather, paper industries; cleaning compositions; wetting agents in acid and alkaline cleaning product; insecticide and herbicide emulsifiers.
70–80†	10–18	Wetters and detergents at high temperatures and strong electrolyte and acid concentrations. Fat, oil, wax emulsifiers; synthetic latex stabilisers.
80–95	18–90	Synthetic latex stabilisers; emulsifiers for polymers; dyeing aids; lime-soap dispersants.

* Completely water miscible.
† Do not cloud out in boiling water.

8.8.4.2. Alkylphenol phosphate esters. Phenol and alkylphenols form esters with phosphoric acid:

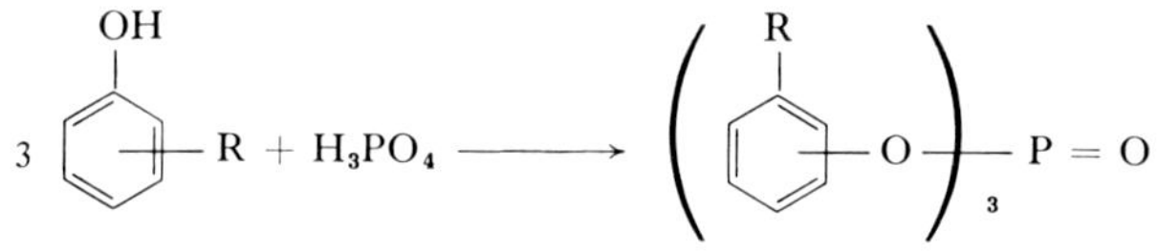

Over 30 million lb of phenol or alkylphenol equivalents were used in this area in 1969. Table VIII.16 gives an approximate breakdown of phenol/alkylphenol phosphate ester production over the years. There appears to be very little growth in this market at present and prices, at least as far as the tricresyl phosphates are concerned, are static or decreasing. About half of these phosphate esters enter the plasticiser[81, 82] field; the phenyl phosphates are used in cellulose acetate, the others for polyvinyl chloride. These substances also give a measure of flame retardance to the plasticised compositions. The balance of these phosphate esters are used in specialty lubricants and in motor fuel additives. This latter application may now be a growth area because of the recent accent on lead pollution due to the tetraethyllead antiknock additives in gasoline.

Plasticisers function essentially as solvents for polymeric systems and thereby allow the fabrication of soft and pliable articles from formerly hard, brittle polymeric materials. Polyvinyl chloride (PVC) is an excellent example. Pure PVC is a hard, brittle solid which is difficult to fabricate because its softening-flow temperature is very close to the decomposition temperature of the polymer. By incorporation of 20–50 per cent of such phosphate plasticisers a soft, pliable material is obtained which has reasonable fabrication characteristics and results in products with good physical properties. Additionally, these phosphate plasticisers act as flame retardants.

Techniques have now been developed which enable unplasticised PVC to be made into film, sheet, moulded products, etc. and the proportion of PVC polymer which is plasticised is falling.

TABLE VIII.16

Production of Various Phenol/Alkylphenol Phosphate Esters in the United States in millions of lb per year

Year	$C_6H_5OP(=O)(OC_6H_4CH_3)_2$	$(CH_3C_6H_4O)_3P = O$*	$(C_6H_5O)_3P = O$†
1958	7·2	26·5	8·0
9	10·5	31·4	8·4
1960	12·0	32·5	9·5
1	13·0	24·6	9·4
2	15·7	27·5	10·4
3	11·3	30·7	10·0
4	16·1	32·4	9·0
1965	19·7	34·8	—
6	20·0	39·8	8·8
7	18·2	42·9	8·7
9‡	20·0	n.a.	8·0

* About one-half is used as plasticiser in PVC, remainder as motor fuel and lubricant additive; price range 31–34 cents/lb.

† Much for cellulose acetate film market.

‡ 29·0 million lb phenol equivalent in 1969.

Frequently these phosphate esters are used in mixtures with dioctylphthalate (DOP). DOP with PVC alone results in flammable products while the addition of a phosphate ester restores the original flame resistance of PVC. Phosphate ester-plasticised PVC coatings on metals give the metal high corrosion resistance.

Alkylphenol esters of phosphoric acid are added to gasoline in order to modify engine deposits. These esters decrease the tendency of the gasoline combustion deposits to glow and initiate premature ignition of the fuel—air mixture. In addition, spark plug fouling is inhibited. Such fouling tends to drain off current and cause misfiring. These phosphate esters are also added to the engine oil to function as an anti-wear agent (tricresyl phosphate). These additives modify the oil-metal interface by absorption or reaction. Such thin layers prevent corrosion and wear under severe load and high temperatures. About 1 per cent phosphate ester is employed.[83, 84]

8.8.4.3. Antioxidants Alkylphenols are widely employed as antioxidants. Most often they are prepared from phenol or phenol derivatives via alkylation with an olefin:

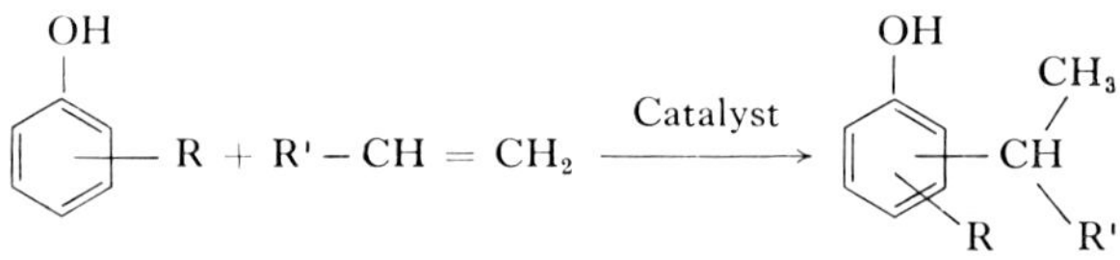

These products are used in the food industry, as motor fuel additives, and in plastics and rubber compositions.[74, 85, 86, 87, 88] In these applications they are always used in very small concentrations (0·1 per cent). Table VIII.17 gives the consumption of alkylphenols in the rubber industry. Considerably greater quantities of phenolic resins (phenol-formaldehyde polymers) and alkyl phosphites are used than alkylphenols.

TABLE VIII.17

Use of Antioxidants in the Rubber Industry in the United States in millions of lb

Year	*Alkylphenols*	*Phenolics and phosphites*
1956	1·6	—
7	2·3	14
8	2·5	11
9	6·0	17
1960	7·1	15·5
1	7·4	14·5
2	8·1	18
3	7·2	17
4	9·2	20
1965	11·4	30
6	12·2	33
7	10·0	31
8	13·2	41

The action of oxidising agents, here oxygen and/or ozone (O_3), can have profound effects on polymers used by man. Four principal areas of use of alkyl phenol-type antioxidants are:

(1) Rubber articles (tyres, etc.)
(2) Plastic products (films, molded objects).
(3) Foodstuffs (notably fat-containing).
(4) Gasoline and lubricating oil additives.

These areas of usage have in common the protection of organic polymers with built-in or pendant vinyl, aryl, alkyl, etc., groups. Such substituents on a hydrocarbon backbone tend to activate

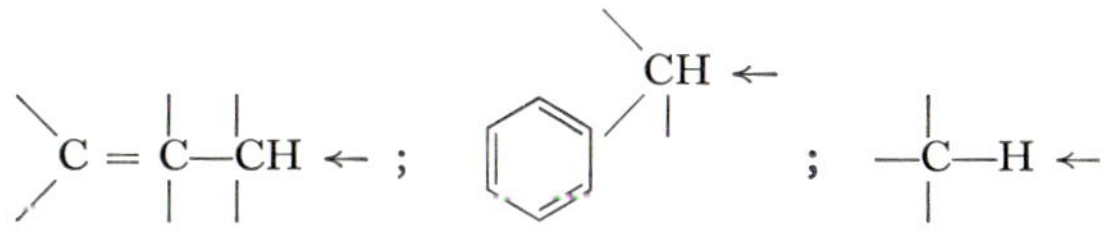

allylic benzylic tert.-hydrogen

containing centres, the C–H bonds of which are now particularly susceptible to attack by oxygen or free radicals derived from peroxides. Such reactions lead to chain scission, degradation of polymer molecular weight and hence loss of desirable physical properties (tensile strength and modulus) and finally to a distintegration of the article. Very small quantities of certain chemicals, amongst which alkyl phenols rank high, inhibit this oxidative degradation.

Antioxidants are often sold as mixtures; one component serves one particular purpose, the other has another use or a synergistic effect on the first. For example, the synergistic pair alkylphenol-alkyl aromatic amine each capture free radicals with ease and interrupt oxidative chain reactions by forming nonradical products. Citric acid, on the other hand, complexes strongly with metal ions (copper, nickel, iron) so that these cannot participate catalytically in peroxide decomposition reactions. Therefore, three-component mixtures of alkylphenol, aromatic amine, and citric acid are in widespread use to stabilise hydrocarbon polymer compositions towards oxidative attack.

Other additives are antiozonants. These specifically inhibit the deleterious effect of ozone (O_3) on rubber articles notably tyres. Another general class of antioxidants are stabilisers. These temporarily inhibit oxidative reactions which could take place in polymer latexes or vinyl monomers during storage or certain processing steps. The effects of these substances should be temporary or they should be readily removable (for example hydroquinone is used as stabiliser for styrene monomer). The choice of an appropriate antioxidant or combination is highly dependent on the substance(s) to be protected and on the conditions of use.

Hindered phenols function as antioxidants by capturing reactive free radicals and convert them to substantially unreactive, non-radical product(s). The reaction of oxygen and a hydrocarbon leads to such radicals; first oxygen or a peroxide, in the initation step, interacts with the carbon-hydrogen bond, by hydrogen abstraction to yield radicals,

$$RH + O_2 \longrightarrow R\cdot + \cdot O_2H$$

$$ROOH \longrightarrow RO\cdot + \cdot OH$$

Propagation steps are:

$$R\cdot + O_2 \longrightarrow RO_2\cdot$$

$$RO_2\cdot + RH \longrightarrow ROOH + R\cdot$$

Various chain transfer and other propagation reactions then lead to degradation of the polymer structure. The antioxidants of this alkylphenol class react exceptionally readily with such radical-intermediates to yield products which are oxidised and non-radical in nature. In effect they function as chain terminators. Thus 2,6-di-tert.-butyl-p-cresol consumes radicals to yield a stable quinone-peroxide structure and a hydroperoxide.

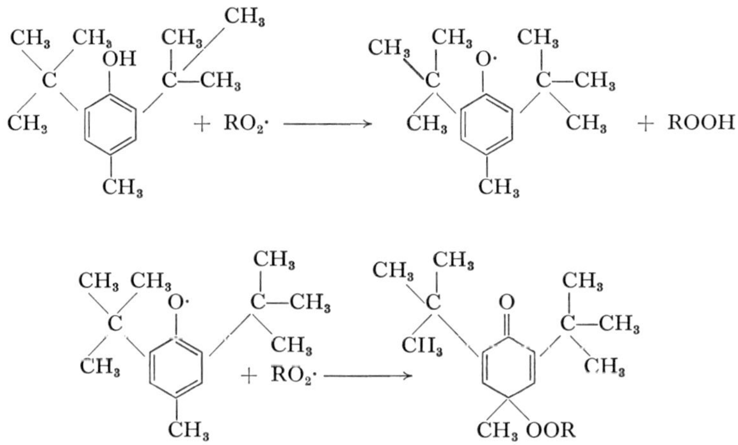

When the alkyl groups ortho to the phenolic hydroxide are strongly branched and there are electron donors on the ring (e.g. methyl, methyl ether), then this anti-oxidant activity is maximised. If metal ions are present in the system, then a chelating agent specific for the metal ion is added in order to prevent the metal ion-catalysed decomposition of these peroxides to radical products. Frequently, appropriate biphenols can act in this manner, 2,2′methylene bis(6,6′-tert.-butyl phenol) is such a chemical:

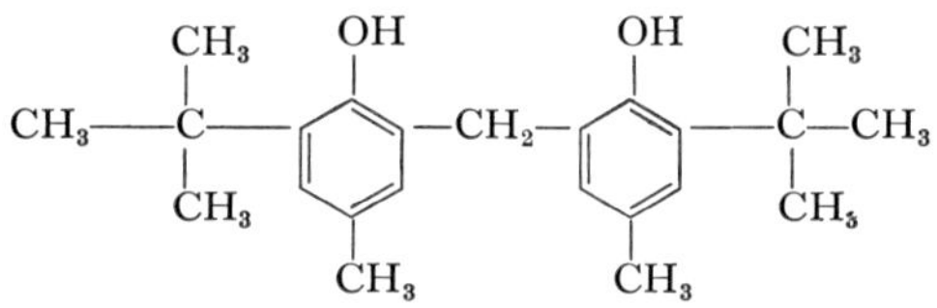

Some classes of phenol-based antioxidants are the following:

(1) Hydroquinone derivatives such as,

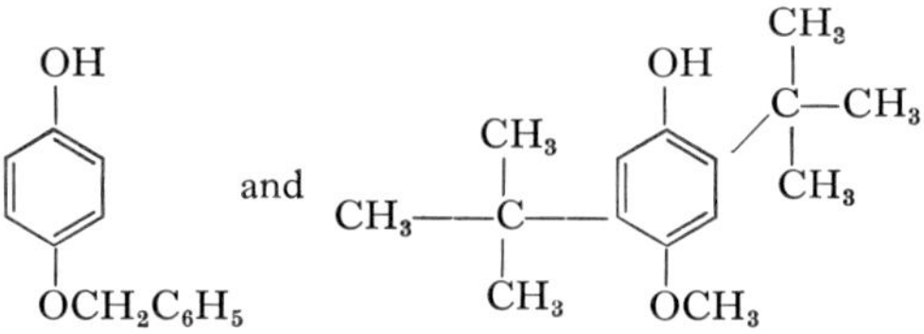

are used as rubber and latex stabilisers, and in food products. Hydroquinone itself is used as a vinyl monomer stabiliser.

(2) 2,4,6-Trialkylphenols such as,

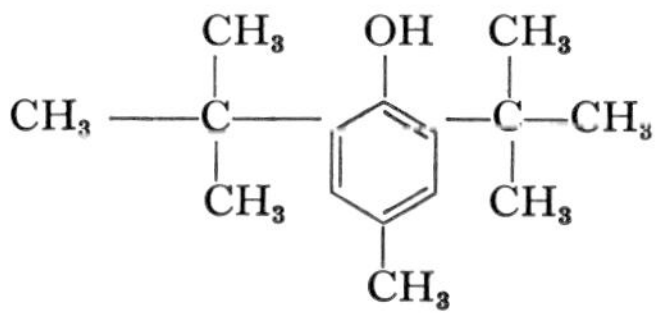

are used in rubber compositions, polyethylene, polypropylene and polyvinylchloride.

(3) Bisphenols,

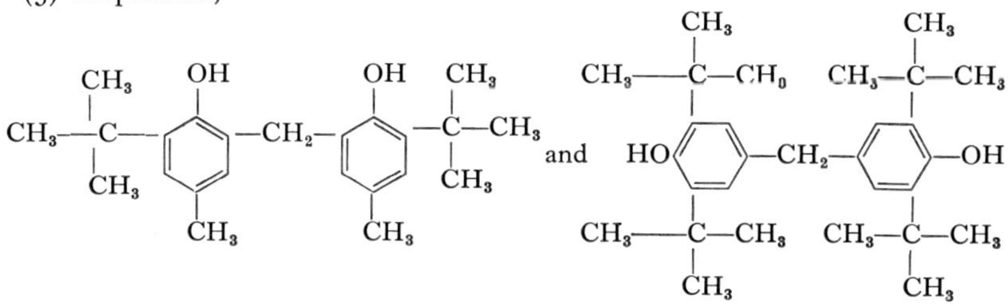

are widely used, the former is an excellent copper ion chelating agent.

(4) Catechol and derivatives,

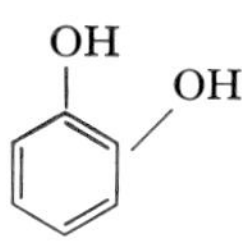

are used in foods and as animal feed stabilisers.

Condensation products of phenol, alkylphenols, and styrene, etc. are cheap and useful in various rubber and polyethylene, polypropylene compositions. Most of these compounds have trade names and manufacturer's numbers. Catalogues of this information and the chemical structures have been published.[86, 87]

Antioxidants are used widely in stabilising foodstuffs, notably fat-containing articles. Naturally occuring chemicals can be employed for this purpose (propylgallate, citric acid, ascorbic acid, etc.), or the synergistic pair butylated hydroxy toluene and butylated anisole are widely used:

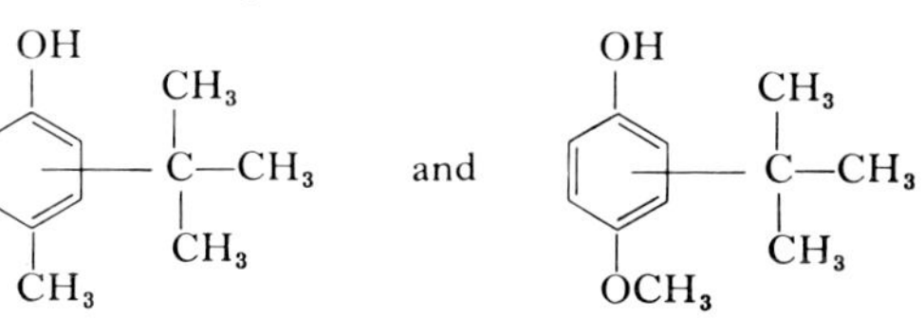

Surprisingly small quantities (0·001–0·05 per cent) of this pair along with a metal ion chelating agent (citric acid) are highly effective.

Raw gasoline, particularly blends containing olefins from catalytically cracked feed stock, have to be stabilised against polymer build-up. Such polymerised gums foul carburator jets and quickly bring even the highest horsepower car to a stop. A combination of metal complexor (to chelate metal corrosion products notably copper) and antioxidants serve this purpose. The latter are usually a synergistic mixture of ortho alkylphenol and a phenylene diamine derivative. About 2–3 lb of antioxidant per 1,000 bbl. (42 US gal. each) are used at present. This is much lower than the 15 lb per 1,000 bbl used ten to fifteen years ago when raw gasolines had a much higher component of cracked feedstock. Diesel oil also contains such antioxidants.[83]

A modern, high performance lubricating oil is a combination of many ingredients other than the hydrocarbon oil itself, each of which serves a particular purpose. The alkylphenol phosphate esters have already been discussed, as were detergents which are designed to keep the metal parts free of gum deposits. Pour point depressents prevent paraffins from crystallising, viscosity index improvers regulate the viscosity of an oil, rust inhibitors are present as well as oxidation inhibitors. For the latter, combinations are again chosen: metal chelating agents along with synergistic pairs of alkylphenols and aromatic amines. These are added in relatively high concentrations, 0·5–1 per cent.[84]

8.9. BISPHENOL-A

by F. M. Konrad

8.9.1. Historical

Bisphenol-A[2,2-bis(4 hydroxyphenyl)propane or 4,4'isopropylidene diphenol] is a major phenol derivative that was first synthesised in 1891 by Dianin[89] from the acid catalysed condensation of phenol and acetone. This compound is sometimes referred to as Dian in Europe while in England, in particular, it is usually known as diphenylolpropane. Bisphenol-A is used principally as a raw material for resins and polymers and its commercial development has closely followed the growth of the plastics industry. In 1923, bisphenol-A was produced in Germany at the Chemische Werke Albert as an intermediate for light coloured phenolic resins.[90]

A significant increase in production capacity and product quality took place during the period 1940–60 to satisfy the needs of a rapidly expanding epoxy resin market. Epoxy resins are still the major consumer for bisphenol-A.

Processes for high purity bisphenol-A developed with the advent of polycarbonate in 1958. These and other linear polymers required bisphenol-A with organic impurity levels of less than 0·5 per cent, and with excellent colour and thermal stability.

Worldwide, annual bisphenol-A production capacity was over 700,000,000 lb by the end of 1972. Although the synthesis and properties of over one hundred other bisphenols have been reported in the literature[91] only bisphenol-A has attained extensive industrial use.

8.9.2. Preparation

Bisphenol-A is prepared by the acid catalysed condensation of phenol and acetone:

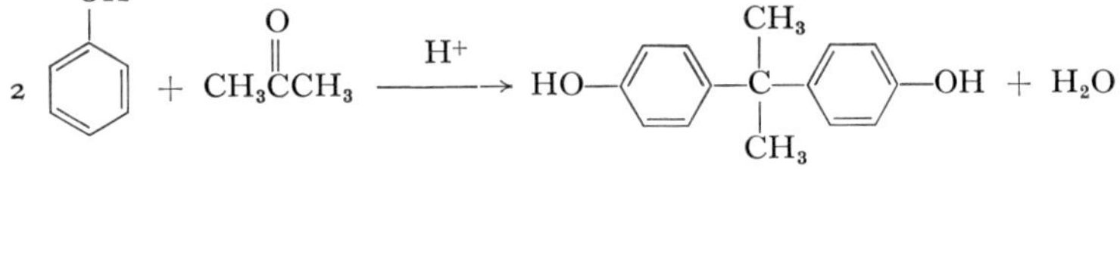

Strong acids such as hydrochloric acid, 70 per cent sulphuric acid, or sulphonated polystyrene-divinylbenzene cation exchange resins are preferred commercial catalysts. Catalysis may also be effected with other hydrogen halides, boron trifluoride, aluminum chloride, phosgene, phosphorus pentoxide, or with mixtures of acetic and sulphuric acids. There is no commercial interest in the alkaline catalysed reaction. Alkaline catalysis requires high temperature and long reaction time, and has the added disadvantage of promoting oxidation of phenol and bisphenol-A.

The extent of reaction is limited by the formation of by-product water. Large excesses of phenol and acid are used to nullify the effect of water. Addition of calcium chloride to the reaction has been used to reduce the inhibiting effect of water.[92]

The reaction rate is accelerated by the addition of catalytic amounts of divalent sulphur compounds such as alkyl mercaptans, thioglycollic acid, thiophenols, mercapto alcohols, and mercaptoalkylsulphonic acids.[93, 94]

Williamson[95] studied reaction promotion with alkylmercaptans of various chain lengths, and concluded that the high product yield obtained with methyl mercaptan was in excess of that expected on comparison with yields obtained under the same conditions with the higher analogues (Table VIII.18). It was also observed that hydrogen sulphide was ineffective. The reaction was carried out at 65°C and with 1 p.s.i.g. hydrogen chloride. The reaction time was 30 mins and 10:1 phenol-acetone molar ratio used.

TABLE VIII.18

Yields from Alkylmercaptan Promoters

Sulphur compound	*Promoter conc. % of theoretical yield of bisphenol A*	*Yield of 2,2 bis(4-hydroxyphenyl propane)* (%)
None	—	12·0
Hydrogen sulphide	1	10·0
Methyl mercaptan	0·5	98·8
Ethyl mercaptan	1·0	70·0
Butyl mercaptan	1·0	30·0

Sulphonated styrene-divinylbenzene ion-exchange resin catalysts with a portion of the sulphuric acid groups converted to mercaptan functionality have been reported as more effective catalysts than the unmodified exchanger.[96–98]

The reaction mechanism has been elucidated by the studies of v. Braum, and by those of Leibnitz and Naumann.[99, 100] In strong acid acetone is protonated to the stable carbonium ion:

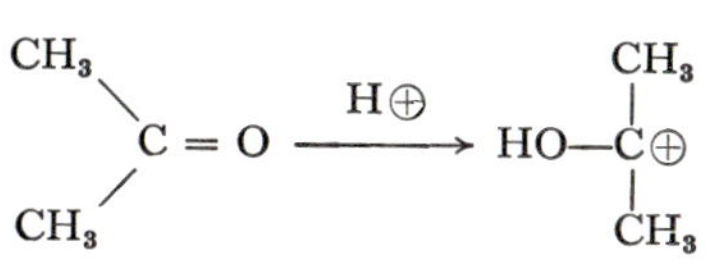

which adds to the limiting quinonoid structure of phenol to yield a protonated carbinol:

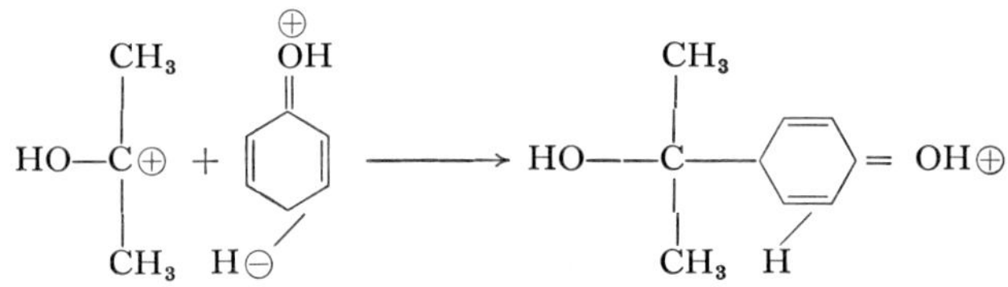

The carbinol rearranges to release water and yields protonated isopropenylphenol:

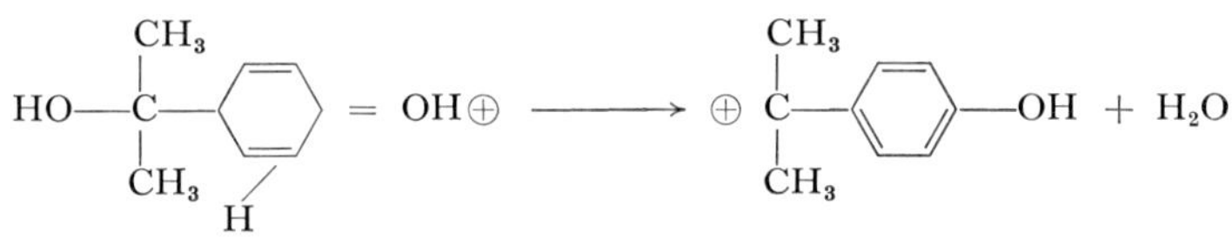

The isopropenylphenol adds to a second phenol molecule to yield bisphenol-A and releases the proton:

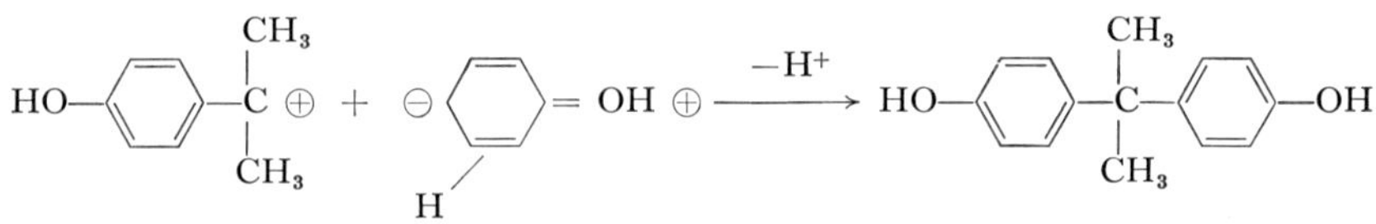

The overall reaction is exothermic and liberates 200 Btu for each pound of bisphenol formed.

Several side products are also formed in addition to the 4,4′ isomer. The major impurity is 2,4′-isopropylidene diphenol formed by the addition of *p*-isopropenylphenol to the phenol ortho position:

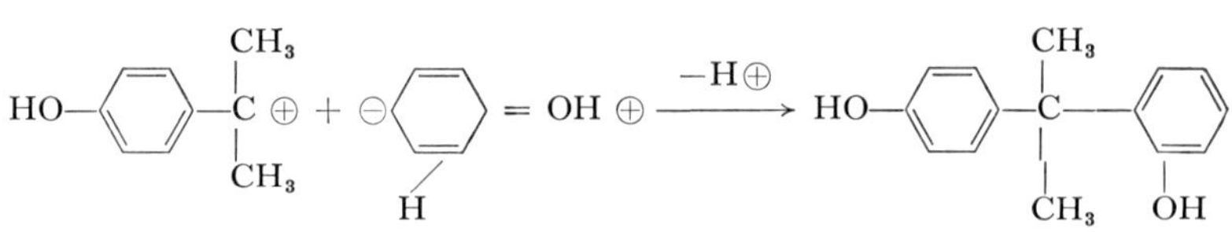

Small amounts of trisphenol are also formed by the addition to both the para and ortho positions of phenol:

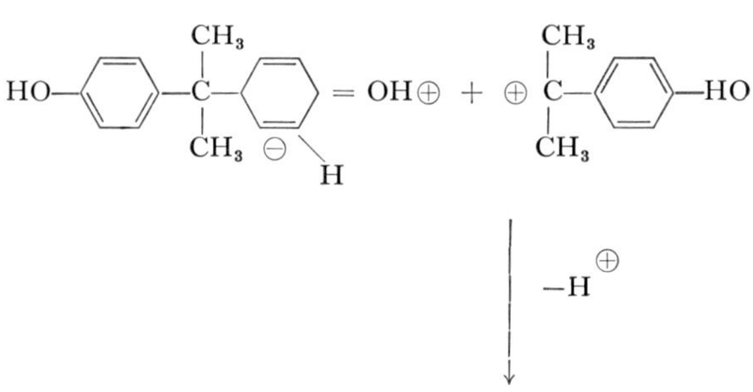

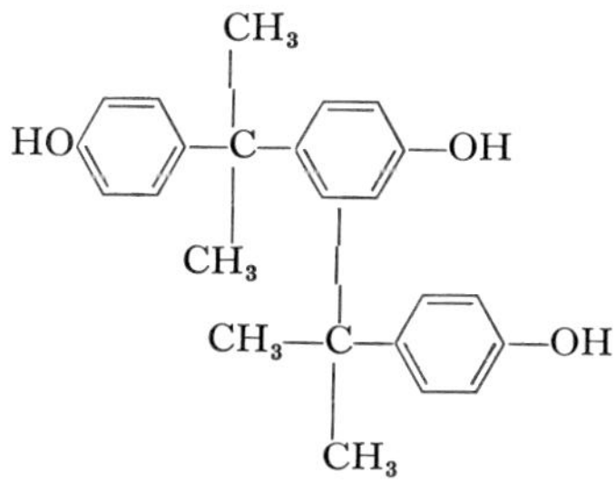

Small amounts of the chroman derivative (Dianin's Compound) arises as a result of the reaction of phenol with mesityl oxide:

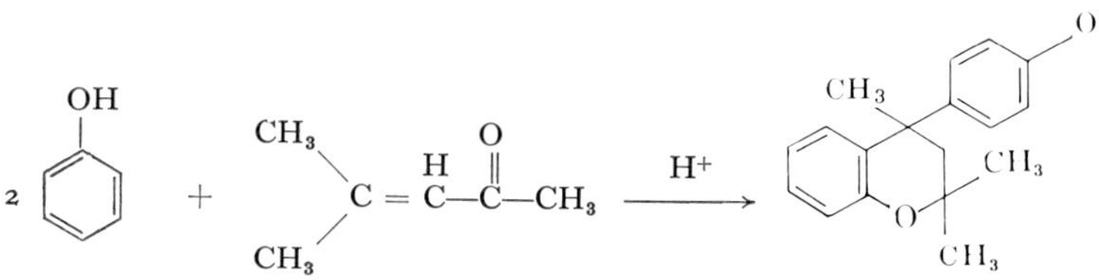

Trace amounts of mesityl oxide have been identified in the phenol acetone reaction stream.

Formation of impurities is supressed by using a large excess of phenol (at least two fold) and low reaction temperature—preferably below 70°C. Crude reaction products containing less than 2 per cent impurities were obtained by running the reaction at conditions (<60°C) to cause crystallisation of the bisphenol-A as an addition compound with 1 mole phenol as rapidly as formed by the reaction.[101, 102]

Acetone is used invariably in the manufacture of bisphenol-A because of its availability at low cost and high purity. Extensive studies have been made with potential acetone substitutes aimed primarily at eliminating the problems associated with by-product water.

Good yields (80 per cent) of bisphenol-A have been reported from the addition reaction of phenol and methylacetylene catalysed with either boron trifluoride[103] or a strong cation exchange resin:[104]

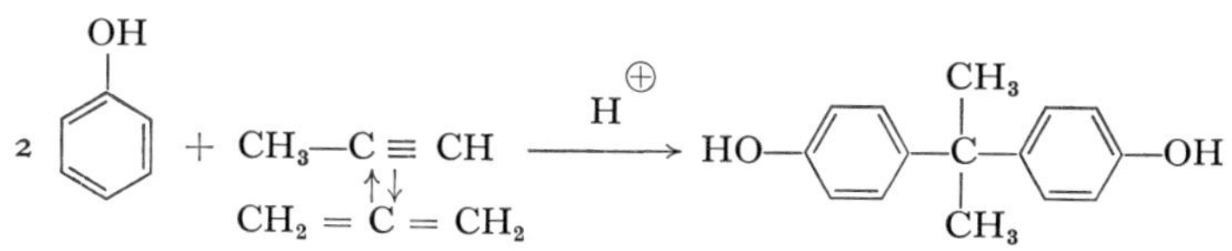

This is an excellent route to bisphenol-A. However, high purity methylacetylene is not commercially available at costs competitive with acetone.

Preparation of bisphenol-A from isopropenylacetate,[105] 2-chloropropene and 2,2'bis(alkylthiopropane)[106] have also been patented but never developed commercially.

The acid catalysed decomposition of cumene hydroperoxide used to manufacture phenol and acetone offers another potential route to bisphenol-A. The hydroperoxide diluted with hydrocarbon solvent and additional phenol to provide more favourable stoichiometry was reacted in the presence of hydrogen chloride gas at 20°C to give a 65 per cent yield of crude bisphenol-A.[107]

Near quantitative yields of bisphenol-A can be obtained by the acid catalysed addition of *p*-isopropenylphenol to phenol at low temperature:

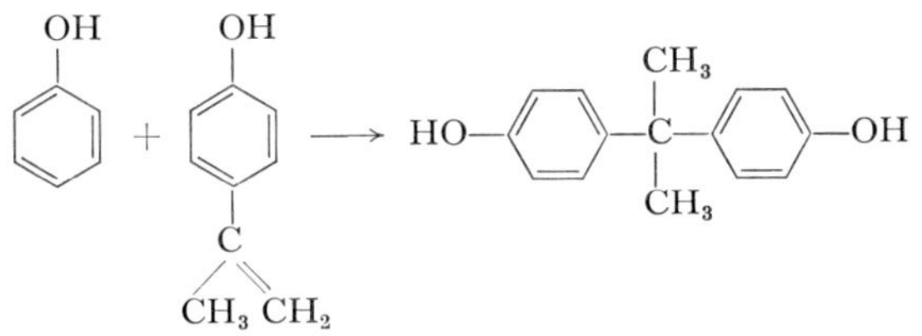

For example the reaction of 135 g *p*-isopropenylphenol with 188 g phenol diluted with 500 cc toluene and catalysed with 10 cc of 10 per cent *p*-toluene sulphonic acid in methanol was worked up to yield 220 g bisphenol-A.[108] *p*-Isopropenylphenol readily undergoes polymerisation to a linear dimer at room temperature:

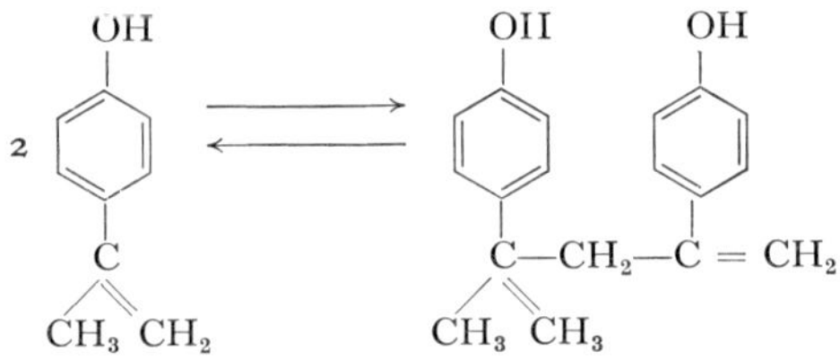

Under acidic conditions the dimer reverts to the monomeric form. Thus high yield of bisphenol-A may also be obtained from the dimeric form via the above reactions.[109]

8.9.3. Physical Properties and Toxicology

Some of the physical properties of bisphenol-A are listed in Table VIII.19.

Bisphenol-A is usually sold under two general specifications, (1) epoxy grade containing up to 5 per cent impurities (phenol, 2,4′ isopropylidene diphenol, trisphenols and chromans) and (2) polymer grade with an assay >99·5 per cent. Typical specifications are shown in Table VIII.20.

APHA colour of the melt and the colour of caustic solutions are also used to specify polymer grade product. Some consumers also specify limitations on specific and total organic impurity content. These impurities are measured by gas chromatography on silylated derivatives.[112]

Toxicologically bisphenol-A is unlike the lower alkyl phenols and is only moderately irritating to humans. Dust, vapours, or solutions can cause eye injury, and skin or lung irritation. Accidental exposure to the skin should be washed immediately with soap and water. Eye contact should be immediately treated with a copious amount of clean water. Medical attention should be obtained immediately. Proper ventilation and personnel protective measures should be taken where dusty conditions exist or where accidental contact may occur .

TABLE VIII.19

Physical Properties of Bisphenol-A

Appearance	White crystals, flakes or prills
Odour	Mild phenolic
Specific gravity (25/25 °C)	1·195
Bulk density, flakes (lb/ft³)	36–42
Melting point (°C)	157·0
Flash point, Cleveland open cup (ASTM D92)	405 °F

Vapour pressure (mm Hg)	*Temperature*[110] (°C)
1	193·0
5	224·2
10	240·8
20	255·5
40	273·0
60	282·9
100	297·0
200	317·5
400	339·0
760	360·5

Solubility (*g/100 g solvent*) *at room temperature*[111]	
Ethyl ether	127
Methylene chloride	·86
Acetone	108·6
n-Hexane	·011
Ethanol	150·5
Benzene	·61
Cyclohexane	·002
Trichloroethylene	·077
Toluene	·34
Acetic acid	21·6
Chlorobenzene	·12

TABLE VIII.20

Bisphenol-A Specifications

	Epoxy grade	*Polymer grade*
Min. melting point (°C) (dry)	155·0	156·5
Max. APHA colour (50·0 g/70 ml methanol)	100	25
Max. phenol content (%)	00·20	0·10
Max. moisture content (%)	0·15	0·15
Max. iron content (p.p.m.)	1·5	1·0
Max. ash content (%)	·02	·02

The oral LD_{50} is 4·04 g/kg of body weight for male albino rats and 2·23 g/km of body weight for rabbits.[113] Thus, only injestion of large quantities of bisphenol-A would be harmful to humans. In the event of such an occurence the bisphenol-A should be removed immediately by induced vomitting, or gastric lavage. Medical assistance should be obtained immediately.

Bisphenol-A may be shipped in bags, fiberpak drums or cardboard boxes, or it may be transported in bulk form with hopper rail cars or trucks. Aluminum equipment is recommended for bulk transfer and storage. Bisphenol-A should be stored in a dry air or nitrogen atmosphere at ambient temperatures.

Bulk transfers may be made by gravity pneumatic conveying (preferably vacuum). Bisphenol-A particles are friable and tend to form dust on collision with other particles or with the transfer conduit walls. Under certain conditions bisphenol-A dust forms explosive mixtures with air.[114] Thus equipment design criteria and operating practices for the bulk handling of bisphenol-A should be common to those for the safe handling of explosive organic dusts. The design should include devices to prevent static electrical discharge or frictional heat accumulation within the system, rapid pressure relief systems for storage vessels, and suitable lockouts to prevent circumvention of safe operating practices by operator errors.

8.9.4. Industrial Processes

Acid catalysed condensation of phenol and acetone is used in all of the commercial routes to bisphenol-A. The variety of processes used by bisphenol-A manufacturers differ primarily in product recovery and purification methods. Two general product grades are manufactured. A regular purity grade suitable for the manufacture of epoxy resins is usually produced as a residue product from the crude reaction effluent. The high purity grade used for polycarbonate resin manufacture is produced by refining the crude with distillation and/or crystallisation techniques.

8.9.4.1. Epoxy grade process Fig. 8.8 depicts a flow diagram for producing a crude residue product. Phenol and acetone at a 20:1 mole ratio along with 1·4 wt per cent hydrogen chloride and 0·2 wt per cent methyl mercaptan are pumped to the first of two continuous stirred tank reactors. The reactants have a mean residence time of 15 minutes at 50°C. The reactors are glass lined and provide cooling to remove heat of reaction. The primary reactor effluent is pumped to a second reactor where a second acetone addition is made to reduce the overall phenol acetone mole ratio to 10:1. The secondary reaction temperature is allowed to rise to 62°C to prevent product crystallisation. A 30-minute residence time is provided by the second reaction.[115] The reactor effluent is fractionated in an evaporator to remove methyl mercaptan, water, acid, and some phenol overhead. Methyl mercaptan and hydrochloric acid are recovered from the overhead by distillation and recycled to the process. The water-rich bottoms from this column are purged from the system. The deacidified mixture of phenol and bisphenol-A is passed through an evaporator to remove the phenol which is recycled. A steam stripper is used to reduce the product phenol content to specification levels. The final product is flaked and packed in shipping containers. Phenol in the water purge and steam stripper overhead is recovered by solvent extraction or distillation. This process yields a product with a 154°C melting point that contains 3–5 per cent impurities. A somewhat less pure product is obtained with a single stage reaction system. Thin-film short residence-time evaporators are desirable to dephenolate the heat sensitive product. In this process, raw material efficiencies are near 100 per cent, when phenol in the waste water streams is recovered and recycled. About 20 lb of hydrogen chloride are consumed per 1,000 lb product.

8.9.4.2. Union Carbide Process In the Union Carbide Process, sulphonated styrene-divinylbenzene cation exchange resin is used to catalyse the phenol acetone condensation reaction.[116, 117] The cation exchange resin is catalytically active only in the anhydrous form (phenol swollen) and thus bisphenol-A output is limited by the water of reaction. Fig. 8.9 outlines a simplified flow sheet for the process.

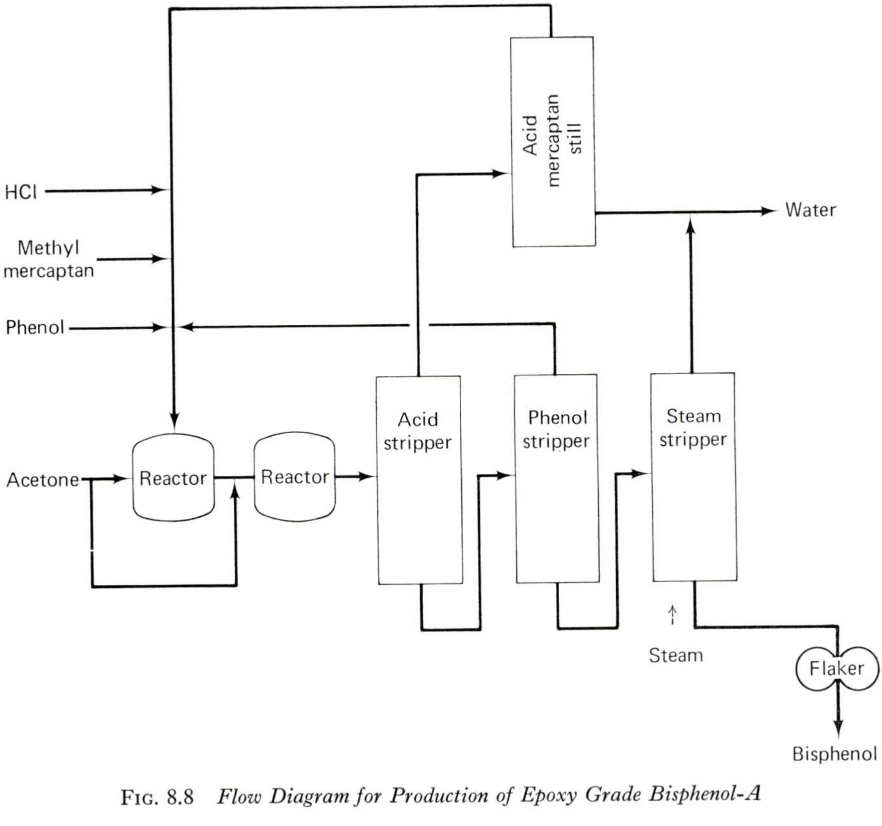

FIG. 8.8 *Flow Diagram for Production of Epoxy Grade Bisphenol-A*

Recycled phenol, containing 10–15 per cent dissolved bisphenol-A and impurities, and 5 per cent acetone is pumped through the catalyst bed at 70°C to provide a 1 hour residence time and to effect about 50 per cent conversion of the acetone. The reactor effluent is passed through an evaporator where unreacted acetone, water and some phenol are removed at 140°C and 200 mm Hg. The base product from the

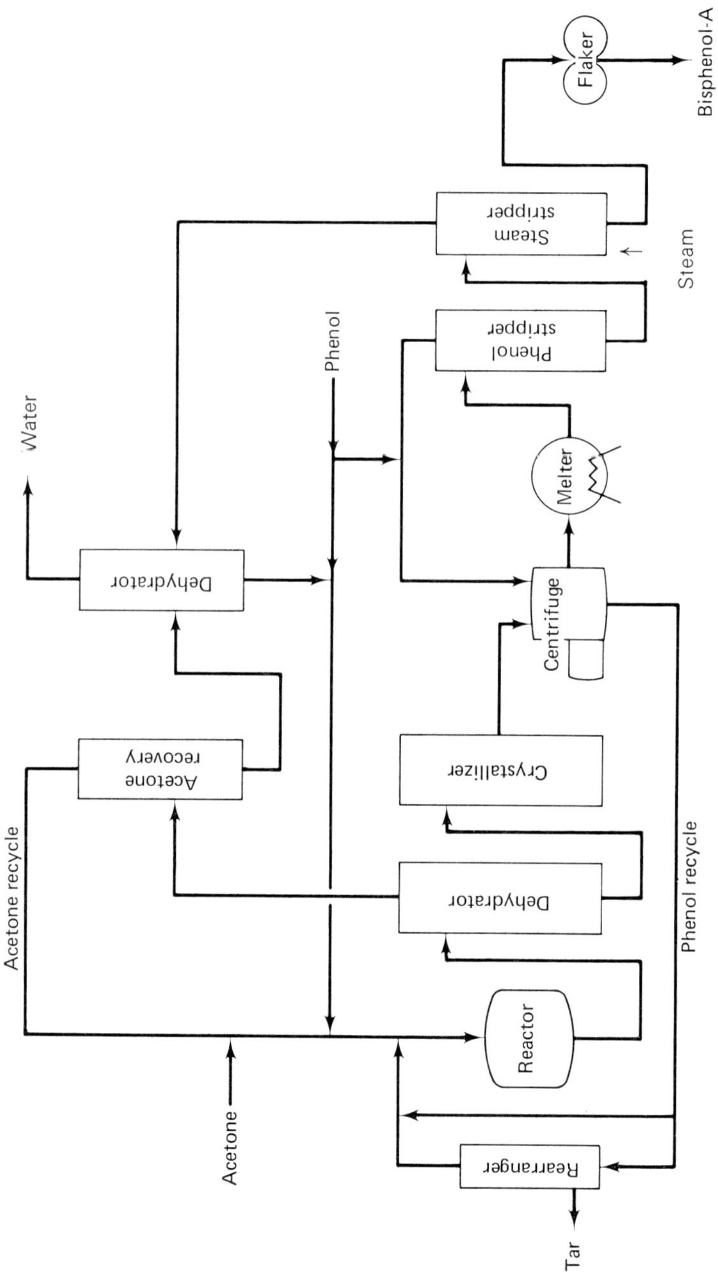

FIG. 8.9 *Union Carbide Process*

evaporator is cooled to 40°C in a crystalliser to produce a slurry of bisphenol-A phenol adduct crystals. These crystals are filtered and washed with pure phenol. The washed adduct crystals are melted and passed through an evaporator and steam stripper where phenol is removed at 180°C under reduced pressure. The finished product is flaked for packaging.

The dehydrator evaporator overheads are refined in two distillation columns to recover phenol and acetone, and to purge water from the system. The crystalliser mother liquor is recycled to the reactor. A fraction of the mother liquor stream is recycled through a rearrangement unit to reform bisphenol-A from reversible impurities and to purge resinous products from the system.

This process can produce either epoxy or polymer grade product by adjusting the amount of phenol wash used for adduct crystal purification. The process operates with raw material efficiencies above 95 per cent and with a low cost catalyst (<2 lb catalyst per ton of product) free of corrosion problems.

8.9.4.3. Hooker process The Hooker process[118] is designed to produce a high purity product by a continuous two-step purification process—distillation and crystallisation (Fig. 8.10). A crude product is formed by the hydrochloric-acid-catalysed condensation of acetone with excess phenol. Water, acid, and some phenol are separated from the crude in a distillation column. The overhead from this column forms two liquid layers on condensation and cooling that are separated in a decanter. The organic layer is refluxed to the distillation column and the aqueous layer is refined in a second column to recover acid and purge water.

The acid free reaction product is distilled under reduced pressure to remove the bulk of the phenol. The residue is then heated in a second still to 240°C and 2 mm Hg to remove residual phenol and the ortho para isomer. This overhead and the prior phenol fraction are recycled to the reactor. Bisphenol-A in the isomer-still residue is separated from high boilers and tars in a thin-film evaporator[119] at 260°C and 2 mm Hg.

The hot liquid bisphenol-A overhead is mixed with benzene under pressure and is then released to a crystalliser to form a slurry of pure crystals in benzene. In this step decomposition products and other impurities that codistill with the product are extracted by the benzene. The crystals are centrifuged and dried. Oxalic acid, terephthalic acid, calcium and magnesium acid phsophates at 0·1 to 0·3 wt per cent may be used to stabilise bisphenol and minimise tar formation during distillation at 250–290°C.[120]

Benzene filtrate from the centrifugation is distilled to recover the pure solvent for recycling. Residues from this distillation contain phenol, *p*-isopropenylphenol monomer and polymer and are recycled to the reactor to reform bisphenol-A.

Tarry residue discarded from the thin-film bisphenol-A evaporator is the only raw material loss.

8.9.4.4. Water/phenol solvent crystallisation purification This process[121] uses large amounts of water to effect recovery of pure bisphenol-A. Crude bisphenol-A is produced by a conventional hydrochloric acid/mercaptan catalysed condensation of 10 moles of phenol with 1 of acetone. This crude is quenched with water and enough

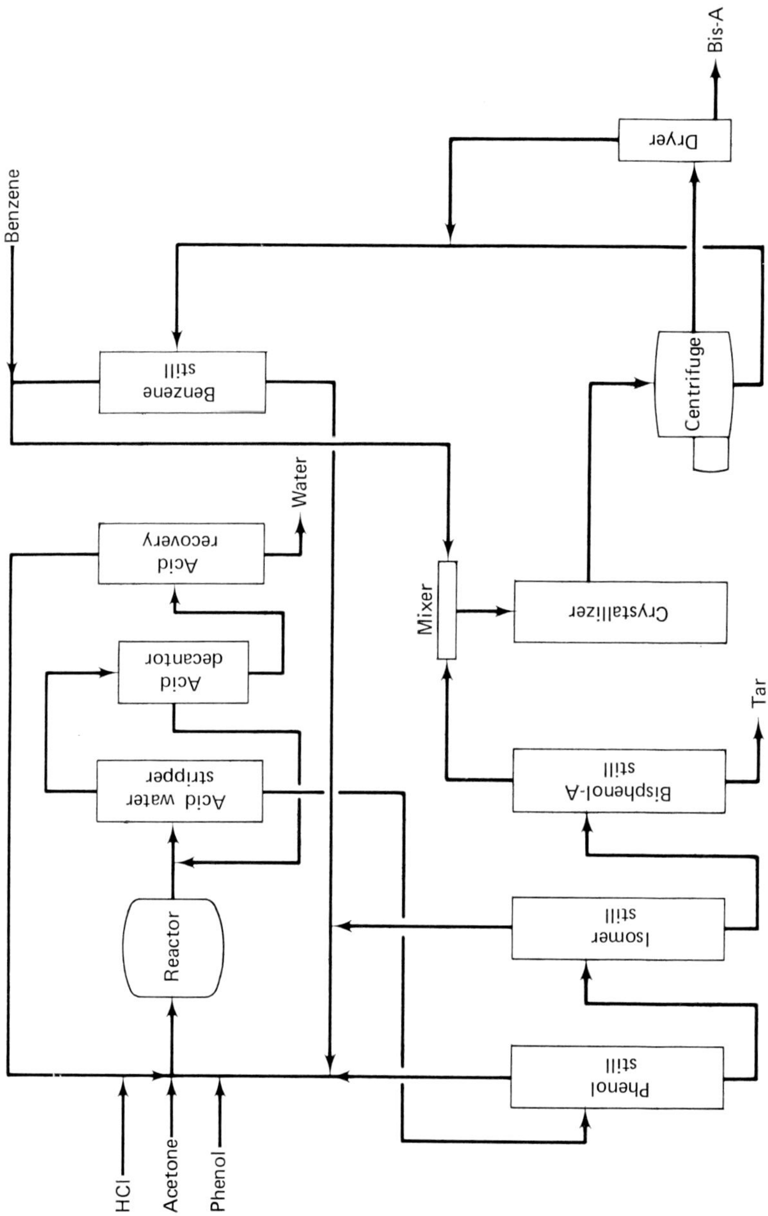

FIG. 8.10 *Hooker Bisphenol-A Process*

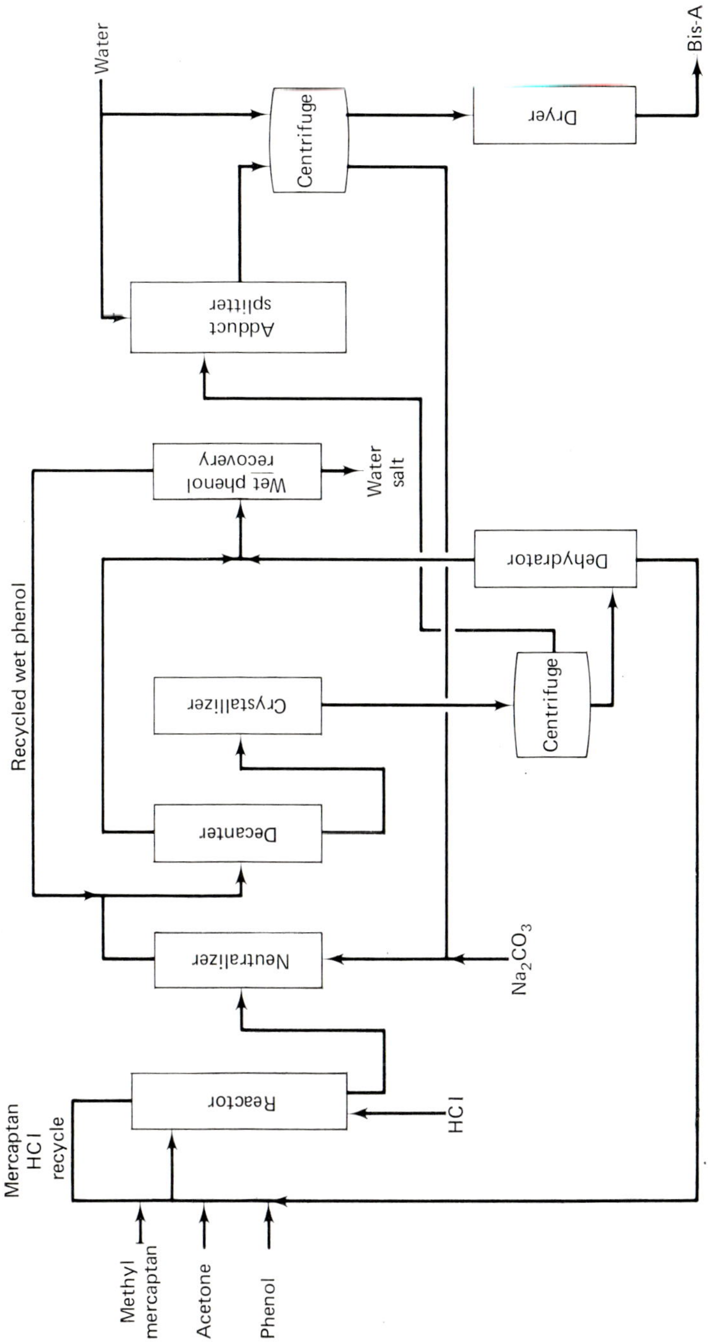

FIG. 8.11 *Purification of Bisphenol-A from Phenol Water System*

caustic to neutralise the acid catalyst and results in a solution containing about 70 per cent phenolics. The solution is cooled to 50°C to form two liquid phases that are separated with a decanter. The aqueous phase contains neutralisation salts and some phenol and is transferred to a phenol recovery unit. The wet organic layer is cooled to 25°C to effect crystallisation of the bisphenol-A-phenol adduct. Phenolic impurities remained dissolved in the wet phenol mother liquor. The crystals are centrifuged from the slurry and transferred to a vessel where they are contacted with 10 vols of water at 45°C to decompose the adduct, dissolve the phenol, and precipitate pure bisphenol-A crystals. These crystals are filtered, water washed, and dried to yield a high purity product. Aqueous mother liquor from the adduct splitting operation is used to quench the reaction mixture. Filtrate from the adduct filtration is dehydrated and returned to the reactor.

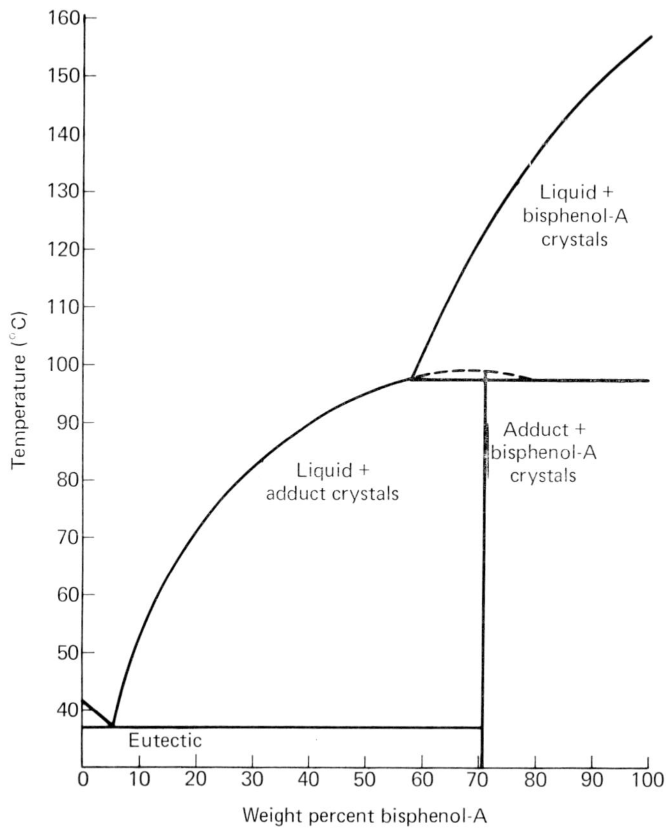

FIG. 8.12 *Melting Point Diagram Phenol-Bisphenol-A*

Treatment of thermally sensitive bisphenol under mild conditions is a major advantage to this process. Another advantage is that all phenolic streams including process impurities are recycled in the process. However, a purge stream would be required to remove irreversible side products from the phenolic mother liquor. The process also requires that large amounts of water, 10 lb/lb of product must be dephenolated.

8.9.4.5. Bisphenol-A purification by melt crystallisation Exceptionally high purity bisphenol-A is produced by melt recrystallisation of either the final product or the phenol adduct intermediate.[122] Bisphenol-A and phenol form a peritectic and eutectic system (Fig. 8.12). The equimolar complex contains about 30 wt per cent phenol and has an incongruent freezing point. On cooling a molten mass of phenol and bisphenol-A mixtures containing 5–58 per cent bisphenol-A yield equimolar adduct crystals; those containing 58–100 per cent bisphenol-A yield bisphenol A crystals.

A crystal slurry feed at 70°C composed of 50 per cent phenol and 50 per cent bisphenol-A is provided to a piston or pulse column crystalliser.[123] This is usually the deacified and concentrated product from the conventional acid-catalysed phenol acetone condensation (Fig. 8.13). The adduct crystal concentration in the feed is 50 per cent. The feed enters the top of column at 70°C and is forced downward with the down stroke of the piston. A portion of the liquor containing the impurities is forced out through screens at the middle of the column. Crystals forced down to the column base are melted with a heating column. A portion of this pure melt is also forced out of the column on the piston down stroke.

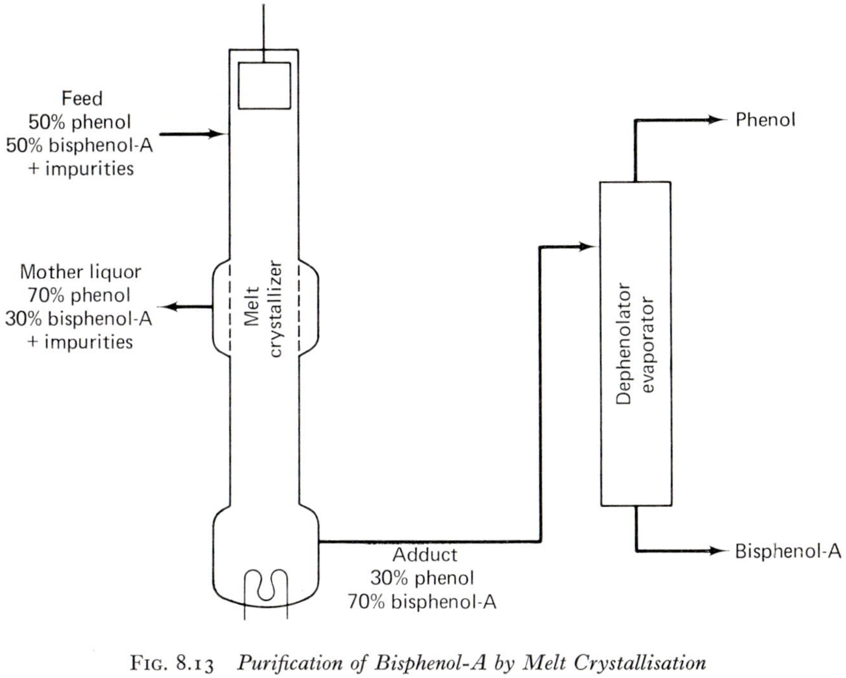

FIG. 8.13 *Purification of Bisphenol-A by Melt Crystallisation*

When the piston is reversed a portion of the pure melt travels up the column to provide a countercurrent wash for the crystals. A new portion of feed also enters the column on the up stroke to provide a continuous flow.

A one to one yield of filtrate and melted adduct product is obtained for a 70°C filtrate temperature and a 140°C product temperature. Phenol is removed from this intermediate product by distillation. A product assay of 99·9 per cent plus by gas chromatography is claimed for this process.

High purity bisphenol-A can be produced directly from a feed containing 15 per cent phenol with a mother liquor temperature of 110°C and a 180°C product temperature by column crystallisation.

8.9.5. Reactions

Bisphenol-A reaction chemistry is similar to phenol. Commercially important reactions involve both the hydroxyl group and the benzene ring. Bisphenol-A also undergoes cleavage to phenol and *p*-isopropenylphenol.

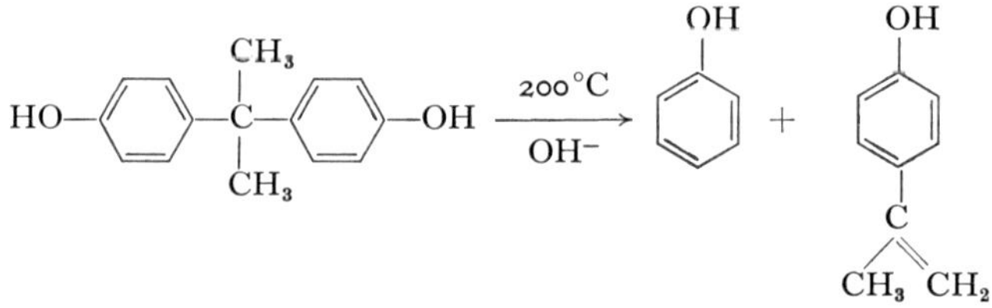

The alkenylphenol polymerises readily to a stable dimer. Good yields are obtained under alkaline conditions. When phenol is selectively removed higher molecular weight species are formed by the reaction of alkenylphenols with bisphenol-A and higher homologs.

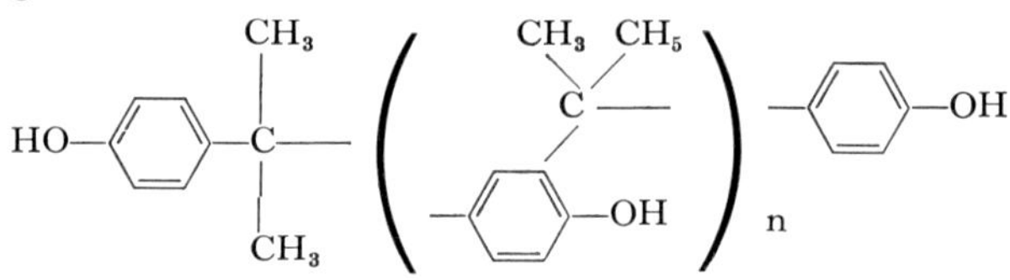

The reaction involving cleavage to phenol and *p*-isopropenylphenol takes place at temperatures as low as 150°C with strong acid catalysts and leads to cyclic products and resins:[124]

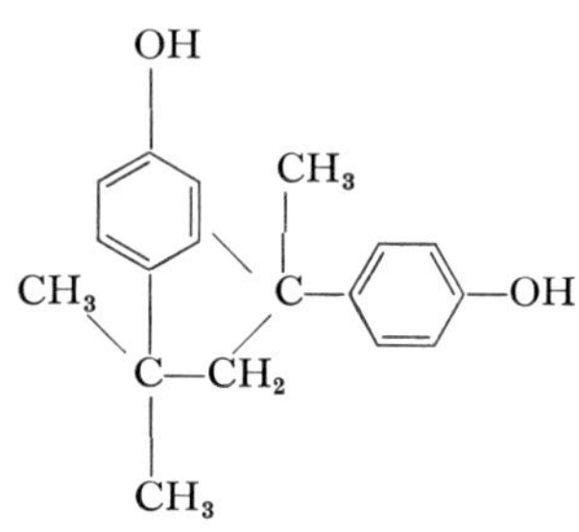

When the phenol is distilled away as rapidly as formed acidic cleavage at high temperature, 250–300 °C, lead to yields of 1·3 to 1·5 moles phenol per mole bisphenol-A.[125]

Alkaline hydrolytic cleavage of bisphenol-A at high temperatures gives high yields of phenol and acetone:[126]

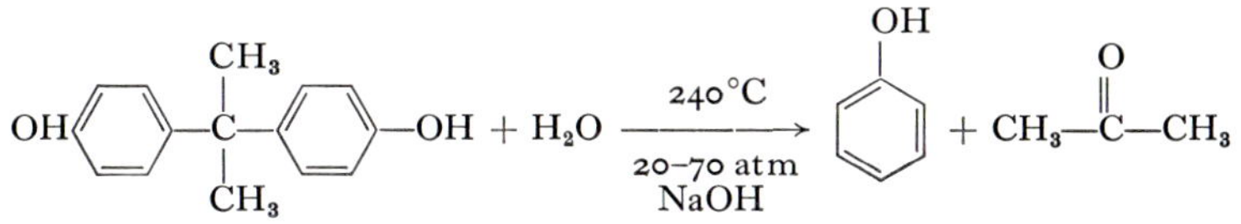

Reductive cleavage with acidic heavy-metal catalyst yields phenol and isopropyl phenol:[127]

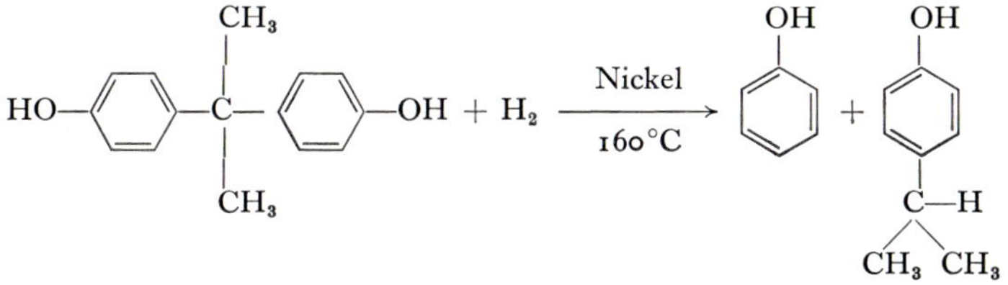

Oxidative cleavage of bisphenol with perchloric acid results in an oxidative-reductive disproportionation which yields *p*-isopropylphenol, quinones and resinous products:

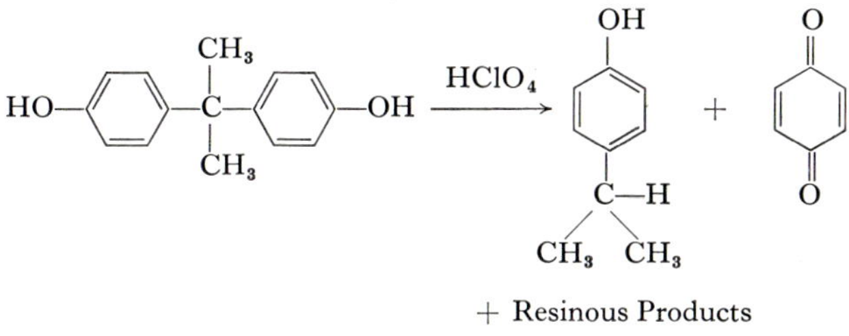

2,2bis(3,5-dibromo-4-hydroxyphenyl)propane, an important flame retardant for epoxies and polycarbonates, is produced by bromination of bisphenol-A in an ethanol-water solvent mixture:[128]

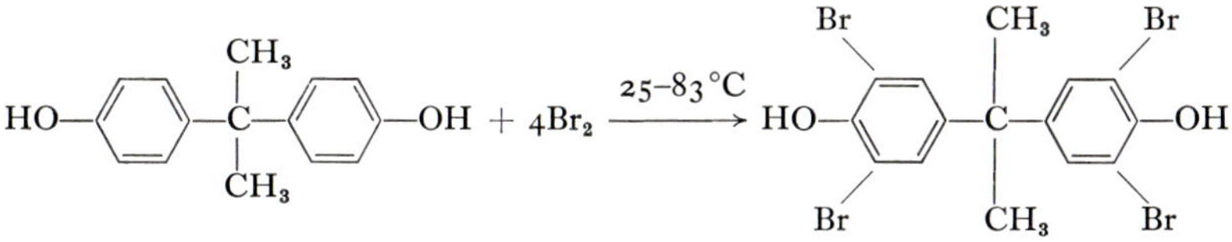

Alkylation of bisphenol-A is difficult as normal alkylating conditions, acid catalysts and temperatures of 100–150°C, lead to bisphenol-A decomposition.[129]

Alkyl derivatives are best prepared by condensation of the corresponding alkyl phenol with acetone:

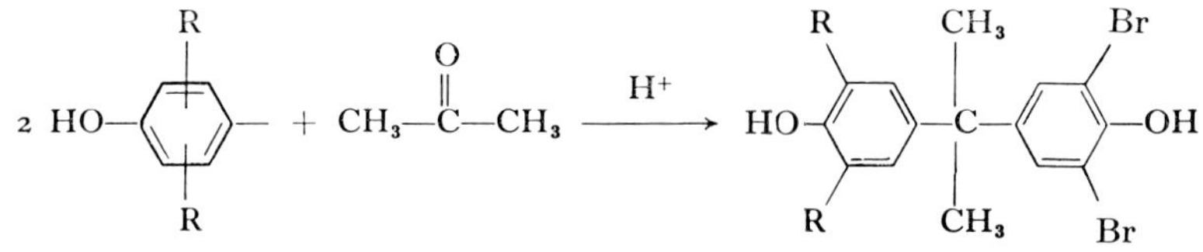

Novolak and resole type phenolic resins can be produced by the reaction of bisphenol-A with formaldehyde.

Hydrogenation of bisphenol-A leads to the corresponding cyclohexane derivative which is used in the manufacture of polyester resins.

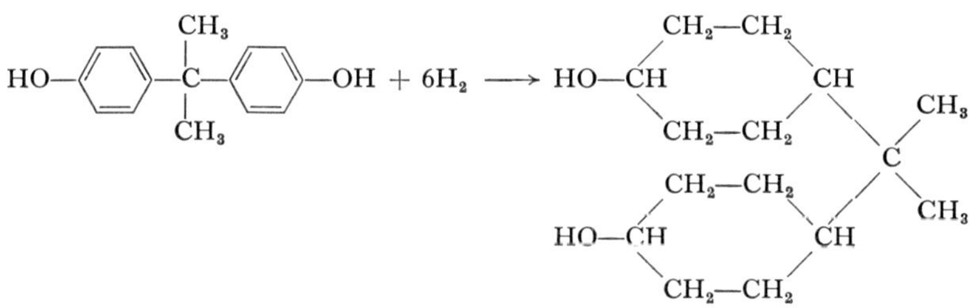

8.9.6. Uses

8.9.6.1. Epoxy resins Epoxy resins are the largest single use for bisphenol-A and are produced by a two-step reaction with epichlorohydrin:

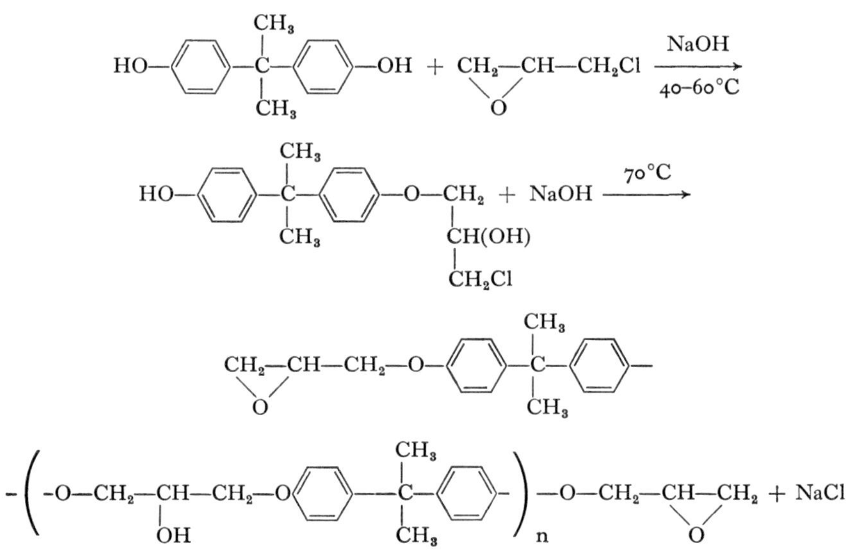

These resins are produced in a range of molecular weights from n = 0 for the diglycidyl ether to n = 10 for some of the solid coatings resins.[130]

Phenoxy resins, poly(hydroxy ethers)

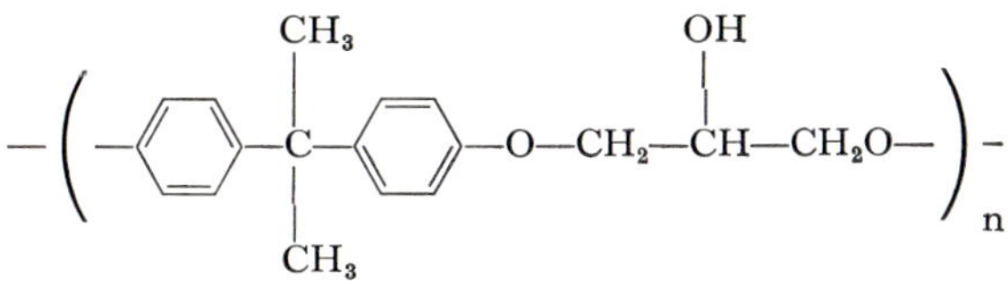

with molecular weights of 25,000 to 30,000 are produced by using essentially stoichiometric amounts of bisphenol-A and epichlorohydrin with a phenolic terminator.[131]

A high melting commercial aromatic polyether is produced by the reaction of bisphenol-A with bis(4-chlorophenyl) sulphone:[132]

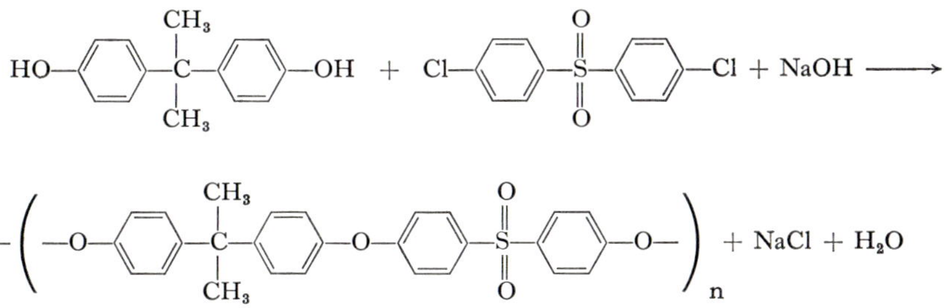

8.9.6.2. Polycarbonates Polycarbonates are prepared by the direct phosgenation of bisphenol-A, by reaction with bischlorocarbonates, or by ester exchange with diphenylcarbonate:

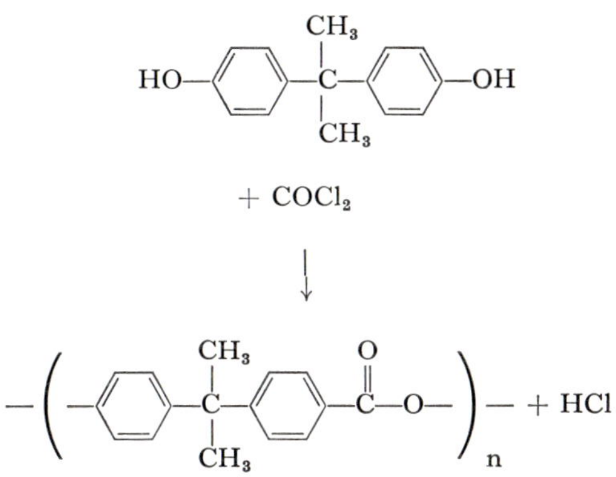

Pyridine or caustic is used to take up the acid. Methylene chloride is used as the solvent for the interfacial condensation of phosgene with the sodium salt of bisphenol-A.[135]

High melting aromatic polyesters are prepared by the reaction of the sodium salt of bisphenol-A with terephthaloyl chloride by interfacial polycondensation.

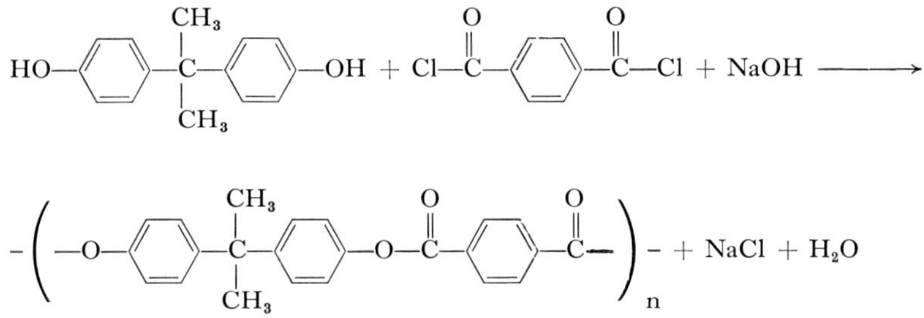

8.9.7. Economics and Markets

Worldwide consumption of bisphenol-A for 1970 is estimated at 470 million lb. Over two-thirds of the output was used in the manufacture of epoxy resins. The remainder was consumed in polycarbonate polymers and other applications (Table VIII.21).

TABLE VIII.21

Estimated Worldwide Bisphenol-A Consumption

	1970 Consumption (*Millions of Pounds*)	
Epoxy	320	68%
Polycarbonate	105	22%
Other:		
Antioxidants, Colour stabilisers, Polyesters, Polysulphones, Phenoxy resins, Flame retardants, Hydrogenation	45	10%
	470	100%

A list of bisphenol-A manufacturers is given in Table VIII.22.

Production of bisphenol-A in the United States for the decade 1960–70 showed a compounded annual growth rate of 14 per cent (Fig. 8.14, Table VIII.23) with 200 million lb produced in 1970 The growth rate was fairly steady over the entire period and continuation of this rate would put 1975 estimated production at 380 million lb. This rate represents 88 per cent of the total United States capacity that will be available by the end of 1972 (Table VIII.24) Thus, U S plants will operate well below capacity for the next few years The same will also be true in Europe where Shell has announced expansion of bisphenol-A facilities to 120 million lb/year at Pernis, the Netherlands and 44 million lb/year at Stanlow, U K [134]

TABLE VIII.22

Bisphenol-A Producers

France	Progil
Netherlands	Shell Nederland Chemie
	Ketjen
Germany	Farbenfabriken Bayer
Japan	Mitsui Toatsu
	Honshu Chemical
United Kingdom	Shell Chemical
	Graesser
United States	Dow Chemical
	General Electric
	Monsanto Chemical
	Shell Chemical
	Union Carbide

TABLE VIII.23

Production of Bisphenol-A in United States[133]

Year	*Production (millions of pounds)*
1960	50·5
1961	45·8
1962	62·1
1963	72·4
1964	92·9
1965	97·2
1966	126·4
1967	130·4
1968	142·3
1969	182·1
1970	200·7

TABLE VIII.24

Production Capacity in the United States

	Capacity (millions of pounds/year)
Dow Chemical	100
General Electric	50
Monsanto	30
Shell Chemical	150
Union Carbide	100
Total	430

Growth in bisphenol-A consumption is expected to continue in epoxy resins and polycarbonates. Over the past decade (1960–70) epoxy resins grew at an annually compounded rate of 9·3 per cent. At the same time polycarbonate production increased at a 55 per cent rate. During the last five years (1965–70) these rates dropped to 7·0 per cent for epoxies and 22 per cent for polycarbonates [133]

Epoxy resins are used in a wide range of applications such as casting, potting and encapsulation of electrical components, adhesives for a wide range of commercial applications, as the binder for lamination of paper, fibre glass cloth, wood, mica, etc., and for the production of tough chemical resistant coatings. The end-use distribution for epoxy resins in the United States for 1971 is shown in Table VIII.25.

TABLE VIII.25

Epoxy Resin End-Use Pattern[135]

Bonding and adhesives	12%
Coatings	43%
Reinforced plastics	14%
Tooling, casting and moulding	7%
Other	24%

Polycarbonate resins have exceptional high impact strength coupled with good dimensional stability, transparency, heat resistance, and electrical properties. They are used in a wide range of applications where these properties can be put to advantage in spite of the relatively high cost. End-use pattern for polycarbonates in the United States are shown in Table VIII.26.

TABLE VIII.26

End-Use for Polycarbonate Resins in United States for 1972[135]

Appliances	16%
Building and glazing	11%
Communications and electronics	29%
Lighting and signs	7%
Recreation	16%
Transportation	13%
Other	8%

Quoted prices for carlots (minimum single shipments of 70,000 lb) at 30·5 ct/lb and less than carlots at 32·5 ct/lb in the United States have held constant over the past ten years.[136] However, the value of large shipments (unit value) which is indicative of contract prices has declined steadily at about 4 per cent per year as a result of the general erosion of chemical prices and excessive production capacity.[133] This is expected to continue for the next few years until demand approaches supply (Fig. 8.14). In practice recent shortages in the availability of phenol have altered the picture.

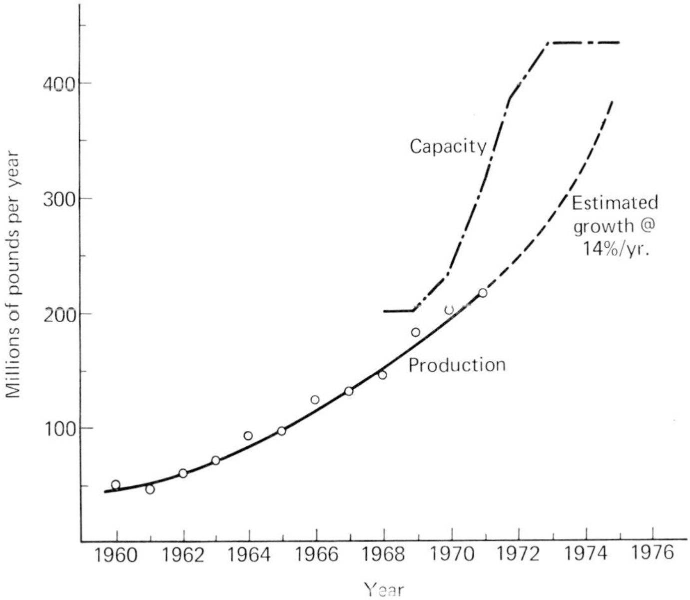

FIG. 8.14 *Bisphenol-A Production and Capacity in the United States*

8.10. PHENOLIC RESINS by F. M. Konrad

8.10.1. Historical

The term phenolic resins is used to denote a variety of phenol formaldehyde reaction products that have a wide range of properties and are used as a binder for composite products in the building, electrical, automotive, and appliance industries. These resins constitute the greatest single use for synthetic phenol.

Peria reported the first phenolic resinous product obtained by heating *o*-hydroxybenzylalcohol (saligenin) derived from a natural glucoside.[138] About thirty years later Baeyer[139] recognised that the phenol-aldehyde reaction was a general one and produced resinous substances by the acid catalysed condensation of benzaldehyde with pyrogallic acid and resorcinol, and by the condensation of phenol with acetaldehyde and methylene diacetate. Michail[140] observed that alkaline catalysis also yield resinous products from phenols and aldehydes. Isolation of crystalline products from these reactions was difficult and discouraged extensive studies by early investigators.

With the advent of commercial formaldehyde in 1890, Kleeberg[141] investigated hydrochloric acid-catalysed phenol-formaldehyde reactions and obtained infusible products that were insoluble in organic solvents, acids, and alkalies. Paal and Serminger[142] found that similar products were obtained on heating saligenin with ammonia. Procedures for the preparation and isolation of saligenin from the phenol-formaldehyde reaction had been worked out by Lederer[143] and Manasse.[144]

During the next two decades, 1890–1910, several patents were issued for the preparation of phenolic resins. Smith[145] and Luft[146] attempted to moderate the reaction and produce fusible products by dilution with solvents. Blumer[147] prepared a shellac substitute from the reaction product of 2 moles of phenol, 2 moles of formaldehyde and 1 mole of tartaric acid. Fayolle[148] attempted to produce a gutta-percha substitute by extending the condensation products with glycerol, pitch and oils.

Baekeland started his studies on the phenol formaldehyde condensation reaction in 1905 which led to the first production of commercial resins in 1909.[149] He separated the process for the conversion of raw materials to useful end products into three stages, 'A', 'B' and 'C'. Baekeland described the 'A-stage' as an initial condensation product, liquid or solid, that is soluble in polar solvents and may be remelted. This product on heating advances to a 'B-stage' which has limited solubility and fusibility but may be moulded under heat and pressure to the end product 'C-stage'. The final 'C-stage' product is insoluble in organic solvents and does not melt at temperatures up to 300°C. Baekeland recognised the importance of using limited amounts of catalyst for reaction control and the need for careful control of the reaction stoichiometry. He observed that reactive 'A-stage' resin (resoles) may be produced with excess formaldehyde and a base catalyst and that acid catalysis with excess formaldehyde led to an unmanageable, infusible product. However, when phenol was in excess, acid catalysis produced a thermoplastic 'novolak' resin. Most important his 'heat and pressure' patent[150] described the conditions necessary for rapid resin curing with containment of the volatile by-products during moulding of useful articles.

In 1912 Aylsworth[151] developed the hexamethylenetetramine-novolak resin curing system that has been the basis for the major portion of the phenolic moulding and bonding compounds used today.

In 1910 Baekeland formed the General Bakelite Company for the manufacture of phenolic resins and compounds and dominated the industry until his patents expired. Today's worldwide industry has a large number of manufacturers, many in captive businesses that produce nearly two thousand million pounds annually.

8.10.2. Chemistry

Phenolic resins are mixtures of various size methylene linked polynuclear compounds with species of up to 20,000 mol wt (e.g. n = 130).

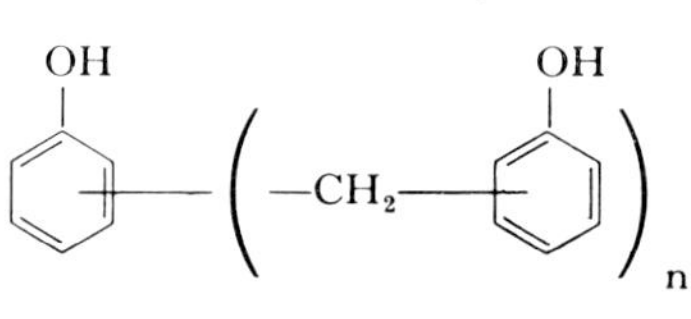

Resins prepared with formaldehyde and alkaline catalysts under mild reaction conditions also contain heat reactive methylol groups (hydroxybenzyl) added to the phenolic ring.

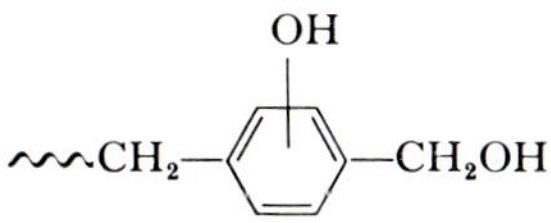

Methylol phenols formed by the addition of formaldehyde to the phenolic ring at the ortho and para positions condense with additional phenol to yield methylene derivatives and water. Repetition of this sequence leads to chains of up to 20 units in length. Formation of methylol phenols via an aldol type mechanism first proposed by Manasse[144] and later favoured by several investigators,[152, 153] involves the reaction of formaldehyde with the phenolic anion.

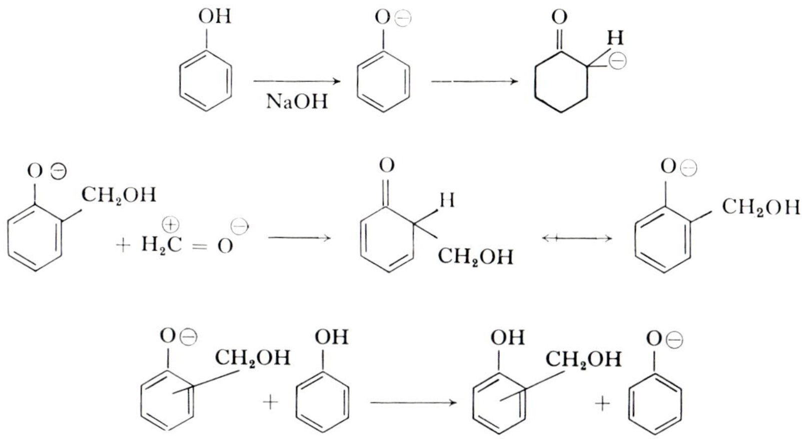

Walker[154] proposed that the reaction takes place via formation of a hemiformal intermediate which rearranges to the methylol phenols.

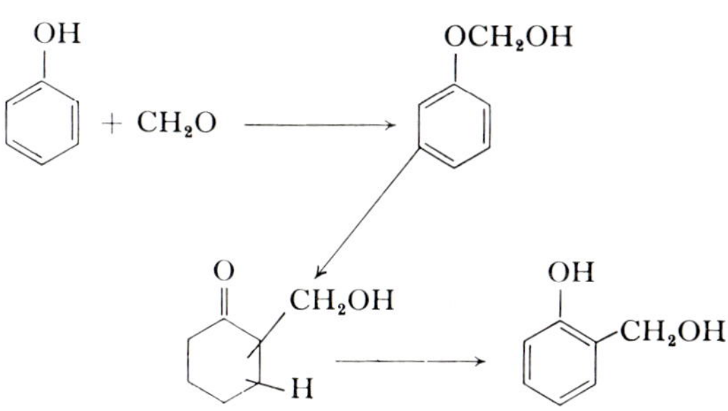

Phenol hemiformals are formed on the addition of anhydrous formaldehyde monomer to phenol. Significant concentrations of hydroxybenzyl hemiformals have been observed in alkaline catalysed phenol formaldehyde reactions by proton magnetic resonance.[155]

The para position is more active than either ortho, however, ortho substitution predominates as there are twice as many available positions. Enhancement of the activity of residual phenol ring positions due to addition to the first and second methylol group has been reported by several investigators.[156] Freeman and Lewis[157] studied the alkaline catalysed reaction using three moles of formaldehyde to one mole phenol in dilute aqueous solution at 30°C. They measured the change in concentration of all possible methylolphenols with time by paper chromatography and were able to assign relative reactivities to the open ring positions of phenol, monomethylol phenols, and the dimethylol phenols which showed that most of the methylol phenols were more reactive than phenol (Table VIII.27). Thus, alkaline catalysed reactions of equimolar quantities of phenol and formaldehyde under mild conditions lead to significant yields of dimethylol and trimethylolphenols and leave a large amount of unreacted phenol although all of the formaldehyde has been consumed.

TABLE VIII.27

Reactivity of Phenol and Methylolphenol Ring Positions

Compound	*Position*	*Relative reactivity*
p-Methylolphenol	Ortho	1·0
Phenol	Ortho	1·4
Phenol	Para	1·7
o-Methylolphenol	Para	2·0
p-Methylolphenol	Ortho	2·3
o,p-Dimethylolphenol	Ortho	2·4
o,o-Dimethylolphenol	Para	11·1

The overall alkaline catalysed phenol-formaldehyde reaction is second order, dependent on the concentrations of phenolic anion and formaldehyde, and has an activation energy of 22 Kcal/mole. De Jong and De Jonge[153] studied the reaction rate at 70–130°C over a wide pH range and observed that the reaction rate increased directly proportional to the hydroxyl ion concentration with a minimum rate at pH5 to a maximum rate at one mole caustic per mole of phenol.

Investigations by Zavitsas[758, 159] which accounted for the concentrations of the ionic methylolphenols and the monomeric formaldehyde-benzylhemiformal equilibrium in addition to the concentration of phenate ion also showed that the reaction conformed to second order kinetics that were in good agreement with the earlier work of Freeman and Lewis.[157]

Although ammonia, primary and secondary amines are used extensively as catalysts for the production of commercial resoles, their role as catalysts has not been clearly elucidated. Primary and secondary amines react with phenol and formaldehyde to form Mannich bases that have been considered as the active catalyst. However, amine catalysis does produce heat reactive resins with methylolphenol structures.[160]

Resoles are not prepared by acid catalysis as methylolphenols are unstable and condense to methylene structures under acidic conditions. However, small amounts of methylolphenols observed during acid catalysed reactions suggest that these compounds may function as reactive intermediates in this system.[161, 162] Reaction of protonated formaldehyde with the phenol quinonoid structure has been proposed as a mechanism.[152]

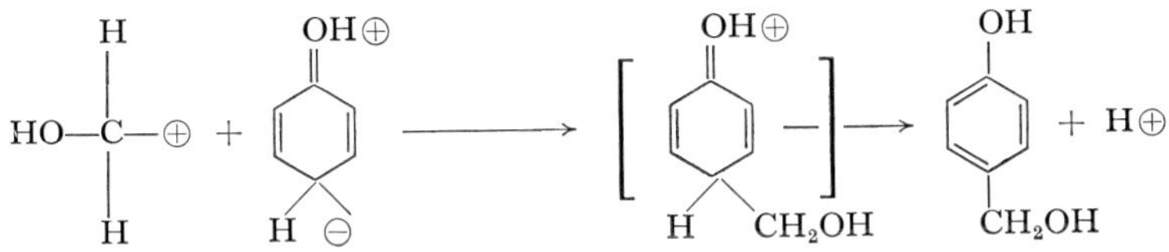

However, acid conditions favour elimination of water by the ionic methylol quinone to yield the reactive quinone-methide which rapidly adds a second phenol (cf. equation (c) on p. 422).

Methylolphenols condense with phenol to form the three isomeric dihydroxydiphenylmethanes:

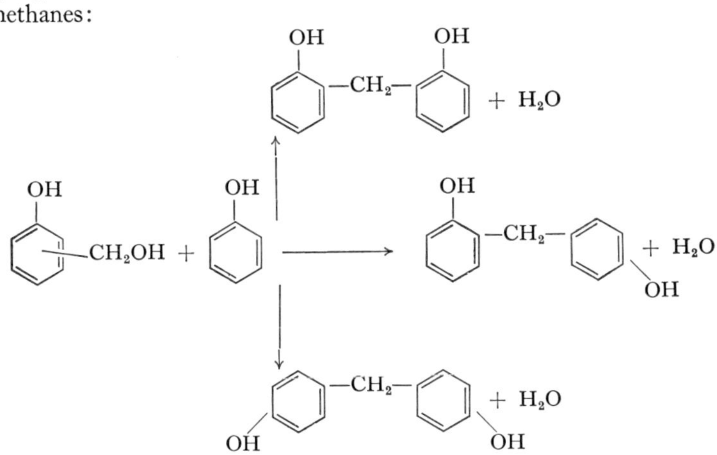

Methylolphenols also self condense to form methylene derivatives which ultimately yield resinous products.

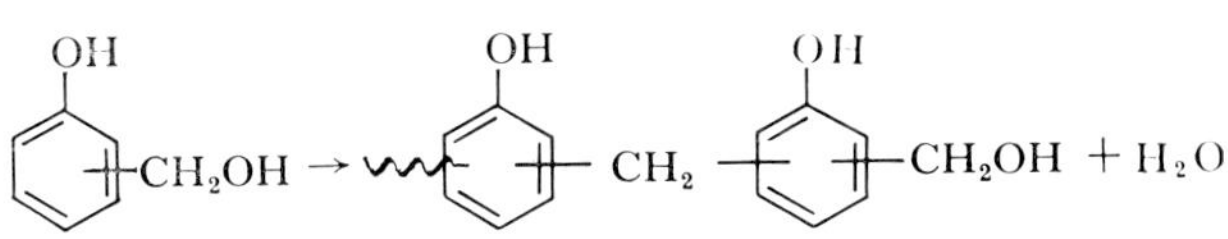

Condensation through the methylol groups with the elimination of water and formaldehyde has also been observed.[157]

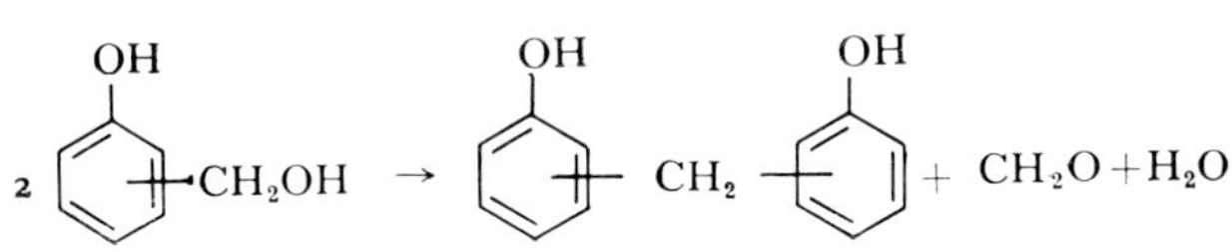

Dihydroxydiphenylmethanes are the initial products in the acid catalysed phenol formaldehyde reaction and may be prepared substantially free of higher molecular weight species by using a large excess of phenol. The reaction mechanism is analogous to that for the phenol-ketone reaction to bisphenols. In Hultzsch's mechanism[163] protonated formaldehyde adds to the quinonoid phenol structure to form a methylol quinone intermediate (equation (a) which elimates water to form the reactive quinone-methide (equation (b)). The quinone-methide reacts immediately with a second phenol molecule to form the dihydroxydiphenyl methane and release the proton (equation (c)).

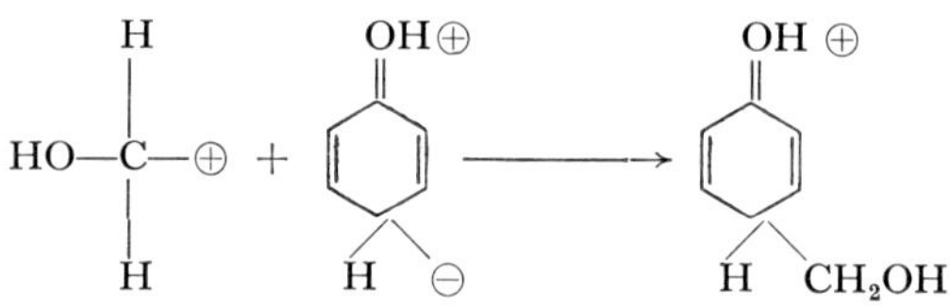

(a)

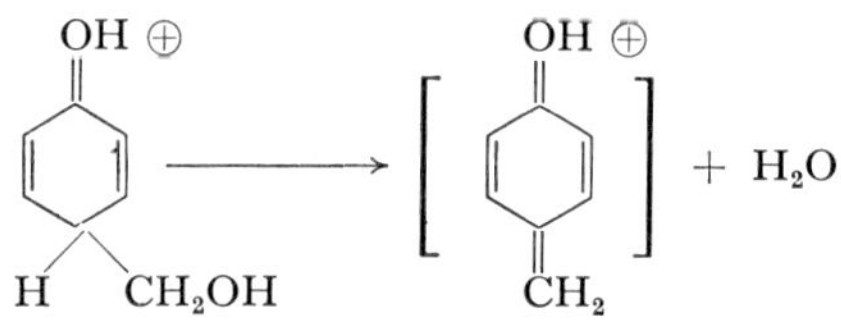

(b)

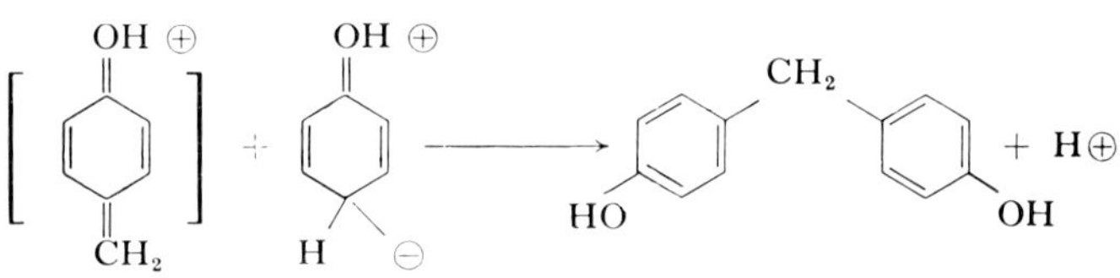

(c)

Repetition of this reaction sequence with product species terminal rings leads to chains containing up to twenty phenolic rings. Substitution takes place at the ortho and para ring positions. Strong acids such as phenolsulphonic or hydrochloric acid promote para substitution. Catalysts with positive, bivalent metal ions such as manganese, zinc, cadmium, magnesium, cobalt and calcium favour ortho substitution[164] and are used to prepare rapid curing novolak resins.[165] Fraser *et al.*, proposed a chelated complex as the reactive species for yielding the highly ortho directed derivatives.

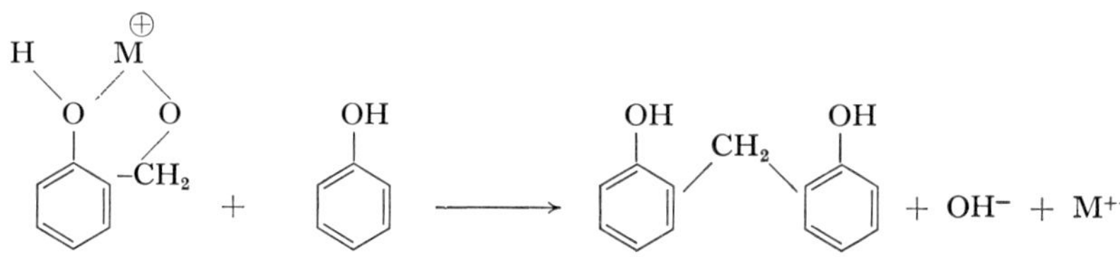

The overall acid-catalysed condensation reaction rate is limited by the rate of formation of the methylol intermediate. The condensation is highly exothermic and liberates 21 Kcal/g mole formaldehyde. A reaction kinetic study by Jones[166] at 80°C and 100°C and at pH 1 to 3 showed that the rate is second order and is dependent on the hydrogen ion concentration.

Dihydroxydibenzyl ethers are formed on heating methylolphenols in neutral or acidic conditions and, under certain conditions, take part in the resole curing process.

On further heating particularly in the presence of acid the dibenzyl ethers yield methylene-bridged phenols and formaldehyde.

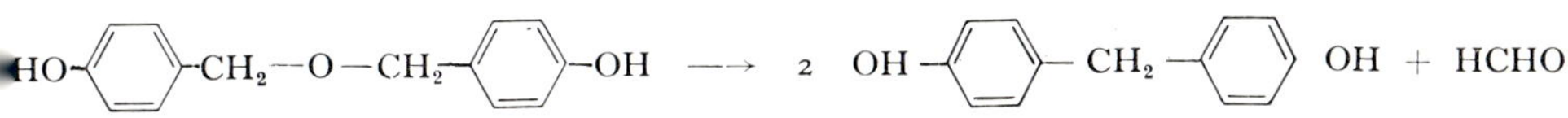

Cyclic 1,3-benzodioxanes are a side product in the normal phenol formaldehyde condensation.

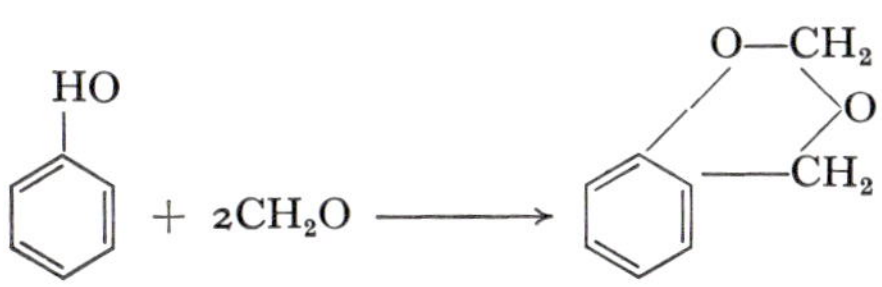

p-Substituted phenols enhance this cyclic formation. Significant yields of 4-alkyl-1,3-benzodioxane are obtained in the preparation of alkylphenol novolak resins.

8.10.3. Resin Structure

The novolak resin structure is made up of phenolic nuclei connected at ortho and para positions with a methylene bridge. The ratio of ortho to para substitution is a function of catalyst type. Reactions catalysed with strong acids yield resins with a large amount of ortho-para internal groupings and 50–80 per cent para substituted end groups. The higher molecular weight species also contain 5–20 per cent tri-substituted phenolic rings giving branched chains. Resin reactions carried out in the 4 to 9 pH range with such catalysts as lead acetate, zinc oxide or lime yield structures with 50–90 per cent ortho-ortho substitution. These resins have only minor amounts, if any, branched chains.

The number average molecular weight of commercial novolaks varies from 200, for a product that is largely a mixture of the isomeric dihydroxydiphenylmethanes, to 800 for the highly viscous resins. The product molecular weight is dependent upon the ratio of formaldehyde to phenol. Fusible novolaks can be produced at ratios as high as 0·90. Higher ratios approaching unity result in gelation. Molecular weight distribution determinations of novolaks by fractionation and gel permeation chromatographic procedures[167] show species with 'n' values of zero to $n = 100$ (Fig. 8.15) and dispersion values (ratio of weight average to number average molecular weight)of 1.5 to 4.0.

The *p*-substituted alkyl and arylphenols yield linear resin chains and a less dispersed molecular weight distribution than resins from trifunctional phenols.

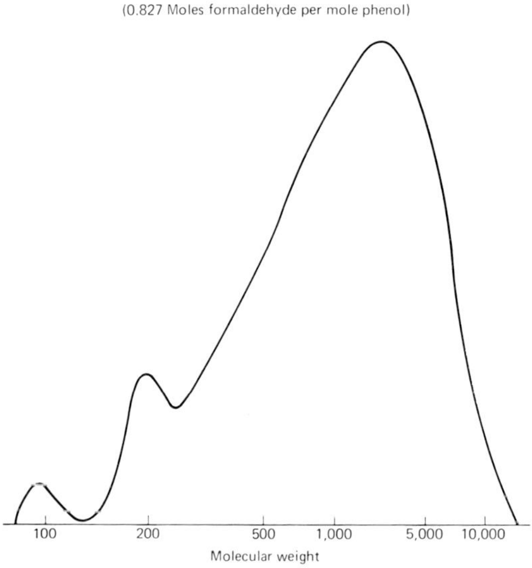

FIG. 8.15 *Molecular Weight Distribution of an Oxalic Acid Catalysed Phenol Formaldehyde Novolak by Gel Permeation Chromatography (0·827 moles formaldehyde per mole phenol)*

The relationship of resin melt viscosity to molecular weight (Fig. 8.16) was studied at various temperatures by Jones.[168] The rapid decrease in activation energy (50–16 Kcal/g mole) observed with increase in temperature of 120–200°C was attributed to a reduction in hydrogen bonding. Dienes[169] reported activation energy values of 70–50 Kcal/g mole for the temperature range 60–100°C and also observed a strong sensitivity of melt viscosity measurements to resin moisture content. For example, at 60°C an appreciable viscosity reduction was effected by 1·5 per cent dissolved moisture.

Resole resins have more complex structures than novolaks and contain methylolphenol, dibenzylethers and benzylhemiformals, in addition to methylene. These structures,

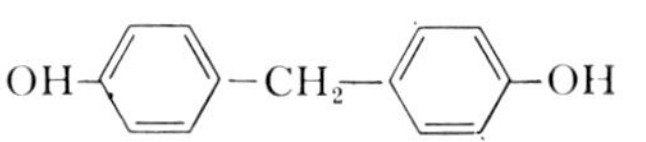 ☐ Methylene

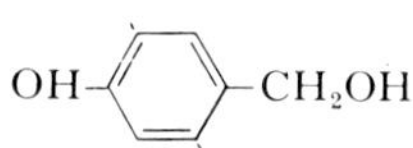 ☐ Methylol

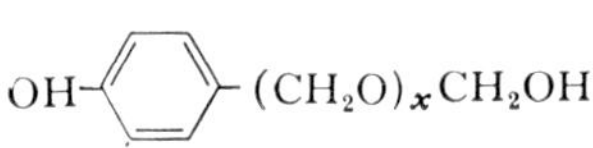 Benzylhemiformal

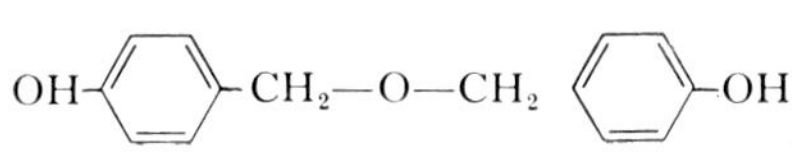 Dibenzylether

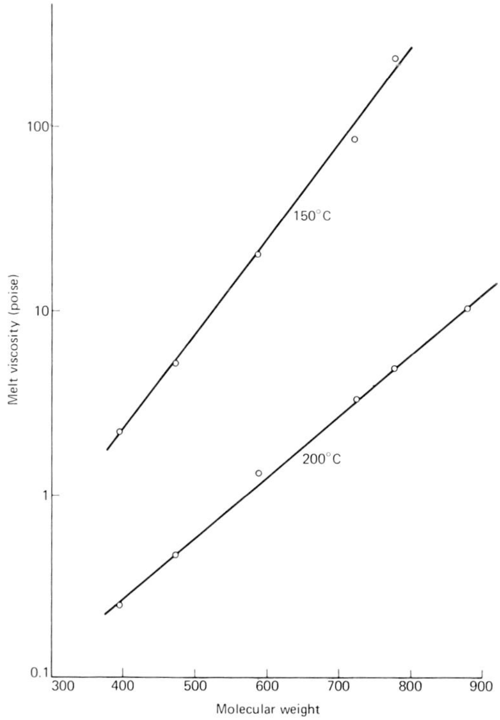

FIG. 8.16 *Melt Viscosity of Novolak Resins as a Function of Molecular Weight*

have been detected in commercial type resoles by proton magnetic resonance studies.[155] The 'labile formaldehyde' structures are favoured by high formaldehyde-phenol ratios and mild reaction conditions. Water dilutable liquid resoles reacted at 50°C with two to three moles of formaldehyde per mole of phenol are essentially mixtures of mononuclear products. Resoles used for laminating varnishes and molding compounds are more advanced and contain a methylene structure with number average molecular weights up to 1,000.

The viscosity of resoles varies widely from thin syrupy water-dilutable types to the solid grindable resins that are extremely viscous at elevated temperatures (110°C).

Resole resins contain the functionality to effect cure to an insoluble, infusible, highly cross-linked product through the reaction of methylol groups to stable methylene links. Water is eliminated during this reaction. The rate of cure is dependent on temperature, resin acidity and structure, free phenol and formaldehyde contents, and the resin moisture content. Solvents, lubricants, catalysts and fillers used with the resin in end-product formulations also effect the rate of cure. A one to one mole ratio of formaldehyde to phenol is the minimum functionality required to effect the curing reaction. Maximum cure is attained with a stoichiometric ratio of 2·5 moles of formaldehyde to one mole of phenol.

Hexamethylenetetramine (hexa) is used almost exclusively to provide the additional formaldehyde functionality needed to cure novolaks. Resoles and paraform have also been used to a limited extent. Hexa concentrations above 6 per cent are required to effect cure and 10–15 per cent produce maximum results. Excessive hexa is detrimental to good end-product performance.

The relative cure rate of resins may be compared via the time required to attain the infusible gel state when a sample is heated on a hot plate. Cure rates for the three isomeric dihydroxydiphenylmethanes with 15 per cent hexa at 160°C is as follows[170]:

2,2′Dihydroxydiphenylmethane	60 sec
2,4′Dihydroxydiphenylmethane	240 sec
4,4′Dihydroxydiphenylmethane	180 sec

Small amounts of water, free phenol or acid catalyst accelerate the curing reaction.

8.10.4. Resin Reactions

Both the phenolic hydroxyl and the open ring positions are used for combining phenolic resins with other substances in several applications.

Halogens react readily under mild conditions at open ortho and para positions on the phenolic ring and will replace the methylol hydroxyl.

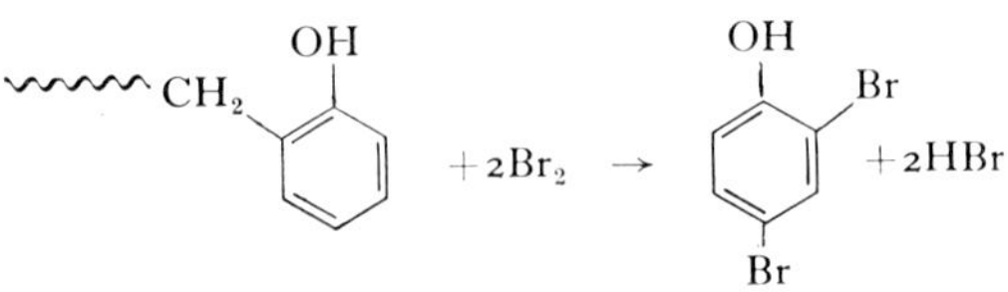

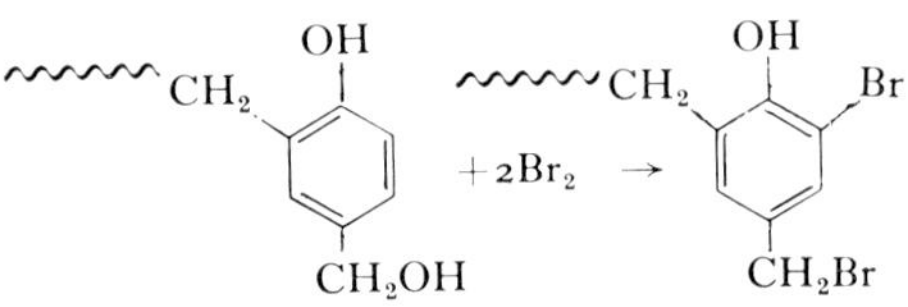

Under more vigorous conditions, halogens will substitute in the meta position and will attack the methylene bridge and ultimately cause cleavage.[171]

Novolak resins are readily sulphonated with sulphuric acid. However, sulphonated cation-exchange phenolic resins are usually prepared by the reaction of a mixture of phenolsulphonic acid and phenol with formaldehyde.

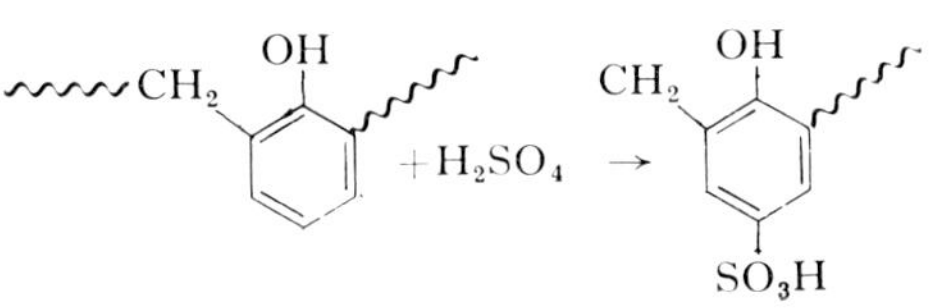

Both the phenolic hydroxyl and the methylol hydroxyl are readily esterified with acid anhydrides.

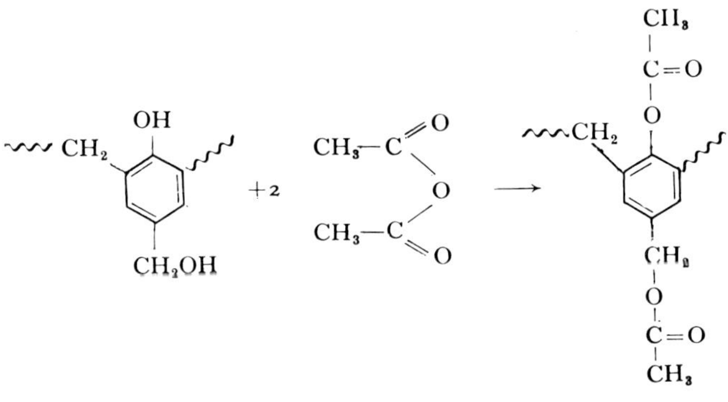

The phenolic hydroxyl on methyldphenols may be alkylated exclusively with alkyl halides in alkaline media.

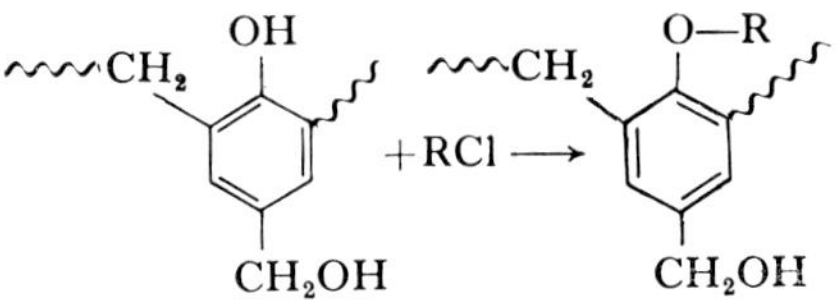

Alkylchlorosilanes react with both the phenolic hydroxyl and the methylol group.

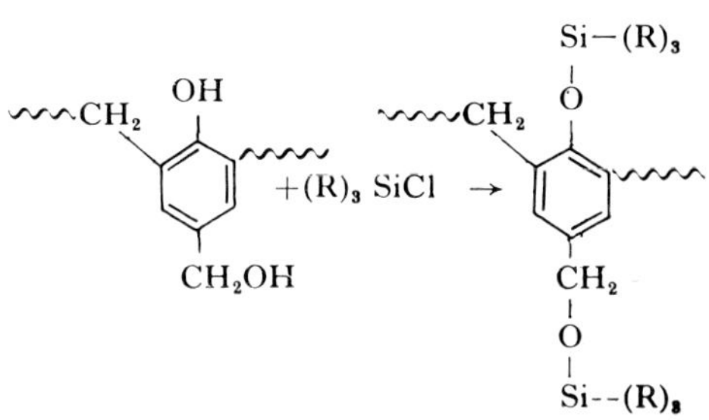

These reactions have been used in analytical procedures to elucidate resin structure.

Alkoxylation of methylol phenols with aliphtaic alcohols has been used to render resoles compatible with oils.[172]

Surface active agents have been prepared from phenolic resins by etherification of the phenolic hydroxyl with ethylene or propylene oxide.

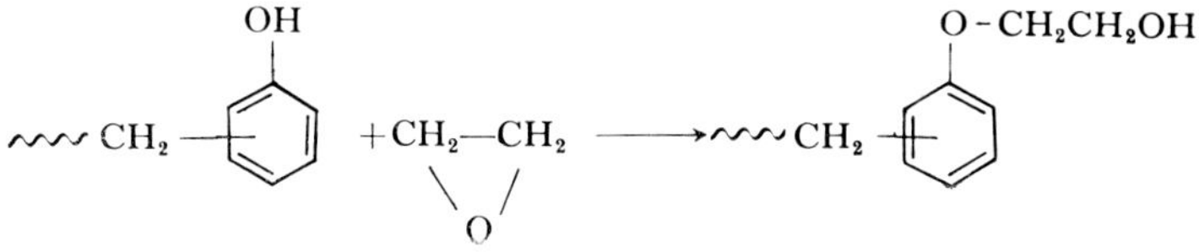

Epoxidation of novolaks with epichlorohydrin is an important industrial process. The product may be liquid or a grindable solid and this depends on the molecular weight of the novolak used. The two-step reaction is the same as that for the production of epoxy resins from bisphenol-A where an excess of epichlorohydrin is used to minimise side reactions with the secondary aliphatic hydroxyl group.

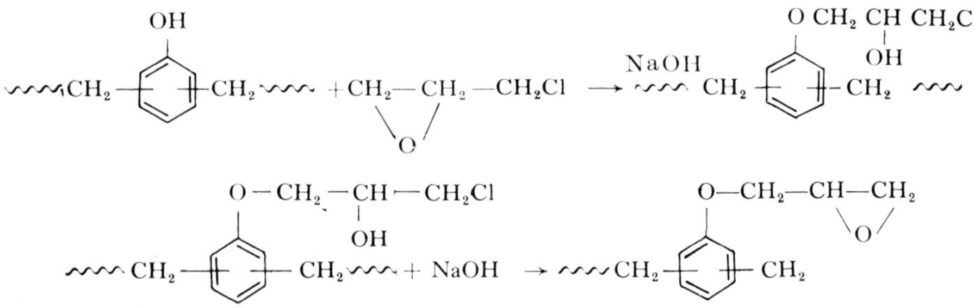

Novolaks are also used as the hardener for curing epoxy resins.

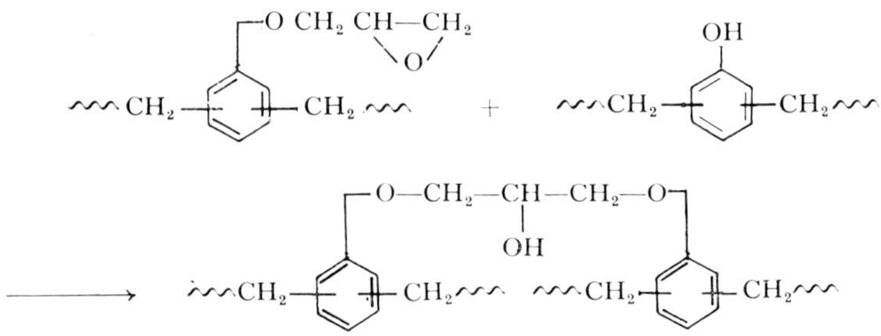

Oil soluble alkylphenol resoles are used to vulcanise butyl rubber. Van der Meer[173] proposes a quinone methide as the reactive intermediate.

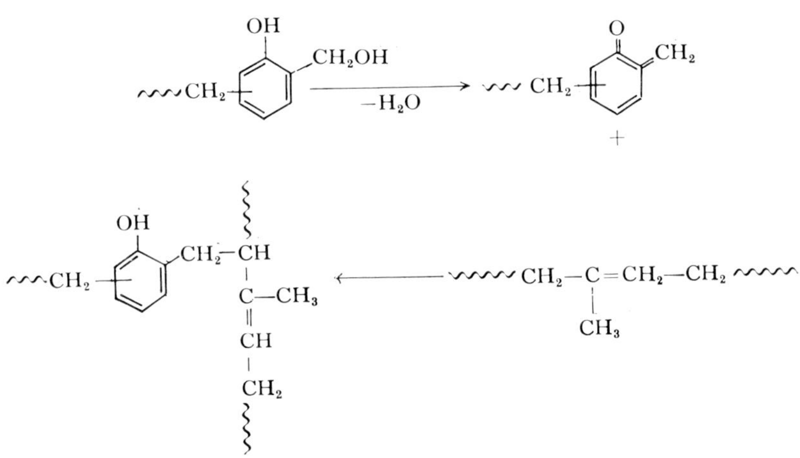

Oil soluble alkylphenol novolak and resole resins are used extensively to prepare varnishes by cooking them with unsaturated drying oils. The loss of unsaturation and phenolic hydroxyls are indications that a reaction has taken place. A clear understanding of the mechanism is not available. The reaction of *o*-methylolphenols to form chromans may take place with resoles but does not explain the behaviour of novolaks. Castle's[174] proposal for alkylation of the phenolic hydroxyl with the aliphatic unsaturation is consistent for both resoles and novolaks.

8.10.5. Manufacturing Processes

8.10.5.1. Novolaks Novalok resins are manufactured batchwise in kettles which range in size from less than 100 gallons to a maximum of 10,000 gallons and are usually fabricated from stainless steel of Hastelloy particularly where strong acid catalysts are used or where light coloured resins are desired. Phenol and formaldehyde are charged to the reactor from heated weigh tanks. The kettle is fitted with an anchor-type or turbine agitator electrically driven through a gear reducer and usually wired for two speeds because of the high end point torque and the large viscosity change during the process cycle. The kettle is jacketed, and in the case of larger vessels, internal coils are provided, for heating and cooling. A shell and tube heat exchanger is used for reflux cooling during the vigorous exothermic reaction stages and for distillate condensation during the dehydration stage. A vacuum receiver is used to collect the distillate. The kettle should also be provided with a vent line capable of carrying the total vapour load generated by the reaction exotherm within the vessel working pressure limits. The unit is instrumented to measure the batch temperature at one or more locations, the kettle pressure, the power input to the agitator, distillate flow rate, and pressure and flow of steam and cooling water.

Facilities for solidification of the product are varied—cooling pans, drums or belt flaker, and a cooling floor have been used.

Novolaks are made with excess phenol that have charge mole ratios of 0·7–0·85 moles formaldehyde per mole of phenol. Formaldehyde is usually charged as an aqueous solution, 'formalin,' which has a concentration of 37–46 per cent and may contain small amounts of methanol.

Phenol and formalin are charged to weigh tanks and then dropped into the resin kettle and are mixed together. The batch temperature is 50–60°C at this point. The catalyst is charged, and mixed into the batch, which is then sampled and tested for catalyst concentration. When catalyst concentration has been assured, the charge is heated with steam or tempered water to about 80°C where upon the heat of reaction becomes sufficiently great for unaided temperature rise. Heat is discontinued and a combination of cooling water and refluxing is used to control the reaction temperature at 80–120°C. With an oxalic acid catalyst the total reaction period is 3 to 6 hr and steam is required to maintain the batch temperature after the first hour of reaction.

Sulphuric acid catalysed reactions are very vigorous and can not be controlled by the sequence described above. In this case the phenol and catalyst are charged and heated to 80°C. Formalin is then metered at a rate to produce a controllable exotherm. Using this procedure the reaction requires 30–90 min.

During the reaction period the batch separates into two liquid phases—aqueous and resinous. At reaction completion the aqueous layer contains less than 0·5 per cent formaldehyde and the catalyst is then neutralised with alkali. Oxalic acid catalysed resins need not be neutralised as this acid decomposes later in the resin dehydration.

The water is removed by atmospheric distillation to a batch temperature of 150°C. Additional water and free phenol can be removed by further heating to 160°C and application of vacuum. The distillation endpoint is prescribed by a melting point or melt flow specification.

The finished resin is solidified by discharging into cooling pans, or to a belt or drum flaker, or it may be discharged directly on a clean floor and then broken up with jackhammers after cooling.

When the resin is used in solution the hot resin is dissolved in solvent and then discharged from the kettle to drums or tanks.

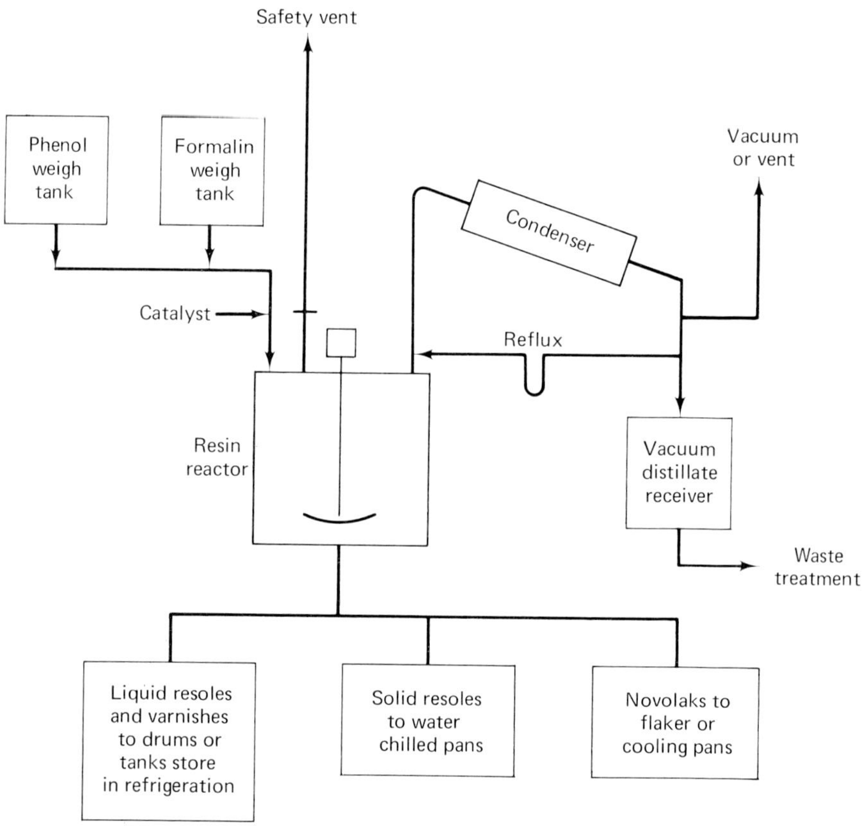

FIG. 8.17 *Schematic Diagram of Phenolic Resin Process*

The cold aqueous distillate separates into two layers—a lower organic phase of phenol and 1,3-benzodioxane and an upper aqueous layer containing dissolved phenol and a small amount of formaldehyde. Although these streams have little economic value, they do present a serious waste disposal problem. In large facilities these distillates may be collected and treated by an extraction or adsorption process to recover the bulk of the phenol and reduce the organic content of the waste stream to a level tolerated by biological waste treatment facilities.

8.10.5.2. Fast curing novolaks Highly ortho substituted, fast curing novolaks are prepared with zinc oxide or acetate or lime catalysts and a large excess of phenol. When consumption of the formaldehyde is complete the bulk of the water is removed by distillation to 130°C at atmospheric pressure. The batch is then held at this temperature to complete condensation of the methylol phenols. The batch is then vacuum distilled to remove the residual moisture and the excess phenol. Premature removal of excess phenol can result in gelation.

8.10.5.3. Liquid resoles Liquid resins are manufactured in the same equipment used for novolak resins. These resins are prepared by reacting 1·5–2·5 moles formaldehyde per mole of phenol below 80°C. After reaction is complete the resins are only partially dehydrated under reduced pressure and are sold as aqueous solutions. These resins continue to react slowly at room temperature and must be used immediately or stored under refrigeration.

8.10.5.4. Laminating varnish Laminating resins are prepared by an alkaline catalysed reaction of one to two moles formaldehyde with one mole of phenol or a phenol alkylphenol mixture. After reaction and dehydration the resin is dissolved to about 60 per cent solids in a solvent. In some cases the resins are modified with rosin, hydrocarbon resins or plasticisers.

8.10.5.5. Solid resoles Solid resoles are manufactured in small kettles—500–2,000 gallon capacity—with scraper blade agitators and a large diameter bottom port for rapid discharge. These resins are reacted to just short of the infusible stage and are difficult to control. The resins are prepared by reacting 1·0–1·2 moles formaldehyde per mole of phenol for 30–60 min at 80°C. The resin is dehydrated to 100–110°C under reduced pressure to a predetermined cure time. When dehydration is complete the resin is discharged rapidly to water cooled trays to prevent further reaction.

8.10.5.6. Substituted phenol resins Resole and novolak alkyl and aryl phenol resins are manufactured by processes similar to those for straight phenolics. In some cases pressure is used to accelerate the more sluggish reactions and higher finishing temperatures may require Dowtherm or electrically heated kettles.

8.10.6. Phenolic Resin Applications

8.10.6.1. Moulding compounds Moulding compounds are prepared by fluxing powdered phenolic resin, filler, lubricant, colourant and hexamethylenetetramine together and then regrinding the flux to the desired particle size. A typical formulation for a general purpose black compound is as follows:

Novolak resin	43%
Wood flour	48%
Hexamethylenetetramine	6%
Lime	1%
Stearic acid	1%
Negrosine	1%

There are four types of moulding compounds, (1) general purpose, (2) high impact grade, where glass fibre, cotton flock, macerated fabric or rubber is used as part or all of the filler, (3) heat resistant grade, where an asbestos filler is used, and (4) electrical grade which uses mica or glass fibre fillers.

Moulding compounds are formulated in various flow ranges for moulding in compression, transfer and injection equipment.

8.10.6.2. Laminates Phenolic laminates are made by impregnating a continuous web of paper or cotton canvas, fibre glass or asbestos cloth with a liquid resole or a resole dissolved in solvent. The web is passed through squeeze rolls and through an oven to evaporate the solvent and partially cure the resin. The web is then cut into sheets which are stacked in a press and laminated together at 190°C and 500–2,000 p.s.i. Decorative laminates use a top design sheet impregnated with a melamine resin for surface hardness and clarity. The electrical and mechanical properties of the many types and grades have been described by Carswell.[175]

8.10.6.3. Foundry Phenolic resin is used as the binder for sand shell moulds used in metal casting. The sand particles are coated with a novolak and hexa and lubricant mixture (3–4 per cent of the sand weight). The shell mould is formed by coating a hot pattern with the sand to fuse the mixture to the desired shape. The shell is removed and baked at 425–525°C for 1–2 min and is then ready to receive the metal casting.

8.10.6.4. Coatings High molecular weight resoles dissolved in alcohol are used as solvent and chemical – resistant coatings in cans, drums, tanks and processing equipment for the food industry. These resins are applied as thin films that are baked on the substrate at 160–200°C. The resins may be modified with polyamides or poly-(vinylformal) to improve the film flexibility.[176]

Oil soluble coatings resins are prepared by the reaction of *p*-*t*-butylphenol, *p*-phenylphenol, *p*-*t*-amylphenol or *p*-octylphenol with formaldehyde. Oil solubility is enhanced by the oleophylic substituent group and the linear chain structure. In some resins bisphenol-A is added to increase functionality. Ortho substituted phenols are undesirable because of poor colour stability. Straight phenolic resins modified with rosin are also oil soluble.

Varnishes are prepared by cooking 6–50 gallons of tung, dehydrated castor, oiticica, or linseed oil per 100 lb of resin at 200–300°C. Coatings with a wide range of properties and economics can be prepared from the many combinations of resin and oil types.

8.10.6.5. Abrasives Phenolic resin is the major bonding agent for high speed grinding wheels. The abrasive, aluminium oxide or silicon carbide is first wetted with a liquid resole and is then coated with a mixture of powdered novolak resin, hexamethylenetetramine and filler. The total resin content is 7–15 per cent of the abrasive. The resin abrasive mix is compacted to a preform by room temperature moulding that is oven cured by slowly raising the temperature to a maximum of 365°F.

Sandpaper and emery cloth are made by bonding the abrasive to heavy paper or cloth with a liquid resole followed by oven curing.

8.10.6.6. Friction products Phenolic bonding systems have been used with asbestos fibres to manufacture brake linings and clutch facings. These materials are produced by either an extrusion or moulding process with resole or novolak resins applied as solutions or dry powder coating. Some resins are prereacted with drying oils to improve product flexibility.

8.10.6.7. Thermal insulation Phenolic resin is used to bond glass and mineral wool fibres to produce mats of thermal insulation. In one process water-dilutable resole at 10–15 per cent solids is sprayed on the hot glass fibres as they emerge from the forming process. The sensible heat of the fibres evaporates the water and leaves a resin film. The fibres are formed into a continuous mat on a belt which is oven cured to form the bond. In a second process the fibres are coated with a powdered resin for subsequent bonding.

8.10.6.8. Wood products Liquid resoles are used as bonding agents for the production of exterior grade plywood. The sheets are roller-coated with a resin wood-flour mixture, stacked cross grain, and cured at 270–300°F and 200–500 p.s.i. for 5 min. Phenolics are also used to bond wood chips, shavings, sawdust and wood flour in the moulding of wood particleboard.

8.10.6.9. Miscellaneous applications Phenolic foam is produced by mixing a highly methylated liquid resole resin with a volatile solvent, surfactant and a strong acid catalyst. The vigorous exotherm causes simultaneous rapid curing and boiling of the solvent to yield a rigid open-celled structure. This foam has found widespread use as a wet base for holding cut flowers.[177]

Alkylphenol resole resins modified with magnesium oxide are used to provide tack, stability and strength to neoprene adhesives.

8.10.7. Economics

The phenolic resin business started in 1910 has had a continuous growth pattern up to the present time. In recent years this has become a mature business in the United States and growth has followed the population trend at 5 per cent compounded annually (Table VIII.28).

TABLE VIII.28

Growth of Phenolic Resin Production in the United States[178]

Year	*US production (millions of pounds)*
1930	18·3
1940	93·4
1950	451
1960	651
1970	1186

The distribution of phenolic resin production according to various market areas in the United States for 1971 is shown in Table VIII.29. The building and electrical industries consumed over 50 per cent of the total production.

TABLE VIII.29

Marketing Areas for Phenolic Resins[179]

	Millions of pounds	*% of total*
Building	447	42
Electrical	136	13
Furniture	64	6
Transportation	50	5
Appliances	46	4
Houseware	25	2
Closures	12	1
Other	287	27
	1,067	

A breakdown of the recent end use pattern for phenolics is given in TableVIII.30

TABLE VIII.30

Phenolic Resin Production End Use Distribution[178]

Application	*1965 (million lbs)*	*1970 (million lbs)*
Moulding	274	334
Laminating	133	128
Abrasive	27	30
Friction	32	47
Thermal insulation	121	133
Foundry	69	95
Plywood	118	199
Wood bonding	32	58
Protective coatings	35	38
Other	49	120
Total	920	1,182

Phenolic resin prices in the United States vary over a wide range and depend considerably on the product solids content and raw material costs. Plywood bonding resins are near 10 cents per pound whereas some of the substituted phenolic coatings resins cost 50–75 cents per pound. The unit value for all phenolic resins in 1970 was 20 cents per pound.[178]

8.11. SALICYLIC ACID, ASPIRIN, AND OTHER PHARMACEUTICALS

by **J. R. Brooks**

8.11.1 Introduction

Quite a number of pharmaceuticals are simple benzene derivatives and a selection of these are reviewed below. Most of these remedies have been known for many years and the chemistry is straightforward. Nevertheless, some attempt has been made to bring out the important points. The discussion deals mainly with salicylic acid and aspirin since these compounds represent by far the greatest tonnage in pharmaceutical manufacturing. The remaining section briefly summarises a selection of other simple pharmaceuticals.

8.11.2. Salicylic Acid

8.11.2.1. Occurrence in nature A review article describing the plant materials from which salicylic acid was first isolated has been published.[180] The article also discusses the medical uses of salicylic acid and of aspirin and briefly recounts some of the ideas which have been advanced to explain the pharmacological properties of aspirin.

Salicylic acid occurs as a glycoside 'salicin' in willow bark. This material has been used for the treatment of malaria from the middle of the eighteenth century and the antipyretic effects in rheumatic fever were noted about 100 years later.[181] Salicin has been converted to salicylic acid by hydrolysis and oxidation:

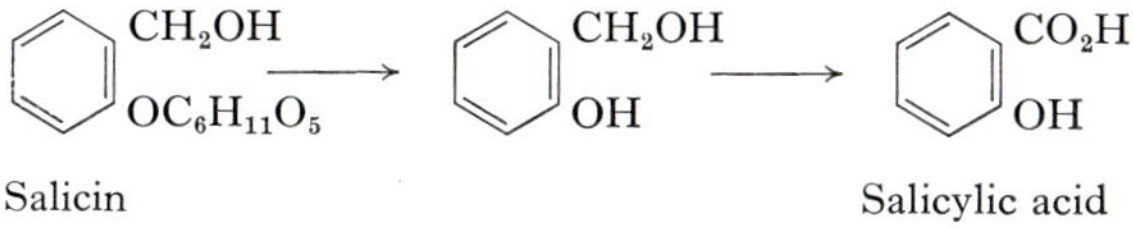

Salicylaldehyde (2-hydroxybenzaldehyde) occurs in the flowers of Meadowsweet (Spirea) and can be extracted and oxidised to salicylic acid. The methyl ester of salicylic acid (Oil of Wintergreen) is also of natural occurrence and can be hydrolysed to the free acid.

8.11.2.2. Synthetic production Salicylic acid was first prepared by Piria in 1838 who treated salicylaldehyde with potassium hydroxide.[182] Today, the method most frequently used on the commercial scale is the Kolbe-Schmidt reaction whereby sodium phenate is combined directly with carbon dioxide. The process was first described by Kolbe and Lautermann[183] but in their process, half the phenol added was recovered unchanged and quantities of disodium salicylate were produced. Schmidt modified the method in 1884 by admitting carbon dioxide gas to cold, dry sodium phenate and subsequently heating under pressure at 120–140°C.[184] The modification gave near quantitative conversion of starting material. More recently it was found that still higher temperatures favour formation of the para isomer, 4-hydroxybenzoic acid.[185]

Current practice is first to prepare sodium phenate from phenol and 50 per cent aqueous sodium hydroxide.[186] This product is dried, heated to 100°C, and carbon dioxide passed in at 5–6 atm pressure. After adsorption of the calculated quantity of gas, the temperature is raised to 140°C to complete the reaction. Acidification of an aqueous solution of the crude product gives the free acid which is crystallised from water.

The reaction can be considered to occur by the following mechanism:

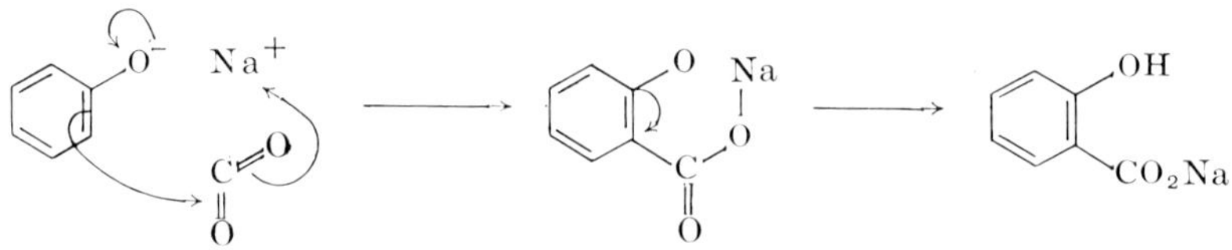

Exact details have been the subject of some controversy in the past. Lindsey and his collaborators have made an extensive study of the reaction[187] and the presence of certain intermediates has been inferred from infra-red studies.[185] Such work has shown that the alkali metal cation migrates from the phenoxide anion to a molecule of carbon dioxide in the first stage of the reaction.[189] The so-formed carboxylate anion absorbs in the infra-red region at 1600 cm^{-1} whereas the un-ionised carboxyl group absorbs between 1660 and 1690 cm^{-1}. These and other similar observations lead to the following postulation as to the mechanism of the reaction:

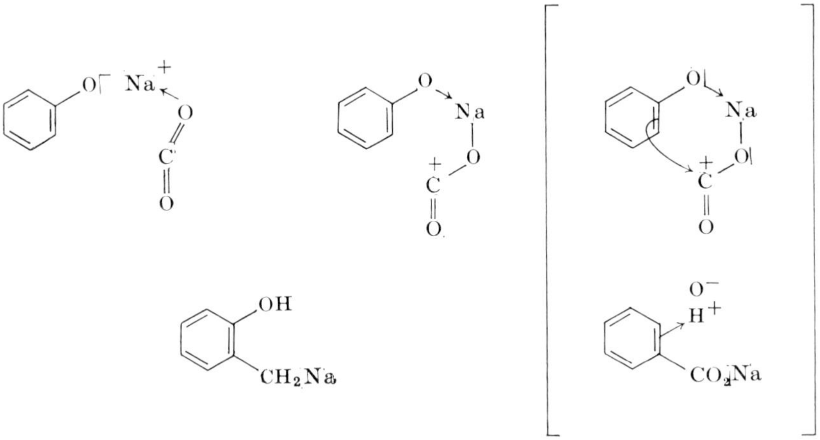

π-complex

An alternative commercial preparation of salicylic acid has been suggested.[190] The technique involves the electrolytic oxidation of *o*-cresol and is claimed to give better yields at lower cost than equivalent commercial methods of oxidation. During electrolysis, oxygen is liberated from the anode (which is made of iron or nickel) and this serves as direct oxidant. In a typical preparation, *o*-cresol (25 kg), potassium hydroxide (20 kg), and sodium hydroxide (20 kg) are heated with sufficient water to cause fusion at 200°C. Current is passed at 4–6 amps./sq. decimeter and the product is formed in near quantitative yield.

8.11.2.3. Chemical properties Salicylic acid is a white crystalline solid melting under pressure at 157–159°C. The solid is stable in air, but sublimes at 76°C. This property has been used as a method of purification. At temperatures in excess of 200°C, decomposition to phenol and carbon dioxide occurs. Salicylic acid is slightly soluble in hot water (8·4 per cent at 100°C) but is soluble in most organic solvents (e.g. 30 per cent in chloroform or ether at 25°C). Solutions give the expected violet colour with ferric chloride. Salts with heavy metals are insoluble.

Sodium salicylate is usually prepared for pharmaceutical use from salicylic acid and sodium bicarbonate. It is very soluble in water (50 per cent at 25°C) and should give a clear solution showing the absence of free acid. There should be no efferves-cence with mineral acids which would indicate the presence of unreacted bicarbonate.

The following salicylate derivatives are the more common:

Salts	*Esters*
Aluminium	Methyl
Ammonium	Ethyl
Calcium	Acetyl
Lead	Benzyl
Lithium	Amyl
Magnesium	Phenyl
Sodium	Salicyl

In addition, 4-amino-2-hydroxybenzoic acid is used in the treatment of tuber-culosis (see later), salicylamide is used as an analgesic and salicylanilide as a fungi-cide. Methylenedisalicylic acid (the 5-5′-isomer) is used in the formulation of alkyd resins to give quick drying properties and a hard gloss finish.

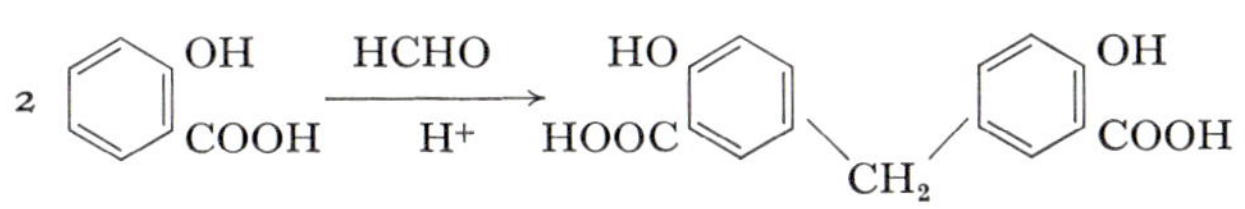

methylenedisalicylic acid

8.11.3. Acetylsalicylic Acid

Acetylsalicylic acid or aspirin is probably the most frequently encountered phar-maceutical substance. It has been reported that in the United States of America in 1968 enough acetylsalicylic acid was synthesised to make 46,000,000,000 aspirin tablets of normal strength.[191] Further, a small survey in Portsmouth, England sug-gested that 2·2 per cent of all men and 8·8 per cent of all women took aspirin every day.[192]

8.11.3.1. Production Salicylic acid was first acetylated by Gerhardt[193] but the first acetylations on the commercial scale were probably carried out by Hofmann in 1893.

Despite prior knowledge of the compound, a patent was subsequently granted to the German Bayer Company on the grounds that this product was 'purer' than had hitherto been known. Whereas Gerhardt had used acetyl chloride for the acetylation, Hofmann reacted acetic anhydride with salicylic acid.[194] Lederer later employed sulphuric acid as catalyst for the reaction which could then be carried out at lower temperature.[195] Typical quantities are given in the following description.[196] Acetic anhydride (250 kg), acetic acid (300 kg), and salicylic acid (250 kg) are placed in a 1,500 litre jacketed vessel and heated to 60 °C. At this point, sulphuric acid (1 kg) dissolved in acetic acid (2 kg) is added. The temperature is then raised to 90 °C for 30 minutes. Subsequent cooling to 20 °C causes crystallisation of the product which, after processing the mother liquors, is obtained in yields approaching 100 per cent.

The product was formerly recrystallised from benzene or carbon tetrachloride but traces of acetic acid frequently remained and gave rise to an unpleasant taste. Also the crystal form was fluffy and not well suited to tabletting. Dioxane may be a more suitable crystallising solvent since these disadvantages are largely eliminated and higher recoveries can be obtained.[197]

Another method of preparation which has not apparently been used commercially is the reaction of keten with salicylic acid.[198] Unfortunately, further reaction of keten with the acid carbonyl group gives mixed anhydrides as by-products.

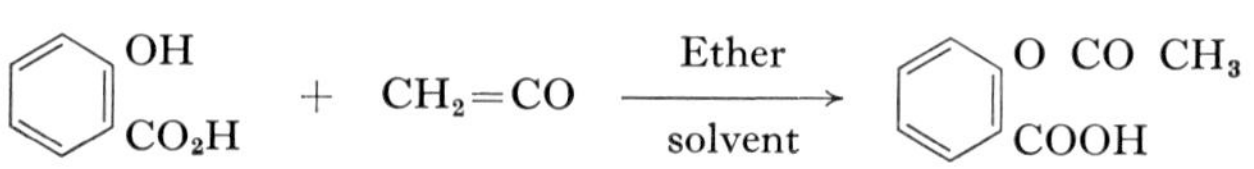

An advantage of the method is that no water is produced in the reaction and evaporation of the solvent leaves crystals of acetylsalicylic acid directly. Salicylic acid and *p*-toluenesulphonylchloride, together with sodium acetate, is said to give a good yield of acetylsalicylic acid together with by-products which are water soluble but again the method does not seem to be used commercially.[199]

8.11.3.2. Chemical properties Acetylsalicylic acid is almost insoluble in water (0·3 per cent at 25 °C). A water-ethanol solution should give no colour with ferric chloride indicating the absence of unacetylated salicylic acid. There should not be any residual acetic acid in the product. The melting point of acetylsalicylic acid depends on the method of heating but is generally described as being 135 °C or higher. Acetylsalicylic acid is stable in dry air but hydrolyses slowly in the presence of moisture. The salicylic acid formed on hydrolysis is irritating to the stomach wall so it is essential to store pharmaceutical grades of aspirin under controlled conditions. Sodium and calcium salts of acetylsalicylic acid have a much greater degree of water solubility and find use in 'soluble' aspirin preparations which are currently preferred by the medical profession. Care is needed during the manufacture of these salts in order to avoid hydrolysis of the acetyl group.

8.11.4. Sulphonamides

Sulphanilamide[200] was first synthesised as a dye-stuff intermediate but it was a number of years later that the antibacterial activity of this and related compounds was discovered.[201] The usual commercial synthesis involves the following four stages.[202]

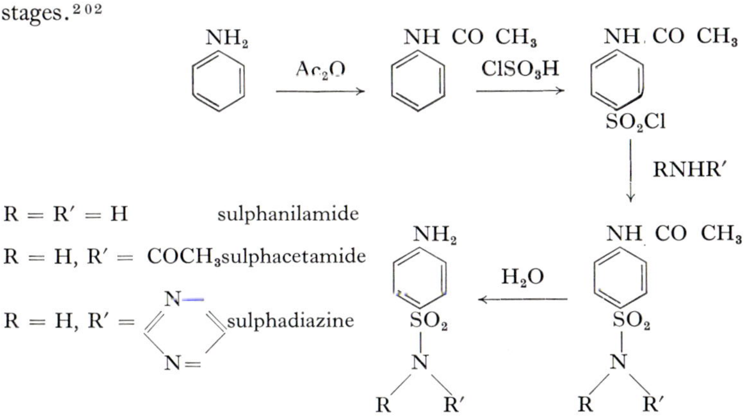

The primary aromatic amino group is essential for activity,[203] but compounds which can be converted to the amine in the body are acceptable, e.g. succinylsulphathiazole:

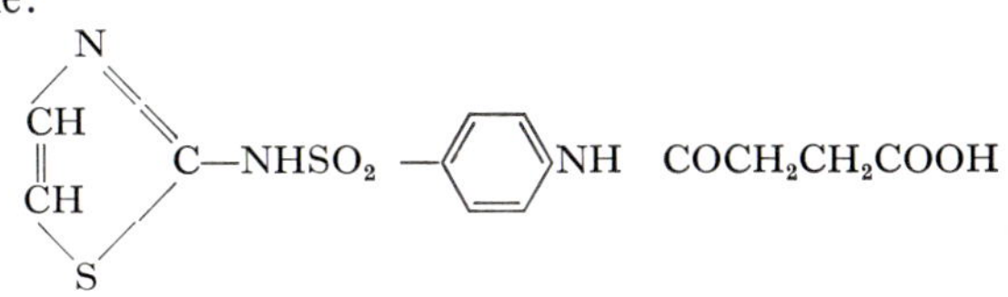

Such compounds have significantly different absorption properties from the unsubstituted analogues. For example, sulphathiazole is almost completely absorbed from the gastrointestinal tract but succinylsulphathiazole is only absorbed to the extent of 10 per cent. Consequently the major use of the latter compound is in the reduction of intestinal micro-organism levels.

8.11.5. Phenols and Antiseptics

The chemistry of phenol and other hydroxybenzenes is described elsewhere (see pp. 346 ff). Some of the more commonly used antiseptics and disinfectants are phenol itself, cresol, resorcinol, hexylresorcinol, thymol, 4-chlorophenol, chloramine and hexachlorophene.

Halogenation of phenol increases the antiseptic activity. Di- and tri- halogen compounds are more active than the mono-halogen derivatives. Alkylation of phenol (giving 2-, 3-, and 4-methylphenol) also increases the antiseptic activity slightly. Resorcinol (1,3-dihydroxybenzene) is somewhat less bactericidal than phenol. Thymol is a useful fungicide and antibacterial and is a common constituent of mouthwashes.

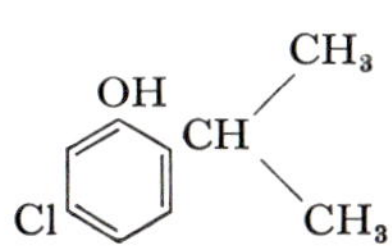

Thymol

It is of natural occurrence (Oil of Thyme). Being phenolic, it can be extracted with caustic alkali and is purified by final distillation.

Hexachlorophene is an important antibacterial but is too toxic for internal use. It is used extensively in deodorant preparations. A common synthesis is from 2,4,5-trichlorophenol and formaldehyde.

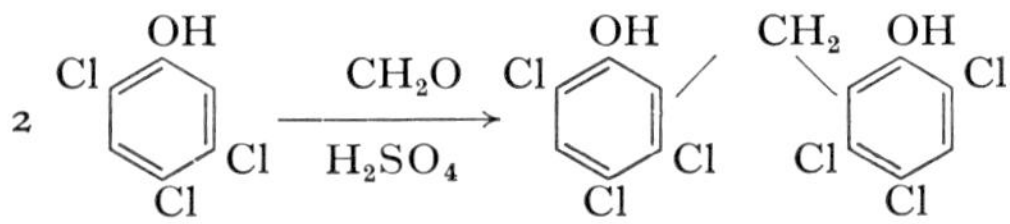

It is insoluble in water so is usually formulated in alcoholic vehicles.

Recent work emanating from the USA has given rise to concern over the safety of hexachloroprene. Under certain conditions, significant absorption can occur through the skin surface. In view of possible damage to the nervous system, drug safety authorities are keeping the use of hexachloroprene under review even though the risk appears to be of a low order.

8.11.6. Mild Analgesics

A number of aniline derivatives have been used as mild analgesics. The three most common are acetanilide, phenacetin and paracetamol.

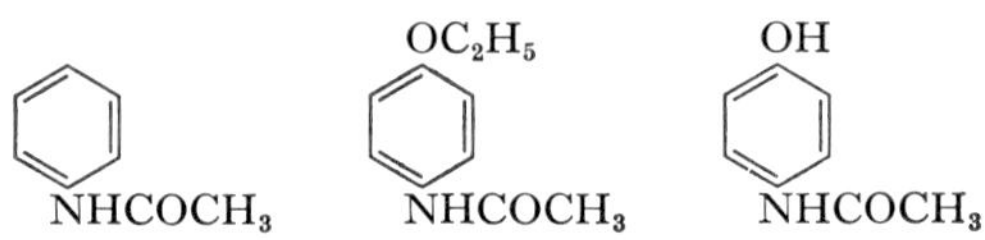

Acetanilide Phenacetin Paracetamol

Aniline itself has the highest activity in the series but is too toxic for human use. Acetylated derivatives are less toxic but probably owe their activity to hydrolysis in the body. Only paracetamol is still in current use.

8.11.7. p-Aminosalicylic Acid

This compound is used in the treatment of tuberculosis. It can be prepared from 4-nitro-2-aminotoluene,[204] or by the reduction of 4-nitrosalicylic acid with sodium sulphide. Decarboxylation readily occurs in aqueous solution with the formation of 3-aminophenol. The sodium salt of the acid is more stable.

p-aminosalicylic acid (PAS)

REFERENCES

[1] Laurent, A., *Ann.*, 1847, **43**, 203.
[2] Hunt, J., *Jahresber. Fortschr. Chem.*, 1849, 391.
[3] Lister, J., *Lancet*, 1867, **2**, 95, 353, 668.
[4] Falkson, R., *Arch. klin. Chir. Lagenbecks*, 1881, **26**, 204.
[5] Deichmann, W. B. and Keplinger, M. L., '*Industrial Hygiene and Toxicity*', 2nd edn., Chapter 33, 1963.
[6] Sax, N. I., '*Dangerous Properties of Industrial Materials*', 3rd edn., 1968.
[7] Dreisbach, R. R., '*Physical Properties of Chemical Compounds*', A C S Advances in Chemistry Series, no. 15, 1955, p. 273.
[8] c.f. The Pharmacopeia of the USA, 17th edn., p. 455, 1965. Mack Printing Co., Easton Pa.
[9] Hickinbottom, W. J., '*Chemistry of Carbon Compounds*', Vol. 3A, Chapter 8, 1954.
[10] Fieser, L. F. and Fieser, M., '*Advanced Organic Chemistry*', Chapter 22, 1961.
[11] Robertson, P. W., de la Mare, P. B. D. and Swedlund, B. E., *J.C.S.*, 1953, p. 782.
[12] Cheronis, N. D. and Ma, T. S., '*Organic Functional Group Analysis*', 1964, p. 448., Interscience Publishers, New York.
[13] Gould, D. F., '*Phenolic Resins*', (Reinhold), 1959.
[14] Martin, R. W., '*Phenolic Resin* ', (John Wiley), 1956.
[15] Lubs, H. A., '*The Chemistry of Synthetic Dyes and Pigments*', A C S Monograph No. 127, Chapter 3, 1955.
[16] Drake, N. L., *Org. React.*, 1942, **1**, 105.
[17] Brannon, J. L., Union Carbide Corp., personal communication.
[18] Fieser, L. F. and Fieser, M., '*Advanced Organic Chemistry*', p. 720.
[19] Gottesman, R. T., Chin D., '*Encyclopaedia of Chemical Technology*', ed. Kirk-Othmer, 2nd edn., Vol. 17, p. 739. Interscience Publishers, New York.
[20] Chiwala, A. and Pailer, M., *J. Prakt. Chem.*, 1939, **260**, 45.
[21] Wacek, A. and Fieldler, R., *Monatsh.*, 1949, **80**, 170.
[22] Peterkin, A. R., *Ind. Eng. Chem.*, 1918, **10**, 738.
[23] Anon., *Chem. Week.*, 20 September 1969, p. 133.
[24] Johnson, R. N., '*Encyclopaedia of Chemical Technology.*', ed. Kirk-Othmer, 2nd edn., Vol. 16, p. 272. Interscience Publishers, New York.
[25] Anon., *Chem. Week*, 20 September 1969, p. 133.
[26] Yen, Y. C., '*Phenol*', Chem. Econ. Programme, Stanford Research Institute, 1967.
[27] Sittig, M. '*Alcohols, Polyols and Phenols Manufacture and Derivatives*' (Noyes, 1968), p. 96; cf. Anon., *Chem. Week*, 4 November, 1967 p. 28.
[28] Krieble, R. H., Holmes, J. B. and Wilcoxon, B. H., *FIAT Final Report*, No. 768, PB-76032, p. 12. Hobart Publishing Co. Washington, D.C.
[29] Kenyon, R. L. and Boehmer, N., *Ind. Eng. Chem.*, 1950, **42**, 1446.
[30] England, S. W., Aries, R. S. and Othmer, D. F., *Ind. Eng. Chem.*, 1953, **45**, 189, 193.
[31] Barbet, P. A., USP 1,459,081 (1923).
[32] Bull, H., USP 1,247,499 (1917).
[33] Dennis, L. M., USP's 1,211,923 (1917), 1,212,612 (1917), 1,228414 (1917), 1,229,593 (1917).
[34] Guyst, A., *Chimie et Industrie*, 1919, **2**, 879.
[35] Tyre, D., USP 1,210,725;6 (1917).
[36] Groggins, P. H., '*Unit Processes in Organic Synthesis*', 5th edn., p. 377 (1958). McGraw-Hill.
[37] Prahl, W. H., *Chem. Eng. News*, 1953, **31**, 4178.
[38] Groggins, P. H. *loc. supra*, p. 222.
[39] Reichle, W. T., *J. Catalysis*, 1970, **17**, 297.
[40] Wittig, G., *Angew. Chem.*, 1965, **77**, 752.
[41] Heanley, H., *Chem. Rev.*, 1962, **62**, 81.
[42] Krieble, R. H., *loc. supra*, p. 1.

[43] Stoesser, W. C., Sommerfield, E. H. and Widiger, A. H., *Ind. Eng. Chem. Products Res., Dev.*, 1971, **10**, 76.
[44] Dalaman, G. W. and Newman, F. W., *J.A.C.S.*, 1968, **90**, 1601.
[45] Bottini, A. T. and Roberts, J. D., *J.A.C.S.*, 1957, **79**, 1458.
[46] Luettringhaus, A. and Ambros, D., *Ber.*, 1956, **89**, 463.
[47] Anon., *Chem. Eng. News*, 18 January 1971, p. 11.
[48] Hay, A. S. and Blanchard, H. S., *Can. J. Chem.*, 1965, **43**, 1306.
[49] Kaeding, W. W., *J. Org. Chem.*, 1961, **26**, 3144.
[50] Strojny, E. J., *J. Org. Chem.*, 1969, **34**, 3685.
[51] Seubold, F. H., Jr. and Vaughan, W. E., *J.A.C.S.*, 1953, **75**, 3790.
[52] J. D. Doedens, '*Encyclopaedia of Chemical Technology*', ed. Kirk-Othmer, 2nd edn., Vol. 5, p. 325.
[53] Anon., *Chem. Eng. News*, 22 March 1971, pp. 12, 39 and 78.
[54] Smith, F. B. and Livak, J. E., USP 2,107, 650.
[55] Holleman, H., *Rec. Trav. Chim.*, 1911, **30**, 81–2.
[56] Lister, J., *Lancet*, 1867, **2**, 1396, 1402.
[57] Doedens, J. D.,'*Encylopaedia of Chemical Technology*', ed. Kirk-Othmer, 2nd edn., Vol. 5, p. 327. Interscience Publishers, New York.
[58] Bacon, R. G. R. and Rennison, S. C., *J.C.S.*, 1969c, 312.
[59] Weingarten, H., *J. Org. Chem.*, 1964, **29**, 3624.
[60] Hancock, E. G. and Weeks, G. A. (Turner and Newall), BP 585,376 (1944).
[61] *Herbicide Handbook*, Weed Science Society of America, 1967.
[62] Plimmer, J. R., '*Encyclopaedia of Chemical Technology*', ed. Kirk-Othmer, 2nd edn., Vol. 22, p. 174.
[63] Suter, C. M., *Chem. Rev.*, 1941, **28**, 219.
[64] National Institute of Health, *Phenol and its derivatives: the relation between their chemical constitution and their effect on the oganism*", Bulletin no. 190.
[65] Preston, R. W. G. and Reed, H. W. B., '*Encyclopaedia of Chemical Technology*', ed. Kirk-Othmer, 2nd edn., Vol. 1, p. 901.
[66] Rosenwald, R. M., '*Encyclopaedia of Chemical Technology*', ed. Kirk-Othmer, 2nd edn., Vol. 1, p. 894.
[67] Weinreich, W., *Ind. Eng. Chem.*, 1943, **35**, 264.
[68] Stevens, D. R., *Ind. Eng. Chem.*, 1943, **35**, 655.
[69] Dutch Patent Appl. 6,410,238.
[70] Knapp, G. G., Coffield, T. H., Napolitano, J. P., Orloff, H. D. and Worrel, C. J., *Proc. 7th World Petroleum Congress*, 1967, **5**, 403.
[71] Dutch Patent Appl. 6,600,159.
[72] Hay, A. S., *J. Poly Sci.*, 1962, **58**, 581.
[73] Endes, G. F. and Kwiatek, J., *J. Poly. Sci.*, 1962, **58**, 593.
[74] '*Autoxidation and Antioxidants*', Vol. 1, p. 206, ed. W. G. Lundberg, 1961.
[75] Lister, J., *Lancet*, 1867, **2**, 1389.
[76] Symes, W. F. and Dawson, C. R., *J.A.C.S.*, 1954, **76**, 2959.
[77] Lister, J., *Lancet*, 1867, **2**, 1407.
[78] Schoenfeldt, N., '*Surface Active Ethylene Oxide Adducts*', 1969.
[79] Stevens, C. E., '*Encyclopaedia of Chemical Technology*', ed. Kirk-Othmer, 2nd edn., Vol. 19, pp. 507 and 532.
[80] Schick, M. J., '*Surfactant Science*', Vol. I, 1967.
[81] Buttery, D. N., '*Plasticisers*', 1947.
[82] Doolittle, A. K., '*The Technology of Solvents and Plasticisers*', 1954.
[83] Lane, J. C., '*Encyclopaedia of Chemical Technology*', ed. Kirk-Othmer, 2nd edn., Vol. 10, p. 476; 477.
[84] Lee, R. E. Jr. and Booster, E. R., '*Encyclopaedia of Chemical Technology*', ed. Kirk-Othmer, 2nd edn., Vol. 12, p. 581; 574.
[85] Maasen, G. C., Fawcett, R. J. and Connell, W. R., 'Antioxidants' in '*Encyclopaedia of Polymer Science and Technology*', Vol. 2, 1965, p. 171.
[86] Bayer, G., '*Antioxidants*' (Noves Data Corp., 1971).

[87] Dugan, L. R., '*Encyclopaedia of Chemical Technology*', ed. Kirk-Othmer, 2nd edn., Vol. 2, p. 588.
[88] Levin, P. I. and Mikhailov, V. V., *Russ. Chem. Rev.*, 1970, **39**, 799 (English transl.)
[89] Dianin, A., *J. Russ. Phys. Chem. Ges.*, 1891, **1**, 488.
[90] Greth, A., *Chem. Ztg., Chem. App.*, 1967, **91**, 357.
[91] Martin, R. W., '*The Chemistry of Phenolic Resins*', 1956.
[92] Alpine Chemische AG, BP 941,995 (1963).
[93] Perkins, R. P. and Bryner, F. (Dow Chemical Co.), USP 2,259,242 (1942).
[94] Jansen, J. E. (Goodrich, B. F.), USP 2,468,982 (1949).
[95] Williamson, G. T. (Shell Development Co.), USP 2,730,552) (1956).
[96] Wagner, R. B. (Hercules Powder Co.), USP 3,172, 916 (1968).
[97] McNutt, B. W. and Gammill, B. B. (Dow Chemical Co.), USP 3,394,089 (1968).
[98] Apel, F. N., Conte, L. B. Jr. and Bender, H. L. (Union Carbide), USP 3,049,568 (1968).
[99] Braum, J. V., *Ann.*, 1929, **472**, 1.
[100] Leibnitz, E. and Naumann, K., *Chem. Techn.*, 1951, **3**, 5.
[101] Meyer, K. H. and Schnell, H. (F. F. Bayer), GP 1,027, 205 (1958).
[102] Rappert, H. Schnell, H. (F. F. Bayer), Belg. P 480,053 (1961).
[103] J. P. Henry (Union Carbide), USP 2,884, 462 (1959)
[104] Conte, L. B. Jr. and Apel, F. N. (Union Carbide), USP 1,161,284 (1964).
[105] Hoaglin, R. I., Plummer, C. W. and Schultze, H. C. (Union Carbide), USP 2,858,343.
[106] Schwarger, C. G., Ballard, S. A. and Luten, D. B., Jr. (Shell Development Co.), USP's 2,602,821–2.
[107] N. V. de Bataafsche Petroleum Maats., Dutch P 85,028 (1957).
[108] Krimm, H., Ruppert, H. and Schnell, H. (F. F. Bayer), Belg.P 611,184 (1961).
[109] Farnham, A. G. (Union Carbide), BP 3,288,864 (1956).
[110] Stull, D. R., *Ind. Eng. Chem.*, 1947, **39**, 517.
[111] Hooker Chem. Corp., BBP 1,039,233.
[112] Brydia, L. E., *Anal. Chem.*, 1968, **40**, 2212.
[113] *Product information bulletin for Bisphenol A* (Union Carbide).
[114] Yowell, R. L., *Chem. Eng. Prog.*, 1968, **64**, 58.
[115] Williamson, G. T. (Shell Development Co.), USP 2,775,620 (1956).
[116] Farnham, A. G., Apel, F. N. and Bender, H. L. (Union Carbide), USP 3,242,219 (1967).
[117] Apel, F. N., Farevaag, P. and Bender, H. L. (Union Carbide), USP 3,049,569 (1962).
[118] Prahl, W. H., Lederman, S. J. and Lichtblau, E. I. (Hooker Chem. Corp.), USP 3,073,868.
[119] Prahl, W. H., Lederman, S. J. and Lichtblau, E. I. (Hooker Chem. Corp.), USP's 3,219,390, 3,290,390.
[120] F. F. Bayer, BP 890,432 (1962).
[121] Luten, D. B., Jr. (Shell Development Co.), 2,791,616 (1957).
[122] F. F. Bayer, Belg.P 738,962 (1969).
[123] Marxil, S. J. and Kolmer, S. J., *Chem. Eng. Prog.*, 1963, **59**, 60.
[124] Farnham, A. G. (Union Carbide), USP 3,288,864 (1966).
[125] Smith, W. E. and Neumann, F. W. (Dow Chemical Co.) USP 3,466,337.
[126] Meyer, K. H. and Schnell, H. (F. F. Bayer), GP 1,094,759.
[127] Jordon, H. (Schering-Kohlbaum), GP 467,640 (1928).
[128] Hennis, H. E., *Ind. Eng. Chem., Prod. Res. and Dev.*, 1963, **2**, 140.
[129] Pospisil, J. and Kohorec, J., Czech.P 129,886 (1968).
[130] Lee, H. and Neville, K., '*Epoxy Resins*', 1957.
[131] '*Encyclopaedia of Polymer Science and Technology*, Vol. 10, 1969.
[132] '*Encyclopaedia of Polymer Science and Technology*', Vol. 11.
[133] US Tariff Commission, '*Synthetic Organic Chemicals, US Production and Sales*' (1960-8, Prelim. reports 1969–70).
[134] Anon., *Chem. Eng. News.*, 16 August 1971.
[135] *Modern Plastics*, January 1972.
[136] *Oil Paint amd Drug Reporter*, July 1960–February 1972.

[137] Author's estimate.
[138] Peria, R., *Ann.*, 1843, **48**, 75.
[139] Baeyer, A., *Ber.*, 1872, **5**, 1094.
[140] Michail, A., *Am. Chem. J.*, 1883/4, **5**, 338.
[141] Kleeberg, W., *Ann.* 1891, **263**, 283.
[142] Paal, C. and Serminger, H., 1894, **27**, 1799.
[143] Lederer, L., *J. Prakt. Chem.*, 1894 (2), **50**, 224.
[144] Manasse, *Ber.*, 1894, **27**, 2409.
[145] Smith, A., USP 643,012 (1900).
[146] Luft, A., USP 753,728 (1903).
[147] Blumer, L. BP 12,880/1902.
[148] Fayolle, E. H., FP 335,584 (1903).
[149] Baekeland, L. H., *Ind. Eng. Chem.*, 1909, **1**, 149.
[150] Baekeland, L. H., USP's 939,966, 942,852 (1909).
[151] Aylsworth, J. W., USP 1,020,593 (1912).
[152] Hultzsch, K., *Angew. Chem.*, 1948, A**60**, 179–184.
[153] De Jong, J. I. and De Jonge, J., *Rec. Trav. Chim.*, 1953, **72**, 497.
[154] Walker, J. F., '*Formaldehyde*', 1964, pp. 304–344.
[155] Woodbrey, J. C., Higgenbottom, H. P. and Culbertson, H. J., *J. Polym. Sci.*, 1965, A**3**(**3**), 1079.
[156] Martin, R. W., '*The Chemistry of Phenolic Resins*', 1956, Chapter 2.
[157] Freeman, J. H. and Lewis, C. W., *J.A.C.S.*, 1954, **76**, 2080–87.
[158] Zavitsas, A. A., *Am. Chem. Soc.*, Div. of Org. Coatings and Plastics, Preprints, 1966, **26** (1), 93–106.
[159] Zavitsas, A. A. and Beaulien, R. D., *Am. Chem. Soc.*, Div. of Org. Coatings and Plastics, Preprints, 1967, **27** (1), 100–113.
[160] Martin, R. W., *J.A.C.S.*, 1952, **74**, 4208.
[161] Reese, J., *Angew. Chem.*, 1954, **66**, 170.
[162] Finn, S. R., James J. W. and Stander, C. J. S., *Chem. and Ind.*, 1954, p. 188.
[163] Hultzsch, K., '*Chemie der Phenolharze*', 1950.
[164] Fraser, D. A., Hall, R. W. and Raum, A. L. J., *J. Appl. Chem.*, 1957, **7**, 676.
[165] Bender, H. L. and Farnham, A. G. (Union Carbide), USP 2,475,587 (1949).
[166] Jones, T. T., *J.S.C.I.*, 1946, **65**, 264.
[167] Gardikes, J. J. and Konrad, F. M., *Am. Chem. Soc.*, Div. Org. Coatings and Plastics, Preprints, 1966, **26** (a), 131X137.
[168] Jones, T. T., *J. Appl. Chem.*, 1952, **2**, 134.
[169] Dienes, G. J., *J. Colloid Sci.*, 1949, **4**, 257.
[170] Bender, H. L., *Modern Plastics*, March 1954.
[171] Amvers, K. and Rietz, *Ann.*, 1908, **356**, 152.
[172] Greth, A. *Angew. Chem.*, 1938, **51**, 719.
[173] van der Meer, S., *Mededial Rubber*, Stichting, Paper 48, May 1944.
[174] Castle, R., *Oil and Colour J.*, 1951, **120**, 493.
[175] Carswell, T. S., '*Phenoplasts; their structure, properties and chemical technology*', 1947.
[176] Purman, D. E. and Floyd, D. E. (General Mills), USP 2,891,023 (1959).
[177] Smithers, V. L., USP 2,753,277 (1956).
[178] US Tariff Comm., *Synthetic Organic Chemicals, US Production and Sales*, 1930, 40, 50, 60, 65 and 70.
[179] *Modern Plastics*, January 1972, p. 179.
[180] Collier, H. O. J., *Sci. Amer.*, 1963, **209**, 97.
[181] Leroux, H., *J. Chim. Med.*, 1829, **6**, 341.
[182] Piria, *Compt. Rend.*, 1838, **6**, 338.
[183] Kolbe, H. and Lautermann, E., *Ann. Chim.*, 1860, **115**, 157.
[184] Schmidt, R., GP's 29,939 and 31,240 (1884).
[184] Alexander, E. R., '*Ionic Organic Reactions*', (Wiley, New York, 1950), p. 197.
[186] Faith, W. L., Keyes, D. B. and Clarke, R. L., '*Industrial Chemicals*', (Wiley, New York, 1950), p. 526.

[187] Lindesey, A. S., *Chem. Rev.*, 1953, **57**, 583; Hunt, S. E., Jones, J. I., Lindsey, A. S., Killoh, D. C. and Turner, H. S., *J. Chem. Soc.*, 1958, 3152.
[188] Halse, J. L., Jones, J. I. and Lindsey, A. S., *J. Chem. Soc.*, 1954, 3145.
[189] Hales, J. L., Jones, J. I. and Lindsey, A. S., *Chem. and Ind.*, 1954, p. 49.
[190] BP 103, 739.
[191] O'Brien, J. R., *Brit. Med. J.*, 1970, **3**, 284.
[192] O'Brien, J. R., *Lancet*, 1970, p. 1055.
[193] Gerhardt, C. F., *Ann. Chim. Phys.*, 1853, **37**, 322.
[194] BP 27,088 (1898).
[195] USP 671,769 (1901).
[196] Bercy, A. *La Nature*, 1936, **2977**, 462.
[197] USP 2,209,019 (1938).
[198] USP 1,604,472 (1926).
[199] USP 749,980 (1904).
[200] Gelmo, P., *J. prakt. Chem.*, 1908, **77**, 369.
[201] Domagk, G., *Dtsch. Med. Wschr.*, 1935, **61**, 250, 892.
[202] Northey, E. H., '*The Sulphonamides and Allied Products*', (Reinhold, New York, 1948).
[203] Woods, D. D., *Brit. J. Exptl. Path.*, 1940, **21**, 74.
[204] Weiss, E. and Gardner, J. S., *J. Amer. Pharm. Ass.*, 1949, **38**, 9.

CHAPTER 9

DETERGENT ALKYLATE

by K. Jones

9.1 INTRODUCTION

ONE of the most important industrial derivatives of benzene is undoubtedly the so-called *Detergent Alkylate.* This is a monoalkylbenzene (I) in which the saturated alkyl group R contains about twelve carbon atoms. All commercially important detergent alkylates are mixtures of isomeric alkylbenzenes of a range of chain lengths. The alkylate is sulphonated to give the corresponding sodium alkylbenzene sulphonate (II).

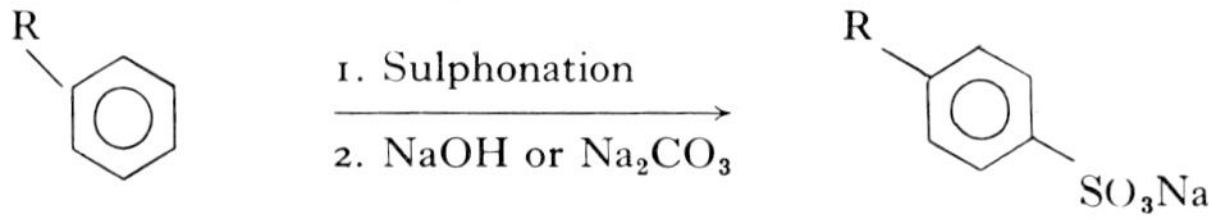

I II

Alkylbenzene sulphonates are the most important synthetic surface active agents (surfactants) currently in use and most modern fabric washing and dishwashing products contain them.

N. K. Adam[1] demonstrated in 1923 that long-chain *n*-alkylbenzene sulphonates showed properties similar to soaps but fourteen years elapsed before the first commercial alkylbenzene sulphonate appeared. This was I. G. Farben's 'Igepal N A' which was made by chlorination of an *n*-paraffin-rich Fischer-Tropsch product (of average chain length C_{14}) to give a chloroparaffin, reaction of the chloroparaffin with benzene under Friedel-Crafts conditions, and sulphonation of the resultant alkylbenzene. More than 800 tons of this material was made in 1943.[2] In the late 1940's and the 1950's, a spectacular growth in the use of alkylbenzene sulphonates occurred; by 1962, 210,000 tons were used in the USA alone.[3]

The materials used in this period had branched alkyl chains. Although they were cheap and performed well, they were biologically 'hard' and the slow rate of their biological breakdown in the environment (biodegradation) resulted in unpleasant foams on rivers and problems in sewage installations in areas of high population density. This has led to their progressive replacement by biologically 'soft' linear alkylbenzene sulphonates which have straight-chain alkyl groups and are much more rapidly degraded than the branched materials. In 1969, H. A. W. Hill *et al.*[4] estimated the world demand for detergent alkylates as 800,000 tons, two-thirds being linear and one-third branched.

It seems certain that alkylbenzene sulphonates will continue to be industrially important for some years to come although they are likely to be increasingly challenged by newer anionic surfactants such as olefin sulphonates.[5]

9.2. TETRAPROPYLENE-BASED BRANCHED-CHAIN ALKYLBENZENE SULPHONATES (ABS)

Branched-chain alkylbenzene sulphonates, which were the most important surfactants in the 1950's and the first half of the 1960's were prepared by converting propylene to a predominantly C_{12} branched-chain olefin mixture, alkylating benzene with this under Friedel-Crafts conditions and sulphonating the resultant alkylbenzene:

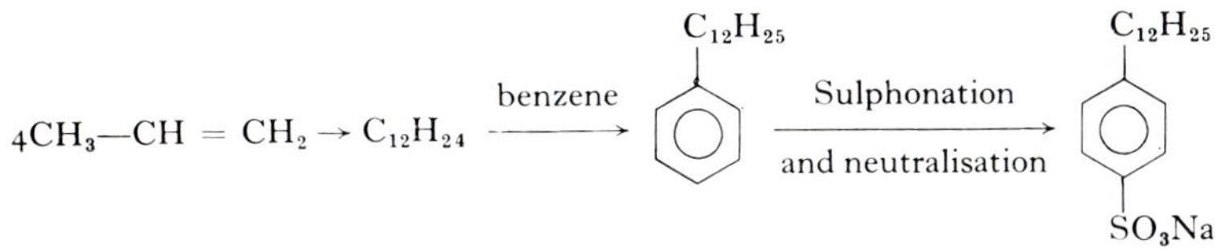

9.2.1. Production of Branched-Chain Alkylbenzenes

9.2.1.1. Tetrapropylene manufacture The growth of the manufacture of high-octane gasoline during World War II resulted in the production of large quantities of by-product propylene and, to utilise this, methods were developed to oligomerise it to medium- and long-chain olefins. The oligomerisation, which is carried out with acid catalysts, usually phosphoric acid supported by Kieselguhr,[6] is highly complex. The scheme on p. 448 exemplifies some of the more important reactions which occur (addition of carbonium ions to olefins, hydride and alkyl shifts and fragmentations) and shows how C_{11}, C_{12} and C_{15} olefins (respectively VI, IV and VII), as well as C_4 fragments (V) can arise. Clearly innumerable reactions analogous to those given in this scheme can and do occur and the product is, therefore, an extremely complex mixture of highly-branched olefins of a range of chain lengths.

The oligomerisation can be run to produce predominantly the tetramer, often termed propylene tetramer or tetrapropylene which is a suitable raw material for providing the side chain of a detergent alkylate.

9.2.1.2. Alkylation of benzene with tetrapropylene The tetrapropylene is treated with benzene under Friedel-Crafts conditions. Process conditions vary quite widely but are largely determined by the catalyst used. Alkylation temperature (usually 20–50°C) residence time (from a few minutes to 1 hr) and tetrapropylene to benzene mole ratio (usually *ca.* 1 to 5–10) are the most important parameters determining alkylation efficiency.[7] Fig. 9.1 shows schematically the elements of most of the industrial processes.

Both anhydrous hydrogen fluoride and aluminium chloride have been used as catalysts for the industrial scale manufacture. Hydrogen fluoride is used in large excess and recycled whereas aluminium chloride is used in small amounts with a promoter (usually water or hydrogen chloride). Like the reaction producing the

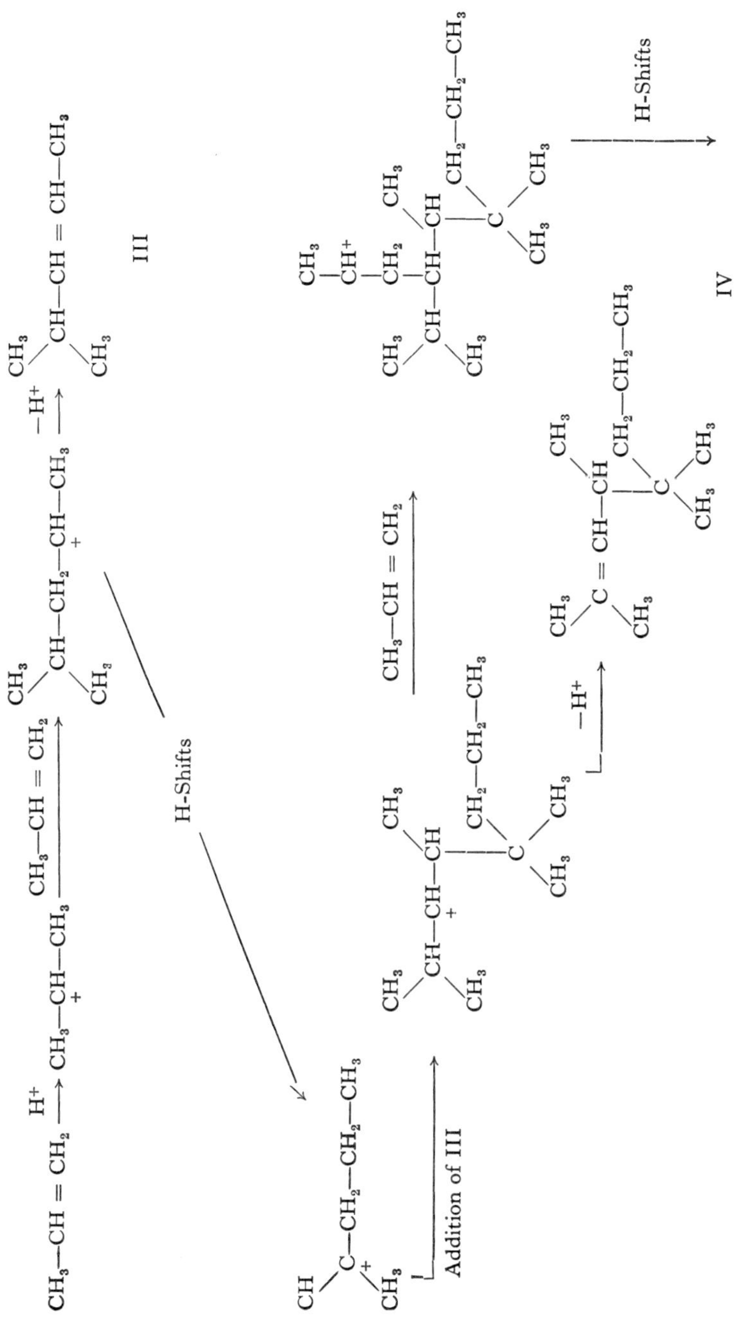
H-Shifts
Addition of III
H-Shifts
III
IV

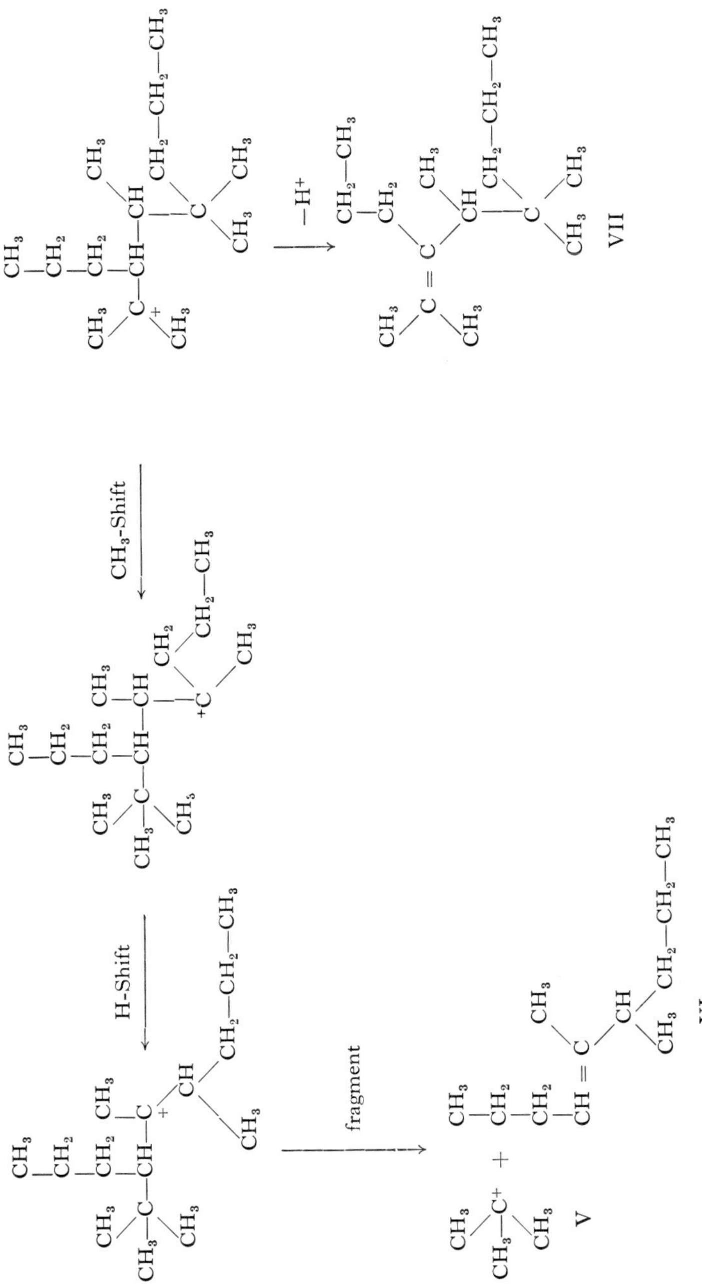

VII

V

VI

tetramer, the alkylation involves carbonium ion intermediates which can rearrange, fragment and polymerise and the system is further complicated by intermolecular hydride transfers which give alkanes.[8] The considerable excess of benzene used in practice facilitates the required mono-alkylation and suppresses the side reactions, although these still occur to an appreciable extent.

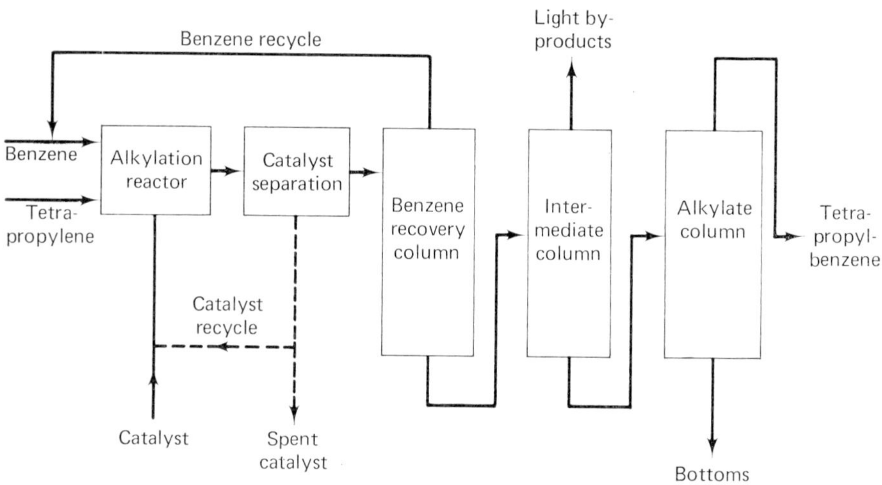

FIG. 9.1 *Idealised Process for the Production of Tetrapropylbenzene*

After removal of the catalyst, the alkylation product is fed into a first distillation unit where the excess benzene is separated off and recycled. The bottoms from this unit are fed into a second distillation unit which separates off a 'light by-products' fraction (usually 5–15 per cent (w/w) of the product tetrapropylbenzene). This consists predominantly of short-chain alkylbenzenes (formed by fragmentation of the propylene tetramer and alkylation of the resultant shorter-chain olefins with benzene) and branched-chain alkanes containing close to twelve carbon atoms (formed by intermolecular hydride transfer reactions). The bottoms from this second distillation are fed to a third distillation unit which separates the product tetrapropylbenzene from less volatile material (usually 9–17 per cent (w/w) of the product tetrapropylbenzene). This bottom fraction consists of di-(tetrapropyl)-benzene, higher polymeric hydrocarbons derived from the tetrapropylene by reactions like those given in the scheme on p. 448 and long-chain alkylbenzenes formed by, for example, tetrapropylene dimerisation and alkylation of benzene with the resultant long-chain olefins.

The yield of the desired tetrapropylbenzene is usually in the range 70–80 per cent.[8,9] Because the tetrapropylene raw material is itself a very complex mixture and the alkylation introduces even more complication, the product is, as shown in Fig. 9.2, a highly complex mixture.

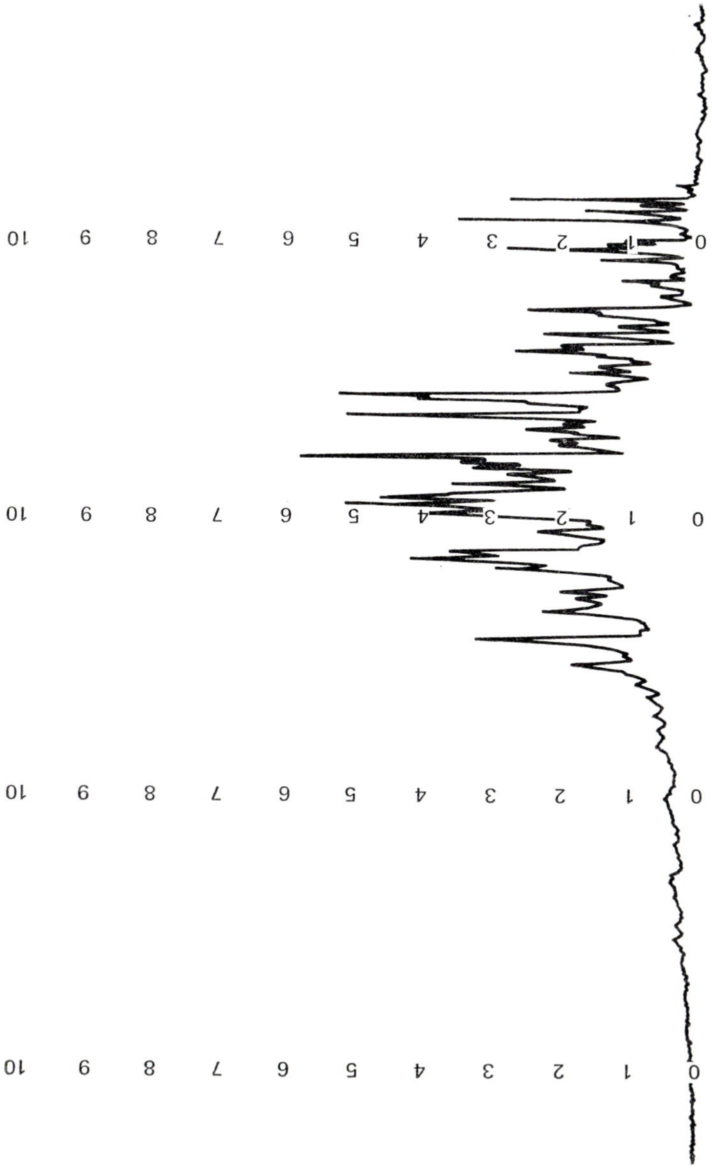

FIG. 9.2 *Gas Liquid Chromatogram of a Typical Tetrapropylbenzene*

The major impurities in the tetrapropylbenzene are branched alkanes of intermediate chain length which cause a problem if present at too high a level, as they cannot be sulphonated. They finish up in the sulphonate product and have a deleterious effect on its foaming properties.

9.2.2. Sulphonation

Detergent alkylates are sulphonated commercially with either oleum or sulphur trioxide.

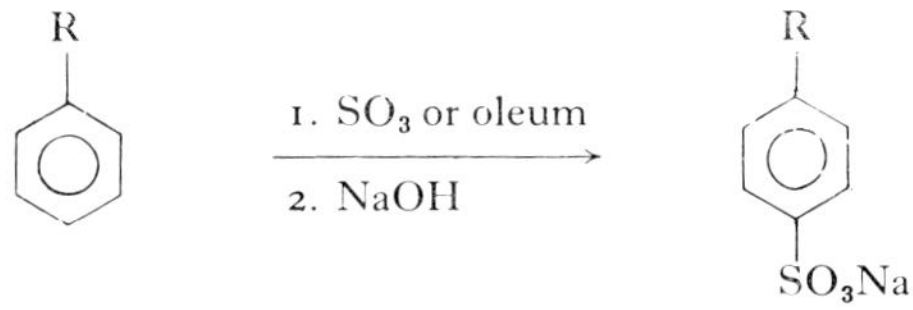

VIII IX

Work on pure alkylbenzenes of known structure[10] established that the orientation of sulphonation depends on the nature of R in VIII. If R is straight-chain alkyl, 80 per cent of the product is *para* sulphonate (IX). If R contains a single methyl branching group on the carbon atom attached to the benzene ring, then 90 per cent of the product is *para* sulphonate. This increase in the extent of *para* substitution is considered to be due to the suppression of *ortho* substitution by steric hindrance. If the single branching group is larger than methyl, the proportion of *para* substitution is higher than 90 per cent and, with tetrapropylbenzene, most of which has *two* alkyl groups on the carbon atom carrying the benzene ring, virtually all the sulphonate is *para*.

9.2.2.1 By oleum For industrial manufacture, batchwise oleum sulphonation procedures were first developed but these were largely superseded by continuous processes. These divide into three main stages—sulphonation, spent acid separation, and neutralisation. In the sulphonation stage, the detergent alkylate is rapidly mixed with an approximately equal weight of 20–25 per cent oleum and the (considerable) heat of reaction is removed so as to keep the reaction temperature at about 50°C. The sulphonation mixture is then passed to a holding section, residence time of the order 20 min., to allow the sulphonation to go to completion (usually >98 per cent alkylate conversion). From this holding section, the sulphonation mixture passes into the spent acid separation system where sufficient water is added to dilute the remaining sulphuric acid to 70–80 per cent. The diluted reaction mixture separates into two phases. The lower spent acid phase, which contains the majority of the sulphuric acid and only traces of the product sulphonic acid, is discarded. The upper phase, containing approx. 85–90 per cent alkylbenzene sulphonic acid and 7–10 per cent sulphuric acid is fed into the final stage where it is neutralised with aqueous sodium hydroxide or sodium carbonate to give a paste containing about 30–50 per cent sodium alkylbenzene sulphonate, 0·3 per cent unsulphonated organic materials and 5–9 per cent sodium sulphate, of suitable colour and odour for processing into finished detergent products. If colours are poor, this can usually be rectified by a mild hypochlorite bleach. Various methods (e.g. treatment of the paste with sodium hypochlorite, sodium borohydride or hydroxylamine hydrochloride[11] or treatment of the tetrapropylbenzene with hydrogen fluoride[12]) have been proposed for improving the product odour (ascribed to ketone and diene by-products[13]).

9.2.2.2. By sulphur trioxide Sulphur trioxide, the second major sulphonating agent used for making alkylbenzene sulphonates, has the important advantage that it gives a product essentially free from sodium sulphate and does not give by-product spent acid. The reaction of detergent alkylate with this reagent is extremely vigorous and, to produce a product of acceptable colour, it is necessary to dilute the sulphur trioxide and avoid contact of the sulphonic acid product with excess sulphur trioxide. Another major problem is to remove the considerable heat of reaction from the viscous sulphonation mixture.

In industrial sulphur trioxide sulphonations, a slight excess (usually *ca.* 5 mole per cent) of vaporised sulphur trioxide diluted with dry air (as low as 2–4 per cent SO_3 by volume) is contacted with the vigorously agitated detergent alkylate. Numerous processes have been evolved to achieve the good gas-liquid contacting and heat transfer needed,[14] and in the most successful sulphonators, the sulphur trioxide/air mixture is passed co-currently over a film of the alkylbenzene flowing over a heat-exchanger surface.[15]

As in oleum sulphonation, the product sulphonic acid is neutralised with aqueous sodium hydroxide or sodium carbonate to give a paste containing 30–50 per cent sodium alkylbenzene sulphonate substantially free from sodium sulphate. With such systems it is possible to produce colours and conversions comparable to those given by oleum sulphonation and sodium tetrapropylbenzene sulphonates produced by sulphur trioxide sulphonation are usually almost odourless.[11]

9.2.3. Major uses

The ability to produce from sodium tetrapropylbenzene sulphonate (usually abbreviated 'ABS') a range of highly efficient products specially designed to fit into a number of different markets where previously only soap-based products (with some deficiencies when used in hard water), or indeed no products at all, had been available, resulted in the spectacular growth in the use of this surfactant after World War II.

By far the most important use of ABS was as the 'detergent active' in domestic heavy-duty fabric-washing powders. The formulation of these varied from country to country depending on indigenous washing practices and washing machine types. They usually contained up to 20 per cent ABS, up to 50 per cent sodium tripolyphosphate builder,* sodium sulphate, sodium silicate, minor amounts of optical brightening agents, antiredeposition agents, and possibly also other surfactants (e.g. alkyl sulphates, soaps), lather boosters (usually long-chain ethanolamides), sodium

* A successful fabric-washing product must function efficiently in hard water. This is not possible if free hardness ions (Ca^{++}, Mg^{++}) are present as these interfere substantially with the detergency process. 'Builders' are materials which remove these ions from solution—in this case by sequestration.

perborate (bleaching agent), etc. Completely new types of domestic washing products which contained appreciable amounts of ABS, namely dishwashing liquids and light-duty liquids designed to give mild washing conditions for sensitive fabrics were also developed and marketed in this period. These also contributed significantly to the growth in the use of ABS.

As ABS is a synthetic product, it was possible to produce materials with different chain-length distributions, investigate how performance depends on chain length, and hence specify the best ABS for a particular application. This was done in certain cases, e.g. Colgate[3] showed that 'pentadecyltridecylbenzene sulphonate' (PDB–TDB) was a much better foamer than 'dodecylbenzene sulphonate' (DDB). The distribution of these two ABS types is shown in Fig. 9.3.

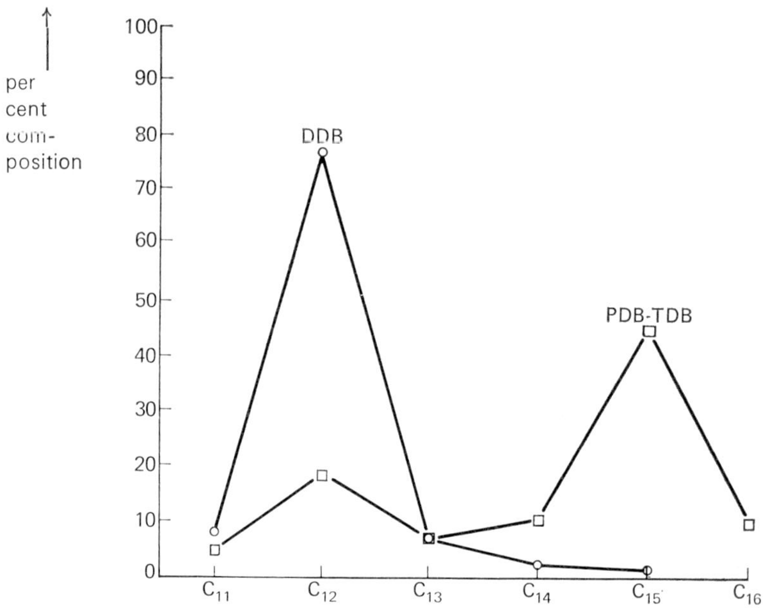

FIG 9.3 *Chain-length Distribution of Dodecylbenzene and Pentadecyltridecylbenzene* (see Rubinfeld J. and Cross H. D., *Soap Chem. Specialities* 1967 **43**, 41)

At a given chain length, performance also undoubtedly depends on the structure of the chain but, because of the complexity of tetrapropylene-based ABS and the virtual impossibility of controlling the propylene oligomerisation or the alkylation of benzene with tetrapropylene to give a particular desired isomer distribution, this was never followed up to generate a 'super ABS'. Colgate[40] did, however, develop a method for the pre-screening and prediction of the performance of commercial ABS's which was based on the determination of the chain length distribution and a 'branching parameter' of the alkylate by mass-spectral analysis prior to sulphonation.

In addition to these major applications, ABS was used as a detergent and emulsifying agent in a variety of smaller applications such as in herbicide and insecticide formulations, textile finishing, emulsion polymerisation, cutting oils and solvents.

9.2.4. Biodegradation[16]

It was recognised in the middle 1950's[17, 18, 19] that the degree of branching of the hydrophobic hydrocarbon chain has a very considerable effect on the rate of biological breakdown (biodegradation) of surfactants. A single methyl branch on an otherwise straight chain has only a small effect but a quaternary carbon atom in the chain can make the molecule very resistant to biological breakdown.

Work by Swisher[20] and by Kölbel *et al.*[21] on model compounds showed that compounds with a quaternary carbon atom at the end of the chain, such as X, are particularly resistant to biological breakdown but that materials containing a quarternary group adjacent to the benzene ring, such as XI are quite readily broken down.

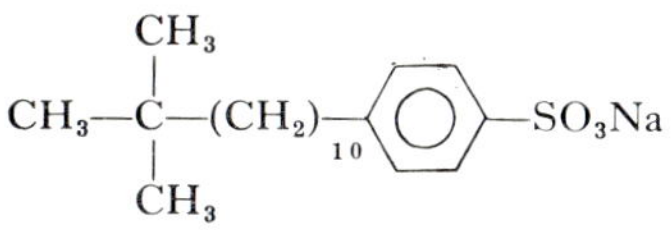

X

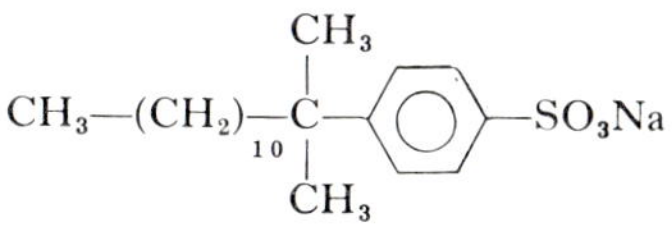

XI

As was described on p. 450, tetrapropylene-based ABS is a very complex mixture of materials, many of which have quarternary carbon atoms in their highly branched side chains. Consequently ABS is resistant to biological breakdown and, as its usage increased, the result was a build-up of the surfactant in natural waters to levels which caused persistent foams. These were aesthetically unpleasant and, more important, in some areas of the world they interfered with sewage treatment and the normal uptake of oxygen by natural waters. As a result, it was clear by the late 1950's that despite its cheapness and very good technical performance, the life of hard propylene-based ABS was limited, particularly in densely populated areas. Clearly, biodegradable alternatives had to be developed.

9.3. BIOLOGICALLY SOFT LINEAR ALKYLBENZENE SULPHONATES (LAS)

Following the considerable technical success of hard ABS, alkylbenzene sulphonates with straight side chains provided the obvious starting point in the search for biologically soft anionic detergents. It was reasonable to expect that such materials could be substituted simply into existing ABS-based products, and that at least some of the processes used for the production of ABS could be adapted for the production of linear alkylbenzene sulphonates (usually abbreviated 'LAS'). To check whether the use of LAS would indeed give the required improvement in biodegradation, an ambitious and very significant field trial was mounted in Luton, (population *ca.* 110,000), England.[22] In late 1958, without any prior publicity, the major detergent manufacturers replaced their hard ABS based products sold in this area with products based on one of the first LAS materials available in commercial quantities and the effect of this replacement on the level of anionic detergent in Luton's treated sewage was monitored up to the end of 1961. A very significant decrease in the detergent level was observed but the decrease was rather less than would have been expected on the basis of laboratory assessment of the biodegradation of LAS types.[17, 18, 19] This was attributed to slow release of adsorbed and occluded ABS from the sewage system as well as the purchase of materials outside the area and failure to achieve a complete switch from ABS to LAS in Luton's retail outlets. The overall conclusion was, nevertheless, that the use of LAS would indeed bring about the required improvement in surfactant biodegradation. This was generally supported by other field tests and the chemical industry was, accordingly, faced with the problem of generating efficient routes to LAS and, as described below, a number of effective routes were developed and rapidly commercialised.

The changeover from ABS to LAS got under way rapidly after the Luton experiment. By mid 1963, 60–70 per cent of the United Kingdom market had switched to biodegradable material and the changeover was effectively complete by 1966. In the meantime, West Germany had already made the change, complying with legislation which became effective in late 1964, and the largest market of all, the USA, had switched to LAS by mid 1965. By the end of 1971, ABS had, in addition, been replaced in the majority of the rest of Europe, Canada, South Africa and Australia.

9.3.1. Production of Linear Alkylbenzenes

The first problem facing a potential linear alkylate manufacturer is how to obtain in large tonnage a basic raw material containing a long hydrocarbon chain which can be alkylated onto benzene. Two such materials, *sec*-monochloroparaffins and olefins have in fact been used.

9.3.1.1. Production of n*-paraffins* *n*-Paraffins are the most important primary raw materials used for LAS manufacture. C_{10-15} homologues, suitable for surfactant manufacture, occur in the kerosene and gas oil fractions of petroleum and two methods have been used to separate these on a commercial scale. The more widely-employed method uses *molecular sieves*. These are calcium zeolites and have a structure within which are cavities and 'windows' of approximately 5Å diameter which provide access to these cavities. *n*-Paraffins (diameter *ca.* 4·9Å) can enter these cavities whereas branched paraffins (diameter roughly 5·6Å or greater) cannot pass through the windows and are hence rejected. After purging the unadsorbed materials (with, for example, N_2 or a low-boiling hydrocarbon) the adsorbed *n*-paraffins can be recovered by increasing the temperature over the sieve bed, decreasing the pressure over the sieve bed or, more usually, by displacement with a low boiling hydrocarbon such as *n*-pentane or *n*-hexane which can be readily removed from the detergent range *n*-paraffin by distillation and recycled. A number of companies have developed processes for the production of *n*-paraffins based on this technology, notably Union Carbide ('Isosiv' process)[23], Texaco ('TSF' process),[24] Universal Oil Products ('Molex' process),[25], BP[26] and Esso ('Ensorb' process).[27] These all operate according to the basic principles described above and differ mainly in the method used to recover the adsorbed *n*-paraffins. With the exception of the liquid-phase 'Molex' process, all are vapour-phase operations. The processes are capable of very large-scale operation and are highly efficient; 85–98 per cent recoveries of *n*-paraffins of better than 97 per cent purity are claimed.

The second method employed for the separation of detergent-range *n*-paraffins from petroleum uses *urea clathration*. If urea is allowed to crystallise in the presence of *n*-paraffins, a clathrate-type adduct is formed in which the *n*-paraffin is contained in a channel running through a hexagonal urea lattice. This channel will not accommodate a branched hydrocarbon chain but a straight chain of more than seven atoms is sufficient to enable the clathrate to form. The stability of the adduct increases with increasing length of the straight-chain residue involved in the adduction. Hence, *n*-paraffins can be separated from mixtures containing branched-chain hydrocarbons of similar chain length. However, those branched-chain hydrocarbons with the branching close to the end of the chain, which consequently contain fairly long straight-chain alkyl groups, also tend to form adducts and this limits the final *n*-paraffin purity which can be obtained.

Several companies hold patents covering processes for the isolation of *n*-paraffins by this method. The major technical problem is to cause the adduct to form and crystallise rapidly in a physical state such that it can be readily separated by filtration. This is achieved by the use of various organic solvents, e.g. methanol, methyl ethyl ketone, dichloromethane, as 'promoters' and by careful choice of the urea/oil contacting conditions. The Edeleanu G.m.b.H. process is described in some detail in the literature.[23] In this, the *n*-paraffin-containing petroleum fraction, dissolved in dichloromethane, is contacted with 76 per cent (w/w) aqueous urea at 80°C. The solid granular urea /*n*-paraffin adduct which forms is separated, washed with solvent and then decomposed by the addition of water and heating. The resultant *n*-paraffin is phase-separated from the aqueous urea, which is recycled. Finally, the last traces of solvent are evaporated from the *n*-paraffin. To give high purity *n*-paraffins (95 per cent is claimed), two adduction stages are used.

Urea processes have been operated by Deutsche Erdöl AG in Germany and by Shell in the Netherlands. The method has not, however, been applied as widely as molecular sieve separation.

9.3.1.2. Conversion of n-*paraffins to alkylating agents* *n*-Paraffins do not react directly with benzene. They must first be converted into derivatives containing functional groups which can be used as sites for the attachment of a benzene ring. One way to achieve this, namely chlorination to *sec*-monochloroparaffins, had already been applied industrially and the other obvious derivative is the corresponding internal olefin.

sec-Monochloroparaffins (e.g. XII) are produced by direct chlorination of the *n*-paraffin as exemplified for *n*- dodecane in the following equation:

$$CH_3(CH_2)_{10}CH_3 + Cl_2 \longrightarrow CH_3(CH_2)_3\underset{\displaystyle Cl}{\underset{|}{C}}H(CH_2)_6CH_3 + HCl$$

XII

(plus all other possible *sec* substitution products)

The reaction is a free-radical chain process and substitution of chlorine occurs to approximately the same extent at each of the —CH_2— groups in the paraffin chain. The terminal methyl groups are less reactive and, in consequence, the product consists of *ca.* 90 per cent *sec*-monochloroparaffin and 10 per cent primary chloride.[29] The substitution of a single hydrogen atom by chlorine in a paraffin chain leaves the reactivity of the majority of the remaining hydrogen atoms essentially unchanged. As a result di- and poly-chlorination can and does occur. This is unsatisfactory for economic reasons as it uses chlorine wastefully and, even more important, undesirable impurities are formed when a linear alkylbenzene is produced by the alkylation of benzene with a monochloroparaffin containing di- or poly-chlorinated impurities (see pages 462 and 465). To suppress polychlorination the paraffin chlorination is, therefore, usually only taken to *ca.* 20–30 per cent conversion. This gives approx. 85–90 per cent yield of monochloroparaffin. The chlorination is normally carried out in the liquid phase at temperatures within the range 60–130°C, or at lower temperatures if photochemically or catalytically initiated. Quite a wide range of process conditions seem to be used industrially. Chevron's two-stage chlorination[30] is probably fairly typical.

The *ca.* 25 per cent solution of *sec*-monochloroparaffin in unreacted *n*-paraffin, produced by the chlorination process, is used as such in the alkylation (see p. 463). This is because it is much easier to separate the paraffin/alkylbenzene mixture by distillation than it is to separate the paraffin/chloroparaffin mixture.

Internal olefins (e.g. XIII) are produced from *n*-paraffins in two ways. The first consists of chlorination to the monochloride, as described above, followed by catalytic dehydrochlorination as exemplified in the following equation:

$$CH_3(CH_2)_{10}CH_3 \xrightarrow{Cl_2} CH_3(CH_2)_3\overset{\overset{\large Cl}{|}}{C}H(CH_2)_6CH_3 \xrightarrow{-HCl} CH_3(CH_2)_2CH=CH(CH_2)_6CH_3 \quad \text{XIII}$$

$$+\ CH_3(CH_2)_3CH=CH(CH_2)_5CH_3$$

Because the starting material for the dehydrochlorination contains all the possible mono-chloro isomers, the product olefin is also a mixture of all the possible olefin isomers.

A very wide range of catalysts (alkalis, alkaline earths, neutral salts, metallic alloys) and process conditions (usually liquid phase, temperature in the range 150–500°C) has been specified in the patent literature.[31] The dehydrochlorination of pure *sec*-monochloroparaffins at 200–250°C over iron catalysts gives almost quantitative yields of olefins. Random *sec*-dichloroparaffins, on the other hand, only give *ca.* 50 per cent of the theoretically possible diolefins, substantial amounts of tarry by-products and cyclic mono-olefins being formed.[50] The by-product HCl is claimed to be very pure [32] and it has been reported [33] that Chemische Werke Hüls use it for vinyl chloride manufacture.

The second route from paraffins to internal olefins uses catalytic dehydrogenation. Such a process is a crucial part of Universal Oil Products' (UOP) linear alkylbenzene process announced in 1966.[25a] The UOP process was claimed to convert detergent-range *n*-paraffins to internal olefins (all possible internal isomers in substantially equal amounts with a small amount of α-olefin) with a selectivity of approximately 95 per cent, the major by-products being conjugated dienes. The process conditions were not disclosed, but merely stated to be relatively mild conditions of temperature and pressure with a new long-lived catalyst. The high selectivity to mono-olefin would seem to imply that the *n*-paraffin conversion is quite low. In patents published after the announcement of their process,[34] UOP covered a wide range of dehydrogenation catalysts involving alumina, noble metals, alkali metals, alkaline earth metals and/or compounds of these. Other workers have patented modified zeolites and copper molybdate.[55]

In principle, the dehydrogenation process, which gives no by-product problems, is more elegant than the chlorination/dehydrochlorination approach, the economics of which depend very considerably on the value of the by-product HCl.

9.3.1.3. Terminal olefins from paraffin wax An important large-scale source of substantially straight-chain olefins is the cracking of long-chain paraffin waxes. This is usually done in the presence of steam in a furnace at 500–600°C with a residence time of up to 10 sec. The wax conversion is of the order 20–50 per cent with a yield of around 50 per cent of liquid olefins of a wide range of chain lengths.[35] This mixture is fractionated to give a C_{10-15} cut suitable for detergent alkylate manufacture. Shell and Chevron are major producers of cracked-wax olefins and a substantial proportion of the LAS used in Europe since the mid-1960's has been made from such materials.

9.3.2. Chemistry of Alkylation

9.3.2.1. Correlation of structure with performance The value of an alkylation process is determined not only by its overall economics but also by the sulphonation characteristics of the detergent alkylate which it produces and the technical performance of the LAS from this sulphonation. Before dealing with alkylation methods, it is therefore appropriate to consider how LAS performance depends on alkylbenzene structure. In contrast to the situation with ABS, the number of isomers of LAS of a particular chain length is very small—e.g. there are only six possible straight-chain dodecylbenzene-*para*-sulphonate isomers ranging from the 1-phenyl isomer (XIV) to the 6-phenyl isomer (XV).

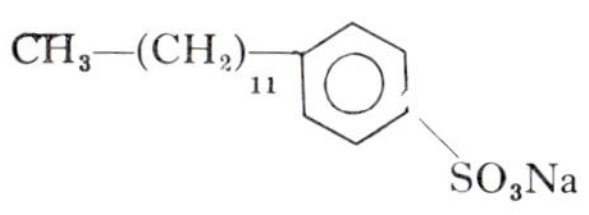

XIV

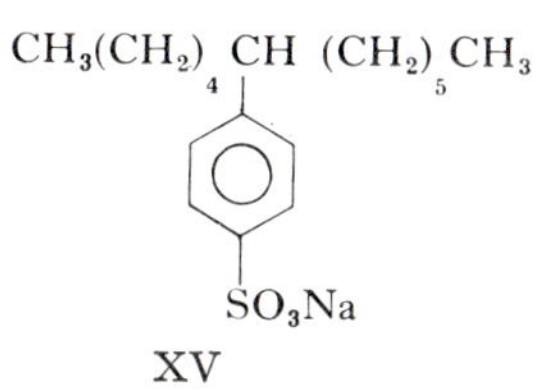

XV

Because of this, the synthesis of a range of pure materials as well as the isolation of concentrates of various structural types from commercial or potentially commercial linear alkylates was possible and hence the performance (detergency and foaming) of LAS isomers could be correlated with structure. A number of groups carried out such studies in 1954–67[10, 36–44] and it emerged that the correlations were far from simple. The ranking order of the LAS isomers for optimum performance was found to depend to a considerable extent on the formulation used for the tests and the test conditions. It was clear, however, that performance was determined both by chain length and phenyl isomer distribution. It was generally agreed that, at a given chain-length, the 'internal' isomers (i.e. those with the phenyl group close to the middle of the chain) give better foaming than the 'end' isomers (i.e. those with the phenyl group close to the end of the chain) but the situation was less clear-cut for detergency. Rubinfeld and Cross[44] (Colgate) concluded that, for optimum performance in heavy duty formulations, the alkylate should have a molecular weight of 258–266 with not less than 70 per cent of the alkylate side chains being C_{13} and C_{14}, 20 per cent C_{12}, and not more than 5 per cent each of C_{11} and C_{15}. The 5- and 6-phenyl content should be greater than 40 per cent with less than 20 per cent 2-phenyl material. The same authors recommended for light duty formulations an alkylate of molecular weight 231–241 with at least 50 per cent C_{10} and C_{11} chains, at least 25 per cent C_{12} and not less than 95 per cent in the C_{10-12} range. The specification for 2-phenyl isomers was put at 15–25 per cent. These alkylates were claimed to give LAS of equal performance to ABS in their respective applications. Most studies arrived at optimum chain lengths fairly close to C_{12} and, in line with the Colgate recommendations, the tendency has been to use rather lower molecular weight alkylate in light duty products and somewhat higher molecular weight material in heavy duty powders.[4]

One other property which depends on the position of the phenyl group on the alkyl chain is biodegradability; the end phenyl isomers degrade somewhat more rapidly than the internal isomers.[71]

9.3.2.2. Factors affecting the isomer distribution in linear alkylbenzenes The chain length of an alkylate is clearly controlled by the chain length of the intermediate (chloroparaffin or olefin) used in the alkylation. The isomer distribution, on the other hand, is controlled by the choice of catalyst, alkylation conditions and process design. Consequently, within certain limits, it is possible for the chain length and isomer distribution of alkylates to be controlled to meet the particular needs of detergent manufacturers.

The same Friedel-Crafts catalysts which were used for tetrapropylbenzene manufacture, anhydrous hydrogen fluoride and aluminium chloride, also serve for the production of linear alkylbenzenes. Olefins can be processed with either catalyst but *sec*-monochloroparaffins do not react with benzene in the presence of hydrogen fluoride under conditions suitable for alkylate manufacture. Because the composition of the product linear alkylbenzenes is a good deal simpler than that of the tetrapropylene-based alkylate, detailed product analysis has been possible by mass spectrometry and gas liquid chromatography and, as a result, much more is known about the detailed chemistry of these alkylation processes than about tetrapropylene alkylation. Friedel-Crafts alkylations are known to proceed *via* intermediates with carbonium ion character[45] and these are capable of undergoing re-arrangement. Accordingly, when benzene is alkylated with a long-chain olefin or *sec*-monochloroparaffin, the distribution of isomers in the alkylating agent, the rate of rearrangement of the carbonium-intermediate formed prior to alkylation, the rate of the alkylation of benzene by this intermediate and the rate of rearrangement of the product alkylbenzene can all influence the isomer distribution in the product alkylbenzene.

An extensive study of the aluminium chloride catalysed and hydrogen fluoride catalysed alkylations of benzene with long-chain olefins has been carried out by H. R. Alul *et al.*[46–49] With promoted $AlCl_3$ at 35–37°C,[46] virtually the same product distribution is given by alkylation of benzene with 1-dodecene or *trans*-6-dodecene, i.e. 32 per cent 2-phenyl-, 19–21 per cent 3-phenyl-, 17 per cent 4-phenyl-, 30–32 per cent 5- and 6-phenyldodecane. In batch alkylations catalysed with aluminium chloride an induction period is observed during which the so-called 'red oil' forms.[63] In Alul's experiments, this was minimised by adding red oil from a previous alkylation, run under the same conditions, as part of the catalyst.

2-Phenyldecane, included in the *trans*-6-dodecene alkylation reaction to diagnose for product isomerisation, was 61·5 per cent rearranged to substantially equal amounts of 3-, 4- and 5-phenyldecane. Corresponding alkylations carried out at 0–5°C (30 min. reaction time) gave quite different distributions (44 per cent 2-phenyl-, 22 per cent 3-phenyl-, 14 per cent 4-phenyl- and 20 per cent 5- and 6-phenyldodecane from 1-dodecene compared to 18 per cent 2-phenyl-, 12 per cent 3-phenyl-, 17 per cent 4-phenyl- and 53 per cent 5- and 6-phenyldodecane from *trans*-6-dodecene) and the marker 2-phenyldecane remained unchanged—i.e. product isomerisation did *not* occur. Thus, at the higher temperature the phenyl isomer distribution was determined by isomerisation of the product alkylbenzenes to an equilibrium composition whereas at the lower temperature, the distribution was determined by the balance between the rates of alkylation and isomerisation of the carbonium ion type intermediate prior to alkylation.

It was also shown that isomer-distributions intermediate between the extreme cases given above could be obtained by modification of the catalyst activity as well as by variation of reaction conditions.

Consequently, the alkylation of benzene with *n*-paraffin-derived random internal olefins or *sec*-monochloroparaffins with an active $AlCl_3$ catalyst at temperatures above 35°C gives around 32 per cent 2-phenylalkanes and, because of the random functional group distribution in these mixed paraffin derivatives, manipulation of the alkylation conditions or catalyst activity cannot greatly affect the 2-phenyl content.

Hydrogen fluoride alkylations differ from aluminium chloride alkylations in a number of important ways.[47, 54] Firstly, the reaction is normally conducted with a large excess of the catalyst which is recovered and re-used and secondly, the product phenylalkanes do not isomerise. Hence the isomer distribution is governed almost exclusively by the balance between the rate of isomerisation of the carbonium ion intermediate and the rate of alkylation. Alul[47] showed that at 55°C with an olefin: benzene: HF mole-ratio of 1:8.7:16.6 (where a separate catalyst phase exists), 1-dodecene and *trans*-6-dodecene give different isomer distributions (25 per cent 2-phenyl-, 17 per cent 3-phenyl-, 17 per cent 4-phenyl- and 41 per cent 5- and 6-phenyldodecane from 1-dodecene as against 15 per cent 2-phenyl-, 15 per cent 3-phenyl-, 17 per cent 4-phenyl- and 54 per cent 5- and 6-phenyldodecane from *trans*-6-dodecene). Apparently, under these conditions, carbonium ion rearrangement is extensive but not complete. At 0°C, both olefins still gave different distributions but with substantially less 2-phenyldodecane and substantially more 5- and 6-phenyldodecane in the products. In alkylations with a separate HF-phase where the benzene and olefin were diluted with a 10-fold amount of hexane, 1-dodecene and *trans*-6-dodecene gave the same product distribution. At 55°C the product contained 14 per cent 2-phenyl-, 15 per cent 3-phenyl-, 17 per cent 4-phenyl- and 54 per cent 5- and 6-phenyldodecane; at 0°C, even less of the 2-phenyl isomer (11 per cent) was formed. It was concluded that the dilution with hexane had slowed down the alkylation sufficiently to allow the carbonium ion isomerisation to go to equilibrium. It is important to note that, with a separate catalyst phase present, even under conditions where the pre-alkylation isomerisation was incomplete, the hydrogen fluoride alkylation gave a lower proportion of 2-penylalkine than was given by the aluminium chloride catalysed alkylation under all the conditions reported by Alul.[46]

9.3.2.3. By-products Industrial linear alkylbenzenes contain cyclic by-products to which a number of authors have ascribed alkylindane (XVI) and alky tetralin (XVII) structures.[25a, 32, 50, 51] Recent evidence indicates that these are mostly, if not completely, tetralins.[52, 53]

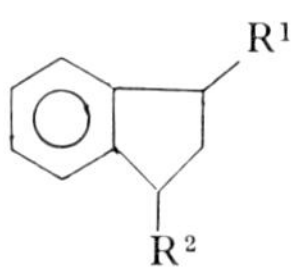

XVI

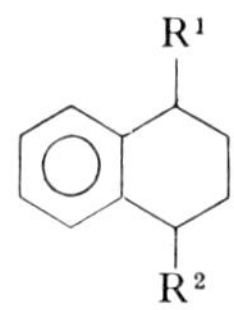

XVII

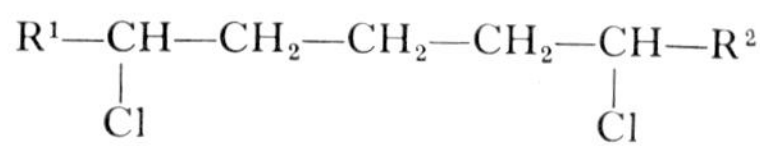

XVIII

These by-products are derived from difunctional impurities, *e.g.* XVIII which would be expected to react with benzene *via* both functional groups, with rearrangement of the intermediate carbonium ions, to give XVII.[32, 50] Small amounts of paraffins are also produced in alkylations involving active aluminium chloride catalysts, presumably by abstraction of hydride ion from the product phenylalkane by the carbonium ion type intermediate; this has been suggested as another possible source of the cyclic by-products[46]:

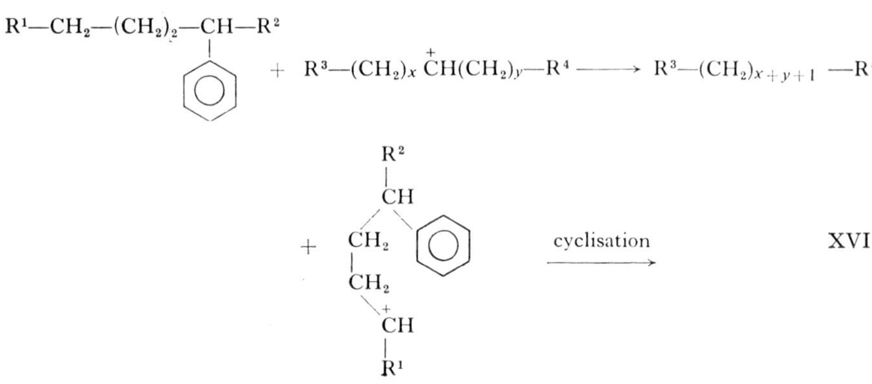

Other by-products are diphenylalkanes (from difunctional alkylating agents such as XVIII), dialkylbenzenes[54] (produced by alkylation of the product monoalkylbenzenes), and traces of polynuclear aromatics (usually considered to be anthracenes). A small amount of 'light alkylate' is also formed, presumably by reactions similar to those which occur during propylene tetramer alkylation. The amount of light alkylate usually produced is considerably less than is normal in the probuction of branched alkylate.

9.3.3. Industrial alkylation processes

The production of linear alkylbenzenes is in principle similar to that of branched alkylate. The main features of the majority of alkylation processes are shown in Fig. 9.4.

Within this general scheme, there are numerous variants, several of which have additional steps (e.g. multistage alkylation, purification of recycle streams, purification of product alkylate). One important difference between the production of linear alkylate from *n*-paraffins and the production of branched alkylate is the necessity of a paraffin recycle step in the former. As explained previously, this is because of the need to operate at low conversion to avoid the production of polyfunctional by-products when producing the alkylating agent (chloroparaffin or olefin), and the impossibility of separating these alkylating agents from the unconverted paraffin by distillation.

9.3.3.1. The sec-*monochloroparaffin/aluminium chloride route*[56]

The alkylating agent in this process, *ca.* 25 per cent monochloroparaffin in *n*-paraffin from the chlorination unit, is fed together with benzene (in 5–15 molar excess to suppress dialkylation) and aluminium chloride (usually 2–5 per cent (w/w) based on chloroparaffin) to the alkylation reactor. The alkylation can be carried out in a multistage process where the 'red oil' catalyst phase is partially recycled.[30] Residence times vary quite widely but are generally within the range 4–300 min., depending on the activity of the catalyst, process design, etc. Reaction temperatures usually fall in the range 40–85°C. At these temperatures with active catalysts, the product phenylalkanes rearrange and hence the process normally gives a product with a high (>30 per cent) 2-phenyl isomer content. The fact that phenylalkanes rearrange under these conditions has been employed to advantage in two variations of the basic process. The first of these variants[57] makes use of the fact that, at a particular chain length, the end phenyl-isomers are less volatile than the internal isomers. Hence, the high 2-phenyl product is separated by distillation into an alkylate of low 2-phenyl content and a (less-volatile) fraction enriched in the 2-phenyl isomer. This high 2-phenyl-fraction is recycled to the alkylation unit where it re-equilibrates. The net result is, therefore, the production of a low 2-phenyl alkylate at the expense of extra distillation and recycle costs.

In the second variant[58] the heavy alkylate is distilled to give a fraction, rich in diphenylalkanes, which is recycled to the alkylation reactor. This is claimed to suppress diphenylalkane formation.

The reaction mixture from the alkylation is phase-separated to give catalyst sludge

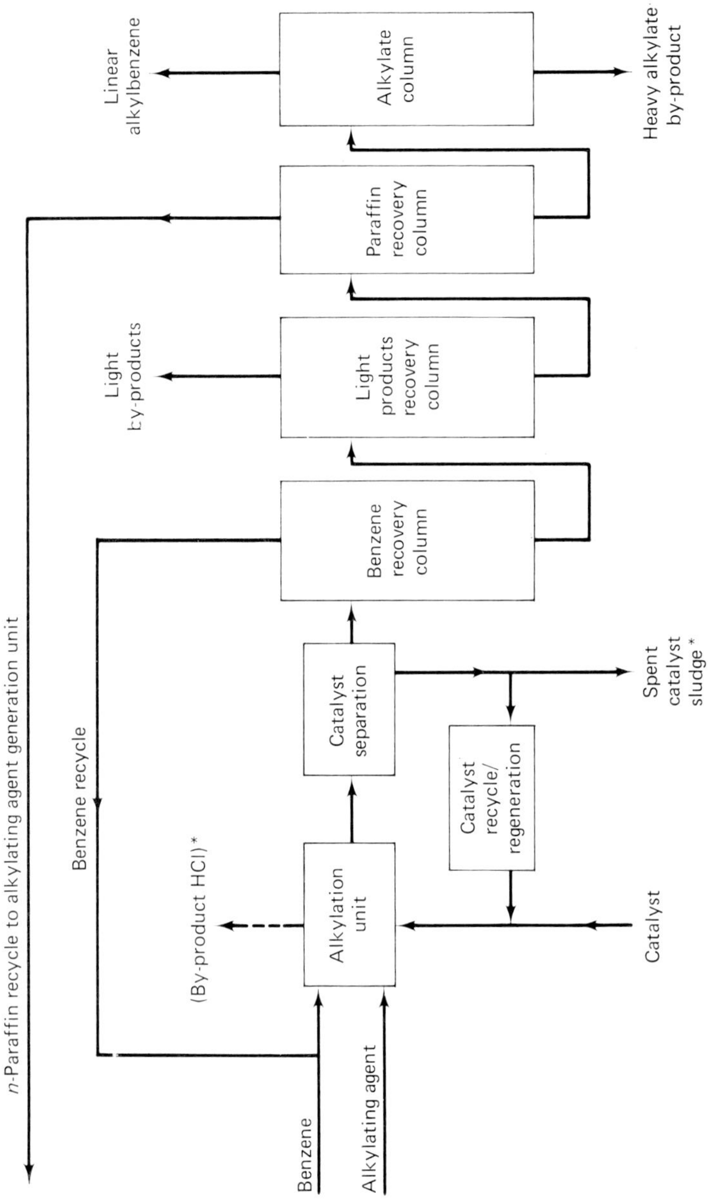

*$AlCl_3$ alkylations only

FIG. 9.4 *Idealised Process for the Production of Linear Alkylbenzene*

which is either discarded or recycled, depending on the process design, and an organic phase containing the product alkylbenzene together with the excess benzene, *n*-paraffin (introduced with the chloroparaffin) and by-products. Any residual catalyst sludge is removed from the organic phase by, for example, treatment with soda ash,[59] ammonia[60] or alumina.[61] The organic phase is then passed to the benzene recovery distillation unit, where the excess benzene is distilled off and azeotropically dried for recycling. The bottoms from this column are fed into a second distillation unit where any 'light alkylate' is removed. The next column separates out the *n*-paraffin, which is recycled to the chlorination unit after purification by, for example, treatment with oleum.[62]

The crude alkylate from the paraffin recovery unit is fed into the final distillation column where it is separated into the product alkylbenzene and heavy alkylate (diphenylalkanes, dialkylbenzenes, etc.). Various purification techniques, designed primarily to improve the colour and odour of the sulphonates derived from the alkylbenzenes, have been patented e.g. treatment with molecular sieves or urea to remove any residual paraffins,[64] treatment with activated carbon,[65] hydrogenation to remove polynuclear aromatics,[65, 66] sulphuric acid treatment either alone[67] or in conjunction with hydrogen peroxide,[68] or treatment with calcium oxide to remove residual organic chlorine.[69] The yield of monoalkylbenzene is determined by a number of factors, the most important being the paraffin conversion in the chlorination, the benzene to chloroparaffin ratio in the alkylation and the alkylation conditions. Increased paraffin conversion, with consequent increase in di- and poly-chlorination, produces more di- and polyphenylalkanes and cyclic by-products, and a decrease in the benzene to chloroparaffin ratio increases the amount of dialkylbenzene formed. Higher reaction temperatures also tend to increase the amount of cyclic by-products. These materials, which have boiling points close to the corresponding open-chain materials and which in consequence normally remain in the product alkylbenzene are generally considered to be undesirable. They are sulphonated readily but their sulphonates are inferior in some ways to the open-chain linear alkylbenzene sulphonates of corresponding chain-length.[50] Hence, a decrease in the paraffin conversion in the chlorination and an increase in the benzene to chloroparaffin ratio increase the yield and quality of the monoalkyl benzene. Unfortunately, changes in this direction result in larger recycle streams and hence increase the processing costs. Accordingly, alkylate manufacturers have to choose the best compromise between yield, quality and cost.

It is not easy to obtain information on actual industrial yields, but data given by H. D. Wulf[50] probably give a reasonable indication of what can be obtained under commercially realistic conditions. This author reports a 79 per cent yield of alkylate produced from an alkylation with a 26 per cent chlorinated *n*-paraffin at 45°C, and benzene to chloroparaffin ratio of 8:1 in a continuous process. The product contained 29 per cent 2-phenylalkane 5 per cent tetralin-type cyclic by-products and 20 per cent (w/w) heavy alkylate (based on product alkylbenzene) was formed. In broad agreement with this, a Societa Italiana Resina SpA (SIR) patent[62] describes the production of a C_{10-14} linear alkylbenzene in 82 per cent yield by alkylation with a 29 per cent chlorinated *n*-paraffin at 60°C with a benzene to chloroparaffin ratio of *ca.* 5:1. SIR's chloroparaffin/aluminium chloride process has been described in some detail in the literature.[59] A Chevron Research Co. patent[30] reports a higher yield of alkylbenzene (93 per cent based on *sec*-monochloroparaffin) and the formation of less bottoms (11 per cent (w/w) based on product alkylbenzene).

9.3.3.2. The olefin/hydrogen fluoride route In this case, the alkylating agent, which is either *ca.* 20–35 per cent random olefin in *n*-paraffin, or 1-olefin, which may also contain paraffins, from wax cracking, is fed together with excess benzene and excess hydrogen fluoride (up to 20 moles per mole of olefin) to the alkylation system. Residence times vary quite widely, usually within the range 5–90 min. Reaction temperatures are usually in the range 4–70°C. These conditions approximate to the 'hexane dilution' conditions described by Alul (see p. 462) and thus the product normally contains less than 20 per cent 2-phenyl isomer. Mitsubishi[70] have patented a variant in which the reaction mixture is held at 40–70°C for up to 90 min. This is claimed to give an increase in the 2-phenyl-isomer content of the product, resulting in especially good biodegradability of the derived sulphonate.

The catalyst is separated and recycled and the organic portion of the alkylation product is processed as described in the previous section for the aluminium chloride catalysed process. The product purification techniques described in the previous section can be used for the products of hydrogen fluoride alkylation but it is more usual to subject the alkylate to a further treatment with substantially anhydrous hydrogen fluoride to remove any residual olefins, non-linear alkylbenzenes, or anthracenes, which might be present.[72]

Yields are very high, usually of the order 90 per cent, and smaller quantities of by-products are formed than is the case during the aluminium chloride catalysed process. UOP's process, which is described in some detail in the literature,[25a] gives a 90 per cent yield of a C_{10-14} alkylate containing only 1·6 per cent tetralin-type cyclic by-products and with a high proportion of internal isomers (15·5 per cent 2-phenyl, 23·6 per cent 3-phenyl, 20·8 per cent 4-phenyl, 24·1 per cent 5-phenyl and 16·0 per cent 6-phenyl). Other by-products of the UOP process are (based on alkylate produced) 6 per cent heavy alkylate, 2 per cent polymer, 1·6 per cent light liquid and 0·7 per cent light gases.

The olefin/hydrogen fluoride route is thus reported to give higher yields and less by-products than the aluminium chloride routes. The former route (which only requires normal carbon steel equipment) has also been claimed to be less demanding on materials of construction than the latter which, according to Wulf (Chemische Werke Hüls)[50] needs special materials (e.g. Hastalloy C) to combat the corrosiveness of aluminium chloride complexes.

SIR, on the other hand, claim to have solved this corrosion problem by careful control of raw materials and process conditions.[59] Their *sec*-monochloroparaffin/aluminium chloride alkylation process, which uses carbon steel in the alkylation section, showed no corrosion problems during three years operation.

The industrially important routes to linear alkylates are summarised in Fig. 9.5.

The choice between these various alternatives depends very much on the overall environment within which the process has to operate. For example, both chlorination/aluminium chloride alkylation and chlorination/dehydrochlorination/hydrogen fluoride alkylation are at a disadvantage in situations where no use exists for the

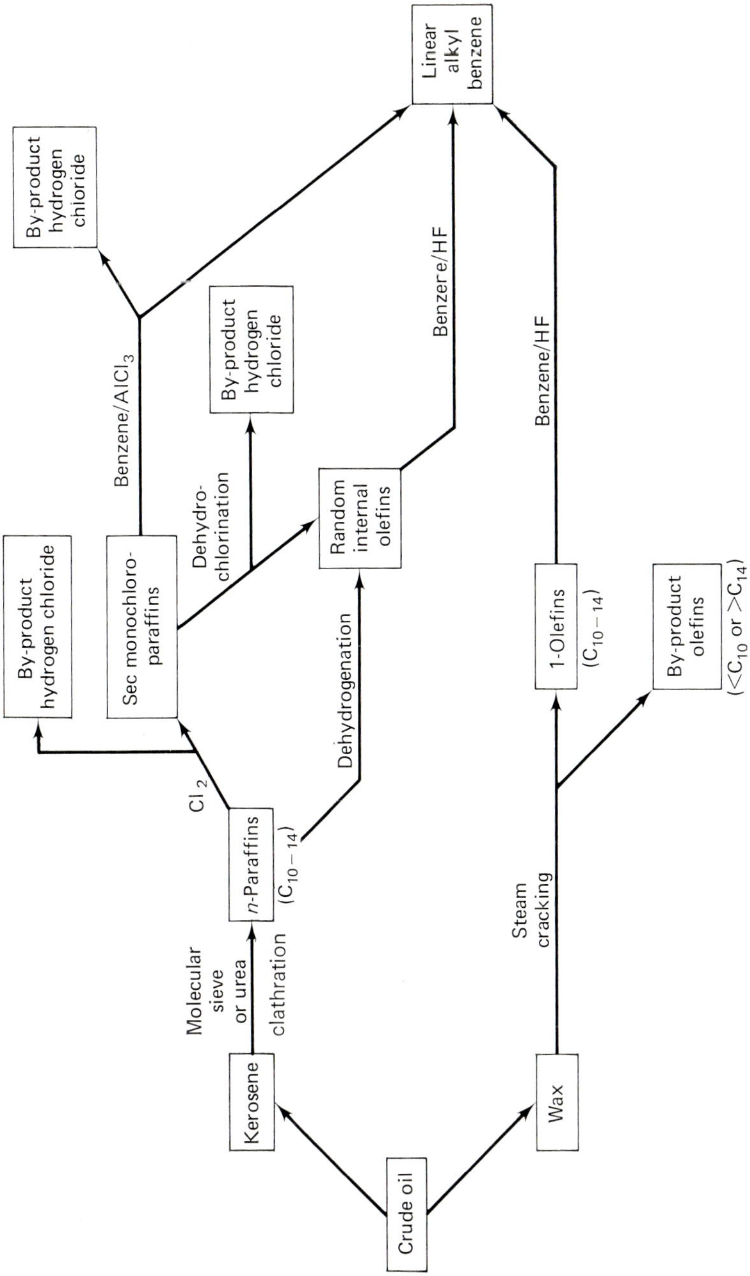

FIG. 9.5 *Routes to Linear Alkylbenzene*

by-product hydrogen chloride, and the cracked-wax olefin/hydrogen fluoride route is best suited to a situation where the by-product olefins can be profitably used.

Plant sizes vary quite considerably, most falling in the range 20–70,000 tons per annum. According to Barilli *et al.* (SIR),[59] total operating costs begin to rise quite steeply below 40,000 tons per annum but particular market and raw materials situations are also important in determining plant size as is evidenced by the fact that quite a number of companies have been reported to operate plants in the 20–30,000 tons per annum range.

9.3.4. Sulphonation

The methods developed and used or the sulphonation of branched chain alkylbenzenes (see p. 452) are used for the sulphonation of linear alkylbenzenes. In changing over from the production of ABS to the production of LAS, some relatively minor modifications in process operation were necessary (e.g. slight changes in the alkylate to sulphonating agent ratio or, in the oleum process, to the amount of water added to give phase separation of the sulphuric acid) but these did not usually necessitate plant modifications. Hence, as far as the sulphonation was concerned, the change from ABS to LAS did not present any significant problems.

9.3.5. Major uses

The replacement of ABS with LAS in heavy and light duty detergent formulations presented some initial problems but these problems were relatively minor. By the middle and late 1960's the producing companies had refined their processes and a range of good quality linear alkylates was available to the detergent manufacturer. From these, LAS could be produced and formulated to give very good quality heavy and light duty products and the growth in the use of these products in the retail market was substantially unaffected by the changeover from ABS to LAS. These products still account for the vast majority of LAS currently used.

LAS is an effective substitute for ABS in the majority of its numerous industrial applications; although not well documented, the changeover to the biodegradable material in the majority of these uses is now presumably substantially complete in the major world markets.

9.4. PRODUCTION AND CONSUMPTION

The world-wide status of detergent alkylate was reviewed in 1969 by H. A. W. Hill[4] who estimated world consumption (excluding Communist Bloc countries) for 1969 at just short of 800,000 tons per annum (65 per cent LAS, 35 per cent ABS) with Western Europe (39 per cent), the USA (26 per cent) and Japan (10 per cent) accounting for 75 per cent of the total. World production capacity was estimated at just over 1,000,000 tons per annum (62 per cent LAS). A growth rate of 2–3 per cent per annum was predicted for the US demand up to 1975 and it was concluded that US capacity existing in 1969 would cope with this up to 1975.

A more rapid rate of growth was predicted for Europe by R. Cociancich and J. Vad.[73] For 1970, these authors estimated the Western European demand at 485,000 tons per annum (68 per cent LAS) (cf. Hill's estimate of 310,000 tons per annum for

1969) and the demand from the Commecon countries at 82,000 tons per annum (76 per cent LAS). They predicted growth to 679,000 tons per annum (98 per cent LAS) for Western Europe and 180,000 tons per annum (92 per cent LAS) for the Commecon countries by 1975. This represents a growth rate of 7 per cent for Western Europe and 17 per cent for the Commecon countries.

Just what the true rate of growth in the use of LAS will be is clearly impossible to say. Probably the only certain conclusion is that LAS is unlikely to lose its current position as the most cost effective anionic surfactant available and that it will continue to be important for a number of years to come.

The following table lists the world's major linear alkylate producers.

Area	*Company**	*References*
USA	Allied Chemical Corp.[a]	74, 75, 76
	Atlantic Refining Co.	75
	Chevron[a]	75, 76, 77
	Continental Oil	75, 76
	Monsanto Co.	76, 78, 79
	Phillips Petroleum[a]	75
	Witfield Chemical Corp.[b]	75, 76, 80
	Union Carbide Corp.	75, 76, 81, 82
W. Europe	Chemische Werke Hüls AG (Germany)	83
	Liquichimica Augusta SpA (Sicily)	84
	Petroquimica Espanola SA (Spain)[c]	77, 85, 86
	Rheinpreussen AG (Germany)[f]	80, 83, 88
	Shell (UK, France)	83, 92
	Societa Italiana Resina SpA (Sardinia)	59, 87
	Wibarco (Germany)[d]	76, 86
Japan	Mitsubishi Petrochemical Co.	89, 90
	Nissan/Conoco[e]	76, 89, 90
	Nippon Sekiyu Detergent Co.	91
	Sumitomo Chemical/Kao Soap	89, 90

a. Production stopped 1968.
b. Joint venture by Witco Chemical Co. and Richmond Oil.
c. Joint venture by Continental Oil Co. and Compania Espanola de Petroleos SA.
d. Joint venture by Wintershall AG, Elektro-Chemie Ibbenbüren GmbH and Atlantic Richfield Co.
e. Joint venture by Nissan Chemical Industries and Continental Oil.
f. Now Deutsche Texaco AG.

* The companies listed in this table have been reported either as actual producers of linear alkylbenzenes or as installing plant to produce alkylbenzenes. As the majority of the reports come from trade journals, the accuracy of the information cannot be guaranteed.

REFERENCES

[1] Adam, N. K., *Proc. Roy. Soc.*, A, 1923, **103**, 676.
[2] BIO S Final Report, No. 421, 15.
[3] Rubinfeld, J. and Cross, H. D., *Soap Chem. Specialities*, 1967, **43**, 41.
[4] Hill, H. A. W., Butler, I. H. and Moffett, J. G., Jr., *Soap Chem. Specialities*, 1969, **45**, 37.
[5] de Witt, W. J., *J. Am. Oil Chemists' Soc.*, 1972, **49**, 361.
[6] Sittig, M., 'Practical Detergent Manufacture', *Chem. Proc. Review*, No. 27, (Noyes Development Corp., 1968, 6.
[7] Feighner, G. C., *J. Am. Oil Chemists' Soc.*, 1958, **35**, 520.
[8] Sharrah, M. L. and Feighner, G. C., *Ind. Eng. Chem.*, 1954, **46**, 248.
[9] Jones, E. K., *Advan. catalysis*, 1958, **10**, 186.
[10] Gray, F. W., *et al.*, *J. Org. Chem.*, 1955, **20**, 511; 1961, **26**, 209.
[11] Seifert, W. K., *Tenside*, 1965, **2**, 216
[12] Universal Oil Products Co., USP 3,501,543.
[13] Seifert, W. K., *Tenside*, 1965, **2**, 182.
[14] Sittig, M., 'Practical Detergent Manufacture', *Chem. Proc. Review*, No. 27, (Noyes Development Corp., 1968, 191.
[15] Chemithon Corp., USP 3,259,645; Stepan Chem. Co., USP 3,169,142; Allied Chem. Corp., USP 3,328,460.
[16] Swisher, R. D., '*Surfactant Biodegradation*', (Marcel Dekker Inc., New York, 1970 covers this complex subject comprehensively.
[17] Bogan, R. H. and Sawyer, C. N., *Sewage Ind. Wastes*, 1955, **27**, 917.
[18] Sawyer, C. N., Bogan, R. H. and Simpson, J. R., *Ind. Eng. Chem.*, 1956, **48**, 236.
[19] Hammerton, C., *J. Appl. Chem.*, 1955, **5**, 517; *Proc. Soc. Water Treat. Exam.*, 1956, **5**, 145.
[20] Swisher, R. D., *J. Water Pollution Control Federation*, 1963, **35**, 877; *Soap Chem. Specialities*, 1963, **39**, 47; *J. Am. Oil Chemists' Soc.*, 1963, **40**, 648; *ibid.*, 1967, **44**, 717.
[21] Kölbel, H., Kurzendörfer, P. and Werner, C., *Tenside*, 1967, **4**, 33.
[22] Ref. 16, p. 194.
[23] *Hydrocarbon Processing*, 1965, **44**, 232.
[24] Cooper, D. E. *et al.*, *Chem. Eng. Progr.*, 1966, **62**, 69; Franz, W. F. *et al.*, *Petrol. Refiner*, 1959, **38**, 125.
[25] (a) Bloch, H. S., *Proc. Symposium Normal Paraffins* (Manchester, England, 1966); supplement published with *European Chem. News*, 2 December 1966, 46. (b) *European Chemical News Large Plants Supplement*, 10 September 1965.
[26] British Petroleum, BP 976,406.
[27] Esso Res. Develop., BP 938,822.
[28] Fritz, B., *Proc. Symposium Normal Paraffins* (Manchester, England, 1966); supplement published with *European Chem. News*, 2 December 1966, 29.
[29] Hurley, D. J. *et al.*, *Ind. Eng. Chem.*, *Prod. Res. & Dev.*, 1965, **4**, 22.
[30] Chevron Res. Co., USP 3,355,508.
[31] Chevron Res. Co., USP's 3,277,204; 3,277,205; Chem. Werke Hüls AG, USP 3,341,615.
[32] Wulf, H. D., *Proc. Symposium Normal Paraffins* (Manchester, England, 1966), supplement published with *European Chem. News*, 1966, 2 December, 41.
[33] *European Chem. News*, 1963, 24 May, 28.
[34] Universal Oil Products Co., USP's 3,484,497 and 3,484,498; BP 1,193,486; Spanish P. 346,405.
[35] Paulis, B., *Proc. Symposium Normal Paraffins* (Manchester, England, 1966); supplement published with *European Chem. News*, 2 December, 1966, 33.
[36] Baumgartner, F. N., *Ind. Eng. Chem.*, 1954, **46**, 1349.
[37] Gray, F. W. *et al.*, *J. Am. Oil Chemists' Soc.*, 1965, **42**, 998.
[38] Tjepkema, J. J. *et al.*, *Proc. 5th World Petrol. Congr.*, New York, 1959, **4**, 237.
[39] Foster, D. J. and Fields, R. R., *Soap Chem. Specialities*, 1964, **40**, 49.
[40] Rubinfeld, J., Emery, E. M. and Cross, H. D., *J. Am. Oil Chemists' Soc.*, 1964, **41**, 822.
[41] Sweeney, W. A. and Olsen A. C., *J. Am. Oil Chemists' Soc.*, 1964, **41**, 815.

[42] Rubinfeld, J., Emery, E. M. and Cross, H. D., *Ind. Eng. Chem., Prod. Res. Develop.*, 1965, **4**, 33.
[43] Anstett, R. M., Munger, P. A. and Rubinfeld, J., *J. Am. Oil Chemists' Soc.*, 1966, **43**, 25.
[44] Rubinfeld, J. and Cross, H. D., *Soap Chem. Specialities*, 1967, **43**, 41.
[45] Patinkin, S. H. and Friedman, B. S. '*Friedel-Crafts and Related Reactions*', ed. G. A. Olah (Interscience, New York, 1964), Vol. II, Chapter 14.
[46] Alul, H. R., *J. Org. Chem.*, 1968, **33**, 1522.
[47] Alul, H. R. and McEwan, G. J., *J. Org. Chem.*, 1967, **32**, 3365.
[48] Alul, H. R., *Ind. Eng. Chem., Prod. Res. Develop.*, 1968, **7**, 7.
[49] Alul, H. R. and McEwan, G. J., *J. Org. Chem.*, 1972, **37**, 3323.
[50] Wulf, H. D. *et al.*, *Fette, Seifen, Anstrichmittel*, 1967, **69**, 32.
[51] Konishi, Kazuo and Inoue, Hiroh, *Chem. Abstr.*, 1970, **72**, 74528m.
[52] Khalaf, A. A. and Roberts, R. M., *J. Org. Chem.*, 1966, **31**, 89.
[53] Flanagan, P. W., Hammings, M. C. and Evens, F. M., *J. Am. Oil Chemists' Soc.*, 1967, **44**, 30.
[54] Olsen, A. C., *Ind. Eng. Chem.*, 1960, **52**, 833.
[55] Monsanto Co., BP 1,045,698.
[56] Sittig, M., 'Practical Detergent Manufacture', *Chem. Proc. Review*, No. 27 (Noyes Development Corp., 1968), 121.
[57] Chevron Res. Co., USP 3,433,846; Continental Oil Co., USP 3,391,210; Allied Chem. Co., Dutch P., 65,15532.
[58] Continental Oil Co., Dutch P., 66,12047.
[59] Barilli, F. and di Fiore, L., *European Chemical News Large Plants Supplement*, 27 September 1968, 44.
[60] Continental Oil Co., BP 1,236,357; USP 3,499,054.
[61] Continental Oil Co., Jap.P 24,325/70.
[62] Societa Italiana Resine SpA, GP (open to public inspection) 1,935,750.
[63] Nelson, K. L. and Brown, H. C., '*Chemistry of Petroleum Hydrocarbon's*, III, (Reinhold Pub. Corp., New York, 1955), Chapter 56.
[64] Societa Italiana Resine SpA, BP 1,165,423.
[65] Atlantic Richfield Co., USP 3,646,238.
[66] Continental Oil Co., Canadian P. 907,649.
[67] Continental Oil Co., USP 3,446,867.
[68] Esso Res. Eng. Co., Canadian P. 800,632.
[69] Gulf Res. Develop Co., USP 3,497,567.
[70] Mitsubishi Petrochemical Co., Canadian P. 872,839.
[71] Ref. 16, p. 207.
[72] Universal Oil Products Co., USP's 3,501,543; 3,501,544; Chemische Werke Hüls AG BP 1,288,335; Shell Int. Res., GP 1,618,911.
[73] Cociancich, R. (SIR) and Vad J. (United Chemical Works, Budapest, Hungary), reported in *Chem. Eng. News*, 16 November 1970, 28.
[74] *Chem. Eng. News*, 23 March 1964, 24.
[75] *Oil, Paint, Drug Reporter*, 21 March 1966.
[76] *Oil, Paint, Drug Reporter*, 28 July 1969.
[77] *European Chem. News*, 6 September, 1968, 6.
[78] *Chem. Eng.*, 20 July 1964, 184.
[79] *Chem. Age*, 13 June 1964, 950.
[80] *European Chem. News*, 22 November, 1963, 26.
[81] *Soap Chem. Specialities*, 1964, **40**, 42.
[82] *European Chem. News*, 6 September 1963, 29.
[83] *European Chem. News*, 23 May 1969, 6.
[84] *European Chem. News*, 26 February 1971, 10.
[85] *Chem. Week*, 25 February 1970, 44.
[86] *Soap Chem. Specialities*, 1970, **46**, 120.
[87] *Oil Gas International*, 1968, **8**, 106.
[88] *Chem. Trade J. Chem. Engineer*, 21 February 1964, 265.

[89] *European Chem. News*, 4 March 1966, 32.
[90] *Chem. Week*, 19 March 1966, 43.
[91] *Detergent Age*, September 1968, 44.
[92] '*New Plants* **1970**', supplement published with *European Chem. News*, 26 February 1971, 34.

CHAPTER 10

NITRATION

by S. T. Swinfin

10.1. NITROBENZENE

10.1.1. Nitration Mechanism

Nitration is one of the most important chemical reactions commercially. In the liquid phase reaction between an aromatic compound and a nitrating agent, e.g. nitric acid, a nitro group is introduced primarily by an ionic mechanism. The nitrating species is the positively charged nitronium ion NO_2^+ which is formed by ionisation of nitric acid.

$$2HNO_3 \rightleftharpoons H_2NO_3^+ + NO_3^-$$

$$H_2NO_3^+ \rightleftharpoons NO_2^+ + H_2O$$

This ionisation is depressed by the presence of water, which is normally produced during the nitration. For this reason a common nitrating agent is 'mixed acid', which is a mixture of fuming nitric acid and sulphuric acid monohydrate. The sulphuric acid, formerly thought to remove water formed during the reaction, acts as a catalyst to increase the formation of nitronium ion.

$$HNO_3 + 2H_2SO_4 \rightleftharpoons NO_2^+ + H_3O^+ + 2HSO_4^-$$

Other strong acids promote the ionisation of nitric acid, including hydrogen fluoride, boron trifluoride and perchloric acid.[1]

The nitronium ion reacts with aromatic compounds by electrophilic attack to form unstable positively charged complexes.

$$RH + NO_2^+ \rightleftharpoons R^+\genfrac{}{}{0pt}{}{\diagup H}{\diagdown NO_2}$$

The rate of formation is slow but the complex decomposes by a fast reaction in the presence of anions, e.g. HSO_4^-, to give a molecule of the nitro aromatic and regeneration of the promoting agent.

$$R^+\genfrac{}{}{0pt}{}{\diagup H}{\diagdown NO_2} + HSO_4^- \longrightarrow RNO_2 + H_2SO_4$$

Such reactions are highly exothermic. The nitro group introduced lowers the electron density of the ring and considerably more severe conditions are needed for the dinitration or trinitration steps in the case of benzene itself. In effect it is possible to stop at the mononitration stage, with formation of traces only of dinitro compounds. The more severe conditions used for further nitration, namely more concentrated mixed acids and higher temperatures, cause more significant side reactions to occur, and a number of by-products may be formed as the result of oxidation.

Where the ring is activated to electrophilic attack, as in the case of phenol, nitration is readily accomplished and a mixture of nitric acid with a small amount of nitrous acid may be used. The reactive species is the nitrosonium ion, NO^+ which attacks the aromatic compound to form a nitroso derivative. Oxidation with nitric acid gives the nitro derivative and regeneration of nitrous acid.

$$HNO_3 + HNO_2 \rightleftharpoons NO^+ + NO_3^- + H_2O$$

$$NO^+ + RH \rightleftharpoons \left[R^+ \begin{matrix} H \\ \\ NO \end{matrix} \right] \longrightarrow RNO + H^+$$

$$RNO + HNO_3 \longrightarrow RNO_2 + HNO_2$$

A more recent technique[2] is the use of stable nitronium salts, e.g. nitronium tetrafluoroborate ($NO_2^+BF_4^-$) in the nitration of aromatic compounds.

Although the role of the nitronium ion in mixed acid nitrations has remained unchallenged, nitrations by nitric acid in organic solvents have been considered to involve less reactive electrophiles such as the nitric acidium ion.[3] Discussion on this point continues.[4]

10.1.2. Nitration: The Unit Process

The process of nitration may be carried out on a batch or on a continuous basis. The majority of lower tonnage nitro-aromatics are still made on a batch basis in standard reactors flexible in use for a range of products. The materials of construction are cast iron or stainless steel, with glass lined mild steel reactors being increasingly used as multi-purpose equipment. The bulk storage and distribution system for nitric acid is in stainless steel; mixed acids can be handled in mild steel. The nitration reaction is very exothermic liberating up to 30 Kcal/g mol and efficient means of cooling are required either by jacket or internal coils.

The nitration reaction has obvious hazards and the reactor must be provided with a full range of safety control equipment. Such equipment must be designed to fail safe in the event of its own breakdown. The following controls are normally provided for a batch nitrator.

(a) Agitation interlock with nitrating acid feed. Agitation is monitored continuously, e.g. electrically by measurement of agitator load in order to detect any failure, the acid feed then being shut off.

(b) Acid feed interlock with services. Failure of power, cooling medium or any other service shuts off acid feed.

(c) Temperature control. Deviation from preset temperature gradients causes the acid feed to be cut off, and may be set to give automatic application of full cooling.

A large diameter vent with appropriate bursting disc should be incorporated in any batch nitrator. Additionally, it is preferable to carry out a nitration reaction with the manhole of the vessel covered only by an unfastened lightweight stainless steel cover on the principle that a frothing out is less damaging than a pressure rupture of the reactor. The problems of batch nitration have been reviewed in two recent papers.[5]

It is probable that the two most common causes of nitration incidents are failure of agitation[6] and excessive reaction temperature, the latter case sometimes being associated with incorrect reactant concentration. In the case of agitation failure, the uncontrolled addition of nitrating acid can lead to a layering effect with an apparently normal reaction temperature, except at the interface. Subsequent restarting of agitation can cause a violent explosion. The explosion is not normally a detonation but is a runaway thermal decomposition producing pressures of up to 3,000 lb/sq. in. at such a speed that a small diameter vent will be ineffectual.

Deviation from correct reaction temperature in either direction is dangerous. If the temperature is too low the reaction slows down and a dangerous excess of nitrating acid can build up. Eventually the quantity of unreacted material is so large that a violent reaction develops. The dangers of exceeding certain key reaction temperatures are well known. It is only more recently that the dangers of holding a reaction mass at some lower, supposedly safe, temperature have been recognised.

Thermal analysis of reaction masses by differential scanning calorimetry or other technique shows that the majority of aromatic nitrations give strong exothermal decompositions at temperatures in the range 160–200°C, although there may be some evidence, e.g. slight gassing, that some decomposition takes place at lower temperatures. It has been found that with deviation from normal process concentrations, e.g. higher concentration of aromatic compounds, or incorrect sulphuric acid charge, the strongly exothermal decomposition can start at a much lower temperature and even within normal processing temperature limits. The danger of holding material for long periods at moderately high temperatures consists in the fact that a slow exotherm can raise the temperature of the mass to the point at which violent decomposition takes place.

Any thermal analysis study of an existing or proposed process for nitration should therefore include:

(a) A close study of the nitration mass under exact process conditions.
(b) A study of the reaction under altered conditions of reactant concentration which might conceivably occur due to process error.
(c) The determination of the maximum safe temperature for prolonged storage e.g. 24 hours. Here, even if there is no intention of storage in the process, it is possible that delay may arise due to equipment failure.
(d) An examination of the effect that vessel or coil failure might have, permitting brine or other cooling medium to enter the reaction mass.

The precautions stressed in the case of nitration reactions apply equally to sulphonations (and chlorosulphonations) of certain simple mononitro-aromatic compounds where a similar hazard exists. Examples where the hazard is well known are the sulphonations of nitrobenzene,[7] *p*-nitrotoluene and *p*-chloronitrobenzene.

Higher nitrated derivatives provide the hazard of detonation especially in the case of trinitro derivatives. A further hazard concerns aromatic compound—nitric acid systems where there is a significant explosion or detonation potential. The detonation potential of the system nitrobenzene—nitric acid—water has been studied.[8] Nitric acid is a powerful oxidising agent and there is a risk that, if the temperature is allowed to increase, a nitration reaction can change to an oxidation—reduction reaction.

10.1.3. Batch and Continuous Processing

Nitrations of intermediates in the lower tonnage range are still carried out in batch reactors, usually of 2,500–5,000 l capacity. Larger tonnage intermediates are manufactured by a continuous process. In its simplest form this may be a small atomated batch reactor where efficient agitation and cooling permit a cycle time of minutes rather than hours, and the reactor operates on a very short repetitive time cycle.

A fully continuous process utilises flow through a series of agitated reactors. Agitation is efficient and the heat evolved during the reaction is removed by heat exchangers or jackets. Organic and acid phases are separated and nitrating acid is recycled. The fully nitrated aromatic compound is finally washed to remove traces of acid. More recent developments include nitration in tubular reactors[9] where agitation is provided by turbulent flow.

An obvious advantage of continuous processing units in nitration from the safety point of view is that the quantities of reactants present at any one time are much smaller than is the case with a batch reactor. The control of highly exothermic reactions on a continuous basis requires a precise knowledge of the chemical kinetics of the reaction and especially the chemical engineering aspects of mass transfer, heat removal, agitation efficiency and residence time, since the reactions may be effectively controlled by these factors on the large scale. The problems of relating the theoretical approach to the kinetics of aromatic nitrations to large scale practice have been studied.[10] The course of a continuous nitration reaction requires close control, with monitoring of the process variables and of the reaction products. It is essential that there should be no chance of unconsumed reactants collecting in secondary processing equipment.

As an aid to equipment design vapour pressure data for nitric and sulphuric acids have been measured.[11]

10.1.4. Nitrobenzene

10.1.4.1. Manufacture. The commercial manufacture of nitrobenzene is by nitration of benzene by a batch or continuous process. The nitrating acid is a mixed acid of approximate composition 30 per cent HNO_3, 60 per cent H_2SO_4, 10 per cent H_2O and is used in only slight excess over theory. The nitration temperature is approximately 50°C and efficient cooling is required to remove the heat generated by the reaction. On completion, the nitrobenzene is separated from the spent acid and can be used directly for aniline manufacture by iron reduction. The nitrobenzene is washed with water and sodium carbonate solution for the technical washed grade; final purification is by distillation. The spent acid is purified from traces of starting materials, concentrated and re-used.

Recent developments in continuous nitration have included plant scale manufacture of nitrobenzene in a tubular reactor, achieving very high conversion rates.[9]

10.1.4.2. Properties Typical specifications for the Technical and Distilled Grades of nitrobenzene are as follows:–[12]

Specifications technical grade	
Appearance	Pale yellow oil with a characteristic odour. It may be slightly hazy owing to the presence of small globules of free water.
Water content	0·5 per cent (maximum)
Specific gravity (15·5/15·5°C)	1·206–1·209
Dinitrobenzene content	0·1 per cent (maximum)
Low-boiling impurities (benzene + aliphatic hydrocarbons)	0·25 per cent (maximum)
Sulphur-containing impurities (CS_2 + nitrothiophen + elementary sulphur)	2·5 p.p.m. (maximum, as sulphur)
Distilled grade (*mirbane oil*)	
Appearance	Clear yellow oil free from visible water and boiling at approx. 211°C at 760 mm/Hg pressure
Distillation range (5–95 ml)	0·5°C (maximum)
Crystallising point	5·5°C (minimum)
Specific gravity (15·5/15·5°C)	1·208–1·211
Aniline content	0·0075 per cent (maximum)
Additional physical properties (not part of the specification)	
Density (20°C)	75 lb/cu. ft (1·20 g/cm^3)
Vapour pressure:	
Temp. (°C)	85 100 140 161 181 203 209
Pressure (mm Hg)	10 20 100 200 350 600 700
Flash point (closed cup)	190°F
Viscosity: Centipoises	1·8 1·2 0·55 0·28
Temp. °C	(25) (50) (100) (150)
Solubility in water: Temp. (°C)	8·8 14·7 30·8
per cent nitrobenzene	0·19 0·22 0·27
Solubility of water in nitrobenzene:	
Temp. (°C)	8·8 38·0 58·8 65·2 65·3
per cent water	0·174 0·194 0·40 0·71 1·5
	65·3°C = critical solution temperature
Heat of combustion	739 Kcal/g mol
Heat of vaporisation	79·1 cal/g at 210°C
Autoignition temperature	495°C

Nitrobenzene is miscible with most organic solvents, and is only slightly soluble in water. It is a good solvent for aluminium chloride and due to its relative inertness is frequently used as solvent in the Friedel-Crafts reaction. Nitrobenzene is highly toxic on inhalation or by skin absorption giving rise to cyanosis. It is a moderate fire hazard.

10.1.4.3. Reactions Nitrobenzene undergoes sulphonation, halogenation[13] and nitration by conventional means, but the ring is deactivated compared with benzene and reaction rates are therefore slower. Substitution takes place primarily in the meta-position in the case of electrophilic attack. However under conditions of nucleophilic attack the nitro group acts as an ortho-para-directing group, as in the case of alkaline fusion with potassium hydroxide which gives the *o*-nitrophenate from which *o*-nitrophenol is obtained on subsequent acidification. As a consequence of the ring deactivation Friedel-Crafts acylation does not take place, unlike the case of benzene itself.

The most important reaction of nitrobenzene is its reduction to aniline, a very widely used intermediate in organic synthesis. The reduction of nitrobenzene gives a wide range of products under varying conditions.

Vigorous reducing conditions with iron (or tin) and hydrochloric acid, titanous chloride and catalytic hydrogenation give aniline.

Phenylhydroxylamine is obtained by reduction with zinc and ammonium chloride, rearrangement in acid solution then yielding *p*-aminophenol.

Zinc and sodium hydroxide give hydrazobenzene as the main product, which on rearrangement with mineral acid gives benzidine. Air oxidation of hydrazobenzene gives azobenzene which may also be obtained by reduction of nitrobenzene with stannous chloride and a base, or with zinc dust and sodium hydroxide in methanol.

Nitrobenzene is reduced to aniline by the action of carbon monoxide at 360 atmospheres pressure in the presence of potassium hydroxide.[14] Aromatic compounds, including heterocyclic compounds, have been arylated by treatment with nitrobenzene in a tubular reactor at 400–600°C.[15] Nitrobenzene has been reduced in good yield to hydrazobenzene by treatment with a diol or polyol, e.g. ethylene glycol, in the presence of 2-hydroxy-1,4-naphthoquinone as catalyst.[16] Nitrosobenzene (and other reduction products of nitrobenzene) have been reduced with alcohols in the presence of a catalyst to give a mixture of products including aniline, primary alcohols and aldehydes.[17] The direct conversion of aromatic nitro compounds, e.g. nitrobenzene to isocyanates with carbon monoxide requires high pressures and a catalyst system consisting of a noble metal and a Lewis acid.[18] Methyl-*N*-phenylurethane has been prepared by the catalysed reaction of nitrobenzene and methanol with carbon monoxide under pressure.[19]

10.1.4.4. Uses The most important use of nitrobenzene is in the manufacture of aniline which is used in the preparation of a wide range of dyestuffs, rubber chemicals, pharmaceuticals and in general organic synthesis. Other significant uses are the manufacture of dye intermediates such as metanilic acid and benzidine. Nitrobenzene is used as an inert solvent in the Friedel-Crafts reaction and in nuclear chlorination reactions with sulphuryl chloride, e.g. the preparation of dichloroindanthrone, and in the chlorination of aniline hydrochloride to give 2,4,6-trichloroaniline. It also finds use as an inert solvent in the depolymerisation of rubber whereby crepe rubber is heated in nitrobenzene under reflux. The liquid depolymerised rubber is used in adhesives and moulding applications. Nitrobenzene is used as an oxidising agent in the Skraup reaction and in the manufacture of the solvent black, Nigrosine. It has also been used as an oxidising agent in the manufacture of vanillin from lignin and in the synthesis of vanillin from guaiacol.

10.1.5. Nitrobenzene-3-sulphonic Acid

Nitrobenzene-3-sulphonic acid is manufactured by sulphonating[20] nitrobenzene in oleum at temperatures up to 105°C. It is essential not to exceed this temperature since higher temperatures have been known to result in explosion.[7] The sulphonator must be equipped with appropriate safety controls including automatic application of full cooling on exceeding a pre-set temperature. The sulphonation mass is quenched, neutralised with lime, and filtered from gypsum. The calcium salt is converted with sodium carbonate into the sodium salt of nitrobenzene-3-sulphonic acid which is used for obtaining resist effects in the printing of vat dyes. It is also used as a mild oxidising agent in anthraquinone synthesis and in the Skraup synthesis of quinolines. In conjunction with sodium cyanide it finds application in metal stripping. It is used in the manufacture of metanilic acid by Béchamp reduction.

10.1.6. Monochloronitrobenzenes and Derivatives

The monochloronitrobenzenes are very widely used starting materials for the manufacture of a large number of dyestuffs intermediates, rubber chemicals, photochemicals and pharmaceutical products. The chlorine atom in the ortho- and para-isomers is activated and undergoes ready reaction by nucleophilic attack giving, for example, phenols and secondary amines.

The commercial manufacture of the chloronitrobenzenes is by the nitration of chlorobenzene with mixed acid. The product contains 35 per cent *o*-chloronitrobenzene and 65 per cent *p*-chloronitrobenzene, with a trace of the meta - isomer. The isomers are separated by a combination of crystallisation and fractional distillation. An alternative procedure is the addition of 1,4-dichlorobenzene to resolve the eutectic mixture and to permit separation of the isomers.[21] The chloronitrobenzenes are considered to be hazardous due to their toxicity on inhalation or by skin absorption. Currently the para-isomer finds the wider application. The kinetics and mechanism of the nitration of halogenobenzenes including chlorobenzene have been studied.[22]

10.1.6.1. o-*Chloronitrobenzene* *o*-Chloronitrobenzene has m.p. 35·5°C and is conveniently stored in bulk in the molten form with distribution by heated pipe-line, thus reducing the toxicity hazard at the handling stage. On further nitration of *o*-chloronitrobenzene the main product is 1-chloro-2,4-dinitrobenzene; sulphonation gives 4-chloro-3-nitrobenzene-1-sulphonic acid. The activated chlorine atom is displaced by nucleophilic attack and hydrolysis with aqueous sodium hydroxide at high temperature yields *o*-nitrophenol. Ammonolysis with aqueous ammonia at high temperature and pressure gives *o*-nitroaniline, while *o*-nitroanisole is obtained by the action of aqueous methanol and potassium hydroxide at high temperature.

o-Chloronitrobenzene reacts with sodium disulphide solution to give 2,2′-dinitrodiphenyldisulphide; the monosulphide is obtained with sodium sulphide. Reduction of *o*-chloronitrobenzene with sodium hydrosulphide yields the sodium salt of *o*-aminothiophenol which on reaction with carbon disulphide gives 2-mercaptothiazole.

The following products are normally derived from *o*-chloronitrobenz ene as starting material:

o- Nitroaniline
o-Phenylenediamine
o-Aminophenol
o-Nitroanisole
o-Dianisidine
o-Anisidine
4-Nitro-*o*-anisidine
5-Nitro-*o*-anisidine
o-Chloroaniline
o-Phenetidine
3,3′-Dichlorobenzidine

10.1.6.2. p-*Chloronitrobenzene.* The reactivity of the chlorine atom is similar to that of the ortho-isomer and ammonolysis with aqueous ammonia at 180°C and 40 atmospheres pressure gives *p*-nitroaniline. *p*-Nitrochlorobenzene is widely used in the synthesis of dyestuffs intermediates, pharmaceutical compounds and rubber chemicals, especially in the manufacture of *N*,*N*′-dialkylated *p*-phenylenediamines by condensation with an alkylamine, followed by reductive alkylation. 4,4′-Dinitrodiphenyl sulphide may be prepared from *p*-nitrochlorobenzene and sodium sulphide in dimethylformamide.[23]

A similar reaction but with selective reduction of one nitro group is the first stage in the synthesis of 4,4′-diaminodiphenyl sulphone, which is used in the treatment of leprosy.

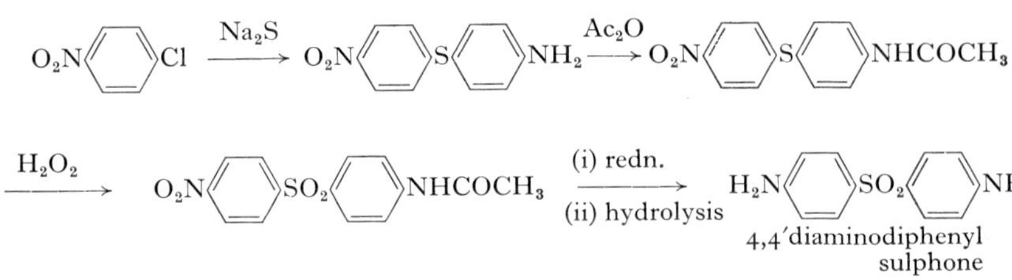

Intermediates normally manufactured from *p*-chloronitrobenzene include the following:

p-Nitroaniline
p-Phenylenediamine
p-Chloroaniline
2-Chloro-*p*-anisidine
3-Nitro-*p*-anisidine

p-Anisidine
p-Nitrophenol
p-Aminophenol
1-Chloro-2,4-dinitrobenzene
p-Nitrophenetole
4-Chloronitrobenzene-3-sulphonic acid
4-Nitroaniline-2-sulphonic acid
1-Chloro-2,4-dinitrobenzene-6-sulphonic acid
p-Phenylenediamine-2-sulphonic acid
4-Chloroaniline-2-sulphonic acid

10.1.6.3. m-*Chloronitrobenzene* *m*-Chloronitrobenzene occurs in trace amounts, up to 1 per cent, in the nitration of chlorobenzene and may also be prepared by chlorination of nitrobenzene. Unlike the ortho- and para-isomers there is no activation of the chlorine atom.

10.1.6.4. 2,5-Dichloro-1-nitrobenzene 2,5-Dichloro-1-nitrobenzene is manufactured by nitration of *p*-dichlorobenzene. 2,5-Dichloro-1-nitrobenzene is important for the manufacture of a number of dyestuff intermediates especially 2,5-dichloroaniline, 4-chloro-2-nitroaniline and 2-amino-4-chlorophenol.

10.1.6.5. 4-Chloronitrobenzene-3-sulphonic acid 4-Chloronitrobenzene-3-sulphonic acid is obtained by the sulphonation of *p*-chloronitrobenzene with oleum. The sulphonation is regarded as hazardous and excessive temperatures have resulted in explosion. This sulphonic acid is used for the preparation of intermediates, mainly for azo dyestuffs. Ammonolysis with aqueous ammonia under pressure gives the ammonium salt of 4-nitroaniline-2-sulphonic acid. Nitration affords the intermediate 1-chloro-2,4-dinitrobenzene-6-sulphonic acid. Iron reduction leads to 4-chloroaniline-3-sulphonic acid.

10.2 ANILINE

Aniline, the simplest primary aromatic amine, is a most important starting material in organic synthesis. Major end uses include the manufacture of rubber chemicals, dyes of the azo, azine and other categories, dyestuff intermediates, heterocyclic compounds, pharmaceutical products and photographic chemicals.

10.2.1. Manufacture

The original commercial method of manufacture was by Béchamp reduction of nitrobenzene with iron and a small amount of hydrochloric acid. This process is now obsolete. The main current commercial process is the continuous vapour phase hydrogenation of nitrobenzene. Operating temperatures are in the range 250–350°C and catalyst systems suggested include copper compounds, metal sulphides, e.g. nickel sulphide, and palladium/charcoal. Some catalyst systems and their respective advantages have been described.[24, 25] Efficiency of conversion is high and the product is much purer than that from the iron reduction process.

An alternative commercial method is the ammonolysis of chlorobenzene with aqueous ammonia at 210°C and 60–70 atmospheres pressure using a cuprous salt catalyst. A recent process of commercial potential involves the reaction of phenol with ammonia at high temperature and pressure using a silica-alumina catalyst. A general method of preparing aromatic amines from phenolic compounds is also claimed.[26]

10.2.2. Properties and Reactions

10.2.2.1. Physical properties Aniline when pure is a colourless oil; the technical grade is pale yellow-brown, darkening on exposure to air.

Boiling point (760 mm Hg)	184·4°C
Crystallising point	−6·0°C
Density, d_4^{20}	1·022
Refractive index, n_D^{20}	1·586
Viscosity at 20°C, centipoise	4·43
Flash point (closed cup)	76°C

Aniline is a toxic material giving rise to cyanosis. Absorption is rapid both through the skin and by inhalation of vapour.

10.2.2.2. Diazotisation Aniline readily undergoes diazotisation with nitrous acid at 0–5°C to give benzenediazonium salts, used extensively in the preparation of the commercially important azo dyestuffs. Benzenediazonium chloride has reasonable stability at 0°C; the stability is however greatly enhanced in a number of derivatives, e.g. with an alkyl, aryl or nitro group in the para-position. The aniline sulphonic acids also give diazo compounds of enhanced stability. A variety of factors need to be taken into account with regard to the effect of substituents on diazo decomposition.

A benzenediazonium salt will react with aniline to give diazoaminobenzene, which rearranges in weakly acid media to *p*-aminoazobenzene.

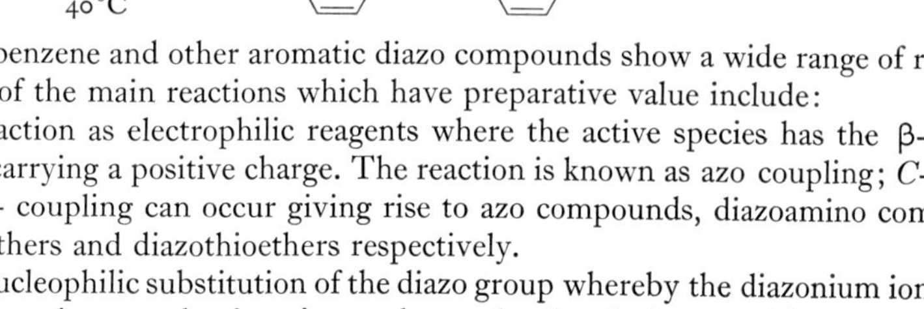

Diazobenzene and other aromatic diazo compounds show a wide range of reactions. Some of the main reactions which have preparative value include:

(i) Reaction as electrophilic reagents where the active species has the β-nitrogen atom carrying a positive charge. The reaction is known as azo coupling; *C*-, *N*-, *O*- and *S*- coupling can occur giving rise to azo compounds, diazoamino compounds, diazoethers and diazothioethers respectively.

(ii) Nucleophilic substitution of the diazo group whereby the diazonium ion decomposes to give a carbonium ion and a molecule of nitrogen. The carbonium ion interacts rapidly with nucleophilic agents.

(iii) Radical decompositions as in the Gomberg-Hey reaction leading to biphenyls. Similar reactions where loss of nitrogen takes place include the Sandmeyer, Gattermann and Meerwein reactions, catalysis with copper compounds normally being required.
(iv) Reduction reactions, especially the hydrogenation of the diazo group giving rise to arylhydrazines on reduction with sodium sulphite. Replacement of the diazo group by hydrogen normally takes place on heating the diazo compound in dimethyl formamide.

The reactions of aromatic diazo compounds have been reviewed by Zollinger.[2]

10.2.2.3. Sulphonation and nitration Sulphonation of aniline with oleum leads to sulphanilic acid, and only traces of the ortho- and meta-isomers. The commercial manufacture of sulphanilic acid is by baking aniline sulphate. Rearrangement then gives substitution almost exclusively in the para-position.

The low temperature nitration of aniline leads to a mixture of ortho-, meta- and para-nitroanilines.[28] However the reaction is accompanied by oxidation and tar formation, especially at higher temperatures, and the process is not used commercially. If the amino group is protected by acetylation the ortho- and para-derivatives are obtained on nitration. When a substituted aniline, e.g. *p*-chloroaniline is dissolved in a large excess of sulphuric acid mono-hydrate nitration occurs primarily meta- to the amino group since the positively charged anilinium ion is deactivating and meta-directing.

10.2.2.4. Halogenation The aromatic ring in aniline is activated by the amino group and halogenation is facile. Excess bromine added to an aqueous solution of aniline hydrochloride readily gives 2,4,6-tribromo-aniline. *p*-Bromoacetanilide is the main product of the bromination of acetanilide. The chloroanilines are usuallly prepared by the reduction of the appropriate chloronitrobenzene.

10.2.2.5. Oxidation Oxidation of crude aniline with hot potassium bichromate and sulphuric acid by Perkin in 1856 yielded Mauveine, the first synthetic dye. Soon afterwards the oxidation dye Aniline Black was discovered by Lightfoot in 1863, and because of its cheapness and good fastness properties is still used.

Oxidation of aniline is a complex process and involves many intermediates. Oxidation with potassium bichromate or manganese dioxide and sulphuric acid leads to *p*-benzoquinone. Hydrogen peroxide or persulphuric acid give a variety of products including phenylhydroxylamine, azobenzene, azoxybenzene and nitrobenzene. With potassium permanganate azobenzene and phenylquinonediimine are formed. Quinonediimines are capable of polymerisation to give dark coloured polymers similar to Aniline Black.

10.2.2.6. Reduction The catalytic hydrogenation of aniline leads to cyclohexylamine or dicyclohexylamine according to the conditions and catalyst employed.[29]

10.2.2.7. Acylation Acylanilides are readily formed by the reaction between aniline and acid chlorides. They are also formed with carboxylic acids in the presence of a catalyst, and with acid anhydrides.

The reaction of aniline with carbon dioxide at high temperature and pressure yields diphenylurea (carbanilide). The reaction with phosgene in the presence of aqueous sodium carbonate or sodium hydroxide also gives diphenylurea. Reaction with phosgene in the vapour phase yields phenyl isocyanate. Diphenylthiourea (thiocarbanilide) is formed by the reaction with carbon disulphide. Unsymmetrical thioureas may be prepared by condensation of an isothiocyanate with an amine in aqueous media.[30]

10.2.2.8. N-*alkylation and arylation* Reaction of aniline with alcohols yields mono- and dialkylanilines according to the excess of alcohol used, and the catalyst system.

In the manufacture of monoalkylanilines a small proportion of dialkyl derivative is present and is difficult to separate. The more vigorous conditions used in dialkylation may give rise to some ring alkylation. Aniline condenses with hydroquinone in the presence of an acid catalyst to give initially *p*-anilinophenol and finally *N,N'*-diphenyl-p-phenylenediamine. *N*-alkyl anilines are formed by reaction with appropriate alkylhalides, and also by reductive alkylation with aldehydes and ketones in the presence of hydrogen and a catalyst. This latter procedure is useful for the manufacture of monoalkylated anilines.

On heating aniline with aniline hydrochloride under pressure, diphenylamine is formed. The commercial process for the manufacture of the *N*-phenylnaphthylamines is the reaction between aniline and 1- or 2-naphthol in the presence of an acid catalyst, e.g. sulphanilic acid.

10.2.2.9. C-*Alkylation.* Aniline reacts with a number of olefins and unsaturated compounds in the presence of catalysts to form *C*-alkyl derivatives, e.g. with ethylene under pressure in the presence of aluminium anilide a mixture of 2-ethyl and 2,6-diethylaniline is obtained. Migration of an *N*-alkyl group can also occur on heating with a suitable catalyst to give a *C*-alkylaniline.

10.2.2.10. Condensation and cyclisation reactions Condensation of aniline with aldehydes yields Schiff bases or anils. With benzaldehyde, benzalaniline is the product.

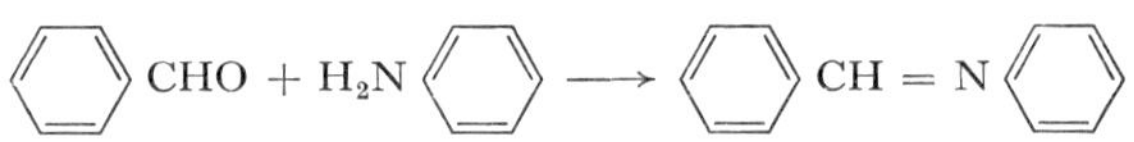

Substituted indoles are formed from aniline by heating with α-halo or α-hydroxy-ketones (Bischler's synthesis).

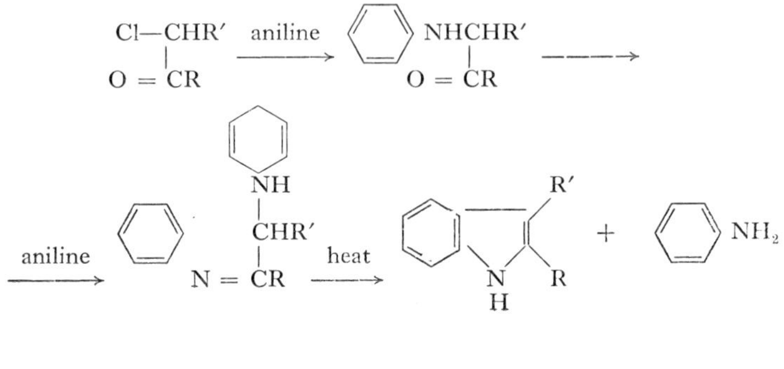

Substituted oxindoles may be made from aniline by a Friedel-Craft cyclisation of an α-chloroacetanilide (Stollé synthesis).

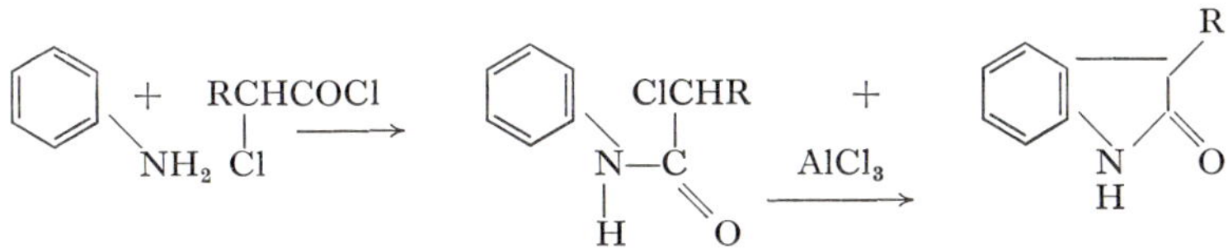

In the Skraup synthesis of quinoline, aniline is heated with glycerol and concentrated sulphuric acid in the presence of an oxidising agent, e.g. nitrobenzene, and with ferrous sulphate as a moderator. Addition of boric acid frequently gives improved yields in the Skraup reaction. The effective condensation agent is acrolein, formed by dehydration of the glycerol.

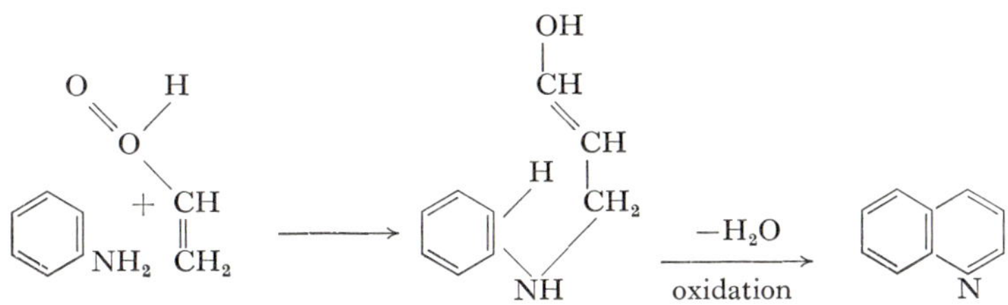

The Skraup reaction is strongly exothermic and requires extensive controls in the commercial manufacture in order to be carried out in safety. If a Skraup reaction goes out of control, a violent exothermic decomposition may take place.

The closely related Döbner–von Miller synthesis involves reaction of aniline, or other primary aromatic amine, with an aldehyde in the presence of hydrochloric acid.

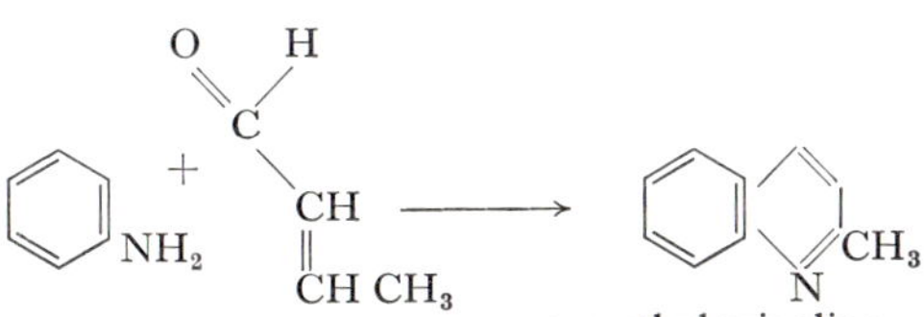

2-methylquinoline

2-Methylquinoline (quinaldine) is used in the manufacture of dyestuffs of the Quinoline Yellow type. The mechanism of the Döbner–von Miller reaction has been examined.[31]

Substituted quinolines may be obtained by two variations of the same basic reaction with ketones or keto esters. Aniline condenses with ethyl acetoacetate at medium temperatures to give an anil, which may be cyclised to the quinoline by dropping into a hot inert solvent. (Conradt–Limpach synthesis.) If however aniline is refluxed with the keto-ester, an anilide is formed which is cyclised by heating in concentrated sulphuric acid (Knorr synthesis).

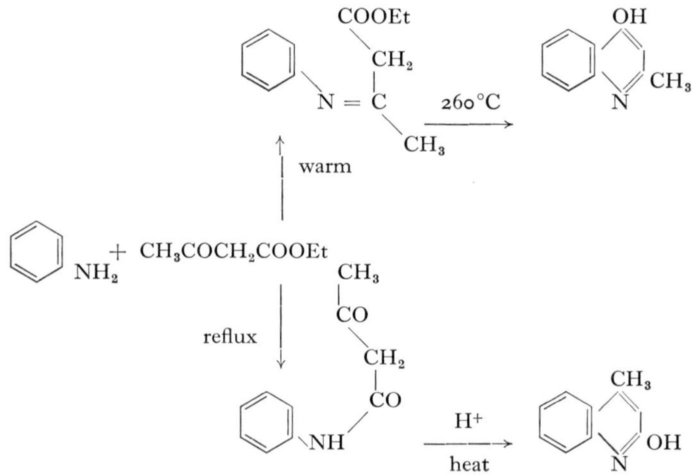

4-Substituted quinolines may be obtained by condensation of ketones with *o*-substituted anilines by the Friedlander synthesis. Quinoline-4-carboxylic acids are prepared by the Döbner reaction; an example is the condensation of benzaldehyde, aniline and pyruvic acid in alcoholic solution.

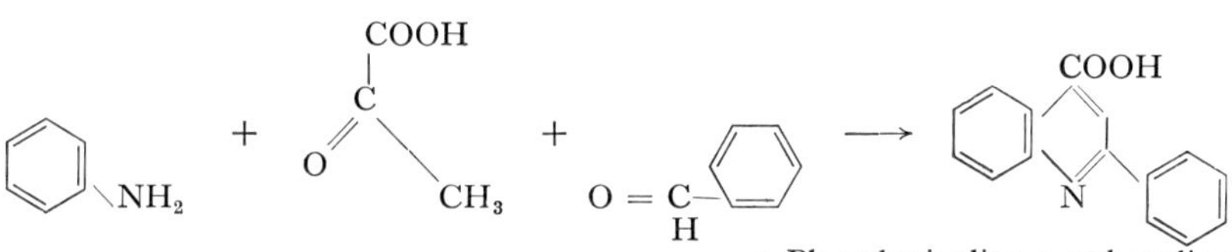

2-Phenylquinoline-4-carboxylic acid

Piperidones may be obtained by reaction of aniline with unsaturated ketones. The condensation of aniline, trichloroacetaldehyde and hydroxylamine yields isatin.

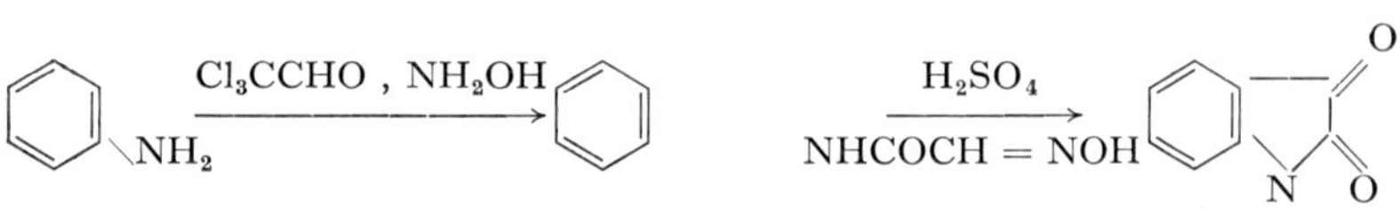

Isatin

Some of the reactions of aniline have been reviewed.[32]

10.2.3. Derivatives of Aniline

10.2.3.1. N-*Acyl derivatives* Acetanilide is manufactured from aniline and acetic anhydride, or by refluxing together aniline and glacial acetic acid. The pure material m.p. 114°C, b.p. 304°C is moderately soluble in water and soluble in polar organic solvents. It is only weakly basic, and it is readily hydrolysed by both acids and alkalies. The acetylamino group is strongly para-directing, or ortho- if the para-position is blocked. Chlorine or bromine give the *p*-halo derivative when reacted with acetanilide in aqueous acetic acid solution. Nitration with mixed acid yields mainly *p*-nitroacetanilide, and this is one of the commercial routes to *p*-nitroaniline. Nitration with acetyl nitrate or nitronium tetra fluoroborate gives predominantly ortho-substitution.[33] It is reported that acetanilide is readily sulphonated with concentrated sulphuric acid to gives *N*-acetylsulphanilic acid[34]. The action of chlorosulphonic acid gives *N*-acetylsulphanilyl chloride which is a key intermediate in the manufacture of sulphonamides. It condenses readily with amines, e.g. with 2-aminothiazole, which on hydrolysis yields sulphathiazole. The amidation of *N*-acetylsulphanilyl chloride with ammonia, followed by hydrolysis gives sulphanilamide.

Acetoacetanilide $C_6H_5NHCO\ CH_2\ COCH_3$, is a white crystalline solid, m.p. 85°C, manufactured from aniline and diketene. It is widely used in the synthesis of dye intermediates, e.g. pyrazolones, and in the manufacture of the Benzidine Yellow pigments.

Phenyl isocyanate has m.p. −30°C, b.p. 165°C. It is prepared by reaction of aniline with phosgene in the vapour phase or in an organic solvent. This reaction is used in the manufacture of the commercially important isocyanates (q.v.). Phenyl isocyanate reacts with primary alcohols to form urethanes, and with primary aromatic amines to form disubstituted ureas. With methylaniline 1,3-diphenyluretidone is given. Amides react slowly with phenyl isocyanate at 100°C to form the acyl urea.

$$C_6H_5NCO + RCONH_2 \longrightarrow C_6H_5NHCONHCOR$$

Phenyl isocyanate reacts with nitrosobenzene to give azobenzene with elimination of carbon dioxide.

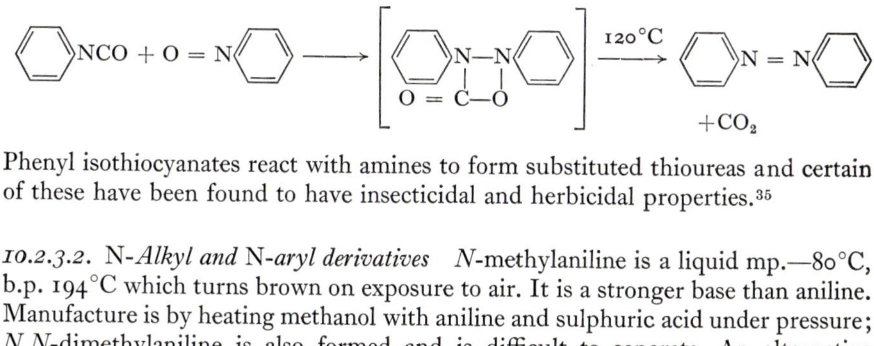

Phenyl isothiocyanates react with amines to form substituted thioureas and certain of these have been found to have insecticidal and herbicidal properties.[35]

10.2.3.2. N-*Alkyl and* N-*aryl derivatives* *N*-methylaniline is a liquid mp.—80°C, b.p. 194°C which turns brown on exposure to air. It is a stronger base than aniline. Manufacture is by heating methanol with aniline and sulphuric acid under pressure; *N*,*N*-dimethylaniline is also formed and is difficult to separate. An alternative method is to condense aniline with formaldehyde, then reduce with zinc and sodium hydroxide. Formylation of *N*-methylaniline with hexamethylenetetramine gives a high yield of *p*-dimethylaminobenzaldehyde.[36]

N-Ethylaniline is a colourless liquid which darkens on exposure to air; m.p.—63°C, b.p. 205°3. It is made by heating aniline hydrochloride with ethanol under pressure. The substituted urea obtained by phosgenation of *N*-ethylaniline is a stabiliser for explosives.

N,*N*-Dimethylaniline is a liquid m.p. 2°C, b.p. 193°C manufactured by heating aniline with excess methanol under pressure in the presence of an acid catalyst. It is employed in organic synthesis and in the manufacture of triarylmethane and basic dyestuffs. Nitrosation of dimethylaniline gives the *p*-nitroso derivative.

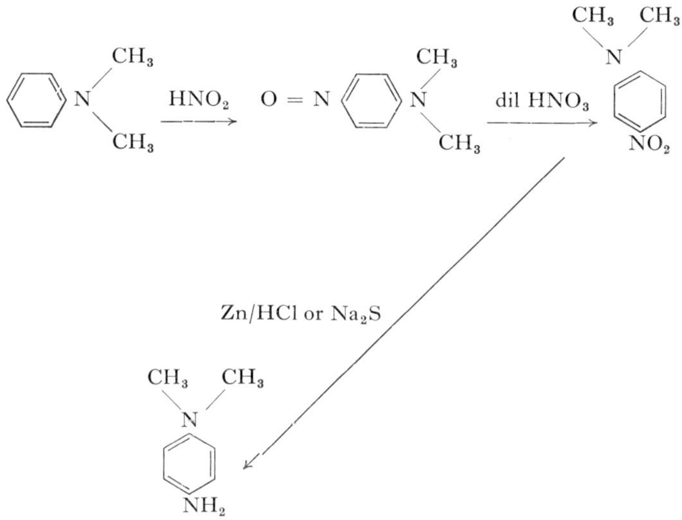

A route to *p*-amino-*N*,*N*-dimethylaniline is claimed in a recent patent.[37]

N-Ethyl-*N*-phenylbenzylamine (ethylbenzylaniline) is a liquid b.p. 310°C, and is manufactured by heating together benzyl chloride and *N*-ethylaniline in a solvent. It is used in the manufacture of triarylmethane dyes. Sulphonation occurs in the 3-position with oleum, and the sulphonic acid is an intermediate for triarylmethane and azo dyes.

Diphenylamine is a white crystalline solid m.p. 54°C, b.p. 302°C, insoluble in water, soluble in organic solvents; it is only weakly basic. For reaction in aqueous media, it is normally used in the form of a fine stabilised dispersion of the molten material. Manufacture is by the condensation of aniline in the presence of a small amount of a mineral acid catalyst at approx. 300°C. Ammonia is liberated and the crude diphenylamine-aniline mixture is separated by fractional distillation. Diphenylamine is also made commercially by the catalytic reaction of chlorobenzene with aniline at high temperature and pressure and by the vapour phase catalytic condensation of aniline. Impurities from the vapour phase reaction may be removed by treatment with orthophosphoric acid which forms salt-like compounds with the primary aromatic amines and tertiary heterocyclic amines present as by-products.[38]

The *o*-amino derivative of diphenylamine gives carbazole on treatment with nitrous acid.

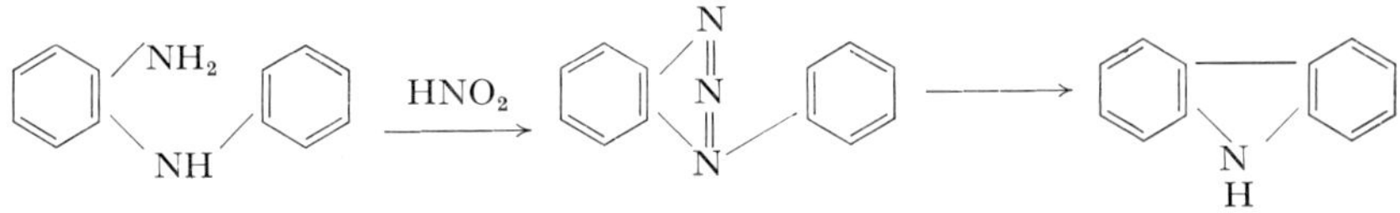

carbazole

Carbazole occurs as a by-product in certain catalytic preparations of diphenylamine. Diphenylamine is used in the manufacture of azo dyes, e.g. CI Acid Yellow 36 (13065) (Metanilic acid → diphenylamine). It finds use as a stabiliser for explosives since it combines with nitrogen oxides and thereby inhibits degradation. It is used in the manufacture of phenothiazine by reaction with sulphur, and of *N*-nitrosodiphenylamine, a widely used rubber vulcanisation inhibitor or retarder, by reaction with nitrous acid. The diphenylamine-acetone reaction product is a rubber antioxidant.

N-Phenyl-1-naphthylamine is manufactured by the reaction of aniline with 1-naphthol in the presence of an acid catalyst, e.g. sulphanilic acid. *N*-Phenyl-2-naphthylamine is prepared similarly, and both diarylamines are useful as rubber antioxidants. They have a small usage as dyestuffs intermediates.

8-Anilinonaphthalene-1-sulphonic acid (Phenyl Peri acid) is made by heating together aniline and 8-aminonaphthalene-1-sulphonic acid (Peri acid). It finds applicastion in the manufacture of blue and black azo dyes, CI Acid Blue 113 (26360) being important. A number of *N*-aryl derivatives of naphthylamine sulphonic acids are prepared by analogous methods and are azo dyestuff intermediates. Phenyl-J acid (6-anilino-1-naphthol-3-sulphonic acid) is however made by the Bucherer reaction. It is an important end coupling component in azo direct cotton dyes and leads to improved wash-fastness.

N-Phenylanthranilic acid may be prepared by the action of aniline upon *o*-chlorobenzoic acid in the presence of anhydrous potassium carbonate and copper oxide. It has been used as an indicator in ceric sulphate titrations.

10.2.4. Aniline Sulphonic Acids

The three aniline monosulphonic acids are all used commercially; the polysulphonic acids are far less important.

10.2.4.1. Orthanilic acid (aniline-2-sulphonic acid) This is a white crystalline solid which decomposes on heating and is slightly soluble in water. The sodium salt is very soluble in water. Sulphonation of aniline under a variety of conditions gives only traces of orthanilic acid and it is necessary to use a rather indirect route from *o*-chloronitrobenzene for its manufacture. Reaction of *o*-chloronitrobenzene with sodium polysulphide solution yields 2,2′-dinitrodiphenyldisulphide which on oxidative chlorination gives 2 moles of 2-nitrobenzenesulphonyl chloride, hydrolysis and iron reduction finally affording orthanilic acid.

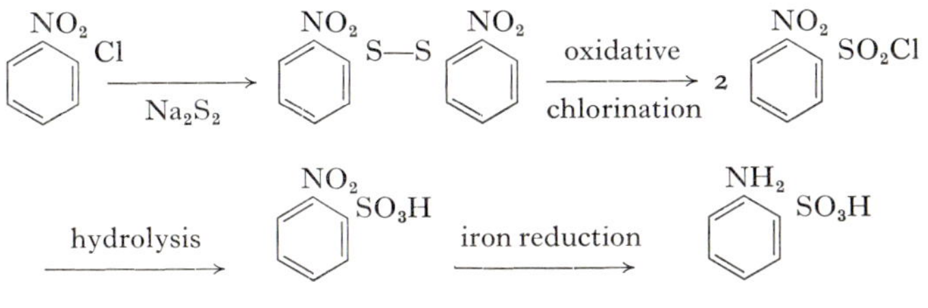

Orthanilic acid occurs in a few specialised azo dyes. It rearranges to give sulphanilic acid on heating in sulphuric acid.[39] The mechanism has been shown to be intermolecular.

10.2.4.2. Metanilic acid (aniline-3-sulphonic acid) This is a white crystalline product when pure. It decomposes on heating and is only moderately soluble in water. Nitrobenzene is sulphonated with oleum to give nitrobenzene-m-sulphonic acid from which metanilic acid is obtained by Béchamp reduction with iron. Reduction with alkaline sulphide has also been examined.[40] Metanilic acid is widely used in the manufacture of azo dyestuffs, although less extensively than sulphanilic acid. It is also used in the manufacture of reactive dyes and optical brightening agents. A typical azo dye incorporating metanilic acid is CI Acid Blue 113 which has the formula Metanilic acid—(1)→ 1-naphthylamine—(2)→ Phenyl Peri acid. The diazo compound derived from metanilic acid is active and will couple under acidic conditions with certain aromatic amines, especially 1-naphthylamine.

10.2.4.3. Sulphanilic acid (aniline 4-sulphonic acid) This is a white crystalline solid which decomposes on heating. It is only slightly soluble in water. The sodium salt is very soluble in water. The normal method of manufacture is to prepare aniline sulphate from equimolecular proportions of aniline and sulphuric acid and to bake at approx. 200°C. The reaction mechanism formerly postulated was the initial formation of phenylsulphamic acid followed by migration of the sulphonic acid group into the ring in the ortho position giving orthanilic acid, then rearrangement to sulphanilic acid. There is however no evidence of formation of orthanilic acid in the reaction, nor in the rearrangement of phenylsulphamic acid itself at high temperatures. At lower sulphonation temperatures orthanilic acid may be present as an alternative product rather than an intermediate product.[41] Sulphanilic acid is the most widely used of the three sulphonic acids. It is an intermediate in the manufacture of many azo dyestuffs. The crude product from the baking is purified by clarification of the sodium salt solution and, if a low aniline content is essential, by recrystallisation.

Sulphanilic acid is readily diazotised in the freshly precipitated free acid form and the diazo compound is of moderate stability. It is more stable than diazobenzene. The enhanced stability of the sulphonic acid derivatives of simple aromatic amines is a common feature in the benzene, and naphthalene series. Sulphanilic acid is a component in 91 dyes whose constitutions are listed in the Colour Index. Typical examples are Tartrazine, CI Food Yellow 4 (15985) which is sulphanilic acid diazotised and coupled to 1-(4-sulphophenyl)-5-pyrazolone-3-carboxylic acid, and Orange II, CI Acid Orange 7 (15510). This latter azo dyestuff for wool was discovered by Z. Roussin in 1876 and is still widely used due to its cheapness and fairly good lightfastness. It is made by diazotising sulphanilic acid and coupling with 2-naphthol.

10.2.5. Nitroanilines

The three isomeric mono nitroanilines are important intermediates in organic synthesis, especially in dyestuffs manufacture.

10.2.5.1. o-*Nitroaniline* This is a deep yellow crystalline solid, m.p. 72°C, b.p. 225°C, soluble in hot water. It is manufactured by the ammonolysis of *o*-chloronitrobenzene in an autoclave at 175°C and 40 atmospheres pressure. In a continuous process rather higher temperatures and pressures are employed. *o*-Nitroaniline is used to a limited extent as a dyestuffs intermediate, and is CI Azoic Diazo Component 6 (37025). Its main use is in the manufacture of *o*-phenylenediamine by sulphide reduction. It undergoes ready diazotisation and forms a fairly stable diazo compound. It does not readily give Schiff bases by condensation with aromatic primary amines.[42] *N*-Cycloalkenic derivatives of *o*-nitroaniline have been prepared from *o*-chloronitrobenzene as starting material,[43] and serve as insecticides and intermediates.

10.2.5.2. m-*Nitroaniline* This is a yellow crystalline compound, m.p. 114°C, b.p. 286°C. The normal method of manufacture is the partial sulphide reduction of *m*-dinitrobenzene. *m*-Nitroaniline is CI Azoic Diazo Component 7 (37030) and is used as a dyestuff intermediate.

10.2.5.3. p-*Nitroaniline* This is a yellow crystalline compound, m.p. 148°C, b.p. 286°C, only slightly soluble in water but very soluble in hydrochloric acid. Manufacture is by the ammonolysis of *p*-chloronitrobenzene under pressure in a batch or a continuous process. The chlorine atom is slightly less reactive than is the case with *o*-chloronitrobenzene. An alternative method is the nitration of acetanilide, followed by hydrolysis. *p*-Nitroaniline is CI Azoic Diazo component 37 (37035). It has extensive use as an intermediate for dyestuffs and is used to a much greater extent than the ortho- and meta- derivatives for this purpose. It is a component in 108 dyestuffs quoted in the Colour Index. An important use is in the manufacture of *p*-phenylenediamine by iron or sulphide reduction. *p*-Nitrodimethylaniline is obtained by treatment of *p*-nitroaniline with formaldehyde and formic acid.[44]

It has recently been reported that an explosion occurred during the manufacture of nitroaniline from the corresponding chloronitrobenzene.[45]

10.2.5.4. p-*Nitroaniline-2-sulphonic acid* This is used in the manufacture of a few azo dyes. It is made by ammonolysis of 1-chloro-4-nitrobenzene-2-sulphonic acid under pressure.

10.2.5.5. 2,4-*Dinitroaniline* This is made by the action of aqueous ammonia at the boil on 2,4-dinitrochlorobenzene under slight pressure. This ammonolysis proceeds much more readily on account of the greater electron withdrawing power of the two nitro groups. The ammonolysis has also been carried out with ammonia in boiling ethylene glycol.[46] 2,4-Dinitroaniline is only feebly basic and requires special conditions for diazotisation, i.e. the use of nitrosylsulphuric acid.

It is used in the manufacture of azo pigments and of the dyestuffs intermediate 6-chloro-2,4-dinitroaniline.

10.2.5.6. 2,4,6-*Trinitroaniline* (*picramide*) This is a yellow compound m.p.188–190°C which explodes on further heating. Manufacture is by nitration of *p*-nitroaniline. An alternative method is to react picric acid with phosphorus pentachloride to form 2,4,6-trinitrochlorobenzene followed by ammonlysis. This illustrates the ease of replacement of substituents in the 1-position as a result of the combined electronic effect of the three nitro groups. A process for a reliable means of diazotisation of picramide has been claimed.[47]

10.2.5.7 Chloronitroanilines A number of hazards are involved in the manufacture and diazotisation of chloronitroanilines. In an attempted laboratory preparation of 4-chloro-2,6-dinitroaniline by direct nitration of 4-chloro-2-nitroaniline, a runaway reaction developed under the conditions used.[48] The product was found to contain a chlorodinitrophenyl diazonium oxide. Diazotisation of chlorodinitroaniline with nitrosylsulphuric acid has resulted in an explosion under certain conditions.[49]

10.2.6. Alkoxy Derivatives of Aniline

Alkoxy derivatives of aniline, e.g. the anisidines and phenetidines are widely used in organic synthesis and in the manufacture of dyestuffs and pharmaceutical compounds.

10.2.6.1. o-*Anisidine* (o-*methoxyaniline*) This is a colourless liquid, rapidly turning brown on exposure to air, b.p. 218°C. It is prepared by the action of methanol and sodium hydroxide on *o*-chloronitrobenzene, followed by reduction. It may be nitrated directly by preparation of crystalline *o*-anisidine nitrate with dilute nitric acid; the crystals are added to 95 per cent sulphuric acid at 0–5°C giving 5-nitro-*o*-anisidine, CI Azoic Diazo Component 13 (37130).

10.2.6.2. p-*Anisidine* *p*-Anisidine is a white solid, m.p. 57°C. It is prepared in a similar manner to *o*-anisidine, starting with *p*-chloronitrobenzene. Both isomers are widely used in the manufacture of azo dyes and azoic intermediates.

10.2.6.3. o-*Phenetidine* (o-*ethoxyaniline*) *and* p-*Phenetidine* These are prepared by the reaction of aqueous ethanol and sodium hydroxide upon *o*-chloronitrobenzene and *p*-chloronitrobenzene respectively, followed by reduction. *o*-Phenetidine is used in the manufacture of CI Azoic Coupling Component 14. *p*-Phenetidine is used in the manufacture of triphenylmethane dyes; condensation with 3-hydroxy-2-naphthoic acid gives CI Azoic Coupling Component 30. *p*-Phenetidine is used in the manufacture of pharmaceutical products especially phenacetin (acetophenetidine). An alternative route for the manufacture of phenacetin from aniline has been investigated.[50] The condensation product of acetone and *p*-phenetidine in the presence of hydrochloric acid is a resinous polymer of 6-ethoxy-2,2,4-trimethyl-1,2-dihydroquinoline and is used as an antioxidant for rubber.

10.2.6.4. p-*Cresidine* (*5-methyl-2-methoxyaniline*) This is a colourless solid when pure, m.p. 52°C. It is made by nitration of *p*-cresol followed by methylation and reduction. Active diazonium salts will couple with cresidine in the 4-position under acidic conditions, giving intermediates used in certain direct dyes for cotton.

10.2.7. Chloroanilines

The chloroanilines are the most important commercially of the halogenated anilines. The normal method of manufacture is the reduction of the corresponding chloronitrobenzene either by the Béchamp method with iron, or by catalytic hydrogenation. A variety of catalyst systems have been examined for hydrogenation of chloronitrobenzenes including platinum[51] and copper chromite.[52] Use of rhenium[53] in conjunction with palladium gave less dehalogenation during reduction. Nickel-carbon however[51] gave significant dehalogenation. Use may also be made of the fact that in the corresponding bromochlorobenzene, the bromine substituent is significantly more labile. Selective ammonolysis of *p*-bromochlorobenzene with aqueous ammonia at 120°C and a cuprous salt catalyst yields *p*-chloroaniline. Acetanilide reacts with sodium hypochlorite and mineral acid to give *p*-chloroacetanilide which yields *p*-chloroaniline on hydrolysis. Some synthesis of new halogenated anilines have been described.[54]

10.2.7.1. o-*Chloroaniline* This is a colourless liquid m.p. −2°C, b.p. 209°C insoluble in water but soluble in hydrochloric acid. The hydrochloride is CI Azoic Diazo Component 44 (37000). *o*-Chloroaniline has been prepared by reduction of *o*-chloronitrobenzene with carbon monoxide in the presence of a catalyst comprising a cupric salt and an aliphatic amine.[55]

10.2.7.2. m-*Chloroaniline* This is a colourless liquid m.p. −10°C, b.p. 230°C, insoluble in water. The hydrochloride is CI Azoic Ciazo Component 2 (37005) and is used in pigment manufacture.

10.2.7.3. p-*Chloroaniline* This is a colourless solid m.p. 71°C, b.p. 232°C, soluble in hot water and in hydrochloric acid. It is used in the manufacture of azoic coupling components and azo and vat dyestuffs. In the manufacture of CI Vat Yellow 5 (56005) *p*-chloroaniline is condensed with *p*-quinone in the presence of manganese acetate, then oxidised with sodium bichromate or hydrogen peroxide.

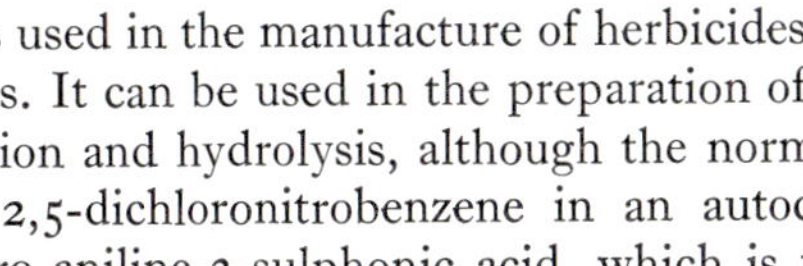

C.I. Vat Yellow 5

p-Chloroaniline is used in the manufacture of herbicides of the carbamate and substituted urea types. It can be used in the preparation of 4-chloro-2-nitroaniline by acetylation, nitration and hydrolysis, although the normal industrial method is by ammonolysis of 2,5-dichloronitrobenzene in an autoclave. The sulphonic acid derivative 4-chloro-aniline-3-sulphonic acid, which is used in a few azo dyes, is made by iron reduction of 4-chloronitrobenzene-3-sulphonic acid.

10.2.7.4. 2,5-*Dichloroaniline* This is manufactured by nitration of *p*-dichlorobenzene, followed by iron reduction. It is insoluble in water but soluble in hot dilute hydrochloric acid. It is CI Azoic Diazo Component 3 (37010) and is used in the manufacture of pigments and azo dyestuffs.

10.2.7.5. 2,6-Dichloroaniline This has been prepared by treatment of sulphanilamide with hydrogen peroxide and hydrochloric acid giving 3,5-dichlorosulphanilamide followed by desulphonation with 70 per cent sulphuric acid at 140–160°C.[56] 2,6-Dichloroaniline has been oxidised to 2,2′,6,6′-tetrachloroazobenzene by treatment with $Pb(OAc)_4$.[57]

10.2.7.6. 3,4-Dichloroaniline This is used in the manufacture of herbicides. Reaction with phosgene leads to the isocyanate, which on condensation with *o*-methylhydroxylamine gives the substituted urea herbicide.

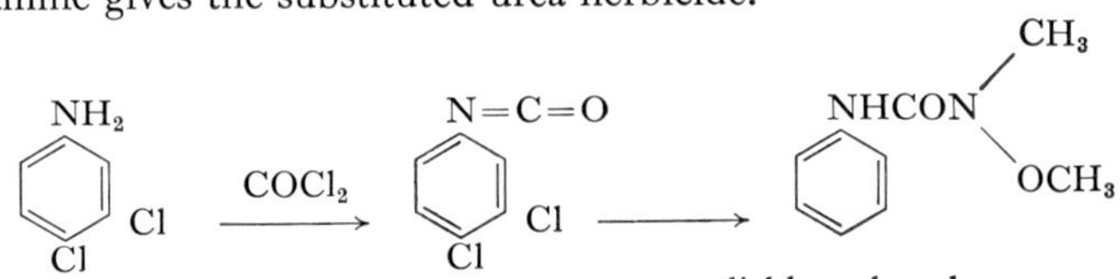

3,4-dichlorophenyl
-1-methoxy-1-methylurea

A method for continuous production of 3,4-dichloroaniline has been claimed.[58] The normal preparation is by catalytic hydrogenation of 3,4-dichloro-1-nitrobenzene.[59]

10.2.7.7. Fluorinated aniline derivatives There is extensive mention of fluorine containing azo dyestuffs in the patent literature. In particular the presence of the trifluoromethyl group is said to result in improved light fastness. The diazo component Fast Orange GGD salt is 1-amino-3,5-bis(trifluoromethyl)-benzene and is made from *m*-xylene by side chain chlorination without carrier followed by halogen replacement with hydrogen fluoride, nitration and reduction.

10.3 ANILINE DYES

10.3.1. Dye Categorisation

The obsolete term Aniline Dye is derived from the early synthetic dyes which were oxidation products of aniline. Although most of these colours are no longer used commercially, aniline itself is still of importance as an intermediate in dyestuffs manufacture.

Dye categorisation is not straight forward. The oldest division into soluble dyes and insoluble pigments, while simple, is hardly valid today since so much overlapping occurs. e.g. Vat dyes are insoluble and can constitute very acceptable pigments, but are applied to the fibre by a temporary solubilisation prodecure during dyeing. The disperse dyes used for the dyeing of many synthetic fibres have low solubility in water, but are in a very fine state of division.

The Colour Index, which is a publication of the Society of Dyers and Colourists and the American Association of Textile Chemists and Colourists, categorises dyes and pigments in two ways. The first classification is according to application where the common factor in each group is the technique of application to a particular substrate. The second classification is according to chemical constitution. These two classifications must be taken in conjunction since dyes of similar chemical structure may have widely differing dyeing characteristics, while a single application class can contain dyes drawn from a range of chemical classes. Synthetic dyestuffs are divided into eighteen classes by method of application and into twenty-five classes on the basis of chemical constitution.

A similar, but simplified, categorisation is used by the U S Tariff Commission for their figures of U S A production of synthetic dyes, and these give a useful indication of the relative importance of the different application and chemical classes.

10.3.2. Production of Synthetic Dyes

Benezenoid Dyes: US production, by Chemical Class, 1969[60]

	Tons
Anthraquinone	23,514
Azo	34,286
Azoic	3,489
Cyanine	233
Indigoid	2,368
Methine	1,365
Nitro	1,020
Oxazine	161
Phthalocyanine	897
Quinoline	731
Stilbene	18,595
Sulphur	8,410
Thiazole	238
Triarylmethane	3,363
Xanthene	626
All other	9,661
Total	108,957

The two largest categories are anthraquinone and azo dyes. The continued strength of azo dyes depends on newer dyes of improved fastness properties, and also on reactive dyes, many of which incorporate azo chromophoric systems.

Benzenoid Dyes US Production by Class of Application 1969[60]

	Tons
Acid	11,130
Azoic (dyes and components)	3,489
Basic	6,752
Direct	17,164
Disperse	11,548
Fibre-reactive	1,101
Fluorescent brightening agents	18,041
Food, drug and cosmetic colours	1,934
Mordant	1,022
Solvent	5,076
Sulphur	8,410
Vat	23,036
All other	254
Total	108,957

The large production of fluorescent brightening agents is obvious; vat, direct, disperse and acid dyes are important groups.

The size and importance of the British dyestuffs industry is indicated by the production and export figures published by the Organisation for Economic Cooperation and Development (OECD). The following table is reproduced from the OECD report, 'The Chemical Industry, 1969–70'.

Production of Dyestuffs in 1968 and 1969

	1968		*1969*	
	Tons	*Million dollars total value*	*Tons*	*Million dollars total value*
Germany	85,312	319·5	98,992	361·2
Belgium	1,790	4·7	2,040	5·6
Spain	11,850	11·8	12,120	12·1
France	19,770	63·4	23,311	69·3
Italy	11,000	22·4	11,000	23·2
Netherlands	2,721	5·3	2,832	5·3
Portugal	284	0·3	251	0·3
United Kingdom	40,600	129·0	43,200	139·1
Switzerland	26,000	145·0	28,000	154·1
Total	199,327	701·4	221,746	770·2
USA	113,000	474·6	n.a.[60]	n.a.
Japan	58,566	117·8	66,154	134·3

The dyestuffs exports are also shown in the OECD report and the table below indicates exports for the three major producing countries in Western Europe.

Country	*Exports tons*	*Million dollars total value*
Germany	67,184	245·8
Switzerland	35,002	not available
United Kingdom	27,115	73·2

10.3.3. Current Trends in Dyestuffs Production

The majority of synthetic dyes are used in the dyeing of natural and synthetic fibres; the remainder, approx. 30 per cent in quantity, being used in the colouration of paper, leather and other substrates. The number of commercially available dyes is very large, and this diversity is a result of the many different types of material to be dyed and the different properties required of the dyestuff, in relation to the cost for a chosen strength of shade.

The increase in production of synthetic fibres in recent years has caused a corresponding increase in the requirements of the appropriate categories of dyestuffs, especially disperse dyes. Selected members of the acid dyes are suitable for nylon. Some cationic dyes of only moderate light fastness show enhanced fastness properties when applied to polyacrylonitrile, causing renewed interest in this class. Many of the older dyestuffs with poor fastness properties have decreased in production. Sulphur dyes as a group have lost some importance since, although cheap, their light fastness is poor. Mordant dyes have also lost ground due to replacement by more modern dyes, and azoic dyes have to a certain extent been replaced by reactive dyes on cellulosic fibres.

Another trend is the improvement of the convenience of usage of dyes by preparation in liquid or rapidly soluble forms for minimum preparation time and speedier time cycle in automated dyeing machinery.

The extensive chemistry of dyes and their intermediates is reviewed in a number of recent books.[61, 62, 63]

10.4. RUBBER CHEMICALS

Natural thermoplastic rubber has been known for many hundreds of years and the birth of the modern rubber industry was the discovery of vulcanisation by Goodyear in 1839. Early vulcanisation accelerators were inorganic and it was the discovery of organic accelerators by Oenslager in 1906 that initiated the research leading to the current wide range of organic rubber additives. A rather arbitrary distinction is made between rubber chemicals, which are normally used only in small quantities, and other compounding materials including pigments, fillers, softeners and inorganic accelerators.

The two most important categories of rubber chemicals are:

(1) Vulcanisation accelerators.

(2) Antidegradants, including antioxidants and antiozonants.

Aniline is an important intermediate in the manufacture of rubber chemicals in both categories. Additionally, many antidegradants are *N*-substituted derivatives of *p*-phenylenediamine. This section, therefore, indicates primarily those rubber chemicals which are ultimately derived from nitrobenzene.

10.4.1. Vulcanisation Accelerators

The rate of reaction between rubber and sulphur is slow, even at high temperatures. The first organic accelerator used was aniline which gave a more rapid cure, and improved properties in the final product. The use of aniline posed a toxicity hazard and, amongst its simple derivatives, thiocarbanilide (*N*,*N*′-diphenylthiourea) was found to be an improvement. Thiocarbanilide is produced by the reaction between aniline and carbon disulphide, and it is still used as an accelerator in a few special cases.

The main types of accelerators currently in use are as follows:

Aldehyde-amine reaction products, made by treating amines, e.g. aniline or *p*-toluidine with aldehydes in the presence of an organic acid catalyst. A Schiff base is formed initially and polymerisation then takes place.

Aryl-guanidines, e.g. 1,3-diphenylguanidine made by the reaction of cyanogen chloride with hot aqueous aniline.

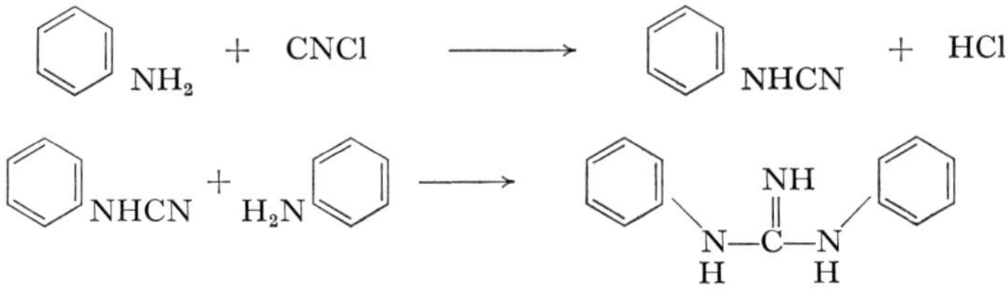

1,3-diphenylguanidine

Dithiocarbamates, prepared by reaction of carbon disulphide with primary and secondary amines.

Mercaptobenzothiazoles, the most important group of vulcanisation accelerators; e.g. in the USA thiazole type vulcanisation accelerators constitutes approximately 80 per cent of all production by weight. 2-Mercaptobenzothiazole is manufactured by heating together aniline, carbon disulphide and sulphur in an autoclave.

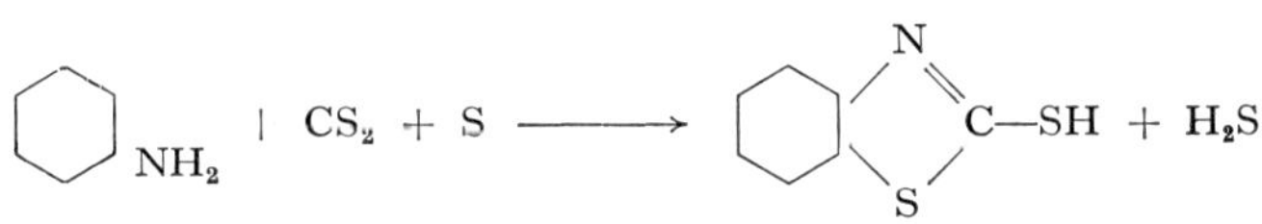

An alternative synthetic route is from *o*-chloronitrobenzene. Oxidation of 2-mercaptobenzothiazole with nitrous acid or chlorine yields 2,2-dithiobisbenzothiazole, an important accelerator.

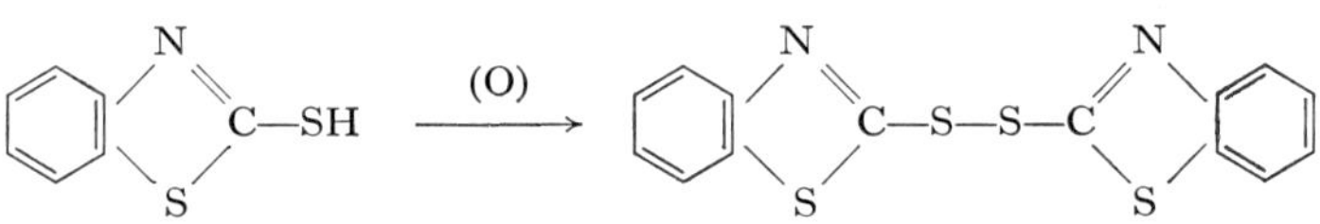

Sulphenamides are widely used accelerators. The general reaction involved in their manufacture is as follows:

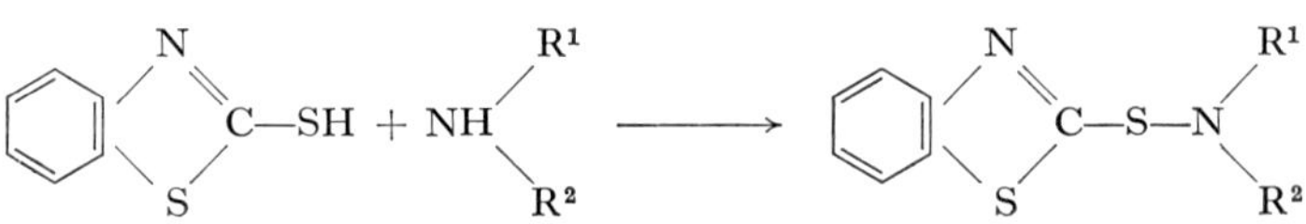

Normally, R^1 and R^2 are H, alkyl or cyclo-alkyl groups and a typical example of these accelerators is *N t*-butylbenzothiazole-2-sulphenamide.

Two preparative routes are available.

(1) By oxidative condensation of 2-mercaptobenzothiazole with the amine in the presence of sodium hypochlorite.

(2) By preparation of the monochloramine of the amine and subsequent reaction with the sodium salt of 2-mercaptobenzothiazole in non-aqueous media.

Retarders Vulcanisation accelerators increase the speed of reaction and thus reduce the curing time. Different accelerators have a different degree of latitude in processing and it is sometimes necessary to increase this latitude, and to prevent premature vulcanisation or 'scorching' from occurring at the compounding stage, because of the rise in temperature caused by frictional forces during blending. Increased latitude may be obtained by the use of retarders, e.g. benzoic acid, maleic anhydride and *N*-nitrosodiphenylamine, the latter being made by nitrosation of diphenylamine hydrochloride with sodium nitrite.

10.4.2. Control of Rubber Deterioration

Rubber deteriorates with age primarily due to oxidation, and a loss of tensile strength and elastomeric properties occurs. Attack by ozone leads to rapid surface cracking and degradation. Intermittent or regular mechanical flexing, radiation, light and heat all accentuate oxidative degradation, while the presence of certain heavy metals such as copper has a pronounced catalytic effect. The mechanism of this oxidative degradation has been extensively studied and is probably a radical chain mechanism initiated by peroxide formation. One of the ways in which antioxidants can work is to provide potential termination steps whereby peroxide intermediates are decomposed without the formation of further free radicals.

Natural rubber latex has a certain natural protection provided by complex phenolic constituents. Antioxidants are therefore only added at the vulcanisation stage. Synthetic rubbers require stabilisation throughout all stages of processing.

The main categories of antioxidants are as follows:

10.4.2.1. Arylamine condensation type The reaction between aniline, or a substituted aniline with ketones, especially acetone, or aldehydes in the presence of an acid catalyst such as hydrochloric acid yields polymers which are often resinous in nature. A typical example is the condensation product formed from aniline and acetone in ths presence of hydrochloric acid, which contains polymers of 2,2,4-trimethyl-1,2-dihydroquinoline.

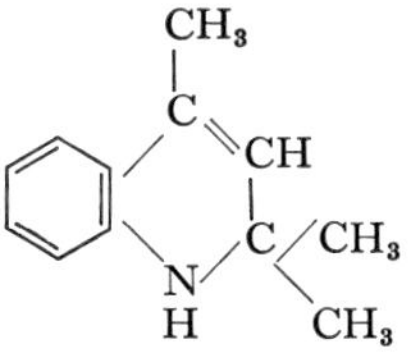

10.4.2.2. Substituted p-*phenylenediamines* *N*,*N'*-Diphenyl-p-phenylenediamine is prepared by condensation of hydroquinone with two moles of aniline in the presence of an acid catalyst. A ring substituted aniline can be used to obtain appropriate alkyl derivatives. When p-phenylenediamine is condensed with two moles of 2-naphthol, *N*,*N'*-di-β-naphthyl-p-phenylenediamine is obtained.

Dialkyl *p*-phenylenediamines may be prepared by condensation of *p*-phenylenediamine with a ketone to form the dianil. Catalytic hydrogenation then yields the required product. These derivatives are very effective antiozonants. The unsymmetrical *p*-phenylenediamines are also effective in protection against ozone attack. An example is *N*-phenyl-*N'*-isopropyl-*p*-phenylenediamine which is manufactured from diphenylamine. Diphenylamine is nitrosated and reduced by hydrogenation to *p*-aminodiphenylamine, which, on reductive alkylation with acetone and hydrogen in the presence of a catalyst, yields the substituted diamine.

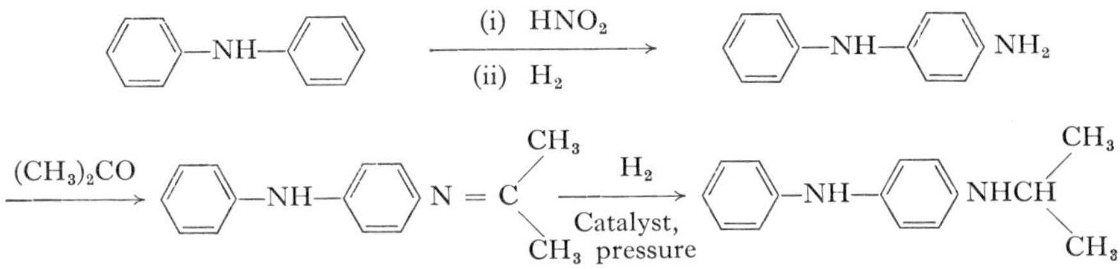

N,N'-Dialkylated *p*-phenylenediamines are made from *p*-chloronitrobenzene by reaction with an alkylamine, followed by reduction and reductive alkylation.

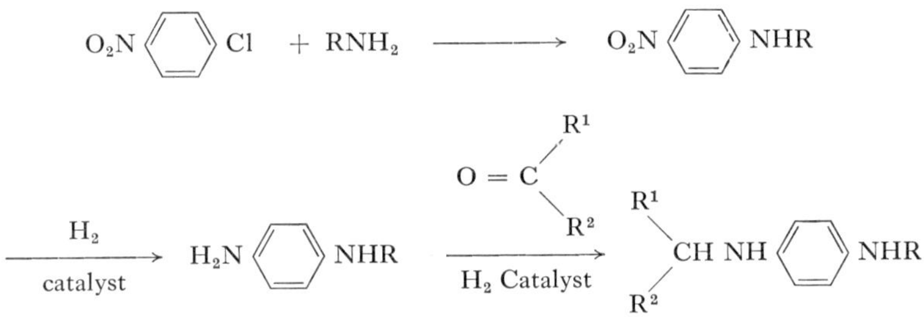

The *p*-phenylenediamines are effective antidegradants, but most of them cause some staining of the rubber. For pale shades of rubber, non-staining antioxidants are used. These are generally of the alkylated phenol, substituted bisphenol or substituted thiobisphenol types.

10.4.2.3. Miscellaneous antidegradants *N*-Phenyl-1-naphthylamine and *N*-phenyl-2-naphthylamine are stabilisers and antioxidants. Their manufacture is by condensation of the appropriate naphthol with aniline in the presence of an acid catalyst.

Mercaptobenzimidazole is particularly effective against metal catalysed degradation. It is prepared by the reaction of aqueous *o*-phenylenediamine with carbon disulphide.

The synthesis of a number of novel *p*-phenylenediamine derivatives has been reported.[64]

10.5 CYCLOHEXYLAMINE

10.5.1. Manufacture

The normal commercial manufacture of cyclohexylamine is by liquid phase catalytic hydrogenation of aniline under pressure. Alternative synthetic routes are the catalytic ammonolysis of cyclohexyl chloride or of cyclohexanol, or by the reduction of nitrocyclohexane. A process for the catalytic ammonolysis of cyclohexanone in the presence of hydrogen has been claimed.[65]

10.5.2. Properties and Uses

Cyclohexylamine is a colourless liquid m.p. −18°C, b.p. 135°C, miscible with water in all propertions. The flash point is 22°C.

Cyclohexylamine is a reactive primary amine whose reactions are mainly typical of an aliphatic amine, forming derivatives which are widely used in the manufacture of rubber chemicals, corrosion inhibitors, plastic and chemicals for textile uses.

A major use of cyclohexylamine has been in the manufacture of sodium (and calcium) cyclohexylsulphamate, the so called cyclamates, formerly widely used as synthetic sweetening agents.

The use of the cyclamates has now been abandoned in many countries. The original method of manufacture was by chlorosulphonation of cyclohexylamine in the presence of excess amine to form the amine salt, followed by treatment with sodium hydroxide yielding sodium cyclohexylsulphamate. More recently preparation has been by sulphonation of cyclohexylamine with sulphamic acid or with sulphur trioxide.

Derivatives of cyclohexylamine find use as antioxidants and antiozonants in the rubber industry and the sulphenamide formed by oxidative condensation of 2-mercaptobenzothiazole with cyclohexylamine is a useful vulcanisation accelerator.

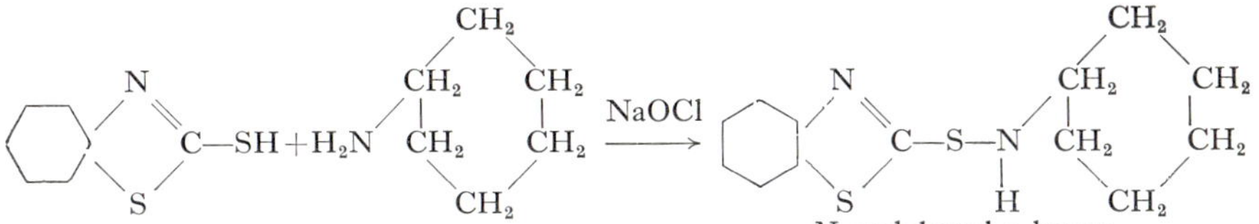

N-cyclohexyl-2-benzothiazolesulphenamide

Derivatives are ferrous metal corrosion inhibitors, and cyclohexylamine itself may be added to boiler feed water in steam raising plant to limit corrosion and scaling. The action is probably to maintain a slight alkalinity, and since cyclohexylamine forms an azeotrope with water the protection is extended to condensers and further equipment. Cyclohexylamine is used to a limited extent as a dyestuffs intermediate.

10.5.3. Dicyclohexylamine

This is a clear colourless liquid m.p. 0°C, b.p. 256°C, slightly soluble in water. It is a strongly basic secondary amine and readily yields *N*-substituted derivatives. Manufacture is by the vapour phase catalytic hydrogenation of aniline.

Like cyclohexylamine, dicyclohexylamine and its derivatives are used as corrosion inhibitors. Solid derivatives of slight volatility are used in the packaging of ferrous metal articles as a protection against atmospheric corrosion. Dicyclohexylamine also gives a sulphenamide used as a vulcanisation accelerator. Salts and metal complexes are used in the textile and in the paint and varnish industry.

10.6 DINITROBENZENE

10.6.1. Manufacture and Uses

Nitrobenzene is further nitrated with mixed acid of composition HNO_3, 33; H_2SO_4, 67 parts by weight. The product is mainly *m*-dinitrobenzene with small amounts of the ortho- and para-isomers. The crude material may be used direct, or, if purification is required, following treatment with sodium sulphite solution, which converts the ortho- and para-isomers into the soluble nitrobenzenesulphonic acid which may be removed by washing with water.

Pure *m*-dinitrobenzene is a colourless crystalline solid m.p. 90°C, b.p. 297°C. It is used in the manufacture of *m*-nitroaniline by partial sulphide reduction, and of *m*-phenylenediamine by Béchamp reduction with iron powder. The further nitration to give 1,3,5-trinitrobenzene, with a stronger mixed acid, is very slow and excessive oxidation occurs.

The isomeric dinitrobenzenes are of little importance. *p*-Dinitrobenzene is a colourless compound, m.p. 173°C, of similar properties to *m*-dinitrobenzene. It is formed by the action of hydrogen peroxide in trifluoroacetic acid on *p*-nitroaniline.

The properties of *o*-dinitrobenzene, m.p. 117°C are also similar to those of the meta-isomer, except that one nitro group is labile and easily replaced by $-NH_2$ or $-OH$ on treatment with ammonia or sodium hydroxide respectively. The electrolytic reduction of *o*- and *p*-dinitrobenzene and their derivatives has been described, and methods of synthesis given.[66]

10.6.2. 1-Chloro-2,4-dinitrobenzene

The commercial manufacture is by dinitration of chlorobenzene or by the nitration of *p*-chloronitrobenzene. 1-Chloro-2,4-dinitrobenzene is a pale yellow solid m.p. 51°C; b.p. 315°C. It is a severe skin irritant and causes dermatitis. Persons who have become sensitised to this material react thereafter to minute traces. Inhalation of vapour and contact with the skin must be totally avoided. Due to its relatively low melting point the material may be conveniently stored and distributed in the molten state; this has the advantage of minimising handling and exposure.

The chlorine atom is even more active than is the case with the chloronitrobenzenes, and displacement by nucleophilic attack normally takes place at about 100°C; pressure is not usually required. 1-chloro-2,4-dinitrobenzene reacts readily with aniline to give 2,4-dinitrodiphenylamine and with aqueous ammonia to give 2,4-dinitroaniline.

The following intermediates are normally manufactured from 1-chloro-2,4-dinitrobenzene:

2,4-Dinitroaniline
4-Nitro-*o*-anisidine
4-Chloro-*m*-phenylenediamine
4-Methoxy-*m*-phenylenediamine
Picric acid
2,4-Dinitrobenzene-1-sulphonic acid

Additional uses include the manufacture of intermediates for sulphur dyestuffs, azo disperse dyes, pigments and rubber chemicals. The vulcanisation accelerator 2(2,4-dinitrophenyl)-2-thiobenzothiazole is prepared by reaction of an aqueous solution of the sodium salt of 2-mercaptobenzothiazole with 1-chloro-2,4-dinitrobenzene.

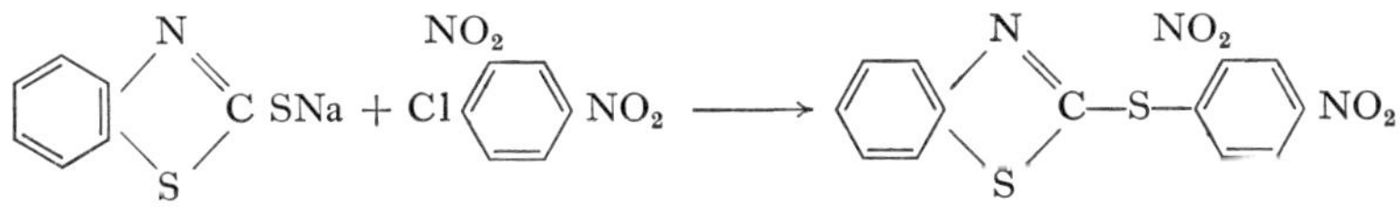

10.6.3. 1-Chloro-2,4-dinitrobenzene-6-sulphonic acid

This intermediate is manufactured by sulphonation of *p*-chloronitrobenzene with oleum, followed by nitration with mixed acid.

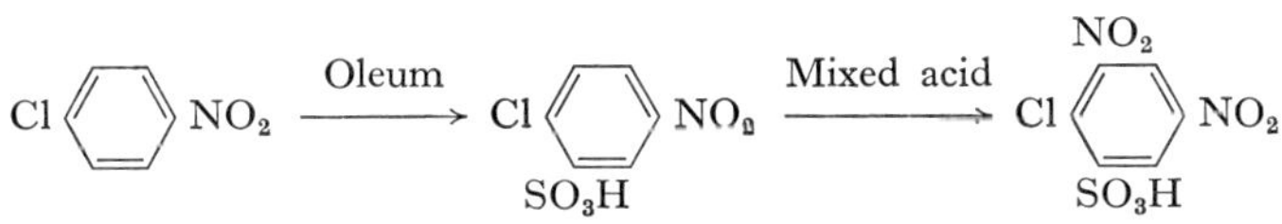

Both reactions are regarded as hazardous at elevated temperatures, and especially above 130°C. The main use of 1-chloro-2,4-dinitrobenzene-6-sulphonic acid is in the preparation of the dyestuffs intermediate 2-amino-4-nitrophenol-6-sulphonic acid by alkaline hydrolysis followed by partial reduction with sulphide. This intermediate is widely used in the manufacture of azo mordant and pre-metallised dyestuffs.

10.7. EXPLOSIVES

A number of derivatives of dinitrobenzene and trinitrobenzene have been used as explosives and detonating agents although some of these have been rendered obsolete by the development of explosives of greater power or stability. Since several derivatives of dinitrobenzene are used in routine organic synthesis, especially in the manufacture of dyestuffs, it follows that appropriate safety precautions are essential in their handling and processing.

10.7.1. 1,3,5-Trinitrobenzene

The further nitration of *m*-dinitrobenzene to give the symmetrical trinitrobenzene is difficult and requires a mixture of nitric acid and oleum at approx. 120°C. Yields are poor and there is excessive by-product formation. An alternative route therefore is from 2,4,6-trinitrotoluene by oxidation to trinitrobenzoic acid, followed by decarboxylation.

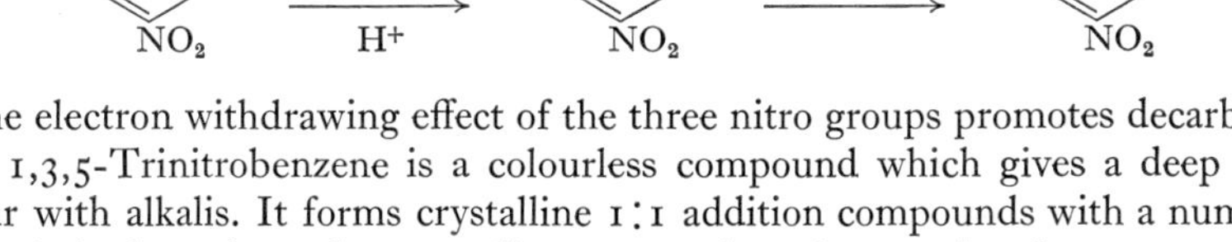

The electron withdrawing effect of the three nitro groups promotes decarboxylation. 1,3,5-Trinitrobenzene is a colourless compound which gives a deep yellow colour with alkalis. It forms crystalline 1:1 addition compounds with a number of aromatic hydrocarbons, heterocyclic compounds and aromatic ethers.

In its explosive properties it is slightly superior to TNT and it is a very stable explosive. It is no longer of importance due to its high cost of manufacture.

10.7.2. Diazodinitrophenol

2-Diazo-4,6-dinitrophenol is probably a cyclic diazo oxide and is manufactured as follows: Picric acid is partially reduced with aqueous sodium hydrosulphide and the picramic acid so formed is diazotised, and the diazo oxide isolated.

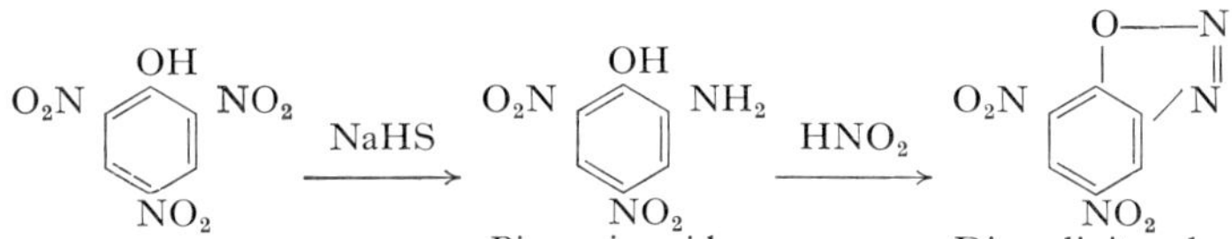

Picramic acid Diazodinitrophenol

Diazodinitrophenol is used commercially as a detonating agent since it is stable on storage, and possesses a high initiating value.

Picramic acid is used in the manufacture of several azo dyestuffs by diazotisation and coupling to appropriate coupling components, e.g. *m*-phenylenediamine.

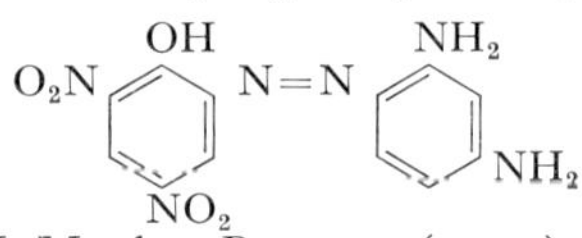

C.I. Mordant Brown 12 (11290)

Due to the explosive nature of the diazo compound, stringent precautions must be taken during manufacture, especially the scrupulous cleaning out of any equipment used for the diazotisation process.

10.7.3. Picric Acid: 2,4,6-Trinitrophenol

The direct nitration of phenol is a very vigorous reaction and difficult to control. Alternative preparative routes are by the hydrolysis of 1-chloro-2,4-dinitrobenzene giving 2,4-dinitrophenol followed by nitration, or from benzene with nitric acid and a mercuric nitrate catalyst. Picric acid is a yellow crystalline compound formerly used to a small extent as a dyestuff. It affords crystalline addition compounds with aromatic compounds e.g. with naphthalene. The combined electronic effect of the three nitro groups leads to easy replacement of the substituents in the 1-position and treatment with phosphorous pentachloride followed by ammonolysis with aqueous ammonia gives 2,4,6-trinitroaniline.

Picric acid has been used extensively in the past in high explosive compositions. It has now been replaced by more modern explosives.

10.7.4. Tetryl: *N*-Methyl-*N*,2,4,6-tetranitroaniline

1-Chloro-2,4-dinitrobenzene is reacted with methylamine giving *N*-methyl-2,4-dinitroaniline, which is nitrated to the tetranitro derivative.

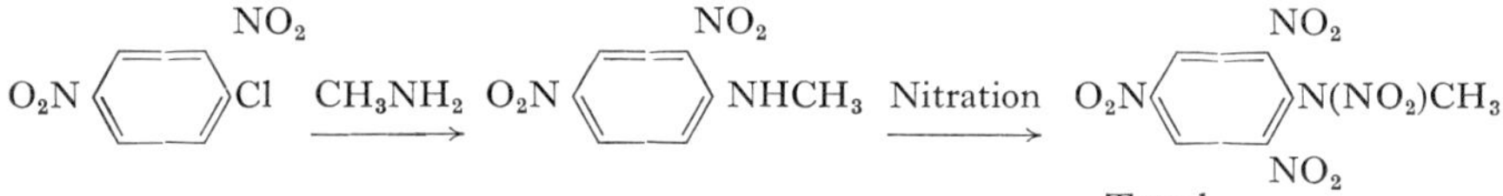

Tetryl

Tetryl has a very high explosive value and is used currently.

10.7.5. Related Explosives

Dinitrobenzene, although of relatively low explosive value, has been used in the past as a substitute for trinitrotoluene, in times of shortage, for high explosive mixtures. It is no longer used due to the availability of more powerful explosives.

Lead styphnate (the lead salt of 2,4,6-trinitroresorcinol) and also lead 2,4- and 2,6-dinitroresorcinates are used as initiating agents in priming compositions.

1,3-Diamino-2,4,6-trinitrobenzene is a heat resistant explosive of stability appropriate to withstand conditions encountered in space flight.

Certain difluoramines have potential use as high energy propellent ingredients, e.g. that prepared from 2,4,6-trinitroaniline.

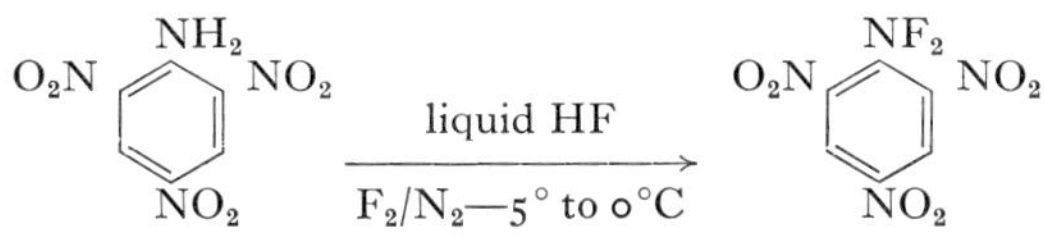

Chemistry in the technology of explosives and propellents has recently been reviewed.[67]

10.8. THE PHENYLENEDIAMINES

10.8.1. *o*-Phenylenediamine

10.8.1.1. Manufacture The normal manufacturing route is as follows.[68] Ammonolysis of *o*-chloronitrobenzene in a cast-iron autoclave under pressure yields *o*-nitroaniline. Reduction of the crude reaction product with sodium sulphide under pressure gives *o*-phenylenediamine, which may be purified by vacuum distillation.

An alternative process is the high pressure amination of *o*-dichlorobenzene with ammonia, containing only small amounts of water, in the presence of copper catalysts at approx. 200°C. The process is analogous to that for the manufacture of aniline from chlorobenzene.

A convenient laboratory preparation is by the reduction of *o*-nitroaniline with zinc dust and alkali.[69]

10.8.1.2. Properties and reactions *o*-Phenylenediamine is a colourless crystalline solid, darkening on exposure to air, soluble in water m.p. 102°C, b.p. 257°C (148°C at 20 mm Hg.). It differs from the isomeric phenylenediamines in its reaction with nitrous acid. Nitrous acid in the presence of glacial acetic acid or hydrochloric acid gives a rapid conversion to 1,2,3-benzotriazole. It has been proposed[70] that a monodiazonium salt is the intermediate stage in this conversion.

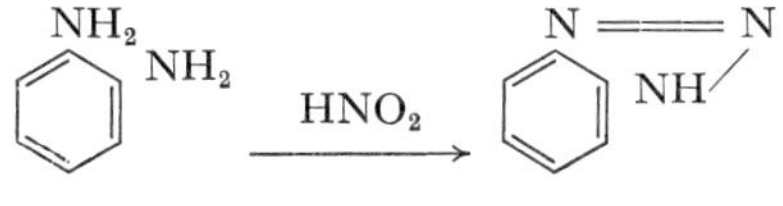

Benzotriazole is an important additive for a wide range of products as an inhibitor of copper corrosion. It is incorporated in antifreeze preparatfons, lubricating oils, polishes and detergents.

Oxidation of *o*-phenylenediamine with ferric chloride gives a diaminophenazine.

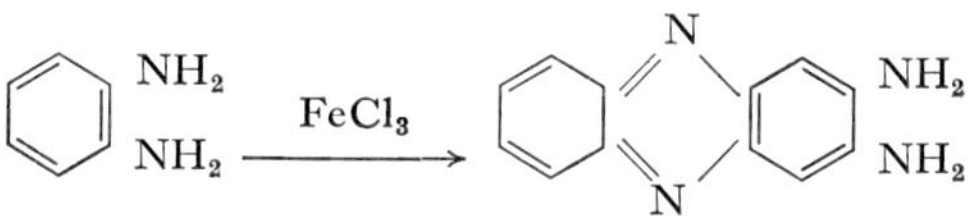

Oxidation with lead dioxide yields *o*-quinonediimine, which on warming gives the diaminophenazine.

A general reaction of *o*-phenylenediamine with 1,2-dicarbonyl compounds such as glyoxal produces quinoxalines.[71]

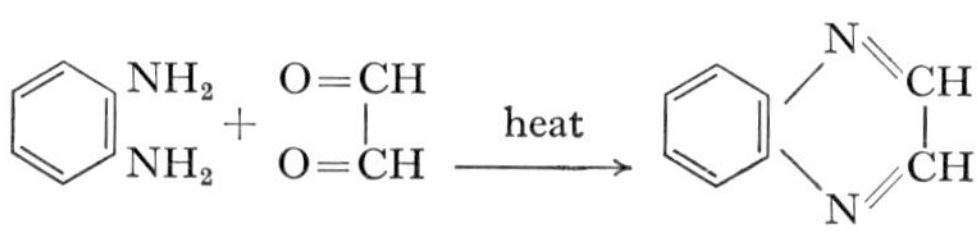

quinoxaline

Another general reaction is with *o*-quinones.

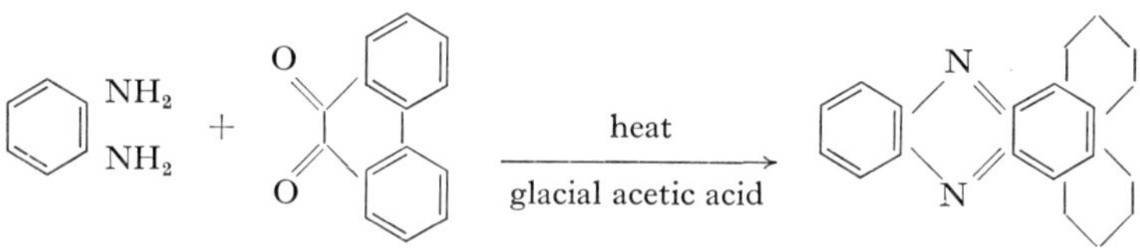

This reaction with 9,10-phenanthraquinone is a means of characterising *o*-diamines. The reactions of *o*-phenylenediamine giving rise to 6-membered rings have been reviewed.[72, 73]

Phenazine (dibenzopyrazine) may be obtained by the condensation of *o*-phenylenediamine with 1,2-dihydroxybenzene (catechol).

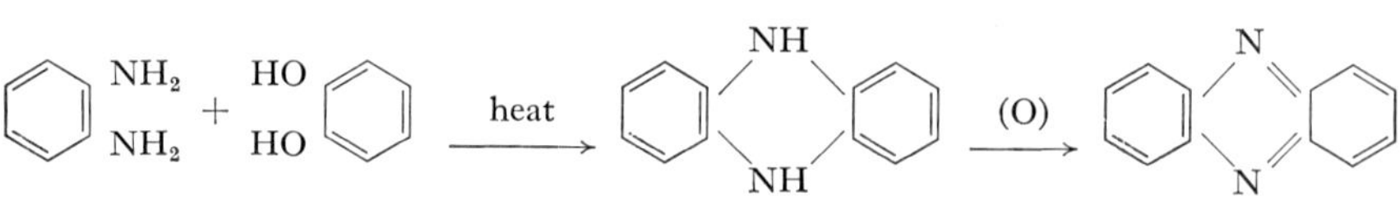

Phenazine

The reaction with alloxan gives alloxazine (2,4-dihydroxybenzopteridine).

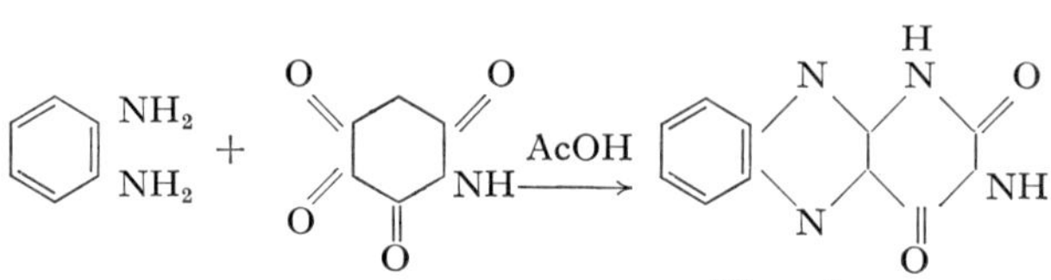

Alloxazine

Oxidation of *o*-phenylenediamine by a Skraup reaction in the presence of glycerol, arsenic acid and dilute sulphuric acid gives 1,10-phenanthroline, which may be used as a Redox indicator.

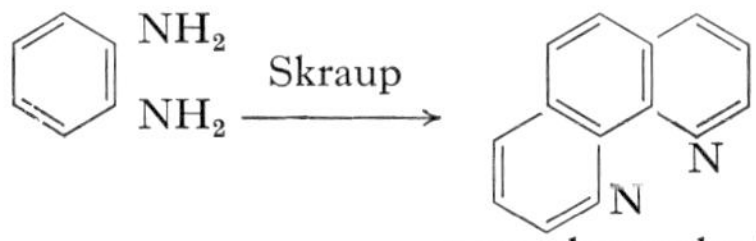

1,10-phenanthroline

Reaction of aqueous *o*-phenylenediamine with carbon disulphide gives 2-mercaptobenzimidazole. The analogous reaction with carbon dioxide under pressure gives a benzimidazolinone. Formation of such derivatives of orthoaromatic diamines is claimed as a technique for their removal in the purification of diamine mixtures.[74]

Condensation of *o*-phenylenediamine with carboxylic acids leads to 2-alkylbenzimidazoles. Aldehydes give rise to disubstituted benzimidazoles. Condensation is carried out in hydrochloric acid at 100°C.

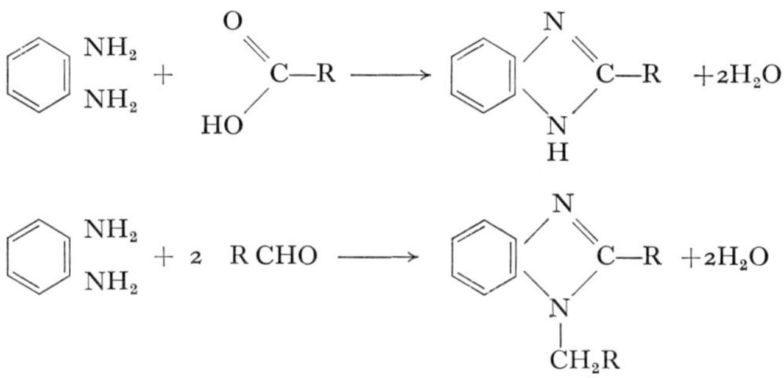

This reaction serves to characterise aldehydes and carboxylic acids.

The perinone pigment CI Vat Orange 7 (71105) is derived from naphthalene-1,4,5,8-tetracarboxylic acid and *o*-phenylenediamine.

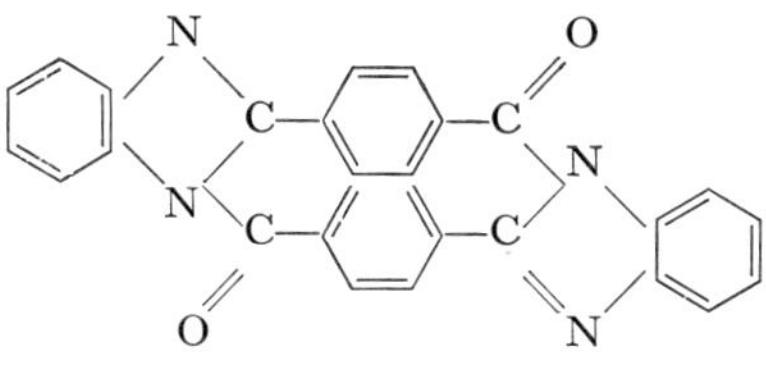

This is a high grade pigment of good fastness properties. A recent patent[75] claims the condensation of benzophenone-3,3′,4,4′-tetracarboxylic dianhydride with *o*-phenylenediamine to give a yellow perinone dye.

N-alkyl and *N*-aryl derivatives of *o*-phenylenediamine may be prepared, e.g. the *N*,*N*-dimethyl derivative is obtained from *N*,*N*-dimethyl-*o*-nitroaniline by reduction with zinc dust in aqueous ammonium chloride.

10.8.1.3. Uses *o*-Phenylenediamine is widely used in organic synthesis. A main use is in the manufacture of the antioxidant, benzotriazole. Other uses include the manufacture of rubber chemicals, dyestuffs, pharmaceutical products and veterinary products.

10.8.2. *m*-Phenylenediamine

10.8.2.1. Manufacture *m*-Phenylenediamine is made by the reduction of *m*-dinitrobenzene with iron and hydrochloric acid, then purified by vacuum distillation.

10.8.2.2. Properties and reactions When *m*-phenylenediamine is treated with nitrous acid a mixture of brown polyazo compounds is formed. Bismarck Brown G (CI Basic Brown 1, 21,000) was the first azo dyestuff made and was discovered by C. Martius in 1863. It is still of some importance as a paper dye. The reaction is very sensitive and has been used for the determination of nitrous acid in low concentration. The normal preparation of Bismarck Brown G is to treat 3 moles of *m*-phenylenediamine with 2 moles of nitrous acid. One constituent is:

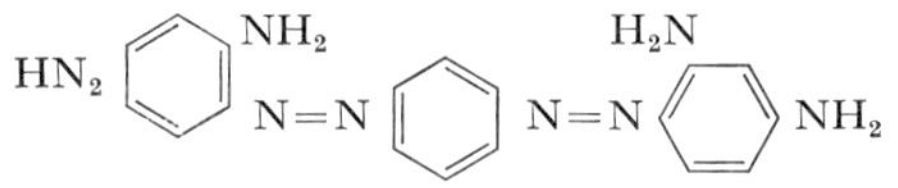

Tetrazotisation of *m*-phenylenediamine without a simultaneous coupling reaction taking place requires special conditions, e.g. use of concentrated hydrochloric acid with excess nitrous acid, or tetrazotisation with nitrosylsulphuric acid in acetic acid. If however substituents are present in the 4- and 6- positions, tetrazotisation proceeds normally.

m-Phenylenediamine couples readily with diazonium salts. Coupling takes place mainly in the 4-position with one equivalent of the diazo compound. Excess of diazo component in acidic solution gives the 2,4-disazo compound, while strongly alkaline solution, or the presence of pyridine, favours formation of the 4,6-disazo compound.[76]

m-Phenylenediamine occurs as end component in certain polyazo direct dyes for cotton and it is also used as a coupling component for dyes diazotised on the fibre (CI Developer 11). The 4-nitro derivative is an oxidation base (CI 76030) and is used in the dyeing of fur and hair. It may be prepared by the acetylation of *m*-phenylenediamine, to protect the amino groups, followed by nitration and hydrolysis. A recent patent[77] has claimed the preparation of resorcinol by sulphuric acid hydrolysis of *m*-phenylenediamine.

m-Phenylenediamine may be converted into the diisocyanate by the normal procedure, by reaction with phosgene, then heating to eliminate HCl. There is considerable by-product formation in the reaction.

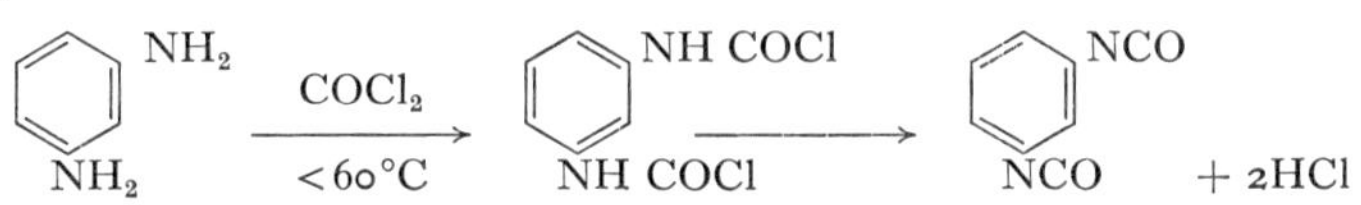

Condensation of such diisocyanates with dihydric alcohols gives straight chain polyurethanes. With higher alcohols, cross-linked polymers are obtained.

10.8.3. *p*-Phenylenediamine

10.8.3.1. Manufacture *p*-Phenylenediamine is normally made by Béchamp reduction of *p*-nitroaniline, which is manufactured either by ammonolysis of *p*-chloronitrobenzene or by nitration of acetanilide followed by hydrolysis.

10.8.3.2. Properties and reactions A colourless crystalline solid, rapidly turning dark on exposure to air, soluble in water, m.p. 140°C, b.p. 267°C. On treatment with nitrous acid under normal diazotisation conditions *p*-phenylenediamine gives a mixture of diazo and tetrazo derivatives. Treatment with nitrosylsulphuric acid in phosphoric/sulphuric acid mixture gives the tetrazo compound. The tetrazo compound couples in a stepwise fashion since one diazonium group is very active. The monoazo diazonium salt then couples with normal activity.[78]

Oxidation of *p*-phenylenediamine with manganese dioxide or sodium bichromate in the presence of sulphuric acid gives *p*-benzoquinone.

Milder oxidation, e.g. potassium ferricyanide in aqueous ammonia, gives *p*-quinonediimine which is unstable and reacts further to give a complex compound known as Bandrowski's base, a type of intermediate formed in the dyeing of hair and fur.

The 2-nitro derivative of *p*-phenylenediamine is prepared by acetylation, nitration and hydrolysis. It is important as an intermediate and a constituent of dyes for hair and fur as CI Oxidation Base 22 (76070). *N*-Alkyl and *N*-aryl derivatives of *p*-phenylenediamine are widely used rubber antioxidants (q.v.) and are made by condensation reactions of *p*-phenylenediamine or *p*-benzoquinone. Substituted diphenylamines may be prepared by condensation of *p*-phenylenediamine with chloronitrobenzenes; these compounds find use as intermediates for azo dyestuffs.

10.8.3.3. Uses *p*-Phenylenediamine and its derivatives are widely used for the dyeing of hair and fur, and as intermediates for the manufacture of dyestuffs of the azo, azine, sulphur and cyanine categories. *N*-Substituted *p*-phenylenediamines are used as antioxidants and antiozonants for rubber.

10.8.4. Phenylenediamines as Oxidation Bases

The phenylenediamines and many of their derivatives are listed in the colour index as oxidation bases.

o-Phenylenediamine	CI Oxidation base 16 (76010)
m-Phenylenediamine	CI Oxidation base (76025)
p-Phenylenediamine	CI Oxidation base 10 (76060)

These oxidation dyes are widely used to give permanent colouration to hair and fur. In general the ortho- and para-diamines and their derivatives are the primary intermediates; meta-diamines, aminophenols, naphthols and polyhydroxy compounds are involved in the secondary condensation reactions to give a wide range of shades. The dyes are formed in two steps.

(1) Oxidation of the base to form a quinonediimine.

(2) Condensation to give an azine dye.

p-Phenylenediamine itself can cause dermatitis and its use is prohibited in some countries. The processes involved in the oxidative hair dyeing procedure have been reviewed.[79] The kinetics and mechanism of oxidative coupling of *p*-phenylenediamines have been discussed.[80] The process of formation of cyanine dyes by such oxidative coupling procedures is of great importance in colour photography. An analogous oxidative coupling reaction applied to hydrazones instead of diamines has led to novel azo dyes not accessible by normal methods.[81]

10.8.5. Other Derivatives

10.8.5.1. Acetyl derivatives The monoacetyl derivatives of the meta- and para-diamines, i.e. *m*-aminoacetanilide and *p*-aminoacetanilide are manufactured by iron reduction of the corresponding nitroacetanilide. The aminoacetanilides are used as dyestuffs intermediates. *p*-Aminoacetanilide occurs as a diazo component in several direct colours for cotton. The condensation product with 1-amino-4-bromo-2-anthraquinonesulphonic acid is CI Acid Blue 40 (62125).

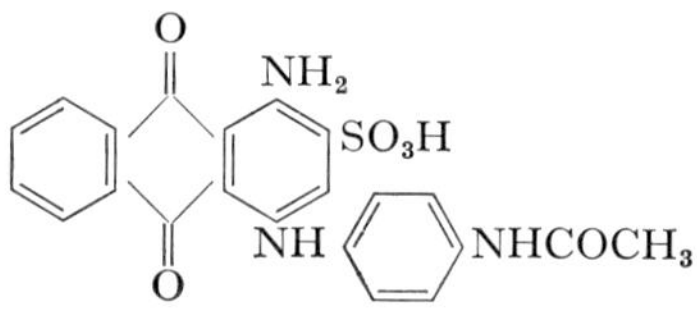

This colour has good fastness properties, especially to light.

10.8.5.2. The phenylenediamine sulphonic acids 1,3-Phenylenediamine-4-sulphonic acid and 1,4-phenylenediamine-2-sulphonic acid are used as dyestuffs intermediates. 1,3-Phenylenediamine-4-sulphonic acid is manufactured by iron reduction of 1,3-dinitrobenzene-4-sulphonic acid. Iron reduction of 4-nitroaniline-2-sulphonic acid yields 1,4-phenylenediamine-2-sulphonic acid. These two intermediates are widely used in the manufacture of azo and reactive dyes. Together with their acetyl derivatives they afford a series of useful blue dyes for wool and polyamide fibres by condensation with bromoanthraquinone derivatives.

The phenylenediamines, and to a lesser extent their sulphonic acid derivatives, darken rapidly on exposure to air and become brown, purple or even black in colour. This oxidative degradation is catalysed by certain impurities in the product and in general a purer preparation has greater stability on storage. Purification may be effected at the diamine preparation stage or at the dinitro precursor stage. A technique has been claimed[82] for treatment of *m*-dinitrobenzene with sodium sulphite, then with water and detergents for subsequent use in the preparation of *m*-phenylenediamine which is almost colourless, and stable during storage.

10.9 AMINOPHENOLS

The aminophenols are widely used in the manufacture of dyestuffs of the azo, azine and sulphur dye categories. The ortho- and para-isomers are readily oxidised and they find application as oxidation bases in fur dyeing and as photographic developers. Derivatives of *o*-aminophenol containing nitro-, chloro, sulphonamide and sulphonic acid groups are especially important in the production of metal complex azo dyestuffs.

10.9.1. *o*-Aminophenols

o-Aminophenol when pure is a white crystalline solid m.p. 170°C rapidly darkening on exposure to air. Like the other unsulphonated aminophenols it is a dermatitis hazard and skin irritant and exposure must be avoided. Manufacture is from *o*-chloronitrobenzene by caustic hydrolysis to *o*-nitrophenol, followed by reduction by either the Béchamp process or with sodium sulphide.

Mild acylation gives the *N*-acyl derivative, e.g. treatment with acetic anhydride gives *o*-hydroxy acetanilide. More vigorous conditions lead to ths *O*,*N*-diacetyl derivative, or 2-methylbenzoxazole. Benzoxazole is obtained by phosgenation of *o*-aminophenol: this technique is used to protect the amino group prior to nitration. The reaction of *o*-aminophenols with acid anhydrides on heating to give substituted benzoxazoles is a general one.

When *o*-aminophenol is heated in the presence of hydrochloric acid, followed by oxidation, phenoxazine is obtained. Phenoxazines also result from the reaction of *o*-aminophenol with *o*-dihydroxy compounds.

2-Amino-4-nitrophenol is an orange crystalline solid m.p. 142°C (anhydrous form). The commercial preparation is from 1,chloro-2,4-dinitrobenzene by caustic hydrolysis to 2,4-dinitrophenol followed by partial reduction with sodium hydrosulphide. The alkaline sulphide reduction of 2,4-dinitrophenol has been studied under various conditions.[83] 2-Amino-4-nitrophenol is used in the manufacture of mordant and pre-metallised azo dyestuffs.

2-Amino-5-nitrophenol is a yellow brown crystalline solid manufactured by nitration of benzoxazole followed by separation of the isomers and caustic hydrolysis. It is used in the manufacture of azo dyes. The diazo compound obtained from 2-amino-5-nitrophenol is dangerous since the heavy orange crystals are very susceptible to friction and mechanical shock and can explode even when wet.

2-Amino-4-nitrophenol-6-sulphonic acid is manufactured by partial reduction of 2,4-dinitrophenol-6-sulphonic acid with sodium hydrosulphide. The dinitrophenol sulphonic acid is obtained by hydrolysis of 1-chloro-2,4-dinitrophenol-6-sulphonic acid. This intermediate is widely used in the manufacture of azo dyestuffs.

2-Amino-4-chlorophenol is manufactured by alkaline hydrolysis of 1,4-dichloro-2-nitrobenzene to give 4-chloro-2-nitrophenol followed by Béchamp reduction. The intermediate is used in the manufacture of azo mordant dyestuffs.

2-Aminophenol-4-sulphonic acid is a colourless crystalline solid, soluble in water. The manufacturing route is from 4-chloro-3-nitrobenzene sulphonic acid by hydrolysis with sodium hydroxide followed by iron reduction. This intermediate is used in the manufacture of a number of azo dyestuffs.

2-Amino-6-nitrophenol-4-sulphonic is manufactured by partial sulphide reduction of 2,6-dinitrophenol-4-sulphonic acid, itself prepared by nitration of phenol-4-sulphonic acid. It is used in azo dyestuffs manufacture.

Other Derivatives. *o*-Aminophenols including sulphonamide and sulphone groups as substituents but without sulphonic acid substituents are extensively used in the preparation of neutral dyeing 1:2 metal-azo complexes.

10.9.2. *m*-Aminophenols

m-Aminophenol is a colourless crystalline material, m.p. 122–123°C, and is fairly stable to oxidation in air, unlike the ortho- and para-isomers. *m*-Aminophenol is manufactured by caustic fusion of metanilic acid, and purified by vacuum distillation.

Alternative methods are by reduction of *m*-nitrophenol with iron and hydrochloric acid, or with sodium sulphide.

m-Aminophenol gives *N*-acyl derivatives with acid anhydrides. The Skraup reaction with glycerol leads to 7-hydroxyquinoline. The condensation of *m*-aminophenol with aniline hydrochloride at 200°C forms *m*-hydroxydiphenylamine.

The main use of *m*-aminophenol is in the manufacture of *p*-aminosalicylic acid (used in the treatment of tuberculosis) by treatment with carbon dioxide under pressure in the Kolbé synthesis. As C.I. oxidation base 7 (76545), *m*-aminophenol is used in fur dyeing.

3-Alkylaminophenols are widely used in the manufacture of basic dyestuffs of the safranine, xanthene, rhodamine and oxazine classes.

m-Diethylaminophenol is the most important of these derivatives and is prepared by ethylation of metanilic acid to give *N*,*N*-diethylmetanilic acid, followed by caustic fusion. It is used in the manufacture of Rhodamine B, C.I. Basic Violet 10 (45170) by fusing with phthalic anhydride.

10.9.3. *p*-Aminophenols

p-Aminophenol is a colourless crystalline solid m.p. 189°C darkening on exposure to air. It forms salts with acids and bases. *p*-Aminophenol is manufactured from *p*-chloronitrobenzene by hydrolysis, then Béchamp reduction. Alternative routes include the catalytic hydrogenation[84] of *p*-nitrophenol, and the rearrangement of phenylhydroxylamine with mineral acid.

p-Aminophenol is extensively used in the manufacture of dyestuffs of many categories. It has important use as a photographic developer. Acylation of *p*-aminophenol in the presence of sodium or potassium iodide has recently been claimed.[85] The competing oxidation reactions during the diazotisation of *p*-aminophenol have been examined.[86]

A number of derivatives of *p*-aminophenol have analgesic or antipyretic properties, the best known being Phenacetin, *p*-ethoxyacetanilide. This is normally prepared by acetylation of *p*-aminophenol with acetic anhydride followed by ethylation of the hydroxyl group with sodium ethoxide—ethyl iodide. An alternative route is by ethylation of *p*-nitrophenol with diethylsulphate forming *p*-nitrophenetole, followed by reduction to *p*-phenetidine and acetylation to Phenacetin.

p-Methylaminophenol is the widely used photographic developer, Metol. In common with other aminophenols it can be a cause of dermatitis. Commercially the method of manufacture involves the reaction of chloroacetic acid with *p*-aminophenol in the presence of alkali to give *N*-(*p*-hydroxyphenyl) glycine which on elimination of carbon dioxide yields *p*-methylaminophenol.

2,4-Diaminophenol is normally handled as the hydrochloride and is the photographic developer amidol. It is manufactured from 1-chloro-2,4-dinitrobenzene by alkaline hydrolysis, followed by Béchamp reduction.

10.10. RELATED PRODUCTS

This section includes a number of organic intermediates derived from, or related to, nitrobenzene and aniline, and includes hydroquinone, benzidine, phenylhydrazine and the pyrazolones. Certain production figures for nitrobenzene and related products are quoted.

10.10.1. Hydroquinone (see also chapter 12, p. 562)

Hydroquinone is a white crystalline solid m.p. 172°C, soluble in water and in polar solvents. It is readily oxidised to quinone and is widely used as a reducing agent, especially as a photographic developer, and as a corrosion inhibitor.

It is manufactured by the oxidation of aniline to yield quinone, followed by reduction to hydroquinone. In the laboratory the oxidation of aniline is carried out with potassium bichromate and sulphuric acid; industrially manganese dioxide, in the form of pyrolusite ore, is used with sulphuric acid. The equation may be written:

$$2\ C_6H_5NH_2 + 4MnO_2 + 5H_2SO_4 \longrightarrow 2$$

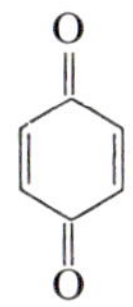

$$+ (NH_4)_2SO_4 + 4\ MnSO_4 + 4\ H_2O$$

The reaction is exothermic and cooling is required. The quinone is removed by volatilisation in steam and reduced to hydroquinone by treatment with iron powder and water. The crude product is purified by vacuum distillation and recrystallisation. An alternative reduction procedure with sulphur dioxide has the disadvantage of producing significant amounts of hydroquinone-2-sulphonic acid as a by-product by an addition reaction. Hydroquinone, and substituted hydroquinones have been made[87] by the catalysed reaction between carbon monoxide and acetylene or a substituted acetylene at high pressure.

Hydroquinone shows phenolic character and forms mono- and diesters and ethers. In the case of halogenation and sulphonation normal ring substitution takes place. Hydroquinone is readily oxidised to quinone and also forms an equimolecular complex with quinone known as quinhydrone which crystallises as green lustrous prisms, and forms the basis of the quinhydrone electrode. Condensation of hydroquinone with phthalic anhydride gives 1,4-dihydroxyanthraquinone (quinizarin).

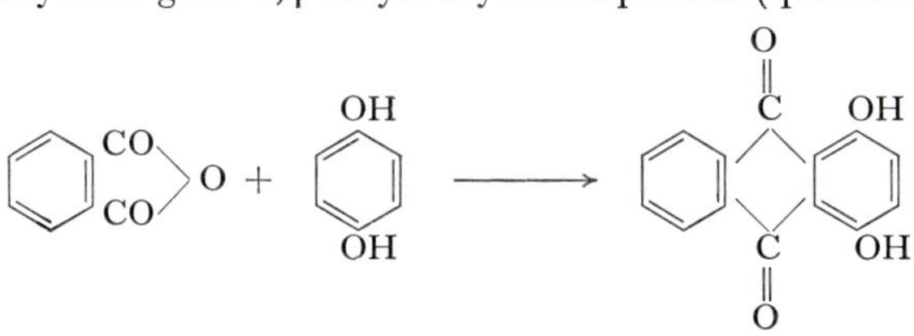

The main use of hydroquinone is as a photographic developer. During development the rate of reduction of the exposed silver halide grains is more rapid than that of the unexposed grains. Sodium sulphite is added to the developer solution to suppress autoxidation of the hydroquinone. Hydroquinone can be used in conjunction with other photographic developers, e.g. derivatives of *p*-aminophenol. Hydroquinone and its derivatives are widely used as antioxidants in fats and oils. Alkyl derivatives are antioxidants for vegetable fats and stabilisers for rubber and synthetic polymers.

10.10.2. Quinones

The quinones are widely used in organic synthesis. They often arise as products or by-products in the oxidation of aromatic amines, *p*-aminophenols and *p*-diamines. The parent quinone, *p*-benzoquinone is manufactured by oxidation of aniline. It forms yellow prismatic crystals m.p. 116°C and is slightly soluble in water. It is olefinic in character and readily forms addition compounds. Addition of hydrogen chloride gives 2-chloro-hydroquinone; sodium bisulphite yields hydroquinone-2-sulphonic acid. An example of addition of a diene to the 2-methyl derivative is afforded by the preparation of vitamin k.

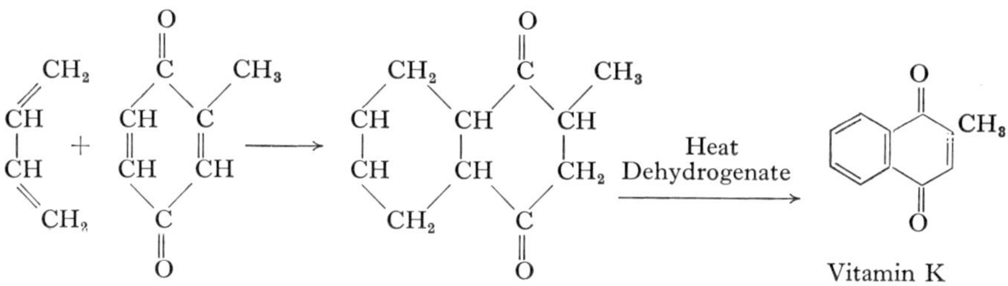

Quinone reacts with 2,4-dinitrophenylhydrazone giving rise to a tautomeric azo compound—hydrazone equilibrium.

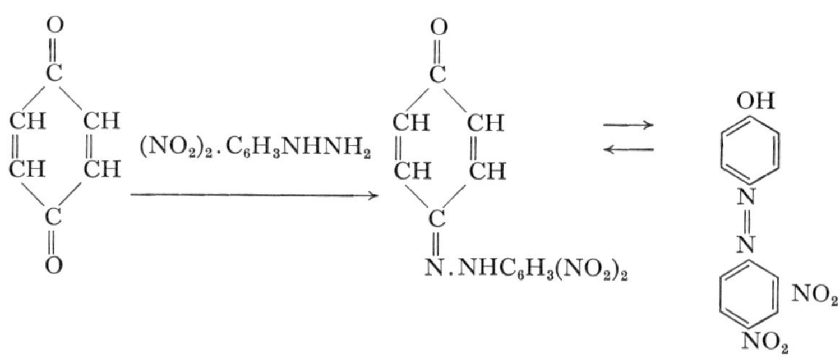

Diazonium salts give a substitution reaction.

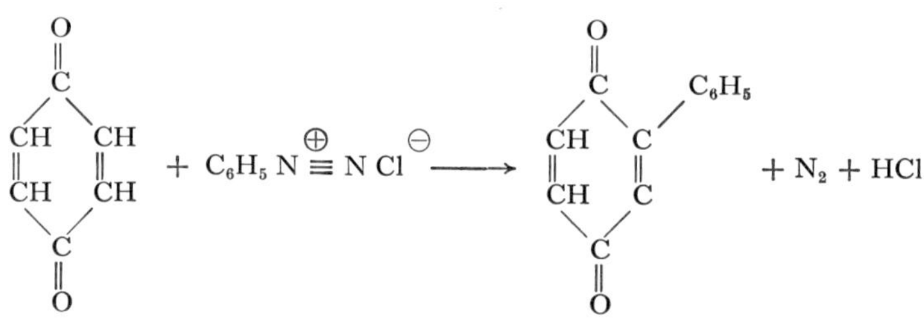

Certain derivatives of quinone, e.g. 2-chloroquinone and 2,5-dichloroquinone are used as bactericides.

2,3,5,6-Tetrachloroquinone (chloranil) is formed by the action of potassium chlorate and hydrochloric acid on many benzenoid compounds. Commercial manufacture is by the chlorination of 2,4,6-trichlorophenol in sulphuric acid monohydrate and chlorosulphonic acid at approx. 90°C.

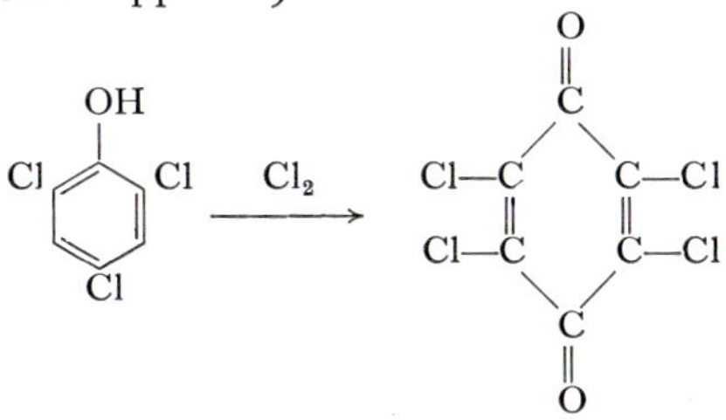

Chloranil is used in the manufacture of dyestuffs in the dioxazine series and for other derivatives of quinone. The dioxazine dye C.I. Direct Blue 106 (51300) is manufactured by condensing chloranil with 2 moles of 4-aminodiphenylamine-2-sulphonic acid, then cyclising with oleum.

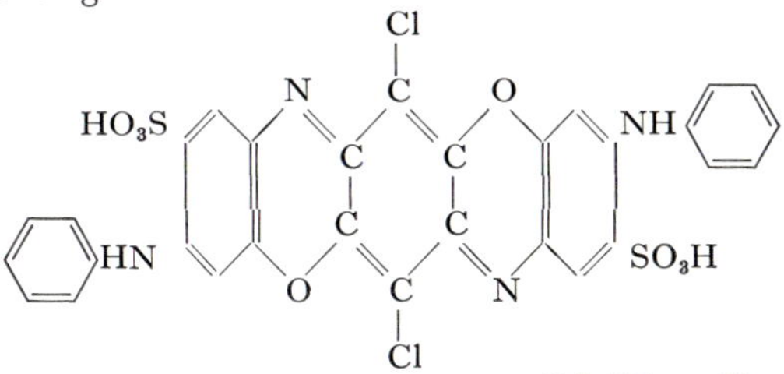

C.I. Direct Blue 106

10.10.3. Benzidine

Benzidine, 4,4′-diaminobiphenyl is a characteristic aromatic amine. It affords salts with mineral acids and undergoes normal acylation and alkylation reactions. Methylation yields tetramethylbenzidine. Halogenation and nitration of the *N*,*N*′-diacetyl derivative occur ortho to the amino group. When benzidine sulphate in sulphuric acid is nitrated, substitution takes place meta- to the amino group. Benzidine is easily tetrazotised, and one diazo group then couples very readily to give a monoazo diazo compound; the activity of the second diazo group is considerably reduced so that the formation of unsymmetrical disazo dyes can be accomplished. Many of these are important direct dyes for cotton. Benzidine and its derivatives are also used in the manufacture of pigments.

Many metallic ions form complexes with benzidine which gives rise to methods for their detection and determination.

10.10.3.1. Manufacture Because of the carcinogenic nature of benzidine, extensive precautions have to be taken today in manufacture. Essentially the manufacture consists in the reduction of nitrobenzene to hydroazobenzene, which is rearranged to benzidine in acid solution. The by-products at the initial reduction stage are azoxybenzene and azobenzene; a small amount of diphenyline (2,4′-diaminobiphenyl is formed at the rearrangement stage.

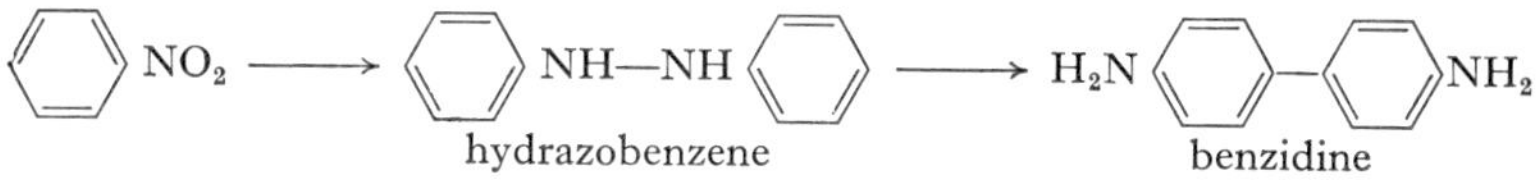

The reduction stage may be carried out by a number of different techniques:

(1) Zinc dust reduction in the presence of sodium hydroxide at 100°C with solvent naphtha as a moderator.

(2) Initial reduction to azobenzene with sodium amalgam, then reduction with zinc dust and sodium hydroxide.

(3) Electrolytic reduction.

(4) Reduction with iron powder and caustic alkali.

The hydrazobenzene formed by the above methods is rearranged in hydrochloric or sulphuric acid and the benzidine so formed is filtered off as the dihydrochloride or sulphate. For dyestuffs manufacture the salt is converted into the tetrazo compound which is regarded as non-carcinogenic, although precautions are still taken in its handling. The tetrazo solution is reasonably stable when stored cold (0°C or lower) and may be transported by road tanker. The free base is formed by treatment of the benzidine salts with sodium carbonate. The mechanism of the benzidine rearrangement has been extensively studied, especially by Ingold[88] and his school. It has been reported[89] that thermal decomposition of substituted hydrazobenzenes can lead to the corresponding anilines and azobenzenes.

10.10.3.2. Benzidine and carcinogenic activity Benzidine shows carcinogenic activity and is known to be a cause of papilloma of the bladder.[90] Other derivatives and related compounds such as ortho- and meta-tolidine, dianisidine and 3,3′-dichlorobenzidine are suspected of some cacinogenic activity. In all cases exposure to these toxic amines is minimised by appropriate plant design using automated closed systems and remote control. The tetrazo compounds derived from the above amines and also their sulphonic acid derivatives are regarded as being without serious hazard.

It is reported that certain plants in Western Europe have recently (March 1971) abandoned the manufacture and use of benzidine.

In the UK use of benzidine and a number of other substances suspected of carcinogenic activity is controlled by the Carcinogenic Substances Regulations 1967.

10.10.3.3. Substituted benzidines A series of substituted benzidines is prepared in an analogous way, making use of the benzidine rearrangement. The tolidines are prepared from the corresponding nitrotoluenes. *o*-Chloronitrobenzene yields 3,3′-dichlorobenzidine and *o*-dianisidine is obtained from *o*-methoxynitrobenzene. *o*-Dianisidine is a component of a number of useful direct blues for cotton and paper. Benzidine-2,2′-disulphonic acid is derived from *m*-nitrobenzene sulphonic acid and is used for a few azo dyes.

3,3′-Dichlorobenzidine finds extensive use in a number of pigments of which Benzidine Yellow, C.I. Pigment Yellow 12, 21090, is of particular importance. Benzidine Yellow is 3,3′dichlorobenzidine tetrazotised and coupled to 2 moles of acetoacetanilide. A more recent chloroderivative is 2,5,2′,5′-tetrachlorobenzidine which is manufactured from 2,5-dichloronitrobenzene, and this gives rise to yellow pigments of improved fastness properties. Some of the Azo condensation pigments

(Cromophtals) of CIBA-GEIGY are condensation products of benzidine derivatives and related diamines and these have outstanding fastness properties.

10.10.4. Phenylhydrazine

Phenylhydrazine forms colourless crystals m.p. 23°C which turn brown on exposure to air. It is basic in character, readily forming salts. It is toxic by skin absorption. Manufacture is by the reduction of a benzene diazonium salt either electrolytically or with a neutral solution of sodium bisulphite.

Phenylhydrazine

In the laboratory, reduction may be effected with stannous chloride. Phenylhydrazine is a powerful reducing agent. On oxidation it gives aniline and ammonia; mild oxidation can give benzene diazonium salts. It forms phenylhydrazones with carbonyl compounds.

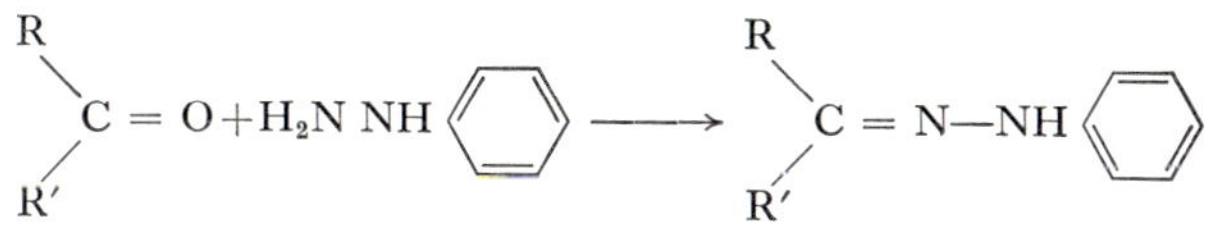

In the Fischer Indole synthesis the phenylhydrazone of an aldehyde, ketone or ketonic acid is converted to the indole by heating with an acid catalyst, e.g. zinc chloride, sulphuric acid or boron trifluoride.

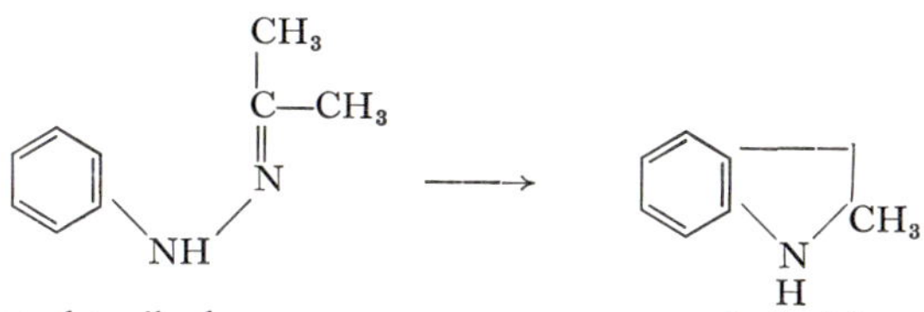

acetone phenylhydrazone 2-methylindole

The reaction between phenylhydrazine and sugars yields an osazone. Both osazones and phenylhydrazones have been used for characterisation purposes. Condensation with β-keto esters yields pyrazolones, e.g. with ethyl acetoacetate:

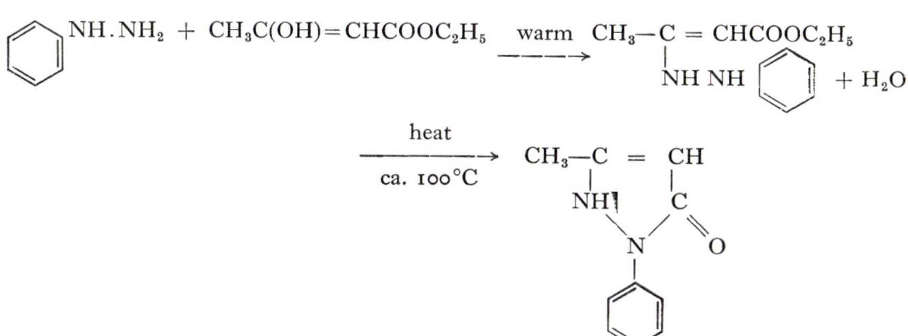

1-Phenyl-3-methyl-5-pyrazolone

10.10.5. Pyrazolones

Pyrazolones are widely used as dyestuffs intermediates in the manufacture of yellow dyes for wool and polyamide fibres, as colour couplers and in the synthesis of pharmaceutical compounds. The general method of manufacture is by the reaction of a β-keto ester with a substituted hydrazone, prepared in situ by sodium sulphite reduction of the appropriate diazonium salt, followed by ring closure on heating. The reaction may be written:

An example of a pyrazolone yellow which is a food colour and a dyestuff for wool is Tartrazine C.I. Acid Yellow 23, 19140.

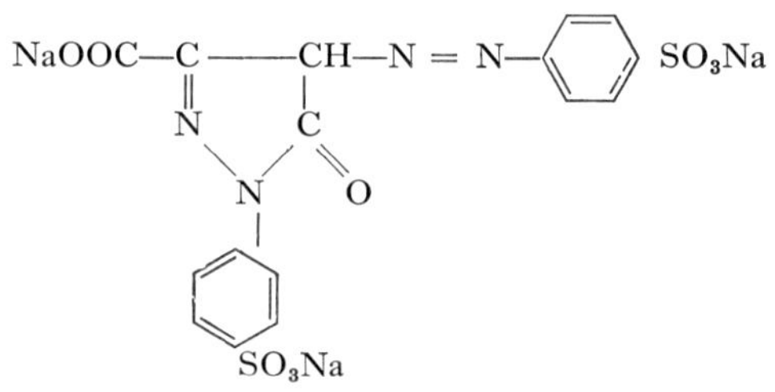

Tartrazine is manufactured by diazotisation of sulphanilic acid and coupling to 1-(4′-sulphophenyl)-5-pyrazolone-3-carboxylic acid. The pyrazolone yellows are characterised by good fastness to light.

The analgesic antipyrine is obtained from 1-phenyl-3-methyl-5-pyrazolone by methylation with dimethyl sulphate in sodium hydroxide and methanol.

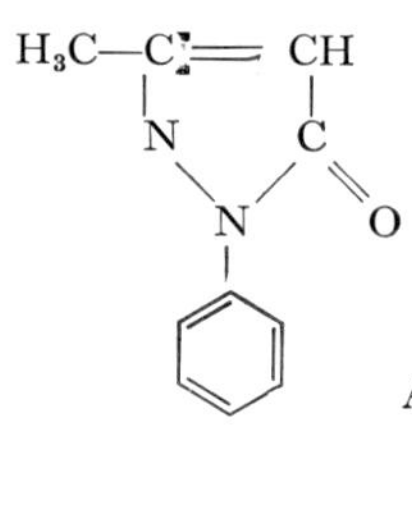

Antipyrine

10.10.6. Arylhydrazines

Unsymmetrical diarylhydrazines may be prepared from the corresponding *N*-nitrosamine by reduction.

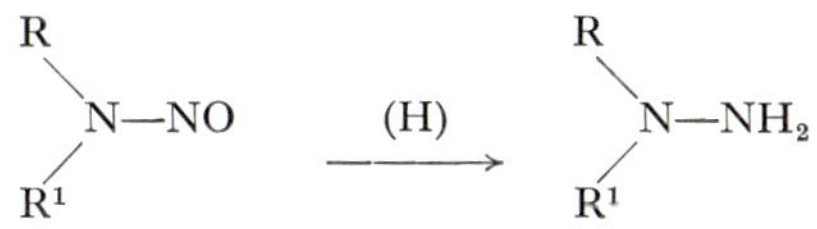

Symmetrical diaryl hydrazines rearrange with acid (see benzidine rearrangement). Tetraaryl hydrazines may be obtained by oxidation of diarylamines with potassium permanganate.

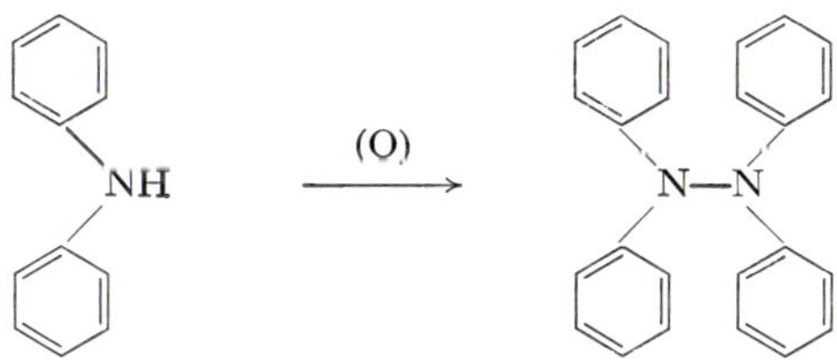

Tetraarylhydrazines are weak bases and give deeply coloured free radicals in solution.

10.10.7. Isocyanates

The industrial preparation of isocyanates is by the reaction of the corresponding amine with phosgene in an organic solvent. Typically a solution of the amine in an aromatic solvent, e.g. xylene, is reacted with a solution of phosgene in the same solvent at a temperature below 60°C. The temperature is gradually raised with further phosgene added to complete the reaction. Side reactions are usually significant.

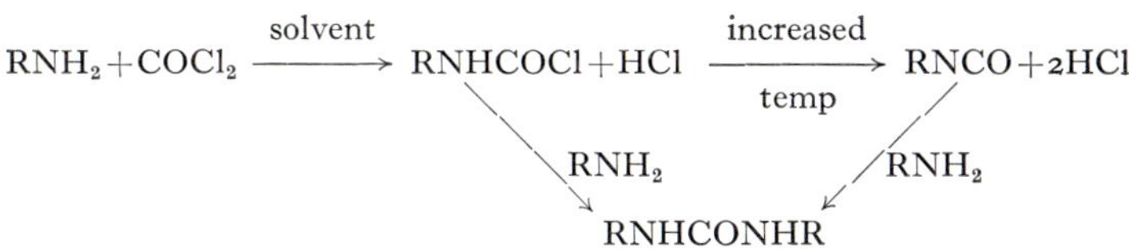

The most important reaction of isocyanates is that with alcohols. With a primary alcohol a urethane is given:

$$RNCO + R'OH \longrightarrow RNH\overset{\overset{\displaystyle O}{\|}}{C}OR'$$

The reaction between a diisocyanate and a dihydric alcohol leads to a straight chain polymer, while polyhydric alcohols give cross linked polymers. The polyurethanes are widely used in foams, coatings and elastomers.

Hydrolysis of an isocyanate with water gives the corresponding urea.

$$2RNCO + H_2O \rightarrow RNHCONHR + CO_2$$

Primary amines give disubstituted ureas; secondary amines react similarly although less readily. Amides react slowly at elevated temperatures to give the acyl urea.

$$RNCO + R'CONH_2 \rightarrow RNHCONHCOR'$$

Carboxylic acids normally yield amides with evolution of CO_2.

$$RNCO + R'COOH \rightarrow RNHCOR' + CO_2$$

The Friedel-Crafts reaction give the amide:

In general, reaction occurs with active hydrogen according to the scheme:

$$RNCO + HY \longrightarrow RNH\overset{\overset{\displaystyle O}{\|}}{C}Y$$

Isocyanates also undergo self addition to form dimers, trimers and polymers. On heating an isocyanate in the presence of a catalyst, a carbodiimide may be formed.

$$RNCO \rightarrow R - N = C = N - R + CO_2$$

Since isocyanates give such a variety of reactions, there is a high probability of side reactions occurring in any system involving their use, especially at elevated temperatures. Choice of catalyst and reaction conditions are particularly important in the commercial manufacture of polyurethanes.

Isocyanates are skin and eye irritants and the vapours are particularly irritating to the respiratory tract. They can give rise to sensitisation.

The following benzenoid isocyanates find commercial use:

Dianisidine diisocyanate
Phenyl isocyanate
m-Phenylenediamine diisocyanate
2,5-Dichlorophenyldiisocyanate
Diphenylmethane-4,4′-diisocyanate (MDI).

MDI is particularly important in polymer formation and is prepared from 4,4′-diaminodiphenylmethane, which is itself manufactured by heating together aniline hydrochloride and formaldehyde.

10.10.8. Production Figures for Cyclic Chemicals

Some idea of the relative importance of synthetic organic chemicals manufactured from benzene derivatives and especially from nitrobenzene and aniline can be obtained from the production and sales figures produced by the United States Tariff Commission.

The production of nitrobenzene and aniline has shown a steady increase in the USA during the years 1960–9.

Production in Tons

Year	*1960*	*1963*	*1964*	*1965*	*1966*	*1967*	*1968*	*1969*
Nitrobenzene	73,622	99,778	108,507	127,161	148,259	157,715	180,503	219,748
Aniline	54,541	70,147	77,070	88,699	108,411	102,311	119,491	151,526

The following production figures are taken from the United States Tariff Commission Production and Sales report 1967, and from the preliminary report 1969.

	1969 *production* *Tons*
Cyclohexylamine	5,250*
o-Chloronitrobenzene	15,520*
p-Chloronitrobenzene	46,040*
N,*N*-Dimethylaniline	5,810†
Diphenylamine	15,760
Hydroquinone	5,600
Rubber antioxidants	68,770
Vulcanisation accelerators (cyclic)	40,210

* 1967 production.
† 1969 sales.

REFERENCES

[1] Coombes, Moodie and Schofield, *J.C.S.*, 1968, 800.
[2] Kuhn, S. J. and Olah, G. A., *J.A.C.S.*, 1961, **83**, 4564.
[3] Olah, G. A. and Kuhn, S. J., *J.A.C.S.*, 1962, **84**, 3684.
[4] Christy, P. H., Ridd J. H. and Stears, N. D., *J.C.S.*, 1970, 797.
[5] Albright and Hanson, *Loss Prev.*, 1969, **3**, 26; Ventrone, *Loss Prev.*, 1969, **3**, 38.
[6] Fritz, *Loss Prev.*, 1969, **3**, 41.
[7] *Quarterly Safety Summary* ((MCA Case History No. 1482), 1968, **39**, 155, p. 24.
[8] Dolah, Van, *Loss Prev.*, 1969, **3**, 32.
[9] Albright, *Chem. Eng.*, 1966, **73**, 161.
[10] Hanson, Marsland and Wilson, *Chem. and Ind.*, 1966, 675; Albright, *Chem. Eng.*, 25 April 1966, p. 169.
[11] Staples, Procopio and Su, *Chem. Eng.*, 16 November 1970, p. 113.
[12] Data Sheet, ICI.
[13] Zetkin, Panchenkov, Zakharov and Kolesnikov, *Zh. Prikl. Khim.*, 1968, **41** (2), 328.
[14] Standard Oil Com., USP 3,293,295.
[15] Fields and Meyerson, *J.A.C.S.*, 1967, **89**, 724.
[16] Montecatini, Neth. Patent 66, 04118.
[17] *Chem. Abs.*, 1967, **66**, 37540.
[18] Hardy and Bennett, *Tetrahedron Lett.*, 1967 (1), 961.
[19] ICI, Neth. Patent 66, 09601.
[20] Mittal and Mahajan, *Indian Chem. Mfr.*, 1969, **7**, 27.
[21] Dikshit and Chivate, *Chem. Eng. Sci.*, 1970, **25**, 311.
[22] Coombes, Crout, Hoggett, Moodie and Schofield, *J.C.S.*, 1970, 347.
[23] Gitis, Ivanov, Kaminskaya, Udris and Nikdaev, *Nauch. Tr. Tul. Gos. Pedagog. Inst.*, 1968, No. 1, 109.
[24] Greenfield, *Trans. N.Y. Acad. Sci.*, 1969, **31**, 1038.
[25] Romanian P. 50,538 (1968).
[26] Halcon International Inc., Belg.P 744,936 and 740,758.
[27] Zollinger, *Azo and Diazo Chemistry*, tr. Nursten (Interscience Publisher Inc., New York, 1963), p. 137.

[28] Hartshorn and Ridd, *J.C.S.*, 1968, 1063.
[29] ICI, B. P. 630,859; BASF, DRP 805,518.
[30] Evans and Milligan, *Aust. J. Chem.*, 1967, **20**, 185.
[31] Forrest, Dauphinee and Miles, *Can. J. Chem.*, 1969, **47**, 2121.
[32] Amin, *Colourage*, 1968, 15 (4), 49.
[33] Lynch, Chen and Wigfield, *Can. J. Chem.*, 1968, **46**, 1141.
[34] Barnett and O'Connor, *Chem. and Ind.*, 1970, 1172.
[35] CIBA, Neth. Patent 66, 04132.
[36] Pol.P 57,927; through *Chem. Abs.*, 1970, 72, 31448.
[37] Montecataini Edison, BP 1,208,118.
[38] ICIANZ, Australian P 253,616.
[39] Spillane and Scott, *Tetrahedron*, 1968, **24**, 5011.
[40] Hashimoto and Sunamoto, *Bull. Chem. Soc. Jap.*, 1968, **41**, 499.
[41] Vrba and Allan, *Tetrahedron Letters*, 1968, No. 43, 4507.
[42] Marshall, Sears and Smith, *J.C.S.*, 1970, 2144.
[43] Universal Oil Products Co., USP 3,291,831.
[44] USSRP 224,502; through *Chem. Abs.*, 1969, 70, 19775.
[45] Vincent, *Chem. Eng. Progr.*, 1971, 67 (6), 51.
[46] USSRP 192,775; through *Chem. Abs.*, 1968, 69, 26982.
[47] ICI, BP 1,220,448.
[48] *Quarterly Safety Summary*, 39, 1968, 156, p. 36.
[49] *Quarterly Safety Summary*, 41, 1970, 163, p. 28.
[50] *Chem. Abs.*, 1968, **68**, 104,659.
[51] *Chem. Abs.*, 1970, **73**, 45026: 45027: 45028.
[52] *Chem. Abs.*, 1970, **73**, 55729.
[53] USSRP 235,039; through *Chem. Abs.*, 1970, 72, 43096.
[54] Godfrey and Thrift, *J.C.S.*, 1967, 400.
[55] Koppers Co. Inc., USP 3,290,377.
[56] E.Ger.P 64,061; through *Chem. Abs.*, 1969, **70**, 77557.
[57] *Chem. Abs.*, 1969, **71**, 90970.
[58] Universal Oil Products Co., FP 1,503,153.
[59] E. I. du Pont de Nemours and Co., USP 3,291,832.
[60] US Tariff Commission, *US Production and Sales of Dyes*, 1969, (preliminary).
[61] Abrahart, *Dyes and their Intermediates* (Pergamon, Oxford, 1968).
[62] Allen, *Colour Chemistry* (Nelson, London, 1971).
[63] Venkataraman, *The Chemistry of Synthetic Dyes* (Academic Press, New York), Vols. I and II 1952, Vol. III 1970, Vol. IV 1971, Vols. V and VI in preparation.
[64] Oberster, Farhat, Kibler, Cook, Setoodeh and Smith, *Can. J. Chem.*, 1967, **45**, 195.
[65] Quaker Oats Co., FP 1,446,554.
[66] Tallec, *Ann. Chim.*, 1968, **3**, 155.
[67] Dunstan, *Chemistry in Britain*, 1971, **7**, 62.
[68] BIOS Final Report, 1153, p. 261.
[69] *Organic Syntheses* (John Wiley and Sons Inc., New York, 1943), Coll. Vol. II, p. 501.
[70] Kroupa and Matrka, *Collect. Czech. Chem. Commun.*, 1970, **35**, 2187.
[71] *Organic Syntheses* (John Wiley and Sons Inc., New York, 1963), Coll. Vol. IV.
[72] *Condensed Pyridazine and Pyrazine Rings*, ed. Simpson (Interscience Publishers Inc., New York, 1953).
[73] *Phenazines*, eds. Swan and Felton (Interscience Publishers Inc., New York, 1957).
[74] Mobay Chemical Co., USP 3,414,619.
[75] BASF, FP 1,564,232.
[76] Zollinger, *Azo and Diazo Chemistry*, tr. Nursten (Interscience Publishers Inc., New York, 1961), p. 259.
[77] Koppers Co. Inc., S. African P 68 07,318.
[78] Zollinger, *Azo and Diazo Chemistry*, tr. Nursten (Interscience Publishers Inc., New York, 1963), p. 263.
[79] Corbett, *Proc. Joint Conf. Cosmet. Sci.*, 1968, 159.

[80] Tong and Glesmann, *J.A.C.S.*, 1968, **90**, 5164.
[81] Hünig, *J. Chem. Education*, 1969, **46**, 734.
[82] Allied Chemical Corp., USP 3,086,063.
[83] *Chem. Abs.*, 1970, **73**, 87562.
[84] Sumitomo Chemical Co. Ltd., Jap.P 69 02, 701.
[85] Universal Oil Products Co., USP 3,409,670 (1968).
[86] Matrka, Sagner, Chmatal and Zverina, *Chem. Prum.*, 1970, **20**, 375.
[87] BASF, Neth. Patent 66, 02880.
[88] Banthorpe, Hughes and Ingold, *J.C.S.*, 1964, 2864, Banthorpe, Ingold and O'Sullivan, *J.C.S.*, 1968, 624.
[89] *Chem. Abstracts*, 1970, **73**, 87554.
[90] Scott, *Carcinogenic and chronic Toxic Hazards of Aromatic Amines* (Elsevier, 1962).

CHAPTER 11

MALEIC ANHYDRIDE

by R. O. Kerr

11.1 INTRODUCTION

About 90 per cent of the maleic anhydride at present produced is derived from benzene oxidation. Other raw materials, butene for example, have been tried but far so have gained little importance. The future relative abundance of benzene and butenes as well as improvements in butene oxidation technology will help determine whether butenes can account for a greater percentage of the feedstock. The growth of maleic anhydride consumption, however, will probably maintain a continuous demand for benzene as a raw material even if butenes gain a third of the market. About 600 to 800 million pounds of benzene is presently being used to produce 500 to 600 million pounds of maleic anhydride, world wide.

Maleic anhydride has beeen known for over 150 years. It was isolated byVauquelin in 1817 from malic acid by dehydrative distillation.

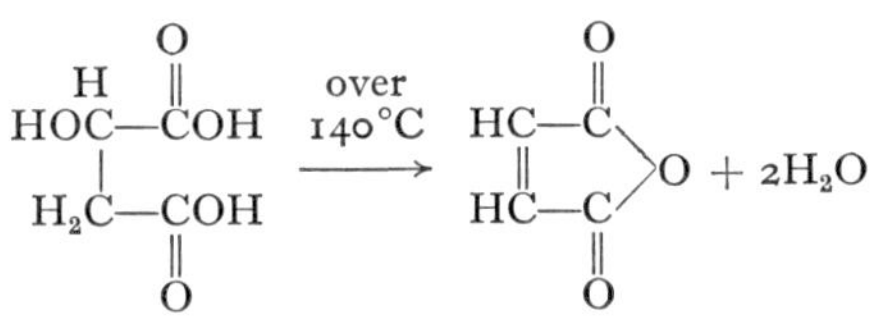

Kempf in 1905 oxidised benzoquinone to maleic acid.

J. M. Weiss and L. R. Downs, while attempting to convert benzene to phenol, catalytically, found that maleic acid was formed as a by-product in yields of 4 per cent. They further developed the system and were granted several patents beginning in 1918.[1, 2, 3, 4, 5, 6, 7] This inital by-product from the catalytic oxidation of benzene has grown considerably since 1930 to over 500 million pounds per year annually. The techniques employed by Weiss and Downs are still being used in a slightly modified form.

Other potential routes to maleic anhydride (and maleic acid) are continuously being investigated. By-product maleic anhydride from phthalic anhydride production might account for 6 per cent of the total maleic anhydride produced while 3 to 4 per cent more is obtained from oxidation of butenes.

This chapter also covers the geometric isomer of maleic acid, fumaric acid.

11.2 PRODUCTION OF MALEIC ANHYDRIDE

11.2.1. Oxidation of Benzene

The benzene-maleic anhydride catalyst has been developed over the past fifty years with numerous papers and patents on the subject being published. The yield of maleic anhydride has been increased to 67–72 mole per cent; that is, for every 100 moles of benzene passed over the catalyst, 70 moles are converted to maleic anhydride. Three to eight moles remain unreacted, while the benzene completely oxidised to $CO+CO_2$ accounts for 15 to 25 moles of benzene. Minor amounts of phenols, carbonyls and carboxylic acids are formed (3–5 mole per cent):

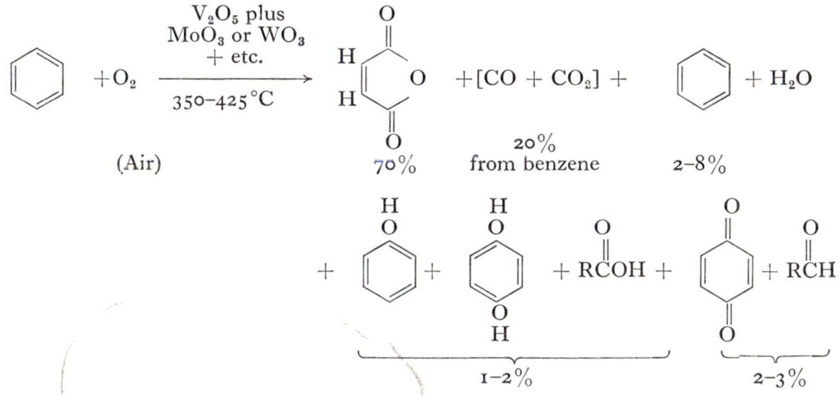

The reaction is generally carried out in fixed bed reactors. The heat released in the oxidation is removed by heat exchanging fluids, such as the eutectic salt, $NaNO_2$–KNO_3. Mercury has been used but, besides being dense, is very poisonous and should be avoided.

The benzene concentration varies from 1 to 1·4 mole per cent in air while its throughput ranges usually from 60 to 130 g C_6H_6 per litre catalyst hour. The success of the naphthalene-phthalic anhydride fluid bed process has raised some interest in applying thus to maleic anhydride. Benzene-maleic anhydride fluid bed systems would have to meet the high yield of the fixed bed and also have an acceptable attrition rate. Use of a fluid bed when butenes are the raw material has been reported. The greater simplicity of the butene catalyst and a cheaper butylene stream, based on fuel value, might make such a system practical in a favourable situation.

The present catalyst for the oxidation of benzene to maleic anhydride is reported to contain $V_2O_5+MoO_3$. This basic composition can give yields of 45 to 55 mole per cent. Additives to the $V_2O_5+MoO_3$ catalyst bring the yield of maleic up to 70 mole per cent. A limited survey of recent literature on high yielding catalysts is given in Table XI.1. The optimum additives then seem to be Ag^+, Co^{+3}, Na^+, Li^+ or TiO_2, WO_3, and P_2O_5.

Benzene is converted to maleic anhydride in yields of 68 to 72 mole per cent or 85 to 91 weight per cent. These are initial yields. The maleic anhydride yield-time trend is related to catalyst composition, benzene throughput and concentrations, extent of benzene conversion, and reactor-heat-removal efficiency. Catalyst life varies from 12 to 36 months under normal conditions with no malfunctions. The catalyst is removed after a 15 to 30 per cent decrease in yield or if a pressure build-up reduces the throughput by 10 to 20 per cent. Catalyst turnaround is often carried out in conjunction with other maintenance repairs.

The yield of maleic from benzene should continue to improve to a level of 73 to 78 mole per cent or 92 to 98 weight per cent. The large volumes of maleic anhydride now being produced require a highly efficient catalyst not only to conserve benzene but to reduce the by-products produced. The maleic anhydride quality produced by the catalyst must be such that the overall yields of 92 to 96 per cent can be maintained in the recovery section.

TABLE XI.1

Benzene-Maleic Anhydride Catalysts

Country/Company	*Ref.*	*Catalyst*	*Yield Mole %*	*Observations*
USA				
Am. Cyanamid	8	V_2O_5 MoO_3 Ag Ce	60–65	
Allied Chem.	9	V_2O_5 MoO_3 Ag_3PO_4 UO_2	70–73	410–460°C On Alundum
Chempatents	10	V_2O_5 MoO_3 Co_2O_3 Li_2O CaO	63–67	On Alundum
Hommel	11	V_2O_5 SnO_2 TiO_2 ZrO	45–55	400–500°C
Monsanto	12	V_2O_5 MoO_3 TiO_2	55–62	Fixed or Fluid Bed
Pittsburgh Chem.	13	V_2O_5 MoO_3	60	
PPG	14	V_2O_5 MoO_3 Co_2O_3 NaCl	65	380–420° On Alundum
Scientific Design	15	V_2O_5 MoO_3 Ni_2O_7 NaCl P_2O_5	68–71	360–400° On Alundum
German				
Badische Anilin	16	V_2O_5 MoO_3 TiO_2 P_2O_5	70	
	17	V_2O_5 WO_3 TiO_2 P_2O_5	70	380–420° ·6–1% C_6H_6
French				
Saint-Gobain	18	V_2O_5 MoO_3 SnO_2 TiO_2 CeO_2 UO_3	64	1·3% C_6H_6
Russian				
I. I. Ioffe *et al*	19	V_2O_5 MoO_3 TiO_2 P_2O_5	60	400°
I. A. Moskaleva	20	V_2O_5 MoO_3 Co_2O_3 Na_2O	63–65	On Alundum
Indian				
B. S. Trehan	21	V_2O_5 MoO_3 Li^+ Na^+	70	Fluid Bed, On Quartz
S. K. Bhattacharyya	22	V_2O_5 MoO_3 WO_3 CoO TiO_2 UO_2	65	
Japan				
Oriental High Pressure	23	V_2O_5 MoO_3 TiO_2	55	
Nippon Cat. Chem.	24	V_2O_5 MoO_3 SnO_2	60	
Takeda Chem. Ind.	25	V_2O_5 MoO_3 Ag^+ Na^+ P_2O_5	70	1% C_6H_6 On Alundum

11.2.2. Oxidation of Butenes

Butene having the same number of carbons per molecule as maleic anhydride would seem to be the more logical feed stock for the production of maleic anhydride. Its molecular weight of 56 as compared to 78 for benzene would seemingly give it the advantage along with its usually lower price per pound. The reaction is as follows:

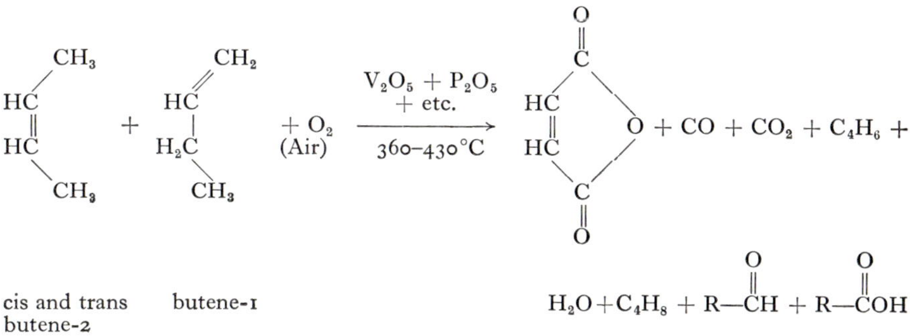

cis and trans butene-2 butene-1

J. H. James[26] was able to oxidise butenes over a molybdenum trioxide catalyst to give a mixture of aldehydes, acids and some maleic anhydride. Union Carbide[27] increased the yield of maleic anhydride using a $Sn(VO_3)_4$ catalyst. Other catalysts patented for use in the oxidation of butenes are listed in Table XI.2.

The usable yields then from butene range from 45 to 52 mole per cent, or 78 to 91 weight per cent. Again, as with benzene-maleic catalyst, these are initial yields. The by-products produced from butene oxidation vary greatly from those of benzene oxidation. While maleic acid dehydration can be carried out in normal fashion, the final purification has to take into account these impurities.

Whether butene or benzene is used as a raw material depends on the relative abundance of these two raw materials and the usually greater capital cost of a butene-maleic anhydride plant. A plant can be readily designed to operate on both feed-stocks with little added costs. The throughput with benzene would be about 50 per cent greater, since both the reactor section and refining section limit the butene capacity. Further work on an improved butene catalyst is required to bring the yield up to 60 mol per cent at high butene throughputs of 90 to 130 g C_4H_8/l cat. hour. The by-product acrolein must also be reduced, otherwise an afterburner must be used.

11.2.3. Maleic Anhydride from Other Sources

Maleic anhydride by-product from phthalic anhydride production is scrubbed out of the stack gases to avoid pollution. This maleic acid can be refined in an adjoining maleic plant or else isomerised to fumaric acid avoiding the capital costs of a dehydrator-refining section.

Butadiene, an intermediate in the oxidation of butene, can be oxidised over the butene-maleic anhydride catalyst in yields of 55 to 62 mole per cent (108 weight per cent). Generally pure butadiene is priced at three to four times the value of butenes, and therefore it would not be available except as a mixed stream with butenes where its presence would give slightly higher yields of maleic anhydride on oxidation.

Recent developments indicate some interest in oxidation of butane to maleic anhydride. The feed costs should run somewhat cheaper than butene. The reported yields range from 30–55 mole per cent maleic anhydride. The butane throughout is low, 60–80% of the benzene factor. The operating temperatures seem to be on the high fringe area.

Numerous papers list other materials as raw materials for maleic anhydride such as furfural, crotonaldehyde, furan, etc., but the cost of maleic anhydride has been reduced to a point which makes the use of these more complex starting materials non-viable.

TABLE XI.2

Butene-Maleic Anhydride Catalyst

Country/Company	*Ref.*	*Catalyst composition*	*Yield mole %*	*Observations*
USA				
Union Carbide	27	$Sn(VO_3)_4+MoO_3+TiO_2$	24	
Std. Oil Development	28	$V_2O_5+MoO_3+SnO+TiO_2$	40	Used Butadiene + Steam
Du Pont	29	$MoO_3+Co_2O_3+B_2O_3$, P_2O_5, V_2O_5	27	Fluid Bed
		$MoO_3+WO_3+TiO_2$	28	
Atlantic Ref.	30	$V_2O_5+P_2O_5$	32	400–400°C Fixed Bed
Petro-Tex	31	$V_2O_5+P_2O_5+$Li or Cu or Cb	45–53	390–460 Fixed Bed
	40	$V_2O_5+P_2O_5$	46–50	
Great Britain				
Distillers	32	$MoO_3+P_2O_5$	30	
ICI	33	$V_2O_5+P_2O_5$	47–48	
French				
Soc. d. Electro	34	$V_2O_5+P_2O_5+Li_3PO_4$	42	
German				
Farb. Hoechst	35	$V_2O_5+P_2O_5$	47	
Badische Anilin	36	$V_2O_5+P_2O_5+TiO_2$	51	380°C
	41	$V_2O_5+WO_3+NiO+P_2O_5+TiO_2$	51·5	410°
Japan				
M. Kitahara	37	$V_2O_5+MoO_3+P_2O_5$	25	
Russia				
Y. Kernos	38	$V_2O_5+P_2O_5+CaO$	47	
T. N. Shakhtakhtinskii	39	$Co_2O_3+MoO_3$	40	Fluid Bed

11.2.4. Maleic Anhydride Isolation and Purification

The maleic anhydride reactor product must be isolated and purified. Usually, the reaction gas stream, after passing through waste heat boilers, is cooled to 55–65°C.

About 40 to 60 per cent of the maleic anhydride can be condensed out of the reactor product stream, reducing the maleic anhydride concentration from a range of 0·6–1·0 mole per cent to 0·35–0·45 per cent. The maleic anhydride remaining in the exit gas from the partial condenser is scrubbed out by water in an absorption tower. Other methods of isolation such as subliming the anhydride or inert solvent absorption have been proposed but have not been adopted generally.

The aqueous maleic acid solution (35–40 weight per cent) is dehydrated by an azeotroping solvent (xylene,[42] anisole or maleic anhydride) to reconvert the acid back to the anhydride. This dehydration can also be carried out in a wiped film evaporator or by a fluid bed of ThO_2 or SiO_2.[43] Nonane can be used as a solvent allowing the maleic anhydride to go overhead effecting a prior purification. Chlorinated solvents such as *o*-dichlorobenzene and tetrachloroethane, while effective, present corrosion problems.

The dehydrator product is combined with the maleic anhydride isolated in the partial condensers, and distilled. A batch distillation is usually carried out to reduce the impurity range to 35–200 p.p.m., giving a freezing point of 52·6 to 52·8°C with a 10 Hazen initial colour. The crude maleic anhydride can be treated in various ways to remove the impurities. Some of the suggested treatments are as follows:

Subjection of the crude to 200°C for 2+ hours.

Addition of the following compounds usually with heat treatment:

(a) 0·1 per cent H_2SO_4 at 100°C
(b) NaOH
(c) Fumaric acid at 200°C
(d) $NaHSO_4$
(e) B_2O_3
(f) H_3PO_4
(*g*) $NaBO_3.4H_2O$.

Generally, however the alkali metals and ammonia should be avoided. Na^+ salts apparently can cause decarboxylation of the anhydride and polymerisation of the acrylic anhydride at high temperatures and pressures. Thus in prudent operations, Na^+ compounds are avoided in maleic plants.

At the present time, 50 to 90 per cent of a plant production can be sold in the molten state. This avoids the costs of solidification, bagging and remelting. Handling of this toxic material is also reduced when shipped molten. The molten product usually must have sufficient thermal stability, since it is kept in the molten state for considerable periods of time at temperatures of 60 to 80°C.

Maleic anhydride and maleic acid are toxic and are irritants to the skin and mucous membranes. However, these chemicals in any form can be safely stored, handled, or used if the physical, chemical, and hazardous properties are fully understood and the necessary precautions, including the use of proper safeguards and personal protective equipment, are observed.

11.3. FUMARIC ACID

Fumaric acid is the trans and more stable form of butenedioic acid:

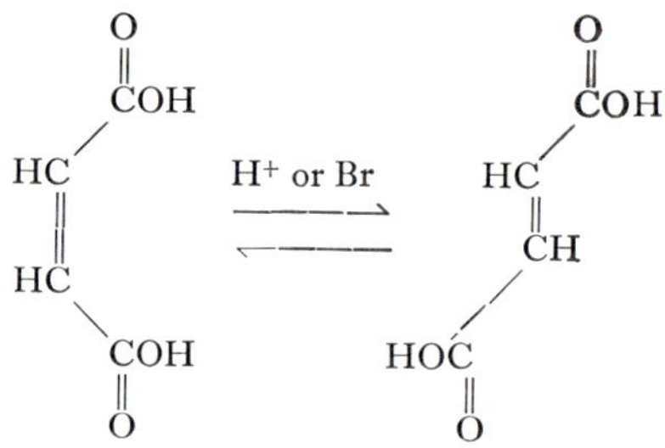

11.3.1. Production

Most of the fumaric acid produced is made from maleic anhydride since maleic anhydride is available in highly purified form by distillation. Using the pure maleic anhydride reduces the amount of isomerisation catalyst required and produces a better grade of fumaric acid. Isolation then of fumaric acid is a simple matter of cooling the reaction mixture, filtration, washing the filter cake, drying and bagging. No further purification is required. Fumaric acid can be produced at times of low maleic anhydride demands since the fumaric acid has a shelf life of several months, while maleic storage either liquid or solid is more limited. Catalysts useful in the isomerisation of maleic to fumaric acid are numerous, e.g.

$KSCN$
NH_2CSNH_2
$KBrO_2$
$NH_4Br + HgAc_2$
$NH_4Br + (NH_4)S_2O_8$[44]
HCl
HNO_3
HCl in acetic acid.

The $NH_4Br + (NH_4)_2S_2O_8$ system can operate at 80 to 100°C in water and the conversion is essentially instantaneous. Maleic acid must generally be added in increments to the isomerisation reactor to control the heat release.

Impure maleic acid from benzene oxidation or by-product from phthalic anhydride production can be used to produce fumaric acid. The impure acid solutions are usually purified by treatment with activated charcoal. The colourless maleic acid solution is then treated in the usual manner. The fumaric acid formed from these impure acid solutions generally does not have the purity found in fumaric acid produced from purified maleic anhydride. Where the crude maleic acid is used for fumaric acid, the maleic reactors can be operated at high conversion levels in order to obtain a light yellowish acid. However, high conversion levels either reduce the catalyst life or benzene throughput. Reactor tubes will also have a shorter life at high hot spot temperatures.

Almost all major producers of maleic anhydride can convert 20 per cent of their maleic anhydride to fumaric acid. The conversion of maleic anhydride to fumaric acid results in an increase in molecular weight from 98 to 116, but solubility factors and handling losses result in a 100 weight per cent yield. The present US capacity of fumaric acid is 95 million pounds divided among six producers as follows:

Allied Chemical	15 million lb/year
Monsanto	25 million lb/year
C. Pfizer	25 million lb/year
Petro-Tex	10 million lb/year
US Steel	10 million lb/year
Tenneco	10 million lb/year

With a greater portion of maleic anhydride being shipped in the molten state where the two isomers are equivalent, maleic anhydride is favoured due to its lower costs. Fumaric acid will be used only where it cannot be displaced by maleic anhydride, as in food acidulents, special unsaturated polyesters, etc.

11.4. PROPERTIES OF MALEIC ANHYDRIDE, MALEIC ACID AND FUMARIC ACID

Maleic acid, maleic anhydride and fumaric acid being polyfunctional molecules react as dibasic acids and as an activated carbon-carbon double bond. This cis structure of maleic acid allows intra-molecular interaction of the acid group accounting for the high acidity of the maleic acid and the ease of formation of the anhydride. The stability of this cyclic anhydride can be observed by its stability over the V_2O_5–MoO_3 catalyst at 410–450°C. Fumaric acid having the trans configuration behaves like a normal dicarboxylic acid having isolated acid functions; however, there is a conjugative effect.

Maleic acid can be determined by titration, plotting ml of NaOH added vs. pH. Two end points are indicated, one at pH 3·7 and the second at pH 9·5. Maleic acid and anhydride can also be determined by gaseous chromatographic analysis using teflon powder packing (du Pont No. 6, 20–40 mesh) with 710 Dow Corning Silicone fluid. The 8–16 ft column is operated at 150 to 180°C. Maleic acid dehydrates to the anhydride so both water and anhydride are observed. Several impurities in maleic anhydride can be determined by this method. Commercial maleic anhydride generally shows impurities in the range of 35 to 200 p.p.m. by chromatographic peak analysis excluding any water.

Both maleic and fumaric acids react with bases to form salts. The monosodium salts are rather insoluble while the disodium salts have greater solubility. The calcium salts have low solubilities by comparison.
These solubilities are as follows:

	Solubilities in g/100 g H_2O	
	25°C	*100°C*
Sodium acid maleate	6·7 g	288 g
Sodium acid fumarate	6·9	30
Sodium maleate· 1/2 H_2O	96	
Sodium fumarate	23	
Calcium maleate·H_2O	2·5	
Calcium fumarate·$3H_2O$	1·5	

Some of the properties of maleic anhydride, maleic acid and fumaric acid are listed in Table XI.3 which follows:

TABLE XI.3

Properties of Maleic Anhydride, Maleic Acid and Fumaric Acid

Property	*Maleic anhydride*	*Maleic acid*	*Fumaric acid*
Structure formula	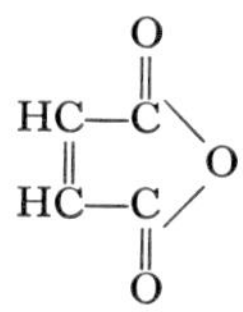	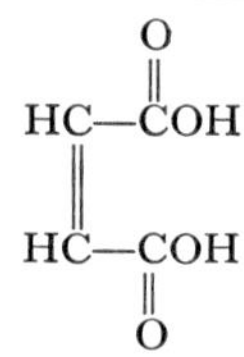	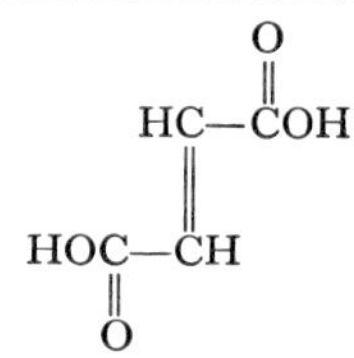
Emperial formula	$C_4H_2O_3$	$C_4H_4O_4$	$C_4H_4O_4$
Molecular weight	98·06	116·07	116·07
Melting point	52·85 °C	130·5 °C	284–300 °C
Decomposition point		138 °C–H_2O	Over 200 °C
Sp. gr. 20/20 °C solid	1·48	1·59	1·635
Boiling point	210 °C	—	290 °C
Vapour pressure	2 mm at 50 °C 25 mm at 100 °C 140 mm at 150 °C		
Ionisation contents			
25 °C K_1		$1·0 \times 10^{-2}$	$9·3 \times 10^{-4}$
K_2		$5·5 \times 10^{-7}$	$3·4 \times 10^{-5}$
Heat of formation, solid kcal/mole	−112·2	−188·9	−193·8
Heat of combustion, solid kcal/mole	−334	−324·5	−319
Specific heat—Solid	0·29		
Liquid	0·40		
Heat of vaporisation, kcal/mole	13·1		
Heat of fusion, kcal/mole	3·26		
Heat of hydrolysis, kcal/mole	−8·33		
Heat of neutralisation, kcal/mole	30·3		
Solubilities (g/100 ml at 25 °C)			
Toluene	23·4		
o-Xylene	17	·01	·03
Benzene	44	·024	·003
Acetone	178	26	1·7
Ethanol	Reacts	70	5·5
H_2O 25 °C	Reacts	78·8	0·7
100 °C		393	9

Maleic anhydride, maleic acid and fumaric acid can enter into a multitude of chemical reactions. Because these are cis and trans isomers of butenedioic acid, their reactions are similar in nature except where the steric configuration leads to a special reaction. The diacids react with alcohols to form diesters or with glycols to produce polyesters:

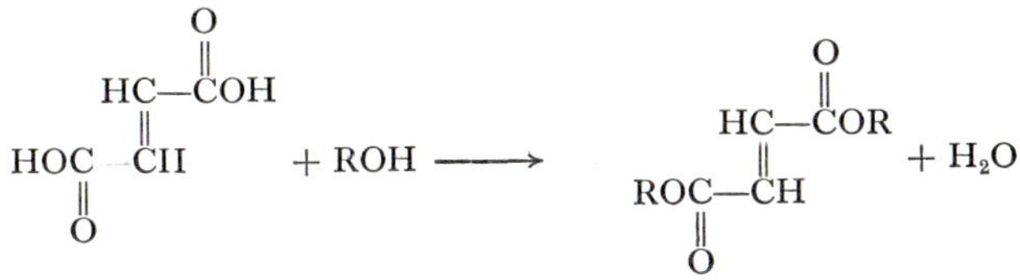

cis or trans

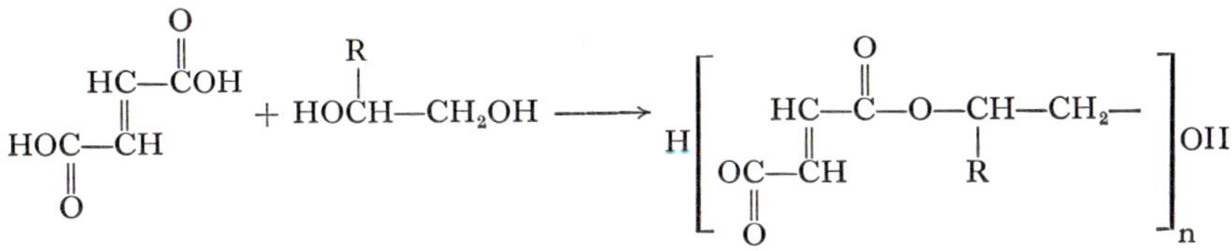

The acid groups react with amines to form the amide. Maleic acid reacts with hydrazine to form maleic hydrazide.

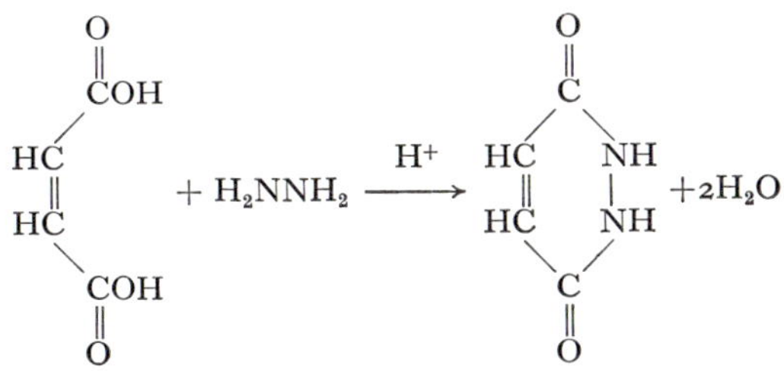

Fumaric acid forms the dichloride with thionyl chloride. Maleic anhydride or fumaroyl chloride react in the presence of a Friedel-Crafts catalyst with aromatic hydrocarbons to form the aryl ketones.

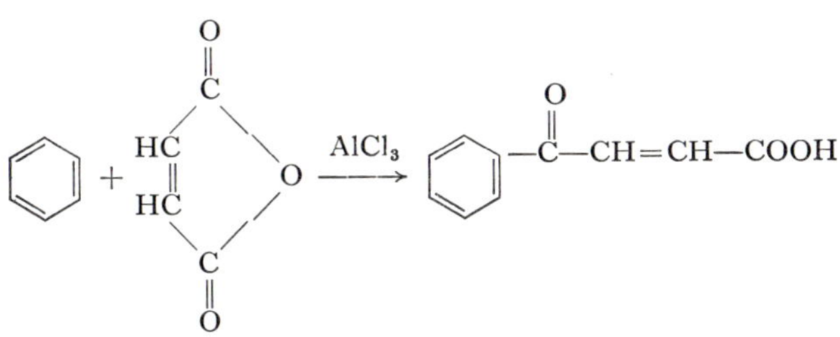

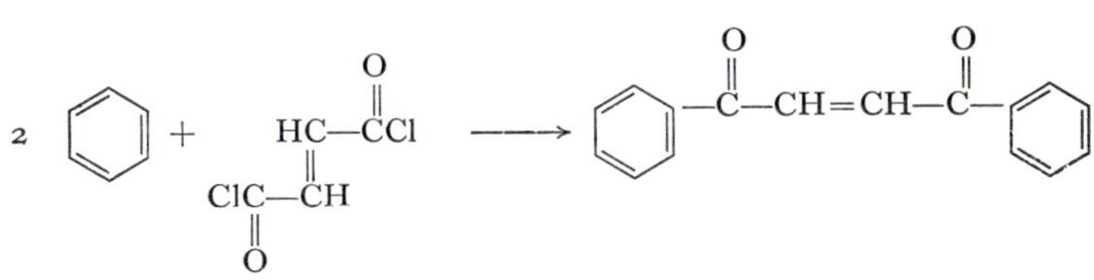

The activated carbon–carbon double bond can react to give the epoxide, dihalide mono-halide, the hydrated malic acid and hydrogenerated to succinic acid.

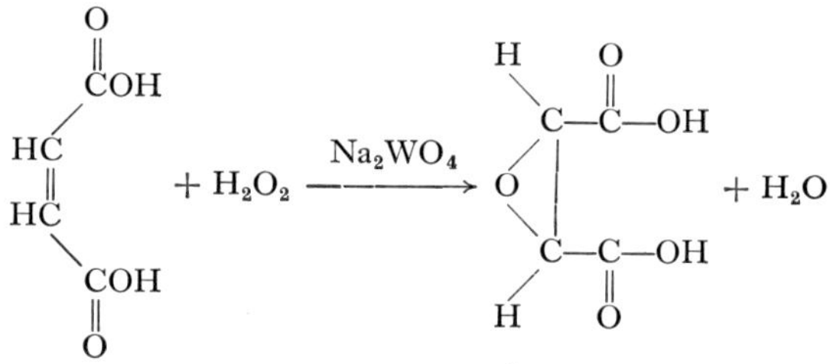

cis-epoxysuccinic acid

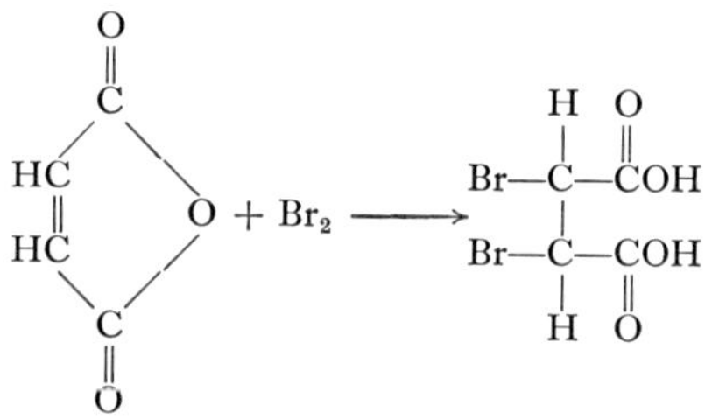

dibromosuccinic acid

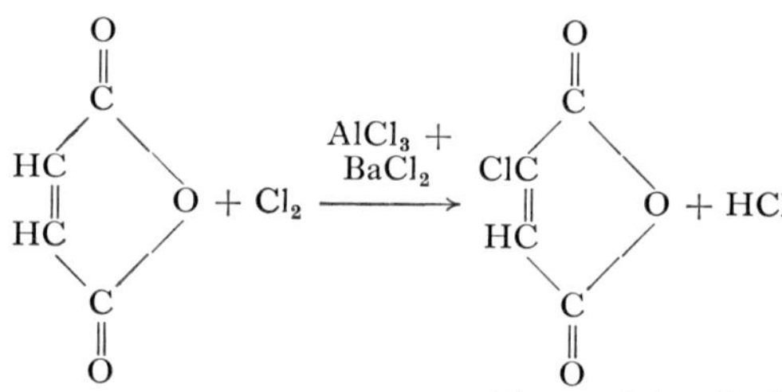

chloro maleic anhydride

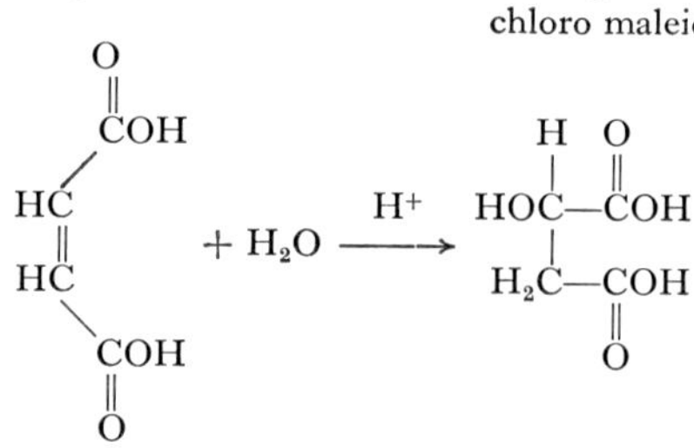

malic acid

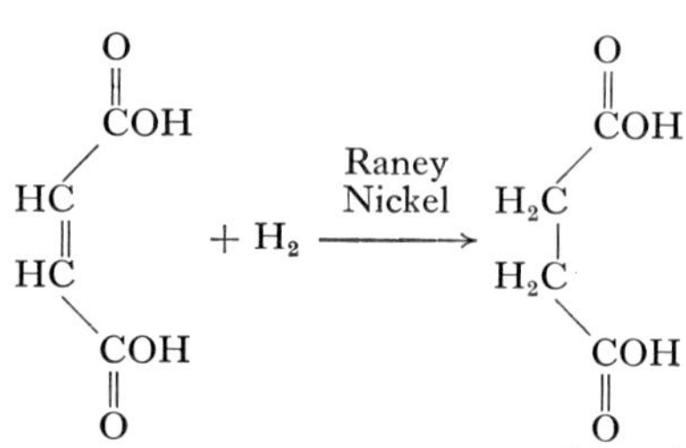

succinic acid

Olefins react with maleic anhydride at 200°C to give an allyl succinic anhydride.

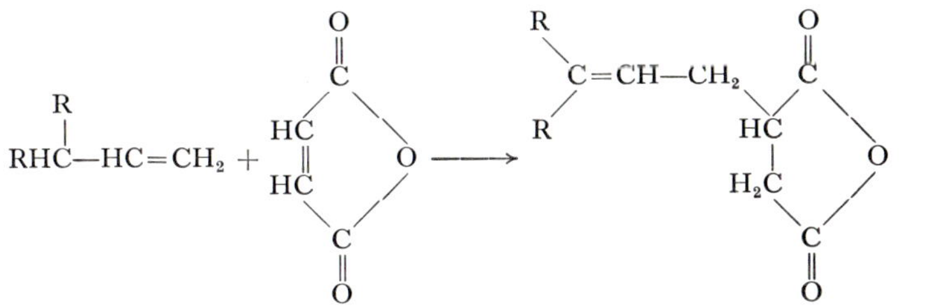

Conjugated diolefins react with maleic to give the Diels-Alder addition compound.

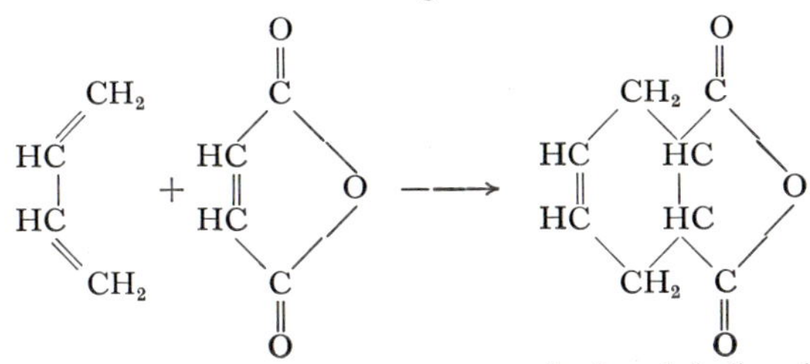

tetra-hydro-phthalic anhydride

The olefin and conjugated diolefin reactions are involved in the reaction of maleic anhydride with unsaturated oils and rosins in making alkyd resins.

Maleic anhydride can be co-polymerised with other vinyl compounds such as styrene, ethylene, methyl vinyl ether, etc. Maleic anhydride dimerises to 1, 2, 3, 4 cyclobutane tetracarboxylic dianhydride in presence of U.V. light. Homopolymerisation of maleic anhydride has been developed recently using free radical initiators like acetyl peroxide at 60 to 80°C in *o*-dichlorobenzene. No large usage for these homopolymers has however been developed. The rather extreme conditions lead to a dark product which requires purification.

Thus the many reactions of maleic anhydride, maleic acid and fumaric acid lead to many commerical applications for these compounds. As old usages diminish in volume, new ones increase the total demand.

11.5. STATISTICS—PRODUCTION AND CONSUMPTION

11.5.1. Production Capacities

The production of maleic anhydride has almost doubled every five years since 1940:

	Production (USA)
1940	4·5
1945	7·9
1950	16
1955	44
1960	90
1965	123
1970	215

During 1971 the USA production should be between 225-235 million pounds. The decrease in growth rate in 1971 was due to the changing economy. The worldwide 1971 production should be 600 million pounds. The recent (1969–70) shortage of maleic anhydride due to the increase in production of unsaturated polyesters, acidulents, lubricants and copolymers brought about large plant expansions which should give enough capacity to last until 1974. The 1971 capacity (USA) was 297 million pounds while the 1972 expansions should make a total of 362 million pounds available. The following lists indicate the USA plants followed by an estimation of world wide capacity.

	Capacity million lb/year 1971	*1972*
USA		
Allied Chemical, West Va.	20	20
Koppers (Am. Cyanamid) Pa.	36	36
Monsanto, Mo	100	105
Petro-Tex, Texas	50	50
Reichhold, N.J. and Illinois	26	86
Tenneco, N.J.	25	25
US Steel (Pitts. Coke), Pa.	40	40
USA	297	362
England	60	86
France	50	50
Belgium	28	28
Italy	117	128
Canada	14	14
Brazil	1	1
Japan	120	220
Austria	22	22
Spain	20	20
West Germany	144	156
Total	576	725
Eastern Europe (estimated)	100	100
Worldwide Total	973	1,187

11.5.2. Maleic Anhydride Consumption

In 1971, USA consumption by application was substantially as follows:

Unsaturated polyesters	48%
Fumaric acid	17%
Agricultural chemicals	10%
Oil additives	7%
Coatings	5%
Plasticisers	5%
Other	6%
Export	2%
	100%

11.5.3. Fumaric Acid Consumption

The production of fumaric acid has only doubled from 1960 to 1970 from 23 million pounds to 45 million pounds in 1970. It had been at the 46 to 48 million pound level since 1966. The distribution of the usage is as follows, given in per cent of total:

	1962	*1968*
Food acidulants	15	35
Synthetic hard resins	40	26
Paper size	10	15
Alkyd resins	20	3
Dialkyl esters		4
Miscellaneous	15	17
	100	100

In the past ten years the price of fumaric acid has increased relative to maleic anhydride now that the major portion of maleic anhydride is being shipped in the molten form. The amount of fumaric acid used in alkyd resins has decreased due to the increased usage of latex paints. Because of Government clearance, fumaric acid is being used in larger volumes as a food acidulent.

11.6. SOME MAJOR APPLICATIONS

11.6.1. Unsaturated Polyesters Resins

11.6.1.1. Introduction and outlook Roughly 50 per cent of the total maleic anhydride produced is consumed in the production of unsaturated polyester resins. Not only is maleic anhydride used in these resins but also fumaric acid, tetrahydrophthalic anhydride, chlorendic anhydride all of which are derived from maleic anhydride.

Polyesters were produced as early as the 1840s. A phthalic anhydride glycerine polyester was manufactured as a paint resin in 1916. Carothers, at Du Pont, further developed the polyesters of ethylene glycol and maleic or fumaric acid. Unsaturated polyesters were fully developed in the early 1940s by a series of advancements. Ellis[45] incorporated unsaturated monomers with these unsaturated polyesters allowing a faster cure rate. In 1942, U S Rubber Company discovered the compatibility of these unsaturated polyesters with fibreglass. These reinforced resins were used as radar domes during World War II. The first fibreglass-unsaturated polyester resin boat was produced about 1946. By 1954, 50 million pounds of unsaturated polyester resins were being produced. The rate has increased yearly thereafter until 1971.:

Year	*Pounds of unsaturated polyester resins produced in the U S A million lb/year*
1954	50
1959	180
1964	230
1969	595
1970	590
1971	660

About 25 per cent of the unsaturated polyester resins produced is used in boat hulls. Their biggest asset is that these fibreglass boats can outlive two engines with absolutely no upkeep except for a patch or two after hitting a rock or the dock. New construction features in the 1970–1 boats have eliminated the wood trusses and beams with fibreglass members.

Cars and trucks account for 14 per cent of the market while construction uses about 12 per cent. New uses like gasoline storage tanks, bathroom units, help continue the increase. For long term growth fire retardant resins need to become increasingly more available. By 1975–6 it is estimated that industry will be using 1,000–1,200 million pounds of unsaturated polyester resins if new developments continue to maintain its competitiveness over reinforced thermoplastics.

11.6.1.2. Unsaturated polyester resins compositions Unsaturated polyester resins contain an active carbon-carbon double bond in their linear polyester chain. Generally maleic anhydride and phthalic anhydride in a mole ratio to 1 to 1–3 are condensed with propylene glycol and/or epoxide. Phthalic anhydride and maleic anhydride are reacted with a slight excess of glycol or epoxide at 100–150°C over a two or three hour period. The mixture is slowly heated to 200°C for four hours under nitrogen or CO_2. Water is taken off to shift the equilibrium. The polyester is then cooled to 150°, and an inhibitor such as hydroquinone in dibutyl phthalate is added. 25 to 60 per cent by weight of styrene is added to the unsaturated polyester resin. The resin is then ready to be crosslinked by free radical initiators such as benzoyl peroxide, methyl ethyl ketone peroxide, etc. This final step is carried out between room temperature and 70°C.

The esterification step can be conducted under vacuum to reduce the temperature. Additives and inhibitors can be used to increase the rates and improve the colour.

Even though maleic anhydride is used, about 60 to 90 per cent is isomerised to the fumarate polyester by the high temperatures and catalyst. The fumarate ester gives greater hardness and a higher heat distortion temperature. The trans structure reacts with the styrene to give a greater degree of crosslinking units, i.e. the styrene-free radical tends to react with the unsaturated fumarate which in turn reacts with another styrene molecule, giving the styrene linking bridge between the polyester chain. Maleates in the polyester gives the resin greater impact strength by being less reactive toward the styrene which tends to fewer crosslinking units and a greater degree of homo-polymerisation of the styrene. Thus one can control the type of resin produced by the reaction conditions during the esterification or by using different initial amounts of fumaric acid.

Isophthalic acid can be used in place of phthalic acid to give greater strength and a higher heat distortion temperature. It also improves the chemical resistance. Adipic acid can be used to give a more flexible polyester. Chlorendic anhydride and tetrabromo phthalic anhydride are used to give a fire retardant polyester.

Diethylene and dipropylene glycols are added for greater flexibility. Propylene glycol is the main glycol used. Ethylene glycol is used with propylene glycol to reduce the cost. Hydrogenated bisphenol-A is used to impart greater chemical resistance toward acid bases and oxidising agents.

The reactions can be formulated as follows:

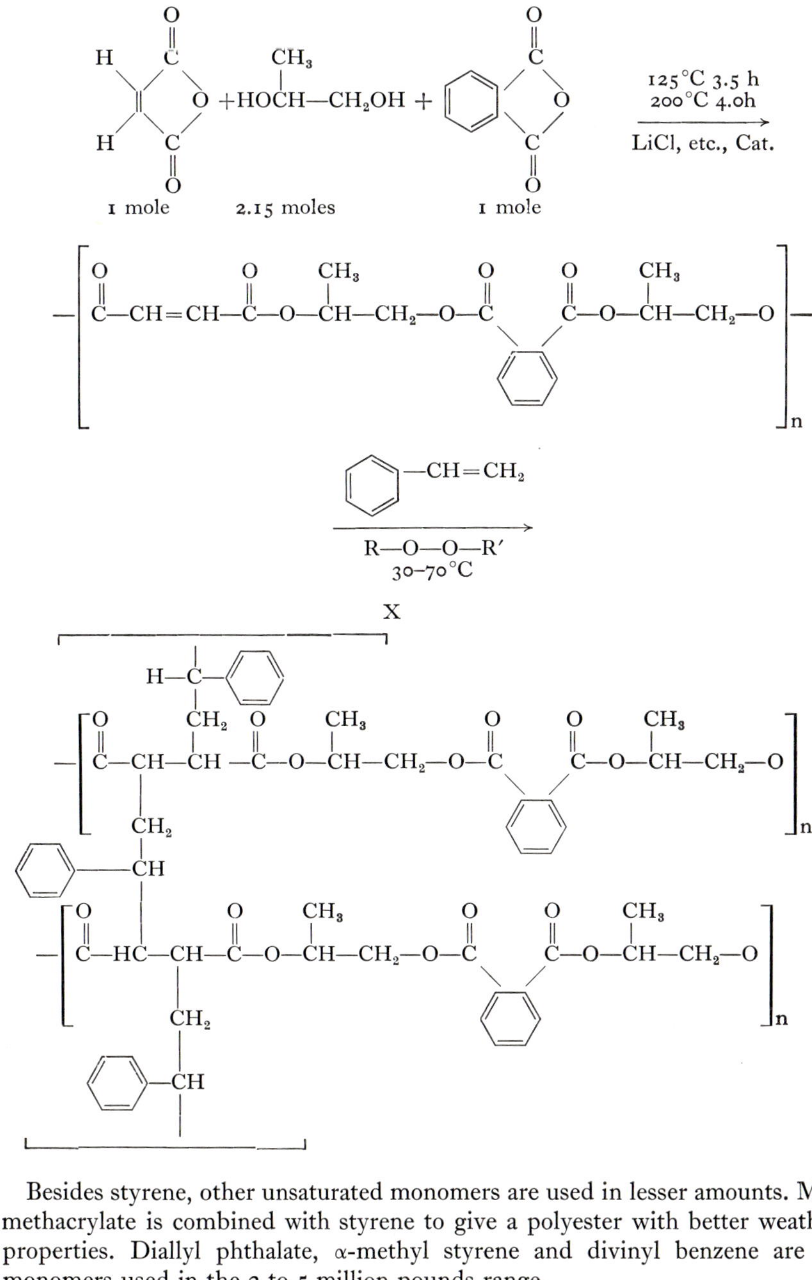

Besides styrene, other unsaturated monomers are used in lesser amounts. Methyl methacrylate is combined with styrene to give a polyester with better weathering properties. Diallyl phthalate, α-methyl styrene and divinyl benzene are other monomers used in the 2 to 5 million pounds range.

11.6.1.3. End use of unsaturated polyesters The distribution of the various usage for unsaturated polyester resins in the USA in 1970 is given below as percentage of the total:

Unsaturated polyester resins in reinforced plastics

Boats, etc.	22%
Transportation	12%
Construction	12%
Tanks, etc.	9%
Electrical	5%
Miscellaneous	20%
	80%

Unsaturated polyester resins in non-reinforced plastics

Castings (Furniture, marble, etc.)	11%
Putty (auto repairs, etc.)	4%
Miscellaneous	5%
	20%

The three largest uses of these unsaturated resins, boats, transportation and construction, account for approximately 50 per cent of the total. During times of reduced spending or changing economy, the decrease in this usage plateaus the production level for two to three years.

11.6.1.4. Producing companies and countries There are about 35 major companies producing these unsaturated polyester resins in the U S A : W. R. Grace, P P G and Reichhold are believed to be the largest. Almost every country in the world produces some unsaturated polyester resins. This is due to the varied usage and formulations and also to the short shelf life of the resins. The equipment is of low capital value and can be used for various other processes.

11.6.2. Alkyd resins

The use of alkyd resins in paints has been falling off in sales due to the replacement of the oil base paints by water base latex paints. The production of maleic alkyd resins was 110 million pounds in 1968 but only 70 MM lb in 1970. The use of maleic anhydride in surface coatings will decrease unless new formulations are developed. The present maleinised oil base paint seemingly has a longer useful life but the water base paints are easier for the novice to apply.

Maleic anhydride is added to the oil base alkyds to react with the conjugated and nonconjugated olefins to give the Diels-Alder or succinate derivatives.

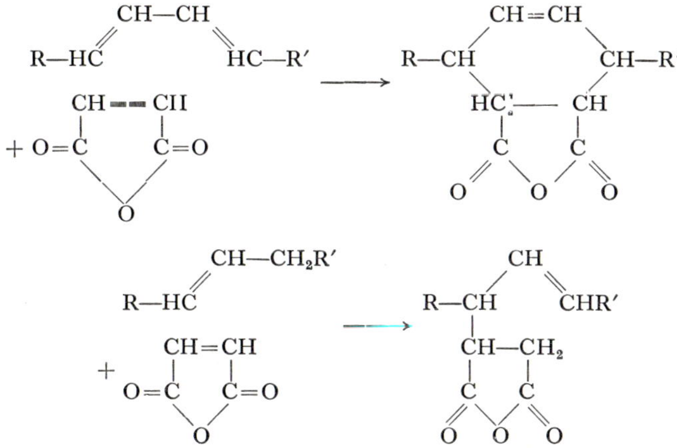

Maleic anhydride or fumaric acid thus adds active sites to the hydrocarbon chains which can react with glycols or glycerine to increase the viscosity. It gives a better coating hardness. These reactions also give better colour retention by reducing the number of double bonds available to react with air.

Maleic anhydride is used to modify rosins in the same Diels-Alder reaction:

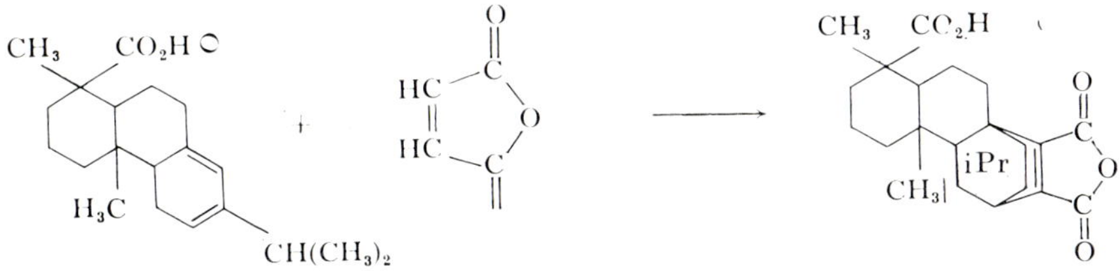

Converting the monobasic levo-pimaric acid to the tri-basic adduct gives greater crosslinking ability. The market for rosin modified maleics is likely to remain at 7 million pounds or drop slightly. New developments will be required to increase the use of maleic anhydride in surface coatings. The present water base systems seemingly need a good crosslinking agent to improve adhesion.

11.6.3. Co-Polymers and Polymer Additives

Maleic anhydride acts as a vinyl monomer when copolymerised with other ethylenic monomers like styrene, vinyl ethers and ethylene. For example, maleic anhydride and styrene give the alternating polymer:

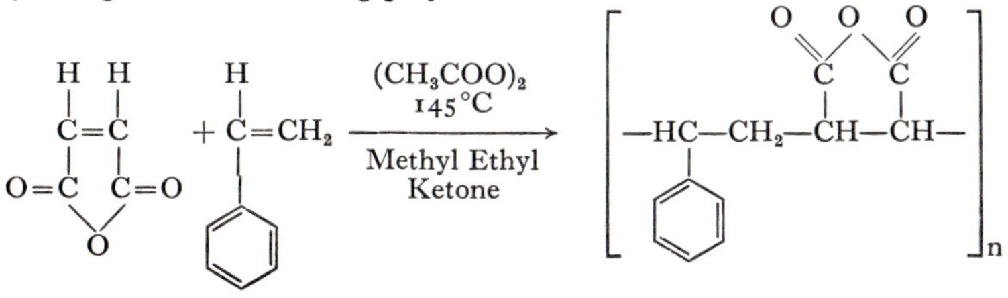

Copolymers of methyl vinyl ether and maleic are produced by General Anilin and Film. It is produced in several molecular weights, and is used as a thickener in solvents, a stabiliser and complexing agent in hair sprays, detergents and textile additives.

The styrene-maleic anhydride copolymer is produced by Arco (Sinclair Petro. Div.) in molecular weight ranges of 700 to 2,000. Monsanto makes a similar polymer converting it to the partial ester and sodium salt. These can be used in latex paints adding adhesion and gloss. The polymer is also used in floor polishes, hair sprays, etc.

The ethylene-maleic anhydride copolymer is also made by Monsanto. The vinyl chloride, vinyl acetate and some tripolymers are also available.

Maleic anhydride was homopolymerised first by radiation and later by peroxides such as peracetic acid at 60 to 80°C.

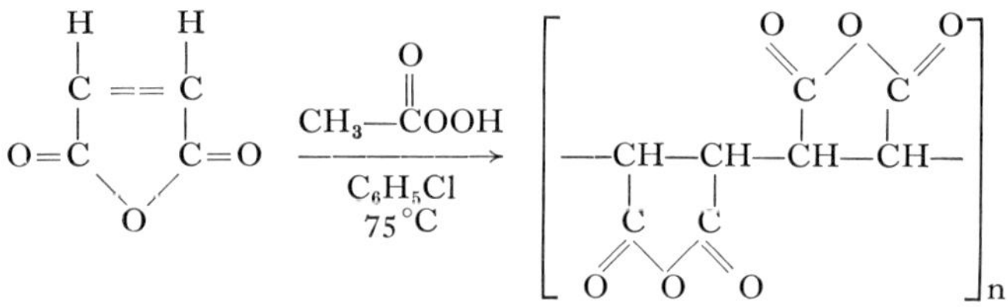

No large usage has been developed so far as is known. The product is generally dark. It can be converted to a lighter colour by hydrolysis, charcoal treatment and bleaching.

Maleic anhydride can be incorporated into polymers already formed by milling the anhydride with the polymer—polyethylene, polypropylene, polybutene, polyisoprene, polyvinyl chloride—and treating with a free radical initiator. This grafting increases the adhesiveness of polyethylene where it can be made to adhere to paper.[48] About 2 to 10 per cent maleic anhydride can be added to the polymers.

11.6.4. Alkyl Succinates

The alkyl succinates are being used as lubricating oil additives acting as a detergent and corrosion inhibitor after further treatment. Olefins react with maleic anhydride at 170 to 250°C. The reactions once started can become quite exothermic. The reaction has been studied by using carbon 14-tagged propylene. The results would indicate that the vinyl carbon is attached to the maleic group.[47]

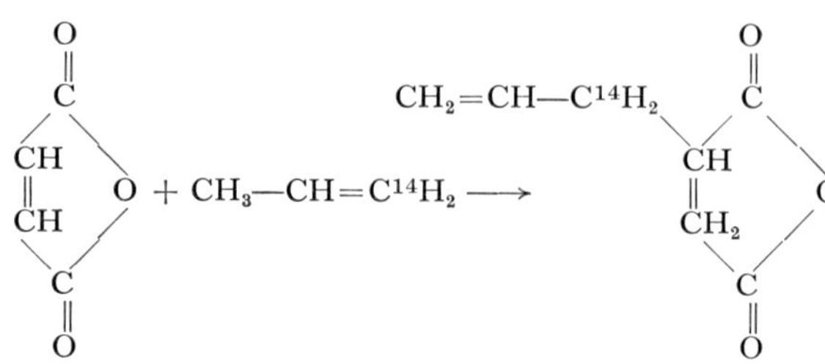

The olefin group can range from C_3 to C_{30} and higher. Ethyl benzene reacts at the alpha position. Polyisobutene can also be used. Non-conjugated dienes can react to form the disuccinate. There is an increasing demand for maleic anhydride for this market.

11.6.5. Other Maleic Derivatives

Maleic hydrazide developed in the 1950s is used as a plant growth retardant. Its use will increase since basic patents are expiring.

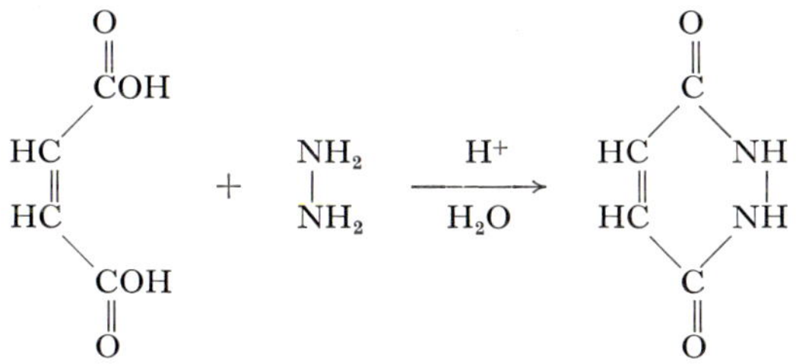

The reaction is readily carried out. The two reactants are mixed, reacted and the product isolated by filtration.

Propyl-N, N-diethyl succinamate is used as an insect repellent.

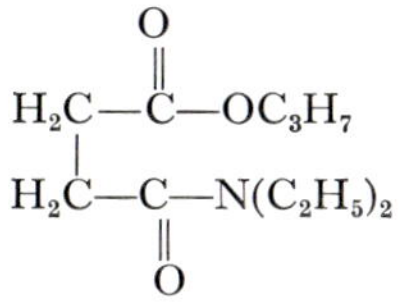

Malic acid can be prepared by the hydration of maleic acid:

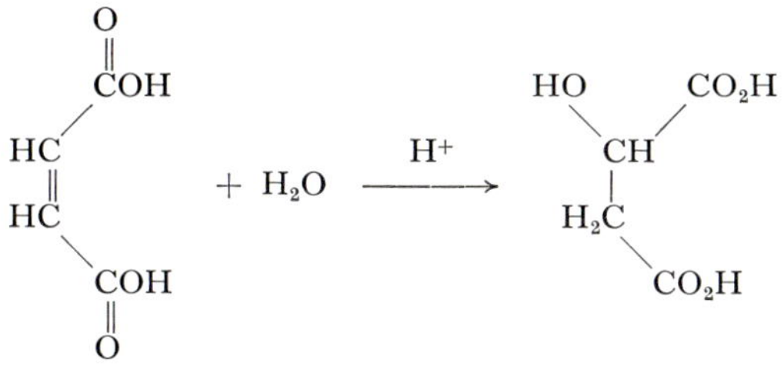

There are about four worldwide producers of malic acid having a total capacity of over 30 million pounds. Allied has a 10 million pound plant in West Virginia. A plant is located in England and two more are found in Japan.

Tetrahydrophthalic anhydride (THPA), the addition product of butadiene and maleic anhydride, is being produced at Petro-Tex in the USA. It is also prepared in Italy, Japan and Germany. Its production might run from 1 to 2 million pounds worldwide. It can be hydrogenated to the hexahydrophthalic (HHPA) anhydride. HHPA with THPA forms a low melting eutectic useful as the curing agent for epoxides. THPA can be epoxidised to form a multifunctional compound. THPA is also incorporated in some special polyesters. The isolated double bond is reactive toward oxygen forming a hard film, but if used in too high a concentration, it will cause the film to turn yellow on aging.

Succinic anhydride, butyrolactone and tetrahydrofuran can be made from maleic anhydride by hydrogenation.

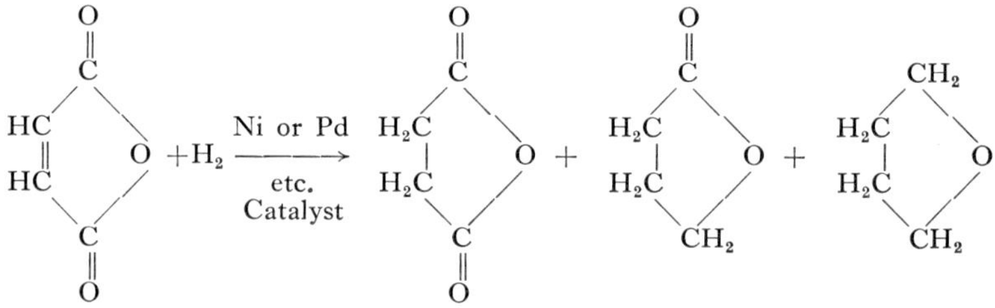

The product distribution depends on the conditions of hydrogenation—catalyst, temperature, pressure and time. A commercial plant is being built in Japan for tetrahydrofuran. A 26 per cent loss in weight occurs on hydrogenation to tetrahydrofuran, so the raw material costs would be 12 to 20c/lb, depending on the current maleic price.

Two chlorinated derivatives of maleic which are available are chlorendic anhydride (Hooker and Velsicol) and Chloran TM (UOP). Chloran TM gives a polyester with greater light stability.

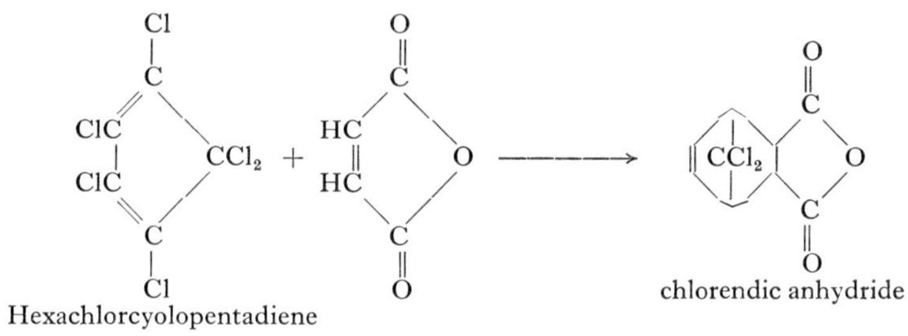

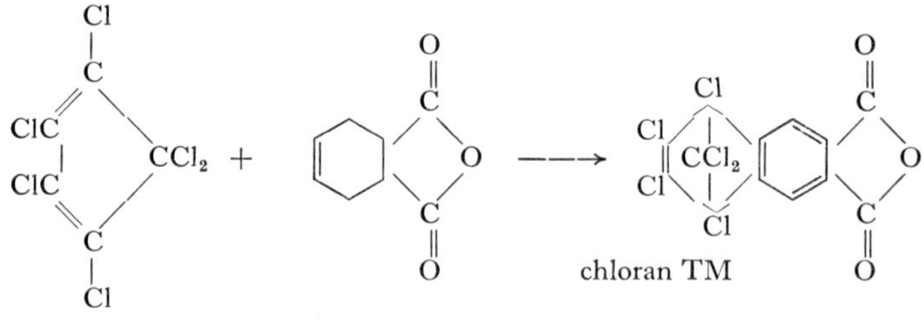

Other chlorinated and brominated derivatives of maleic anhydride are also being used.

REFERENCES

[1] Weiss, J. M. and Downs, C. R., USP 1,318,631 (1919).
[2] *Ibid*, USP 1,318,632 (1919).
[3] *Ibid*, USP 1,318,633 (1919).
[4] *Ibid*, Canadian P 192,766 (1919).

[5] Barrett Co., BP 122,167 (1918).
[6] Barrett Co., GP 475,808 (1921).
[7] Weiss, J. M. and Downs, C. R., *J. Ind. Eng. Chem.*, 1920, **12**, 228.
[8] American Cyanamid, USP 2,885,409 (1959).
[9] Allied Chemical, USP 3,277,017 (1966); USP 3,417,108 (1968).
[10] Chempatents, USP 2,777,860 (1957).
[11] Hommel Co., USP 2,698,306 (1954).
[12] Monsanto, USP 3,106,569 (1963).
[13] Pittsburgh Chem., USP 3,005,831 (1961).
[14] PPG, USP 3,163,613 (1964).
[15] Scientific Design, USP 2,967,185 (1961).
[16] Badische Anilin, GP 1,115,241 (1961).
[17] Badische Anilin, Belg.P 677,270 (1966); USP 3,538,122 (1970).
[18] Prod. Chim. Pech. Saint-Gobain, FP 1,372,476 (1964).
[19] Ioffe, I. I., *Zhur. Fiz. Khim.*, 1961, **35**, 2348; CA 1962, **56**, 5437.
[20] Moskaleva, I. A., USSRP 173,717 (1965); CA 1966, **64**, 8043.
[21] Trehan, B. S., Ind.P 74,450 (1962); CA 1963, **59**, 9808.
[22] Bhattacharyya, S. K., *J. Appl. Chem.* (London), 1958, **8**, 728; CA 1959, **53**, 14,974.
[23] Oriental High Pressure, Ind., Jap.P 3,522 (1958); CA 1959, **53**, 5549.
[24] Nippon Cat. Chem., Jap.P 5,418 (1957); CA 1958, **52**, 10,163.
[25] Takeda Chem. Ind. USP 3,535,346 (1970).
[26] James, J. H., USP 1,588,836 (1926).
[27] Union Carbide, USP 2,097,904 (1937).
[28] Std. Oil Development, USP 2,260,409 (1941); USP 2,537,568 (1951).
[29] DuPont, USP 2,625,519 (1953); USP 2,605,238 (1952).
[30] Atlantic Refining, USP 2,719,853 (1955); USP 2,773,838 (1956).
[31] Petro-Tex, USP 3,255,211–3 (1966).
[32] Distillers, BP 693,904 (1953); CA 1954, **48**, 8448.
[33] ICI, FP 1,382,523 (1964); BP 1,086,498 (1967).
[34] Soc. d. Electro, FP 1,291,659 (1962); CA 1963, **58**, 1356a.
[35] Farb. Hoechst, BP 988,447 (1965).
[36] Badische Anilin, USP 3,478,063 (1969); USP 3,538,122 (1970).
[37] Masao Kitahara, Rikagaku Kenkyusho Hokoku (Tokyo) 1958, **34**, 283–93; CA 1960, **54**, 5458.
[38] Kernos, Y. D., USSRP 145,571 (1962); CA 1962, **57**, 13625i.
[39] Shakhtakhtinskii, T. N., *Azerb. Khim. Zh.*, 1965, (3), 80–3; CA 1966, **64**, 1951.
[40] Petro-Tex, BP 1,088,696 (1967).
[41] Badische Anilin, German Patent Appl. P. 1468814.3.
[42] Stone and Webster, USP 2,683,110 (1954).
[43] Badische Anilin, GP 1,176,126 (1964); CA 1964, **61**, 13200.
[44] Scientific Design, USP 3,262,972 (1966).
[45] Lawrence, J. R., '*Polyester Resins*' (Reinhold, New York, 1960).
[46] Hicks, J. S., '*Low Pressure Lamination of Plastics*' (Reinhold, New York, 1947).
[47] Sublett, B. J. and Bowman, N. S., *J. Org. Chem.*, 1961, **26**, 2594.
[48] Clarke M. A. (Union Carbide), BP 1,059,973 (1967); CA 1967, **66**, 116355.

CHAPTER 12

OTHER INDUSTRIAL USES OF BENZENE

By R. A. Fraser

THIS chapter is intended to cover the major uses of benzene not considered in the previous chapters. Although the title emphasises 'industrial' uses, it would be inappropriate, in a book of this nature, to omit a usage which was at one time the major outlet for hydrocarbons, namely as a constitutent of motor fuels. This is emphasised by the fact that the forerunner of the British textbook '*Benzoles, Production and Uses*'[1] was entitled '*Motor Benzole, Its Production and Use*'.[2]

12.1. MOTOR FUELS

The development of the benzene industry ancillary to coal carbonisation has been discussed in Chapter 1, where the growing usage in Great Britain and Germany of refined benzole in motor fuel is described. The British usage in this field reached its zenith in 1949, when 235,000 tons of motor benzole were produced.[3] In Germany, the highest figure was reached in 1941 when about 420,000 tons were made.[4] Subsequent developments in the European chemical industry increased the demand for pure benzene as a raw material, thereby reducing the availability of benzene as a fuel, which has also been affected by the substitution of natural gas for coal gas. In 1970 the output of this grade had fallen to 13,540 tons in the UK and about 180,000 tons in Western Germany.[5]

The benzene content of motor benzole varies with the type of carbonisation employed, the method of recovery and the market for toluene and xylenes. Benzole from high-temperature processes such as coke ovens may contain upwards of 70 per cent benzene, while those from vertical retorts or oil gasification may contain only 45 per cent benzene with a high content of non-aromatic hydrocarbons.[6]

TABLE XII.1

Compound	*Blending octane number*	
	Research	*Motor*
Benzene	99	91
Toluene	124	112
o-Xylene	120	103
m-Xylene	145	124
p-Xylene	146	127

The advantages of the aromatic hydrocarbons as constituents of motor fuel are associated with their high calorific value on a volumetric basis, their high antiknock value as measured by the blending octane number, and their favourable vapour-pressure characteristics when mixed with non-aromatic hydrocarbons. The octane numbers of some hydrocarbons are quoted in Table XII.1, which gives the blending octane ratings in 20 per cent vol/vol concentration in a mixture of 60 per cent vol/vol *iso*-octane and 40 per cent vol/vol *n*-heptane.[7]

Benzene has also been used as a blending agent in motor fuel containing alcohol, to ensure solubility of the alcohol in the hydrocarbon constituents of the fuel.[8]

Owing to the relative economics in the United States of motor fuel from petroleum and benzole from coal, usage of motor benzole in that country has always been very small, although aromatics from reforming processes are widely used (see Chapter 3). In order to maintain the anti-knock rating of motor fuels at an acceptable value economically, lead tetraethyl and tetramethyl are also used in about 98 per cent of American motor fuels, but concern has recently been expressed over the effects of these additives on air pollution and on exhaust-emission control devices. It has therefore been proposed in the United States that the lead content of motor fuels should be drastically reduced by 1975. This would mean that the average aromatic content might have to be increased from about 20 to 40 per cent to maintain the anti-knock value. Such an eventuality could have a profound effect on the availability and price of benzene for chemical synthesis.[9, 10]

12.2. BENZENE AS A SOLVENT

As described in Chapter 1, some of the earliest applications of the aromatic hydrocarbons were as solvents, particularly in the rubber and surface coating industries. The increasing availability of petroleum solvents, coupled with the more recent awareness of the toxic properties of benzene, have considerably reduced the use of benzene as such for solvent purposes (see Chapter 5).

In 1965, for instance, the Paintmakers' Association adopted a resolution that neat benzene should not be used as an ingredient of any product of the paint industry. Furthermore, benzene that might be introduced as an impurity in toluene or other solvent should be limited to not more than 1 per cent of the final product.[11] The higher aromatic hydrocarbons are however, still widely used, in view of their superior solvent properties, compared with non-aromatic hydrocarbons. Mellan has given details of these properties, together with specifications and applications of aromatic solvents.[12]

In certain industrial processes, where the emission of vapours can be closely controlled to avoid the build-up of toxic concentrations of benzene vapour in the atmosphere, benzene continues to be used for its solvent properties. Important examples are the solvent extraction processes developed in the petroleum industry for the refining of lubricating oils. These are directed to two ends: the removal of wax and the improvement of the viscosity index, which measures the rate of change of viscosity with temperature. In the removal of wax, the oil is diluted with a suitable solvent and chilled to precipitate the wax, which is separated by filtration or centrifuging, the solvent being recovered by distillation. Mixtures of benzene and toluene with acetone or methyl ethyl ketone, or ethylene dichloride and benzene, are commonly used. Improvement in the viscosity index may be achieved by extracting the oil with a mixture of liquid sulphur dioxide with about 25 per cent in volume of benzene. This preferentially removes the high boiling constituents having a high proportion of aromatic rings in the molecule, which change their viscosity rapidly on heating.[13] Other industries in which benzene has been used successfully as a solvent are in the production of polyethylene by the ICI high-pressure process, in the preparation of rubber cements, in the manufacture of rubber hydrochloride (Pliofilm) and in the making of artificial leathercloth. The grade of benzene used in these processes is often that known as 'Industrial Benzene' or in the UK as '90's Benzene', which may contain up to 10 per cent toluene. In the preparation of nuclear fuels a solution of thiophenyl trifluoracetone in benzene has been used to extract plutonium from dilute nitric acid solution in the form of a benzene-soluble chelate.[14]

An application of benzene analogous to its use as a solvent is that of an azeotrope-former for the removal of water, as in the purification of ethanol[15] and in the esterification of higher alcohols with phthalic anhydride.

12.3. BENZENE SULPHONIC ACIDS

The main application of the benzene sulphonic acids is in the manufacture of phenols (see Chapter 8), although for phenol itself other routes have almost entirely replaced that in which benzene monosulphonic acid is used.

In the sulphonation of benzene one of three possible reagents may be used: sulphuric acid, sulphur trioxide, or oleum. Their relative advantages have been discussed by Gilbert.[16] It is possible to use the Strecker reaction, in which an aromatic chloro compound is treated with sodium sulphite but, so far as is known, this reaction has not been applied commercially to benzene.

12.3.1. Benzene Monosulphonic Acid[17]

Whichever reagent is employed it is generally accepted that the effective sulphonating agent is sulphur trioxide:

$$C_6H_6+SO_3 \rightarrow C_6H_5SO_3H$$

The rate of this reaction has been studied by Crooks and White,[18] who found that at moderate temperatures the reaction with concentrated sulphuric acid readily occurs, but the time for completion is relatively long, owing to the limited solubility of benzene in sulphuric acid and the dilution of the latter with the water formed in the reaction. Certain catalysts, such as a mixture of sodium sulphate and vanadium pentoxide, notably increase the reaction rate,[19] which is also improved by increase in pressure and temperature.[20] The main by-product of the reaction is diphenyl sulphone and a number of patents have appeared covering the addition of various substances to repress this sulphone formation.[21]

In a typical batch process of manufacture, the sulphonation is carried out in a cast-iron vessel equipped with an efficient agitator, vented condenser and a jacket for heating by means of steam or hot oil. For preparing the mono-sulphonic acid, 66° Beaumé sulphuric acid (95 per cent wt/wt H_2SO_4) is used, the temperature being gradually raised to 170°C. The water formed in the reaction is removed azeotropically by passing benzene vapour through the reaction mixture. The water must be removed at such a rate as to maintain the acid concentration above 78 per cent wt/wt throughout the reaction. A development of this process using a cascade system, with a series of sulphonation vessels, has been described by Kenyon and Boehmer.[22]

As with most chemical processes, continuous operation is now preferred and the application of this principle to the sulphonation of benzene was described some thirty years ago in a Russian paper.[23] Their process conserves sulphuric acid by continuously adding liquid sulphur trioxide to maintain the acid concentration at 100 per cent, the benzene sulphonic acid being removed as a solution in the excess benzene present. The same authors have also described a process in which the benzene is vaporised before passing to the sulphonator, which is fed continuously with sulphuric acid. The reaction mixture is then passed into a plate column, where it reacts further with an ascending stream of fresh benzene vapour. A seven-stage reaction, operating with a molecular ratio of benzene to acid of 10:1 at a temperature of 180°C has a residence time of 1 hour and it is claimed that the continuous process has ten times the capacity of a batch plant covering the same area.[24]

A combination of liquid- and vapour-phase sulphonation is also used in a process using oleum. A presulphonator is charged with benzene and 9·5 per cent oleum is added at such a rate as to avoid undue rise of temperature. The reaction mixture containing excess acid is then pumped to further sulphonators where benzene vapour is fed in until the desired residual acid level is reached, following which the mass flows to vessels in which excess benzene and water are removed by distillation, the benzene being scrubbed with sodium hydroxide solution, washed, dried and returned to the process.[25] A high-pressure process, in which oleum up to 25 per cent concentration, is used in the liquid-phase at 190–225°C and 150–425 p.s.i. has been patented. It is claimed that a yield of 80–95 per cent benzene sulphonic acid is obtained without the necessity of removal of water.[26]

Sulphur trioxide has several advantages as a sulphonating agent. It remains liquid when stored at 34°C, but at room temperatures even a trace of moisture converts it to solid polymers unsuitable for sulphonation. It can, however, be stabilised by the addition of small quantities of certain derivatives of boron, silicon or sulphur. A solution of sulphur trioxide in liquid sulphur dioxide is sometimes used for sulphonation since the low boiling point of the latter allows for autofrigeration during the reaction, as in the two-stage process in which benzene is allowed to react with a solution, containing one part of sulphur trioxide in eight parts of sulphur dioxide, at −9°C. The liquid sulphur dioxide is allowed to evaporate until a temperature of 19–20°C is reached; a further quantity of benzene is then added and the temperature allowed to rise to 50–70°C, sulphur dioxide being removed in a stream of air. Yields of approximately 95 per cent benzene sulphonic acid are claimed.[27]

For purification the benzene sulphonic acid may be extracted from the reaction mass with benzene,[28] crystallised out at −10°C,[29] or extracted with ethylene dichloride.[30] Use can also be made of the fact that the calcium salt is soluble in water, unlike calcium sulphate.[31] Anhydrous benzene sulphonic acid melts at 65–6°C and has density at 80°C of 1·30.

12.3.2. Benzene Disulphonic Acids and other Derivatives

The important benzene disulphonic acids are the *meta* acid and the *para* acid. The former is obtained by reacting the monosulphonic acid with oleum at 30–85°C for about 7 hours.[32] The *para* acid cannot be made by direct sulphonation but the sodium salt is prepared by heating sodium benzene sulphonate at 300°C under pressure with a heavy metal catalyst.[33]

Diphenyl sulphone is a by-product of the sulphonation of benzene but a process for its specific preparation has been patented. Preheated benzene vapour is passed at 265 to 270°C into the base of a reaction column maintained at 260°C down which 95 per cent sulphuric acid is flowing.[34]

Benzene sulphonyl chloride may be prepared by reacting sodium benzene sulphonate with phosphorus pentachloride or by treating benzene with chlorosulphonic acid.

12.3.3. Applications

The prime use for benzene sulphonic acid was for the synthesis of phenol and in 1961 about 200,000 tons was used in the United States for this purpose, as well as about 250,000 tons in the rest of the world.[17] Although largely replaced by other phenol processes, notably that in which cumene is an intermediate, the sulphonation process still holds its own where there is a market for the by-product sodium sulphite.[35] Resorcinol is made from *meta*-benzene disulphonic acid (see p. 560). As starting materials in synthesis, some of the benzene sulphonic acid derivatives, such as the sulphonyl chlorides, are important owing to their reactivity, especially as intermediates for dyestuffs. The acid itself is used as an accelerator in resin formation and is also employed in certain condensation processes as in the reaction of phenols with terpenes and phenols with primary aromatic amines. The acid is also used in the dehydration of alcohols and castor oil and in the esterification of fatty acids with alcohols.

12.3.4. Statistics

Owing to the limited number of manufacturers of the benzene sulphonic acids no production figures are available. They are made in the United Kingdom by Albright and Wilson Ltd. (Marchon) and Midland Yorkshire Tar Distillers Ltd., and in the United States by Koppers Co. Inc., Organic Materials Division, Nease Chemical Co. Inc. and United States Pipe and Foundry Co.

12.4. HALOGENATED COMPOUNDS OF BENZENE

The most important halogenated compounds of benzene are those containing chlorine and of these monochlorobenzene and *ortho* and *para*-dichlorobenzenes have a large industrial application.[36] Monochlorobenzene was in fact the first heavy organic chemical produced on a large scale, being made at Widnes by the United Alkali Co. in 1909.[37] It acquired major importance during World War I, when it was used for making phenol, required as an intermediate for picric acid. In 1915 the Dow Chemical Co. in America started to make monochlorobenzene as an outlet for their chlorine, no longer used for making bleach solutions.

12.4.1. Manufacture

As there is a demand for the dichlorobenzenes as well as monochlorobenzene, the chlorination of benzene is usually operated as a three-product process, and can be applied batchwise or as a continuous process but, if the maximum yield of the monochlorinated product is required, the former is favoured. In either case, to avoid undesirable by-products and low yields, benzene free from high-boiling hydrocarbons and non-aromatics should be used. Freedom from water is also essential; to achieve this the benzene should be dried by azeotropic distillation or by treatment with a drying agent such as silica gel or alumina. The chlorine is scrubbed with concentrated sulphuric acid. The catalyst commonly used is ferric chloride in a concentration of about 0·1 per cent and this may be prepared *in situ* by the preliminary addition of iron turnings to the benzene in the reaction vessel.[38, 39] Alternatively metallic aluminium[39] or the chlorides of aluminium or tin may be used.[40] The use of fuller's earth as a catalyst has been favoured in Germany and is claimed to reduce the formation of dichlorobenzenes, but in this case glass-lined plant must be used, since traces of ferric chloride adsorbed on the fuller's earth tend to nullify the catalytic effect.[41] With metallic chloride catalysts iron or mild steel vessels may be used but the cooling coils and chlorine inlet pipes are of lead or lead-covered. Chlorine is added at such a rate as to maintain the reaction temperature between 40° and 70°; excessive temperatures lead to polychlorination, which leads in the ultimate to complete formation of hexachlorobenzene at 230°C under pressure. The progress of the reaction is followed by density measurement and on completion the product is washed with caustic soda solution and distilled. The by-product hydrogen chloride is recovered by washing the vent gases from the reaction with a refrigerated solvent such as *o*-dichlorobenzene.

In a typical continuous process a series of externally cooled tubes made of corrosion-resistant material such as stainless steel, is used. These tubes contain a catalyst such as iron Raschig rings, and flows of benzene and chlorine are maintained with a large excess of the former to avoid excessive temperature rise which lead to uncontrolled reactions. Temperatures below 40°C reduce the formation of dichlorobenzenes and this possibility is also reduced by rapid removal of the mono product from the reaction zone. Unreacted benzene is separated from the product by fractionation and returned to chlorination. An efficiently operated process can yield a product containing up to 95 per cent monochlorobenzene.[42]

The continuous chlorination procedure lends itself well to the oxychlorination process, in which a mixture of hydrogen chloride and air or oxygen is used instead of chlorine. This was developed by Raschig at Heidelberg in 1928 and a number of variations have since been patented.[43, 44] This process, which was first used on a plant scale at Ludwigshafen in 1931, received great impetus from the development of the processes of Dow and Raschig for making phenol from chlorobenzene, since the hydrogen chloride evolved in the hydrolysis can be returned as a reactant to the chlorination stage. This appeal, however, has less importance following the rapid development of the cumene process for phenol (see Chapter 8).

In the usual technical application of the Raschig process a mixture of benzene vapour, air, steam, and hydrogen chloride is preheated and contacted at 220–260°C and atmospheric pressure with a catalyst, which may consist of cupric oxide with a promoter on silica gel. In this process also temperature control is very important, to avoid combustion of the benzene. In any case, only 10 per cent of the benzene is reacted at each pass, to avoid the formation of higher chlorinated products. Penner, of the Frontier Chemical Co., has described a high-pressure oxychlorination process, operating at 234–315°C and 40–80 p.s.i.g., giving a greater conversion of benzene per pass.[44] It has been stated that the chlorides of sodium and potassium have a poisoning effect on copper catalysts, although lithium chloride may be used as a promoter.[45] Other continuous processes which have been patented include chlorination of benzene vapour in a quartz tube at 400–500°C[46] and the use of nitrosyl chloride.[47]

12.4.1.1. Dichlorobenzenes A mixture of dichlorobenzenes is obtained as a high-boiling residue from the distillation of the product from the chlorination of benzene which contains about 2–3 lb *para*- to 1 lb *ortho*-dichlorobenzene. Further quantities may be produced by the extended chlorination of monochlorobenzene in the presence of ferric chloride at 150–190°C.[49]

Production techniques devolve upon the separation of the *ortho* and *para* isomers; in technical chlorination, very little *meta* isomer is formed. The separation can be achieved by fractional distillation or by crystallisation of the *para* compound (see Table XII.3). The latter, however, is preferred and the last traces of the *ortho* isomer can be removed from the crystals of *para* by washing the latter with methanol, which is removed from the solid mass by heating to 100°C under vacuum.[36]

Although there is an increasing market for *ortho* the *para*-isomer is at present in greater demand and processes for increasing the yield of this isomer have been patented. If the distillation residue from the chlorination process is treated with chlorosulphonic acid at 15°C the *ortho* compound is selectively converted to the sulphonic acid from which the *para* compound may be easily separated by distillation.[50] An ingenious method of separation depends on the slower rate of further chlorination of the *ortho* isomer. Crude dichlorobenzenes are chlorinated in the presence of ferric chloride and fractionated when the *para* compound is easily separated from 1,2,4-trichlorobenzene (b.p. 210°C).[51] Orientation of the chlorination of benzene in the direction of the *para* isomer by the use of suitable catalysts such as benzene-sulphonic acid has been patented.[52]

Meta-dichlorobenzene is usually obtained from one of the other isomers by an isomerisation process, such as treatment with aluminium chloride and hydrogen chloride at 120°C and 650 p.s.i.[53]

12.4.1.2. Higher chlorobenzenes The higher chlorobenzenes can be obtained from the chlorination residues of benzene, but hexachlorobenzene can be made by the reaction of hexachlorocyclohexane (BHC, see page 557) with chlorine[54, 55] or with air at 450–550° in a fluidised bed of cupric chloride.[56]

12.4.2. Properties

The main properties of the chlorobenzenes are given in Table XII.3.

TABLE XII.3

Properties of Chlorobenzenes

Compound	*Mono*	*Di* O	*Di* m	*Di* p	*Tri* *1, 2, 4*	*Tetra* *1, 2, 3, 4*	*Hexa*
Molecular weight	112·56	147·01	147·01	147·01	181·45	215·90	284·79
Density	$1{\cdot}107_{4}^{20}$	$1{\cdot}305_{4}^{20}$	$1{\cdot}288_{4}^{20}$	$1{\cdot}458_{4}^{21}$	$1{\cdot}454_{4}^{20}$	—	$1{\cdot}5691^{23{\cdot}6}$
Melting point, °C	−45·2	−17·6	−24·7	53	17·1		
Boiling point, °C	132·1	179·0	173·0	173·7	213·5	254	322 (sublimes)

12.4.3. Applications

The main uses for monochlorobenzenes were given in 1965 as follows:[57]

Phenol	60 per cent
DDT	25 per cent
Aniline + misc.	15 per cent

It is probable that since that date the proportion used for making phenol has diminished but its use for making aniline by ammonolysis is increasing. A developing usage is that for conversion to *para*-nitrochlorobenzene, an intermediate for the manufacture of phenylenediamines.

Ortho-dichlorobenzene is used largely for organic synthesis and also as a high-boiling solvent. US applications have been given as:[58]

Organic synthesis	46 per cent
Miscellaneous solvent uses	24 per cent
TDI process solvent	18 per cent
Dyestuffs	8 per cent
Others	4 per cent

A large usage in the solvent field is that for the removal of carbonaceous and lead deposits from internal combustion engines.[59, 60] It has also been used as a heat transfer medium and as a coolant for high-energy magnetic coils.[61] It is probable that the application with the highest growth rate is that as a solvent in the manufacture of toluene di-isocyanate (TDI) from toluene 2,4-diamine and phosgene. TDI is an intermediate for the urethane polymers. It has also been used in the sulphonation of anthraquinone to give dyestuff intermediates.[62] Another recent application is as a swelling agent in the dyeing of polyester fibres.

About 70 per cent of the *para*-dichlorobenzene sold is used in moth repellants and air deodorisers, the latter being a rapidly growing market.[63, 64] Other uses are mainly in chemical synthesis and dyestuffs, an example being the manufacture of the anti-diabetic drug, acetohexamide.[65]

Trichlorobenzenes are used as a mixture of isomers as a solvent with a boiling point of about 215°C for degreasing and in the application of oil-soluble dyes. They are also used for the extermination of termites.

Developments in agricultural chemicals have stimulated interest in the tetrachlorobenzenes, which are used in the production of 2,4,5-trichlorophenoxyacetic acid, a herbicide especially effective in woody plants,[66] and for making Tecnazene (2,3,5,6-tetrachloronitrobenzene), which delays sprouting in potato tubers.[67]

Hexachlorobenzene is an important raw material for making pentachlorophenol used as a wood preservative and in seed dressings.[68]

12.4.4. Statistics

United States production of principal chlorobenzenes is given in Table XII.4.[69]

TABLE XII.4

US Production of Chlorobenzenes

Year	*Production, thou. tons*		
	Monochlorobenzene	o-*DCB*	p-*DCB*
1966	263	24	30
1967	219	23	30
1968	262	27	32
1969	269	31	23
1970	216	30	31
1971	183	24	31
1972	179	—	—

The main producers of monochlorobenzenes in the United States are Dow, Hooker and Montrose, although about fourteen other manufacturers are involved. The main producers of *ortho*-dichlorobenzene are Dow, Monsanto and Standard Chlorine. Statistics for other countries are not available, owing to the limited number of producers, although there is a major output in the United Kingdom, Western Germany, France and Japan. The only producer in the United Kingdom is ICI Ltd. and annual production of monochlorobenzene is probably of the order of 5,000 tons per year. It has been estimated that the US demand for *ortho*dichlorobenzene will rise from about 34,000 tons in 1970 to 45,000 tons in 1975, the doubtful factor being the demand for chlorinated pesticides.[68]

12.5. PESTICIDES

It is appropriate to consider in this section pesticides derived from benzene by chlorination, mainly dichloro-diphenyl-trichloroethane (DDT) and hexachlorocyclohexane (BHC).

12.5.1. DDT

The structure of DDT, which is strictly described as 1,1,1,-trichloro-22bis(*p*-dichlorophenyl)ethane is shown in Fig. 12.1.

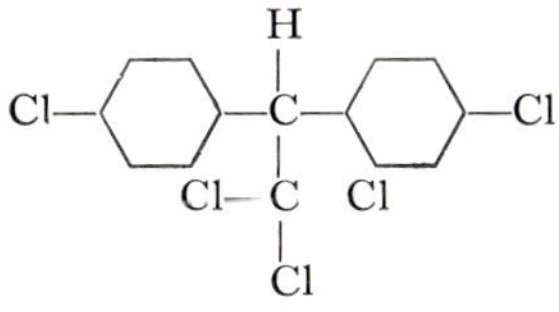

FIG. 12.1

12.5.1.1. Production It was first synthesised by Zeidler in 1874[70] but its insecticidal properties awaited discovery by Müller of the Geigy organisation in 1939.[71] The usual process for the manufacture of DDT involves the condensation of monochlorobenzene with chloral in the presence of sulphuric acid:

$$C_6H_5Cl + CCl_3CHO \longrightarrow (C_6H_4Cl)_2CHCCl_3 + H_2O$$

Ethanol is chlorinated to chloral and the product reacted with monochlorobenzene in the presence of 100 per cent sulphuric acid or oleum, at 15–30°C, using a glass-lined reactor. The spent acid is withdrawn and the product washed and neutralised with soda ash, following which it is dried by heating, which also removes any unreacted chlorobenzene.[72, 73]

12.5.1.2. Properties Technical DDT is a white, amorphous powder containing between 65 and 80 per cent *para para'*-dichloro-diphenyl-trichloro-ethane, together with up to fourteen other chlorinated compounds. It should have a setting point of not less than 89°C and contain between 9·5 and 11·0 per cent hydrolysable chlorine. It is soluble to the extent of 4 g/100 ml in refined kerosine and as much as 60 g/100 ml in xylole.[73]

12.5.1.3. Uses DDT was the first organic insecticide to be manufactured and used on a large scale. It acts both as a contact and a stomach poison towards a large number of insects and is therefore effective against insects that merely come into contact with it without necessarily eating it. The only important class of insects not affected are spiders, but of plant pests, the black aphid, red spider, snail, and woolly aphis are not affected.[74] DDT is usually applied as a dust or as a spray of an aqueous emulsion or suspension containing 0·1–5·0 per cent DDT. It may also be applied in solution as an aerosol.

Mellanby has described how the insecticidal properties of DDT were discovered and applied in World War II, first to the elimination of lice and then to malaria control by its effect on the anopheles mosquito.[76] It had also the great advantage of relatively low toxicity to man, about the same as that of aspirin.

The widespread use of DDT to reduce the ravages of typhus and malaria, in wartime and the period immediately afterwards, placed a strain on the production capacity but, when supplies became more generally available, new uses became evident. This was notable in the case of house-flies, particularly around refuse tips and farms. It was also used successfully to control other domestic pests, such as cockroaches, ants and bedbugs.

After a few years, control of insects, particularly those carrying disease, became less effective and this was found to be due to the development of strains resistant to DDT.[77] Some attempt was made to overcome this disadvantage by the incorporation in the insecticide formulation of other chemicals, such as *N*,*N*-di-n-butyl-p-chlorobenzene sulphonamide.[78] Alternative insecticidal chemicals have also been developed, such as methoxychloro (dimethoxyldiphenyltrichloroethane) and Gix (*p*,*p*′-difluorodiphenyltrichloroethane), as well as numerous other products not derived from benzene. It has also been found that after suspension of a DDT campaign for periods of about a year, the insect population may again become susceptible.

A more serious defect of the use of DDT, and indeed of chlorinated hydrocarbon insecticides generally, has been the discovery that their residues become concentrated in fatty substances, such as the tissues of insects such as caterpillars and earthworms, which are eaten by birds. These may then accumulate a toxic dose of the insecticide. These effects, however, have been particularly evident where massive applications of DDT, up to 25 lb per acre, have been used. Mellanby describes a careful study made of an application on the Cannock Chase area of 1 lb per acre to eliminate pin looper caterpillars where no bird casualties were reported and only low increases in DDT contents of their tissues were found. He also points out that fish are particularly susceptible to DDT, probably due to the take-up from the large amounts of water passed through their gills during respiration, but DDD (1,1-dichloro-2,2,bis p-chlorophenyl ethane) has a much lower toxicity to fish and warm-blooded animals than DDT. A well-established case of poisoning of birds, following a treatment of a lake in California for midge control, appeared to be due to concentration of DDD first in plankton, then in fish and finally in the birds.[76] There has also been some concern about the accumulation of DDT residues in the soil, but recent work at the University of Wisconsin has suggested that other compounds naturally present in the soil may have given false positive results for DDT.[79] In spite of all these fears, it has recently been stated by the United Nations Food and Agriculture Organisation that the continued use of DDT is essential for the maintenance of world food supplies.[80]

12.5.1.4. Statistics Production of DDT in the United States is given in Table XII.5.

TABLE XII.5

US Production of DDT

Year	*Thousand tons*
1966	64·0
1967	46·7
1968	63·1
1969	55·8
1970	26·5

According to a recent American report, the only US producers in 1970 were Lebanon and Montrose with a total annual capacity of about 45,000 tons; demand in 1970 was about 32,000 tons having declined by 8·2 per cent per year since 1960, but demand was expected to remain at this level through 1975, the price at the end of 1970 being 22 cents per lb.[79]

12.5.2. Benzene Hexachloride

Benzene hexachloride (BHC) is more correctly described as 1,2,3,4,5,6-hexachlorocyclohexane and was first prepared by Faraday in 1825 although, as with DDT, its insecticidal properties were not discovered until a long period has elapsed, in this case until 1942 when, under the wartime stimulated demand for insecticides, its value was shown by ICI workers.

12.5.2.1. Production The BHC molecule may theoretically exist in sixteen stereoisomeric forms, but only four have been isolated and of these the γ-isomer is the most toxic and hence useful as an insecticide. Modifications to the manufacturing process have therefore been devoted to obtaining the maximum concentration of this isomer.[83]

Fundamentally the production of BHC consists in the chlorination of benzene in the absence of a substitution catalyst (such as ferric chloride) and under the influence of actinic light (of wavelength 2500–4500Å) or in the presence of organic peroxides. The process is well fitted to continuous operation[84] and an excess of benzene to chlorine over the theoretical ratio is maintained to prevent precipitation of the less soluble isomers, the BHC concentration in the reactor being 12–15 per cent. The reaction is exothermic and refrigerated equipment is necessary to obtain the maximum yield, the preferred range for maximum γ-isomer production being $-30°C$ to $0°C$.[85] The low chlorine concentration favours addition rather than substitution; this is also aided by a low residence time. The importance of eliminating addition catalysts such as ferric chloride from the reaction medium led at first to the use of expensive glass-lined or nickel equipment, but Twiehaus found that the reaction can be carried out in ferrous equipment without the occurrence of undue substitution, provided that the water content of the reaction mixture is maintained at less than 50 p.p.m.[86] Reaction additives that have been claimed to increase the proportion of γ-isomer is the product include alkyl sulphates,[87] acetic anhydride or acetyl chloride,[88] and hydroxyhexyl hydroperoxide, cyclohexyl hydroperoxide or tetraisopropyl tin.[89] Careful control of reaction temperature and chlorine concentration, together with the use of selected solvents of high polarity such as acetic anhydride has also been described,[90] as has the replacement of air in the reaction vessel by nitrogen or carbon dioxide.[91] In order to avoid colour and oiliness in the product it is important to have a minimum content of toluene in the benzene used.[92]

The reaction product is concentrated by distilling off the unreacted benzene. The liquid crude BHC is pumped to a crystalliser where some of the α and β-isomers crystallise. After filtration, the filtrate is steam distilled and the molten bottoms cast in metal pans, cooled, ground to the required size and packed in fibre drums for shipment.[93]

The crude product usually contains 10–15 per cent γ-isomer and this may be increased to about 25 per cent by the purification process described and is then known as 'Gammexane' (really an ICI trade name). While suitable for certain purposes, such as the control of boll weevil in cotton, this product has a pronounced odour, which has led to the development of a concentrated product made by further crystallisation to contain not less than 99 per cent γ-isomer.[94] This is known as Lindane. The melting point of this pure γ-isomer is 112·5°C.

12.5.2.2. Uses Crude BHC is widely used as a soil poison, for grasshopper control and against cotton pests. It has found particular value when used against locusts in Africa and the Middle East. Swarms of these insects, amounting to 100,000 tons in weight, often travel 1,000 miles in a few weeks and have a daily food consumption of the order of their own weight of plant material. Baiting operations against locust hoppers have given kills of up to 34 kg of hopper per gram of BHC used.[95] As already mentioned, it is unsuitable for use in food crops owing to its musty and persistent odour. Lindane, however, has not this disadvantage and has a toxicity for insects five to ten times that of DDT. It controls more species than any other insecticide due to its action as both a stomach and contact poison, as well as the activity of its vapour. It also has a lower dermal and oral toxicity than DDT to warm-blooded animals.[96] It has found an increasing usage to combat insects that have developed resistance to DDT. Up to the present BHC does not appear to have been particularly dangerous to wild life, possibly because its unpalatability prevents dangerous amounts being consumed, but widespread usage, as by aerial spraying, might be more harmful.[97]

12.5.2.3. Statistics Little information is available owing to the paucity of manufacturers, but American sales in 1966 amounted to about 4,000 tons.[81]

12.5.3. Other Pesticides

Benzene enters to a lesser extent into the manufacture of certain other pesticides. The waste isomers from BHC production can be converted to trichlorobenzene, an intermediate for trichlorophenol, which is used for making hexachlorophone, a germicide. Tetrachlorobenzenes can also be recovered and the 1,2,4,5-isomer is useful as a starting material for 2,4,5-trichlorophenoxyacetic acid, a herbicide and defoliant. Parathion, the first of the organiophosphorus insecticides, is *O*-*O*-diethyl-p-nitrophenyl thiophosphate. It is a very effective wide-range insecticide, but owing to its high toxicity to humans, it is being replaced by other compounds, such as Malathion, which is not a benzene derivative. Certain phenyl derivatives such as Sulphone (*p*-chlorophenyl phenyl sulphone) are used as acaracides against mites and ticks, and azobenzene has been used as an ovicide for red spiders in greenhouses. Compounds such as dimethyl phthalate and 2-phenylcyclohexanol are widely used as insect repellants and chlorobenzenes are used in the production of mothproofing agents such as Eulan CN and Mitin FF.

12.6. DIHYDRIC PHENOLS

The three dihydric phenols, catechol, resorcinol and hydroquinone, are all of commercial importance.

12.6.1. Catechol

Catechol, *ortho*-dihydroxybenzene, was first prepared by Reinsch in 1839 during the distillation of catechin, a tannin-like substance,[98] which was one of the vegetable dyes used in cotton printing. Since catechol was first obtained by a destructive distillation process, it is also known as pyrocatechol, but the shorter name, catechol, is that now usually employed. Prior to its synthetic production from benzene, it was obtained from its monomethyl ether, guaiacol, which occurs in beechwood tar, by distillation with hydriodic acid.

12.6.1.1. Manufacture In addition to the catechol producted from wood tars and waste sulphite liquors, a considerable amount is now made from benzene by way of the *ortho*-sulphonated or halogenated phenols. The usual process consists of fusing *o*-chlorophenol with an alkali or by heating a mixture of *o*-chlorophenol and 1-phenol-2,4-disulphonic acid under pressure with an alkaline hydroxide and water. After acidification and removal of residual chlorophenol by blowing with steam, the catechol is extracted from the solution with a solvent such as ether and purified by distillation. Much of the industrial development of catechol production has been carried out in Germany. I. G. Farbenindustrie heated *o*-chlorophenol with a mixture of barium and sodium hydroxides, the barium content of the reaction mixture being recovered by precipitation as carbonate. Freshly precipitated lead hydroxide, with cuprous catalyst, could also be used, the low solubility of lead chloride in cold water facilitating its recovery.[99] In another German process, a mixture of one mole of *o*-chlorophenol with three moles of calcium chloride was heated to 225–230°C; the yield was 90–92 per cent of theory based on the chlorophenol.[98]

12.6.1.2. Properties The physical properties of the three dihydric phenols are given in Table XII.6.

TABLE XII.6

Properties of the Dihydric Phenols[101]

Property	*Catechol*	*Resorcinol*	*Hydroquinone*
Density $\frac{15^\circ C}{4^\circ C}$	1·371	1·272	1·324–1·328
Melting point, °C	105	110	172
Boiling point, °C	246	276	285/70 mm
Solubility, g/100 g at 25°C			
Water	31·2	58·4	6·9 mol %
Benzene	0·8	2·2	0·02 mol %
Ethanol	58·2	61·0	36·4 mol %

The chemical properties of catechol are generally those of phenol, but it can react in two tautomeric forms:

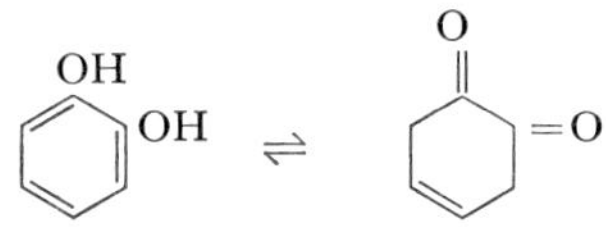

Many of the heavy metal salts are insoluble in water and the solutions of the alkali metal salts readily absorb oxygen, darkening in colour owing to the formation of quinones and polymeric condensation products. It has the most active reducing properties of the three dihydric phenols and can precipitate the finely divided metal from solutions containing a number of the heavy metals. When passed with air over silica gel it oxidises rapidly at room temperature whereas phenols without adjacent hydroxyl groups do not react under these conditions. Ammonolysis yields *o*-aminophenol and its esters undergo the Fries rearrangement to give phenolic ketones. Co-ordination compounds of some heavy metals, such as iron, aluminium, and titanium are readily formed.

12.6.1.3. Applications Owing to its powerful reducing action, catechol finds a major usage as an anti-oxidant, particularly for hydrocarbon oils containing olefines. Similarly, it is used as an anti-skinning agent for paints and varnishes, and as a photographic developer, particularly in formulations for fine-grain development and in colour-cinematography. Its properties also render it useful in such divers applications as fur dyeing and in electroplating baths. A major use of catechol, particularly in the United States, is in the production of 4-*tert*-butyl catechol, which is a powerful anti-oxidant and is used as a polymerisation inhibitor for butadiene and styrene. Condensation products of catechol with various aldehydes form synthetic resins, some of these products being used to improve the stability to light of nylon fabrics

The condensation product with formaldehyde is used as a synthetic tanning agent and other derivatives of catechol have important applications in the synthetic perfume of piperonal (heliotropin) via the aldehyde obtained by treating catechols with chloroform and caustic soda, the aldehyde then being reacted with methylene iodide, also in the presence of caustic soda. Catechol is used as the starting material in the synthesis of epinephrine, the active principle of the medulla of the adrenal gland. A possible application in poultry disinfectants might develop with a reduction in the price of catechol.[100]

12.6.1.4. Statistics Owing to the limited number of manufacturers, statistics for the production and usage of catechol are scanty, but it has recently been stated that, at the present time, the whole of the United States requirements of catechol, amounting to not more than 700 tons per year, are imported.[100] Catechol is produced in the United Kingdom by Coalite and Chemical Products Ltd.

12.6.2. Resorcinol

Resorcinol, *meta*-dihydroxybenzene, was first prepared by Körner in 1868 by fusing *m*-iodophenol with caustic alkali.

12.6.2.1. Manufacture The usual method for the manufacture of resorcinol involves fusion of sodium *m*-benzenedisulphonate with caustic soda. The melt is dissolved in water, acidified with hydrochloric acid and filtered. The filtrate is extracted with amyl alcohol or ether, which is then distilled off and the crude resorcinol purified by vacuum distillation, sublimation, or crystallisation from benzene.[101] The process as operated by I. G. Farbenindustrie at Leverkusen has been described in the reports of the British and American technical missions to Germany.[102] It has been claimed that, by modifying the sulphur trioxide content in the benzene sulphonation stage, an improved yield of resorcinol may be obtained on hydrolysis.[103] Resorcinol can also be made by acid hydrolysis of *m*-phenylenediamine or *m*-aminophenol. The isomerisation of *ortho*- and *para*-substituted dihydric phenols can be used to increase the yield of resorcinol as in a Bayer process, in which the heating of *ortho*- or *para*-chlorophenols or their mixtures with alkaline hydroxides, followed by rapid removal of the volatile products, is claimed to yield a product containing 59 per cent resorcinol and 38 per cent catechol.[103] Not unexpectedly, it has been found that by a process analogous to that using cumene for the production of phenol (see p. 370), good yields of resorcinol and acetone can be obtained by the oxidation of *m*-di-isopropyl benzene; a by-product is *p*-isopropyl phenol.[104]

12.6.2.2. Properties The physical properties of resorcinol are given in Table XII.6.

Its chemical properties are those associated with the presence of two phenolic groups. The resorcinates do not give the free resorcinol on treatment with carbon dioxide and it differs from catechol and hydroquinone in the fact that it does not form a quinone as a result of gentle oxidation. The esters undergo the Fries rearrangement to yield 4-acylresorcinols. It will be seen from Table XII.6 that it is very soluble in water (58·4 g/100 g solution at 20°C) and the solid is very hygroscopic, easily caking in the presence of moisture. Pure resorcinol is a white, crystalline material but this tends to acquire a pink tint on exposure to light and air; this tendency may be reduced by ensuring a minimum content of unsaturated hydrocarbons in the benzene used in the sulphonation stage of manufacture. In order to avoid darkening of the resorcinol, contact with metals other than aluminium and stainless steel should be avoided. Other metals, such as zinc, which does not cause darkening, may nevertheless react with resorcinol and cause contamination. The toxicology of resorcinol has been reviewed by Von Oettingen.[105]

12.6.2.3. Applications As with the other dihydric phenols, resorcinol finds considerable usage as an anti-oxidant and stabiliser, particularly for halogen compounds, and also, together with its derivatives, as a photographic developer, particularly in diazo-type processes.

The antiseptic properties of resorcinol find application in cosmetics and in unguents for the treatment of skin diseases. Derivatives such as resorcinol monoacetate and hexylresorcinol, which are less toxic than the parent compound, also find usage in pharmaceutical preparations.[106] Hexylresorcinol is a very effect anthelmintic and is made by reacting resorcinol with caproic acid and reducing the resultant ketone.

An important derivative of resorcinol is the lead salt of 2,4,6-trinitroresorcinol, known as lead styphnate, which is an explosive compound used in primary detonating charges.[107] In the production of dyestuffs resorcinol is coupled with diazonium salts to give azo dyes and particularly in making fluorescent dyes such as fluorescein which is made by condensing 1 mole of phthalic anhydride with 2 moles of resorcinol in the presence of zinc chloride at 100°C. Fluorescein is a yellow compound giving a green fluorescence; eosin, tetrabromofluorescin, dyes pink with an orange-red fluorescence. Formulations containing resorcinol are used in the dyeing of furs.

Formaldehyde reacts more readily with resorcinol than with phenol to give hard thermosetting resins and these have been particularly developed as adhesives to cure at room temperatures without the addition of acid or alkali, which have proved of great importance in the lamination of timber, paper and cloth, and in the construction of roof trusses and wooden boats. The bonding is proof against water and fungus development and does not have any deleterious effect on the bonded materials. Resorcinol-formaldehyde resins are also used in bonding rubber to textiles, as in the fabrication of tyres and industrial belting.[108] The curing rate and water resistance of phenolic and other resins may be modified by the addition of resorcinol. Epoxy ethers, made from resorcinol tar-still residues, have been suggested as coating and jointing compounds.[109]

Certain derivatives of resorcinol, notably its reaction products with aldehydes and quinones, have been suggested as tanning and cotton mercerisation agents.[110] A recent development in the dyestuffs field has been that of optical brighteners, used on fibres, and in soaps and plastics, as well as in suntan lotions as screening agents for ultra-violet light. One of the most important of these compounds is β-methylumbelliferone(7-hydroxy-4-methylcoumarin):

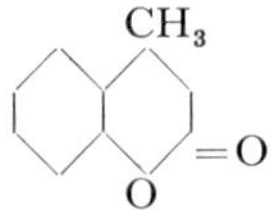

This can be made by condensing resorcinol with acetoacetic ester in the presence of sulphuric acid or phosphorus pentoxide.

12.6.2.4. Statistics As with the other dihydroxyphenols, production statistics for resorcinol are not available. United States producers are the Koppers Co. and United States Pipe and Foundry Co. It has been made in Britain by L. B. Holliday and Co. Ltd.

12.6.3. Hydroquinone

Hydroquinone, *para*-dihydroxybenzene, was first obtained by Pelletier and Caventon in 1820 by the dry distillation of quinic acid and its structure was established in 1844 by Wöhler.[110] Glucosides of hydroquinone are widely distributed in the leaves of plants, an example being arbutin, found in cranberry leaves.

12.6.3.1. Manufacture The conventional method for the production of hydroquinone involves two stages. In the first stage aniline is oxidised with manganese dioxide (pyrolusite) or sodium dichromate in sulphuric acid to quinone, which is reduced in the second stage by iron powder in water to give hydroquinone. In a typical process the reaction vessel is charged with dilute sulphuric acid and about 25 per cent of pyrolusite, which should be pure and of suitable physical texture; the mixture is cooled to about 10°C before addition of the aniline sulphate solution, and the remainder of the pyrolusite added to maintain an excess of 10–20 per cent over the theoretical requirement. Emphasis is generally laid on the necessity to maintain good cooling of the reaction mixture, but one patent describes a process in which the reaction is carried out at 20–70°C and the quinone removed by volatilisation at reduced pressure.[111] The quinone reduction may be carried out continuously by subjecting the crude quinone product to a counter current flow of steam in a column filled with coke. This is then passed into another column carrying a suspension of iron dust in water. The proportion of iron to water is adjusted so that a nearly saturated solution of hydroquinone is obtained (about 9 g per 100 g of solution at 28°C). It has been claimed that improved yields and a more pure product can be obtained when the iron-dust suspension is held at 60–65°C and a somewhat reduced pressure.[113] The reduction mixture is filtered to remove iron oxides and hydroxides, and the filtrate passed to crystallisers. In the I. G. Farbenindustrie, Wolfen, plant the hydroquinone was purified by vacuum distillation.[114] The final product is decolorised using active carbon, sodium bisulphite or zinc dust. Yields are about 80 per cent theory on the aniline.

Other possible methods for the production of hydroquinone include the hydrolysis of *para*-hydrogenated phenols with aqueous alkali hydroxide solution,[115] electrolytic oxidation of benzene in the presence of sulphuric acid, using a lead peroxide anode and lead cathode, followed by reduction of the quinone thus formed,[116] and treatment of *para*-dialkylbenzenes with hydrogen peroxide, followed by decomposition of the peroxides with acid.[117] Hydroquinone may also be prepared by a variation of the cumene-phenol process, in which esters of 4-(α,α-dialkylmethyl) phenols are oxidised to hydroperoxides, which are then decomposed with acid.[118] According to a Du Pont patent, both hydroquinone and quinhydrone may be produced by passing a mixture of acetylene and carbon monoxide at high pressure over a ruthenium or rhodium catalyst.[119]

12.6.3.2. Properties The physical properties of hydroquinone are given in Table XII.6.

Chemically it behaves as a phenol, giving ethers and esters, as well as mono- or diphenoxides with strong bases. Ring substitution easily yields halides, sulphonates, and alkyl derivatives.

The most important technical applications of hydroquinone depend upon its facile oxidation to quinone by a reversible reaction; furthermore hydroquinone and quinone form an equimolar complex called quinhydrone, which is the basis of the well-known electrode for measuring hydrogen-ion activities since the redox potential depends on the pH value of the surrounding medium. In the initial reaction of hydroquinone in aqueous solution with oxygen, the quinone and hydrogen peroxide formed have an autocatalytic effect on the reaction. Quinone acceptors such as sodium sulphite are employed for stabilising alkaline solutions of hydroquinone used as photographic developers. The darkening of old solutions of hydroquinone is due to the formation of condensation products with quinoid structures by continued oxidation. The ready reaction of hydroquinone with free peroxy radicals has led to its extensive use as an antioxidant.

12.6.3.3. Applications The principal use of hydroquinone is as a photographic developer, since it reduces light-exposed grains of silver halide in the emulsion at a rate considerably faster than that at which it reacts with the unexposed grains:

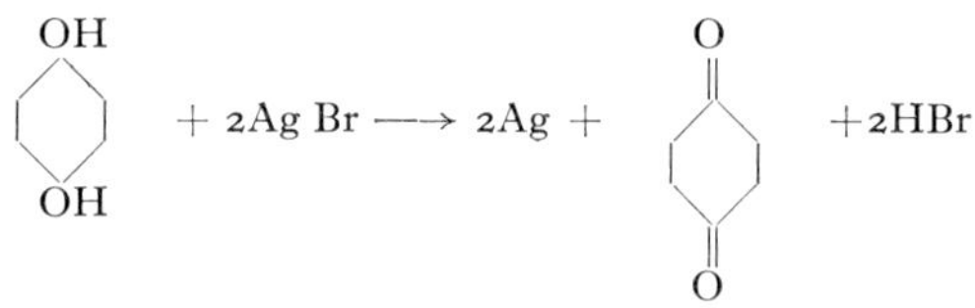

As already stated, sodium sulphite is usually added to hydroquinone developer solutions to prevent autoxidation. Hydroquinone produces high-density negatives and images of high contrast compared with the amine-type developers, but is very sensitive to bromide concentration in the developing solution; for this reason mixtures of hydroquinone and *para*-methylaminophenol sulphate are now frequently used.[111]

Another major application of hydroquinone is its use as an inhibitor of autoxidation in products ranging from foodstuffs such as fats and lard to linseed oil, furfural and isopropyl ether. Large quantities are used for this purpose in preventing the premature polymerisation of styrene in storage.[120] It is also used as gum inhibitor in motor fuels, particularly those containing tetraethyl lead.[121] Its anti-oxidant properties lead to its use as an anti-skinning agent for paints and as an inhibitor of the oxidation of vitamin A in fish-liver oils and of carotene in cottonseed oil and butter. It reduces the rate of deterioration on exposure to light of cotton fabrics treated with copper naphthenate as a rotproofing agent.[122] Considerable interest has been shown in the United States in the possible replacement of bisphenol-A in epoxy resins by hydroquinone, but this would depend on the development of a production process more economical than those currently used.[123]

Hydroquinone derivatives of industrial interest include the alkoxy hydroquinones, of which 2,5-diethoxyhydroquinone is particularly effective for inhibiting the polymerisation of styrene.[124] Condensation of hydroquinone with phthalic anhydride gives quinizarin, a dyestuff intermediate, and with acetoacetic ester gives 6-hydroxy-4-methyl-coumarin.

12.6.3.4. Statistics Statistical information on hydroquinone is scanty, but United States capacity in 1967 has been given as 8,000–9,000 tons per year, with demand of about 7,000 tons, the growth rate being expected to increase from 5 per cent per year to 7–10 per cent per year in the early 1970's.[123] A new plant has recently been announced in France to be built by Rhône-Poulenc with a total capacity of 18,000 tons per year of hydroquinone and catechol.[125] Hydroquinone is made in the United Kingdom by Kodak Ltd., Chemical Division.

Production in USA in 1971 was 6500 tons.

12.7. ANTHRAQUINONE

By reason of its use as a dyestuffs intermediate anthraquinone is an important derivative of benzene. It was first obtained from anthracene by Laurent in 1835, but its commercial importance dated from the discovery by Graebe in 1869 that alizarin was dihydroxyanthraquinone.[126]

12.7.1. Manufacture

The original method of production for anthraquinone involved the oxidation of anthracene with nitric or chromic acid.

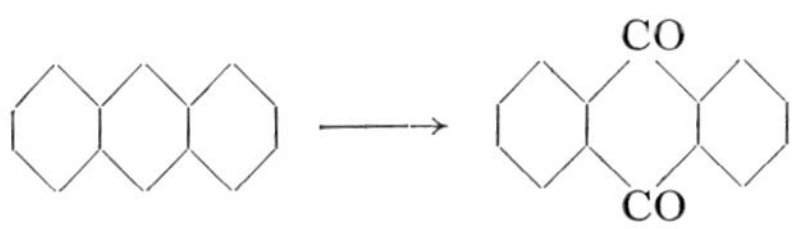

The use of chromic acid was found to be particularly attractive, since the by-product chromium sulphate could be used for the production of chrome green or as a chrome tanning agent. A more satisfactory method of oxidation, however, was found to be the continuous catalytic process, using a ferro-vanadate catalyst. The commercial success of this process depends on the availability of anthracene from high-temperature coal tar and as this became in short supply, notably in the United States, an alternative method of manufacture based on benzene and xylene has been developed. *Ortho*-xylene is oxidised to phthalic anhydride, which is reacted with benzene in the presence of aluminium chloride to give *ortho*-benzoylbenzoic acid; treatment of this acid with concentrated sulphuric acid gives a high yield of anthraquinone.

This process may be carried out by two methods. In the solvent method, a large excess of benzene is charged into a cast-iron jacketed reactor containing aluminium chloride equipped with a condenser and scrubber system to remove hydrogen chloride. Phthalic anhydride is then added with good agitation of the slurry and cooling applied to maintain the reaction temperature between 50 and 60°C. When reaction is complete the mass is transferred to an acid-proof reactor containing dilute sulphuric acid, thus decomposing the aluminium chloride complex of *ortho*-benzoylbenzoic acid. After separation of the aqueous solution of aluminium salts, the benzene layer is extracted with soda solution, which on acidification precipitates the *ortho*-benzoylbenzoic acid. This is filtered off, washed and treated with strong sulphuric acid or oleum to give anthraquinone in about 95 per cent yield.[127]

In the ball mill method, the reactants are present in practically stoichiometric quantities, thereby diminishing the tendency to form diphenylphthalide as a by-product.[128] At the end of the reaction the product is discharged as the dry aluminium chloride complex of *o*-benzoylbenzoic acid, which can then be treated directly with 5–20 per cent oleum to give anthraquinone. No solvent recovery is required and the hydrogen chloride emitted during the reaction is dry and therefore less corrosive. Yields are about the same as in the solvent process.[129]

For dyestuffs production the purity of the anthraquinone obtained in the processes given above is sufficiently high to be used directly but for certain specialised uses, as in electrical capacitors, an extremely high purity is required but this can be obtained by sublimation.

12.7.2. Properties

The physical properties of anthraquinone as given in Table XII.7.

TABLE XII.7

Physical Properties of Anthraquinone

Density $^{20}_{4}$	1·419–1·438
Melting point, °C	286
Boiling point, °C	379–381 (sublimes)
Solubility in toluene at 100°C g/100 g	6

Chemically anthraquinone is a relatively inert compound and many reactions characteristic of quinone take place only with difficulty. Nitration, however, occurs so readily that it is difficult to obtain a pure nitro product free from isomers and more highly nitrated derivatives. Halogenation also occurs readily but sulphonation requires energetic and carefully controlled conditions. With oleum of 20–25 per cent strength at temperatures of 140–150°C, sulphonation occurs mainly at the 2-position. Higher temperatures result in the formation of disulphonates as well as oxidation products, hence it is preferred to use stronger oleum (40 per cent), when mixtures of 2,6 and 2,7-disulphonic acids are formed.[126]

The substituted anthraquinones are of great importance in the dyestuffs industry. 2-Chloroanthraquinone is obtained from phthalic anhydride and chlorobenzene and 2-methylanthraquinone from toluene, but an important reaction for the general production of chloroanthraquinones is that in which the sulphonic acid group is exchanged for chlorine by means of sodium chlorate in hot hydrochloric acid. The aminoanthraquinones may be produced by reaction of the potassium salts of the sulphonic acids or of the chloroanthraquinone with ammonia.[130]

12.7.3. Applications

As indicated above, the major use of anthraquinone is as an intermediate in the dyestuffs industry, particularly in the production of mordant and vat dyes, such as the indanthrene blues and flavanthrones.[131]

Substituted anthraquinones are finding increasing usage in non-electrolytic processes for the production of hydrogen peroxide such as that developed by the Becco Chemical Division of FMC, in which 2-ethyl or 2-*tert*-butyl anthraquinone is dissolved in an organic solvent, together with a suitable catalyst and hydrogenated at 40°C. The catalyst is removed by filtration and the anthrahydroquinone solution passed through a series of air or oxygen oxidisers, where it is reconverted to the quinone and hydrogen peroxide, the yield of the latter being 90 per cent of theory. This process is also being used by BASF and the Degussa Co. at Frankfurt.[132]

12.8. POLYPHENYLS

From the commercial standpoint the most important polyphenyls are diphenyl (or biphenyl) and terphenyl.

Diphenyl was first reported by Fittig in 1862 and was made in quantity by Berthelot in 1867 by passing benzene vapours through a red-hot tube.[133]

12.8.1. Manufacture

Although the production of diphenyl by thermal treatment of benzene appears to be a simple process, great care must be taken to avoid the heterogeneous gas-solid phase reaction by which the benzene is completely cracked to carbon and hydrogen. This reaction is catalysed by heavy metals, particularly iron, nickel and copper.[134] Originally a bath of molten lead was used, through which benzene vapour was bubbled at 750°C.[135] Molten salts such as the chlorides of calcium and sodium have also been used,[136] but later developments have led to the use of electrical resistance heaters made of metals with low catalytic activity, such as nichrome or copper alloys In order to obtain minimum contact time, high gas velocities to obtain good heat-transfer rates are used and tubular resistance elements, with the benzene vapour flowing inside, are favoured.[137] The addition of sulphur compounds, such as carbon disulphide, to the benzene is claimed to repress carbon formation and to increase conversion rates.[138] The addition of steam, acetone, methanol or ethanol also has a similar effect.[139] Dilution of the benzene vapour with superheated steam[140] or with high-temperature combustion gases[141] has been used to inhibit contact of the benzene vapour with the hot surface of the reactor. A number of catalysts have been examined to improve yields in the process, but with such a system it appears to be difficult to maintain highly turbulent flow throughout the reactor. A well-designed thermal reactor, operated to give a condensed product containing 12–15 per cent diphenyl, based on benzene used, gives higher polyphenyls, namely terphenyl, in amounts corresponding to about 20 per cent of the diphenyl yield which, based on benzene consumed, should be 80–85 per cent of theory.[133]

It is possible that the hydrodealkylation process operated in petroleum refineries could become a major source of diphenyl, which also appears as a by-product of direct oxidation processes for the manufacture of phenol from benzene. Diphenyl oxide, is also a by-product from such processes as well as from the Dow chlorobenzene process (see p. 364).

The high-boiling by-products from the manufacture of diphenyl are the only commercial source of terphenyl[142] but, if demand increases, this could be met by recycling diphenyl to the reactor.[143] A typical sample of the high-boiling by-products from the diphenyl process contains 8 per cent *ortho*-terphenyl, 49 per cent *meta*-, 23 per cent *para*- and 20 per cent triphenylene (9,10-benzophenanthrene), together with tetraphenyls etc. This mixture is distilled at 100 mm Hg to give a mixture of the three terphenyl isomers, which are difficult to separate, but this can be achieved to some extent by fractionation at high reflux ratios (more than 10 to 1) or by centrifugation at high temperatures slightly above the melting points of the isomers. Variation in the proportions of the isomers can be achieved by treatment with aluminium chloride under differing conditions.[144]

12.8.2. Properties

The physical properties of diphenyl and the three terphenyls are given in Table XII.9.

TABLE XII.9

Physical Properties of Diphenyl and Terphenyls[133, 142]

	Diphenyl	*Terphenyl*		
		o	*m*	p
Melting point, °C	69	56	87	213
Boiling point, °C	255	320	359	367
Density of solid	0·991	1·022*	1·040*	0·960†
Flash point, °F	235	340	405	410

* Commercial grades, liquid at 100°C, g/ml.
† Commercial grade, liquid at 211°C.

The polyphenyls behave chemically as aromatic hydrocarbons. They may be nitrated and the products reduced to give analogues of aniline and it may be noted that benzidine is 4,4′-diaminodiphenyl. Extensive studies have been reported on the preparation and reactions of *ortho*-terphenyl.[145]

12.8.3. Applications

The main use of diphenyl is as a heat-transfer fluid, but owing to its high melting point it is usually employed in eutectic mixtures with diphenyl oxide such as Dowtherm, marketed by Dow Chemical Co., or Thermex of ICI, which contains 26·5 per cent diphenyl and 73·5 per cent diphenyl oxide. It is capable of withstanding temperatures up to 400°C and has a high heat-transfer coefficient. A similar liquid is marketed by Bayer in Germany as Diphyl.[146] A nuclear power plant has been designed that uses diphenyl as a working fluid for the turbine system. The vapour can be expanded without condensation, eliminating the moisture separation and vapour reheat required by steam turbines.[147] Diphenyl is also used as a paper impregnant for wrapping citrus fruits[148] and as a dye carrier for plastics and synthetic fibres.[149] The chlorinated diphenyls are marketed as Aroclors by Monsanto for heat-transfer fluids and as lubricating oil additives. The sodium salt of 2-hydroxydiphenyl (obtained as a by-product in the alkaline hydrolysis of chlorobenzene) is used as a preservative, germicide and fungicide.

The terphenyls are used namely as high-temperature heat-transfer fluids and are marketed for this purpose by Monsanto as Santowaxes. They have found particular application in nuclear power plants having a very high factor for radiation absorption and decomposition.[150] The best mixture as regards heat-flux has been found to consist of 15 per cent diphenyl, 34 per cent *ortho*-terphenyl, 24 per cent *meta*- and 3 per cent *para*-.[151] It is essential, when used for this purpose, that the polyphenyls should contain not less than 2 p.p.m. free chlorine, to avoid corrosion of zirconium alloy used in nuclear reactors.[142]

A number of hydrogenated derivatives of the polyphenyls such as dicyclohexyl and Monsanto H13–40, a partially hydrogenated mixture of terphenyls, have been developed as solvents and plasticisers. HB–40 has also been used as a high-temperature chain lubricant in bakery ovens. A number of polyphenyl ethers have been prepared and investigated as synthetic lubricants.[152]

Mixtures of chlorinated diphenyls, such as Monsanto's Aroclor 1254, mainly pentachlorodiphenyl, containing 54 per cent chlorine, have been widely used as plasticisers for paints and adhesives, and as pre-resistant hydraulic and transformer fluids, but in 1967 it became apparent that these compounds presented an ecological hazard particularly to fish-eating birds. As a result Monsanto have restricted the sales of the polychlorinated diphenyls to controllable usage in the electrical field where the products are used in enclosed, sealed systems. It is possible, however, that a biodegradable series of similar compounds may be developed.[153, 154]

12.8.4. Statistics

Diphenyl and terphenyl are made only by Dow and Monsanto in the United States and by Monsanto in Britain. Polychloro diphenyls are made by Monsanto in the United States, Britain and Japan, by Bayer in Germany, Prodelec in France and Flix in Spain. There are also plants in Russia, Poland and Czechoslovakia.

REFERENCES

[1] Claxton, G., '*Benzoles, Production and Uses*' (National Benzole and Allied Products Association, London, 1961).

[2] Hoffert, W. H. and Claxton, G., '*Motor Benzole, its Production and Use*' (National Benzole Association, London, 1938), 2nd edition.

[3] Ref. (1), p. 709.

[4] Ref. (1), p. 19.

[5] Information from National Benzole and Allied Products Association.

[6] Ref. (1), p. 381.

[7] Ref. (1), p. 687.

[8] Pleeth, S. J. W., '*Alcohol as a Fuel for Internal Combustion Engines*' (Chapman and Hall Ltd., London, 1949), p. 174.

[9] *Chem. Age*, 1969, **99**(*2631*), 22.

[10] Chopey, N. P., *Chem. Engineering*, 1970, **77**(*10*), 40.

[11] *Paint, Oil and Colour J.*, 1965, **147**, 876.

[12] Mellan, I., '*Industrial Solvents Handbook*' (Noyes Development Corporation, New York, 1970).

[13] Kalichevsky, V A. and Kobe, K. A., *Petroleum Refining with Chemicals* (Elsevier Publishing Co., New York, 1956), p. 359.

[14] Spence, R., *J. Roy. Inst. Chem.*, 1957, **81**, 357.

[15] Pleeth, S. J. W., *loc. cit.*, p. 22.

[16] Gilbert, E. E., *Petroleum Engr.*, 1954, **21**(*2*), C21; '*Encyclopedia of Chemical Technology*' ed. Othmer, Kirk, (Interscience, 1969), **19**, 279.

[17] Quinn, J. F., *Petroleum Engr.*, 1964, **3**, 401.

[18] Crooks, R. C. and White, R. R., *Chem. Eng. Prog.*, 1950, **46**, 249.

[19] Ambler, J. A. and Cotton, W. J., *Ind. Eng. Chem.*, 1920, **12**, 968.

[20] Allied Chemical and Dye Corporation, USP 2,697,117 (1954).

[21] Union Carbide and Chemical Corporation, USP 2,578,823 (1951); Allied Chemical and Dye Corporation, USP 2,704,295 (1955); Monsanto Chemical Co., USP 2,889,360–1 (1959).
[22] Kenyon, R. L. and Boehmer, N., *Ind. Eng. Chem.*, 1950, **42**, 1556.
[23] Plagovski, A. N. and Kagan, S. Z., *Org. Chem. Ind. USSR*, 1940, **27**, 296; CA, 1941, 3485.
[24] Idem, *Khim. Prom.*, 1944, **9**, 5; CA, 1946, 2123.
[25] Smith, F. C., *Southern Power Ind.*, 1944, **62**(*10*), 95.
[26] Allied Chemical and Dye Corporation, USP 2,697,117 (1954).
[27] Tennessee Corporation, USP 2,798,089 (1957).
[28] Dennis, L. M., USP 1,229,593 (1917).
[29] Pennsylvania Coal Products Co., USP 2,125,189 (1938).
[30] Ratcliff, G. A., *Univ. Microfilm Publ.* No. 9798.
[31] Groggins, P. H., '*Unit Processes in Organic Synthesis*' (McGraw-Hill, New York, 1958), 5th edition, p. 312.
[32] Gilbert, E. E. and Jones, E. P., *Ind. Eng. Chem.*, 1951, **43**, 2030.
[33] Imperial Chemical Industries Ltd., BP 834,250–1 (1960).
[34] Stauffer Chemical Co., USP 2,593,001 (1952).
[35] Norris Shreve, R., '*Chemical Process Industries*' (McGraw-Hill, New York, 1967), 3rd edition, p. 676.
[36] Hardie, D. W. F., '*Encyclopedia of Chemical Technology*', ed. Kirk-Othmer (Interscience, New York, 1964), **5**, 253.
[37] Idem, '*A History of the Chemical Industry in Widnes*' (Imperial Chemical Industries Ltd., London, 1950), p. 155.
[38] Sittig, M., *Chem. Proc. Review No.* **10**, Noyes Development Corporation, New York, 1968, p. 169.
[39] Solvay and Co., BP 605,693 (1948).
[40] Monsanto Chemical Co., USP 2,111,866 (1938).
[41] BIOS Final Report 1869, Item 5.
[42] Lummus and Co., USP 2,395,777 and 2,470,366.
[43] F. Raschig GmbH, Ger.P 539,176 and 575,765.
[44] Frontier Chemical Co., USP 3,148,222 (1964).
[45] F. Raschig GmbH, USP 1,963,761 (1934).
[46] Imperial Chemical Industries Ltd., BP 388,818 (1933).
[47] Solvay and Co., USP 2,152,357 (1939).
[48] *Chem. Eng. News*, 1964, **42**(7), 26.
[49] Monsanto Chemical Co., USP 3,029,296 (1962).
[50] Imperial Chemical Industries Ltd., BP 440,205 (1935).
[51] Dow Chemical Co., USP 1,923,419 (1933).
[52] Societe d'Ugines, FP 1,209,222 (1958).
[53] Columbia-Southern Chemical Corporation, USP 2,819,321 (1958).
[54] BASF, GP 973,588 (1960).
[55] Farbwerke Höchst, GP 1,100,610 (1963).
[56] Imperial Chemical Industries Ltd., Can.P 544,274 (1957).
[57] Faith, W. L. *et al.*, '*Industrial Chemicals*' (Reinhold Publishing Corporation, New York, 1965), p. 261.
[58] *Oil, Paint and Drug Reporter*, 8 March 1971, p. 9.
[59] Turco Products Inc., USP 2,964,429 (1960).
[60] *Chem. Eng. News*, 1955, **33**, 412.
[61] Tenner, A. G. and de Klerk, D., *Rev. Sci. Inst.*, 1957, **28**, 206.
[62] E. I. du Pont de Nemours and Co., USP 2,135,346 (1938).
[63] Imperial Chemical Industries Ltd., BP 806,726 (1958).
[64] Progil SA, Belg.P 601,358 (1961).
[65] *Chem. Trade J.*, 1963, **153**, 459.
[66] *Ibid.*, 1951, **129**, 1221.
[67] Brook, M. and Chesters, C. G. C., *Ann. Appl. Biol.*, 1957, **45**, 263.
[68] The Sherwin Williams Co., Can.P 590,736 (1960).

[69] *US Tariff Commission Reports.*
[70] Zeidler, O., *Ber.*, 1874, **7**, 1180.
[71] J. R. Geigy AG, BP 547,871, 547,874 (1942).
[72] *Chem. Engineering*, 1950, **57**(*11*), 204.
[73] West, T. F. and Campbell, G. A., '*DDT and Newer Persistent Insecticides*' (Chapman and Hall Ltd., 1950), 2nd edition, p. 41.
[74] Metcalf, R. L., '*Encyclopedia of Chemical Technology*', ed. Kirk-Othmer (Interscience, New York, 1966), **11**, 677.
[75] Frear, D. E. H., '*Chemistry of Insecticides, Fungicides and Herbicides*' (D. Van Nostrand Co. Inc., New York, 1948), 2nd edition, p. 71.
[76] Mellanby, K., '*Pesticides and Pollution*' (Collins Fontana New Naturalist, 1969), pp. 120–32.
[77] *Chem. Trade J.*, 1955, **136**, 152.
[78] *Chem. Eng. News*, 1962, **40**(27), 21.
[79] *Guardian*, 2 June 1971.
[80] *Ibid*, 21 May 1971.
[81] *US Tariff Commission Reports, Insecticides and Related Products.*
[82] *Oil, Paint and Drug Reporter*, 4 January, 1971, 9.
[83] Sittig, Marshall, '*Pesticide Production Processes*' (Noyes Development Corp., New York, 1967), p. 113.
[84] *Chem. Engineering*, 1954, **61**(*10*), 338.
[85] Ethyl Corporation, USP 2,729,603 (1956).
[86] Twiehaus, H. C. *et al.*, to Columbia-Southern Chemical Corp., USP 2,744,145 (1956).
[87] Columbia-Southern Chemical Corp., USP 2,858,260 (1958).
[88] Idem, USP 2,797,195 (1957).
[89] Langenbeck, W. *et al.*, *Chem. Tech.* (*Berlin*), 1952, **4**, 247.
[90] Bissinger, W. E. *et al.*, *Ind. Eng. Chem.*, 1959, **51**, 523.
[91] Klosa, J., *Arch. Pharm.*, 1953, **286**, 248.
[92] Melidoni, A., *Rev. fac. cienc. quim.* (*Univ. de la Plata*), 1948, **23** (**7**), 75.
[93] Faith, W. L. *et al.*, '*Industrial Chemicals*' (John Wiley, New York, 1965), 3rd edition, p. 136.
[94] Henard, P., *Industrie Chimique*, 1952, **39**, 139.
[95] Rainey, R. C., *Chem. and Ind.*, 1957, 629.
[96] Hensill, G. S., *Soap Sanit. Chemicals*, 1951, **27** (*9*), 135.
[97] Mellanby, K., *op. cit.*, p. 132.
[98] Raff, R. J. and Ettling, B. V., '*Encyclopedia of Chemical Technology*', ed. Kirk-Othmer (Interscience, 1966), **11**, 462.
[99] *US Dep. Com. Office Tech. Serv.*, PB Rept. 63959.
[100] *Oil, Paint and Drug Rep.*, 7 September 1970, p. 11.
[101] Raff, R. J. and Ettling, B. V., '*Encyclopedia of Chemical Technology*', ed. Kirk-Othmer (Interscience, 1966), **11**, 472.
[102] *US Dep. Com. Office Tech. Serv.*, PB Repts. 49272 and 63627; BIOS Reports 939, 1946.
[103] Farbenfabriken Bayer, BP 1,040,563 (1958).
[104] Distillers Co. Ltd., BP 775,813 (1956); 982,512 (1965).
[105] Von Oettingen, W. F., *Nat. Inst. Health* (*US*) *Bull No. 190*, 1949.
[106] Hill, J. H., *J. Am. Med. Assoc.*, 1935, **105**, 100.
[107] Ministry of Technology, USP 3,291,663 (1966).
[108] Glauert, R. A., *Brit. Plastics*, 1947, **19**, 223.
[109] Koppers and Co. Inc., USP 3,133,033 (1964).
[110] Shand, E. W., *Rayon and Synthetic Textiles*, 1950, **31** (*11*), 79.
[111] Raff, R. J. and Ettling, B. V., '*Encyclopedia of Chemical Technology*', ed. Kirk-Othmer (Interscience, 1966), **11**, 483.
[112] B. F. Goodrich Co., USP 2,343,768 (1944).
[113] Eastman Kodak Co., USP 1,998,177 (1935).
[114] BIOS Final Report 773, 1946.

[115] Dow Chemical Co., USP 1,934,656 (1933).
[116] Eastman Kodak Co., USP 2,043,912 (1936). Kanegafucha Kogo KK, Jap.P **172,351** (1946).
[117] Many patents, including Rhone Poulenc, BP 724,224 (1955); Eastman Kodak Co., USP 2,748,172 (1956); Distillers Co. Ltd., BP 910,735 (1962); Universal Oil Products Co., BP 1,158,173 (1969).
[118] Hercules Powder Co., USP 2,799,715 (1957).
[119] E. I. du Pont de Nemours and Co. Inc., USP 3,055,949 (1962).
[120] Frank, R. L. and Adams, C. E., *J. Am. Chem. Soc.*, 1946, **68**, 908.
[121] Standard Oil Development Co., USP 2,461,972 (1949).
[122] Bogaty, H., *Am. Dyestuffs Reptr.*, 1949, **38**, 253.
[123] *Oil, Paint and Drug Rep.*, 12 June 1967, p. 8.
[124] Mathieson Chemical Corporation, USP 2,455,745 (1948).
[125] *Eur. Chem. News New Plants*, 1970, 26 February 1971, p. 31.
[126] Cofrancesco, A. J., '*Encylopedia of Chemical Technology*', ed. Kirk-Othmer (Interscience, 1963), **2**, 431.
[127] Groggins, P. H., *Ind. Eng. Chem.*, 1931, **23**, 152.
[128] Rubidge, C. R. and Qua, N. C., *J. Am. Chem. Soc.*, 1914, **36**, 736.
[129] American Cyanamid Co., USP 2,496,894 (1950).
[130] Rieche, A., '*Outline of Industrial Organic Chemistry*' (Butterworths, 1968), 3rd edition, p. 271.
[131] Shreve, R. N., '*Chemical Process Industries*' (McGraw-Hill), 3rd edition, p. 843.
[132] *Petrol Refiner*, 1957, **36** (*11*), 256; *Chem. Trade J.*, 1958, **143**, 48.
[133] Poffenberger, N., '*Encyclopedia of Chemical Technology*', ed. Kirk-Othmer (Interscience, New York, 1965), **7**, 191.
[134] Zanetti, J. E. and Egloff, G., *Ind. Eng. Chem.*, 1917, **9**, 350.
[135] Monsanto Chemical Co., USP 1,894,266 and 1,894,283 (1933).
[136] Idem., USP 1,891,514 (1932).
[137] For numerous references see ref. 133.
[138] Dow Chemical Co., USP 1,978,069 (1964).
[139] Montecatini Chemical Co., Ital.P 586,113 (1962); J. W. Weiss and G. R. Downs USP 1,322,983 (1919).
[140] Dow Chemical Co., USP 1,938,609 (1933).
[141] Chemical Construction Co., BP 709,035 (1954).
[142] Hubbard, H. L., '*Encyclopedia of Chemical Technology*', ed. Kirk-Othmer (Interscience, New York, 1965), **7**, 194.
[143] Progil, BP 953,140 (1964); Farbenfabriken Bayer AG, BP 957,065 (1964).
[144] Monsanto Chemical Co., USP 2,363,209 (1944); Progil, FP 1,345,021 (1963).
[145] Allen, A. F. H. and Pingert, F. P., *J. Am. Chem. Soc.*, 1942, **64**, 1365, 2630.
[146] *Chem. and Ind.*, 1959, 1458; Szymkowiak, J., *Chem. Ing. Tech.*, 1961, **33**, 243.
[147] *Chem. Engineering*, 1961, **68**, 75.
[148] Farkas, A., USP 2,265,522 (1941); *Chem. Trade J.*, 1958, **143**, 426.
[149] Hendrid, N., *Melliand Textilber.*, 1962, **43**, 258.
[150] Traynard, P. and de Gaudemaris, G., *Rev. Inst. franc. Pétrole*, 1961, **16**, 607; *Chem. Week*, 20 April 1963, p. 69.
[151] Core, T. C. and Sato, K., *US At. Energy Comm.*, 1959, IDO 28007.
[152] Reith, H. *et al.*, *Freiberge Forschungsch.*, 1962, A**250**, 147.
[153] Tinker, J., *New Scientist*, 1 April 1971, p. 16.
[154] Edwards, R., *Chem. and Ind.*, 1971, p. 1340.

SUBJECT INDEX

Notes: All entries carrying (P) after them mean that the compound listed is mentioned *only* in the section on Phase equilibria between Benzene and other substances. This section is pages 116–179.
Catalysts are listed together and not specifically under processes and reactions, but cross-references, e.g. Potassium salts See: catalysts, are included.
Safety precautions are also entered as one section, but some references are also included elsewhere, e.g. 'hazards' under Phenol manufacture.

INDEX OF COMPANIES AND NAMES

Where a name to which reference is made in the text is also contained in the list of bibliographical notes at the end of a chapter, its number in that list of literature references is shown in parenthesis in this index.

Type set by Gloucester Typesetting Co. Ltd.
Printed in Great Britain at
The Pitman Press · Bath